WOOD ENGINEERING

Second Edition

German Gurfinkel

Professor of Civil Engineering
University of Illinois at Urbana-Champaign
Registered Structural Engineer
Ph.D., Fellow ASCE, Member ASAE, ASTM, ACI.

Southern Forest Products Association

KH

KENDALL/HUNT PUBLISHING COMPANY
Dubuque, Iowa

Library of Congress Catalog Card Number: 81–81843

ISBN 0–8403–2476–6

Printed in the United States of America

B 402476 01

Dedicated to Ana, my wife.

Contents

Preface

Wood Engineering is the result of a number of years spent by the author as a teacher, designer and researcher of wood structures. It was written as a textbook for undergraduate and graduate courses in schools of Civil Engineering, Architecture and Agricultural Engineering. The book may also be used for introductory courses in the material properties of wood, designing with wood and for students of engineering technology.

Professional Engineers and Architects will find this book contains material useful in their work—a number of examples of practical structures and a comprehensive appendix, containing the latest tables, monographs and other design aids.

A cursory review of the contents will reveal that the book consists of two distinct parts. Thus, Chapters 1 through 5 discuss the material wood. Natural and physical characteristics of wood, its strength properties, allowable stresses of structural wood, glued-laminated wood and plywood, and the durability of wood structures, are thoroughly studied. Chapters 6 through 11, on the other hand, discuss the specific uses of wood as a structural material. Behavior and design of various fasteners and connections, wood beams and columns, design of buildings and bridges made out of wood and composite wood-concrete-steel design, and a number of miscellaneous wood structures including arches and rigid frames, trusses, foundations and pole-type buildings are discussed.

Throughout the text, a large number of examples illustrate various concepts in behavior, analysis and design of wood structures. Following suggestions by users of the First Edition of the text the examples have been expanded to include the required formulas which now precede numerical calculations. A set of relevant problems (with answers included) is given at the end of most chapters.

Considerable material has been added to the text in its Second Edition to account for major developments in research and design of wood structures that have taken place since the book first appeared in 1973. Thus, the following new subjects are discussed at length: 1) Glulam beams made of mixed species of wood, 2) Moment-curvature and load-deflection relationships in wood beams subjected to transverse loading that varies from zero to ultimate, 3) Deflection of wood beams under service conditions including the effects of shear strains, the passage of time, and variation of the modulus of elasticity of wood, 4) Beam-column behavior including secondary load-deflection (P–Δ) effects, magnification factors and interaction equation for design, 5) Iteration procedure to analyze continuous, composite wood-concrete bridges that accounts for the actual variation of moment of inertia of the cross-sections with bending moment, 6) Design and behavior of glulam-deck bridges, and 7) Cantilever-suspended beam construction including an iteration procedure to determine the optimum position of the necessary hinges to minimize the maximum design moments.

Cooperation by the U.S. Forest Products Laboratory, American Society for Testing and Materials, Koppers Company, American Institute of Timber Construction, National Forest Products Association, American Plywood Association, and Timber Engineering Company, in allowing the use of original photographs, charts and nomographs, is gratefully acknowledged. Credit has been given throughout the text to these contributors, as well as to authors and researchers, in captions and references at the end of each chapter. Any omission is regretted.

The author had, in addition, the benefit of thorough reviews of all chapters by his civil engineering students at the University of Illinois at Urbana-Champaign, who used them as class notes. Many useful suggestions were given by the students.

The author is grateful to the staff of Southern Forest Products Association and to his son Daniel for having read the galleys and the page proofs.

German Gurfinkel

Urbana, Illinois
June 1981

Chapter 1
Characteristics of Wood

1.1 INTRODUCTION

The presentation of the material wood is the object of this chapter. Natural, physical and strength-reducing characteristics of wood are discussed from the point of view of the structural user of the material.

A cross-section of a log and a greatly magnified isometric view of wood are used to discuss the cycle of growth of a tree, and such morphological details as the annual rings of wood cells, the bark, the sapwood and heartwood portions of the log and others.

The presence of water in wood is measured by the moisture content; the effects that changes in moisture content have on the dimensional stability of wood are studied in detail because of their importance. The various specific gravities of wood and their standard methods of determination are discussed at length since the concept is directly related to the strength properties of the material. Finally, the fact that wood is a material of natural origin is recognized in the presence of strength-reducing characteristics such as knots, cross-grain and others. Discussion of these strength-reducing characteristics of wood centers in definition and origin; their influence on strength properties is discussed elsewhere, see Chapter 2.

1.2 NATURAL CHARACTERISTICS

A certain degree of familiarity with the nature of wood is necessary to understand the factors that affect the strength properties and behavior of the material. To this purpose, the following describes briefly some important botanical features and other characteristics of wood.

A cross-section at right angles to the longitudinal axis of a tree is shown in Fig. 1.1. The following well-defined features are easy to recognize as one moves from the outside of the section of the tree to the center: 1) bark, which consists of two parts, namely, the outer (corky, dead) part, A, that varies in thickness with different species and age of trees, and the thin (inner, live) part, B; 2) wood, which in most commercial species may be clearly differentiated between sapwood, D, and heartwood, E; and 3) the pith, F, a small central core darker in color, where primary growth originated. The cambium* layer, C, see Fig. 1.1., is microscopic; it is here, however, that all growth in thickness of bark and wood arises by cell division.

1.2.1 Wood Cells and Annual Rings

A greatly magnified isometric view of the wood structure of an evergreen tree is shown in Fig. 1.2; the three perpendicular axes represent the principal directions of growth of the tree, see also Fig. 2.1. Since growth in diameter of a tree in a temperature region is more rapid during the spring than later in the season, cells produced during this early period are large (2 to 5 mm. long

*Latin for change.

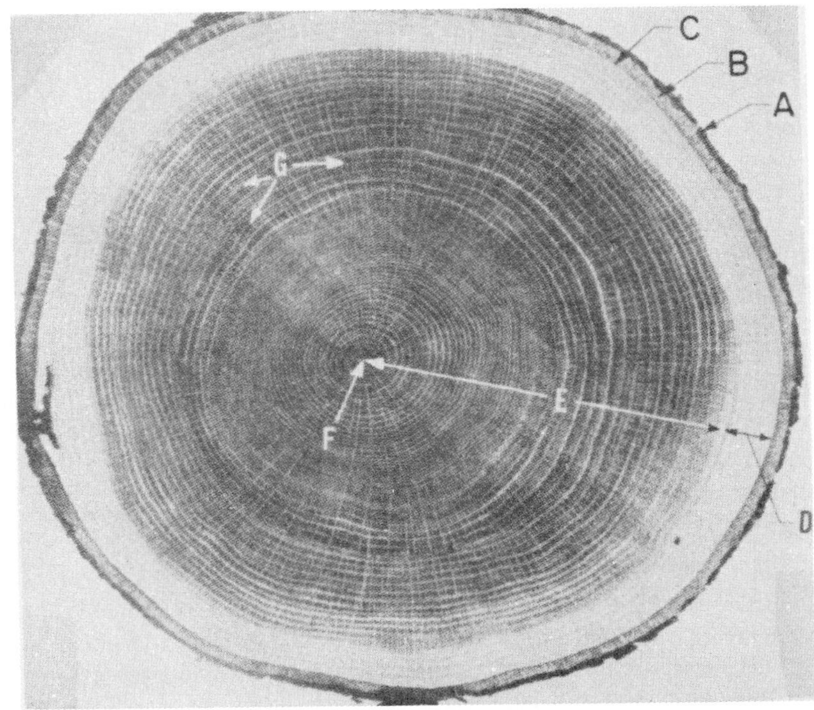

Figure 1.1. The tree trunk: A, outer bark or corky layer is composed of dry dead tissue. Gives general protection against external injuries. B, Inner bark is moist and soft. Carries prepared food from leaves to all growing parts of tree. C, Cambium layer (microscopic) is inside of inner bark and forms wood and bark cells. D, Sapwood is the light-colored wood beneath the bark. Carries sap from roots to leaves. E, Heartwood (inactive) is formed by a gradual change in the sapwood. Gives the tree strength. F, Pith is the soft tissue about which the first wood growth takes place in the newly formed twigs. G, Wood rays connect the various layers from pith to bark for storage and transference of food. From Ref. 1.

and 30 to 70μ† wide, see Ref. 2) and of low density. Their walls are thin (2–3μ) and enclose a large hollow space; earlywood, also known as springwood, is generated during this period. As the growing season advances denser and darker wood is produced; latewood, also known as summerwood, has smaller cells of thicker walls (5–7μ). The difference between the two types of cells is clearly shown in Fig. 1.2. The cycle is repeated every year, resulting in a series of well-defined rings; the age at any cross-section of the trunk may be determined by counting these rings. This, however, is to be considered as only an estimate since in trees in which growth has been interrupted by drought or defoliation by insects, more than one ring may be formed in the same season.

The number of annual rings per inch and the proportion of summerwood are used (1)* to classify some specimens as possessing either: 1) dense grain, not less than six rings per inch and

†A unit of length equal to one thousandth of a millimeter.

*Numbers in parentheses refer to bibliographic items listed at the end of each chapter. Equations are given by two digit numbers appearing in parentheses along side of equation.

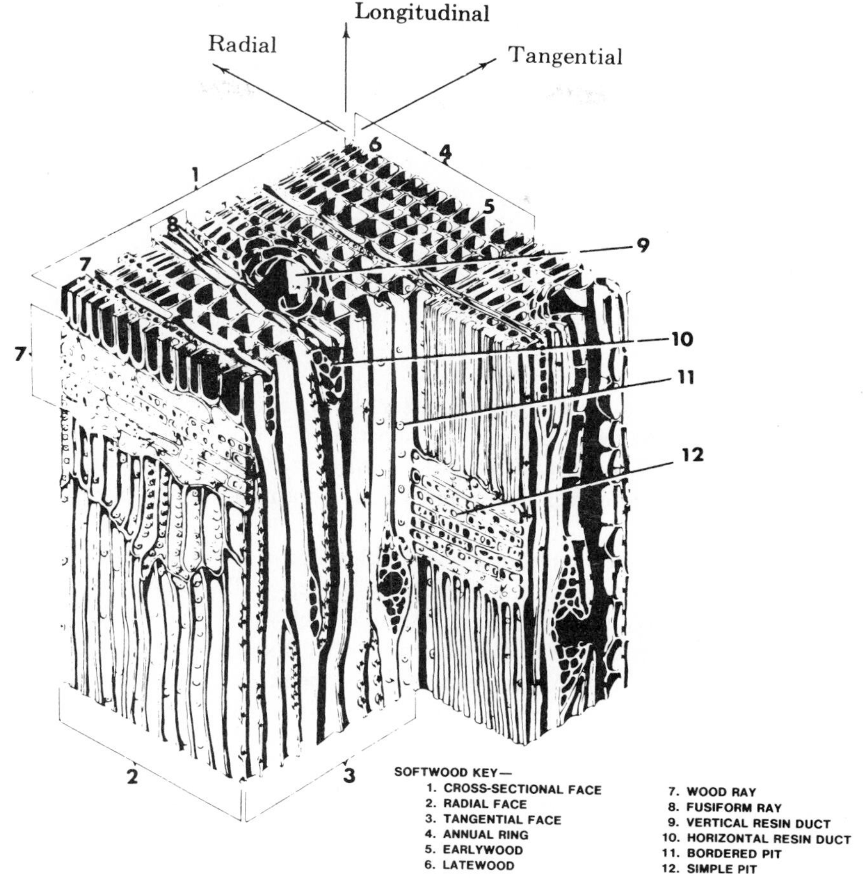

Figure 1.2. Wood structure of a softwood. From Ref. 3.

⅓ or more of summerwood†; 2) close grain, not less than 6 nor more than 30 annual rings per inch; or 3) open grain, which is not restricted as to the rate of growth.

1.2.2 Sapwood and Heartwood

As shown in Fig. 1.1, the main portion of the trunk consists of sapwood and heartwood. The former is a living part of the tree, ranging from 1.5 to 2 in. in radial thickness and participating in the conduction of sap from the roots to the leaves. In addition, it serves as storage of reserve food (4). With the passage of time old portions of sapwood die and are transformed into a zone of inactive cells, the cavities of which are frequently impregnated with various materials. This zone becomes heartwood, which is often easy to distinguish from sapwood because of its darker color.

†Pieces averaging not less than 4 rings per inch are accepted as dense if they average one-half or more summerwood.

The durability of untreated heartwood is greater than that of untreated sapwood when used in exposed conditions; this is due primarily to the infiltrations of materials deposited in the heartwood cells, which reduce the availability of moisture and air necessary for fungal growth. However, for the same reason, sapwood is easier to treat with pressure-impregnated chemicals because of the unoccupied space in its cell cavities. Regarding strength properties and dry unit weight, no consistent difference exists between sapwood and heartwood.

The ratio of cross-sectional surfaces between sapwood and heartwood increases toward the crown of the tree. This is a result of the thickness of sapwood increasing as the diameter of the tree decreases toward the crown. The cross-sectional area of sapwood is fairly constant at all levels of the stem (4); thus, as the diameter of the tree decreases toward the crown, the thickness of sapwood must increase to keep the area constant.

1.2.3. Chemical Composition of Wood

Wood consists fundamentally of organic matter, such as carbon, hydrogen, oxygen and a certain amount of nitrogen. Wood substance, irrespective of species, contains approximately 50 percent carbon, 6 percent hydrogen and 44 percent oxygen together with an insignificant 0.1 percent of nitrogen. In addition to organic matter, wood contains a number of mineral substances in small amounts.

The fundamental elements are combined with each other in wood to form complex chemical substances such as cellulose and lignin†, which form the cell walls. Of the two, cellulose is the most abundant constituent, comprising about 70 percent of wood; paper, explosives, synthetic textiles and plastics are obtained from it. Lignin is the other major (18 to 28 percent) constituent of wood; it is believed (1, 4) of lignin that, by cementing the structural units of wood together in an action similar to that of portland cement in concrete, it is responsible for the characteristic rigidity and hardness of wood. Lignin also contributes to reduce water absorption, thereby increasing the dimensional stability of wood.

In addition to cellulose and lignin, the wood structure contains a small amount (0.2 to 1.0 percent) of ash-forming minerals* which constitute the nutrient plant-food elements of the tree. Not as part of the actual wood structure, but enclosed in the cavities of the cells, are various extractives which contribute to the wood color, odor, taste and even resistance to decay. These extractives, which can be removed or extracted from the wood by neutral solvents such as water, alcohol and others, include tannins, starch, coloring matters, oil, resins, fats and waxes (1).

1.2.4 Wood-Producing Species

Two classes are recognized in the industry: hardwoods, also known as deciduous because they shed their leaves at the end of each growing season, and softwoods, also known as conifers, since all native species bear cones of some kind or another. Softwoods may have scale-like leaves, as the cedars, or needle-like leaves, as the pines; most softwoods are evergreen.

It should be recognized that the terms hardwood and softwood have no direct relation to the strength properties of the wood. Some hardwoods are extremely strong, such as cherry-bark red oak, and hickory or sugar maple; others however, such as cottonwood and aspen, have softer wood

†From Latin "lignum" for wood.

*The term derives from the fact that these minerals are left as ash when the lignin and cellulose are burned.

than softwoods, such as the white pines and true firs. Certain softwoods such as Southern pine and Douglas fir produce some of the strongest available woods.

For specific information on characteristics of commercial woods grown in the United States refer to the Wood Handbook (1) and to Panshin et al (4).

1.3 PHYSICAL CHARACTERISTICS

In the following paragraphs are given the definitions of some important concepts such as moisture content, specific gravity and unit weight of wood. The effects that variations of these quantities have on the strength properties of any given wood are discussed in Chapter 2; the dimensional changes that occur as a result of changes in moisture content or temperature are discussed herein.

A symbolic representation of wood as composed of solid matter, water, and air is shown in Fig. 1.3(a); weight and volume symbols are given at both sides. Various useful definitions and relations which can be derived using this representation are summarized in Fig. 1.3(b) for convenience purposes.

1.3.1 Moisture Content and Dimensional Changes in Wood

The presence of water in wood is a well known fact. Quantitatively, the amount of water in wood is measured by the moisture content, M. For a given piece of wood, moisture content is defined as the percentage weight of water, W_w, to weight of oven-dry wood, see Fig. 1.3.

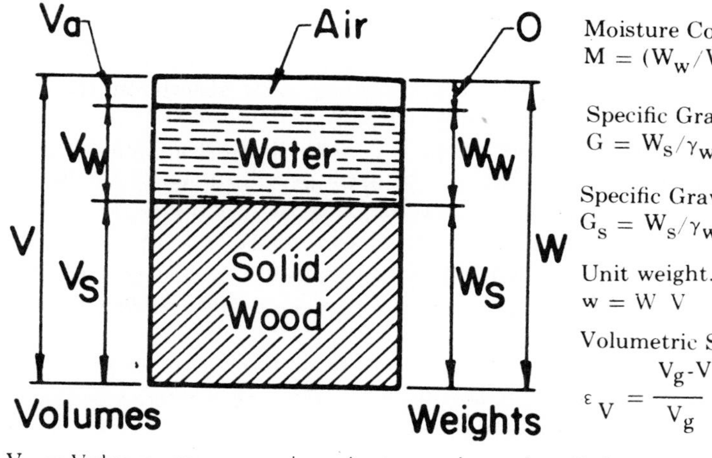

V_g = Volume green; an invariant equal to the volume at a moisture content equal to, or greater than, fiber-saturation point.

V = Volume at a given moisture content. V is less than V_g for a moisture content below fiber-saturation point.

(a) Symbolic Representation

Moisture Content, percent.
$M = (W_w/W_s)\,100$

Specific Gravity.
$G = W_s/\gamma_w V$

Specific Gravity of Wood Substance.
$G_s = W_s/\gamma_w V_s = 1.53$

Unit weight.
$w = W/V$

Volumetric Shrinkage, percent.

$$\varepsilon_V = \frac{V_g - V}{V_g} \times 100,$$

Relations between variables:
$w = \gamma_w G(1 + M/100)$

$\gamma_w = W_w/V_w = 1\ \text{gr/cm}^3$ or $62.4\ \text{lb/ft}^3$

(b) Various Definitions

Figure 1.3. Symbolic representation of wood and various definitions.

Standard determination of moisture content requires that the weight of the specimen, W (in the case of lumber, specimens should be obtained at least two inches from the ends of the pieces), be determined immediately before any drying or reabsorption of moisture occurs. Otherwise, the specimen should be placed in a plastic bag until it can be weighed. The specimen is then kept in an oven heated to a temperature of 103 ± 2 degrees Celsius until it has reached constant weight, W_s; this may take between 12 and 48 hours. The difference between the two weights, $W - W_s$, is the weight of water, W_w, that evaporated in the process; in some woods a small amount of extractives and other volatile matter may be included also. Thus, using this method, the percentage moisture content of a given specimen of wood may be determined from $M = 100(W - W_s)/W_s$.

For a rapid, nondestructive and reliable determination of moisture content use of electronic moisture meters is recommended. Two types are available commercially; namely 1) the resistance meters, based on the effect moisture content has on the resistance of wood to passage of a direct current, and 2) dielectric meters, based on the effect of moisture content on a capacitor in a high-frequency circuit in which the wood serves as the dielectric material. The reliability of these meters, however, is limited to a range in M of 0 to 20 percent using dielectric-type meters but only 6 to 30 percent when using a resistance-type meter. The principal advantages of the electrical method over the oven-drying method are its speed (only a few seconds are required) and the fact that the wood need not be cut or damaged.

Two other procedures can be used for the determination of M, namely: the distillation method and the hygrometric system. Both are as destructive as is the oven-dry method; however, they allow accurate determination of moisture content when the wood contains volatiles other than water, such as pitch, oil-type preservatives or other chemicals. The hygrometric method is limited to wood having a moisture content less than 30 percent. Complete information on standard methods for the determination of moisture content in wood may be obtained from ASTM Standard D2016, see Ref. 5.

The average moisture of green wood varies greatly, not only among the various species but within a given wood. Thus, moisture content is normally higher in the sapwood than in the heartwood. Typical value of average moisture content of green wood are listed in the Wood Handbook (1) for United States species. One finds Douglas-fir Coast-type at M = 37 and 115 for heartwood and sapwood, respectively. Some Southern pines such as Loblolly, Longleaf and Shortleaf are listed at 33 and 110, 31 and 106, and 32 and 122, respectively. As a note of interest, the maximum value for average moisture content of green wood is that of the sapwood of Western red cedar at 249 percent.

Variations of moisture content within a given piece of wood occur as a result of its hygroscopic nature. The natural affinity of wood for water makes it able to absorb it by direct contact or in vapor form from the atmosphere. Since maximum moisture content in wood occurs in the green condition, natural seasoning or drying processes will lower the moisture content of wood to that of the surrounding air. Wood in service will tend to reach an equilibrium-moisture content with the surrounding air. In most covered structures, the equilibrium-moisture content may be about 12 to 15 percent. In northern climates, at the peak of the heating season, the moisture content of wood in an enclosed building may be as low as 5 percent. Because of its hygroscopic nature, wood in service will undergo slight changes in moisture content at all times; daily and seasonal changes are to be expected.

It is important to recognize the two ways in which wood can hold water, namely: in the hollow space between the walls of the cell, and as part of the cell walls. When wood dries out, the water that is first to evaporate is the free water contained in the hollow space between the cell walls. The moisture content at which all the free water has evaporated, but the cell walls are still fully saturated, is known as the fiber-saturation point; this usually occurs at 30 percent moisture content.

Very small dimensional changes take place in wood between the green condition and that corresponding to the fiber saturation point ($M \simeq 30\%$); the volume of any given piece remains constant in this range in spite of the loss of water. It is only after the water in the actual cell walls begins to evaporate that dimensional changes occur. The approximate variation of dimensional change in wood from green to oven-dry moisture content, expressed as a percentage of green dimension, is shown in Fig. 1.4 for Southern pine and Douglas fir. Note that for any given change in moisture content, 1) the tangential change, ϵ_t, is about sixty percent greater than that which occurs in the radial direction, ϵ_R; 2) the longitudinal shrinkage, ϵ_L, is insignificant usually being between 0.1 and 0.2 percent for most species of wood; and 3) the volumetric shrinkage, ϵ_v (as defined in Fig. 1.3) is approximately equal to the sum of the linear dimensional changes. As the increasing moisture content approaches fiber-saturation point, dimensional changes become very small. The combined effects of radial and tangential shrinkage can distort the shape of wood pieces because of the difference in shrinkage and the curvature of annual rings. The major types distortion due to these effects is shown in Fig. 1.5.

It is possible to estimate shrinkage, S_M, from the green condition to moisture content M (where $M < 30\%$) using the following formula:

$$S_M = S_o\left(\frac{30-M}{30}\right) \tag{1.1}$$

where S_o is the total shrinkage from the green to the oven-dry condition. This formula may be used for determination of radial, tangential and volumetric shrinkages. Consider finding the values for Douglas fir at $M = 10$ percent if S_o is 4.8%, 7.6% and 12.4% for radial, tangential and volumetric shrinkages, respectively. Substituting these values in the formula yields S_{10} values of 4.4%, 5.1% and 8.3%, respectively. Average values of S_o for United States woods, and some imported woods from green to oven-dry moisture contents are given in the Wood Handbook (1); a coefficient of variation of approximately 15 percent on these average values can be expected.

1.3.2 Specific Gravity

The specific gravity of wood, G, is defined as the weight of a given volume of oven-dry wood, W_s, divided by the weight of an equal volume of water, $\gamma_w V$, see Fig. 1.3(b). Since V varies with the moisture content of the wood, the definition of specific gravity must be qualified by the volume of wood that has been used in the determination. Specific gravity is frequently determined for: 1) green wood, G_g; 2) air-dry wood, at about 12 percent moisture content, G_a; and 3) oven-dry wood, G_o. In view of the fact that the volume of any given piece of wood decreases with loss of moisture content below fiber-saturation point, see Fig. 1.4, it follows that $V_g > V_a > V_o$ and thus, $G_g < G_a < G_o$.

The specific gravity of wood substance, G_s, is defined as the ratio $W_s/\gamma_w V_s$, and is practically constant for all species; its value is approximately 1.53. Considering this fact, it is apparent that the individual specific gravities defined above are indicative of the amount of wood substance

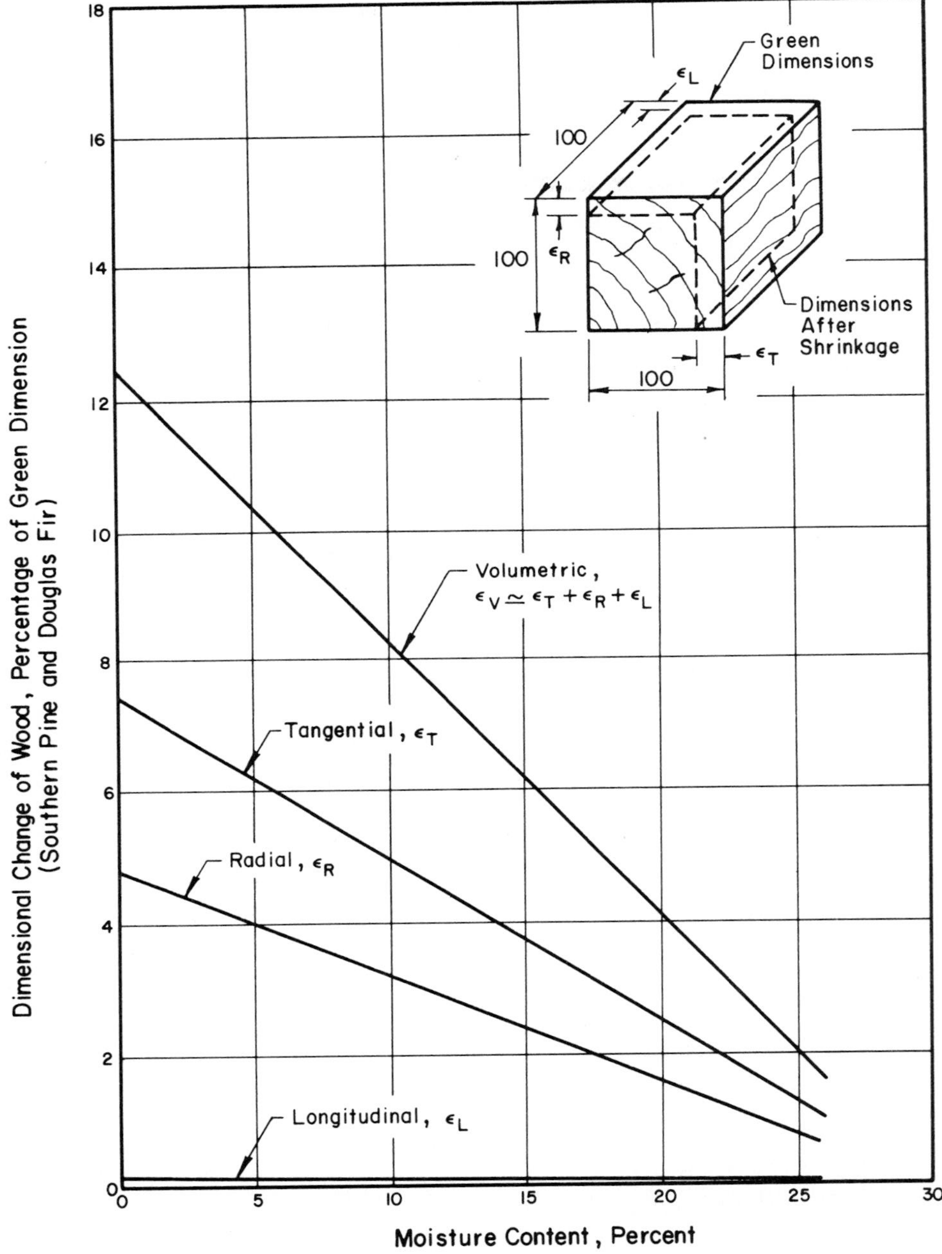

Figure 1.4. Dimensional changes of wood as the moisture content varies between fiber saturation point (M ~ 30%) and the oven dry condition (M=O).

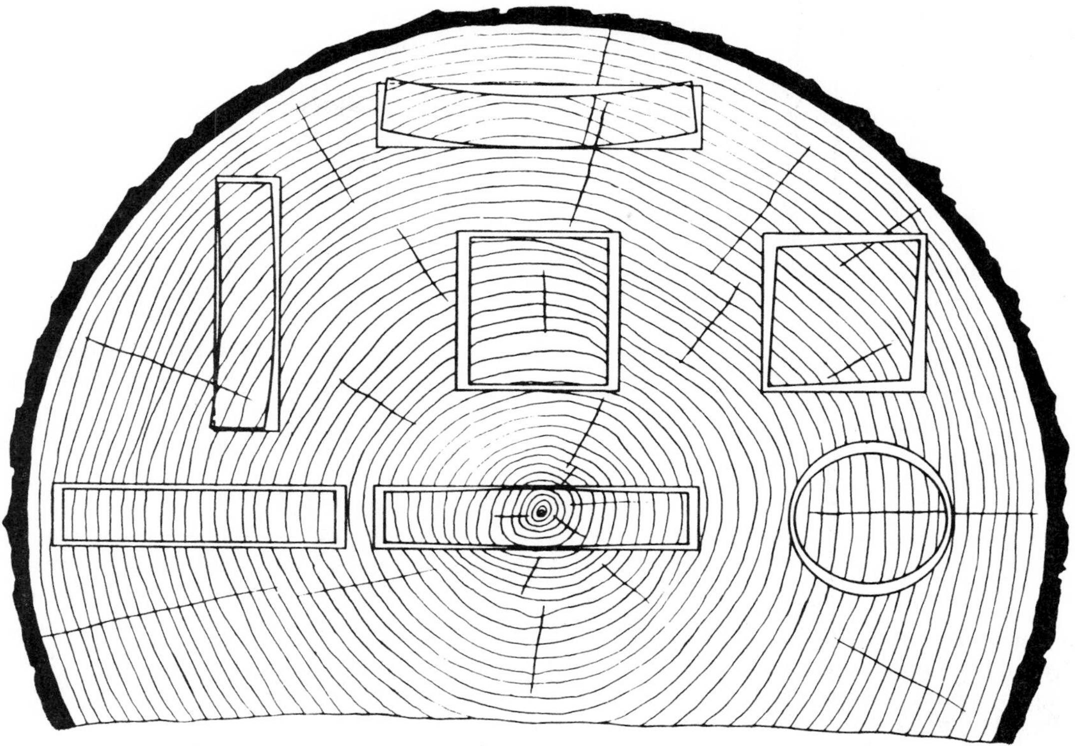

Figure 1.5. Characteristic shrinkage and distortion of flats, squares, and rounds as affected by the direction of the annual rings. Tangential shrinkage is about twice as great as radial. From Ref. 1.

present in any given piece of wood. Since strength properties of wood depend on the actual amount of wood substance present, it is easy to conceive of possible numerical relations between strength and specific gravity; see Section 2.6.

Several standard methods exist (6) for the determination of specific gravity of wood, the most important of which are: A) Volume by measurement; B) Volume by water immersion; C) Flotation tube; and D) Volume by mercury immersion. Methods A, B and D differ mainly in their determination of the volume of wood. Method A can be very accurate if the representative specimen is carefully prepared and regular in shape; however, the volume of irregularly shaped specimens is best determined by immersion in water or mercury, methods B and D. To prevent a change in the moisture content of the specimen when using method B, it is necessary to dip air-dry or oven-dry specimens in hot paraffin wax before making volume determinations. A similar coating may be necessary also for volume determination by mercury immersion; specimens with open pores or voids may entrap mercury and thus result in erroneous measurements. The flotation tube, method C, is an approximate procedure that is very useful because of its simplicity. A prismatic specimen $1 \times 1 \times 10$ in. is allowed to float upright, for a short time, in a slender cylinder filled with water. The specimen is retrieved once a mark of the water level is made on it.

The value of G at the given moisture content is obtained by dividing the immersed length of the specimen by the total length, i.e., 10 in. It is estimated (6) that specific gravity can be obtained to the nearest 0.02 using the flotation tube method.

A chart is available, see Fig. 1.6, to convert the specific gravity at one moisture content to that at some other moisture content. Values of G_a and G_o are read on the left-hand scale; values G_g are plotted on the diagonal lines. The use of the chart is illustrated by means of the following two examples: 1) Determine G_a for a given wood at $M = 12\%$ if $G_g = 0.55$. The solution is shown in dashed lines in Fig. 1.6. Enter the chart at $M = 12\%$, move vertically to the point where this line intersects the $G_g = 0.55$ curve, then move horizontally to the left-hand scale to read $G_{12} = 0.60$; 2) Determine G_a at $M = 15\%$ if $G_o = 0.54$. The solution is also shown in dashed lines in the chart; one obtains $G_{15} = 0.50$.

Most values of the specific gravity G_g of wood fall between 0.35 and 0.60. Extreme values of G_g in commercially important woods grown in this country, according to Wood Handbook (1), are 0.29 for northern white cedar and 0.66 for pignut hickory. Panshin et al (4) show a greater range in U.S. native woods, from 0.21 for corkwood to 1.04 for black ironwood. Worldwide, the range is from 0.04 to 1.40.

1.3.3 Unit Weight

The unit weight of wood, w, at a given moisture content, is defined as the ratio of its weight W to total volume V, see Fig. 1.3. Since V varies with the amount of water in the wood, it is obvious that unit weight depends on moisture content. The maximum value of w is obtained for the fully saturated condition, in which case weight and volume are both maximum. As the wood dries below the fiber-saturation point, both weight and volume decrease, the latter at a slower rate. Thus, w becomes smaller with reducing values of M; as M tends to zero, the unit weight reaches a minimum. The reverse is also true, since w increases with M; this explains why blocks of some woods are buoyant when thrown air-dry into the water, but sink once their air spaces are saturated.

The unit weights of various commercial wood specimens are given in Ref. 1 for 15 and 8 percent moisture contents. At 15 percent moisture content the unit weight of Douglas fir Coast type is 34 lb/ft^3, and that of Southern pine ranges between 36 and 44 lb/ft^3 depending on the given specimens. In general, a unit weight value equal to 40 lb/ft^3 may be used for design purposes. In the case of wood treated with chemical preservatives to prevent destruction due to attacks of decay, insects, marine borers or other causes (see Chapter 5), the unit weight of wood for design purposes is greater and should be taken conservatively as 50 lb/ft^3.

1.3.4 Illustrative Example

The weight and volume of a stick of Pacific silver fir measuring 6.05 × 2.001 × 2.002 in. is determined in the laboratory at various moisture contents. The following results are obtained:

	Weight, W	Volume, V
	gr.	cm^3
Green	201.3	393.8
Air-dry	168.0	360.3
Oven-dry	149.9	332.1

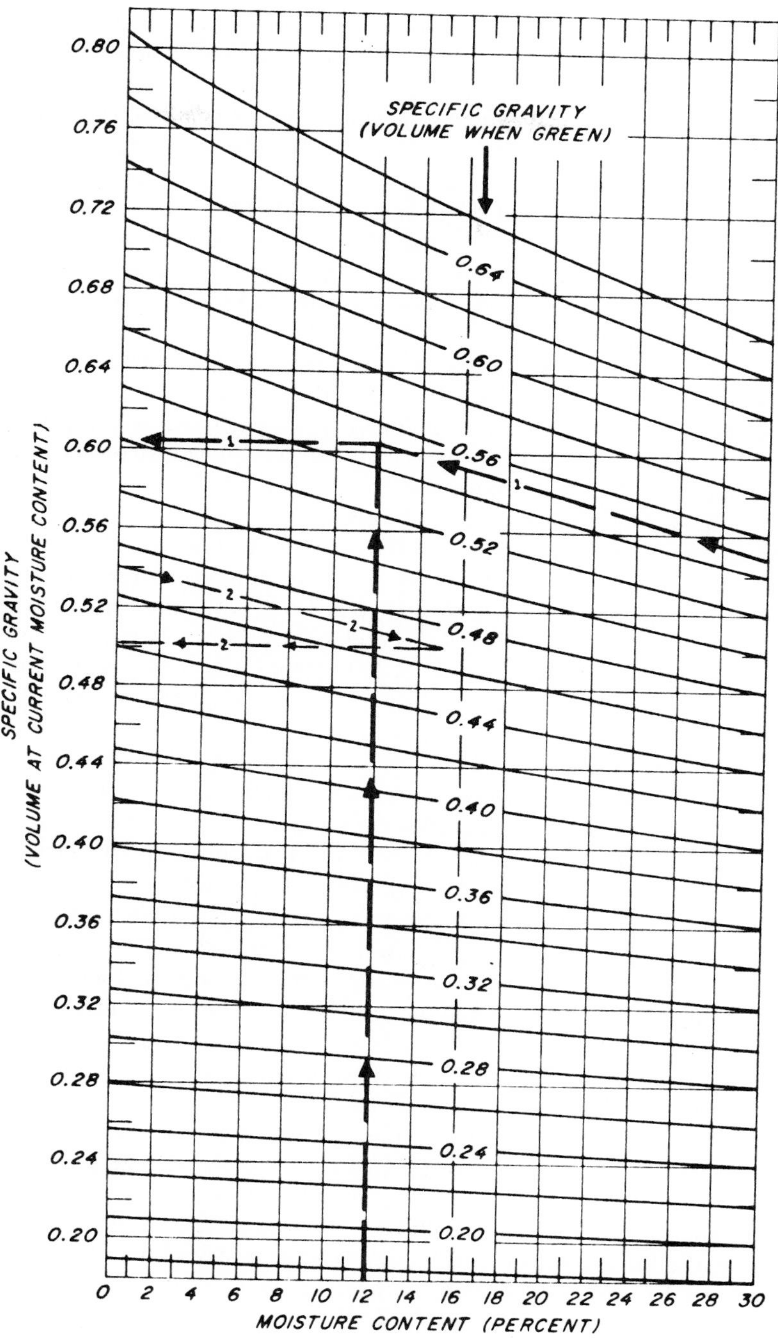

Figure 1.6. Relation of specific gravity and moisture content. From Ref. 1.

It is wished to determine the values of moisture content, specific gravity and unit weight for each case. Also, determine the volumetric shrinkage that occurs between the green-wood and the air-dry and oven-dry conditions.

The weight of actual wood in the stick is given by the weight of the stick in the oven-dry condition, i.e., W_s = 149.9 gr. The weight of water, W_w, in the green and air-dry conditions is the difference between the weight of the stick and W_s; thus, W_w = 51.4 and 18.1 gr., respectively. The corresponding moisture content, M, is given by (51.4/149.9)100 = 34.3% for green wood, and (18.1/149.9) 100 = 12.1% for air-dry wood. M = 0 for oven-dry wood.

Specific gravity of wood, G, is determined as the ratio W_s/V, where V is the volume at the given condition as shown above. Thus, the values of G are: 149.9/393.8 = 0.381 for green wood, 149.9/360.3 = 0.416 for air-dry wood, and 149.9/332.1 = 0.451 for oven-dry wood. The use of the approximate chart shown in Fig. 1.5 for G = 0.38 renders G = 0.41 for M = 12%, and G = 0.43 for M = 0; these values are somewhat lower than the actual results.

The unit weight of wood, w, is given by the ratio W/V. Thus, the unit weight is: 201.3/393.8 = 0.51 gr/cm³ for green wood, 168.0/360.3 = 0.467 gr/cm³ for air-dry wood and 149.9/332.1 = 0.451 gr/cm³ for oven-dry wood. Since 1 gr/cm³ = 62.4 lb/ft³ the above unit weights are easily converted to 31.8, 29.1, and 28.1 lb/ft³, respectively.

Volumetric shrinkage occurs in wood once the moisture content drops below fiber-saturation point. Above that point, the volume of wood remains fairly constant all the way to full saturation. This property is used to define volumetric shrinkage in wood, ϵ_V, as the percentage loss of volume relative to the constant volume of green wood. Total loss in volume for the air-dry and oven-dry conditions is 33.5 and 61.7 cm³; thus, ϵ_V is (33.5/393.8)100 = 8.5% for air-dry wood, and (61.7/393.8)100 = 15.7% for oven-dry wood.

1.3.5 Thermal Expansion

Wood structures change less in dimensions with variations in temperature than steel and concrete structures. The coefficient of expansion for wood along the grain, i.e., in the longitudinal direction, varies between 1.7×10^{-6} to 2.5×10^{-6} per 1°F; this is small when compared to 6.5×10^{-6} and 7.0×10^{-6} per 1°F for steel and concrete, respectively.

Usually, the effects of thermal expansion due to normal temperature changes are negligible compared to dimensional changes (swelling or shrinking) produced by changes in moisture content, see Fig. 1.4. Since longitudinal change is quite small in both cases, it follows that longitudinal motion of wood structures can be neglected for most structural designs; in other words, the need to make provision for expansion or contraction joints in wood structures and buildings is much less than in similar concrete or steel structures. However, in the case of wood-concrete composite structures, such as long bridges consisting of glued-laminated stringers and composite concrete slab, see Section 9.6, it is necessary to consider the restraint against shrinkage of the concrete and motion due to temperature change that is imposed by the wood stringers on the concrete slab. Provisions to resist the shear created at the interface concrete-wood must be made.

In the radial and tangential directions the respective coefficients of expansion of wood are directly proportional to the specific gravity G_o, based on oven-dry volume of the given wood. The ranges in values are: for radial expansion, between 25 and 35 $G_o \times 10^{-6}$ per 1°F; and for tangential expansion, between 32 and 45 $G_o \times 10^{-6}$ per 1°F. Thus, for the usual values of G_o the coefficients of thermal expansion of wood in the transverse directions are 7 to 10 times larger than the coefficient of thermal expansion in the longitudinal direction. For a range of about 0.1 to 0.8 in. oven-dry specific gravity, G_o, the following formulas are given (1)

$$\alpha_r = (32\ G_o + 9.9)\ 10^{-6}/°F$$

and

$$\alpha_t = (33\ G_o + 18.4)\ 10^{-6}/°F \qquad (1.2)$$

Where α_r and α_t are the radial and tangential thermal expansion coefficients, respectively, for oven dry wood. These values may be considered independent of temperature over the temperature range of −60 to 130°F.

Normally, wood in structures contains a certain amount of moisture. As wood temperature increases its thermal expansion will be compensated by shrinkage caused by a loss of moisture content. The net result, unless the wood was very dry initially (less than 4 percent moisture content), would be a negative dimensional change, i.e., a contraction (1). Thus, for wood having a range in moisture content between 8 and 20 percent, heating will result first in expansion followed by a gradual shrinking to a volume smaller than the initial volume, as the wood gradually loses water.

Other thermal properties of wood such as conductivity, specific heat, and diffusivity, of interest mainly in insulation applications, may be found in Ref. 1.

1.3.6 Coefficient of Friction

The coefficient of friction of wood is variable and depends on its moisture content, the corresponding surface roughness and the velocity of relative movement. The values given in Table 1.1 are obtained from Ref. 1 and should vary little with species except for those species which contain abundant oily or waxy extractives.

Lignum vitae may be one such exception. A wood species found in Southern Florida, and other countries in Central and South America, it is characterized by its resin content which may constitute up to about one fourth of the air-dry weight of the heartwood. Coefficients of friction of this wood against an unpolished steel surface are 0.20 and 0.34 for dry and green conditions, respectively.

1.3.7 Miscellaneous Physical Characteristics

Some physical behavior of wood such as: 1) its resistance against chemical action, 2) its electrical properties (conductivity, dielectric constant and dielectric power factor) and 3) the effects of nuclear radiation has been investigated and reported in the technical literature (1). The limited importance of these characteristics to the user of structure wood precludes their discussion in this text.

TABLE 1.1. COEFFICIENTS OF FRICTION OF WOOD, μ.

Condition of Wood	Friction Against	μStatic	μSliding*
Dry	Unpolished steel	0.70	0.70
Green	Unpolished steel	0.40	0.15
Dry, smooth	Dry, smooth wood	0.60	—
Green, smooth	Green, smooth wood	0.83	—

*At a relative movement of 13 feet per second.

1.4 STRENGTH-REDUCING CHARACTERISTICS

This term is used to include all such characteristics of wood that may tend to reduce structural strength. Major natural characteristics in wood are slope of grain and knots; other natural characteristics are compression and tension wood, shakes, brashness, pitch and bark pockets. In addition, seasoning defects, such as checks and warping, occur as a result of improper drying of lumber. In the following paragraphs, these characteristics and defects are studied in detail. Discussion of other defects, less common or unimportant, may be found in Refs. 1 and 4.

1.4.1 Slope of Grain

The tangent of the angle that wood fibers make with the longitudinal axis of a given piece is known as slope of grain. The piece of wood shown in Fig. 1.7 is a case of combined slope of grain; the angle between fibers and longitudinal axis is that between lines AD and AC. The determination of the combined slope of grain, 1/s, consists of measuring the simple slopes of grain 1/x and 1/y in planes ACB and ACE, respectively, followed by use of the radical expression shown in Fig. 1.7. For instance, let $1/x = 1/14$, and $1/y = 1/8$. Then $1/s = \sqrt{(1/14)^2 + (1/8)^2} \simeq 1/7$. Practically, the slopes 1/x and 1/y may be determined by drawing scribe lines on the wood with a needle-pointed scribe on a swivel handle, or by observing the direction of ink lines formed by applying free-flowing ink from a pen. Other devices for measuring slope of grain are given in Ref. 7.

Slope of grain in a piece of wood, also known as cross grain, may be due to natural causes such as spiral grain and the presence of knots, or as a result of sawing. Spiral grain is caused by the fibers growing in a winding or spiral course about the bole of a tree instead of in a vertical course. Distortion of the fibers around knots, see Fig. 1.8, can cause severe slopes of grain, often

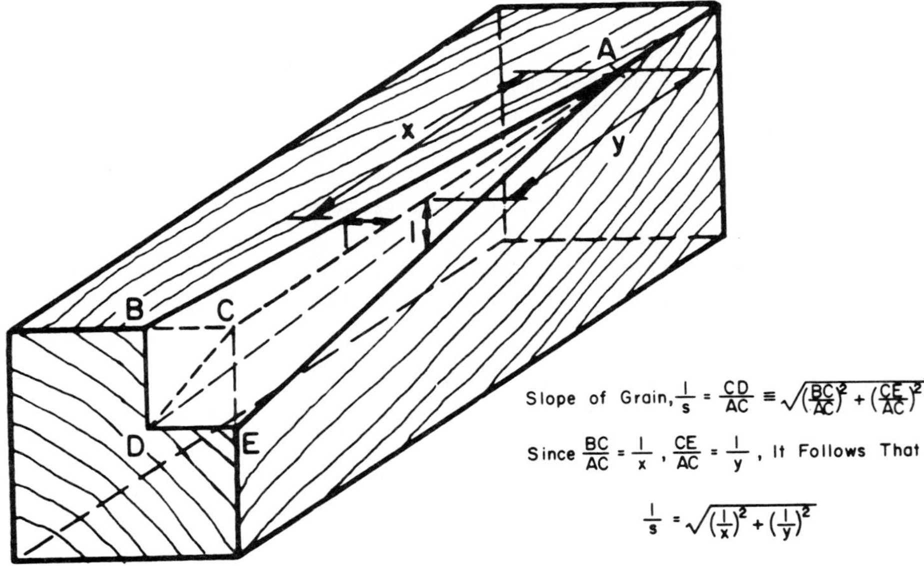

Figure 1.7. Determination of combined slope of grain.

Figure 1.8. View of a knot. From Ref. 8.

as much as 1 in 3. Sawing a piece of lumber parallel to the pith, or at an angle other than parallel with the bark, will cause slope of grain. Whether cross grain is due to natural or industrial causes the presence of slope of grain in a piece of wood reduces its strength, see Section 2.5.

1.4.2 Knots

The presence of knots in commercial wood is by far the characteristic that is easiest to recognize, and of which most users of structural wood are aware. Fundamentally, a knot is a portion of a branch or limb, see Fig. 1.8, which has been surrounded by subsequent growth of the wood of the trunk. While the branch is alive, growth continues simultaneously at both the main trunk of the tree and the branch. This is the cause of intergrown (otherwise known as tight or live) knots in lumber. However, when the branch dies or is pruned, its growth is stopped, although that of the trunk of the living tree is not. This leaves an interior dead stub that, as time goes on, is gradually embedded by new wood of the trunk. The result is a loose knot, otherwise known as a dead or encased knot. The improper handling of loose knots during fabrication of lumber often results in fallen knots, and consequently, in knotholes.

The shape in which a knot appears on a sawed surface depends on the geometry of the cut, see Fig. 1.9. Sawing a knot at right angles to its length results in a round knot; if cut diagonally, an oval knot, and when sawed lengthwise, a spike knot.

The presence of knots is detrimental to the strength of any given structural member. Their actual effect depends on size and the probable structural use of the given piece. The influence of knots on the stress-grading practice of commercial lumber is great, as shown in Section 2.5.

1.4.3 Other Natural Characteristics of Wood

Compression and tension wood are abnormal types of wood that occur in leaning trees where the pith is off-center. Some believe that this reaction wood is created to compensate for higher stresses induced by the eccentricity of the weight (1). Others feel that it is a result of an uneven distribution of growth hormones in the leaning stem (4). Compression wood occurs in softwoods generally; it is denser and harder than normal wood and is characterized by wide annual growth rings and large proportion of summerwood. As its name implies, compression wood is generated in the lower, compression side of the leaning tree. Tension wood, on the contrary, occurs mainly in hardwoods, in the upper, tension side of the leaning tree.

A shake is a separation or crack that runs along the grain, the greater part of which occurs between rings of annual growth. It is believed that this defect is a result of internal stresses created by lateral growth; however, such external causes as bending due to wind action may contribute to its formation.

Brashness is an abnormal condition that causes some pieces of wood to be relatively low in shock resistance for the given species. When subjected to bending, brash wood structural members fail abruptly in the tension side without splintering of the wood fibers and at comparatively small deflections. Brash wood has an unusually light weight and a low proportion of summerwood.

A pitch pocket is an oval opening containing tree resin, which extends parallel to the annual rings. In a transverse cross-section of the tree, a pitch pocket would appear flat toward the pith and curved on the bark side. A bark pocket is simply a small portion of bark partly or wholly enclosed in wood; the defect originating as a result of bird pecks or tunneling by insects.

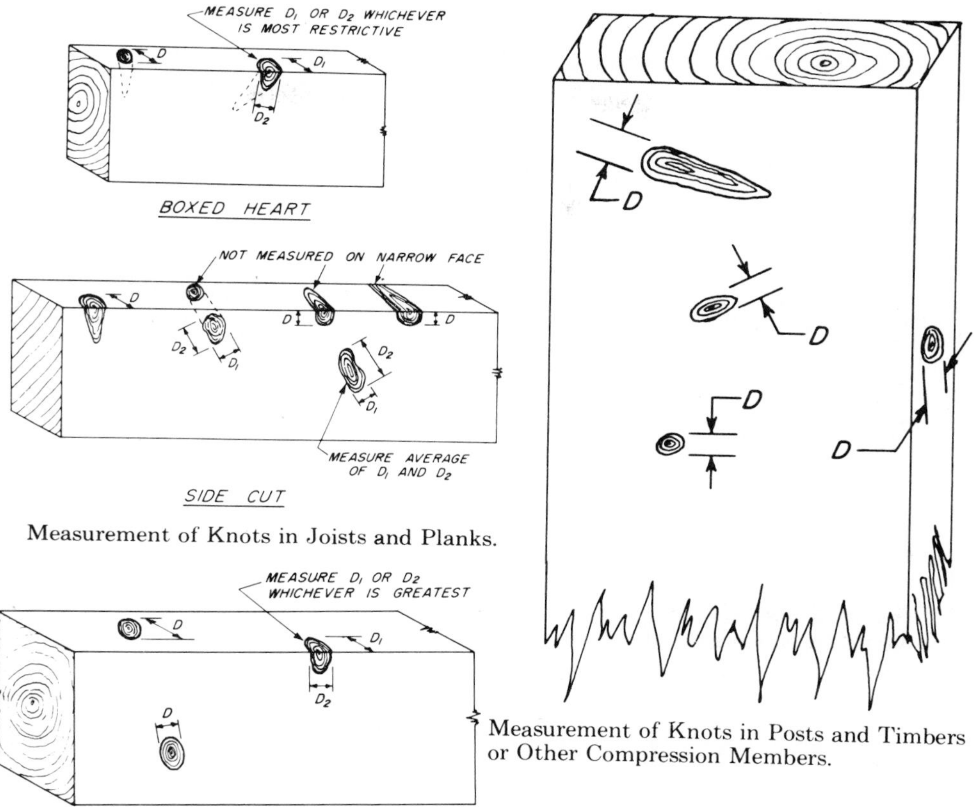

Measurement of Knots in Joists and Planks.

Measurement of Knots in Posts and Timbers or Other Compression Members.

Measurement of Knots in Beams and Stringers.

Figure 1.9. Shapes in which knots appear in various structural members and methods of measurement. From Ref. 9.

1.4.4 Seasoning Defects

Seasoning consists of removing moisture content from green wood to improve its serviceability. In the process, whether air or kiln-dried, wood changes its volumetric dimensions once the moisture content drops below fiber-saturation point, see Fig. 1.4. The differences in dimensional change, i.e., the differential shrinkage between longitudinal, radial and tangential directions create strains in the wood, which may cause checks. The latter are separations along the grain, the greater part of which occurs across the rings of annual growth. The possible appearance of checks can be reduced by providing conditions for slow and uniform evaporation of water in the wood.

Warping a piece of lumber can occur as a result of improper seasoning. Several types of distortion exist such as twisting, cupping and bowing (1). Obviously, there is very little use for warped lumber in structural applications.

PROBLEMS

1. A stick of Pacific silver fir measuring 6.03 × 2.004 × 2.001 in. is weighed and its volume is determined at various moisture contents. The following results are obtained.

	Weight, gr	Vol, cm^3
Green	223.1	392.0
Air-dry	160.9	360.9
Oven-dry	143.5	334.2

Determine the values of moisture content, specific gravity and unit weight (lb/ft^3) for each state. Also, determine the volumetric shrinkage that occurs between the green condition and the air-dry and oven-dry conditions.

Ans.	M, %	G	w, lb/ft^3	ϵ_V, %
Green	55.5	0.366	35.5	0
Air-dry	12.1	0.398	27.8	7.9
Oven-dry	0	0.429	26.8	14.7

2. The theoretical maximum moisture content of wood, M_{max}, assumes that all the void space in wood is occupied by water. Show that $M_{max} = [1/G_g - \gamma_w/\gamma_s]100$, where G_g is the specific gravity of wood based on oven-dry weight and green volume, and γ_w and γ_s are the unit weights of water and wood substance, respectively. Since $\gamma_w/\gamma_s \simeq 1/1.5$, show that $M_{max} \simeq 100\,(1.5 - G_g)/1.5\,G_g$.
3. Measurements in a piece of wood such as that shown in Fig. 1.6 render BC = 2 in., CE = 4 in., AC = 40 in. Determine the combined slope of grain of the given piece.
   Ans. Approximately 1.9.
4. For a given piece of wood $G_g = 0.44$. Determine the approximate values of specific gravity at: a) M = 0%, b) M = 12%, c) M = 24%.
   Ans. (a) 0.50, (b) 0.47, (c) 0.45
5. Show that the unit weight of green wood, w, for given values of specific gravity, G_g, and moisture content, M, is given by $w = G_g\left(1 + \frac{M}{100}\right)\gamma_w$, where γ_w is the unit weight of water.

Draw a table with values of w for a range in M between 30 and 100 percent, in increments of 10 percent, and a range in G_g between 0.3 and 0.7, in increments of 0.1.

Ans.

M, percent	Weight in pounds per cubic foot when G_g is:				
	0.3	0.4	0.5	0.6	0.7
30	24.3	32.4	40.6	48.7	56.8
40	26.2	34.9	43.7	52.4	61.2
50	28.1	37.4	46.8	56.2	65.5
60	30.0	39.9	49.9	59.9	69.9
70	31.8	42.4	53.0	63.6	74.3
80	33.7	44.9	56.2	67.4	78.6
90	35.6	47.4	59.3	71.1	83.0
100	37.4	49.9	62.4	74.9	87.4

Values under the dash line correspond to non-buoyant wood.

REFERENCES

1. Forest Products Laboratory, "Wood Handbook," Agriculture Handbook No. 72, U.S. Department of Agriculture, Washington, D.C., revised August 1974.
2. Karlsen, G. G., et al, "Wooden Structures," translated from the Russian by W. L. Goodman, MIR Publishers, Moscow, 1967.
3. Folger, A. N., "Classroom Demonstrations of Wood Properties," PA-900, Forest Products Laboratory, Madision, Wis., 1969, 41 pp.
4. Panshin, A. J., DeZeeuw, C., and Brown, H. P., "Textbook of Wood Technology," Volume I, Second Edition, 1964, McGraw Hill Book Company, N.Y.
5. ASTM Designation D2016–74, "Standard Methods of Test for Moisture Content of Wood," American Society for Testing and Materials, 1978 Book of ASTM Standards, Part 22, pp. 616–631, 1978.
6. ASTM Designation D2395–69, "Standard Methods of Test for Specific Gravity of Wood and Wood-Base Materials," American Society for Testing and Materials, 1978 Book of ASTM Standards, Part 22, pp. 717–727, 1978.
7. Anderson, E. A., Koehler, A., and Krone, R. H., "Instruments for Rapidly Measuring Slope of Grain in Lumber," Report No. 1592, U.S. Forest Products Laboratory, Madison, Wis., 1945.
8. ASTM Designation D9–76, "Standard Definitions of Terms Relating to Timber," American Society for Testing and Materials, 1978 Book of ASTM Standards, Part 22, pp. 8–22, 1978.
9. ASTM Designation D245–74, "Standard Methods for Establishing Structural Grades and Related Allowable Properties for Visually Graded Lumber," American Society for Testing and Materials, 1978 Book of ASTM Standards, Part 22, pp. 139–162, 1978.

Chapter 2
Strength Properties and Allowable Stresses of Structural Wood

2.1 INTRODUCTION

Design of wood structures requires thorough knowledge of the behavior of wood under load, and the stress levels to which the material can be used safely. Thus, the object of this chapter is twofold. First, it is concerned with the determination of strength properties of clear wood, and second, with their use in obtaining the allowable stresses necessary for design purposes.

The chapter is developed along these lines. Section 2.2 is devoted to the determination of strength properties of clear wood using standard methods, and to the interpretation of the test results. The natural variability of strength properties of wood is recognized in Section 2.3; the concept is used to discuss the methods employed by the wood industry to determine the clear wood strength values of all commercial species. The influence of various factors such as strength-reducing characteristics, moisture content, specific gravity, temperature, duration of loading and fatigue on the strength properties of clear wood is studied in Sections 2.4 to 2.10.

Before allowable stresses can be determined, structural wood must be graded in relation to strength properties, as shown in Section 2.11. Allowable stresses for normal loading condition are obtained in Section 2.12 using the clear wood strength properties of the given species modified by adjusting factors which include time-duration of loading, safety, seasoning effects and others; illustrative examples of the various determinations are given. Allowable stresses for loading conditions other than normal and an example of determination of the governing loading condition for design are given in Section 2.13.

Tables corresponding to this chapter in the appendix provide information of permanent value on strength properties of U.S. commercial woods, on actual sizes and geometric constants for standard dressed sizes of lumber, and the allowable unit stresses for structural lumber.

2.2 STRENGTH PROPERTIES OF CLEAR WOOD

It is necessary to determine the strength properties of wood using clear specimens; this must be done to avoid possible uncertainties in test results that are usually caused by the presence of defects. The principal strength properties of interest to users of structural wood are 1) compression parallel to grain; 2) compression perpendicular to grain; 3) tension parallel to grain; 4) static bending and 5) shear parallel to grain. The first is necessary for the design of columns and other members subjected to compression such as arches and chords of trusses; the second is used in the design of end bearing of beams and connections bearing across the grain. Tension parallel to grain is used in the design of bottom chords and some web members of simply-supported trusses. Bending occurs in all members subjected to the action of transverse loads such as beams and floor planks; shear parallel to grain accompanies bending.

Standard procedures for testing these properties and others are given in ASTM's Designation D143–52 (reapproved 1972), see Ref. 1. It is important to note that, in recognition of the significant influence of temperature and humidity on the strength of wood, these factors must be strictly controlled during testing; thus, it is recommended that the testing room and rooms for preparation of test specimens have some means of temperature and humidity control.

The anisotropic nature of wood, made evident already in the different dimensional changes along the three principal axes, see Fig. 1.4, affects also the various strength properties. The three axes are mutually perpendicular in the longitudinal, radial and tangential directions of the trunk of the tree; they are identified in Fig. 2.1, as L, R and T. The direction of the fibers is given by axis L; and the largest strength of wood is obtained when load-induced strains and stresses follow this direction. This is particularly true of compression parallel to grain as compared to compression perpendicular to grain; also, the modulus of elasticity in the L direction is approximately twenty and ten times larger than the corresponding moduli in the T and R directions, respectively. The same phenomenon is true in the case of shear stresses, as shown later in this chapter.

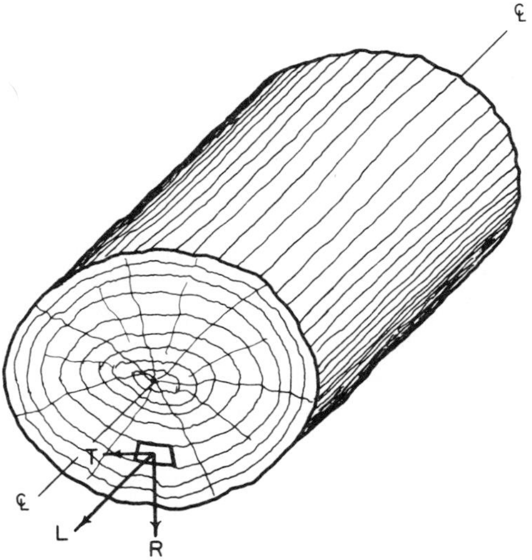

Figure 2.1. Identification of longitudinal, radial and tangential directions in wood.

Following is a discussion of the most important strength properties of wood and the standard methods for their determination. A list of these properties for all U.S. commercial species of wood is given in Tables A.2.1 and A.2.2.

2.2.1 Compression Parallel to Grain

This test is made on 2 × 2 × 8 in. specimens in which special care has been taken to ensure that end-grain surfaces are parallel to each other and normal to the longitudinal axis. The results of the test are usually plotted as a stress-strain diagram, see the top portion of Figs. 7.2 and 7.5. It is easy to see that the relation between f and ϵ is linear for the initial stage of loading; the slope of the line defines the modulus of elasticity of wood in compression parallel to grain. The end of the linear relation between strains and stress defines the proportional limit. Beyond it, the relation becomes non-linear; a smooth curve represents the transition between the proportional limit and ultimate. Values of maximum compression parallel to grain, also known as crushing strength, are listed in the above-referred Tables.

2.2.2 Compression Perpendicular to Grain

To test this property specimens 2 × 2 × 6 in. are loaded across the grain through a metal bearing plate 2 in. in width, placed across the upper surface of the specimen at equal distances from the ends, and normal to the length. It is important that quarter-sawed* specimens be used so that the load will be applied through the bearing plate parallel to a radial surface.

Results of a typical test are shown in Fig. 2.2, where load in pounds is plotted against shortening in inches; since the specimen measured 2.015 × 2.012 × 6.07 in. the figure can be converted to a stress-strain diagram easily if one divides the ordinates by 4 $in.^2$ and the abscissas by 2.01 in. Observation of Fig. 2.2 shows an initial variation that is almost linear and a well-defined proportional limit. Beyond this point the variation becomes non-linear; loads continue increasing with deformation, however, at a smaller rate of change. Ultimate load for compression normal to grain is not well defined. After the flattening and failure of the cell walls, a compaction of the wood occurs with increasing deformations; this is followed by a continuous rise in the resistance of the compressed specimen. A large value of ultimate load is possible; however, at an unacceptable deformation. Thus, values of fiber stress at proportional limit, see Tables A.2.1 and A.2.2, are used to determine allowable stresses in compression perpendicular to grain.

Test results indicate that fiber stress at proportional limit obtained in the standard test procedure is about 50 percent greater than that obtained in loading the entire area of a 2 in. cube. As the ratio of loaded to unloaded area is decreased, further increases in proportional limit stresses are obtained. The explanation lies in the fact that in the partially loaded case, the fibers under the load are subjected to deformations that throw them into tension, the vertical component of which contributes to the support of the load. This increase is taken into account in the determination of safe working stresses (2) for bolted connections loaded normal to grain, see Section 6.2, and for end bearing of wood beams, see Section 7.12.

2.2.3 Compression at an Angle of Load to Grain

Hankinson showed (3) that the strength of wood in compression at an angle of load to grain, θ, lies between a maximum value P for $\theta = 0°$ and a minimum value Q for $\theta = 90°$. The following expression was developed from Hankinson's experimental work:

$$N = \frac{PQ}{P \sin^2\theta + Q \cos^2\theta} \tag{2.1}$$

where: N is the strength in compression at an angle θ or load to grain; and P and Q are the strengths in compression parallel to grain, $\theta = 0°$, and perpendicular to grain, $\theta = 90°$, respec-

*Lumber that has been sawed so that the wide surfaces extend approximately at right angles to the annual growth rings.

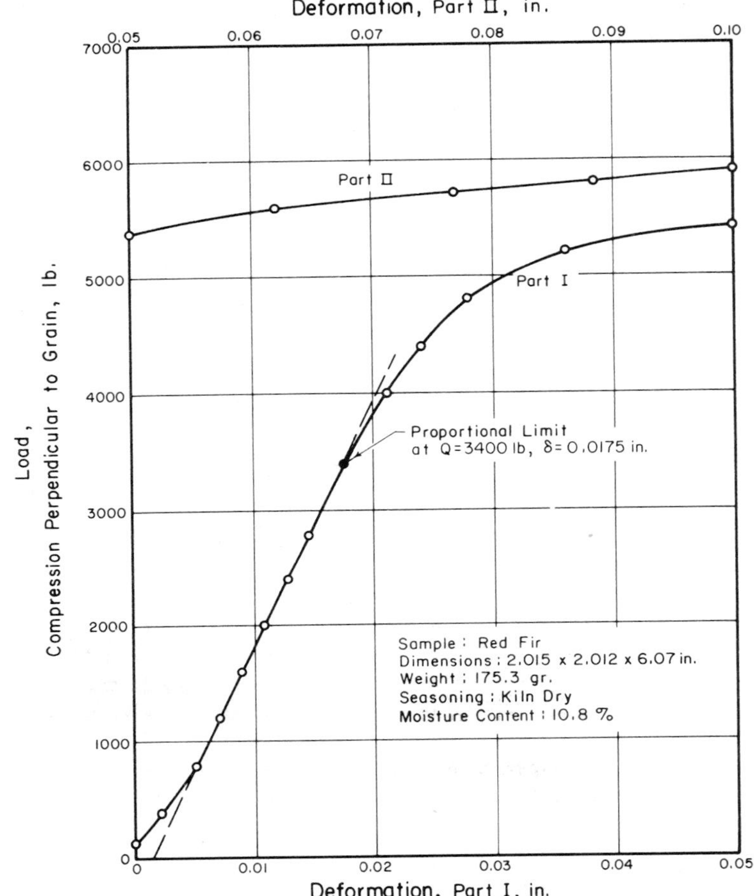

Figure 2.2. Results of a standard test of compression perpendicular to grain. Adapted from Ref. 1.

tively. For the sake of simplified computation it is convenient to use the following version of Hankinson's formula:

$$N = \frac{P}{1 + \left(\frac{P}{Q} - 1\right)\sin^2\theta} \tag{2.2}$$

Examples of use of Hankinson's formula are given in Sections 6.2.2.1 and 6.5.1.

Excellent accuracy may be obtained using a nomograph developed by Scholten, see Graph A.2.3. Hankinson's formula applies not only to strength but also to allowable stresses as well as allowable loads of connections; an example of use of Scholten's nomograph for the latter case is given in the above-referred Graph.

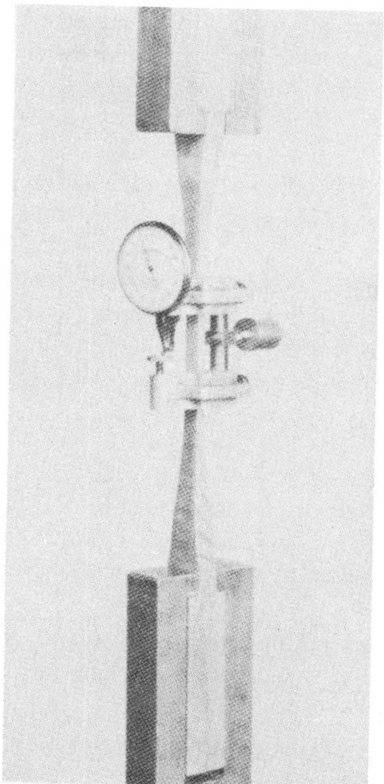

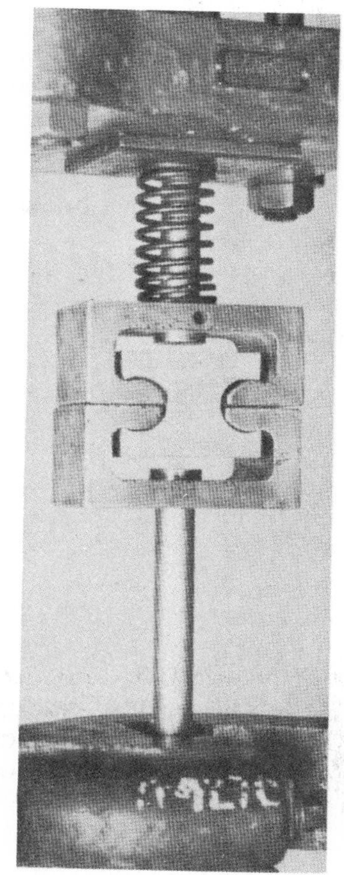

Figure 2.3. Tension-parallel-to-grain test assembly showing grips and use of 2-in. (5-cm) gage length extensometer for measuring deformation. From Ref. 1.

Figure 2.4. Tension-perpendicular-to-grain test assembly. From Ref. 1.

2.2.4 Tension Parallel to Grain

A special specimen and grips are needed, see Fig. 2.3, to test this strength property; in addition, the specimen must be so oriented that the direction of the annual rings at the critical section is normal to the greater cross-sectional dimension.

Test results are shown in the bottom portion of Figs. 7.2 and 7.5. The stress-strain relation for tension parallel to grain is a straight line almost to failure; a slight deviation occurs just before splitting of the fibers. The modulus of elasticity in tension is slightly larger (4) than that obtained in tests of compression parallel to grain; this fact is used in Section 7.3 to explain some inconsistencies of the elastic theory of bending of wood beams.

Only a small amount of data is available for the tensile strength parallel to grain of various species. Recognizing this fact, ASTM's D2555–78, see Ref. 5, recommends that for clear wood strength in tension parallel to grain the clear wood strength for modulus of rupture be used. This is conservative, as shown in Section 7.3; however, the pronounced effect that stress concentrations and slope of grain have upon tensile strength makes it desirable to use conservative values.

2.2.5 Tension Perpendicular to Grain

Seldom are wood structures subjected to tension perpendicular to grain. However, this type of stress may occur in curved members subjected to bending, as in the case of glued-laminated wood arches, see Section 7.14.

Not much data are available, but all tests, see Fig. 2.4, have consistently indicated small strength. This is not surprising when one realizes that the action tends to separate the fibers from each other; only the strength of the lignin, acting as a bonding agent, is available to resist it. It is recommended (5) that tension strength across the grain be taken as only 0.33 times the clear wood strength value for shear parallel to grain.

2.2.6 Static Bending and Modulus of Elasticity

The strength of wood in bending under transverse loads is one of its most important properties because of the major use made in construction of flexural elements for roofs and floors of buildings, and as stringers and girders in bridges. Tests are conducted as shown in Fig. 2.5. Specimens are

Figure 2.5. Static bending test assembly showing method of load application, specimen supported on rollers and laterally adjustable knife edges, and method of measuring deflection at neutral axis by means of yoke and dial attachment. (Adjustable scale mounted on loading head is used to measure increments of deformation beyond the dial capacity.) From Ref. 1.

2 × 2 × 30 in. with an actual 28 in. span; a bearing block of hard maple is used for applying the load.

The results of a typical test are plotted in Fig. 2.6 showing the relation between transverse load and deflection at midspan. The latter are measured by means of a steel yoke and dial attachment to measure only the deflections due to bending; the use of this device results in automatic disregarding of the additional deflection due to local deformation caused by compression across the grain at the supports.

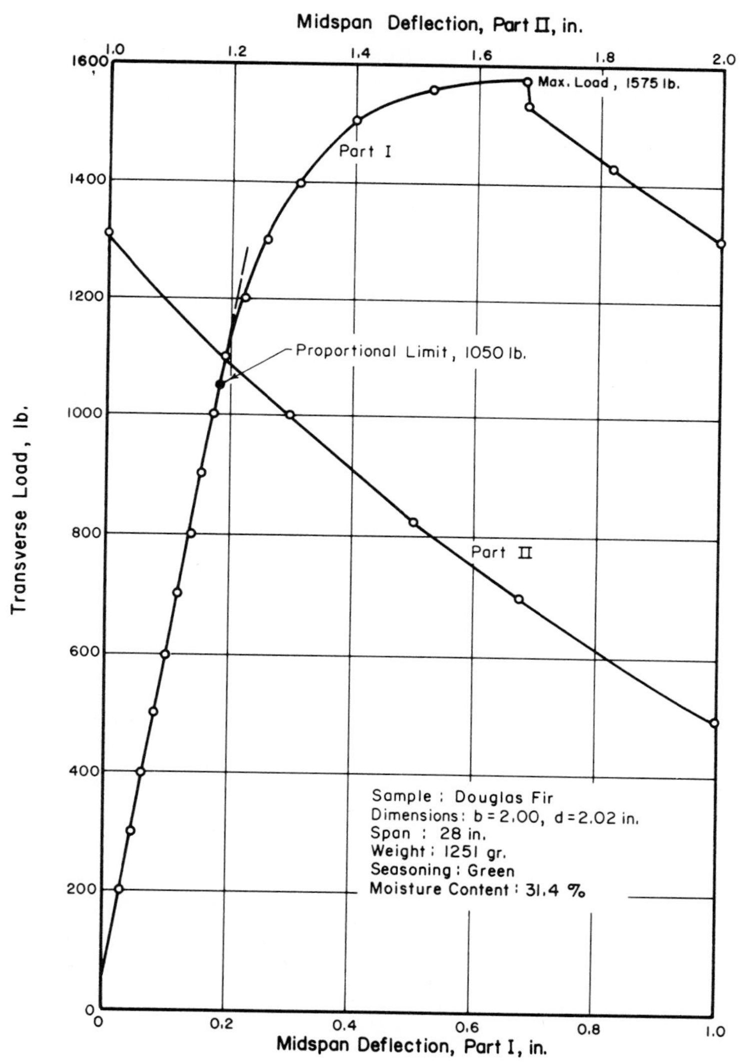

Figure 2.6. Results of a standard test of static bending. Adapted from Ref. 1.

A linear relation between transverse load and midspan deflection is evident for the initial stage of loading. The proportional limit, which marks the end of the linear relation, is clearly shown at a load $P' = 1050$ lb and deflection $\delta = 0.19$ in. The fiber stress at proportional limit is obtained as $(P'L/4)/(bd^2/6)$, where b and d are the width and depth of the section, respectively. The non-linear behavior beyond the proportional limit is continued until the maximum load, $P_u = 1575$ lb, is achieved at $\delta_u = 0.66$ in. deflection. The modulus of rupture, R, is obtained as $(P_uL/4)/(bd^2/6)$. R is a fictitious ultimate stress, see Section 7.3; it is nevertheless used as a good measure of strength. The descending portion of the curve is unstable since the rate of change of load with deformation is negative, i.e., $dP/d\delta < 0$. This is only possible because testing machines apply deformations rather than loads; under gravity loads in actual structures, collapse would occur when the above maximum load is reached. Refer to Sections 7.2 through 7.5 for a thorough discussion of bending of wood beams.

It is possible to determine the apparent modulus of elasticity in bending, E, using the elastic relation between P′ and δ′, where the prime denotes values at proportional limit. For the standard tests shown in Fig. 2.5, the following relation is obtained:

$$E = \frac{P'L^3}{48\ I\delta'} \tag{2.3}$$

where $I = bd^3/12$ is the moment of inertia of the section. The value of E obtained from Eq. 2.3 is smaller than the actual value since δ′ is affected by the deflections due to shear, which are not of negligible importance in wood beams.

The true modulus of elasticity in bending, E′, can be obtained with a variation of the standard test using two concentrated loads, rather than one, as shown in Fig. 2.7. The deflection Δ′ in the

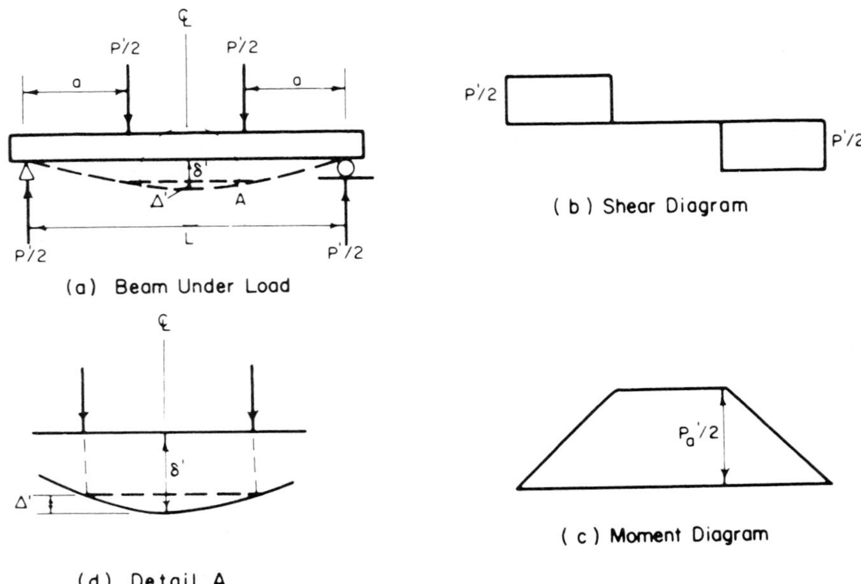

Figure 2.7. Determination of true modulus of elasticity in bending.

span L-2a, see Fig. 2.7d, may be measured directly using the yoke arrangement described before; the supports of the yoke are under the loads. Using elastic analysis it is easy to show that

$$\Delta' = \frac{P'a(L-2a)^2}{16\ E'I} \tag{2.4}$$

where all terms have been defined before and from which the value of E′ may be obtained directly as follows:

$$E' = \frac{P'a(L-2a)^2}{16\ I\Delta'} \tag{2.5}$$

Evidently, the intent of this determination is to disregard the deflections due to shear in the outer portions of the beam. The results can be checked using the total deflection at midspan, δ'. It is easy to show that the value of δ' is given by:

$$\delta' = \frac{P'a}{48\ E'I}(3L^2-4a^2) + \frac{P'a}{2AG}\kappa \tag{2.6}$$

where the first term is the deflection contribution of moment and the second term that of shear; G is the shear modulus of wood, which may be taken as $E'/16$; κ is a constant which for rectangular sections is equal to 6/5 ($\kappa = 10/9$ for circular sections, see Ref. 6); and $A = bd$ is the cross-sectional area of the beam. From Eq. 2.6 it follows that

$$E' = \frac{P'a(3L^2 - 4a^2)}{4bd^3\delta'\left(1 - \dfrac{3P'a}{5bdG\delta'}\right)} \tag{2.7}$$

The value of E′ given by either Eq. 2.5 or Eq. 2.7 is the true modulus of elasticity of wood in bending; it follows that $E' > E$.

Only apparent moduli of elasticity are listed in Tables A.2.1 and A.2.2, as well as in separate Table A.2.5 for design purposes. This allows direct determination of deflection of wood beams using the same simple expressions that are used for steel and concrete structures. The allowance for deflections due to shear in wood beams is in the form of a reduced modulus of elasticity; however, this is only an approximate procedure and for some cases, such as short solid beams or plywood-lumber box or I beams under concentrated transverse loads, it may be necessary to determine the real deflections as given by Eq. 2.6, using the true value of $E' \simeq 1.1E$ for deflections due to bending and $G \simeq E'/16$ for deflections due to shear. A compilation of useful expressions for the determination of the latter is given in Ref. 6 and Fig. 7.23.

2.2.7 Shear Parallel to Grain and Shear Across the Grain

The standard test for the determination of this strength property uses a 2 × 2 × 2½ in. specimen with a ¾ × ½ × 2 in. notch. The load is applied in such a way as to shear the 2 × 2 × ¾ in. step from the rest of the piece; only the maximum load is recorded. The shearing strength is obtained as the ratio of the maximum load, P_u, and the cross-sectional area, bd. For example, in a test of Pacific silver fir (5) it was determined that $P_u = 2770$ lb., $b > 2.016$ in., $d = 2.000$ in.; the shearing strength is $2770/2.016 \times 2.000 = 687$ psi. Tables A.2.1 and A.2.2 give values of clear-wood shear strength for U.S. Commercial woods.

Structural interest in this strength property stems from the fact that shear parallel to grain, known also as horizontal shear, is usually generated in members subjected to bending, see Fig. 2.8.a. Shear parallel to grain occurs in longitudinal-tangential (L-T) or longitudinal-radial (L-R) planes, see Fig. 2.1, with shear stresses parallel to the direction of the fibers. Simultaneously, however, shear stresses of equal intensity are generated across the grain, in R-T planes. Because the shear strength of wood is much larger across the grain than along the grain, a structural designer is seldom concerned with shear strength of wood across the grain. Thus, for design of wood beams, attention is focused only on the actual capacity to resist horizontal shear stresses. A thorough discussion of shear strength of wood beams of structural size may be found in Sections 7.7 and 7.8.

2.2.8 Rolling Shear

Shear stresses that occur in longitudinal-tangential (L-T) or longitudinal-radial (L-R) planes, see Fig. 2.1, and are normal to the direction of the grain, are known as rolling shear stresses. These stresses occur most commonly in plywood panels used in bending, as in the case of floor sheathing, see Fig. 2.8.c. Plywood panels are composed of glued thin plies laid in such a way that the direction of the fibers are at right angles to each other in alternate plies, see Chapter 4. In those plies in which the grain is at 90° with the span, shear stresses are generated, see Fig. 2.8d, which tend to roll one fiber over the other, thus justifying the descriptive title of rolling shear stresses by which they are normally recognized.

The strength of wood against rolling shear stresses is much smaller than the strength against shear stresses parallel to grain. In limited tests in solid wood rolling shear strengths were 10 to 20 percent of the parallel-to-grain shear values (10). Rolling shear strengths were about the same in the L-R and L-T planes.

The low strength of wood against rolling shear stresses is reflected in the design specifications for the industry; it is recommended that the allowable shear stress in rolling shear, i.e., in the

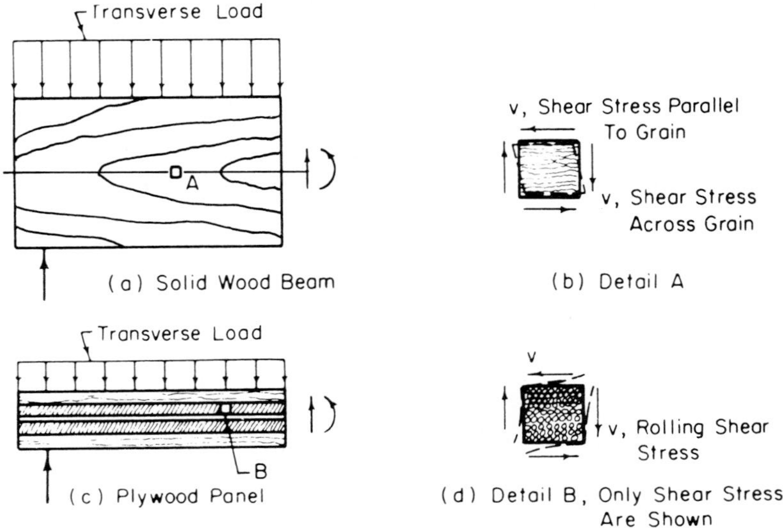

Figure 2.8. Various types of shear stresses affecting wood members.

plane of plies, be taken as 30- percent of the allowable stress in shear in a plane perpendicular to the plies for the lowest species group in the panel, see Table A.4.10. However, it should be mentioned that in actual practice a condition such as shown in Fig. 2.8.c generates small rolling shear stresses, which seldom govern design.

Examples of cases where rolling shear stresses are important and must be checked, occur in the design of plywood stressed-skin panels (for roof or floor decking) see Section 10.3.1, and plywood box beams, see Section 10.3.2.

A standard for testing plywood in rolling shear is given in Ref. 8.

2.2.9 Elastic Properites of Clear Wood

Recognition of wood as an orthotropic elastic material, for analysis of curved members and three-dimensional stress analysis, requires determination of elastic characteristics in its three principal directions, see Figs. 1.2 or 2.1. These directions are longitudinal (L), tangential (T) and radial (R).

Orthotropic elastic theory shows that twelve parameters are needed to describe the elastic behavior of which only nine are independent. Three moduli of elasticity (E_L, E_T, E_R), three moduli of rigidity (G_{LR}, G_{LT}, G_{RT}) and six Poisson's ratios (μ_{LR}, μ_{LT}, μ_{RT}, μ_{RL}, μ_{TR}, μ_{TL}) are needed.

These parameters vary not only between wood species but also within any given species. In addition, they vary with moisture content of the wood and specific gravity. For the purpose of illustration, values of the 12 elastic constants are given in Table 2.1 for seven wood species. Prediction equations to estimate the values of the elastic constants have been obtained by Bodig and Goodman and are given in Ref. 18.

2.3 VARIABILITY OF STRENGTH PROPERTIES OF CLEAR WOOD

The strength properties given in Tables A.2.1 and A.2.2 were obtained using the standard methods of testing discussed in the preceding Section. For any species listed, these properties are the average values of individual tests. Since wood is a natural material which is subject to numerous influences such as moisture, soil conditions and growing space, it may show a decided variation in properties even in clear wood. If a frequency distribution curve is used to represent the variation of a given strength property with number of tests, a curve such as shown in Fig. 2.9 is obtained, where the dots show the sample distribution; it seems reasonable to assume that the Gaussian distribution can be used to determine probabilities. From the sample data, we can compute the sample average $\overline{X}$ and sample variance σ^2, using the following equations:

$$\overline{X} = \frac{\sum_{1}^{n} X_i}{n} \tag{2.8}$$

$$\sigma^2 = \frac{\sum_{1}^{n} (X_i - \overline{X})^2}{n} \tag{2.9}$$

where X_i are individual observations and n is the number of possible observations. These values can be used to model the actual mean and variance in the Gaussian distribution.

TABLE 2.1. ELASTIC CONSTANTS OF VARIOUS SPECIES†

Species	Approximate specific gravity[1]	Approximate moisture content (pct.)	Modulus of elasticity ratios		Ratio of modulus of rigidity to modulus of elasticity			Poisson's ratios					
			E_T/E_L	E_R/E_L	G_{LR}/E_L	G_{LT}/E_L	G_{RT}/E_L	μ_{LR}	μ_{LT}	μ_{RT}	μ_{RL}	μ_{TR}	μ_{TL}
Balsa----------------	0.13	9	0.015	0.046	0.054	0.037	0.005	0.229	0.488	0.665	0.217	0.011	0.007
Birch, yellow ------	.64	13	.050	.078	.074	.068	.017	.426	.451	.697	.447	.033	.023
Douglas-fir---------	.50	12	.050	.068	.064	.078	.007	.292	.449	.390	.287	.020	.022
Spruce, Sitka------	.38	12	.043	.078	.064	.061	.003	.372	.467	.435	.240	.029	.020
Sweetgum----------	.53	11	.050	.115	.089	.061	.021	.325	.403	.682	.297	.037	.020
Walnut, black-----	.59	11	.056	.106	.085	.062	.021	.495	.632	.718	.379	.052	.035
Yellow-poplar-----	.38	11	.043	.092	.075	.069	.011	.318	.392	.703	.329	.029	.017

1. Based on ovendry weight and volume at the moisture content shown.

†From Table 4.1 of Ref. 10

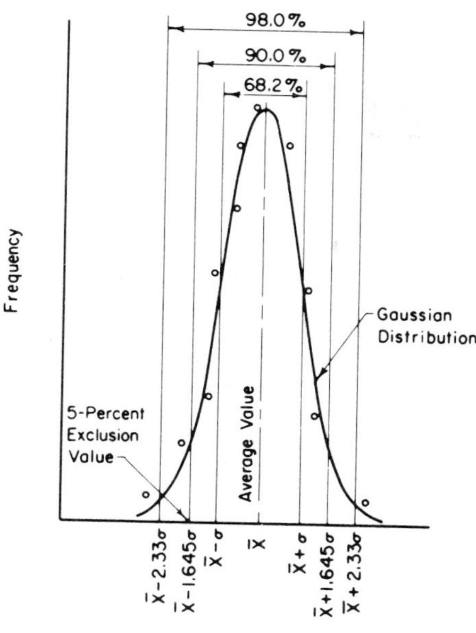

Figure 2.9. Frequency distribution curve of a given strength property of wood.

The average value $\overline{X}$ is the best single estimate of the true population mean. It can be shown that the range $\overline{X} - \sigma$ to $\overline{X} + \sigma$ contains 68.2 percent of the results of the tests, see Fig. 2.9; in other words, 68.2 percent is the probability that any given test will fall in this range. From another point of view, only (½)(100−68.2) = 15.9 of every 100 tests will have a strength smaller than $\overline{X} - \sigma$. Along the same line of reasoning only 5 percent of the test results will fall below $\overline{X} - 1.645\sigma$, and only 1 percent will fall below $\overline{X} - 2.33\sigma$. The value $\overline{X} - 1.645\sigma$ is known as the 5 percent exclusion value (9) and is extensively used in the standard method of establishing clear wood strength values (5), see Section 2.4.

The concept of coefficient of variation is used frequently; it is defined by

$$v = \frac{\sigma}{\overline{X}} \tag{2.10}$$

and since it is dimensionless, it facilitates comparisons between various strength properties. A list of coefficients of variation for a number of strength properties is given in Ref. 10; from this list ASTM Standard D 2555–78, see Ref. 5, extracted the following values: 1) 0.16 for modulus of rupture; 2) 0.22 for modulus of elasticity; 3) 0.18 for maximum crushing strength parallel to grain, 4) 0.14 for maximum shear strength; 5) 0.28 for compression perpendicular to grain strength; and 6) 0.10 for specific gravity.

The use of these concepts in the determination of clear wood strength values is discussed in the following Section.

2.4 DETERMINATION OF CLEAR WOOD STRENGTH VALUES OF COMMERCIAL SPECIES

Before safe working stresses for wood, glued-laminated wood, plywood and other wood products can be determined it is necessary to have clear wood strength properties of all commercially important species. ASTM Designation D 2555–78, see Ref. 5, was established to meet these needs and to provide, in addition, information on factors for consideration in the adjustment of clear wood strength properties to working stresses for design.

Two methods, known as A and B, are presently used for establishing tables of clear wood strength properties for different species and regional subdivisions. Method A provides for the use of the results of surveys of wood density involving extensive sampling of forest trees, in combination with the data obtained from standard strength tests made in accordance with ASTM Designation D 143, see Ref. 1. The average strength properties are obtained from wood density survey data through linear regression equations establishing their relation to specific gravity. Density surveys have been completed only for a limited number of species, namely Douglas fir, White fir, California red fir, Grand fir, Pacific Silver fir, Noble fir, Western hemlock, Western larch, Black cottonwood and Southern pine (see Table A.2.1). Data are not currently (1980) available for the use of method A for all commercial species.

Method B provides for the establishment of tables of strength values, see Table A.2.2, based on standard tests of small clear specimens in the unseasoned condition for use when data from density surveys are not available. A thorough description of method B is given in Ref. 5 and 17. For the purposes of this work, it is considered that discussion of method B is best accomplished by means of the following example.

It is wished to determine the modulus of rupture, R, a clear wood strength property of Southern pine using method B. Southern pine has been selected because it allows application of the general procedure used when administrative and marketing considerations make it necessary or desirable to combine basic groups having relatively similar properties into a single marketing combination. In this case, Longleaf, Slash, Shortleaf and Loblolly pine are marketed as Southern pine. When this occurs, equitable treatment for each species in the combination is assured by using weighting factors, β, based on the standing timber volume of the combination; Table 2.2 gives the volume-weighting factors for this particular example.

TABLE 2.2.* DETERMINATION OF VOLUME-WEIGHTING FACTORS FOR FOUR MAJOR SOUTHERN PINE SPECIES

	Volume 10^6 cu. ft.	Volume-Weighting Factor
Longleaf	5,534	0.1016
Slash	5,017	0.0921
Shortleaf	16,328	0.2996
Loblolly	27,610	0.5067
	⩽=54,489	⩽=1.0000

*Data taken from Table 4 "Standing Timber Volume for Commercially Important Species Grown in the United States," of Ref. 5.

TABLE 2.3.* DETERMINATION OF MODULUS OF RUPTURE, R OF SOUTHERN PINE COEFFICIENT OF VARIATION, v = 0.16†

Species	Average R, $\overline{R}$, ksi	Std. Dev., $\sigma = v\overline{R}$, ksi	5% Exc. Val. = $\overline{R}$-1.645σ, ksi	Vol. weighting Factor, β	Weighted $\overline{R} = \overline{R}\beta$ ksi	Weighted Exc. Val. = ($\overline{R}$-1.645σ) β, ksi	CDF = ($\overline{R}$-A)/σ
Longleaf	8.670	1.387	6.388	0.1016	0.881	0.649	2.17
Slash	8.570	1.371	6.315	0.0921	0.789	0.582	2.12
Loblolly	7.340	1.174	5.409	0.5067	3.719	2.740	1.43
Shortleaf	7.650	1.224	5.637	0.2996	2.292	1.689	1.63
				Weighted Mean =	7.681	A = 5.660	

The lowest CDF = 1.43 for Loblolly is less than 1.48, the minimum allowable value. Thus, the exclusion value assigned to the combination is not the weighted "5% Exclusion Value" A = 5.660 ksi, but may not exceed $\overline{R}$ − 1.48σ = 7.340 − 1.48 × 1.174 = 5.602 ksi.

*Adapted from Ref. 5.

†Other coefficient of variations are 0.22 for modulus of elasticity, 0.18 for maximum crushing strength parallel to grain, 0.14 for maximum shear strength, 0.28 for compression across the grain strength, and 0.10 for specific gravity, see Ref. 5.

The solution is given in Table 2.3. The values of the average moduli of rupture, $\overline{R}$ and standard deviations $\sigma = v\overline{R}$, are obtained from Table A.2.2. The volume-weighted value of $\overline{R}$ is given by the sum of the various $\overline{R}\beta$. Thus, from Table 2.3, the weighted mean value of modulus of rupture is 7.681 ksi. Use of the 5 percent exclusion value for each species, $\overline{R} - 1.645\sigma$, guarantees that only 5 percent of the pieces may have a smaller value of the given strength property, see Fig. 2.9. In the case of a combination of species, an approximate value for the 5 percent exclusion limit can be obtained by computing the volume-weighted average 5 percent exclusion value, $A = \Sigma(\overline{R} - 1.645\sigma)\beta$, for all included species. From Table 2.3, A = 5.660 ksi; only 5 percent of the area under the combined frequency distribution will fall below the value of A. However, before selecting A as the modulus of rupture assigned to the combination it is necessary to check that the composite dispersion factor CDF, a statistical quantity defined as $(\overline{R}\text{-}A)/\sigma$ is not less than 1.48. Observation of Table 2.3 discloses that, for the case of Loblolly pine, CDF = 1.43; thus, the 5 percent exclusion value for the combination Southern pine, as shown in the table, is given by $\overline{R}$-1.48σ where $\overline{R}$ and σ correspond to values of Loblolly pine.

2.5 INFLUENCE OF VARIOUS CHARACTERISTICS ON STRENGTH

The presence of slope of grain in a structural member, see Section 1.4, reduces its strength. This is particularly true in the case of tensile strength. Tension stresses in a piece of wood with cross grain, induced by either direct tensile force or bending, is resisted by one component in the direction of grain and another in a direction normal to grain. The latter tends to separate the fibers causing early splitting of a structural member under direct tension and a reduction in modulus of rupture in a member subjected to bending. Compression parallel to grain is also affected by cross grain; however, at a smaller scale. The strength ratio† of wood members with various grain slopes is given in Table 2.4.

TABLE 2.4.* STRENGTH RATIOS CORRESPONDING TO VARIOUS SLOPES OF GRAIN

	Maximum Strength Ratio, percent	
Slope of Grain	Bending or Tension Parallel to Grain	Compression Parallel to Grain
1 in 6	40	56
1 in 8	53	66
1 in 10	61	74
1 in 12	69	82
1 in 14	74	87
1 in 15	76	100
1 in 16	80	
1 in 18	85	
1 in 20	100	

Adapted from Table 1 of Ref. 11.

Knots reduced the strength of wood principally because of interrupting the direction of grain; localized steep slope-of-grain concentrates around knots. No distinction is made between live knots, dead knots and knotholes in stress-grading of lumber, in spite of the fact that live intergrown knots resist some stresses while dead knots and knotholes are of little value. However, so much greater distortion of grain exists around intergrown knots, than around dead knots, that the overall strength effects are roughly equalized (10).

The effect of a knot on strength depends on the proportion of the cross section of the given piece occupied by the knot, and on its relative location in the piece. Limits on knot sizes are made in accordance with the width of the face in which the knot appears. The location of a knot affects bending strength more because the distribution of stresses, see Fig. 7.1, varies along the depth of a beam. A knot located close to the axis will have less effect on strength than one located close to the edge; this is especially true for the edge subjected to tensile stresses, since the effect of knots on strength is greater in tension than in compression. Knot requirements suitable for tensile stress are applied to both edges. Standard provisions for handling knots during stress-grading are given by ASTM Designation D 245–74, see Ref. 11. Three tables are given for the determination of strength ratio of members ranging in width between 1 and 16 in. of narrow face, and between 2 and 24 in. of wide face; the size of knots varies between ¼ in. and 8½ in. A set of formulas for determining strength ratios corresponding to various knot sizes and width of face is also given in Ref. 11.

Shakes, checks and splits, by their nature, reduce the transverse section necessary to resist horizontal shear, i.e., shear along the grain. The shear strength of members subjected to bending

†See Section 2.11.1.

due to transverse loads, such as joists, planks, beams and stringers is affected by the presence of these defects, see Section 7.8. The strength of columns that are subjected mainly to compression due to longitudinal loads of small eccentricity, and not to transverse loads, is not affected by these defects. Again, standard provisions accounting for the presence of shakes, checks and splits in grading are given in Ref. 11.

2.6 EFFECT OF SPECIFIC GRAVITY ON STRENGTH

The specific gravity of wood gives a reliable measure of the amount of wood substance in any given wood. Since strength properties depend greatly on the actual amount of wood present, it is easy to see why useful relations can be established between strength properties and specific gravity for any given species. As an example, Fig. 2.10 shows the variation with specific gravity of fiber

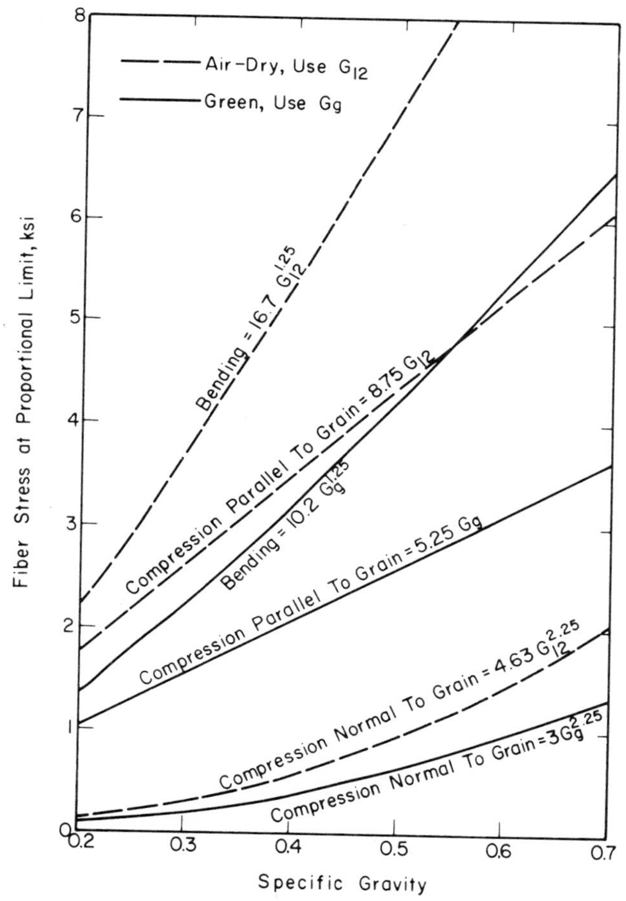

Figure 2.10. Variation of fiber stress at proportional limit with specific gravity of wood for air-dry and green conditions. From data given in Ref. 10.

stress at proportional limit in bending and compression parallel and normal to grain. The relations are given for green wood and air-dry wood at M = 12%, and were obtained at the Forest Products Laboratory (10) from average results of strength tests of more than 160 species. Observation of Fig. 2.10 shows that the given strength property increases with specific gravity for both green and air-dry conditions, and also that it is larger for air-dry wood than for green wood.

Other relations between various strength properties and specific gravity are given in Table 2.5. The user of these expressions should keep in mind that strength values within a species may vary somewhat from the average strength values of different species. Thus, the relations given in Fig. 2.10 and in Table 2.5 might be best regarded as a qualitative indication of trend rather than as a reliable source of information on strength of any given wood.

TABLE 2.5. FUNCTIONS RELATING MECHANICAL PROPERTIES TO SPECIFIC GRAVITY OF CLEAR, STRAIGHT-GRAINED WOOD†

Property		Specific gravity-strength relation[1]	
		Green wood	Air-dry wood (12 pct. moisture content)
Static bending:			
Fiber stress at proportional limit	p.s.i.	$10{,}200G^{1.25}$	$16{,}700G^{1.25}$
Modulus of elasticity	million p.s.i.	$2.36G$	$2.80G$
Modulus of rupture	p.s.i.	$17{,}600G^{1.25}$	$25{,}700G^{1.25}$
Work to maximum load	in.-lb. per cu. in.	$35.6G^{1.75}$	$32.4G^{1.75}$
Total work	in.-lb. per cu. in.	$103G^{2}$	$72.7G^{2}$
Impact bending, height of drop causing complete failure	in.	$114G^{1.75}$	$94.6G^{1.75}$
Compression parallel to grain:			
Fiber stress at proportional limit	p.s.i.	$5{,}250G$	$8{,}750G$
Modulus of elasticity	million p.s.i.	$2.91G$	$3.38G$
Maximum crushing strength	p.s.i.	$6{,}730G$	$12{,}200G$
Compression perpendicular to grain, fiber stress at proportional limit	p.s.i.	$3{,}000G^{2.25}$	$4{,}630G^{2.25}$
Hardness:			
End	lb.	$3{,}740G^{2.25}$	$4{,}800G^{2.25}$
Side	lb.	$3{,}420G^{2.25}$	$3{,}770G^{2.25}$

[1]The properties and values should be read as equations; for example: modulus of rupture for green wood = $17{,}600G^{1.25}$, where *G* represents the specific gravity of ovendry wood, based on the volume at the moisture condition indicated.

†From Table 4.9 of Ref. 10

2.7 EFFECT OF MOISTURE CONTENT ON STRENGTH

The effects of specific gravity, G, on strength, indicate that the strength of wood increases with specific gravity. For any given piece of wood, specific gravity depends on volume which, in turn, is a function of the moisture content, M, see Fig. 1.5. Changes in moisture content affect specific gravity and thus, various strength properties of wood. For example, the drying-out process from a green condition to M = 12% in specimens of clear wood, may double the strength in compression parallel to grain (10); drying out to M = 5% may triple the strength. Actual changes in strength properties do not begin, however, until M is less than that corresponding to the fiber-saturation point. Obviously, this results from the fact that volume and specific gravity are constant during the drying-out process until the fiber-saturation point is reached, at approximately M = 30%. Below this moisture content, as volume of wood decreases, the specific gravity and strength properties of wood increase.

A numerical example of evaluation of change in strength properties with moisture content is given in Fig. 2.11. A linear relation is shown between certain strength property, plotted in a logarithmic scale, and moisture content in the range $0 < M < M_p$. The actual variation in a small range about M_p is also shown by a dash curve. For the range $M > M_p$, strength remains practically constant, as shown by the horizontal line extending to the green wood condition. The value of M_p has been obtained (10) in the laboratory by the intersection of the sloping line with the horizontal line. For most species, M_p = 25 percent; however, low and high values of M_p are 21 percent (Loblolly pine) and 28 percent (Western hemlock) respectively.

Strength properties such as toughness and shock resistance, which depend on the work done by the load during deformation, or the energy absorbed by the member, do not increase with a decrease in moisture content. In fact, these strength properties decrease, since dried wood will not bend as far as green wood before failure, although it will take a greater load.

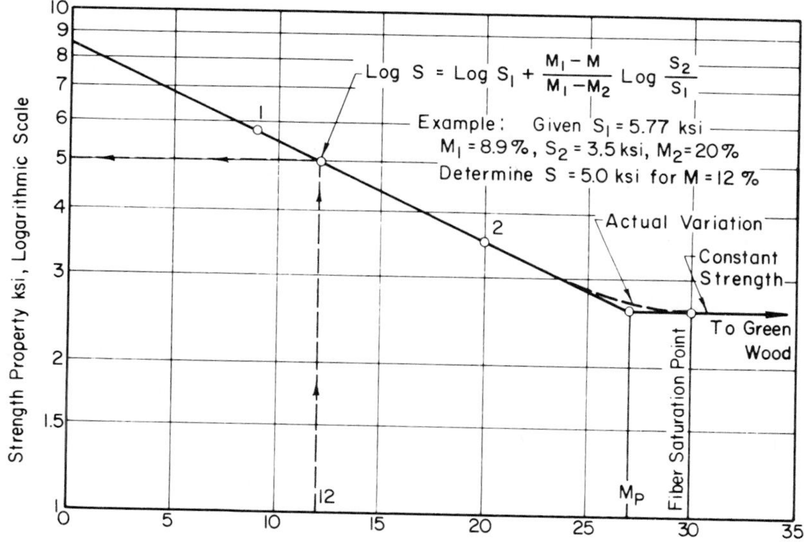

Figure 2.11. Variation of strength with moisture content. Data taken from Ref. 10.

2.8 EFFECT OF TEMPERATURE ON STRENGTH

Strength properties of wood are sensitive to variations in temperature. In general, an inverse relation exists between temperature and strength properties; as temperature increases strength decreases, and when temperature decreases, the strength of wood increases. Under ordinary atmospheric conditions, wood exposed for a limited amount of time to moderate increases in temperatures above normal can be expected to recover all of its original strength when the temperature is reduced to normal (10); tests of air-dry wood exposed to 105°F for a year assert this statement.

The effects of temperature on strength properties depend on the moisture content of the wood; changes are greater as moisture content increases. In the range of 0° to 150°F it is estimated that, for dry wood (M = 12%) an increase or decrease in the strength at 70°F of about ⅓ to ½ percent occurs for each 1°F decrease or increase in temperature, respectively.

Exposure of wood to high temperature for long or permanent periods of time has a definite detrimental influence on strength properties. In ordinary structural service, such as in buildings or bridges, one can expect that wood will seldom be subjected to exposure to high temperatures for a prolonged time. Thus, these effects are considered beyond the scope of this text; for further reference see the Wood Handbook (10).

2.9 EFFECT OF DURATION OF LOADING ON STRENGTH

Consider the case of identical wood specimens loaded with large sustained loads of different values. Failure occurs at different times; the greater the load, the faster failure of the specimen takes place. Also, below a certain load, the specimens do not fail no matter how long the duration of the load. These tests leave no doubt about the influence of the duration of loading on strength. If the results of these tests are plotted, using strength as ordinates and time-to-failure as abscissas, a curve such as shown in Fig. 2.12 is obtained. The asymptotic nature of the curve indicates that, although strength is reduced with duration of loading, a minimum strength exists which is independent of time. This is known as sustained strength; physically it gives the limiting value of stress or load which the specimen can sustain indefinitely without failing.

The difference in behavior between a specimen loaded to $\sigma < \sigma_{sust}$, case I, and a specimen loaded to $\sigma > \sigma_{sust}$, case II, is illustrated in Fig. 2.13. For case I the deformation increases with time beyond the instant, elastic deformation. The increase, however, takes place at a reduced rate of change with time; in other words, in the course of time the deformation approaches a certain limit. For case II deformations increase constantly with time; the behavior may be studied using the rate of change of deformation with time, i.e., the slope of the tangent to the curve, $d\delta/dt$. As shown in Fig. 2.13 $d\delta/dt$ is reduced in portion CD of the curve, remains constant at a small value in portion DE and attains very large values beyond E. Failure occurs rapidly once the accelerated growth of deformations begins. This phenomenon has application in the study of the safety of heavily stressed timbers (12). A deformation continuing to increase, but at a decreasing rate, even after a long period of time does not presage failure. On the other hand, deformation that continues to increase at a uniform rate may be a danger signal, and when the rate of change accelerates, failure may be imminent.

Loads acting on structures are not all sustained indefinitely. As a matter of fact, only the self-weight of the structure and other similar weights are permanent loads. All other actions such as produced by wind, live load, snow, earthquake and impact are repeated loads, that are applied

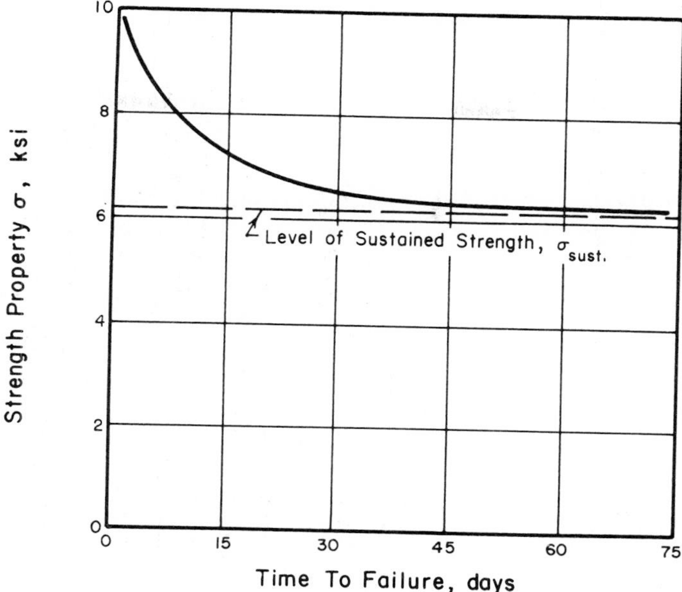

Figure 2.12. Variation of strength with duration of loading.

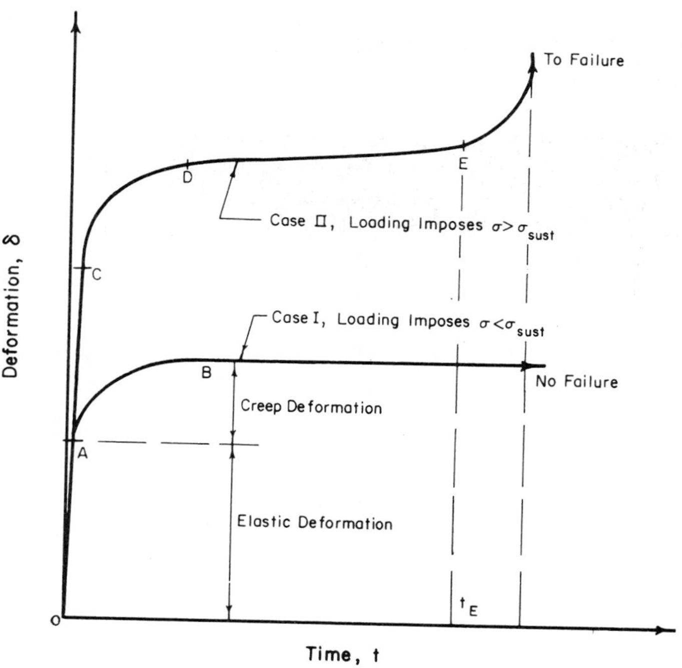

Figure 2.13. Variation of deformation with time for two identical specimens loaded to different stress levels.

for certain periods and are reduced in part, or altogether, at other periods. Long-time loading tests have shown that these loads have a cumulative effect such that the sum of all loaded periods before failure will at least equal the duration of sustained load at the same stress level (12). For design purposes (2) the total duration of the repeated loads is estimated as: 10 years for live load, 2 months for snow load, 7 days for temporary construction loads, 1 day for wind or earthquake load and 1 second for impact loads.

Based on the existing evidence of variation of strength with duration of loading, strength properties determined in tests that last usually from 6 to 8 minutes can be converted to other durations of loading. Strength properties for the so-called normal loading conditions* may be determined by multiplying standard strength properties by the factor 1/1.6; the same can be done for other loading conditions using corresponding factors. Since present design of wood structures is based on service load conditions and not on ultimate, the conversion of strength properties for different durations of loading has seldom any practical significance. However, the concept remains useful for the determination of allowable stresses for different loading conditions as shown in Section 2.13.

2.10 EFFECT OF FATIGUE ON STRENGTH

A wood structure may be subjected to many cycles of loading and unloading of the live load during its lifetime. Some consideration should be given to possible failure by fatigue when the repetitions may be more than 100,000. Experience has shown that the possibility of fatigue failures of wood members is generally less than in those constructed of other materials (12). The fatigue strength for large number of repetitions of stress of fibrous materials, such as wood, is a higher proportion of the static strength than that of crystalline materials like metals. Thus, actual stresses in a wood structure are usually lower than the level of stress which, if repeated, would produce failure during the normal service life of the structure.

Tests performed on Douglas fir and oak in tension parallel to grain for a ratio $\sigma_{min}/\sigma_{max} = 0.10$ (where σ_{min} and σ_{max} are the minimum and maximum value of the repeated stress, respectively) have shown that the fatigue strength, for 30 million cycles, is about 50 percent of the ultimate static strength. A summary of reported results of 30 million-cycle fatigue studies on clear wood specimens is given in Table 2.6. A similar summary is available for 2 million-cycle fatigue studies of bending of wood specimens with small knots (strength ratio between 50 and 90 percent), specimens with 1:12 slope of grain, and specimens with both knots and slope of grain, see Table 2.7. The influence of these strength-reducing characteristics in the fatigue life of the specimens is great; when knots and slope of grain were both present, the specimens had approximately 30 percent of their static strength. Compare this to clear specimens of straight-grained wood which after 2 million cycles of bending still generated 60 percent of the strength of similar specimens tested under static conditions (10).

It is generally considered that the failure danger of wood structures from fatigue in service is minor. It has been concluded that, if maximum repeated stresses do not exceed the proportional limit in bending, fatigue failures due to repeated loads are unlikely. Further research in this area is needed.

*Normal loading conditions assume continuous or cumulative duration of maximum design load of not more than 10 years during the life of a structure and/or 90 percent of this load continuously for the rest of the life of the structure.

TABLE 2.6. A SUMMARY OF REPORTED RESULTS OF FATIGUE STUDIES[1]†

Loading	Conditions	Range ratio (minimum stress ÷ maximum stress)	Fatigue life (million cycles)	Fatigue strength (percent of strength from static test)
Tension parallel to grain...	Clear, air dry	0.10	30	50
Cantilever bending........	Clear, air-dry, solid wood	—1.00	30	30
Simple beam bending	Clear, green	.10	30	60
Rotational bending	Clear, air dry	—1.00	30	28

1. Results from Forest Products Laboratory studies except for rotational bending results (from Fuller, F. B., and Oberg, T. T. 1943. Fatigue Characteristics of Natural and Resin-Impregnated Compressed Laminated Woods. J. Aero. Sci. 10(3):81–85.)

†From Table 4.8 of Ref. 10

TABLE 2.7. FATIGUE OF SPECIMENS WITH KNOTS AND CROSS GRAIN†.

Strength	Specimen	Index of Strength (Pct.)
Static...........................	Clear, straight grained	100
Fatigue (2×10^6 cycles)	Clear, straight grained	60
Do	Small knots, straight grained	50
Do	Clear, 1:12 slope of grain	45
Do	Small knots, 1:12 slope of grain....	30

†From Ref. 10

2.11 GRADING OF STRUCTURAL LUMBER

A requirement for orderly marketing of lumber is the establishment of grades to allow the procurement of any specified quality of lumber in any desired quantity. Structural grades are established in relation to strength properties. They include the effects on strength of: 1) variability of clear wood; 2) specific gravity; 3) seasoning; 4) characteristics such as knots, cross grain and others; 5) duration of stress; and 6) temperature.

Two methods exist for stress-grading structural lumber, namely visual and machine stress-grading. In the latter method each piece of lumber passes through a machine that imposes a transverse deformation on it that allows measurement of the stiffness, EI, of the piece. Since the dimensions are known, the moment of inertia I is available; thus, the modulus of elasticity may be derived easily. In addition, each piece receives a visual inspection for those features that are important for its end usage. The grading machines are adjusted to meet the stiffness requirement and bending fiber-stress characteristics of any given grade. In 1971 machine stress-grading had been in use for about 9 years (13); however, while virtually all U.S. mills use the visual grading technique, only about 12 mills were using machine stress-grading. An increased use of the method is expected in the future.

Visual grading is accomplished from an examination of all four faces and the ends of the piece; the location, size and nature of knots and other characteristics appearing on the surface are evaluated over the entire length. In the following sections, the principles and application of visual stress-grading are discussed in detail.

2.11.1 Strength Ratio

The concept of strength ratio has been created in connection with visual grading of structural lumber. It is defined, for any given piece of lumber, as the hypothetical ratio of its strength to that which it would have if no weakening characteristics were present. Thus, a piece with a strength ratio of 70 percent would be expected to have 70 percent of the strength of the clear piece.

Various rules and tables (11) are given to visual graders for the determination of strength ratios. Using these tables, for instance, stress graders may determine the maximum size of knots and cross grain that a beam with actual section dimensions 7.5 × 15.5 in. may have in a grade with a strength ratio of 70 percent. The following is obtained: 1) maximum size of knot: 2⅛ in. in the 7.5 in. face and 4¼ in. in the 15.5 in. face; and 2) maximum slope of grain: 1 in 12.

2.11.2 Use Classification of Structural Lumber

Visual grading requires evaluation of various factors affecting strength, the effects of which depend on the kind of loading and major stress to which the piece will be subjected. Thus, it is necessary to classify structural lumber according to its size and use. According to ASTM Standard Designation D 245–74, see Ref. 11, four classes are widely used as follows:

Structural Joists and Planks. Pieces of rectangular cross section, 2 to 4 in. in least dimension, graded primarily for strength in bending edgewise or flatwise but also frequently used where tensile or compressive strength is important. Lumber 2 in. in nominal thickness is often placed in grades separate from the thicker joists and planks.

Beams and Stringers. Pieces of rectangular cross section, 5 by 8 in. (nominal dimensions) and larger, graded for strength in bending when loaded on the narrow face.

Posts and Timbers. Pieces of square or nearly square cross section, 5 by 5 in. (nominal dimensions) and larger, graded primarily for use as posts or columns.

Structural Boards. Lumber less than 2 in. (nominal) in thickness and of any width, graded primarily for use where the principal stresses are in axial compression or tension.

The principles of stress-grading permit the assignment of any kind of allowable property to any of the classes of structural lumber, whether graded primarily for that property or not. Thus, use of the various classes of structural lumber, for purposes other than their titles may indicate, is acceptable. There is need for compressive and tensile values for joists and planks if used in the chord or web members of timber trusses. Post and timbers may give services as beams, or be subjected to combined action of axial load and bending. Of course some loss of efficiency results from grading for such general use; however, it offers the advantage of a more simple system of stress grades.

2.11.3 Stress-Grading Agencies

Stress-grading structural lumber found in the U.S. market is graded by agencies certified by the American Lumber Standards Committee and meets the requirements of Standard PS20–70, see Ref. 16. Various regional agencies* such as Southern Pine Inspection Bureau in Florida, West Coast Lumber Inspection Bureau in California, Northeastern Lumber Manufacturers Association in Maine and others, publish rule books that are used for visual grading of structural lumber. Stress-graded lumber is properly marked with the grademark of the certifying agency.

As a result of stress grading it is possible to determine allowable stresses for use in structural design. Such a list of allowable stresses for stress-graded wood is published by National Forest Products Association (2) and appears in Appendix Tables A.2.5.

2.12 PRINCIPLES USED IN THE DETERMINATION OF ALLOWABLE STRESSES

An evaluation of the strength properties of clear wood of any given species, see Section 2.4, is necessary for the determination of allowable stresses of green lumber. The latter are based on normal loading duration and the assumption that each member carries its own load; provisions for repetitive-member use are given in Section 7.6.

Allowable stresses are given in a separate Table A.2.5 for the various U.S. commercial species. The following are listed: 1) extreme fiber in bending, F_b; 2) tension parallel to grain, F_t; 3) horizontal shear, F_v; 4) compression perpendicular to grain, $F_{c\perp}$; 5) compression parallel to grain, F_c; and 6) modulus of elasticity, E. Only the allowable modulus of elasticity is intended to be an average for the species group and other stress grades; the allowable stresses are intended to be less than the stress permissible for 95 percent of the pieces in a species group and stress grade. Using the same statistical criteria as in the determination of strength properties of clear wood, see Table 2.3, the allowable stresses are based also on the concept of a 5 percent exclusion limit. Thus, the strength properties obtained for clear wood, are multiplied by the inverse of the adjustment factors given in Table 2.8 to obtain the respective allowable design properties for clear straight-grained wood under normal duration of loading. These stresses are for clear wood of a given species. Allowable design stresses for any given grade are obtained as the product of the given strength ratio times the clear wood stress, modified if necessary as described below.

TABLE 2.8.* ADJUSTMENT FACTORS[1] TO BE APPLIED TO THE CLEAR WOOD PROPERTIES

	Bending Strength	Modulus of Elasticity in Bending[2]	Tensile Strength Parallel to Grain	Compressive Strength Parallel to Grain	Horizontal Shear Strength[3]	Proportional Limit in Compression Normal to Grain[4]
Softwoods	2.1	0.94	2.1	1.9	4.1	1.5
Hardwoods	2.3	0.94	2.3	2.1	4.5	1.5

*Adapted from Table 9 of Ref. 11.

1. The factors include an adjustment for normal duration of loading and a factor of safety.
2. Adjusted from a span-depth ratio of 14 (standard test) to a span-depth ratio of 21 and an assumed uniform loading.
3. Large adjustment factor reflects additional safety necessary to compensate uncertainty of calculating shear stresses in checked beams using elastic expressions, see Sections 7.8 and 7.9.
4. Notice that proportional limit rather than strength is used. Factor includes an adjustment for average ring position.

*Mailing addresses of these agencies may be found in Table 3.1 of Ref. 18.

2.12.1 Principal Modifications of Allowable Stresses for Design Use

The determination of modulus of rupture, as shown in Table 2.3, is based on 2 in. depth. It has been determined experimentally that, for any given wood, modulus of rupture decreases with increasing depth of section, see Sections 7.4 and 7.5. The modifying factor $(2/d)^{1/9}$, where d is the net surfaced depth in inches, is used to take this strength variation in account (14).

ASTM (11) also provides for percentage increases in allowable stresses due to seasoning effects, for lumber 4 in. or less in thickness; see Table 2.9. In view of the shrinkage that accompanies loss of moisture content, see Fig. 1.6, increases of working stresses for drying are always based on the actual dimensions of the dry pieces. For wood of structural size, thicker than 4 in., the increase of strength due to drying-out is largely offset by possible seasoning degrade such as checks and honeycombing; the latter are checks that occur in the interior of a piece of wood. Only an increase of 10 percent above allowable stress values, for green lumber based on net size at the time of manufacture, (for compression members thicker than 4 in., of all lengths regardless of grade) may be taken for drying regardless of grade. Two percent increase is allowed for modulus of elasticity, providing that the lumber is seasoned to a substantial depth before full load is applied.

TABLE 2.9.* MODIFICATION OF ALLOWABLE STRESSES FOR SEASONING EFFECTS FOR LUMBER 4 IN. AND LESS IN NOMINAL THICKNESS

Property:	Percentage Increase in Allowable Property Above that of Green Lumber When Maximum Moisture Content is: 19 percent	15 percent
Bending	25	35
Modulus of Elasticity	14	20
Tension Parallel to Grain	25	35
Compression Parallel to Grain	50	75
Horizontal Shear	8	13
Compression Across the Grain†	50	50

*From Table 11 of Ref. 11

†Increase is the same for all moisture content below fiber saturation since the outer fibers, which season rapidly, have the greatest effect on this strength property regardless of the extent of seasoning of the inner fibers.

The effects of the passing of time in old timbers may include seasoning, weathering, or chemical change, in addition to the effects from duration of load. In the absence of decay, these additional aging effects are structurally unimportant; old lumber may be given the same working stress values as those for new lumber of equivalent species and grade (12).

The presence of decay, see Section 5.2 is not accepted in most structural grades. However, if confined to knots and not present in wood surrounding them, decay may be permitted in some structural grades.* Special attention given by the designer to such features as drainage and ventilation will help in preventing decay. However, for such applications where decay is likely to occur, it is necessary to use wood treated with chemical preservatives. In this regard, some possible reductions in strength may result from the high temperatures and pressures used for conditioning

*Decay must be confined to knots smaller than or equal to one half the size of permitted sound knots.

of wood at a high moisture content under approved methods of treatment. National Design Specification (2) provides no reduction of allowable stresses to lumber pressure-impregnated by an approved process and preservative; however, for lumber pressure-impregnated with fire-retardant chemicals it requires a 10 percent reduction in allowable stresses.

Occasional exposures to temperatures up to about 150°F, and longer exposures up to about 125°F, are provided for in the allowable stresses. However, a special allowance should be made for lumber subjected to abnormally high temperature, particularly for long periods of time. This is due to the weakening effects of high temperatures applied to wood for a long time (10).

2.12.2 Illustrative Example

This example of stress-grade development from ASTM's Designation D 245, see Ref. 11, is given as typical of present practice. Consider the case of joists and planks for light building construction, 1½ in. thick and 5½ in. wide, at 19 percent moisture content, and of a fictitious softwood species. It is wished to achieve a strength ratio in bending of 60 percent, in compression parallel to grain of 65 percent, and in shear of 75 percent.

The following determinations are required:

1.) Limiting provisions for the given grade and size regarding strength-reducing characteristics.
2.) List of allowable stresses for normal loading conditions.

Table 2.6 gives the limiting characteristics that will provide the required strength ratios; it is based on tables of Ref. 11. Using the tabulated values, the limiting provisions for this grade and size are:

1.) Slope of grain no more than 1 in 10.
2.) Knots on narrow face no larger than ¾ in.
3.) Knots at centerline of wide face no larger than 2⅛ in.
4.) Knots at edge of wide face no larger than 1⅜ in.
5.) Shakes and checks no larger than ½ in.
6.) Splits no longer than 4⅛ in.

TABLE 2.10.* SELECTION OF LIMITING CHARACTERISTICS.

Property	Limiting Characteristic	Strength Ratio, percent
Bending	narrow face knot, ¾ in.	62
	knot on centerline wide face, 2⅜ in.	60
	knot at edge of wide face, 1⅜ in.	60
	slope of grain, 1 in 10	61
Compression strength parallel to grain	knot on any face, 2⅛ in.	65
	slope of grain, 1 in 8	66
Shear	size of shake or check, ½ in.	76
	length of end split, 4⅛ in.	75

*Adapted from Table 12 of Ref. 11.

For this grade the allowable properties for normal loading conditions are developed and given in Table 2.11. For other duration of loading allowable stresses can be obtained as shown in the following section.

TABLE 2.11.* ALLOWABLE STRESSES FOR THE SAMPLE STRESS-GRADE.

Property	Clear Wood Strength, ksi	Adjustment Factor	Strength Ratio ÷ 100	Seasoning Adjustment	Special Factors	Allowable Stresses,[3] ksi
Bending	4.432	1/2.1	0.60	1.25	0.89[2]	1.400
Compression Parallel to Grain	2.174	1/1.9	0.65	1.50		1.100
Horizontal Shear	0.576	1/4.1	0.75	1.08		0.115
Tension Parallel to Grain	4.432	1/2.1	0.60×0.55[1]	1.25		0.850
Modulus of Elasticity	1,304.	1/0.94	1.00	1.14		1,580.
Compression Normal to Grain	0.282	1/1.5	1.00	1.50		0.280

*Adapted from Table 13 of Ref. 11.

1. Strength ratios in tension parallel to grain are 55 percent of the corresponding bending strength ratio.
2. Depth factor $= (2/5.5)^{1/9} = (0.363)^{1/9} = 0.894$.
3. Obtained by multiplying together the 5 preceding columns.

2.13 MODIFICATION OF ALLOWABLE STRESSES FOR VARIOUS LOADING CONDITIONS

Allowable stresses derived by the method given in the preceding section are applicable to normal loading conditions, see Section 2.9. For durations of load other than normal loading all allowable stresses, except modulus of elasticity, may be modified using the information contained in Fig. 2.14. In the latter the values of $\overline{\alpha}$, the ratio of working-stress for a given loading condition to allowable stress for normal loading condition are plotted against the duration of loading. Some values of $\overline{\alpha}$ are as follows: 2.0 for impact,* 1.33 for wind and earthquake, 1.25 for seven days (construction loads), 1.15 for snow and 0.9 for permanent loading conditions. For normal loading conditions $\overline{\alpha} = 1.0$.

Structural members may be subjected to forces induced by a series of loads, the duration of which could be different. It is necessary to study the various combinations that may exist in order to determine the governing loading condition. The latter is such that, when the member is designed to satisfy it, exactly or by a margin, the solution obtained satisfies all other loading conditions by an even greater margin.

*Recent work by Keeton (15) indicates, however, that increases in strength properties may not be as large as this for shock and impact loading. In particular, dynamic strength increases for green specimens subjected to shock and impact loading were insignificant when compared to static values.

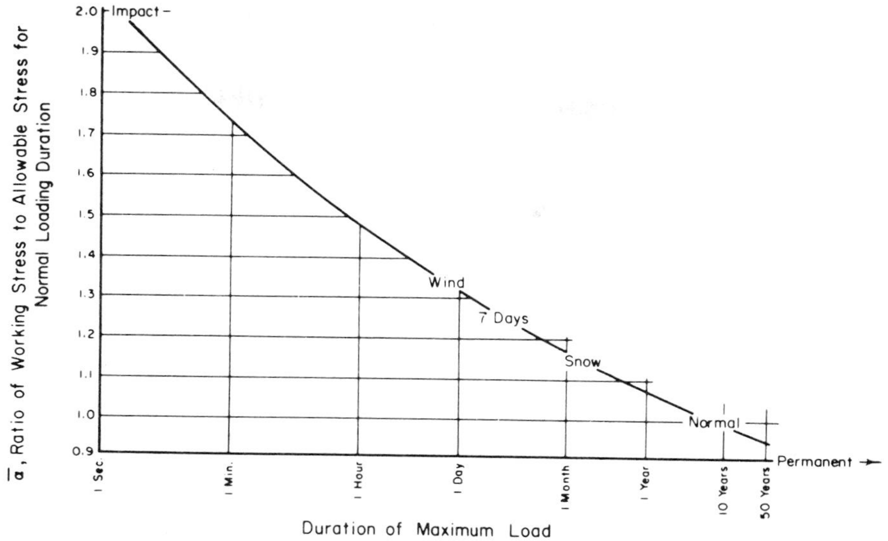

Figure 2.14. Adjustment factors of working stresses for various durations of load.

It is necessary to point out that the allowable stress for a given loading condition is defined by the particular component load of smallest duration. Thus, in the case of a structure subjected to the action of its own weight D, live load L, and wind W, three possible combinations exist, namely, D, D + L and D + L + W; the allowable stresses are 0.9 F, F, and 1.33 F, respectively. The following simple example, in which the governing loading condition of a given member is sought, illustrates this concept.

2.13.1 Illustrative Example

The wood column of a building several stories high is subjected to axial forces created by the following loads:

Permanent, D	20 k
Live, L	40 k
Snow, S	30 k
Wind, W	20 k
Earthquake, E	8 k

Determine the loading condition that governs the design of the column.

For the sake of simplicity it is assumed that the design of the column is accomplished by the determination of its cross-sectional area using the ratio of the acting load and the allowable stress; refer to Chapter 8 for a thorough discussion of design of wood columns.

The results are shown in Table 2.12. The allowable stress F_c for the given wood corresponds to normal loading conditions; the allowable stresses for the various other conditions are obtained using the factors α shown in Fig. 2.14. The governing loading conditions turn out to be No. 3 for which the ratio $P/\bar{\alpha}F_c$, i.e., the required cross-sectional area, is a maximum.

TABLE 2.12. DETERMINATION OF THE GOVERNING LOADING CONDITION FOR THE DESIGN OF A GIVEN COLUMN.

Loading Condition	Load P. kip	Duration	Allowable Stress. $\bar{\alpha}F_c$	Required Area $P/\bar{\alpha}F_c$
1. Permanent, D	20	50 years	0.9 F_c	$22.2/F_c$
2. Normal, D + L	60	10 years	1.0 F_c	$60.0/F_c$
3. Snow, D + L + S	90	2 months	1.15 F_c	$78.2/F_c$**
4. Wind* D + L + (½) S + W	95	1 day	1.33 F_c	$71.4/F_c$
5. Earthquake: D + L + S + E	98	1 day	1.33 F_c	$73.7/F_c$

*Only one half of the snow load is considered likely to be present under the action of the design wind.

**Maximum.

2.14 DIMENSIONS OF STRUCTURAL LUMBER

On September 1, 1970, the U.S. Department of Commerce made effective its National Bureau of Standard PS 20–70 "American Softwood Lumber Standard" (16). This comprehensive voluntary standard was developed by the National Bureau of Standards in cooperation with producers, distributors and users of lumber in this country; it signified a hallmark of standardization and uniformity for the lumber industry.

PS 20–70 classifies structural lumber in various groups. Of immediate interest to structural users of wood is its classification of lumber according to extent of manufacture. The following three types are recognized:

1.) Rough Lumber. Has not been dressed (surfaced) but has been sawed, edged and trimmed at least to the extent of showing saw marks in the wood on the four longitudinal surfaces of each piece for its over-all length.
2.) Dressed (surfaced) Lumber. Has been dressed by a planing machine for the purpose of attaining smoothness of surface and uniformity of size on: one side (S1S), two side, (S2S), one edge (S1E), two edges (S2E), or a combination of sides and edges (S1S1E, S1S2E, S2S1E, S4S).
3.) Worked Lumber. In addition to being dressed, has been matched, ship-lapped, or patterned.

The nominal and minimum-dressed sizes of boards, dimension lumber and timbers are given in Table 2.13. Since lumber used in structural design is almost exclusively dressed lumber, Tables A.2.4 give only the sectional properties of standard dressed (S4S) sizes. Section properties of rough lumber may be obtained using their nominal dimensions.

PROBLEMS

1. A number of tests of compression parallel to grain are made using Douglas fir. The following results are obtained for crushing strength: 4.00, 3.50, 3.80, 3.70, 3.90, 3.60, 3.50, 3.70, 3.90, 4.20 ksi. Disregarding the fact that this is a very small population sample, determine: (a) average value, (b) standard deviation, (c) variability and (d) 5 percent exclusion value.

   Ans. (a) 3.780 ksi, (b) 0.213 ksi, (c) 0.056, (d) 3.430 ksi

TABLE 2.13.* NOMINAL AND MINIMUM-DRESSED SIZES OF BOARDS, DIMENSION AND TIMBERS.

(The thicknesses apply to all widths and all widths to all thicknesses.)

Item	Thicknesses			Face widths		
		Minimum dressed			Minimum dressed	
	Nominal	Dry[1]	Green[1]	Nominal	Dry[1]	Green[1]
		Inches	Inches		Inches	Inches
Boards.	1	¾	25/32	2	1-½	1-9/16
	1-¼	1	1-1/32	3	2-½	2-9/16
	1-½	1-¼	1-9/32	4	3-½	3-9/16
				5	4-½	4-⅝
				6	5-½	5-⅝
				7	6-½	6-⅝
				8	7-¼	7-½
				9	8-¼	8-½
				10	9-¼	9-½
				11	10-¼	10-½
				12	11-¼	11-½
				14	13-¼	13½
				16	15-¼	15-½
Dimension.	2	1-½	1-9/16	2	1-½	1-9/16
	2-½	2	2-1/16	3	2-½	2-9/16
	3	2-½	2-9/16	4	3-½	3-9/16
	3-½	3	3-1/16	5	4-½	4-⅝
				6	5-½	5-⅝
				8	7-¼	7-½
				10	9-¼	9-½
				12	11-¼	11-½
				14	13-¼	13-½
				16	15-¼	15-½
Dimension.	4	3-½	3-9/16	2	1-½	1-9/16
	4-½	4	4-1/16	3	2-½	2-9/16
				4	3-½	3-9/16
				5	4-½	4-⅝
				6	5-½	5-⅝
				8	7-¼	7-½
				10	9-¼	9-½
				12	11-¼	11-½
				14	. . . .	13-½
				16	. . . .	15-½
Timbers.	5 and thicker	. . . .	½ off	5 and wider	. . . .	½ off

*From National Design Specification for Stress-Grade Lumber and Its Fastenings. 1977 Edition by NFPA.

1. Dry lumber is defined as lumber which has been seasoned to a moisture content of 19 percent or less. Green lumber is defined as lumber having a moisture content in excess of 19 percent.

2. Determine the modulus of rupture for a given combination of species, using Method B of ASTM Designation D 2555–78 and the following information:

Species	Average R, ksi	Std. Deviation, ksi	Vol. Factor
D	6.951	1.112	0.25
E	7.202	1.152	0.30
F	6.301	1.008	0.45

Ans. R = 4.81 ksi

3. Determine horizontal shear strength, coefficient of variation v = 0.14, for a given combination of species, using Method B and the following information:

Species	Average Shear Strength, ksi	Vol. Factor
D	1.037	0.1016
E	0.958	0.0921
F	0.850	0.5067
G	0.872	0.2996

Ans. Weighted mean = 0.886 ksi; weighted 5 percent exclusion value = 0.682 ksi; lower CDF = 1.42; exclusion value assigned to combination = 0.674 ksi.

4. Determine the required section modulus, see Section 7.6, for a simply-supported roof rafter of 14 ft. span carrying the following uniformly distributed loads:

Permanent	2,000 lb. total
Snow	8,000 lb. total
Wind	5,000 lb. total

Assume a grade of lumber with an allowable fiber stress of 1,600 psi.

Ans. S = 114 in^3

5. Determine the required cross-sectional area of a tension member to carry the loads given in the following two cases. The allowable tensile stress parallel to grain is 1,600 psi.

(a)	Permanent	4,000 lb.
	Live	8,000 lb.
	Wind	6,000 lb.
(b)	Permanent	10,000 lb.
	Live	20,000 lb.
	Wind	4,000 lb.
	Impact	20,000 lb.

Ans. (a) 8.45 in.^{2}, (b) 18.75 in.^{2}

6. A given species of wood is tested to determine its modulus of rupture. For green wood and wood at 6 percent moisture content, test results were 3,000 psi and 8,000 psi respectively. Determine the modulus of rupture for 12 percent moisture content.

Ans. R = 5870 psi for M_p = 25%

7. For a given species of wood, the unit stress in compression parallel to grain is 1,600 psi and in compression perpendicular to grain is 400 psi. Make a table of values of allowable compressive stress when the angle between the direction of grain and direction of load varies between 0° and 90° in increments of 10°. Determine your results with an accuracy of 10 psi; plot N against θ in a scaled drawing.

   Ans. 1600, 1480, 1180, 920, 720, 580, 490, 440, 410, 400 psi.

8. The statement is usually made that impact loading conditions do not govern design of wood structures. Draw your conclusions regarding this belief after determining the critical condition of loading for the following two structures:

   (a) **Timber bridge** made out of wood deck and stringers. Normal loading conditions: 20-ton truck, considered as a uniformly distributed lane load. Impact load: an additional 30 percent of truck weight. The self-weight of bridge may be estimated as 10-ton, also uniformly distributed.

   (b) **Bomb shelter** made out of timber arches and deck. Normal loading conditions: live load 20 lb./sq. ft. Loading conditions during winter: 30 lb./sq. ft. due to snow. Impact and shock load during bomb raid are estimated at one lb./sq. inch. The self-weight of the structure may be assumed as 100 lb./sq. ft.

   Ans. Impact and shock loading control design (b)

9. Determine the modulus of rupture for Canadian Spruce, using Method B of ASTM Designation D2555–78 and the following information:

Species	Standing Timber Volume, 10^6ft^3	Average R, ksi	Std. Deviation, ksi
Black	111,259	5.870	0.939
Engelmann	18,354	5.660	0.906
Red	2,835	5.880	0.941
Sitka	5,868	5.420	0.867
White	77,073	5.100	0.816

   Ans. Volume weighting factors: 0.5165, 0.0852, 0.0132, 0.0272, 0.3579.
   Weighted Mean = 5.564 ksi, weighted 5-percent exclusion
   Values = 4.100 ksi; lower CDF = 1.23 for White Spruce, exclusion value assigned to combination = 3.89 ksi

REFERENCES

1. ASTM Designation D143–52 (Reapproved 1972) "Standard Methods of Testing Small Clear Specimens of Timber," American Society for Testing and Materials, 1978 Book of ASTM Standards, Part 22, pp. 52–109, 1978.
2. "National Design Specification for Stress-Grade Lumber and Its Fastenings," 1977 Edition. National Forest Products Association, Washington, D.C.
3. Hankinson, R. L., "Investigation of Crushing Strength of Spruce at Varying Angles of Grain," United States Air Service Information Circular No. 259, 1921.
4. Dietz, A. G. H., "Stress-strain Relations in Timber Beams of Douglas Fir," ASTM Bulletin No. 118, October 1942, pp. 19–27.

5. ASTM Designation D 2555–78 "Standard Methods for Establishing Clear Wood Strength Values," American Society for Testing and Materials, 1978 Book of ASTM Standards, Part 22, pp. 734–755, 1978.
6. Roark, R. J., "Formulas for Stress and Strain," McGraw Hill Book Co., Inc., New York, 1954.
7. American Plywood Association "Plywood Design Specifications," 1966. Published by American Plywood Association, Tacoma, Washington.
8. ASTM Designation D 2718–76 "Standard" Method of Testing Plywood in Rolling Shear (Shear in Plane of Plies)," American Society for Testing and Materials, 1978 Book of ASTM Standards, Part 22, pp. 789–794, 1978.
9. ASTM Designation D 2915–74 "Standard Method for Evaluating Allowable Properties for Grades of Structural Lumber," American Society for Testing and Materials, 1978 Book of ASTM Standards.
10. Forest Products Laboratory "Wood Handbook," U.S. Department of Agriculture Handbook No. 72, U.S. Government Printing Office, Washington, D.C., 1955, revised August 1974.
11. ASTM Designation D 245–74 "Standard Methods for Establishing Structural Grades and Related Allowable Properties for Visually Graded Lumber," American Society for Testing and Materials, 1978 Book of ASTM Standards, Part 22, pp. 139–162, 1978.
12. "Duration of Load and Fatigue in Wood Structures: Progress Report of Sub-Committee of the Committee on Timber Structures of the Structural Division," Journal of the Structural Division, ASCE Vol. 83, No. ST5, Proc. Paper 1361, May 1957.
13. Galligan, W. L., and Snodgrass, D. V., "Machine Stress Rated Lumber: Challenge to Design," Journal of the Structural Division, ASCE, Vol. 96, No. ST12, Proc. Paper 7772, Dec. 1970, pp. 2639–2651.
14. Bohannan, Billy, "Effect of Size on Bending Strength of Wood Members," Forest Products Laboratory, U.S. Department of Agriculture, Research Paper, FPL 56, May 1966.
15. Keeton, John R., "Dynamic Properties of Small, Clear Specimens of Structural-Grade Timber," Technical Report R573 Naval Civil Engineering Laboratory, Port Hueneme, California.
16. U.S. Department of Commerce, National Bureau of Standards "American Softwood Lumber Standard," U.S. Government Printing Office, Washington, D.C. 20402.
17. Southern Pine Inspection Bureau, "Determination of Design Values for Southern Pine Stress Grades," SPIB, Pensacola, Fla. 1970.
18. "Wood Structures. A design Guide and Commentary" Compiled by Task Committee on Status-of-the-Art: Wood. Committee on Wood ASCE Structural Division. Published in 1975 by American Society of Civil Engineers, 345 East 47th St., New York, 10017.

Chapter 3
Glued-Laminated Timber

3.1 INTRODUCTION

Glued-laminated timber structural members are made of material glued together from smaller pieces of wood, either in straight or curved form, with the grain of all the laminations essentially parallel to the length of the member; thus, laminated wood is basically different from plywood, see Chapter 4, in which the grain direction of adjacent plies is at right angles.

Laminating standards require that laminations shall not exceed 2 inches in thickness; usually ¾ in. and 1½ in. actual thicknesses are employed. Since modern methods of end joining of lumber allow indefinite extension, the laminations can be any length, as required. In the fabrication of a glued-laminated member, different wood species may be used. However, it is important that their dimensional changes with changes in moisture content, see Fig. 1.4, be similar; to avoid as much differential change as possible, the moisture contents at fabrication time of the various laminations should be within a few percent of each other. This condition should be observed always, even when wood of the same species is being used.

Glued-laminated members may be either horizontally or vertically laminated. A horizontally laminated member is one comprising 4 or more laminations and in which the loads on the member act in a plane perpendicular to the plane of the laminations; it is the most commonly used form. A vertically laminated member is one comprising 3 or more laminations and in which the loads on the member act in a plane parallel to the plane of the laminations. Hereafter in this text, the term glued-laminated member is to be construed as a horizontally laminated member; vertically laminated members are specifically referred to as such.

The present importance of this industry is best illustrated by the buildings, bridges and other structures that have been erected using glued-laminated construction. The Keystone Wye bridge in South Dakota, see Fig. 3.1, is an outstanding example of a well-designed glued-laminated structure. The athletic building of Montana State College, measuring 300 ft. across, was built using a glued-laminated wood dome, see Fig. 3.2; details of another major work of the same kind are shown in Fig. 3.3. Examples of rigid-frame construction and smaller-span arches are shown in Fig. 3.4 and 3.6, respectively. Post and beam construction, typical of warehouses and other buildings consisting of a number of wide bays, is shown in Fig. 3.5. Discussion of analysis and design of a number of these structures is given in Chapters 9 through 11.

3.1.1 SPAN RANGE OF GLUED-LAMINATED STRUCTURES

The structures in Figs. 3.1 through 3.6 illustrate modern use of glued-laminated timber. While theoretically no limit need be established for span length because of structural strength or stiffness, economic considerations of competition with solid-sawn wood at the lower end of the span scale, and with steel and concrete structures for large structures, dictate some practical limits. Table 3.1 indicates the present economic span ranges for the most common glulam structures.

Figure 3.1. The Keystone Wye Bridge in South Dakota. Three glued-laminated arches hinged at the supports and at midspan support glulam columns and stringers carrying the reinforced-concrete deck. The bridge was designed by the department of highways of the state of South Dakota. Photo courtesy of Department of Highways.

Figure 3.2. Interior view of the glued-laminated dome of the athletic building at Montana State College. Photo courtesy of AITC.

Figure 3.3. Details of connections of secondary arches to main arches of glued-laminated dome. North Dakota State Teachers College Fieldhouse. Photo courtesy of AITC.

Figure 3.4. Barrel vault building design for tennis club. Clear heights of 40 feet are necessary for indoor tennis. Photo courtesy Koppers Company, Inc.

Figure 3.5. Typical use of glued-laminated post and beam construction.

Figure 3.6. Modern church design consists of a 167 ft. diameter dome supported by eight pairs of glued-laminated wood arches. Pressure-treated wood-decking on canopied arches spanning between the principal arches forms the roof of the dome. Photo courtesy of Koppers Company, Inc.

TABLE 3.1. PRESENT USE OF GLULAM STRUCTURES.[1]

Type	Range, ft
Beams	20–130
Bowstring Trusses	50–200
Pitched Trusses	50–90
Parallel Chord Trusses	50–150
Arches	30–300
Domes	50–350

1. From Ref. 21

3.2 COMPARISON BETWEEN GLUED-LAMINATED AND SOLID-SAWN CONSTRUCTIONS

The following are significant advantages glued-laminated timber offers:

1. Higher utilization of wood, since lower-grade material can be used for the less highly stressed central laminations in beams without adversely affecting structural integrity. In addition, because fabrication of large glued-laminated structural members is possible from smaller pieces, lumber from smaller trees can be used effectively.
2. Size and length of structural wood members are not limited by length of tree; laminated arches and beams in excess of 200 ft. spans, with sections as deep as 7 ft. have been used.

3. Strength-reducing characteristics of wood can be controlled through quality specification of individual laminations. Since the latter are thin enough to be readily seasoned before fabrication, seasoning checks (that are otherwise associated with large solid-sawn members) are minimized.
4. For dry conditions of use of the finished member, fully seasoned laminations may be used. Thus, the increased strength of seasoned wood can be realized in design.
5. Members of variable depth can be designed and fabricated. Economy can be obtained for variations in cross-section that reflect strength requirements.
6. Architecturally pleasing effects are obtained with glued-laminated design; when combined with proper lighting and other component materials for a given building, a psychological feeling of the warmth of wood and the well-being of occupants or users is easy to obtain.

However, not all are advantages for glued-laminated construction, especially in small structures, for which solid-sawn members may be available for use. When compared to solid-sawn members of equivalent capacity, glued-laminated members may be more expensive for the following reasons:

1. More time is required to cut and season lumber and do the necessary laminating and gluing than is required to cut solid green timbers.
2. The laminating process requires special equipment, plant facilities, adhesives, fabricating skills and quality control that are not needed to produce solid-sawn timbers.

Hence, at the present time, glued-laminated construction does not compete economically with solid-sawn designs for short spans and small size members.

3.3 FABRICATION OF GLUED-LAMINATED MEMBERS

Industrial fabrication of laminated wood members consists of various well-defined steps that include the selection and preparation of lumber for laminating, layout of laminated assemblies and gluing of the finished member. A succinct review of these processes is given herein, after which, the special cases of fabrication of tapered members is discussed. In view of the important role that glue plays in the fabrication of these members, a brief discussion of the various available glues is given first.

3.3.1 Glues

For many years only glues of natural origin were used by the laminating industry. These included animal, vegetable-starch, casein, vegetable-protein and blood-albumin types; casein was the most widely used. Fabricated members were mostly limited to interior use. However, in the past twenty years, the development of synthetic-resin glues has resulted in improved performance that allows use of fabricated products for exterior applications such as bridges and utility poles.* Among the synthetic-resin glues that are frequently used for resorcinol resins and phenol-resorcinol combinations. A thorough description of properties of various glues is beyond the scope of this

*The insulating properties of wood alone would justify the use of glued-laminated poles and cross arms in high-voltage transmission structures. For example, the Long Island Lighting Co. installed in 1970 a line of 138 Kv conductors supported by 110 ft. long octagonal-shaped glued-laminated poles.

text; for further details refer to the work by Freas and Selbo (1) and to a number of standards on adhesives given by ASTM (2).

Since glues are generally chosen on the basis of their durability under service conditions, designers must specify to fabricators whether dry or wet use is intended for the given glued-laminated member. Certain types of glues may be used under any exposure, such as continuous immersion in water or intermittent wetting and drying. These conditions may occur outdoors or in buildings where high humidities are encountered for a considerable time. Within this group the wet-use adhesives such as phenol-resorcinol, resorcinol and melamine glues are required. Casein glue is the representative of the dry-use group and is adequate only under limited exposure or where protection from the elements is provided. Urea adhesives are not permitted for use in structural glued-laminated timber. (Ref. 5.)

3.3.2 Selection and Preparation of Lumber for Laminating

It has been shown by tests that glue joints of high strength, comparable to the shear strength of the wood, can be produced in most commercial species of wood with casein and synthetic-resin glues. It is important, however, that the moisture content of lumber at the time of gluing approximate the average value that the member will attain in service; serious changes in moisture content after gluing induce shrinking or swelling of the wood with resulting stresses and possible checking in the wood or along the glue line. A maximum moisture content of 16 percent (generally averaging 13 percent in practice) at the time of gluing is considered satisfactory for laminated members intended for normal interior use. When the moisture content is 16 percent for the member in use, the moisture content at the time of gluing should not exceed 20 percent. These members will later develop higher moisture contents; however, gluing is accomplished at the specified range since, otherwise, good joints would be more difficult to obtain. For obvious reasons, uniformity of moisture is necessary, differential moisture content between laminations should not exceed 5 percent if internal stresses and checking are to be avoided.

3.3.3 Layout and Gluing Laminated Assemblies

Designers presently specify the quality desired in standard glued-laminated members by means of combination symbols, see Tables A.3.1, 2 and 3. The requirements for the laminations that constitute these members were given by AITC Standards 117–71, 120–71 and 119–71, see Refs. 3, 4 and 20 respectively, but presently (1980) are all contained in AITC Standard 117–79, see Ref. 29. The quality of laminations is specified by zones. The tension laminations demand the best material, especially the outer laminations; the inner zone of the member, i.e., the interior half-depth, can be fabricated with wood of lesser quality. Hence, a plan for the position of laminations in the glued assembly is necessary because of the different grades of the laminations that might be used, and to indicate the position of edge and end joints.

Normally the individual pieces of lumber are end jointed to form full length laminations. These laminations may or may not be resurfaced prior to gluing depending on the thickness tolerance of the lumber used and the cleanliness of the surface for gluing. When plain scarf joints are used, see Fig. 3.7, the end gluing of the individual pieces of lumber may be accomplished at the same time as face joining provided the lumber is clean and is within the specified thickness tolerance.

Successful fabrication of the laminated assembly depends on following the right procedure when gluing; the laminations must be arranged in the correct order in a convenient place for proper feeding to the glue spreader. The following steps are considered necessary:

1. Mixing and spreading the glue properly;
2. Placing the laminations on the jig or gluing bed;
3. Applying adequate pressure uniformly and quickly to avoid initial setting of the glue before application of pressure is completed.

A thorough discussion of these operations with illustrative details of necessary equipment and procedures is beyond the scope of this work; however, a wealth of information may be found in FPL's Technical Bulletin No. 1069, see Ref. 1. Additional discussion is also given in the bulletin on clamping techniques for straight and curved members; curing requirements for assemblies made with various types of glues, including means devised for applying heat to laminated assemblies to facilitate curing of the glue; and quality control instruments and methods. Since the time of publication of Bulletin No. 1069 in 1954, major improvements have taken place.

3.3.4 Members of Variable Depth

The depth of the cross section of a member can be varied along its length to meet the strength requirements. This can be accomplished by varying the number of laminations and by tapering the member to a smooth outline. Thus, single and double-tapered beams are used to provide pitched roofs; tapered arches and rigid frames are also frequently used.

It is recommended (18) that any sawn taper cuts be made on the compression face of tapered beams. It is also recommended that pitched or curved tension faces of beams not be sawn across the grain, but, instead, that the beam be so manufactured that the laminations are parallel to the tension face. Analysis and design of tapered beams is given in Section 7.15.

3.3.5 Dimensions of Glued Laminated Structural Members

Efficient and economical production results when standard lumber sizes are used for the laminates. Nominal 2 in. thick lumber of standard nominal width is used to produce straight members, and curved members where the radius of curvature is within the bending radius limits for the given thickness and species, see Section 7.13; otherwise, nominal 1 in. thick boards are used. Standard practice (26) calls for surfacing nominal 2 in. laminations to a net 1½ in. thickness and nominal 1 in. laminations to a net ¾ in. thickness. Maximum allowable variations on actual thickness of laminations is ± 0.008 in.

To provide a uniformly smooth surface in glulam members and to remove the glue squeeze out, it is necessary to surface the wide faces. This causes the net finished width to be even smaller than the nominal standard width of lumber. Net finished widths of glued-laminated members are: 2¼, 3⅛, 5⅛, 6¾, 8¾, 10¾, 12¼ and 14¼ in.

The glulam industry may fabricate structural members using the following tolerances:

1. width: ±1/16 in of specified width
2. depth: + 1/8 in/ft of specified depth
−1/16 in/ft of specified depth or − 3/32 in., whichever is the larger

3. length: $\pm$ 1/16 in to 20 ft
   $\pm$ 1/16 in/20 ft of specified length
   where length dimensions are not specified or critical, length tolerances do not apply
4. squareness: $\pm$ 1/8 in/ft of specified depth

3.4 STRENGTH OF GLUED-LAMINATED MEMBERS

Consideration of a number of factors is involved in the determination of strength of glued-laminated members. The species and grade of laminations used are of utmost importance; characteristics such as knots affect the strength of the laminations and thus, depending on their position in the member, influence the overall strength. Slope of grain affects strength also, but is taken into account by imposing limitations that make its effects similar in magnitude to those of knots of permitted size. The presence of checks affects horizontal shear as in solid beams; however, their presence is rare in laminations. Shakes have the same effect on shear as checks; thus, material containing shakes is usually excluded or placed in areas of low shear. Efficient use of material is obtained by using wood of the higher grades in the outer laminations and that of lower grades in the inner laminations.

The need for joining the laminations together to obtain the necessary lengths and widths required for fabrication creates a new variable, that of joints. End and edge joints of various types are used in glued-laminated members; because of their importance, a number of tests have been made to determine their effects on strength of the member. Other factors such as depth, time, and moisture content affect strength of glued-laminated members; in addition, overstressing of laminations and creation of radial stresses in curved members are problems that affect glued-laminated construction in particular.

The effects of the variables described above on the strength of glued-laminated members are studied in more detail in the following paragraphs.

3.4.1 Effects of End Joints

End joints of laminations are used to obtain the lengths required for a given member. Various types of end joints are shown in Fig. 3.7, namely, the plain scarf joint, the butt joint and the finger joints.

The glued butt joint, see Fig. 3.7b, is extremely weak and quite variable in tensile strength; hence, no reliability can be placed on the strength of a butt joint. However, butt joints may be used for compression laminations (other than the outer lamination) of a member that is subjected to bending provided that all laminations having such joints, at a given cross section, are disregarded in computing the moment of inertia.

Plain scarf joints may be made to have a considerable proportion of the strength of the lamination in both tension and compression. The slope of the plane of the scarf with respect to the axis of the piece designates the scarf joint; the slope is computed from the ratio of thickness of the lamination to length of scarf, see Fig. 3.7a. The strength of plain scarf joints increases as the slope decreases; joint factors, expressed in percentages of full strength of lamination, are given as follows (1):

Slope of Scarf	Joint Factor, percent
1/12 or flatter	85
1/10	80
1/8	75
1/5	60

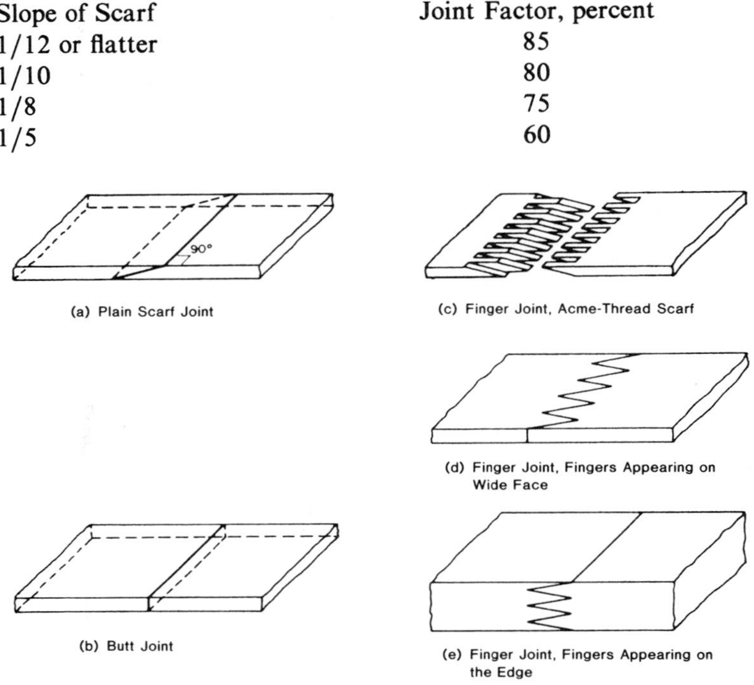

Figure 3.7. Various types of glued end joints for laminations.

The length of a lamination lost due to lapping for a scarf joint, and hence its cost, increases as the slope of the scarf decreases. It is necessary that the tips of scarf joints of adjacent laminations be separated in accordance with the provisions of CS 253–63 (5).

Finger joints of various designs are used instead of scarf joints by most fabricators; Fig. 3.7 shows three types. The strength of finger joints varies greatly with the type and configuration used; however, flat slopes and sharp tips are factors contributing to higher strength. Qualification tests (5) are always necessary to determine joint factors of new designs; once the joint factors are established, allowable stresses for finger joints can be determined in the same way as for scarf joints (5). Finger joints consume less wood than scarf joints and their cost is usually less.

Allowable values of shear stress and compression across the grain need not be reduced because of the presence of end joints in the laminations of a beam.

3.4.2 Effects of Edge Joints

Laminations may consist of two or more pieces of lumber placed edge to edge when the width of the glued-laminated member is greater than that of the lumber available. The edge joints of these pieces need not be glued except when occurring in members subjected to torsion and in members loaded normal to the lamination edge. As an example of the latter case, vertically laminated beams (for which the total depth is obtained by edge gluing laminations together) demand high performance of these joints; the capacity to resist longitudinal shear depends on the edge joints.

Plain-edge joints are recommended (1) over tongue-and-groove and other types because of the ease with which good joints can be made for most species. As compared to plain joints, machine joints facilitate alignment but are more wasteful of material, require special equipment and, because of difficulties in machining, may actually afford less effective gluing area.

3.4.3 Effects of Knots

The presence of knots in the constituent laminations of a glued member affects strength as much as it does in solid-sawn timbers. The number, size, and position, with respect to the neutral axis of the member, of the knots close to the critical section affect bending strength and stiffness. For axially loaded members subjected to either compression or tension, only the total area of knots in the cross section affects strength; the position of the knots is not a factor.

A statistical study at the Forest Products Laboratory (1) showed that a relation existed between bending strength and stiffness and the ratio I_K/I_G, where I_K is the sum of the moments of inertia of the cross-sectional areas of all knots within 6 in. of a single cross section of a beam and I_G is the moment of inertia of the full or gross cross section; both values being computed about the centroidal axis of the full cross section. The results are shown in Fig. 3.8, where curves A and B give the variation of percent of modulus of elasticity and flexural stress for clear wood, respectively, with the ratio I_K/I_G.

The preceding investigation was completed with the development of a method to estimate the value of the ratio I_K/I_G for beams of various grades and combinations of laminations. It is not likely that knots of the maximum size allowed for the given lamination grade will occur in all laminations at the critical section; some degree of dispersion exists that makes the probability of such concentration low. A thorough survey was made of a considerable number of assemblies with randomly chosen laminations from which the distribution of the computed values of the ratios I_K/I_G was studied.* The results of this study are given by Curve B of Fig. 3.9. The ratio of the value of I_K/I_G obtained using the preceding method, to the value of I_K/I_G which would result if the largest knot permissible in the grade were assumed to be in every lamination, is plotted against the number of laminations in the member. Observation of the figure verifies the fact that, as the number of laminations increases the probability decreases that the largest knots will occur in each lamination at the critical cross section. Thus, using Fig. 3.9, the probable value of the ratio I_K/I_G for a given section can be obtained, once the maximum value of I_K/I_G is determined.

Although all current laminating specifications using visually graded lumber are based on the I_K/I_G concept, recent tests have indicated that modification of the I_K/I_G theory is necessary for the deeper beams, see Section 3.5. The laminations in that tension zone must be of higher quality than indicated by the I_K/I_G theory. On the other hand, the quality of the lumber in the compression zone can be somewhat lower grade.

The variations of percent of compressive stress and tensile stress for clear wood with the ratio K/b of knot size, K, to finished lamination width, b, is given by curves A and B of Fig. 3.10, respectively. The ratio K/b and the ratio A_K/A_G (where A_K and A_G are the cross-sectional areas

*The methods used in this investigation are beyond the scope of this work; refer to Appendix A of Ref. 1 for a complete description.

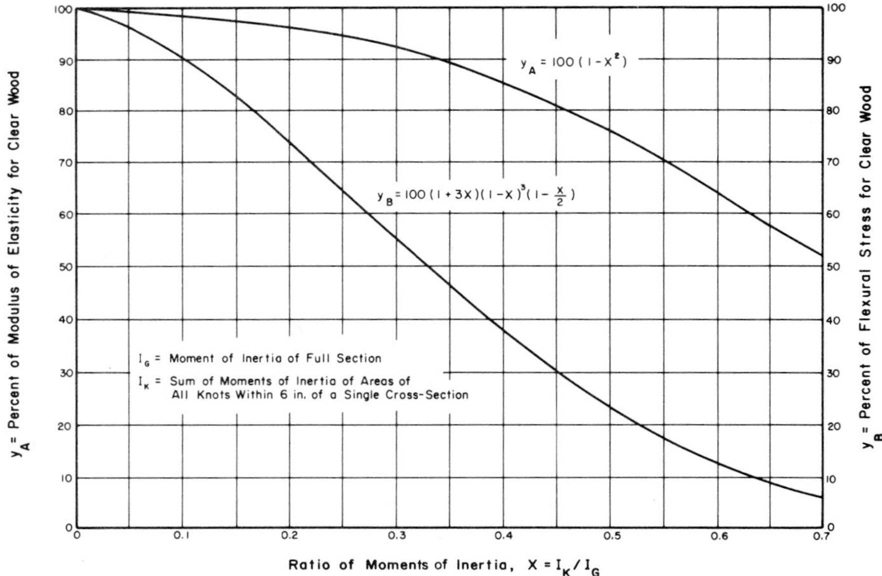

Figure 3.8. Curves relating allowable flexural stress and modulus of elasticity to moment of inertia of areas occupied by knots in laminations. Data from Ref. 1.

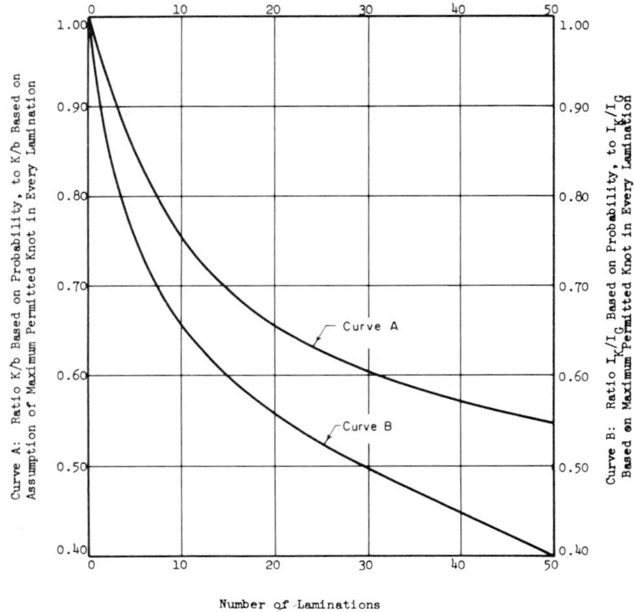

Figure 3.9. Probable values of the ratios K/b and I_K/I_G for a glued-laminated section with a given number of laminations. Data from Ref. 1.

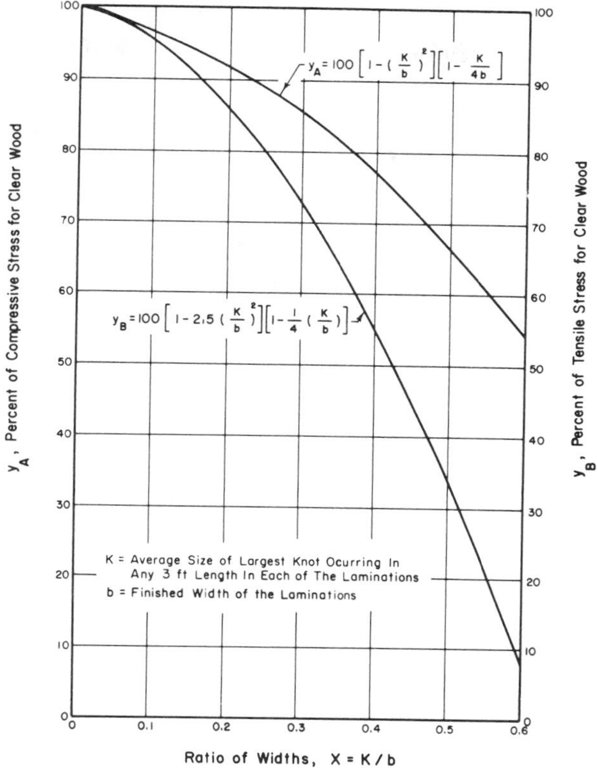

Figure 3.10. Curves relating allowable compressive and tensile stresses to widths occupied by knots in laminations. Data from Ref. 1.

of the knots and the full section, respectively) are identical in sections for which all laminations have an equal thickness. The variation of the ratio K/b based on a probability analysis, see Appendix A of Ref. 1, to the ratio K/b based on the assumption of maximum permitted knot in every lamination at the critical section, is given by curve A of Fig. 3.9. The conclusions and method of solution for this case are similar to those corresponding to the preceding case.

The presence of knots in the laminations of a member does not reduce its strength in shear or in compression across the grain.

Illustrative Example. A glued-laminated beam consists of 15 laminations of 1½ in. thickness of a certain species of wood. Two grades have been used, a better one for the three outer laminations at each side, K/b = 0.25; and another for the nine interior laminations, K/b = 0.5. As before, the quantity K/b expresses the ratio of maximum size of knot to width of the lamination. It is wished to determine: 1) the maximum possible values of the ratios I_K/I_G and K/b; 2) the percentages of flexural stress and modulus of elasticity for clear wood, respectively, that the given member is likely to have; and, 3) the percentages of compressive stress and tensile stress for clear wood, respectively, that the given member is likely to have.

The maximum value of I_K can be determined assuming that a maximum size knot occurs in each lamination at the critical section. For the given section, it is admissible to consider that all laminations have a ratio $K/b = 0.25$, with the central nine laminations having an additional $K/b = 0.25$. The value of I_K can be determined as the sum of two terms: 1) I_K for a section with all laminations having $K/b = 0.25$; and, 2) I_K for the nine interior laminations having an additional $K/b = 0.25$. The first term leads to $I_K = 0.25\ I_G$; the second term, because of the position of the interior laminations, gives $(0.25)\ (9/15)^3\ I_G$. Therefore, the maximum value of I_K for the section is

$$I_K = 0.25\ I_G + (0.25)(9/15)^3\ I_G \tag{3.1}$$

from which the maximum value of the ratio I_K/I_G is determined to be 0.304. Similarly, it is easy to show that the maximum value of K/b for the given section is $0.25 + 0.25\ (9/15) = 0.40$. The maximum value of K/b corresponds to the maximum value of A_K/A_G, the ratio of the total area of knots to the area of the full cross section.

The corresponding value of the ratio I_K/I_G that is most likely to occur statistically can be obtained from curve B of Fig. 3.9, using $n = 15$ laminations. A coefficient equal to 0.6 is obtained; hence,

$$I_K/I_G = 0.6 \times 0.304 = 0.1824 \tag{3.2}$$

Entering Fig. 3.8 with $I_K/I_G = 0.18$ yields $y_A = 97$ and $y_B = 76$, which are the percentages of modulus of elasticity and flexural stress for clear wood, respectively, for the given beam.

Similarly, using $n = 15$ laminations and curve A of Fig. 3.9 a coefficient equal to 0.70 is obtained; hence, $K/b = 0.70 \times 0.40 = 0.28$. Entering Fig. 3.10 with the preceding value of K/b one obtains 87 and 75 percent of compressive stress and tensile stress for clear wood, respectively, as the answers required by item 3 of the problem.

3.4.4 Effects of Slope of Grain

The relations between strength ratio and slope of grain for the component laminations of a given glued member are identical to those listed in Table 2.3. The most stringent demands in beams are obviously for the outer laminations; interior laminations may have larger cross grain if their strength is sufficiently high to resist the stresses imposed by flexure. Flexural stresses are assumed to vary linearly from a maximum at the outermost fiber to zero at the neutral plane, see Section 7.2. Hence, in the case of beams, it is not necessary to use the quality specification for cross grain of the outside laminations for the inside laminations. For instance, if a lamination with 1/16 slope of grain is used for the outside tension lamination, a strength ratio equal to 0.80 is obtained, see Table 2.3. In view of the linear distribution of flexural stresses, the stress at a lamination one-eighth the distance from the outside fiber to the neutral axis is 7/8 that of the stress in the outside fiber; the required strength ratio is $(7/8)(0.80) = 0.70$. From Table 2.3, laminations having slope of grain as large as 1/12 can be used from the neutral axis down to the above referred lamination, without changing the strength ratio of the section. Following the same argument it can be shown that laminations with 1/8 slope of grain can be used between the neutral axis and a point at three-eighths the distance from the outer fibers to the neutral axis.

Normally, a strength ratio is established based on the effects of knots; the slope of grain is limited in such a way that the strength ratio defined by it is equal to, or larger than, the strength ratio resulting from knots. Laminations containing steep slope of grain should not be used (1), because the occurrence of such steep slopes is likely to result in severe warping, twisting and high stresses when such pieces are clamped in gluing.

3.4.5 Effects of Miscellaneous Variables

Tests have shown that the relative strength of glued-laminated wood beams varies with the size of the cross section (6). A decrease in relative strength, as measured by the modulus of rupture, occurs when the depth of the cross section increases. Refer to Sections 7.4 and 7.5 where a full discussion of this effect is given as part of the general presentation of flexural behavior of wood beams.

The factors that affect strength of straight glued-laminated members affect also the strength of curved members. In addition, certain factors are applicable to curved members only. A definite loss of strength with increased curvature of the laminations was obtained in tests of arches by Wilson (7); see Section 7.13 for a discussion of this effect. For any given curvature, loss of strength is greater in 1½ in. laminations than in ¾ in. laminations. Hence, sharply curved members such as knees of rigid frames are usually fabricated with ¾ in. rather than 1½ in. laminations.

When a curved member is subjected to bending moment, stress is induced in a radial direction and thus acts normal to the grain. These radial stresses may be either compression or tension, depending on the curvature of the section and the acting moment, see Section 7.14. Strength of the member can be impaired if longitudinal checking, or a separation of the laminations at the glue line, occur due to excessive radial tension. To prevent this from happening the allowable value for radial tensile stress is taken as a portion of the allowable horizontal shear value and varies according to species (8), see Section 7.14. Cross-bolts, if used, are effective only after separation of laminations has occurred.

The effects of various load durations on strength of glued-laminated members are not different from those discussed in Section 2.13 for solid-sawn wood members; adjustment factors of allowable stresses for various loading conditions are given in Fig. 2.14. The effects of permanent loading on deformations of glued-laminated members are also similar (1) to the effects on solid wood members, see Section 2.9. It is necessary to provide either camber or extra stiffness where deflection under long periods of loading must be limited in amount. This can be done, as shown in Section 7.11, by taking one and one half times the initial deflection due to long-time loading.

The effects of increased moisture content on strength of glued-laminated members can be represented also by the relation given in Fig. 2.11. Design specifications (3,4,8) allow smaller working stresses for wet than for dry conditions of use, see Tables A.3.1 to A.3.3. In addition, it is considered necessary that wood in glued-laminated members intended for exposed use in conditions conducive to decay, see Section 5.2, be made durable by the application of preservative chemicals. Treatment may be given to the lumber before laminating, or to the finished member. Since preservative treatments, see Section 5.8, do not prevent wood from absorbing moisture in service, design of treated glued-laminated members for exposed use is accomplished using allowable stresses corresponding to wet conditions.

3.5 RESEARCH ON FLEXURAL STRENGTH OF GLUED-LAMINATED MEMBERS

The research conducted at Forest Products Laboratory (1,7,9) between the years 1930 and 1954 was mainly on beams 12 in. or less in depth; only a few beams over 30 in. were tested. Design criteria based on the statistical variation of strength with the I_K/I_G ratio, see Section 3.4, were developed and based on these tests. In the early 1960's, tests on beams 31½ in. deep (10) showed strengths lower than those that were predicted using the I_K/I_G theory. Research on prestressed-wood beams (11) showed that the use of a clear, straight-grained, outer lamination in a beam had a pronounced effect on its modulus of rupture; when compared to equivalent beams, this addition caused increases of as much as 32 percent in flexural capacity. It was concluded that a significant improvement in flexural strength of structural glued-laminated beams is possible by increasing the quality of the outer few laminations.

A total of 41 beams was tested in two series of experiments conducted at Forest Products Laboratory in the late 1960's (12,13). Wood was either Southern pine or Douglas fir, the latter in both coast and north-interior types. The majority of the beams were 5¼ in. wide, approximately 24 in. deep and 40 ft. long. There were six beams 9 in. wide, 31.5 in. deep and 50 ft. long. The beams were manufactured using lumber grade combinations with a tested bending stress of 2600 psi; the only change was that one tension lamination meeting stringent AITC 301 requirements* was substituted for the outer tension lamination of each beam.

Failure of the beams was classified into three categories: 1) compression failure, 2) failure through a strength-reducing characteristic in the outer tension lamination, and 3) failure at an end joint in the outer or first interior tension lamination. Only two out of 15 beams in this series failed by compression first; nine of the other 13 beams involved edge knots in the tension lamination; the other four failures involving edge knots also involved end joints, see Fig. 3.11, two in the outer tension laminations and two in the first interior tension lamination. It may be pointed out that the two beams that failed first in compression did so at relatively high loads and that the lowest test loads of the series occurred in beams which failed first in tension. It was concluded that, in general, the use of better quality outer-tension laminations results in significantly stronger glued-laminated beams. In addition, the role of the first interior tension lamination was recognized as a possible source of failure; upgrading of this lamination was strongly recommended. The results of this research were extensively used in writing the 1971 AITC standard specifications for structural glued-laminated timber, see Ref. 3.

3.5.1 Beams with Laminations Positioned by Visual-Stiffness Criteria

The idea of using a stiffness criterion, in addition to visual grading, to classify laminations has been explored in recent years. The thought behind this concept is based on the following considerations. Deflection limitations often govern the design of flexural members, see Section 7.11. Deflection of a given beam is inversely proportional to the stiffness of the section; any attempt to increase the value of E will contribute to reduce the deflection under load. If lumber graded by visual methods would receive consideration of its modulus of elasticity, as an added measure of quality to visual grading, it would permit more accurate assessment of the quality of the lami-

*AITC 301–67 requirements were used for tests reported in Ref. 12 while those of AITC 301–A69 were used for tests reported in Ref. 13.

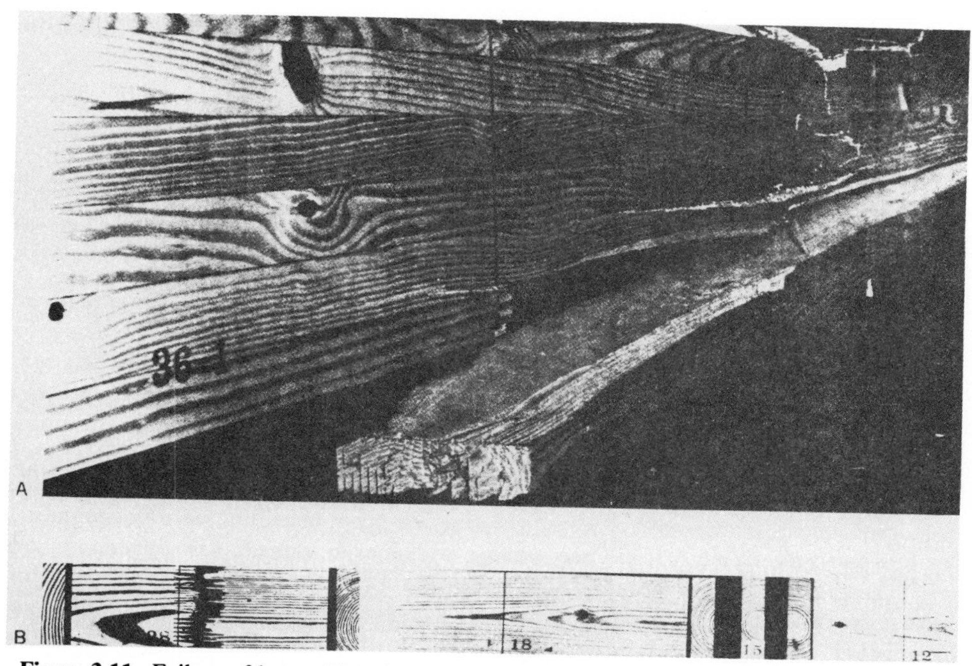

Figure 3.11. Failure of beam 36 at finger joint in tension lamination. Shown are: (A) side view of beam following test, (B) bottom view of tension lamination showing failure and end sections. From Ref. 13.

nations; then, using the stiffest laminations in the outer areas of a beam could produce both stiffer and stronger beams. This idea was explored in the early 1960's (14,15) and re-examined in 1970 (16).

A series of beams 5¼ in. wide, 24⅜ in. deep and 40 ft. long were laminated in such a way that the laminations were assembled in descending order of stiffness from the outside to the inside of the members. The modulus of elasticity of each board was determined using a vibration technique; modulus of elasticity ranged between 2200 ksi for the outer laminations to 1000 ksi for the central lamination. It is easy to show that, theoretically, the equivalent modulus of elasticity of the section is given by $\sum_{1}^{n} E_i I_i / I$, where E_i, I_i are the modulus of elasticity and moment of inertia respectively, of the ith lamination, n is the number of laminations, and I is the gross moment of inertia of the section. This series was compared to another series of similar beams in which modulus of elasticity was not used to locate the laminations during manufacture.

The results of the tests showed that in four of the five beams tested, failure originated at edge knots in the tension laminations, see Fig. 3.12; the other beam failed through a centerline knot in the tension lamination. Once tension failure occurred, compression failure of the top fibers of the beams followed instantly. Hence, failure occurred in a tension-compression sequence precipitated by the presence of knots. Comparison between the beams made using both visual and

Figure 3.12. Failure of beam 32 initiated at an edge knot in the tension lamination (lower center) and propagated through remainder of beam. Beam 32 had a modulus of rupture (adjusted to 12 pct. moisture content) of 6,320 psi and a full-span modulus of elasticity of 1.97 × 10 psi. From Ref. 16.

stiffness criteria for positioning laminations, and those beams made using visual criteria only, showed the first group significantly stiffer. In addition, the average modulus of rupture, see Section 7.3, of the visual-stiffness beams was 14 percent higher than the average for the visual-only beams. However, the minimum values within each group were nearly equal. The lessons learned in this test series were used to set up another series of tests of 54 beams, see Ref. 19. The results of these investigations were used to develop the 1971 Standard Specifications for Structural Glued-Laminated Timber Using "E"-rated and Visually Graded Lumber (4).

3.5.2 Beams Made of Mixed Species of Wood

Two softwoods, Douglas Fir and Southern Pine, have been commonly used for laminating. Many reasons justify this widespread use, including the fact that there has always been an adequate raw material supply. However, in the early 1970's the glulam wood industry, concerned with potentially inadequate supply of Douglas Fir and Southern Pine in the face of rising demand, recognized the need for using other woods in combination for the fabrication of glulam members. In particular, for members such as beams, which are principally subjected to bending, a combination of species (with higher strength species in the outer laminations and lower strength species as inner laminations) is quite feasible and would result in a wider raw material base for laminating (23).

In 1973 initial research in this area by Moody, at the Forest Products Laboratory (23), was devoted to two-species beams. The design approach used in this study was a combination of a transformed-section type analysis and the I_K/I_G concept (with tension lamination modifications) for evaluating knot effects used for single-species beams. Twenty beams, 41 ft. long and using finger-jointed lumber, were manufactured of Douglas Fir and Lodgepole pine. The beams were tested to failure simply-supported on a 38 ft. span with two symmetrically-placed load points 8 ft. apart. The beams were approximately 5 in. wide and 24 in. deep. They were designed in three distinct groups to represent practical levels of design bending stress (1600 f, 2000 f and 2400 f) and in order to detect, had they existed, any gross inconsistencies in the proposed concepts for section design.

Beam failures were generally sudden and complete, with the exception of two beams that failed in the outer compression laminations. Tension failures occurred either through edge knot and associated grain deviation in the outer tension lamination or by end joint failure in the outer or second tension lamination.

The results of the tests substantiated the transformed-section concept. Substituting Lodgepole pine for Douglas-fir inner laminations had little effect on modulus of rupture and modulus of elasticity properties of the beams tested. All strength values exceeded the predicted near-minimum levels, thus supporting the combination of design method (transformed section and I_K/I_G concept) and the other assumptions made. The study further supported previous findings that finger joints controlled the strength of many beams and that a balance between known lamination strength, as controlled by defect, and by finger joints, was approached with AITC's 301–24 grade.

The design method for proportioning glulam sections which resulted from these studies was adopted by ASTM 3737–78 Standard, see Ref. 24. The following example, consisting of the design of a 20-lamination glulam beam using two grades of Douglas Fir for the outer laminations and one grade of Lodgepole pine for the inner laminations illustrates the method.

3.5.2.1 Illustrative Example

Design a 20-lamination Glulam section for which the allowable combination bending stress and modulus of elasticity are: $F_b = 2{,}400$ lb/in^2 and $E = 1{,}800{,}000$ lb/in^2, respectively. Consider the section shown in Fig. 3.13a as a possible solution and determine F_b and E using the standard method given by ASTM D3737–78, see Ref. 24.

As shown in Fig. 3.13a the section is symmetrical. Each half section consists of two L1 and four L2 Douglas Fir outer laminations, and four L3 grade Lodgepole pine inner laminations. Clear wood design stresses in bending and modulus of elasticity of these wood grades are given in Table 3.2. Note the difference in design bending stress and modulus of elasticity for the three grades of wood used in the beam.

Because of the different moduli of elasticity of the various grades of wood the transformed section appears as shown in Fig. 3.13b. The widths of transformed section for the various wood grades are b, $(E_2/E_1)b$ and $(E_3/E_1)b$, respectively, where b is the width of the original section. Note that the latter is transformed to a uniform wood section with constant modulus of elasticity, E_1.

For a linear strain distribution due to bending, the stress distribution appears jagged, see Fig. 3.13c, at levels where adjacent wood has different moduli of elasticity. The maximum bending stress at these levels, namely: f_1, f_2 and f_3, should not exceed the corresponding allowable stresses for each grade of wood. Usually, f_1, the tension stress in the outer lamination controls the cross-section design and is made equal to the allowable design stress; in that case f_2 and f_3 are equal to, or lower than, the corresponding allowable stresses for each grade of wood.

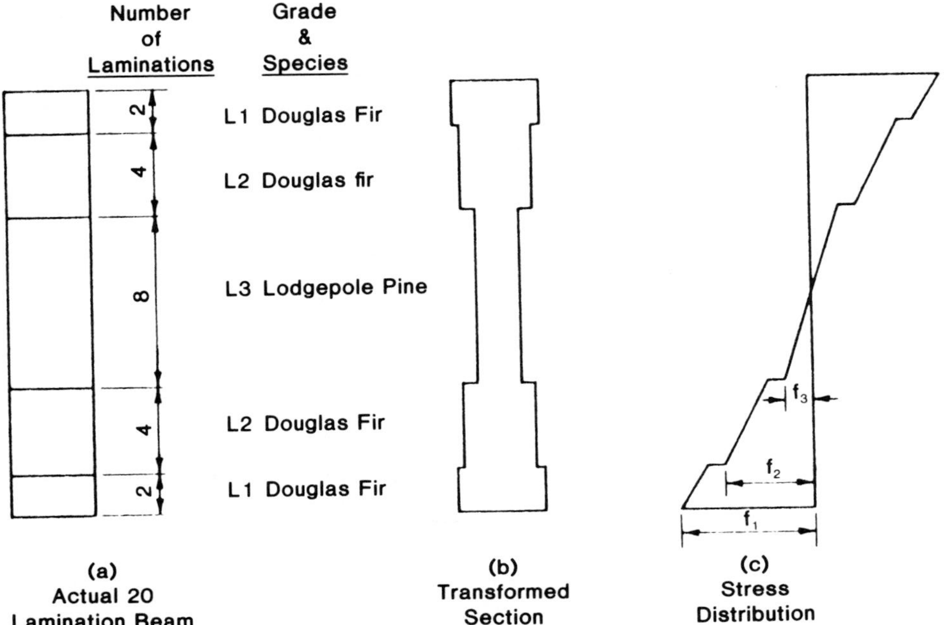

Figure 3.13. Analysis of 20-lamination beam. Taken from ASTM D–3737–78, See Ref. 24.

TABLE 3.2. GLULAM BEAM. CHARACTERISTICS OF VARIOUS WOODS USED.

Wood	Grade	Number of Laminations in Half-Section	Clear-wood Design Stresses: Bending f, psi	Clear-wood Design Stresses: Modulus of Elasticity, E psi
Douglas Fir	L1	2 outer	3,500	2,100,000
Douglas Fir	L2	4 intermediate	3,000	1,800,000
Lodgepole	L3	4 inner	1,933	1,100,000

Practical use of this information by designers requires that the properties (A,I,S) of the actual physical section be used instead of those of the transformed section which require more elaborate calculations. Thus, for use with the section modulus, S, and moment of inertia, I, of the original cross-section, it is required that F_b and E be used instead of f_1 and E_1. The values F_b and E are obtained as follows:

$$F_b = f_1 T$$
$$E = 0.95 E_1 T$$

where T is the transformed-section moment-of-inertia factor. The 0.95 factor applied to the modulus of elasticity should take care of shear deflections for all beams with span-to-depth ratios larger than 14:1.

The calculations necessary to determine the set of allowable stresses f_1, f_2 and f_3 are summarized in Table 3.3. Note that determination of a set of strength ratios, SR, is necessary since allowable stress is obtained by multiplying clear-wood design stress by a corresponding SR. The latter may be calculated as a function of R, the I_K/I_G ratio of the corresponding beam. The whole process is given in Table 3.3 with footnotes containing references and formulas used in the various calculations. In addition, one can find f_2 and f_3 evaluated in the last column of the Table when f_1 = 2640 psi; the fact that f_2 = 1810 psi and f_3 = 550 psi are lower than the corresponding allowable stresses 1890 and 970 psi, indicates that f_1 = 2640 psi is the controlling allowable stress for the section.

TABLE 3.3. GLULAM BEAM.

Determination of Bending Stresses for the Various Woods Used

Section	$R^1 = \frac{I_K}{I_G}$	Strength Ratio[2], SR	Bending Stresses: Allowable[3], psi	Bending Stresses: Actual[4], psi
Whole Beam	0.191	0.753	f_1 = 2,640	f_1 = 2,640
16-lam. interm. beam	0.259	0.629	f_2 = 1,890	f_2 = 1,810 < 1,890, OK
8-lam. inner beam	0.490	0.500	f_3 = 970	f_3 = 550 < 970, OK

1. Using formula 3 from Ref. 24
2. $SR = (1+3R)(1-R)^3(1-R/2)$
3. f = (SR) (clear wood design stress from Table 3.3)
4. From $f_i = f_1 (E_i/E_1)(d_i/d_1)$ for f_1 = 2640 lb/ in², $d_2 = 0.8\ d_1$, $d_3 = 0.4\ d_1$, $E_2/E_1 = 1.8/2.1$ and $E_3/E_1 = 1.1/2.1$

At this point all that remains to be calculated is the transformed-section moment-of-inertia factor, T. It can be shown that for a symmetrical beam section with three stiffness zones, T may be calculated as follows:

$$T = 1 - \left(\frac{d_2}{d_1}\right)^3 \left(1 - \frac{E_2}{E_1}\right) - \left(\frac{d_3}{d_1}\right)^3 \left(\frac{E_2}{E_1} - \frac{E_3}{E_1}\right)$$

For the given section $d_2/d_1 = 0.8$, $d_3/d_1 = 0.4$, $E_2/E_1 = 1.8/2.1$ and $E_3/E_1 = 1.1/2.1$. Substituting these values in the preceding formula yields

$$T = 1 - (0.8)^3 \left(1 - \frac{1.8}{2.1}\right) - (0.4)^3 \left(\frac{1.8}{2.1} - \frac{1.1}{2.1}\right) = 0.906$$

The required values for the allowable combination bending stress, F_b, and modulus of elasticity, E, are given as follows:

$$F_b = f_1\ T = (2{,}640)\ (0.906) = 2392 \text{ psi}$$

$$E = 0.95\ E_1\ T = (0.95)\ (2{,}100{,}000)\ (0.906) = 1{,}807{,}500 \text{ psi}$$

which are finally rounded off to F_b = 2400 psi and E = 1,800,000 lb./in². These were the values sought by the cross-section designer, thus, the solution shown in Fig. 3.13 a is acceptable.

3.5.3 Further Research on Improved Lumber Utilization

Additional studies made by Moody (24) during the period 1974 through 1977 further explored the use of lower grade wood for the design of glulam beams. Three additional possibilities were investigated, namely: 1) reducing grade of lumber on the compression side of beams, 2) using lumber with wane (bark or lack of wood from any cause on edge or corner of a piece) for inner laminations and, 3) extending criteria developed for visually graded lumber to E-rated lumber, i.e., to lumber which has had its modulus of elasticity determined.

A total of 120 beams 20 ft long were manufactured and tested simply-supported over a 19 ft span with 4 ft between loading heads; lateral supports at about 4 ft from each end, with roller contacts to minimize friction force, were provided for lateral stability. The beams were 3⅛ in. wide and either 12 in. or 18 in. deep. The beams failed mostly in tension (108 out of 120), 10 beams failed in compression and only 2 failed in shear. Of the tension-related failures the largest number, 47, occurred because of selected tension-lamination characteristics (edge and centerline knots, low density, etc.). This was followed by 26 failures caused by finger joints alone, 17 by finger joint and other defects and 18 by other tension lamination characteristics (slope of grain, local grain deviations and others).

Analysis of the test results indicated justification for using reduced wood grades in the compression laminations of a glued-laminated beam. Also, the procedure used to design visually-graded glulam beams was found acceptable for design using E-rated laminations. Finally, in beams made with lumber having wane occupying up to one-sixth of the width at either edge, shear weaknesses larger than assumed were not apparent; under dry conditions design of such beams for a shear stress equal to ⅔ of that with wane-free lumber would appear justified. Beams made with wany lumber should not be used unless dry conditions are guaranteed. This results from the fact that wane may introduce problems due to stress concentrations at the gluelines in beams subjected to repeated wetting and drying.

3.5.4 Vertically Glued Laminated Beams

These beams are presently designed following provisions of ASTM Standard D245, see Ref. 27. For vertically glued-laminated beams consisting of three or more laminations, a 15-percent increase over single-member bending stress is permitted based on the same repetitive-member provision familiar to designers of light-frame components such as roof or floor systems. Thus, the design bending stress, F_b, may be taken as 1.15 times the weighted average of the corresponding design stresses of the various component laminations. For example, if laminations occupying two thirds of the width have a 1.5 ksi design bending stress and the remainder have 1.2 ksi, then the design bending stress, F_b, for the section may be taken as:

$$F_b = 1.15\,(1.5 \times \tfrac{2}{3} + 1.2 \times \tfrac{1}{3}) = 1.61 \text{ ksi}$$

A recent study (1979) of these members was performed by Wolfe and Moody at Forest Products Laboratory, see Ref. 27. The work included actual bending tests on a sample of 12 ft long 5⅛ in. deep beams, loaded on an 11 ft long, simply-supported span, with load heads symmetrically placed 22 in. either side of midspan. Some significant conclusions of this study indicated that:

1. The average modulus of elasticity was not affected by the number of laminations.
2. No significant difference was found between the average bending strength (expressed by the modulus of rupture) with 3, 4, or 5 lamination members.
3. Variability in both modulus of elasticity and rupture decreases with the number of laminations. The corresponding coefficients of variations are inversely proportional to the square root of the number of laminations.
4. The commonly used 1.15 repetitive-member factor is conservative for members with two or more laminations, varying from 1.5 for the highest grade to over 2.5 for the lowest grade.
5. Procedures for estimating the near-minimum strength (assumed to be near the statistical 5-percent exclusion value of the beam population) varied in their conservatism depending upon quality and number of laminations.

It is expected that this work will influence design procedures for vertically glued-laminated beams.

3.6 DESIGN OF GLUED-LAMINATED MEMBERS

The behavior and design provisions of glued-laminated members subjected to bending, and combined axial load and bending are discussed in Chapters 7 and 8, respectively. Provisions for design of connections are given in Chapter 6. Examples of design can be found in Chapters 6 through 11.

The provisions of National Design Specification "Design Values for Wood Construction" (28) are given in Tables A.3.1, A.3.2 and A.3.3. The first of these Tables gives design values for normal loading conditions and dry conditions of use for glulam members made out of softwoods (from Western Species or Southern Pine) and stressed primarily in bending. Design values are given for bending about the strong axis (loaded perpendicular to wide faces of laminations) and for bending about the weak axis (loaded parallel to wide faces of the laminations). In addition, design values are given to check the axial-loading term in the design interaction equation (see Chapter 8) since members in this category may include rigid frames with cross-sections that are subjected, not only to bending, but also to a certain amount of thrust. Glulam combinations using both visually-graded species (V) and E-rated species (E) are given. In addition, the designer using Western woods is provided also with information on the wood species from which outer laminations and inner (core) laminations may be obtained for any given combination symbol. Thus, for combination symbol 22F-V1, Table A.3.1 for Western woods provides a structural designer with the following information:

22F-V1 visually-graded western species

DF/WW outer laminations: Douglas Fir-Larch
core laminations: Western Woods or Canadian Softwood Species

Bending About Strong Axis:

F_{bxx} = 2200 psi (tension zone stressed in tension)

F_{bxx} = 1100 psi (compression zone stressed in tension, as in simple beam with a short cantilever)

$F_{c\perp xx}$ = 450 psi (tension face)
$F_{c\perp xx}$ = 385 psi (compression face)
F_{Vxx} = 140 psi
E_{xx} = 1,600,000 psi

Bending About Weak Axis:

F_{byy} = 1050 psi
$F_{c\perp yy}$ = 190 psi
F_{Vyy} = 130 psi
E_{yy} = 1,300,000 psi

Axially Loaded:

F_t = 850 psi
F_c = 1100 psi
E = 1,300,000 psi

Similar information is provided by Table A3.1 for glulam members made out of Southern Pine which, however, is not used in combination with any other wood species. It should be noted, because of its importance, that species classification for fastener design (which is a function of the design stress in compression perpendicular to grain, see Chapter 6) is also listed in footnote 14 of Table A3.1.

Table A.3.2 provides identical information as Table A.3.1, except for the design of glulam softwood timbers subjected primarily to axial tension or compression. These members are frequently encountered in practice as they constitute columns, arches and compression chords of bowstring trusses and other structures as well as tension chords, ties and diagonals in trusses, etc. These members are also designed following combination symbols both for visually graded or E-rated lumber. As in the case of Table A.3.1 one finds the Table divided between Western woods and Southern pine. However, no two-species combinations, not even for Western woods, are used for these members.

Finally, there is Table A3.3 containing design values that are given for glulam members made out of hardwood timber.

More comprehensive tables than these, including design values for California Redwood and end-grain bearing in wood, can be found in AITC 117–79, see Ref. 29, in which the American Institute of Timber Construction gives all laminating combinations of both visually-graded lumber and E-rated lumber in one specification.

In order to increase availability of laminated timbers designers are advised to specify by strength requirements whenever possible. This should allow more glulam manufacturers to make the structural member than would be the case when species and grades are limited. Also, for the purpose of economy designers may wish to employ standard sections used by all fabricators; Table A3.4 lists properties of standard sections ranging in widths between 3⅛ in. and 14¼ in. and in depth between 3¼ in. and 66¾ in. For recommended rules and practices on fabrication, transportation and erection of glued-laminated wood structures refer to the latest edition of Timber Construction Standards (17, 18) published by the American Institute of Timber Construction.

REFERENCES

1. Freas, A. D., and Selbo, M. L., "Fabrication and Design of Glued-Laminated Wood Structural Members," U.S. Forest Products Laboratory Technical Bulletin No. 1069, February 1954.
2. American Society for Testing and Materials, "1978 Annual Book of Standards," Part 22, Wood; Adhesives.
3. American Institute of Timber Construction, "Standard Specifications for Structural Glued-Laminated Timber of Douglas Fir, Western Larch, Southern Pine and California Redwood," AITC, 117–71, October 1971.
4. American Institute of Timber Construction, "Standard Specifications for Structural Glued-Laminated Timber Using 'E' Rated and Visually Graded Lumber of Douglas Fir, Southern Pine, Hem-Fir and Lodgepole Pine," AITC 120–71, August 1971.
5. Commercial Standard CS253, "Structural Glued-Laminated Timber," U.S. Department of Commerce, 1963.
6. Bohannan, B., "Effect of Size on Bending Strength of Wood Members," U.S. Department of Agriculture Research Paper FPL 56, May 1956.
7. Wilson, T. R. C., "The Glued-Laminated Wooden Arch," Technical Bulletin No. 691, U.S. Department of Agriculture, Washington, D.C., October 1939, 123 pp.
8. "National Design Specification for Stress-Grade Lumber and Its Fastenings," 1977 Edition by National Forest Products Association, Washington, D.C.
9. Wilson, T. R. C., and Cottingham, W. S., "Tests of Glued-Laminated Wood Beams and Columns and Development of Principles of Design," Forest Products Laboratory Report R1687, 1952.
10. Bohannan, B., "Flexural Behavior of Large Glued-Laminated Beams," U.S.D.A. Forest Service Research Paper FPL 72, December 1966.
11. Bohannan, B., "Prestressed Laminated Wood Beams," U.S.D.A. Forest Service Research Paper FPL 8, January 1964.
12. Bohannan, B., and Moody, R. C., "Large Glued-Laminated Timber Beams with Two Grades of Tension Laminations," U.S.D.A. Forest Service Research Paper FPL 113, September 1969.
13. Moody, R. C., and Bohannan, B., "Glued-Laminated Beams with AITC 301 A-69 Grade Tension Laminations," U.S.D.A. Forest Service Research Paper FPL 146, October 1970.
14. Koch, Peter, "Strength of Beams with Laminae Located According to Stiffness," Forest Products Journal, Vol. 14, No. 10, pp. 456–460, 1964.
15. Koch, P., and Bohannan, B., "Beam Strength as Affected by Placement of Laminae," Forest Products Journal, Vol, 15, No. 7, pp. 289–295, 1965.
16. Bohannan, B., and Moody, R. C., "Flexural Properties of Glued-Laminated Southern Pine Beams with Laminations Positioned by Visual Stiffness Criteria," U.S.D.A. Forest Service Research Paper FPL 127, February 1970.
17. American Institute of Timber Construction, "Timber Construction Standards," AITC 100–69, Fifth Edition, 1969.
18. American Institute of Timber Construction, "Timber Construction Manual," Second Edition, John Wiley and Sons, New York, 1974.
19. Johnson, J. W., "Design and Test of Large Glued-Laminated Beams Made of Non-destructively Tested Lumber," Forest Research Laboratory, School of Forestry, Oregon State University, Report T-27, November 1971.
20. American Institute of Timber Construction, "Standard Specifications for Hardwood Glued-Laminated Timber," AITC 119–71, August 1971.
21. Wood Structures. A Design Guide and Commentary. Published in 1975 by American Society of Civil Engineers, 345 East 47th St., New York 10017.
22. PS 56–73, Voluntary Product Standard for Structural Glued Laminated Timber, U.S. Department of Commerce.
23. Moody, R. C., "Design Criteria for Large Structural Glued-Laminated Timber Beams Using Mixed Species of Visually Graded Lumber" USDA Forest Service Research Paper FPL 236, 1974.

24. Standard Method for Establishing Stresses for Structural Glued Laminated Timber (Glulam) Manufactured from Visually Graded Lumber ASTM D3737–78, American Society for Testing and Materials, 1979 Book of Standards, Part 22.
25. Moody, R. C., "Improved Utilization of Lumber in Glued Laminated Beams" USDA Forest Service Research Paper FPL 292, 1977.
26. "Standard for Dimensions of Glued Laminated Structural Members," American Institute of Timber Construction Standard AITC 113–75.
27. Wolfe, R. W. and Moody, R. C., "Bending Strength of Vertically Glued-Laminated Beams with One to Five Plies," Forest Products Laboratory Research Paper FPL 333, 1979.
28. "Design Values for Wood Construction," A Supplement to the 1977 Edition of National Design Specification for Wood Construction by National Forest Products Association, April 1980.
29. American Institute of Timber Construction, "Design Standard Specifications for Structural Glued Laminated Timber of Softwood Species," AITC 117–79, June 1979 with changes issued June 1980.

Chapter 4
Plywood

4.1 INTRODUCTION

Plywood has been defined as a glued wood panel that is made up of layers, or plies, with the grain of one or more layers at an angle, usually 90°, with the grain of the others (1). The greater amount of plywood panels are made of an odd number of plies,† of which the outside plies are called faces or face and back, the center ply and inner plies are called core or centers, and the plies immediately below the face and back are called crossbands.†† Plywood panels normally used in construction have plies ranging between 1/12 in. to 1/4 in. thick and have a total panel thickness that varies between 1/4 inch and 1 1/8 inch (3); the number of plies varies between 3 and 7. Plywood panels may consist of plies of different thicknesses, species and grades of wood.

The main advantage of plywood, as compared to solid wood, are its tendency to reduce the difference in strength properties in orthogonal directions, i.e., along the length and width of the panel; greater resistance to checking and splitting; and more dimensional stability, in other words, less change in dimensions with changes in moisture content. The rectangular shape of the plywood panel, because of its wide dimensions (4 × 8 ft. is standard size but at least two mills do manufacture 5 × 12 ft. panels), permits rapid erection and coverage of large areas with consequent savings in labor costs.

In the following is given, first, a description of important structural applications of plywood, followed by general characteristics and manufacturing details of the material. The determination of stiffness and strength properties of plywood is discussed at length and finally, for design purposes, effective section properties and allowable stresses of commercially available plywood panels are given. The provisions of U.S. Product Standard PS1–74 (18), which regulates the plywood industry, and the specifications of the American Plywood Association (4) (5) (6) are used extensively.

4.2 STRUCTURAL APPLICATIONS

More than one half the amount of plywood produced in this country is used in residential construction, the main uses being roof and wall sheathing, sub-floor, underlayment, and a small amount used for soffits (3). About 20 percent of the production is employed in the industrial field, for such uses as crates, pallets and boxcar lining. Next in volume, about 13 percent, is plywood used for wall and roof sheathing of factories and warehouses, offices and institutions with a substantial amount used for concrete forming (7). The rest is used in home workshop projects and agricultural uses such as for pallet bins, bulk storage structures and animal shelters.

†Some ½ in. and ⅝ in. panels are made as "4-ply" with the middle layers consisting of two parallel plies. For design purposes they may be considered a "3-ply."

††Standard definitions of terms relating to veneer and plywood are given by ASTM (2).

Figure 4.1. Tongue and groove panels, 1–⅛ in. thick, 2–4–1 plywood used as combination sub-floor underlayment over supports at 48 in. o.c. Two men may install 1000 sq. ft. in 4 hours. Photo courtesy of American Plywood Association.

A good example of the structural application of plywood is shown in Fig. 4.1 where tongue-and-groove 2–4–1 type, 1 1/8 in. thick panels are used as combination sub-floor underlayment in a building. The panels act as one-way slabs between intermediate supports; the tongue-and-groove edge connections between panels facilitate placement, eliminate blocking, and enhance lateral distribution of loads. The latter is particularly true in the case of horizontal loads induced by violent winds or earthquakes. The Alaskan earthquake of April 1964 proved that plywood-sheathed houses and buildings behaved as very stiff structures; major rigid-body displacements were sustained with minimum structural distortions (8).

The explanation of this unusual strength lies in diaphragm action, which occurs when the total sheathing behaves under in-plane forces as a deep, thin, structural unit in its own plane. Large shear forces can be taken by the plywood sheathing acting in this manner; flexural stresses are taken by ridge and eave chords in an action similar to that of beam flanges. Thus, lateral forces on the structure are carried by diaphragm action to the supporting end walls, which in a similar fashion take them to the foundation, see Fig. 4.2. A diaphragm structure results when a series of such diaphragms (floors, roof and walls) are adequately tied together and to the foundation (3). For a structure subjected to wind or seismic loading the designer may use Table A.4.13 to determine the recommended shear, in pounds per foot, of horizontal plywood diaphragms and vertical shear walls; in addition, Table A.4.3 compares the strength and rigidity of various other types of frame walls with those made of plywood.

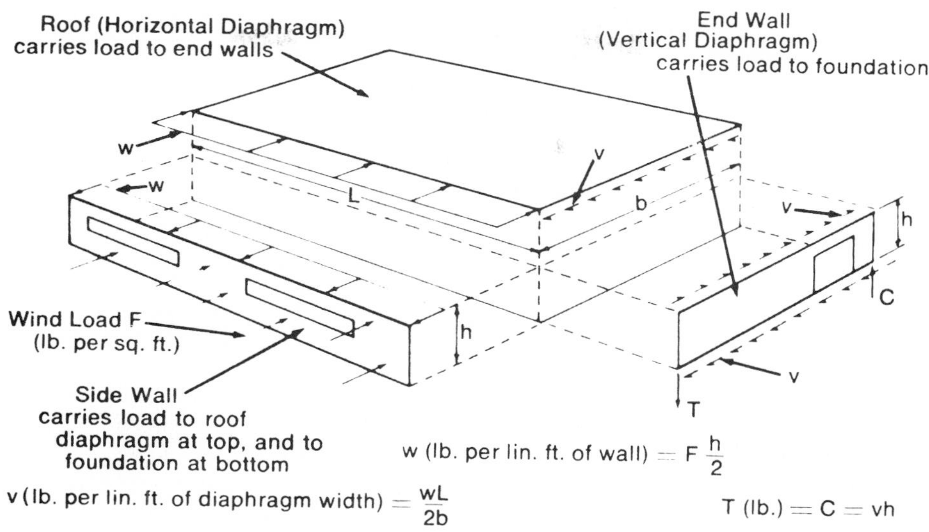

Figure 4.2. Diaphragm action in a one-story building.

The fact that plywood diaphragms are able to resist large shear forces may be extended to the design of folded plate structures (9), see Fig. 4.3, on which these forces are mainly created by gravity in addition to those caused by wind or earthquake. Long, clear spans may be obtained with this type of structure which is ideal to take advantage of the high shear strength and stiffness of plywood panels. Various shapes are possible, namely: rectangular folded plates, see Fig. 4.4; radial folded plates; and space planes, see Fig. 4.5. Analysis of these structures is beyond the scope of this work; reference should be made to existing literature on the subject (9) (10) (11).

Diaphragm action occurs also in plywood-lumber and plywood-glued laminated box beams, see Fig. 4.6, in which lumber on glued-laminated flanges carry most of the bending and plywood webs carry the shear. For a discussion of box beams and an illustrative example of design used in a given building, refer to Section 10.3.2. Similar diaphragm action is used for design of nailed plywood and lumber beams (19) which are frequently used for roofs and garage-door headers. These are lightweight members which are easy to fabricate in the field and can be made as strong and stiff as required.

A case where the plywood panels take flexural stresses and the lumber stringers take shear stresses occurs in flat stressed-skin panels, see Fig. 4.7. There is virtually no-slip of the glued joints between the plywood skins and the framing elements, thus assuring composite action of the panel. These economical structural elements are used for long-span and heavily-loaded floors and roofs and also for walls. Insulating material, if necessary, can be inserted easily during the manufacturing process of the panel. A discussion of design of flat stressed-skin panels can be found in Section 10.3.1 which includes an illustrative example of application to a given building.

(a) Erection of Prefabricated Folded Plate.

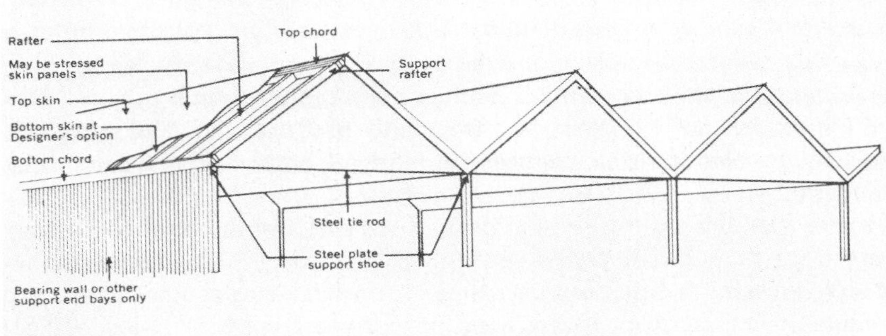

(b) Details of Multiple Folded Plate.

Figure 4.3. Use of plywood panels in the design of folded plates. Courtesy of American Plywood Association.

Figure 4.4. Multiple rectangular folded plates used in modern hotel design. Plates are 40–ft. and 60–ft. long, 2–3/8 in. thick, composed of 5/8 in. plywood covers on 2 × 4, 2 × 6, 2 × 8 chords. The structure was built on site. Photo courtesy of American Plywood Association.

Figure 4.5. Plywood space planes form a clear span area of 32 × 110 ft. for a modern church building in St. Louis. The eighty stressed skin panels used to build this structure were fabricated with 3/8 in. Plywood nail glued to 2 × 4 stringers 24 in. o.c. and 2 × 8 perimeter framing. Photo courtesy of American Plywood Association.

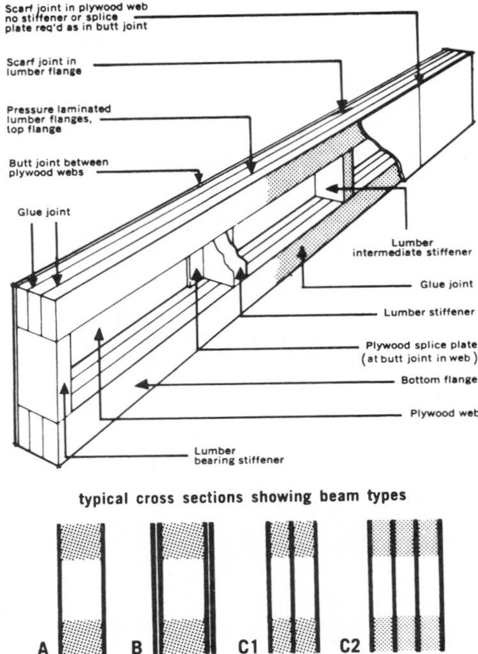

Figure 4.6. Box-beam using plywood webs. From Ref. 6.

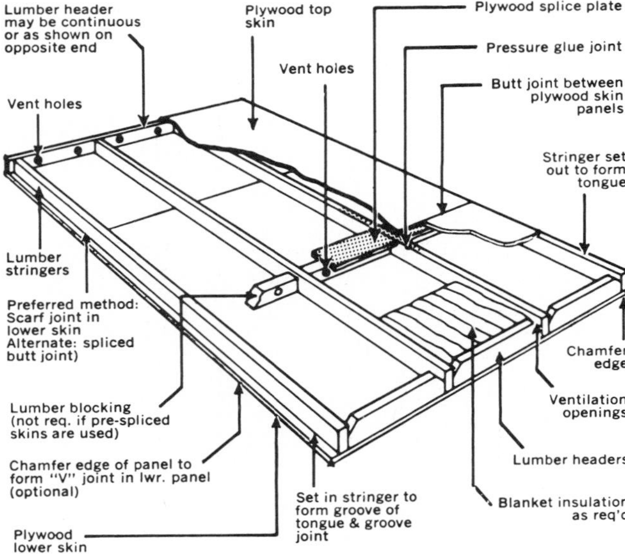

Figure 4.7. Details of stressed-skin panels using plywood for top and bottom skins. From Ref. 6.

Stressed-skin panels need not be flat; curved panels may be used successfully for roof structures, see Fig. 4.8. The shape of these elements allows arching action to develop; since little bending occurs, curved panels with relatively thin cross-sections are capable of spanning long distances between supports. If bending of these elements is to be avoided, tie rods (end buttresses or rigid walls are also acceptable) should be used to take the arch thrust and prevent the transverse opening of the structure. Three types of panels are available, namely: the ribbed panel, the solid core panel and the panel with lightweight sandwich core. Of these, details of the first one is shown

(a) Use of Curved Plywood-Panels for Canopy of Shopping Center.

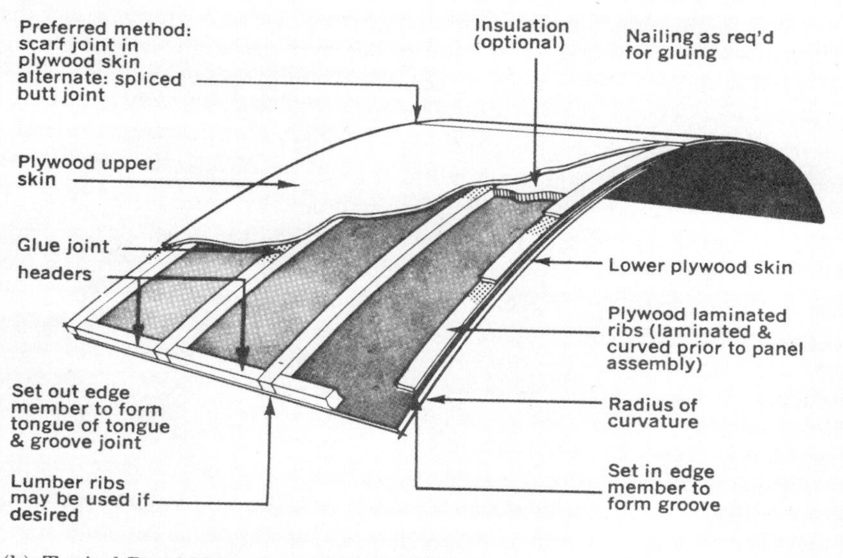

(b) Typical Panel Using Curved Plywood Ribs.

Figure 4.8. Use of curved plywood panels. Courtesy of American Plywood Association.

in Fig. 4.8; the latter uses a resin-impregnated paper honeycomb core sandwiched between top and bottom plywood skins. Minimum radii of bending of plywood panels are given in Table A.4.5.

Examples of these and other applications of plywood, such as nailed gusset plates, various floor and wall systems, domes and hyperbolic paraboloids, are given in Refs. 12 and 13. Analysis and design of plywood-lumber rigid frames are particularly considered in Ref. 14.

4.3 GENERAL CHARACTERISTICS

Plywood is manufactured from more than fifty different species which have been classified into five groups by U.S. Products Standard PS 1–74 (18), see Table A.4.1. Plywood used for structural applications is generally manufactured from the material of softwood trees such as Douglas fir, Western hemlock, Southern pine and the true firs. Plywood made of hardwood trees is mostly used for furniture and decorative paneling; therefore, in the following, only softwood panels are considered.

4.3.1 Identification

Plywood is manufactured in two types, namely: exterior type with waterproof glueline and interior type with highly moisture-resistant glueline. Exterior-type plywood is used for all material exposed to the weather, or whose equilibrium moisture content for other reasons continuously or repeatedly exceeds 18 percent. Veneers in backs and inner plies of the exterior type are of higher grade than those in the interior type; exterior plywood has all veneer of grade C or better.* Interior type plywood should not be specified at all for applications that will be permanently exposed to the weather. However, plywood made with interior veneers and an exterior glue is excellent for sheating where long construction periods are expected, and for some protected exposures where a high moisture level might some time be reached. Because of use of D-grade veneer, this material does not qualify as exterior plywood; also, some panels may develop localized glueline delaminations when permanently exposed to the weather. Such interior plywood with exterior glue has an advantage in addition to increased moisture resistance; it permits use of the same shear stresses as those for exterior plywood because the phenolic-resin glue used strengthens the wood fibers.

Reference is made to Tables A.4.6 and A.4.7 for a guide to appearance grades and engineered grades, respectively, of plywood panels. The group number that appears in many APA-grade trademarks is based on the species used for the face and back of the panel, or the weaker species, if face and back are not the same. A set of two numbers separated by a slash, the Identification Index, appears in the grade trademarks of sheating panels, see Table A.4.7. The number on the left indicates recommended maximum spacing of supports, in inches, when the panel is used for roof decking** with face grain across supports; the number on the right shows maximum recommended spacing for floor supports, in inches, when the panel is used for subflooring*** with face grain across supports. Refer further to APA's guide to identification index on engineered grades in Table A.4.8.

*Veneer grades used in plywood are given in Table A.4.2

**Good for at least 35 psf live load.

***Good for at least 165 psf live load uniformly distributed, and normal concentrated loads such as pianos, home freezers, water heaters, etc.

The Span Index used on STURD-I-FLOOR panels is similar to the Identification Index for sheathing panels, except that the number is the maximum spacing of supports for combination subfloor-underlayment under average residential loading.

The "Index" system, consisting of the Identification and Span Indexes, was established to simplify specification of plywood without resorting to structural design. The indexes indicate performance without the need to refer to species group or panel thickness by giving the allowable span when the face grain is placed across supports. For example, for roof sheathing on a 24-in. span, the user will specify a 24/0 C-D INT-APA panel, which in actuality might be a ⅜-in. Group 1, or ½-in. Group 2, 3 or 4 plywood. So that there is no problem with two different panel thicknesses at the same job it is wise to specify thickness also; thus, the final order might read ⅜ in., 24/0 C-D INT-APA.

Plywood panels of several grades may be modified by the STRUCTURAL designation. Structural applications of plywood, such as box-beams, gusset plates, stressed-skin panels and diaphragms, etc. . . . demand the use of structural-grade panels. STRUCTURAL I is limited to Group 1 species, see Table A.4.1. STRUCTURAL II allows Species Groups 1, 2 and 3. Both are made only with exterior glue, and have some further restrictions, as on knot sizes and repairs.

Panels in grades other than those mentioned above are designated by the panel thickness, by the Veneer classification of face and back veneers, and by the species group of the veneers. PLYFORM is an exception, where Class designates a species mix.

4.3.2 Manufacturing

The following basic steps used in the manufacture of softwood plywood panels are similar for all species. The panels are generally made up of an odd number of veneers or plies, glued together with the grain of adjacent veneers or layers at right angles. Veneer for all panels covered by APA's Plywood Design Specification (4) is peeled, rather than sliced or sawn.

Peeler logs are cut into desired length of "bolts," usually about 8½ ft. long, which are placed in a giant lathe and rotated against a long knife. The latter peels the wood in long, continuous, thin sheets known as veneer. The veneer is conveyed to clippers which cut it to desired widths, after which it is run through dryers where its moisture content is reduced to about two or three percent. Veneers are graded in five groups N, A, B, C or D, as shown in Table A.4.2, according to quality; after grading, the veneer goes to the glue spreaders where adhesive is applied.

The majority of softwood plywood panels is made using a hot-press process, whereby the glue is set in a few minutes by the application of heat and pressure. Upon removal from the press, panels are trimmed to size, and some grades are sanded. Plywood produced by APA mills to conform with U.S. Product Standard PS 1–74 (18) carries a copyrighted trademark on every panel, which assures easy identification and quality control.

4.3.3 Dimensional Stability

Crossbanded products such as plywood panels may be subjected to warping, caused by uneven shrinking and swelling due to moisture changes, unless they are balanced. Balancing is a condition normally attained by arranging plies in similar pairs on each side of the core, thus justifying the use of an odd number of plies. The usual panels with three, five and seven plies, in which plies are placed with the fiber grain at exactly right angles to adjacent plies, remain flat through changes in moisture content.

From an oven-dry condition to fiber-saturation point, plywood shrinks or swells approximately 0.2 to 0.3 of one percent in the plane of the panel (3); the maximum shrinkage or swelling under normal conditions of use is about 0.1 of one percent. Shrinkage in thickness is approximately the same as that of wood, see Section 1.3.

Changes in temperature affecting plywood panels depend on the wood species used. For instance, the coefficients of thermal expansion per 1°F of a 3-ply Douglas fir plywood panel are: 2.3×10^{-6} for a direction parallel to the face grain of the panel; 4.0×10^{-6} in the direction perpendicular to the grain of the face ply; and 15×10^{-6} in the direction of the thickness of the panel. These values are similar to those of solid wood, see Section 1.3.

4.3.4 Mechanical Fasteners

The strength of plywood connections made with mechanical fasteners can be determined using Table A.4.4, which gives ultimate loads carried at failure. A reduction factor, at least equal to four, is necessary to determine allowable loads from the table. The nailing characteristics of plywood are not greatly different from those of solid wood except that the greater resistance of plywood to splitting, when nails are driven near an edge, is a definite advantage (1). Thus, in general, the provisions and suggested design values given in Chapter 6 for mechanical fasteners may be applied to plywood with confidence.

4.3.5 Glued Joints

Plywood used in structural applications such as box beams, stressed-skin panels, gusset plates for lumber rigid frames, etc. . . . are attached normally by means of highly-efficient glued joints. The latter are arranged to transmit loads through shear stresses. Because shear strength across the grain of wood (often called rolling shear strength for the tendency to roll of the wood fibers) is considerably less than parallel to grain, see Section 2.2.8, sufficient area must be provided between plywood and flange members of box beams and plywood and stringers of stressed-skin construction to avoid shearing failure perpendicular to the grain in the face veneer, in the cross-band veneer next to the face veneer, or in the wood member. Thus, design of glued joints using plywood is controlled by rolling-shear strength of the material.

Glued joints should not be designed to transmit load in tension normal to the plane of the plywood sheet because of the low tensile strength of wood in a direction perpendicular to the grain.

4.3.6 Dimensional Tolerances and Squareness of Plywood Panels

The following provisions are given by U.S. Product Standard PS 1–74, see Ref. 18, for the manufacture of plywood panels.

Size:	+ 0.0 in.,—1/16 in. on the specified length and/or width	
Thickness:	Sanded Panels:	$\pm$ 1/64 in. for $t \leq 3/4$ in.
		$\pm$ 0.03 t for $t > 3/4$ in.
	Unsanded, Touch-Sanded and overlaid Panels:	
		$\pm$ 1/32 in. for $t \leq 13/16$ in.
		$\pm$ 0.05 t for $t > 13/16$ in.

Squareness: $1/64$ in./ft. for $L \geq 4$ ft.
$1/16$ in. measured along short direction for $L < 4$ ft.

Straightness: All panels manufactured so that a straight line drawn from one corner to the adjacent corner falls within $1/16$ in. of panel edge.

4.4 STIFFNESS AND STRENGTH

The nature of a strip of plywood is such that its strength and stiffness are smaller than those of a strip of wood of the same size. Plies having their direction of fibers at 90° to the direction of actual stress contribute only a fraction of the strength given by the corresponding areas of the solid wood strip. This also explains why strength properties in the principal orthogonal directions are closer in plywood than in solid wood; in other words, the strength of plywood in a direction normal to the face grain may not be as small as it would be in an equivalent piece of wood.

The following sections give formulas for the determination of stiffness and strength of plywood, in the two principal orthogonal directions, under direct axial stress, pure bending moment and edgewise shear stress.

4.4.1 Modulus of Elasticity in Direct Axial Stress

The modulus of elasticity of a plywood panel in direct axial stress, either tension or compression, can be determined as in Fig. 4.9 which shows a panel consisting of three plies of equal thickness t and width b. The determination is given for the two principal cases of axial stress, namely: parallel to the face grain and perpendicular to the face grain.

Under the action of direct axial force N, the strain distribution is a uniform ϵ over the section. The actual stress distribution is obtained by multiplying strains by corresponding values of modulus of elasticity; thus, stresses in the face and back plies are given by ϵE_L and those in the center ply by ϵE_T, where E_L and E_T are moduli of elasticity of the veneer parallel to grain and perpendicular to grain, respectively.* The stress distribution in the equivalent homogeneous panel is uniform at an intensity equal to the strain times the modulus of elasticity. Equating the resultant of the stress distributions to N leads to the determination first of E_w, the modulus of elasticity of the panel under axial stress parallel to face grain and second of E_x, the modulus of elasticity of the panel under axial stress perpendicular to face grain. The following expressions are obtained in Fig. 4.9:

$$E_w = \frac{1}{3}(2\,E_L + E_T)$$

$$E_x = \frac{1}{3}(E_L + 2\,E_T) \tag{4.1}$$

The general expression (15) for E_w and E_x may be obtained using the procedure given in Fig. 4.9; they are given by Eq. 4.2 as follows:

$$E_w \text{ or } E_x = \frac{1}{h}\sum_{1}^{n} E_i\,h_i \tag{4.2}$$

*If the veneer is rotary cut, the value of E_T is the modulus of elasticity in the tangential direction. For quarter-sliced veneer, the modulus of elasticity in the radial direction E_R, should be substituted for E_T. Values of the ratio E_T/E_L and E_R/E_L are listed in the Wood Handbook (1) for various species; practically, $E_T/E_L = 0.05$ and $E_R/E_L = 0.10$.

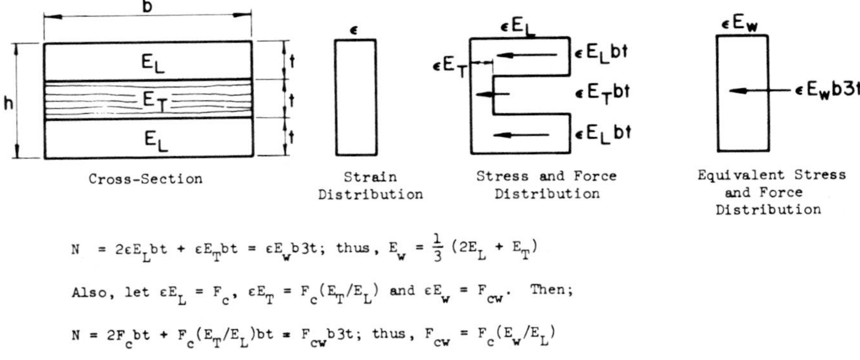

(a) Axial Stress Parallel to Face Grain.

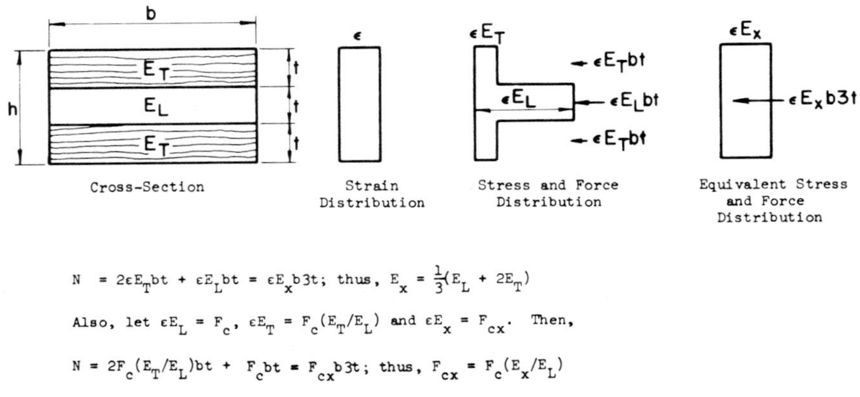

(b) Axial Stress Perpendicular to Face Grain.

Figure 4.9. Determination of moduli of elasticity and strength in direct axial stress of a given plywood panel.

where: E_i is the modulus of elasticity parallel to the applied load of the veneer in ply i; h_i is the thickness of the veneer in ply i; h is the thickness of the plywood; and n is the number of plies. When all plies are of the same thickness and wood species, Eq. 4.2 yields:

$$E_w = \frac{1}{2n}\left[(E_L + E_T)\,n + (E_L - E_T)\right]$$

and (4.3)

$$E_x = \frac{1}{2n}\left[(E_L + E_T)\,n - (E_L - E_T)\right]$$

For n = 3, the preceding formulas reduce to Eqs. 4.1.

The modulus of elasticity in compression, E_θ, at an angle θ to the face-grain direction is given approximately (1) by:

$$\frac{1}{E_\theta} = \frac{1}{E_w} \cos^4\theta + \frac{1}{E_x} \sin^4\theta + \frac{1}{G_{wx}} \sin^2\theta \cos^2\theta \tag{4.4}$$

where G_{wx} is the modulus of rigidity under edgewise shear, and all other terms have been defined before. A formula for the determination of G_{wx} is given by Eq. 4.13.

Under compressive loads applied at an angle to the face grain, plywood elements experience shear strains, in addition to direct compressive shortening. Formulas for the calculation of these effects, as well as the dimensional changes resulting from the effects of Poisson's ratio, are given in Ref. 16.

4.4.2 Strength Under Direct Axial Stress

The strength of a strip of plywood panel subjected to edgewise compressive forces can be determined as shown in parts (a) and (b) of Fig. 4.9. The general expressions are:

$$F_{cw} = F_c\frac{E_w}{E_L}$$

and

$$F_{cx} = F_c\frac{E_x}{E_L} \tag{4.5}$$

where F_{cw} and E_w are the compressive strength and modulus of elasticity, respectively, of plywood parallel to the face grain; F_{cx} and E_x are the compressive strength and modulus of elasticity, respectively, of plywood perpendicular to the face grain; F_c and E_L are the compressive strength and modulus of elasticity, respectively, of the veneer parallel to the grain. If more than one species is used in the longitudinal plies, values for the species having the lowest ratio F_c/E_L should be used in Eq. 4.5.

The strength of a plywood strip in tension parallel or perpendicular to the face grain is taken as the sum of the strength values of only the plies having their grain direction parallel to the applied load, since perpendicular plies contribute essentially nothing to strength. The tensile strength of the plies may be taken as equal to the modulus of rupture of the given wood species (1). The tensile strength parallel to the face grain is designated as F_{tw} and the tensile strength perpendicular to the face grain as F_{tx}.

Along a similar line of thought, APA (4) recommends for design purposes that safe axial load be determined as the corresponding product of cross-sectional area, as listed in Table A.4.9, times allowable stress for the species group of the faces of the given plywood panel, see Table A.4.10. The areas listed are effective areas for tension and compression, based only on those plies whose grain is parallel with the stress.

Other formulas for edgewise compression or tension at an angle θ to the face grain, and compressive buckling stresses, are beyond the scope of this text, see Ref. 1.

4.4.3 Modulus of Elasticity in Bending

As in the case of direct axial stress, modulus of elasticity of a plywood panel in bending depends on the direction of the face grain relative to the plane of bending. The two principal cases occur when the face grain is either parallel or perpendicular to the span. The determination of both moduli of elasticity, for a three-ply panel with equal thickness of plies, is given in Fig. 4.9.

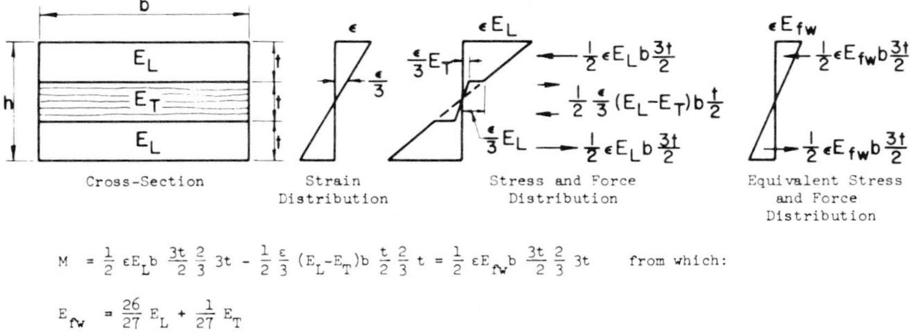

(a) Bending with Face Grain Parallel to Span.

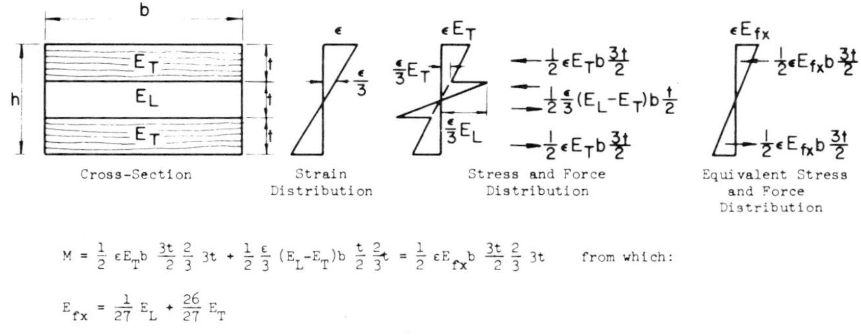

(b) Bending with Face Grain Perpendicular to Span.

Figure 4.10. Determination of moduli of elasticity in bending of a given plywood panel.

A linear strain distribution results, as shown in Fig. 4.10, when the section is subjected to a given bending moment M*. Maximum strains ϵ of opposite sign are obtained in the extreme fibers; zero strain occurs at mid-depth. The actual stress distribution in the section is obtained, assuming a linear-elastic behavior, by multiplying strains by corresponding modulus of elasticity; the figure shows the stress distribution and forces generated at the given panel. The stress distribution in the equivalent homogeneous panel is given by the strain distribution times the equivalent modulus of elasticity E_{fw} or E_{fx} as the case may be, see Fig. 4.10. E_{fw} is the modulus of elasticity of plywood in bending when the face grain is parallel to the span; E_{fx} is the modulus of elasticity of plywood in bending when the face grain is perpendicular to span. Equating the bending moment due to corresponding force distributions, as shown in the figure, yields:

*The reader is referred to Chapter 7 for a full discussion of the bending phenomenon.

$$E_{fw} = \frac{26}{27} E_L + \frac{1}{27} E_T$$

and (4.6)

$$E_{fx} = \frac{1}{27} E_L + \frac{26}{27} E_T$$

where all terms have been defined before.

The general expressions for E_{fw} and E_{fx} may be developed (17) using similar procedures. The following result is obtained:

$$E_{fw} \text{ or } E_{fx} = \frac{1}{I}\sum_{1}^{n} E_i I_i \tag{4.7}$$

where E_{fw} and E_{fx} have been defined before; E_i is the modulus of elasticity of the $_i{}^{th}$ ply in the span direction; I_i is the moment of inertia of the $_i{}^{th}$ ply about the center line of the full cross-section; I is the moment of inertia of the full cross-section about the center line; and n is the number of plies. For instance, in the case of the panel shown in Fig. 4.10, substitution of its properties in Eq. 4.7 renders:

$$E_{fw} = \frac{1}{b\dfrac{(3t)^3}{12}}\left[2\,E_L\left(\frac{bt^3}{12} + bt^3\right) + E_T\,\frac{bt^3}{12}\right]$$

$$E_{fx} = \frac{1}{b\dfrac{(3t)^3}{12}}\left[2\,E_T\left(\frac{bt^3}{12} + bt^3\right) + E_L\,\frac{bt^3}{12}\right] \tag{4.8}$$

from which Eqs. 4.6 can be obtained. When all plies have the same thickness, and are of the same species of wood, Eqs. 4.7 yield the following simple expressions:

$$E_{fw} = \frac{1}{2n^3}\left[(E_L + E_T)\,n^3 + (E_L - E_T)\,(3\,n^2 - 2)\right]$$

$$E_{fx} = \frac{1}{2n^3}\left[(E_L + E_T)\,n^3 - (E_L - E_T)\,(3\,n^2 - 2)\right] \tag{4.9}$$

Again, for $n = 3$, the preceding formulas reduce to Eqs. 4.6.

For the determination of deflections of plywood panels, APA (4) lists effective moduli of elasticity in Table A.4.10; the latter include an allowance for shear deflection. Plywood sheathing is generally used in applications where the loads are considered to be uniformly distributed and where the ratio $1/h$ of span to thickness of panel varies between 30 and 50. Tests of these panels have shown that shear deformation accounts for only a small percentage of the total deflection occurring at these span-depth ratios.

In certain cases, however, where short spans are involved, i.e. $1/h < 20$, deflections computed using the tabulated modulus of elasticity may underestimate the actual deflection. In such cases, the shear deflection should be calculated separately and added to the deflection due to bending; the latter should be calculated from the conventional formulas, using a true modulus of elasticity in bending equal to 1.1 times the tabulated effective modulus of elasticity as in Section 2.2.

4.4.4 Strength in Bending

The ultimate load-carrying capacity of a plywood strip is given by the following expression:

$$M_u = Kr\,\frac{FI}{c} \tag{4.10}$$

where M_u is the resisting moment of the plywood strip; K is an experimental constant determined by Freas (17) as 0.85 for outer-parallel plywood, 1.15 for 3-ply outer-perpendicular plywood and 1.00 for 5-ply or more outer-perpendicular plywood; r is the ratio E_{fw}/E_L for plywood strips having face grain parallel to span and E'_{fx}/E_L for plywood strips having face grain perpendicular to span, where E'_{fx} is the same as E_{fx} except that the outermost ply in tension is neglected; F is the strength of the outermost longitudinal ply and I and c are the moment of inertia of the full cross-section about its centerline, and the distance from the neutral axis to the outer fiber of the outermost longitudinal ply, respectively.

For design purposes, APA (4) recommends the use of the following expression:

$$M = F_bKS \tag{4.11}$$

where M is the allowable service-load bending moment, F_b is the allowable stress in bending for the given panel, see Table A.4.10; and KS is the effective section modulus of the given panel, see Table A.4.9. The values of KS in both directions are obtained from the term KrI/c of Eq. 4.10; an example of determination of section moduli of an actual plywood panel is given in the next Section. The tabulation of effective section moduli for commercial plywood panels in Table A.4.9 makes individual determinations unnecessary and thus, simplifies design considerably.

4.4.5 Modulus of Rigidity

Assume a rectangular plywood panel subjected to equal edgewise shear stresses, τ, in all its perimeter. The radian measure, γ, of the angle-change at one of the corners is called the shear strain; the relation between τ and γ, assuming linear elastic behavior, is the following:

$$> = \gamma\, G_{wx} \tag{4.12}$$

where, G_{wx} is the modulus of rigidity under edgewise shear.

The value of G_{wx} for a plywood panel can be obtained using the same approach as before. The following expression is obtained:

$$G_{wx} = \frac{1}{h}\sum_{1}^{n} G_i\, h_i \tag{4.13}$$

where G_i is the modulus of rigidity of the ith ply*; h_i is the thickness of the ${}_i$th ply; and, h is the total thickness of the plywood panel.

Refer to the Wood Handbook (1) for modulus of rigidity at an angle to the face grain.

*G_i may be either G_{LT} for rotary-cut veneers or G_{LR} for quarter-sliced veneers. Values of G_{LT} and G_{LR} are given in terms of E_L for various species in the Wood Handbook (1). Practically, the following value may be used: $G_{LT}/E = 0.06$ and $G_{LR}/E = 0.075$; also, as shown in Section 2.2, $G \simeq E/16$.

4.4.6 Strength Against Shear

The ultimate strength of plywood elements in shear, F_{swx}, with the shearing forces parallel and perpendicular to the face-grain direction, is given by the following empirical formula (1):

$$F_{swx} = 55 \frac{n-1}{h} + \frac{9}{16h}\sum_{1}^{n} F_{swxi}\, h_i \tag{4.14}$$

where, n is the number of plies, F_{swxi} is the shear strength of the ith ply, and the factor (n-1)/h should not be assigned a value greater than 35. A basic shearstress, equal to $F_{swx}/4$, is suggested for the determination of allowable shearing stress.

As in other cases before, formulas exist for the determination of shear strength of plywood panels when shear is applied at an angle θ to the face grain and for the determination of the critical shear-buckling stress (1). Details of design for grain directions other than parallel or perpendicular to panel edges are given in Ref. 16.

4.5 DESIGN PROPERTIES

The formulas of the preceding Section can be used for calculating the stiffness and strength properties of plywood at proportional limit or ultimate, or for estimating working stresses, depending upon the strength property that is substituted in the formulas for the property of the veneer. Design of plywood structures may be based on these formulas. However, the modern approach is to simplify design by assuming that the material behaves as if it were a homogeneous orthotropic plate; i.e., a plate made of a material with different properties in two orthogonal directions. By using the corrected or "effective" section properties, once a list becomes available to him, the designer need not be concerned with the actual multilayered make-up of the material. The effective section properties are used in conjunction with a list of allowable stresses assigned to the species group of the face ply; this, regardless of actual species used in inner plies, since section properties have been adjusted to compensate for such differing materials (4). Thus, for design purposes, a designer needs only a list of effective section properties and a list of allowable stresses. The former is discussed herein, the latter in the next Section.

In view of the great number of possible layups of plies in the manufacture of panels by various fabricators, effective section properties are calculated on the basis of APA's standard veneer layups as listed in Table A.4.11. This leads to properties parallel to the face grain of the plywood panel that are based on a panel construction giving minimum values in that direction. Properties perpendicular to the face grain are based on a different panel construction, giving minimum values in that direction. Thus, values that are obtained in this manner are conservative, but should not be added in any manner to obtain other properties of the full panel. Illustrative examples of the use of Table A.4.11 are given elsewhere in this Section.

The top portion of Table A.4.9 applies to all panels having veneers from mixed species groups, including most Product Standard grades. The bottom portion applies to panels having all veneers from the same species group. Grades included are Structural I, II and Marine. Each table includes a section giving properties for unsanded plywood, and separate sections giving properties for sanded and touch-sanded panels.*

The various section properties listed in Table A.4.9 are briefly discussed as follows.

*Normally all panels with A or B-grade faces are sanded; all Standard, C-C, and Structural grades are unsanded. "Sanded" section properties should be used for overlaid panels.

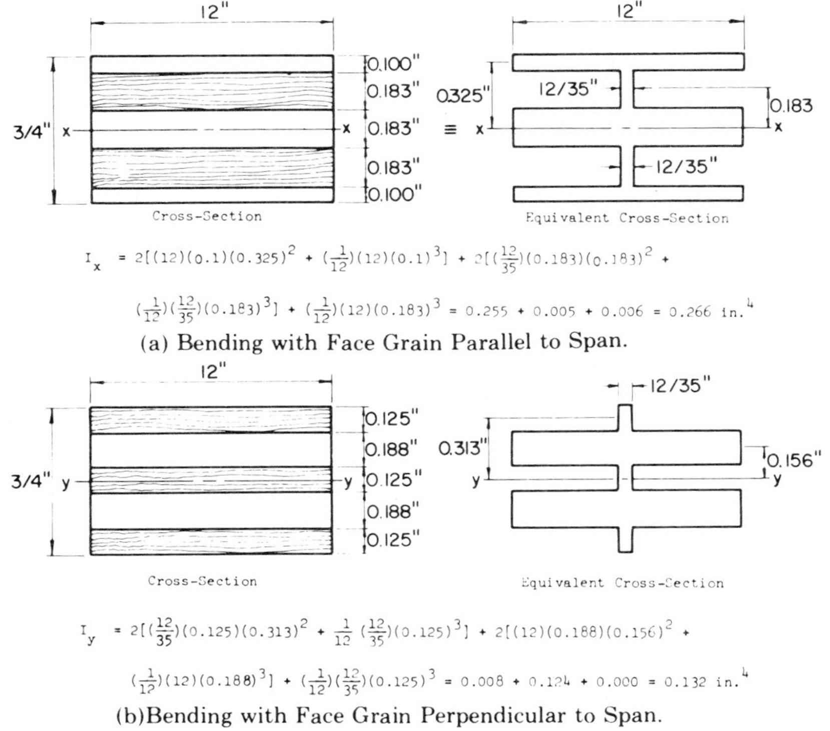

Figure 4.11. Determination of moments of inertia of ¾ in. Unsanded plywood panel with veneer layups used by APA (4) in calculating section properties. See tables A.4.9 and A.4.11.

4.5.1 Thickness for Shear

Column 3 of the table presents effective thicknesses for calculation of shear-through-the-thickness, as in the case of plywood webs of box beams, see Section 10.3.2. These values include provisions to compensate for the reduced effectiveness of inner plies in mixed-species panels and also for the increased shear resistance afforded by exterior glue.

Effective thickness for mixed-species panels, see top portion of the table, tends to be lower than the nominal thickness due to reduced effectiveness of inner plies as compared to the face plies. On the other hand, effective thickness of panels glued with exterior glue is increased due to a strengthening effect of the phenolic resin adhesives. This effect varies with the thickness of the panel and the number of gluelines.

4.5.2 Areas for Tension and Compression

Effective areas are given by columns 4 and 8 of Table A.4.9 for tension and compression parallel or perpendicular to the face grain, respectively. In the calculation of effective areas only those plies whose grain is parallel with the stress are taken into account, since perpendicular plies contribute essentially nothing to tensile and compressive strength.

For tension or compression at 45° to the face grain, allowable stresses from Table A.4.10 may be applied to the full thickness of the panel if all plies are of the same species group, as in STRUCTURAL I. If inner plies are not of the same species group as the faces, an effective thickness can be calculated by taking the thickness of the inner plies as the product of their actual thickness times the ratio of their modulus of elasticity to that of the face plies. This adjusted thickness of inner plies is then added to the thickness of the face and back (4). An approximate value of the effective thickness may be obtained by taking 70 percent of the actual thickness.

4.5.3 Moments of Inertia

The values listed in columns 5 and 9 of Table A.4.9 have been adjusted to account for several variables in such a way that they may be used in conjunction with the modulus of elasticity of the face plies, in either direction, without reference to actual physical make-up of the panel. The following example is presented to clarify this concept.

Consider the ¾ in. unsanded plywood panel shown in Fig. 4.11. Parts (a) and (b) of the figure deal with the determination of the effective moment of inertia of the panel for bending with face grain parallel to span and perpendicular to span, respectively. The veneer layup has been obtained from Table A.4.11; note that the layup is different in both panels to allow for variations in panel veneer constructions by various manufacturers.

The equivalent cross-section of the plywood panel, for each case, is also shown in the figure. The width of the plies in which the grain runs normal to flexural stresses is reduced to 1/35th of the actual width. This compensates for the fact that the modulus of elasticity of wood perpendicular to the grain is considered (4) as 1/35th that of its parallel modulus. The computations for the effective moments of inertia use the equivalent cross-sections. Moment of Inertia, I, may be used only in stiffness calculations and not at all in bending-stress calculations.

4.5.4 Section Moduli

Effective section moduli are given in columns 6 and 10 of Table A.4.9, and are to be used in conjunction with the allowable stresses for the face plies. Adjustments for species of inner plies, direction of inner plies or K factor, see Section 4.4.4, must not be applied to the properties listed; all these factors have been included in the calculations in compliance with the research findings of US FPL-059, see Ref. 17. For instance, the effective section modulus KS_x of the panel shown in Fig. 4.11a is obtained as follows:

$$KS_x = 0.85 \times \frac{0.266}{0.375} = 0.603 \text{ in.}^3$$

where the factor 0.85 reflects the value of K for outer-parallel plywood; $I_x = 0.266$ in.^{4} and $c = 0.375$ in. from Fig. 4.11a.

The effective section modulus perpendicular to the face grain is calculated ignoring the outermost tension ply in accordance with Freas' suggestions (17). The outermost ply on the tension side of a plywood panel stressed perpendicular to face grain adds little strength to the panel. Omission of the outer ply on the tension side results theoretically in a shift in the neutral axis from the center of the depth toward the compression face of the panel. However, the effect of this shift is generally small enough to be neglected, and the neutral axis may be assumed to be at the center of the depth of the section with little error. Thus, for the case of the panel shown in Fig.

4.11b the following values are used in the calculation of the effective section modulus perpendicular to face grain:

$I_y = 0.132 - 0.004 = 0.128$ in.^{4}; $c = \frac{0.125}{2} + 0.188 = 0.251$ in.; and, $K = 1.0$, see Section 4.4.4. Thus:

$$KS_y = 1.00 \times \frac{0.128}{0.251} = 0.509 \text{ in.}^3$$

These values coincide with those listed in the table.

4.5.5 Rolling Shear Constant

Columns 7 and 11 of Table A.4.9 give the rolling shear constant Ib/Q for stress parallel and perpendicular to face grain, respectively. Rolling shear is induced when plywood is subjected to loads perpendicular to the panel; the wood fibers in the ply at right angles to the principal shearing force tend to roll, thus the name. The phenomenon is shown in Figs. 2.8c and d for plywood panel used as sheathing; it should be noted, however, that rolling shear is seldom of any consequence in sheathing applications. Examples of calculation of rolling shear stresses in a stressed-skin panel and a box beam are given in Sections 10.3.1.7 and 10.3.2.6 to which the reader is referred. In this respect, Table A.4.12 is given to assist in the calculation of the necessary constants for the calculation of rolling shear stresses in stressed-skin panels.

The allowable stresses presented in Table A.4.10 apply to most cases of rolling-shear stress, except where the stressed area occurs at the edge of a panel (as in the examples cited above) so that stress concentrations may occur. For applications where such stress concentrations are expected, due to imbalance of the stress on the total area under consideration, it is conventional (4) to reduce allowable design rolling-shear stresses by 50 percent.

4.6 ALLOWABLE STRESSES

A list of allowable stresses for dry, untreated, plywood panels conforming to U.S. Product Standard PS 1 for softwood plywood (construction and industrial) as compiled by APA (4), is given in Table A.4.10. The allowable stresses are divided into three levels which are related to grade. Plywood with exterior glue, and with face and back containing only N, A or C veneers, use level one (S-1) stresses. Plywood of Exterior type or Interior type with exterior glue, and with B, C-plugged or D veneers in either face or back use level two (S-2) stresses. All grades with interior or intermediate glue use level three (S-3) stresses.

The deviation of the various stress levels is based on research that showed the following facts:

1.) Bending, tension and compression stresses depend on the grade of the veneers,
2.) Shear stresses do not depend on veneer grade but do vary with kind of glue, and
3.) Stiffness and bearing strength do not depend either on glue or on veneer grade, but on species group.

Actual stresses computed on the basis of section properties given in Tables A.4.9 shall not exceed the allowable stresses, except as modified for loading conditions, treatment or moisture content. Adjustment factors apply for these variations as follows.

The allowable stresses are for normal loading conditions, see Section 2.9. Conditions of loading other than normal have the same effects on plywood panels as in solid wood structures, see Section 2.13; the values of the coefficient $\bar{\alpha}$, apply here also.

The allowable stresses listed in Table A.4.10 apply also to plywood pressure-impregnated with preservative chemicals, see Section 5.8, in accordance with American Wood Preservers Association AWPA Specifications C-9. However, for plywood pressure-impregnated with fire-retardant chemicals, see Section 5.9, in accordance with AWPA Specification C-27, allowable stresses shall be reduced 1/6th and modulus of elasticity shall be reduced 1/10th.

The allowable stresses listed and the preceding adjustments, apply to plywood used under conditions continuously dry, as in most covered structures. Where equilibrium moisture content in service is 16% or greater, as in most applications that are directly exposed to the weather, the allowable stresses of Table A.4.10 must be modified by factors given in the footnote labeled "Wet or Damp Location." These factors range in value between 61 percent for allowable compression stress to 89 percent for modulus of elasticity. In addition, where equilibrium moisture content is above 18 percent, only exterior-type plywood should be used.

4.6.1 Effects of Joints

Three types of joints are commonly used in plywood construction, the scarf joint, the finger joint and the butt joint. These are similar to the corresponding ones shown in Fig. 3.7, except that butt joints are backed by a glued plywood splice plate on at least one side, having its grain perpendicular to the joint and of a grade and species group equal to the plywood spliced. Also the splicing plate may not be thinner than the panel itself. The necessary length of the splice plate depends on the plywood thickness, see Table A.4.14.

The following provisions (4) apply:

Tension and Flexure. Scarf joints and finger joints 1:8 or flatter shall be considered capable of transmitting full allowable stress; only 75% of the allowable stress shall be considered for scarf joints having a 1:5 slope. Scarfs steeper than 1:5 shall not be used. Finger joints are acceptable at design levels supported by adequate design data. Butt joints as described above, may be considered capable of transmitting the tensile and flexural stresses indicated in Table A.4.14; strength may be taken proportionately lower for shorter splice lengths.

Compression. For scarf joints and finger joints with slopes not steeper than 1:5 and butt joints as described above, see Table A.4.14, 100% of the compressive strength of the panels may be taken. Strength may be reduced proportionately for butt joints with splice-plates lengths shorter than those specified in Table A.4.14.

Shear. Scarf joints, along or across the face grain, with slope 1:8 or flatter, may be designed for 100% of the shear strength of the panels joined. Finger joints are acceptable, at design levels supported by adequate design data. Butt joints as described above (except the minimum glued-plywood splice-plate length need only be equal to 12 times the panel thickness) may be designed for 100% of the strength also; for shorter splice-plate lengths strength may be taken proportionately less.

Combination of Stresses. Joints subjected to more than one type of stress (for example, tension and shear), or to a stress reversal (for example, tension and compression), shall be designed for the most severe case.

For the case of glued-laminated wood used in combination with plywood, the specifications provide joint factors identical to those considered in Section 3.4.

4.6.2 Using Section Property and Allowable Stress Tables

The Section properties and allowable stresses presented in Tables A.4.9 and A.4.10 are to be used with the proper type and grade of plywood produced under U.S. Product Standard PS 1. Because the section properties must represent a wide variety of manufacturing techniques and combinations of species, they are of necessity conservative. This is especially true for the Identification Index and Span Index panels, see key in Table A.4.15. To relate plywood type and grade to Tables A.4.9 and A.4.10 a GUIDE TO USE OF ALLOWABLE STRESS AND SECTION PROPERTY TABLES has been provided for those grades most often used in engineering design, see Table A.4.16.

The following examples from Ref. 4 illustrate the use of these Tables.

Sheathing-Grade Panel.

The "Guide," see Table A.4.16, indicates C-D INT-APA should be used for Interior application and C-C EXT-APA is needed for exterior exposure. Both grades may be modified to the STRUCTURAL category.

For a 32/16 C-D INT-APA panel the "Guide" indicates that section properties from Table A.4.9 Sub-Table 1 should be used in conjunction with stress level three (S-3). Refer to the "Key to Identification Index, Span Index and Species Group," Table A.4.15. The "Key" indicates that a 32/16 Identification Index is available in 1/2″ or 5/8″ thickness. Selecting the 1/2″ thickness indicates the use of Species Group 1. Hence, for a 1/2″ 32/16 C-D INT-APA panel, the following values for stress applied parallel to the face grain are extracted from Tables 1 and 3: I = 0.086, KS = 0.247; Ib/Q = 4.189. Group I stresses in the dry condition for stress level three (S-3) are: E = 1,800,000, F_b = 1650, F_s = 48, all in psi.

Should the panel be changed to 1/2″ 32/16 C-C EXT-APA the same section properties would be used but stress level one (S-1) would be used. Stress level two (S-2) could be used with a C-D INT-APA panel if exterior glue is specified.

If STRUCTURAL I C-D INT-APA is used, the "Guide"; Table A.4.16, indicates that Table A.4.9 Sub-Table 2 section properties should be used along with level two (S-2) stresses. For a 1/2″ STRUCTURAL I C-D INT-APA panel the following section properties are obtained: I = 0.091, KS = 0.318, Ib/Q = 4.497 where stress is applied parallel to the face grain. Should the stress be applied perpendicular to the face grain, the following section properties should be used: I = 0.017, KS = 0.145, Ib/Q = 2.574.

Sanded-Panel.

Plywood produced with an A or B face is generally fully sanded and considered an appearance grade. For a ¾″ Group 3 B-C EXT-APA panel the "Guide," Table A.4.16, indicates that stress level two (S-2) should be used. The "Guide" also shows that the sanded portion of Table A.4.9 Sub-Table 1 should be used to obtain section properties; they are: I = 0.197, KS = 0.452 and Ib/Q = 7.881 for stress applied parallel to the face. The allowable stresses for the wet condition are: F_b = 820, F_s = 44, and E = 1,100,000, all in psi.

Touch-Sanded Panel.

Plywood manufactured as STURD-I-FLOOR, UNDERLAYMENT, C-D Plugged or C-C Plugged is generally touch-sanded. To find the properties and stresses for a 19/32-inch, 20 o.c. STURD-I-FLOOR INT-APA panel (with veneer inner plies), the "Guide," Table A.4.16, indicates level two (S-2) stresses should be used with Table A.4.9 Sub-Table 1 section properties. Also refer to the "Key to Identification Index, Span Index and Species Group," Table A.4.15, which indicates that for 19/32-inch thickness and 20 o.c. Span Index, species Group I stresses should be used. Stresses are therefore: E = 1,800,000, F_b = 1650, F_s = 53, all in psi. Section properties are: I = 0.123, KS = 0.337, Ib/Q = 5.403. If 19/32-inch Group 1 UNDERLAYMENT INT-APA with exterior glue is specified, the same section properties and stresses are used.

REFERENCES

1. Forest Products Laboratory, "Wood Handbook," Agriculture Handbook No. 72, U.S. Department of Agriculture, Washington, D.C. revised August 1974.
2. ASTM Designation: D1038–52 (Reapproved 1970), "Standard Definitions of Terms Relating to Veneer and Plywood," American Society for Testing and Materials, Book of ASTM Standards, Part 22, pp. 333–338, 1978.
3. Countryman, D. R., Carney, J. M., and Welsh, J. L., Jr., "Plywood," Chapter 4 of *Composite Engineering Laminates,* edited by A. G. H. Dietz, MIT Press, 1969.
4. American Plywood Association, "Plywood Design Specifications," revised April 1978 by APA, P.O. Box 11700, Tacoma, Washington, 98411.
5. American Plywood Association, Supplements to Plywood Design Specifications.
   No. 1, Design of Plywood Curved Panels.
   No. 2, Design of Plywood Beams.
   No. 3, Design of Flat Plywood Stressed-Skin Panels.
   No. 4, Design of Flat Plywood Sandwich Panels.
6. American Plywood Association, Specifications No.
   CP-8, Fabrication of Curved Plywood Panels.
   BB-8, Fabrication of Glued-Plywood Beams.
   SS-8, Fabrication of Plywood Stressed-Skin Panels.
   SP-61, Fabrication of Flat Plywood Sandwich Panels.
   FP-62, Fabrication of Plywood Folded Plates.
   All revised 1971, except SP-61 in 1961.
7. Hurd, M. K., "Formwork for Concrete," American Concrete Institute, Committee 622, Special Publication No. 4, ACI, Detroit.
8. American Plywood Association, "Plywood Diaphragm Construction," 14 pp., 1970.
9. Carney, J. M., "Plywood Folded Plates Design and Details," Laboratory Report No. 121, American Plywood Association, 92 pp., and Appendix, 1971.
10. Yitzhaki, D., "The Design of Prismatic and Cylindrical Shell Roofs," North Holland Publishing Company, Amsterdam, 1959.
11. Ramaswamy, G. S., "Design and Construction of Concrete Shell Roofs," McGraw Hill Book Co., N.Y., 1968. See Chapter 12.

12. American Plywood Association, "Plywood Construction Systems for Commercial and Industrial Buildings," November 1970.
13. Countryman, D., "Evolution of the Use of Plywood for Structures," Journal of the Structural Division, ASCE, Vol. 93, No. ST2, Proc. Paper 5166, April 1967, pp. 13–24.
14. American Plywood Association, "Plywood Rigid Frame Design Manual," published by APA, AIA File No. 19F, 1962.
15. Liska, J. A., "Methods of Calculating the Strength and Modulus of Elasticity of Plywood in Compression," U.S. Forest Products Laboratory Report No. 1315, Madison, Wis., Revised 1955.
16. U.S. Dept. of Defense, "Design of Wood Aircraft Structures," ANC Bul. 18, 234 pp., 1951. Issued by Subcommittee on Air Force-Navy-Civil Aircraft Comm. Munitions Board.
17. Freas, A. D., "Bending Strength and Stiffness of Plywood," U.S. Forest Service Research Note No. FPL-059, U.S. Forest Products Laboratory, Sept. 1964.
18. "U.S. Product Standard PS 1–74 for Construction and Industrial Plywood," Product Standards Section, National Bureau of Standards, Washington, D.C.
19. "Nailed Plywood and Lumber Beams for Roofs and Garage Door Headers" by American Plywood Association, Publication Z416D.

Chapter 5
Durability of Wood Structures

5.1 INTRODUCTION

Wood structures will last indefinitely if built properly and when the appropriate amount of maintenance and protection is given. Thus, wood buildings exist today that were built several centuries ago. However, under certain conditions of exposure wood may be attacked by decay-producing fungi, insects or marine borers that will considerably shorten the useful life of the structure. In addition, as with any type of construction, fire must be given consideration as a possible cause of destruction.

Some wood species are more durable than others. For all species, however, durability can be enhanced by simple means in a number of cases. For instance, decay can be prevented by keeping the moisture content of wood below a certain level, as in covered structures, or through the use of proper construction details, or by keeping it in a complete absence of oxygen, as under the water table. The destructive action of insects may be prevented by keeping the wood above ground level and avoiding construction details that favor access of the insects to the wood structure. However, in many cases, such as poles, railroad ties, bridges, piers, wharves, etc., it is not possible to protect wood as described above. For these applications wood is treated with chemical preservatives in which toxic substances are impregnated by pressure or other means into the mass of the wood. A special treatment is required for fire-retardant action. The extensive and successful use of treated wood in this country is evidence of the effectiveness of the protection afforded.

The object herein is the study of the various agents of wood destruction and the preservative methods that are presently available to treat wood.

5.2 DESTRUCTION BY DECAY

In its earliest stage decay appears as a slight discoloration of the infected piece of wood. However, as decay progresses, bore holes and general dissolution of cell walls become conspicuous and marked changes in color, texture and strength properties of wood take place. In an advanced stage of decay, wood may become either soft and spongy (soft rot), or stringy (white rot), or crumbly (brown rot); in addition, strength is reduced drastically and a definite odor develops in the rotting wood. Examples of effects of decay in wood are shown in Figs. 5.1 and 5.2.

Decay occurs in untreated wood which is in direct contact with the soil or in places where moisture collects but does not readily evaporate. Railroad crossties, posts and poles, bases of porch columns and steps of porches are examples of the first situation. The ends of beams or columns fixed into exterior concrete or masonry walls, and the interior wood structure of some buildings such as textile and paper mills, breweries, and cold-storage plants (in which high humidity favors condensation on exposed wood surfaces) are examples of the second situation.

Figure 5.1. Cross-sectional, radial and tangential views of white pocket rot in Port Orford Cedar. Photo courtesy of Forest Products Laboratory.

Figure 5.2. Typical brown rotted wood. The checking across the grain is typical; it is found otherwise only in soft rotted wood. Photo courtesy of Forest Products Laboratory.

The cause of wood decay is a series of parasitic micro-organisms, known as fungi, which derive their food from the wood cells. Starches, sugar, proteins and other cell contents are also utilized as food by the fungi. For the successful development of decay in wood the following three conditions are necessary: 1) moisture content in excess of 19 percent, 2) oxygen in adequate supply, and 3) a favorable temperature.

The fiber-saturation point, see Section 1.3, marks the point at which there is no water in the cell cavities but the cell walls are fully saturated (approximately 25 to 30 percent). This is the minimum moisture content for the onset of decay. Below this level the digestion process of wood by fungi cannot proceed. Kiln-dried wood is sterile as long as its moisture content remains below fiber-saturation point. For the same reason, decay can be arrested in an infected piece of wood for as long as its moisture content remains below fiber-saturation point.

An excess of moisture content, as for example in the case of the submerged portion of a wood pile, inhibits decay by limiting the oxygen required by fungi. The supply of oxygen is also very limited below the surface of the ground, especially in dense soils; wood piles supporting the foundations of buildings and bridges may be considered free from attack by decay if they are fully submerged under the water table. The service life of these piles has no limit.

The range of temperatures between 75° and 90°F is one in which activity by decay-producing fungi reaches a peak. Once the temperature moves away from this range, inhibition of growth occurs gradually. At temperatures below 50°F growth is a small percentage of optimum and at somewhat lower temperatures it stops altogether; the fungi hibernate for as long as necessary until the temperature rises again. When temperature increases above the upper limit of the optimum range, fungal growth stops gradually, at 115°F all species of fungi except one were fully inhibited (1). Tests have shown that complete sterilization requires a 12-hour exposure of wood to steam (moist heat) at 131°F, or dry heat at 221°F.

The durability of untreated wood against attack by fungi depends on the species; within the species, it depends on whether heartwood or sapwood is used. The heartwood of certain species such as Redwood, Cedar and Juniper, among the softwoods, and Black Locust, Black Walnut and Chestnut, among the hardwoods, is quite durable. That of Douglas fir and Southern pine is not as durable, while Spruce, Maple, Ash, Birch and others is considered nondurable (1). Heartwood is more durable than sapwood because of the extractives, see Section 1.2, that are deposited during the life of the tree in the dead cells of the heartwood; the extractives are toxic to the decay-producing fungi. Where the durable species are not used pressure-treated wood should be specified for structural members where decay is likely to occur and can not be prevented by other means. The designer should be responsible for preventing situations where moisture can collect, by providing construction specifications and details in which proper ventilation of wood is guaranteed.

5.3 DESTRUCTION BY INSECTS

Millions of dollars are lost every year in unprotected and untreated wood structures through the continuous, and usually undetected, work of wood-destroying insects. Two types of insects are generally considered, depending on whether they attack wood before or after it is put into use or service. In the first group are ambrosia beetles or timber worms, which make pinholes as large as ¼ in. in diameter in living trees and logs; larger holes up to 1 in. in diameter are made by locust borers, carpenter ants, carpenter bees and other insects. Among the destroyers of wood in service, outstanding damage is caused by termites, powder-post beetles and carpenter ants. Only this group is discussed herein because of their importance to the user of wood.

5.3.1 Termites

Termites are recognized as the major wood destroyers. They may exist anywhere in this country, but are more destructive in the coastal states and the Southwest. It is estimated that their damage to buildings alone amounts to millions of dollars per year.

From an entomological point of view termites are social insects. They live in colonies with a high degree of organization of activities among three castes, namely, reproductives, soldiers and workers. In addition, there is a queen dedicated to laying eggs, a function which results in her body swelling to a size several times larger than either soldiers or workers. The belief that destroying the queen may end the life of the colony is false, since termites have the ability to develop other queens; a colony may continue to prosper indefinitely provided food and moisture are in adequate supply.

The reproductives swarm once or twice a year (usually in early spring and autumn) for mating purposes; when this occurs close to a wood structure it should be considered an unequivocal sign of infestation by termites. The wind takes these insects away during the mating flight. Many perish but those pairs which are able to escape from their natural predators lay the foundations of a new colony by eating their way into wood. The soldiers are dedicated only to the defense of the colony against predators such as ants or termites from other colonies. The workers are the most numerous group; like the soldiers, the workers are sterile, wingless and blind. It is this group that is responsible for all the destruction in the wood since the material provides termites with both shelter and food. Termite burrowing takes place only on the inside of the timber, see Fig. 5.3; these insects are very careful to leave a protective outside shell which they keep intact. This hollowed-out condition may be detected by lack of resonance of the wood when struck with a hammer. Otherwise, since termites avoid open air, very little evidence of infestation of a wood structure exists from the outside.

Termites can be classified in two distinct groups (2) according to their living habits, namely, subterranean and nonsubterranean. It is the first group that is most common in the U.S.; the nonsubterranean termites are confined to a narrow strip of territory along California and the southern border, up the Atlantic coast to Virginia. The subterranean termites enter the wood from the ground. Since these insects are particularly sensitive to lack of moisture, they build conduits, out of soil particles cemented by their own excretions, over masonry or concrete foundations. This enables them to reach and successfully penetrate wood that is above ground. These earthlike conduits are definite indications of infestation of wood by subterranean termites; once burrowing of wood is started, termites follow the grain, eating the softer springwood first. Sapwood is definitely more attractive than heartwood.

The question may be asked whether or not species of wood exist that are naturally resistant to the attack of termites. The answer is positive, but with reservations. The heartwoods of Redwood and Baldcypress are resistant, nevertheless, evidence (1) has shown that even these woods can be damaged. However, when used in combination with other woods in the same structures, termites will bypass Redwood to burrow in the other more attractive woods. The only actual and practical solution to defend wood from attack of termites consists of adequate impregnation with a suitable preservative, see Section 5.7.

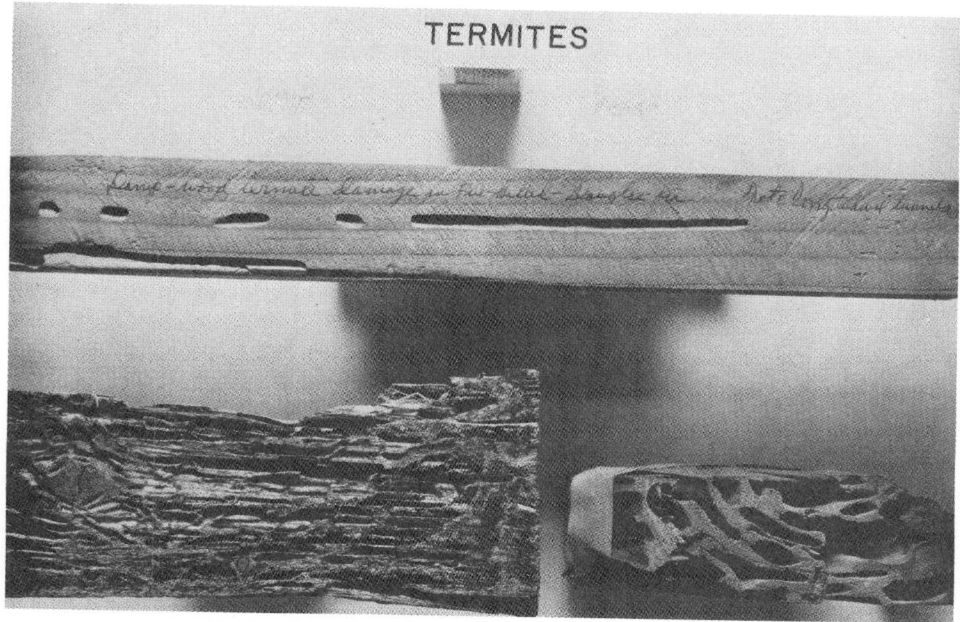

Figure 5.3. Damage in wood due to the action of termites. Photo courtesy of Forest Products Laboratory.

5.3.2 Powder-Post Beetles

The larvae of the Lyctus beetles bore through wood for food and shelter leaving their irregular burrows packed with powdery, undigested remnants of wood. Wood that is infested by these insects releases this powdery residue when moved or jarred; the powder falls out of holes in the wood surface made by emerging winged adults, see Fig. 5.4. The emergence holes and the powdery wood residue that collects in small piles are indications of infestation of wood by powder-post beetles. Damage by these insects can be avoided by proper sanitation measures at the manufacturing plant and prompt manufacture and utilization of wood products.

5.3.3. Carpenter Ants

These insects are true ants, of either black or brown color. They excavate wood for shelter rather than food. Like termites, carpenter ants live in colonies and are also organized in castes. Unlike termites, carpenter ant workers are not as concealed and are often seen outside the wood in which they are burrowing. Typical damage caused by these insects is shown in Fig. 5.5.

5.4 DESTRUCTION BY MARINE BORERS

Wood piles used to support wharves, piers and other structures in harbors or brackish waters are subjected to the action of marine borers; the piles are destroyed rapidly if the wood is untreated.

Two types of marine borers are recognized; namely, mollusks like teredo and martesia, and crustaceans like limnoria and chelura. The mollusks are distantly related to oysters and clams;

Figure 5.4. Damage in wood caused by powder-post beetles. Photo courtesy of Forest Products Laboratory.

Figure 5.5. Damage in wood due to carpenter ants. Notice that the galleries cross the grain of the wood. Photo courtesy of Forest Products Laboratory.

the crustaceans to lobsters and crabs. In addition, mollusks and crustaceans differ in their methods of destroying wood. Mollusks use wood not only as food but also as habitat; crustaceans, unlike mollusks, do not live inside the wood but move about it.

On becoming attached to the surface of the wood pile, the young mollusk larvae begin boring their way into the wood across the grain, using their bivalve shells; once away from the surface, burrowing continues in a more or less longitudinal direction. As a result of this the pile is weakened considerably, see Fig. 5.6, and failure occurs abruptly. Since the entrance holes of the young

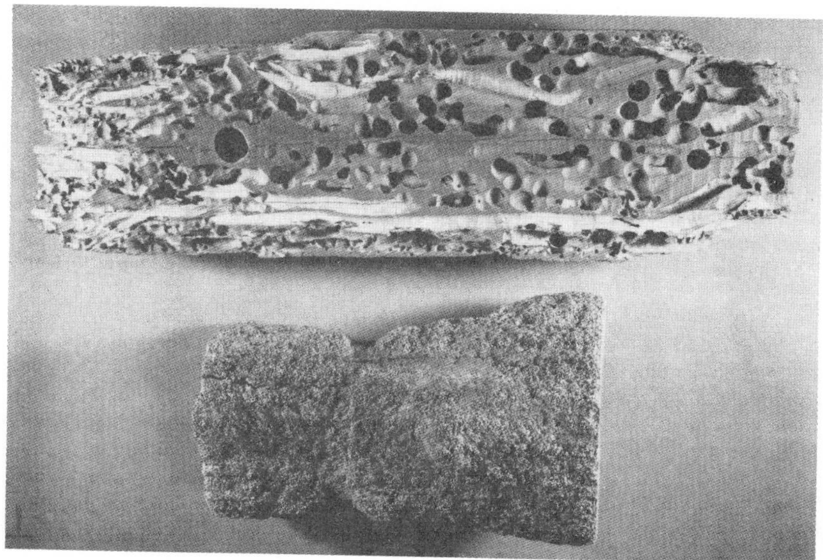

Figure 5.6. Damage caused by marine borers in a wood pile. Photo courtesy of Forest Products Laboratory.

larvae are small, and remain small, infestation by mollusks such as teredo (otherwise known as shipworms) is difficult to detect. Crustaceans, on the contrary, burrow into the timber to make galleries that seldom extend far from the surface of the wood. The infested timber is honeycombed in such a way that the small partitions of wood separating the galleries are broken away by wave action, thereby exposing a new surface of wood to the crustaceans. As destruction is carried deeper, the piles assume an hour-glass type shape, which is easily recognized. Failure occurs as a result of substantial loss in the section of the pile.

No commercial U.S. wood resists the action of marine borers if not treated properly with coal-tar creosote and creosote-coal-tar solutions. Douglas fir and Southern pine piles that are thoroughly treated may have a minimum average life of 15 years, and frequently last much longer. Recent research (3) has shown that treatment with creosote alone is enough against attacks by teredo but leaves wood unprotected against attack by limnoria. However, if creosoting is preceded by a double-diffusion treatment designed to deposit insoluble copper arsenate in the wood, 10-year tests have shown that only very light attack by limnoria occurs. The foregoing results support the case of dual treatment wherein both an insoluble copper-arsenate compound and creosote are injected by pressure. Various other types of protection such as covering the wood piles over the entire surface exposed to borer attack down to below the mud-line with sheet metal, cast-iron pipe, concrete and vitrified-clay have been tried in the past with various degrees of success (4).

5.5 DESTRUCTION BY MECHANICAL WEAR

A wearing away action in wood can be caused by a continuous or repeated motion of a load bearing on it. The typical case is that of railroad ties where mechanical wear occurs under the steel base plate that ties the rail to the wood. It also occurs in factory floors and treads of wood stairs. In the past, untreated wood ties lasted only for a few years, a period that is not long enough to allow mechanical wear to occur to a significant extent. Presently, however, preservative-treated ties resist the action of fungi so successfully that mechanical failure due to wear has become a factor in the determination of useful life of ties.

Preservative treatment increases resistance against mechanical wear indirectly by preventing decay which, in turn, reduces the original hardness of the material.

5.6 DESTRUCTION BY FIRE

Thermal decomposition of wood under increasing temperatures starts with vaporization of moisture and volatilization of extraneous materials. This is followed by emission at considerable force of a mixture of gases containing carbon; burning of these gases in the surrounding air constitutes the flaming phase of combustion. Formation of charcoal (charring), occurs soon thereafter. At the usual temperatures of fires (1400–1600°F) charcoal does not become volatile and thus, remains on the member, see Fig. 5.7. Even a negligible layer of ash on the charcoal is able to prevent it from smoldering; in fires, as a rule, ash remains on the charcoal saving it from burning. The charcoal becomes an insulator to the wood in the interior of the member. However, in the case of fires of great intensity, with temperatures above 2000°F, the charcoal becomes very volatile and burns with a flame.

The loss of load-carrying capacity of wood members under the action of fire is the result of two major causes; namely, the carbonization of the outside portion of the member, and the weakening, due to the rise in temperature, of a thin wood layer that is just below the charcoal,

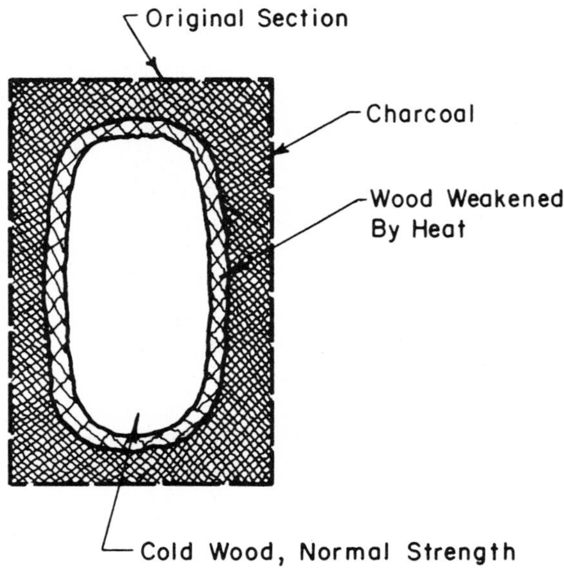

Figure 5.7. Section of a wood beam after being subjected to fire.

Figure 5.8. Glued-laminated rigid frames endure effects of fire.

see Fig. 5.7. Structural failure occurs when the strength of the sound wood is overcome by the loads on the member. In the case of heavy timbers this is delayed by the size of the member; frequently enough, see Fig. 5.8, main structural members are able to survive even the effects of conflagrations.

The natural resistance of wood structures to fire, as related to spread of flame on its surface and to its fuel potential, can be improved in various ways. Fire-retardant treatments are available, see Section 5.9, in which chemicals are added by pressure-impregnation or surface treatment of the wood. Designers of wood structures may also provide built-in fire resistance through protective coverings, good design and construction details, see Section 5.10.

5.7 WOOD PRESERVATIVES

Certain chemical substances are applied to wood to improve its durability against the various destructive agents. Such chemicals, recognized as wood preservatives, have in common some, or all, of the following characteristics:

1.) they are toxic to wood destroyers;
2.) their action is durable;
3.) they are able to penetrate wood;
4.) they are non-harmful to wood or metal;
5.) they are safe and economical to use.

The most important characteristic of a good preservative is its toxicity to wood-destroyers. For this purpose, a preservative must be at least partially soluble in the biological fluids of the attacking organisms to be lethal; some solubility in water is a good indication of this ability (1). As far as duration is concerned the preservative action of any given chemical must be as long as the intended life of the structure; not less than 50 years in most cases. The preservative should not evaporate away from the treated wood, nor should it be changed into other non-toxic chemicals or be leached away when the wood comes in contact with water. It is very important that the preservative be able to penetrate wood for adequate protection; surface applications are easily worn away. Finally, the chemical substances used must be available in industrial amounts, they must be safe to handle by workmen and their cost should be competitive with others.

Three major groups of preservatives exist; namely, oils and oil-borne preservatives, oil solutions of toxic chemicals, and water-borne preservatives (5, 6). In the first group are coal-tar creosote (7) and creosote-coal tar solutions (8) that have been widely used by the wood-preserving industry since the last century. Pentachlorophenol is an extremely toxic chemical that can be dissolved in petroleum oils; it constitutes an example of the second group of preservatives. Water-borne preservatives usually contain salts of two or more of the elements zinc, chromium, copper and arsenic, but may contain others as well. Unless kiln dried after treatment, the swelling action brought about during treatment may cause wood to shrink out of shape.

Coal-tar creosote is a distillate of coal-tar produced by high temperature carbonization of bituminous coal. As defined by the American Wood Preservers Association, creosote consists principally of liquid and solid aromatic hydrocarbons and contains appreciable quantities of tar acids and tar bases; it is heavier than water.

Pentachlorophenol is an industrial chemical compound that is usually borne by petroleum solutions. Its preservative action against fungal decay and insects is comparable to that of coal-tar creosote, but it is ineffective against marine borers; thus, should not be used for protection of wood piles in salt water. An effectiveness against both decay and insects greater than that of either coal-tar creosote or pentachlorophenol petroleum solutions has been obtained by adding 2-percent pentachlorophenol to standard coal-tar creosotes. A table similar to the preceding one is also given in Ref. 6 for minimum net retentions of pentachlorophenol solutions.

Water-borne preservatives penetrate wood very well and leave surfaces clean and odorless. Some are non-leachable and recommended for use in water and on ground contact. A number of water-borne preservatives exist which use various ingredients. Refer to the standards of the American Wood Preservers Association for additional information and to Federal Specification TT-W-571i (6) for minimum net retentions for various wood products.

Recommended minimum net retentions for the various types of preservative treatments are given in Table 5.1.

5.8 WOOD PRESERVATION PROCESSES

There are two major processes used for the application of preservatives to wood; the distinctive difference between the two being that one requires application of pressure beyond normal atmospheric and the other does not. A number of different methods are available that follow fundamentally one process or the other; thus, there is a group of pressure processes, and another of non-pressure processes. The latter is responsible for only a small percentage of the treated wood in this country. Such processes as brush and spray, dipping, steeping, kyanizing, hot and cold bath and others belong to the non-pressure group. Their study is beyond the scope of this work; information on these and other non-pressure processes is given in Refs. 1 and 9.

TABLE 5.1. MINIMUM NET RETENTIONS FOR PRESERVATIVE TREATMENTS.

Water-borne Preservatives (3) See Note (4) for trade names

MATERIAL AND USAGE	CREOSOTE (5)	CREOSOTE-COAL TAR (5)	CREOSOTE-PETROLEUM	PENTACHLOROPHENOL (3)	ACID COPPER CHROMATE (ACC)	AMMONIACAL COPPER ARSENITE (ACA)	CHROMATED COPPER ARSENATE (CCA)	CHROMATED ZINC CHLORIDE (CZC)	FLUOR CHROME ARSENATE PHENOL (FCAP)	AWPA STANDARDS (6)
LUMBER AND TIMBER	Retention by Assay of Treated Wood – lbs./cu. ft.									
Above ground	9	9	9	0.40	0.25	0.25	0.25	0.45	0.25	C2
Ground contact										
Non-structural	9	9	9	0.50	0.50	0.40	0.40	NR[7]	NR	C2
Structural	9	9	9	0.60	NR	0.60	0.60	NR	NR	C14
In salt water	25	25	NR	NR	NR	2.5	2.5	NR	NR	C14
PILES										
Land or fresh water use and foundations	12	12	12	0.60	NR	0.80	0.80	NR	NR	C3
Salt water										
Moderate bore hazard – pholads only	20	20	NR	NR	NR	NR	NR	NR	NR	C18
Severe borer hazard – limnoria tripunctata only	NR	NR	NR	NR	NR	2.5[8] and 1.5	2.5[8] and 1.5	NR	NR	C18
For both pholads and limnoria tripunctata use a dual treatment										
First treatment	–	–	–	–	–	1.0	1.0	–	–	C18
Second treatment	20	20	–	–	–	–	–	–	–	C18
POLES										
Utility										
Normal	7.5	7.5	7.5	0.38	NR	0.60	0.60	NR	NR	C4
Severe service conditions (high incidence of decay and termite attack)	9.0	9.0	9.0	0.45	NR	0.60	0.60	NR	NR	C4
Building poles – structural	9.0[9]	NR	NR	0.45	NR	0.60	0.60	NR	NR	C23
POSTS										
Fence, guide, and sight										
Round, half-round, and quarter-round	8	8	8	0.40	0.50	0.40	0.40	NR	NR	C14
Sawn four sides	10	10	10	0.50	0.62	0.50	0.50	NR	NR	C14
Guardrail and sign (including spacer blocks)										
Round	10	10	10	0.50	NR	0.50	0.50	NR	NR	C14
Sawn four sides	12	12	12	0.60	NR	0.60	0.60	NR	NR	C14

FOOTNOTES:

1. Southern Pine protected from the weather or exposed in a manner not to permit water to stand for any appreciable length of time does not require preservative treatment. Building codes generally require wood floors closer than 18 inches, or wood girden closer than 12 inches, to exposed ground be pressure treated (preservative).

2. AWPA Standard CI applies to each of the treating processes and all types of material.

3. Pentachlorophenol in suitable solvents or water-borne preservatives can provide a clean, paintable, odorless, dry surface. When painting after treatment is intended the processor should be so advised when the order is placed.

TABLE 5.1. *(Continued)*

4. Trade names of water-borne preservatives:

Acid Copper Chromate (ACC)
  Celcure*

Ammoniacal Copper Arsenite (ACA)
  Chemonite*

Chromated Copper Arsenate, Type A (CCA Type A)
  Greensalt

Chromated Copper Arsenate, Type B (CCA Type B)
  Boliden* CCA
  Koppers CCA-B
  Osmose K-33*

Chromated Copper Arsenate, Type C (CCA Type C)
  Chrom-Ar-Cu (CAC)*
  Langwood*
  Osmose K-33*
  Wolman* CCA
  Wolmanac* CCA
  Wood Last* CCA-C

Chromated Zinc Chloride (CZC)

Fluor Chrome Arsenate Phenol (FCAP)
  Osmosalts* (Osmosar*)
  Tanalith
  Wolman* Salts FCAP
  Wolman* Salts FMP

*Reg. U.S. Pat. Off.

5. When these preservatives are specified for materials to be used in salt water the creosote-coal tar shall conform to Standard P2 or P12 and the creosote shall conform to Standard P1 or P13.
6. All recommended retentions are taken from the standards of the American Wood Preservers Association (AWPA).
7. NR—Not Recommended.
8. The assay retentions are based on two assay zones—0 to 0.50 inch and 0.50 to 2.0 inches.
9. Not recommended where cleanliness and freedom from odor are necessary.

For information on pressure treating standards, quality control programs, and other technical assistance write or contact any of the following.

American Wood Preservers Association
7735 Old Georgetown Rd
Bethesda, Md. 20014
(301) 652-3109

Society of American Wood Preservers, Inc.
1401 Wilson Blvd. Suite 205
Arlington, Virginia 22209
(703) 841-1500

American Wood Preservers Institute
1651 Old Meadow Road
McLean, Virginia 22101
(708) 893-4005

Southern Pressure Treaters Association
2920 Knight St., Room 121
Shreveport, Louisiana 71105
(318) 861-2479

It is easy to understand why pressure processes are presently used to such an extent. The penetration of preservative under pressure is deep and uniform, thereby providing a more effective protection to the treated wood. In addition, the process is easily adaptable to large-scale industrial output, thus making it possible to be highly competitive in cost. There are various pressure processes that differ in some details. However, they all have in common that treatment takes place under pressure, in long cylinders of up to 180 ft. in length, the diameter of which varies between 4 and 9 ft. The wood is moved in and out of the tank, and about the treating yard, on special cars riding on steel tracks. Additional plant equipment consists of storage and measuring tanks, pressure and vacuum pumps and steam boilers.

Two general methods are used to apply preservatives under pressure, namely, the full-cell and the empty-cell. The greatest amount of preservative remains in the former process, the wood cells are filled as much as possible. In the empty-cell process a portion of the preservative is forced out of the wood, leaving the wood cells only coated, rather than filled with preservative; thus, deep penetration is possible with limited final retention of preservative. A number of variations exist within the two methods.

Brief descriptions of one of the many full-cell and the empty-cell methods are given as follows. Full-cell processes start with an initial vacuum, 20 in. of mercury, that is applied to the wood for a period of 15 to 60 minutes to draw the air out. After this, hot preservative oil with a temperature of 180 to 200°F is forced in under pressure that may reach a maximum of 200 psi. The pressure is maintained until the desired absorption of preservative is attained by the wood. Once the pressure is released the remaining oil is drained from the tank, after which a final vacuum is applied to remove the surface oil from the wood. In the empty-cell process preliminary air pressure is applied and forced into the wood before the preservative is introduced in the chamber at a higher pressure. This results in air being compressed inside the wood. Once the desired absorption of preservative is attained, the pressure is released and the remaining preservative is drained from the chamber. While this is going on, the air compressed inside the wood expands forcing out a considerable amount of applied preservative. A final vacuum in the chamber contributes to recovery and shortens the dripping period. The penetration of preservative into wood with this method is deep and uniform.

5.8.1 Effects On Strength Properties

Enhancement of the durability of wood against the various agents of destruction should not take place at the expense of a reduction in strength properties. Modern methods of treatment of wood accomplish preservation without weakening the wood.

It has been shown that creosote, as well as coal-tar creosote and petroleum creosote mixtures, are inert to wood; no chemical reaction takes place between these chemicals and the wood. However, experience has shown that a loss of strength can occur during treatment if excessive preconditioning, such as steaming at high temperatures (more than 240°F) and high pressure (more than 200 psi) is used. The combined effect of both, while wood is hot and soft, may cause severe distortions of the piece under treatment. Thus, strict control of temperature and pressure is necessary. Standards of the industry permit a maximum temperature of 210°F, with an average of 180°F, and a maximum pressure of 200 psi, depending on the species and the ease with which the wood takes the treatment. Commonly, however, maximum pressure varies between 150 and 175 psi (9).

The fact that modern preservative treatments do not affect strength properties of wood is reflected in design specifications (10); the allowable unit stresses given for untreated wood apply also to lumber pressure-impregnated by an approved process and preservative.

5.9 FIRE-RETARDANT TREATMENTS

There are two basic fire-retardant treatments available. For new wood construction, pressure impregnation with effective fire-retardant chemicals is the best method. For wood in existing structures, surface applications offer the principal means of increasing resistance against fire.

The methods used for impregnation of wood with fire-retardant chemicals are similar to those described in Section 5.8; the wood is placed within a sealed treating cylinder and the treating solution forced in by means of pressure (9). Among the chemicals commonly used are monoammonium and diammonium phosphate, ammonium sulphate, borax, boric acid and zinc chloride. In addition, fire-retardant chemicals may be used in combination with other preservative salts that are toxic to wood-destroying fungi, thereby extending the protection coverage of the treatment.

Wood that has been pressure-treated with fire-retardant chemicals may be charred or disintegrated with a very limited amount of flaming if subjected to continuous exposure to intense heat from an outside source; however, once the heating is discontinued the charring and flaming ceases. The treatment retards the normal increase in temperature under fire conditions, decreases the rate of flame spread and allows fires to be more easily extinguished. These beneficial fire-retardant properties, however, are not obtained without some effect on other properties of wood. Pressure-impregnation of wood with fire-retardant chemicals has some detrimental effect on its strength properties. This is recognized by present design specifications (10) which require that the allowable unit stresses for lumber pressure-impregnated with fire-retardant chemicals be reduced 10 percent.

Fire-retardant coatings, in contrast to pressure-impregnation methods, are applied by brush. Coatings of fire-retardant paints are applied to obtain the desired insulation of wood from the heat and to prevent free access of the oxygen required by combustion. These coatings of fire-retardant paint incorporate materials which decompose under heat to release gases which puff and swell the paint film to form an intumescent cellular insulating layer (13).

5.10 DETAILING FOR FIRE PROTECTION

Through good design and construction details it is possible to provide a high degree of fire resistance in the wood types of construction, namely heavy timber, ordinary and light-frame construction.

An example of heavy-timber construction is shown in Fig. 5.9. This type of wood construction is used for multistory structures as shown in the figure, with interior wood columns and beams (usually glued-laminated) and exterior masonry walls. Beams and girders are not less than 6 in. thick and not less than 10 in. deep; wood columns are not less than 8 × 8 in. when supporting floor and 8 × 6 in. when supporting roofs only. Floors have a minimum 4 in. thickness; roof decks have a minimum 2 in. thickness (all dimensions given are nominal). Fire resistivity of this type of construction lies in the slow rate of burning of massive wood and the avoidance of concealed spaces in which fire may originate and spread undetected.

Figure 5.9. Inside view of a seven-story asbestos processing mill 126 ft. × 132 ft. in plan, erected in British Columbia, Canada. See Ref. 11.

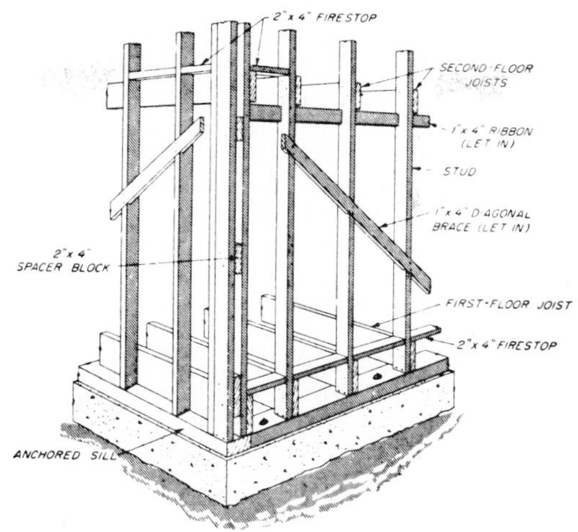

Figure 5.10. Light-frame wood construction with firestop provisions. From Ref. 9.

Ordinary type construction is similar to, but somewhat less massive than, heavy-timber construction. The exterior walls are of masonry or other approved noncombustible material having the required fire resistance and in which the interior structural elements are wood members not less than 2 in. thick and fire protected as required in accordance to use. Fire protection requirements should be as set forth in local building codes.

Light-frame wood structures, as in the case of most residential and most commercial buildings, consist of stud-walls, joisted floors and ceilings, and raftered roofs (12). These buildings do not have the fire resistance provided by heavier wood frames or ordinary wood construction; thus, it is here that attention to good construction details is most important to delay the spread of fire and reduce hazards to occupants. Fire spreads by movement of high-temperature air and gases through open channels, hence, it is important that firestops, see Fig. 5.10, be provided in concealed air spaces to prevent passage of flames up or across a building. Firestops should be provided at exterior walls, at each floor level, and at the level where the roof connects with the wall (9).

REFERENCES

1. Hunt, G. M., and Garratt, G. A., "Wood Preservation," McGraw Hill Book Company, New York, Third Edition, 1967.
2. St. George, R. A., Johnston, H. R., and Kowae, R. J., "Subterranean Termites, their Prevention and Control in Buildings," U.S. Dept. of Agriculture, Home and Garden Bulletin, No. 64. 1940.
3. Baechlet, R. H., Gjovik, L. R., and Roth, H. C., "Marine Tests on Combination-Treated Round and Sawed Specimens," American Wood Preservers' Association, Preprint prepared for the 1970 Annual Meeting in Chicago, Illinois.
4. Chellis, R. D., "Pile Foundation," McGraw Hill Book Company, Inc., New York, Second Edition, 1961.
5. ASTM Designation D 1760–60, "Standard Specification for Pressure Treatment of Timber Products," American Society of Testing and Materials, 1971 Book of ASTM Standards, Part 16, pp. 548–568, July 1971.

6. Federal Specification TT-W-571i, "Wood Preservation: Treating Practices," U.S. Government Printing Office, Washington, D.C., 1969.
7. ASTM Designation D 390–67, "Standard Specification for Coal-Tar Creosote for the Preservative Treatment of Piles, Poles, and Timbers for Land and Fresh Water Use," American Society for Testing and Materials, 1971 Book of ASTM Standards, Part 16, pp. 194–195, July 1971.
8. ASTM Designation D 391–63, "Standard Specification for Creosote-Coal Tar Solution," American Society for Testing and Materials, 1971 Book of ASTM Standards, Part 16, pp. 196–197, July 1971.
9. Forest Products Laboratory, "Wood Handbook," U.S. Dept. of Agriculture Handbook No. 72, U.S. Government Printing Office, Washington, D.C., 1955.
10. "National Design Specification for Stress-Grade Lumber and Its Fastenings," 1977 Edition, National Forest Products Association, Washington, D.C.
11. Gower, L. E., "Laminated Timber Used in High-Rise Mill Building," Civil Engineering Magazine, ASCE, December 1971, pp. 40–42.
12. Anderson, L. O., "Wood-Frame House Construction," U.S. Department of Agriculture, Forest Products Laboratory, Agriculture Handbook No. 73, Revised July 1970. For sale by Superintendent of Documents, U.S. Government Printing Office, Washington, D.C. 20402.
13. Fleischer, H. O., "The Performance of Wood in Fire," Report No. 2202, U.S. Forest Products Laboratory, Madison, Wisconsin, November, 1960.

Chapter 6
Fasteners

6.1 INTRODUCTION

The importance of fasteners in wood engineering can not be overemphasized since wood structures are made up of elements that must be connected together. Thus, connections occur in trusses, rigid frames and arches, post and beam construction and light-frame structures, among others. A good understanding of the behavior and use of fasteners is necessary to the designer of wood structures.

Of the various mechanical fasteners, bolts, lag screws and timber connectors are studied first, because of their particular importance in heavy timber construction. Usually these fasteners are designed to transmit axial forces and shear between connected members. However, rigid connections of prefabricated laminated frames and arches (which may be subjected to the action of axial load, shear and bending moment) can also be obtained using these fasteners. Problems and solutions for rigid connections are discussed herein. Finally, for light-frame construction, fasteners such as nails, wood screws, sheet-metal anchors and connector plates are studied. Tables and graphs that can be used as design aids for the various mechanical fasteners are given in Appendix A.6.

Only mechanical fasteners are described herein since the use of glue to fabricate laminated members and plywood panels is studied in Chapters 3 and 4. Illustrative examples of use of various mechanical fasteners are given throughout Chapter 6 and in some succeeding chapters.

6.2 BOLTS

Common bolts have been used for many years as fasteners in wood construction. Details of wood bridges built in the eighteenth century show bolts used in numerous ways; however, since the development of modern wood connectors, bolted joints are no longer employed as frequently as they once were. The use of bolts in combination with so-called timber connectors is discussed in Section 6.4. In the following, the behavior and design of conventional bolted connections is studied.

Present recommendations for design of bolted joints of National Design Specification (1) are based on experimental research conducted at the U.S. Forest Products Laboratory at Madison, Wis., by Trayer (2) in the early 1930's. In general, two types of tests, see Fig. 6.1, were made in the investigation; in one the applied load acted in a direction parallel to the grain of the wood, and in the other the applied load was perpendicular to the grain. Both wood and metal splice plates were used in the tests. The metal splice plates were ¼ in. thick, and the wood splice plates were one half the thickness of the main member. Only one bolt was used in each test, since the latter were made primarily to determine the bearing strength of wood under bolts of various diameters and lengths.

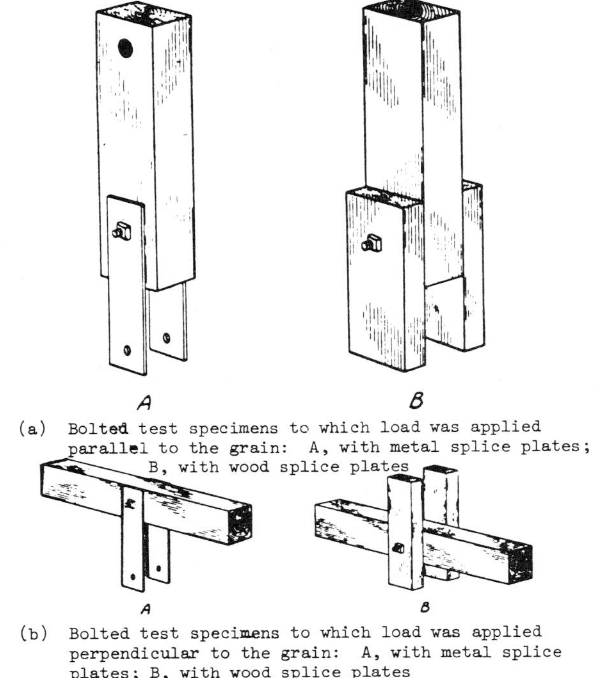

(a) Bolted test specimens to which load was applied parallel to the grain: A, with metal splice plates; B, with wood splice plates

(b) Bolted test specimens to which load was applied perpendicular to the grain: A, with metal splice plates; B, with wood splice plates

Figure 6.1. Tests of bolted connections. From Figs. 2 and 3 of Ref. 2.

In order to eliminate the extremely variable element of friction, the bolt nuts in Trayer's tests were not drawn tight; it was considered that, under service conditions, friction could not be relied upon because of the shrinkage that takes place in the wood. However, it was observed in the tests that, at very high loads, the bolts are usually bent enough to draw the splice plates snuggly against the center piece, thereby introducing considerable friction.

6.2.1 Discussion of Test Results

The results of load-testing a joint, made by connecting ¼ in. metal splice plates to a 2-in. timber by means of a single ½ in. bolt, are shown in Fig. 6.2. The relation between load on the bolt, applied parallel to the grain of the wood, and slip in the joint is given. The first point is on the axis of ordinates and not at the origin, because readings were begun only after a small initial load had been applied to take up all slack in the joint. Successive points fall along a straight line up to a load $P = 2{,}800$ lb., where P is the proportional-limit load. The latter is the load at which the slip in a bolted joint ceases to be proportional to the load. The concept of proportional-limit stress, f_P, is derived from this and is defined as the average stress under the bolt for the proportional-limit load. The value of f_P is given by P/LD, where L and D are the length of bolt in the main member (equal to the actual thickness of the main member) and the diameter of the bolt, respectively. The product LD is the projected bearing area of the bolt. The average stress f_P is not

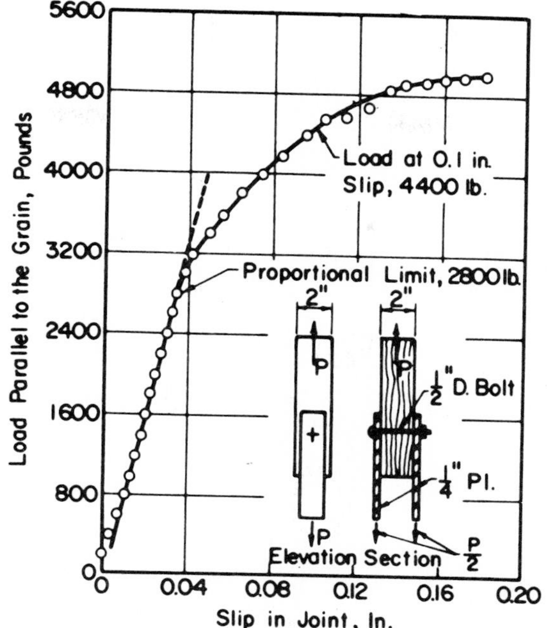

Figure 6.2. Variation of load parallel to grain with slip in the joint for an individual test specimen. From Fig. 4 of Ref. 2.

the actual stress at the proportional-limit load; at the edges of the member the stress is much greater than this value and near the center it is much less. The value of $f_{0.1''}$ is similarly defined, i.e. $f_{0.1''} = P/LD$, where P is the load at 0.1 in. slip, see Fig. 6.2.

Trayer's most significant conclusion was that both f_P and $f_{0.1''}$ drop off gradually as the ratio L/D increases. This is shown in Fig. 6.3 where the variations of f_P and $f_{0.1''}$ are given for L/D ratios between 0 and 13. The average proportional-limit bearing stress under a bolt is expressed as a percentage of the maximum crushing strength of the wood in which that bolt was used. For small values of L/D, up to L/D = 2.5, there is no reduction in the values of f_P. In the range $2.5 < L/D < 5.5$ there is a definite reduction in the value of f_P. Beyond L/D = 5.5 the average proportional-limit stress drops off at the same rate that the L/D ratio increases. In other words, for $L/D > 5.5$, f_P is inversely proportional to L/D. Since $P = f_P LD$ it is easy to show that for this case P is directly proportional to the term D^2 and independent of L. This is shown graphically in Fig. 6.4, where the curve A is identical to the curve for f_P of Fig. 6.3, and the curve B gives the load for a given diameter of bolt. Note that curve B increases uniformly up to L/D = 2.5 and then gradually merges into a horizontal line at L/D = 5.5. Beyond this L/D ratio the load increases no more, remaining constant at the maximum value, in spite of the increasing values of L.

Tests with bolts which subject wood to bearing stresses perpendicular to grain, as compared to tests in bearing parallel to grain, showed similar influence of the L/D ratio. However, a larger

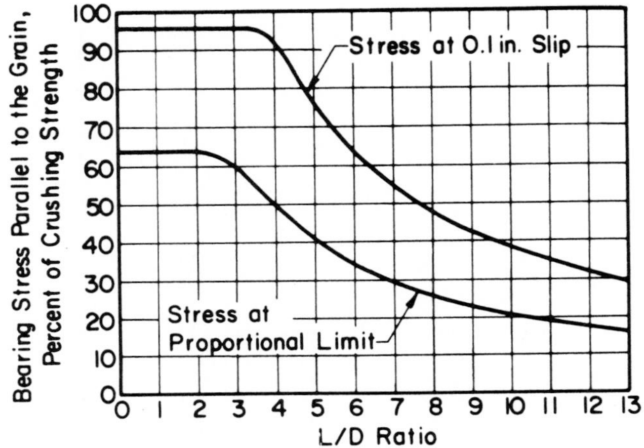

Figure 6.3. Variation with L/D radio of average bolt-bearing stress at proportional limit and at 0.1 in. slip in softwoods. From Ref. 2.

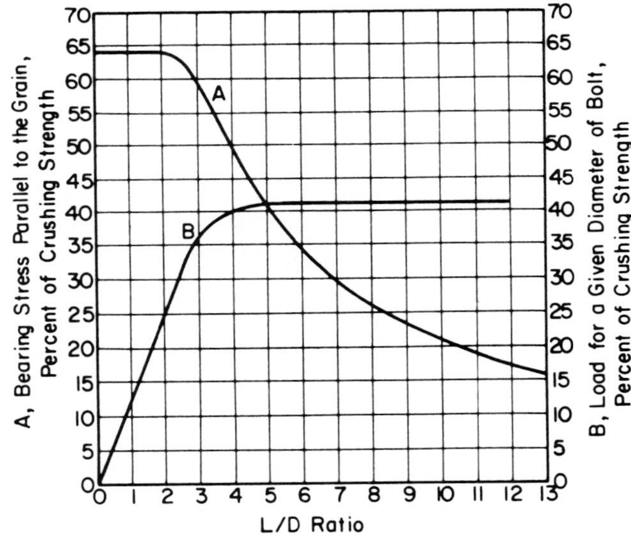

Figure 6.4. Variation with L/D ratio of average proportional limit bolt-bearing stress parallel to grain, curve A, and proportional-limit load, curve B. From Ref. 2.

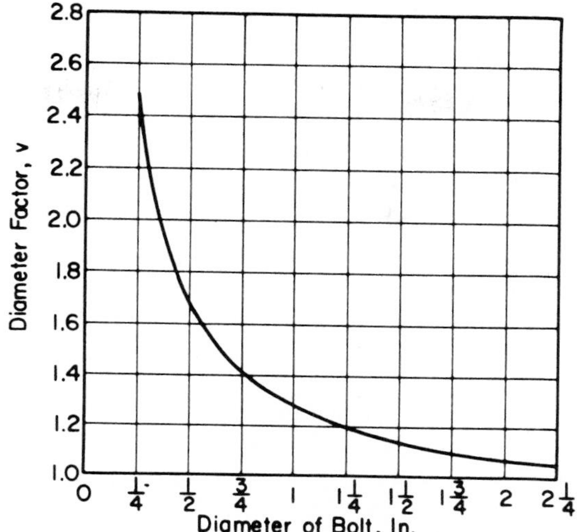

Figure 6.5. Effect of diameter of bolt on capacity of joints for wood subjected to compression across the grain. From Ref. 2.

L/D ratio is necessary before the load drops off, which then takes place at a slightly greater rate. There is a definite influence of the diameter of the bolt, D, on the capacity of the bolt when bearing perpendicular to the grain of wood. This is contrary to the case of wood loaded parallel to grain, for which the average proportional-limit stress is independent of the diameter of the bolt. Trayer's explanation of the influence of the bolt diameter on bearing normal to grain is based on the similarity of this phenomenon to that of a specimen of wood loaded normal to the grain on only a portion of its surface. When load is applied to the specimen, the fibers along the edges of the bearing surface are thrown into bending and tension. This action contributes to supporting the applied load with the result that the observed proportional-limit stress across the grain is higher than it would be if the whole specimen were covered. Since this edge effect remains constant, the percentage increase that it causes on the average calculated bearing stress varies as the width of plate is changed, naturally becoming greater at smaller widths, or smaller diameters as in the case of bolts. As a result of Trayer's tests, the variation of the ratio of proportional-limit stress across the grain under a short bolt to proportional-limit stress across the grain of control specimens, v, with D is shown in Fig. 6.5. The term v is referred to as the diameter factor.

6.2.2 Design of Bolted Connections

6.2.2.1 Basic Stress Method

The results of Trayer's tests were used to determine safe bolt-bearing working stresses. A table of basic bolt-bearing stresses (2) (3) was obtained first for small, clear, green specimens of wood in both compression parallel and perpendicular to grain. The basic bolt-bearing stresses in compression parallel to grain were obtained from the average crushing-strength parallel to grain as modified by duration of loading, increase in strength with drying, variability, and a reduction factor for safety. The basic bolt-bearing stresses in compression perpendicular to grain were obtained as before except the average proportional-limit stress perpendicular to grain was used.

The safe working stress in compression parallel to grain, $F_{\parallel}$ for a bolt of diameter D for a given species, is determined as the product of the corresponding basic stress in compression parallel to grain times a factor r_1; the latter is the percentage of basic stress parallel to grain for the particular L/D ratio of the bolt. Similarly, the safe working stress for the same bolt in compression perpendicular to grain, $F_{\perp}$, is obtained as the product of the corresponding basic stress in compression perpendicular to grain times the factors r_2 and v. The factor r_2 is the percentage of basic stress normal to grain for the particular L/D ratio of the bolt. The variations of r_1 and r_2 with L/D are shown in Fig. 6.6. The factor v has been discussed before, its variation with D is given in Fig. 6.5. For an angle of load to grain θ, other than zero or ninety degrees, Hankinson's formula or Scholten's nomograph, see Section 2.2.3, may be used to determine the allowable load.

The allowable loads P and Q are determined as the product of the corresponding allowable stress, $F_{\parallel}$ and $F_{\perp}$, times the projected area of the bolt, LD.

As an example, determine the allowable loads in compression parallel and perpendicular to grain and at a 45° angle of load to grain, of a ¾ in. bolt in a 3 in.-thick timber having basic stresses of 1,450 psi and 320 psi compression parallel and perpendicular to grain, respectively.

The L/D ratio is 3/3/4 = 4. Hence, from Fig. 6.6: $r_1 = 0.92$, $r_2 = 1.00$; from Fig. 6.5, for D = ¾ in., v = 1.4. Thus, the allowable stresses are:

$$F_{\parallel} = (1{,}450)(0.92) = 1{,}333 \text{ psi}$$

$$F_{\perp} = (320)(1.00)(1.4) = 448 \text{ psi}$$

from which the allowable loads may be determined as follows:

$$\theta = 0°,\ P = (1{,}333)(3/4)(3) = 3{,}000 \text{ lb.}$$

$$\theta = 90°,\ Q = (448)(3/4)(3) = 1{,}010 \text{ lb.}$$

and using Eq. 2.2:

$$\theta = 45°,\ N = \frac{P}{1 + \left(\frac{P}{Q} - 1\right)\sin^2\theta} = \frac{3{,}000}{1 + \left(\frac{3{,}000}{1{,}010} - 1\right)\frac{1}{2}} = 1{,}500 \text{ lb.}$$

6.2.2.2. Actual Design Practice

There is no need for designers to follow the basic stress method when using commercial lumber. Table A.6.1a provides designers directly with allowable loads, in pounds, for one bolt (f_y = 45 ksi) loaded at both ends in double shear for commercial species of wood under normal loading conditions. For loading conditions other than normal, the usual values of the load-duration factor apply. Allowable load for more than one bolt, each of the same or miscellaneous sizes, is the sum of the loads permitted for each bolt. The tabulated loads apply to species, irrespective of grade of lumber used except that some species have different values for dense, medium grain and open grain lumber.

The influence of the coefficients r and v on NDS-tabulated loads can be shown, for example, using Southern pine wood. First, consider the effect of r using a ¾ in. bolt, in compression parallel to grain. The values of the average stress, as L and the ratio L/D are increased, are shown in

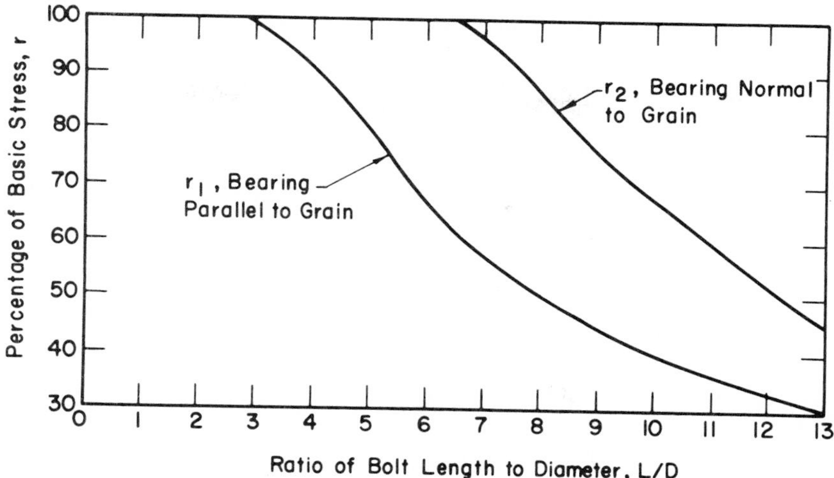

Figure 6.6. Variation of percentage of basic stress, r, with L/D for Southern Pine and Douglas Fir. Data from Ref. 3.

Table 6.1. It is immediately apparent that as L/D increases, the average stress decreases, thereby acknowledging the effect of r. Also, as shown before, beyond a certain value of L/D the allowable load P remains constant. The influence of v on capacity can be shown by comparing the allowable stresses in compression perpendicular to grain for two bolts of different diameter, for instance ½ in. and 1 in., but equal L/D ratio. Table 6.2 shows the results for various L/D ratios. As expected, the average stress under the bolt with the smaller diameter is consistently greater.

TABLE 6.1. USE OF FACTOR r IN NATIONAL DESIGN SPECIFICATION (1).

L in.	L/D	LD in.²	P lb.	P/LD psi
1.5	2.00	1.125	1,420	1,262
2.0	2.67	1.500	1,890	1,260
2.5	3.33	1.875	2,310	1,232
3.0	4.00	2.250	2,630	1,169
3.5	4.67	2.625	2,800	1,067
4.0	5.33	3.000	2,850	950
4.5	6.00	3.375	2,860	847
5.5	7.33	4.125	2,860	693
7.5	10.00	5.625	2,860	508
9.5	12.67	7.125	2,860	401

For design purposes a discussion is necessary on details such as: 1) conditions of lumber, 2) service conditions, 3) side members, 4) critical section, 5) spacing, 6) end distance and 7) edge distance of bolts.

TABLE 6.2. USE OF FACTOR v IN NATIONAL DESIGN SPECIFICATION (1).

L in.	D in.	L/D	LD in.^{2}	Q lb.	Q/LD psi
1.5	½	3	0.75	430	573
3.0	1	3	3.00	1,300	433
2.0	½	4	1.00	570	570
4.0	1	4	4.00	1,730	433
2.5	½	5	1.25	720	575
5.0	1	5	5.00	2,160	433
3.0	½	6	1.50	860	573
6.0	1	6	6.00	2,540	424

Conditions of Lumber. The condition of lumber at the time of fabrication of the bolted joints affects design. Tabulated loads are for bolts in lumber seasoned to a moisture content approximately equal to that to which it will eventually come in service. For lumber installed at, or above, fiber saturation point and which becomes seasoned in place, some limitations exist. This is due to the fact that the wood will shrink substantially across the grain as seasoning takes place. If shrinkage is restrained by splice plates with various rows of bolts, tensile stresses across the grain will occur which will split the member. The collapse of wooden trusses in a building in Portland, Oregon (4), where substantially green lumber was used, is cited by McKaig (5) as having originated because of this phenomenon. NDS (1) requires that for green lumber the full allowable load be used only for: 1) a bolted joint with wood side members having a single bolt and loaded parallel or perpendicular to grain, 2) a single row of bolts loaded parallel to grain, 3) multiple rows of bolts loaded parallel to grain with separate splice plate for each row, and 4) a bolted joint with steel gusset plates having a single row of bolts parallel to grain in each member and loaded parallel or perpendicular to grain. For other arrangements of bolted joints, the allowable loads are 40 percent of the tabulated loads. For lumber partially seasoned when fabricated, adjusted intermediate values may be used. See also Tables A.6.27 and A.6.28.

The weakening effect of pressure-impregnation of wood with fire-retardant chemicals is accounted for in the following way. If the wood is kiln-dried after treatment, 90 percent of the tabulated loads apply. For lumber not kiln dried after treatment, provisions similar to those described above for unseasoned wood apply, except the base is to be taken at 90 percent of the tabulated loads.

Service Conditions. Tabulated loads as adjusted for condition of lumber apply to bolted joints used under conditions continuously dry as in most covered structures. When joints are to be exposed to weather, 75 percent, and where always wet, 67 percent, of the tabulated loads apply.

Side Members. Tabulated bolt loads are for side members of wood. Bearing thrust on side plates is assumed to be parallel to fibers. When steel plates are used for side members, the tabulated loads for parallel-to-grain loading may be increased 25 percent, but no increase can be made for perpendicular-to-grain loads. The thickness of the side members affects capacity of the joint. Also, two-member and multiple-member joints are subjected to different rules for the determination of capacity. Illustrative Example 1 presents all the various rules which the reader is advised to study carefully.

Critical Section. That section of the member, taken at right angles to the direction of the load which gives the maximum stress in the member based on the net area, is the critical section. The required net area is determined by dividing the total load which is transferred through the critical section by the allowable stress in tension parallel to grain, F_t, for tension members, or by the allowable stress in compression parallel to grain, F_c, for compression members, for the species and grade of lumber used.

Spacing of Bolts. The minimum spacing of bolts for parallel-to-grain loading is four times the bolt diameter. For perpendicular-to-grain loading, as in the case of a beam attached to the side of a column, spacing between bolts in a row perpendicular to grain shall be limited by the spacing requirements of the attached member or members (whether of metal or of wood loaded parallel to grain). If the design load is less than the bolt-bearing capacity of side members, the spacing may be reduced proportionately.

The spacing across the grain between rows of bolts is at least 1½ diameters for parallel-to-grain loading. For perpendicular-to-grain loading the spacing is at least 2½ diameters for $L/D = 2$, and 5 diameters for $L/D > 6$. For the range $2 < L/D < 6$, the spacing is obtained by straight-line interpolation. In addition, the spacing between rows of bolts paralleling the member shall not exceed 5 in. unless separate splice plates are used for each row of bolts.

End Distance. The distance from the end of a bolted member to the center of the bolt hole nearest to the end is the end distance. For parallel-to-grain loading the minimum end distance is: in tension, 7 D for softwoods, 5 D for hardwoods; and in compression, 4 D. For perpendicular-to-grain loading, 4 D.

Edge Distance. The distance from the edge of the member to the center of the nearest bolt hole is the edge distance. For parallel-to-grain loading the minimum edge distance is 1.5 D, except that for $L/D > 6$ use at least 1.5 D or one-half the distance between rows of bolts, whichever is the greater. For perpendicular-to-grain loading, the minimum edge margin toward which the load is acting is 4 D and the minimum margin on the opposite edge is 1.5 D.

For other provisions and recommendations refer to the latest edition of NDS (1). The examples that follow show the use of bolted connections in wood design.

6.2.2.3 Illustrative Example

A number of bolted connections of wood members under normal loading conditions are shown in Fig. 6.7. They represent the various cases discussed in NDS (1). It is wished to determine the load P that each connection is allowed to transmit. Assume that the members are of Southern pine medium grain wood surfaced dry and used at 19% maximum moisture content under conditions continuously dry.

The solution is shown in Fig. 6.7. Connection (a) is the typical case for which tabulated loads, see Table A.6.1a, apply directly. Connection (b) with thicker side members has no additional strength, since capacity is determined by the main member. Connection (c) has thin side members which determine its capacity. One can see that using twice the thickness of the thinnest side member to determine P increases the effect of the factor r, thereby penalizing capacity. It can be easily checked that a greater capacity, although not acceptable by NDS, could have been obtained by taking P equal to two times the capacity of each individual side member.

The same is true for connection type (d), where two equal members are joined together. Capacity is obtained by taking one-half the tabulated load of a member with twice the thickness. In the case of members of unequal thickness, see case (e), capacity is determined as before except

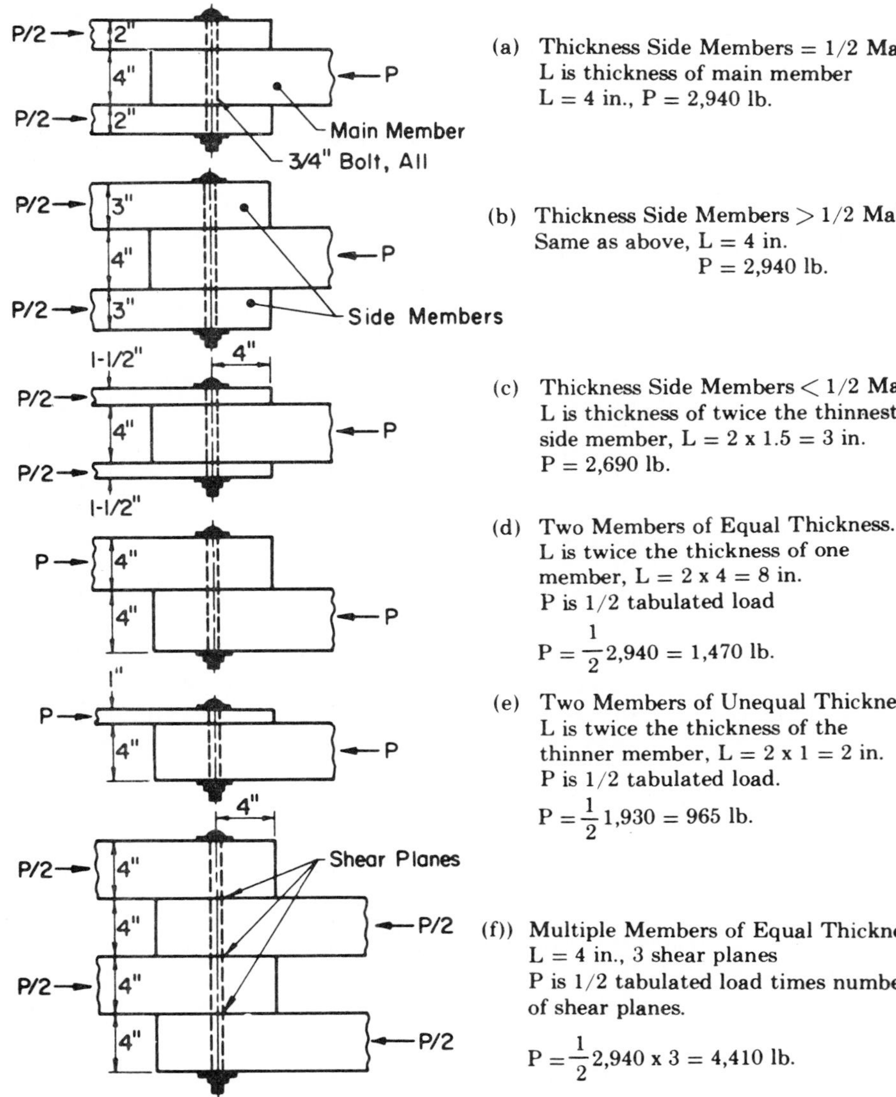

Figure 6.7. Capacity of various bolted joints of Southern Pine for normal loading conditions, determined using Table A.6.1, Group 3.

that the thickness of the thinnest member is taken. The case of multiple-member joints with pieces of equal thickness is shown in Fig. 6.7f. The allowable load depends on the number of shear planes involved; for each plane it is equal to one-half the tabulated load for a piece with a thickness equal to that of the member.

Provisions for design of a joint with four or more members, not of equal thickness, may be found in Table A.6.1.b.

6.2.2.4 Group of Fasteners

The preceding calculations determined the capacity of the single bolt for various possible cases. When two bolts are used, the capacity of the connection doubles as each bolt is subjected to the same load. However, for connections using more than two bolts and, especially in long multiple-bolt joints, an unequal distribution of load among the bolts results upon application of the load. Recent research (19, 20) has shown this fact conclusively.

Consider the multiple-fastener joint shown in Fig. 6.8a used to splice, for instance, the main member of the bottom chord of a truss. The load F is transferred from the main member to the splice plates through 5 bolts. The load in the main member is equal to F to the left of bolt No. 1, and is equal to F-P_1 to the right of bolt No. 1. At other cross sections the load in the main member is reduced by an amount equal to the sum of the bolt loads to the left of the given section. At any cross section the sum of the load carried by the main member and that carried by the splice plates must equal the load on the joint. Thus, as the load in the main member decreases by transfer through the bolts, the splice plates pick up the load by an equal amount. The respective variations of load in the main member and the splice plates are given in Fig. 6.8b, the steps representing the load taken by the bolts. It is noted that the load steps are not of the same intensity; the load steps for the end bolts, 1 and 5, are greater than those corresponding to the center bolts, 2, 3 and 4. The uneven distribution of loads in the bolts is caused by the flexibility of the joints,

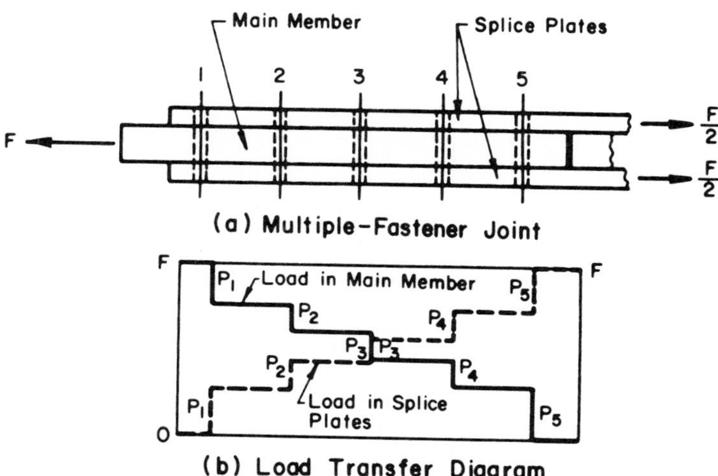

Figure 6.8. Transfer of load in main member to splice plates in a multiple-fastener joint. Adapted from Ref. 19.

i.e. by the elongations that occur under load in the main member and the splice plates. Because of this the shear strain imposed on the bolts and, therefore, the load on the bolts, is a variable quantity that depends on the relative location of the bolt on the multiple-bolt joint.

The relation between the load on the end bolt and the load on the center bolts, as a function of the number of bolts in a given multiple-bolt joint, is given by Cramer (19). The relation depends primarily on the relative stiffness of the main member and splice plates; Fig. 6.9 gives the relation for a case where the stiffness of the main member is equal to that of the sum of the splice plates. The figure shows the variation of bolt load, P, given as a percent of the joint load, T, with the number of bolts in a row, N, of a multiple-bolt joint. Three curves are given to show the actual loads for end bolts and center bolts as well as the theoretical average load, $P = 100\ T/N$, that would hypothetically occur as a result of an even distribution of load among all bolts. It is noted that for the given case, the load at the end bolts decreases as N increases but reaches a lower bound, for $N = 6$, at approximately 30 percent of T. For a greater number of bolts, the load of the end bolts remains constant; the two end bolts together carry over half of the load in multiple-fastener joints. The actual load on the center bolts is always lower than the theoretical average value corresponding to an even distribution of loads.

These results are used by NDS (1) to determine the capacity of a multiple-fastener joint. Included in the Design Provisions that follow are not only bolts, but also lag screws, see Section 6.3, and connectors, see Section 6.4.

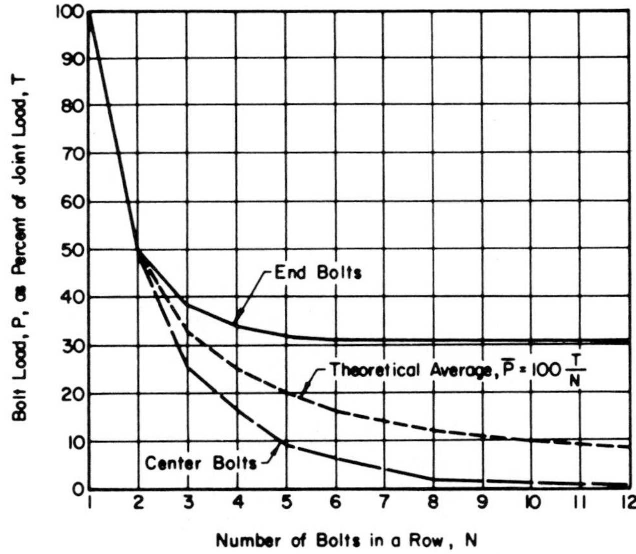

Figure 6.9. Relation between the load on the end and center bolts and the number of bolts in a row, in a given multiple-bolt joint. Adapted from Ref. 19.

Design Provisions

A group of fasteners is defined as one or more rows of fasteners arranged symmetrically with respect to the axis of the load. A row of fasteners consists of the following aligned with the direction of load:

a. Two or more bolts of the same diameter loaded in single or multiple shear; or
b. Two or more connector-units, see Section 6.4, or lag screws of the same type and size, see Section 6.3, loaded in single shear.

When fasteners in adjacent rows are staggered, see Fig. 6.10, the following is used to determine whether the rows act as individual rows or not. If the distance between adjacent rows, a, is greater than one fourth the distance, b, between the closest fasteners in adjacent rows measured parallel to the rows, i.e, if a τ b/4, the adjacent rows are considered as individual rows, see Fig. 6.10a. If a $<$ b/4 the adjacent rows shall be considered as one row for purposes of determining the allowable load on the group. For groups of fasteners having even number of rows, see Fig. 6.10b, this principle applies to each pair. For groups of fasteners having odd numbers of rows, the more conservative interpretation applies, see Fig. 6.10c.

The allowable load, P_r, on a row of fasteners of the same size and type in which the allowable load, P, on each individual fastener is the same is determined by:

$$P_r = N_f \, P \, K \tag{6.1}$$

where N_f = the number of fasteners
K = the modification factor appropriate for the type of side member, member size and number of fasteners in a row, as given in Tables A.6.25 and A.6.26.

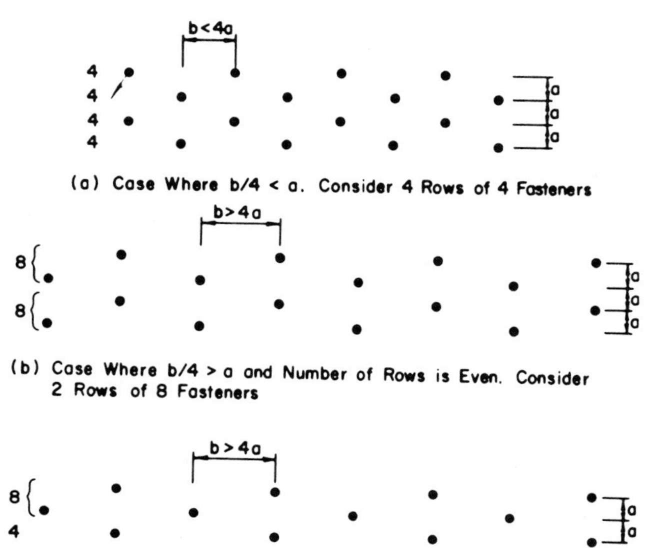

(c) Case Where b/4 > a and Number of Rows is Odd. Consider 1 Row of 8 Fasteners and 1 Row of 4 Fasteners

Figure 6.10. Staggered fasteners in multiple-fastener joints.

For the condition where the allowable load, P, for each individual fastener is not the same, as for example when reductions are applied because of the spacing or end distance, N_fP is expressed as P_s. Thus Eq. 6.1 can be expressed as follows:

$$P_r = P_sK \tag{6.2}$$

where P_s = the summation of the allowable loads on individual fasteners in a row as determined in the relevant requirements of the specification.

In all cases, the allowable load on a group of fasteners is the sum of the allowable loads on the rows in the group.

Tables A.6.25 and A.6.26, for the determination of the modification factor, K, use gross cross-sectional areas (with no reductions for net section) in calculating cross-sectional area ratios. When a member is loaded perpendicular to grain direction, its equivalent cross-sectional area (for calculating cross-sectional area ratios) is the product of the thickness of the member and the overall width of the fastener group, see Fig. 6.11a. When only one row of fasteners is used, the width of the fastener group shall be considered as the minimum parallel-to-grain spacing of the fasteners, see Fig. 6.11b.

The application of the preceding design provisions is illustrated by examples in this chapter as follows: Sections 6.2.2.5 and 6.2.2.6 discuss design of multiple-bolt joints; Section 6.5.2.1 considers a multiple-connector joint. Additional examples may be found in Chapter 10, e.g. at Section 10.2.2, connection of beam to girder, Section 10.3.3, column connection to girder, and Section 10.3.2, connection at midspan.

6.2.2.5 Illustrative Example

A 3 × 6-in. S4S Douglas fir member fabricated at 15% moisture content is spliced with two 2 × 6 side plates by means of six ¾-in. bolts, as shown in Fig. 6.8. The member is part of a structure that is used under cover and for which the moisture content will remain at 15%. The force T in the member is caused by snow-loading conditions. Determine the maximum value of T that the splice may take.

Since the thickness of the side members is greater than one-half that of the main member, the capacity of a bolt can be determined as shown in Fig. 6.7a. Entering Table A.6.1, using group 3, L = 2.5 in., d = ¾ in., yields P = 2,310 lb. The allowable load per bolt for a single bolt has been found. However, since this is a multiple-fastener connection, a modification factor, K, from Table A.6.25, is necessary to determine the actual allowable load per bolt. For the given case: $A_1 = 2.5 \times 5.5 = 13.75$ in², $A_2 = 2 \times 1.5 \times 5.5 = 16.5$ in², and $A_1/A_2 = 0.833$. For A_1 between 12 and 19 and N = 3, Table A.6.25 gives K = 0.95 for $A_1/A_2 = 0.5$, and K = 0.98 for $A_1/A_2 = 1.0$. Interpolating between K = 0.95 and K = 0.98, for A_1/A_2. = 0.833, yields the actual modification factor K = 0.97. The actual allowable load per bolt is then 0.97 × 2310 = 2240 lb. Thus, T = 6 × 2,240 × 1.15 = 15,460 lb., where 1.15 is the load-duration factor corresponding to snow-loading conditions. The end distance, 5¼ in., is seven times the diameter of the bolt as specified for softwoods; the spacing between bolts, 3 in., meets the minimum spacing of four times the bolt diameter specified for softwoods. The spacing between rows of bolts, 2½ in., and the edge distance, 1½ in., are in each case more than the minimum one and one-half bolt diameters, 1⅛ in., specified for parallel-to-grain loading. Hence, the geometry of the splice permits the development of the tabulated capacity.

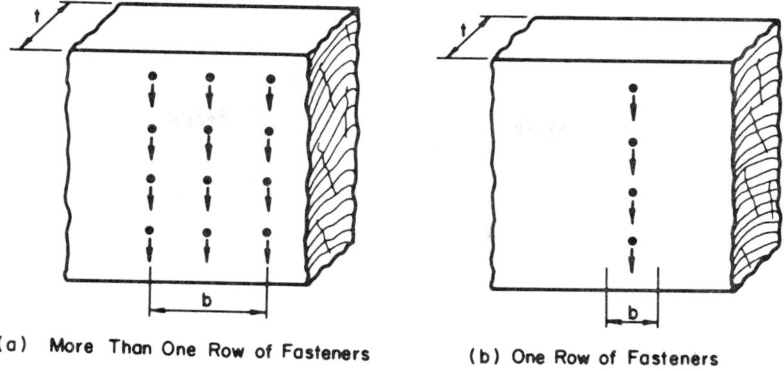

Figure 6.11. Equivalent cross-sectional area, A = bt, for the determination of modification factor, K, in members loaded perpendicular to grain.

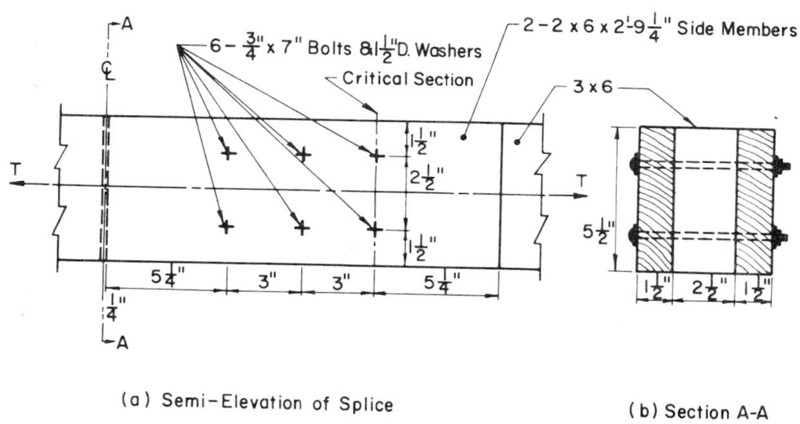

Figure 6.12. Capacity of a bolted splice connection.

Only the critical section remains to be checked. For the given splice, the critical section in the main member occurs where the full load T meets the first two bolts, see Fig. 6.12a; at that section a load T/3 is transmitted to the side members. The net area of the section of the main member is $13.75 - 2 \times 13/16 \times 2.5 = 9.96$ in.². For the given wood, from Table A.2.5 on allowable stresses, $F_t = 1{,}200 \times 1.15 = 1{,}380$ psi. Thus, $T = 1{,}380 \times 9.69 = 13{,}400$ lb. Therefore, the capacity of the splice is limited to 13,400 lb., as determined by the critical section of the main member rather than the bolts. Since this is very close to the capacity of the given bolts, the number and/or the diameter of the latter need not be reduced for a more economical splice.

6.2.2.6 Illustrative Example

A 4 × 4 in. tie is fastened to a 4 × 8 in. beam by two ¼ in. steel side plates and ¾ in. bolts as shown in Figs. 6.13a and b. Wood is surfaced-four-sides Southern pine No. 1, surfaced dry and used at 19% maximum moisture content. The connection transmits a 5,000 lb. load, due to normal

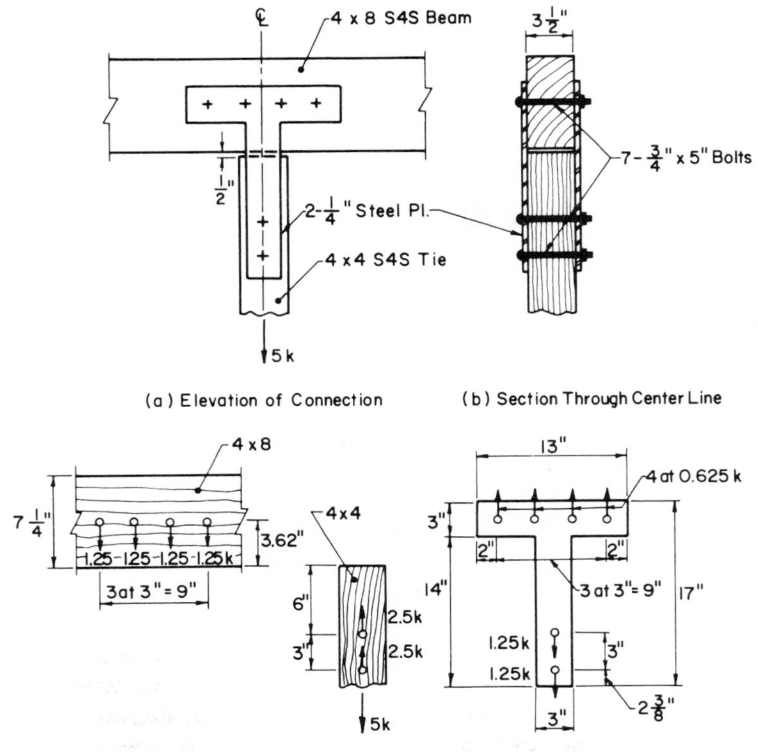

Figure 6.13. Bolted connection of a vertical tie to its supporting beam.

loading conditions, from the suspended vertical tie to the beam. Determine the number of bolts required and check the adequacy of the connection.

The ¾ in. bolts connecting the tie to the steel plates are acting in compression parallel to grain, see Fig. 6.13c, from Table A.6.1, for Group 3 at 3½ in. actual thickness, P = 2,800 lb. An increase of 25 percent is allowed for metal side plates, therefore, P = 1.25 × 2,800 = 3,500 lb. Consider using 2 bolts for the tie. The modification factor for multiple fasteners when N = 2, from Table A.6.26, is K = 1.00 for all values of A_1 and ratios A_1/A_2. Therefore, the capacity of the 2 bolts is 3500 × 2 = 7,000 lb > 5,000 lb; use 2 bolts. The transfer of the load from the steel plate to the beam occurs through bolts acting in compression perpendicular to the grain of the beam, for which Q = 1,260 lb. Consider using 4 bolts for the beam, placed in a single row, as shown in Fig. 6.13a. The provisions for multiple fasteners do not apply in this case because there is only one row of fasteners and it is not aligned with the direction of the load, but is perpendicular to it. The load applied to each bolt is essentially the same, and the total capacity of the 4 bolts is 4 × 1,260 = 5,040 lb > 5,000 lb; use 4 bolts.

The development of the connector forces shown in the free-body diagrams of Fig. 6.13c requires satisfaction of the necessary end and edge distances, as well as the spacing between connectors. For the beam connectors the required loaded-edge margin is four times the bolt diameter, i.e., 4 (¾) = 3 in. < 3.62 in., provided. Placement of the bolts as high as possible above the beam neutral axis is advisable to reduce the interaction between tension-across-the-grain stresses and horizontal shear stresses. The spacing between connectors is a minimum four times the bolt diameter, i.e., 3 in., as provided. The same is true for the connectors at the 4 × 4 in. tie. In the latter, the end distance, 6 in., is a little larger than the recommended seven times the diameter of the bolt, i.e., 5¼ in. Thus, the values of P and Q determined above can be developed by the connection.

The design of the T-shaped steel plate requires provisions of tensile capacity in the vertical leg, and bending and shear capacities in the horizontal portion, see Fig. 6.13c. Assume that the plate is ¼ in. thick and 3 in. wide. The net area in tension is (3–13/16) (¼) = 0.547 in.^{2}; at 20 ksi, this allows a tensile force of 10.94 k > 2.5 k. The critical section in the horizontal portion of the T is at the inside connector hole. Properties of the steel section are A = 0.547 in.^{2}, I = (1/12) (¼) [3^3 − $(13/16)^3$] = 0.552 in.^{4}, S = 0.368 in.^{3}. Thus, at 12.5 ksi and 20 ksi allowable stresses in shear and bending, respectively, the allowable shear force and bending moment are V = (0.547) (12.5) = 6.84 > 0.625 × 2 kip, and M = (0.368) (20) = 7.36 > 0.625 × 3 kip-in. In direct bearing against a ¼ in. plate, a ¾-in. bolt at 22.5 ksi may develop (¾) (¼) (22.5) = 4.22 > 1.25 kip. Since all requirements are met by an ample margin, the steel plate may be considered acceptable; and so is the connection.

6.3 LAG SCREWS

These fasteners are used instead of bolts where nuts on the surface of the wood may be objectionable or where it would be difficult to fasten a bolt. The anchoring effect of the nut is replaced by the threaded end of the bolt. Lag screws are available in lengths from 3 to 16 in. and in diameters from ¼ to 1¼ in. Typical dimensions of standard lag screws for wood are given in Table A.6.2a. The normal use of these fasteners is either in direct withdrawal from the wood, as, for instance, when fastening a vertical tie subjected to tension to the bottom of a beam, or through development of lateral resistance, as, for example, when connecting diagonal braces to piles of a bridge abutment. In addition to this independent action, lag screws may be used instead of bolts with timber connectors for multiple-plate connections, as shown in Section 6.5.

Research conducted at Forest Products Laboratory (3) resulted in the following expression for P_u, the ultimate withdrawal load in pounds per inch of penetration of the threaded part into side grain of the member holding the point:

$$P_u = 8{,}100\, G^{1.5}\, D^{0.75} \tag{6.3}$$

where D is the shank diameter in inches and G is the specific gravity of the wood based on oven-dry weight and volume at 12-percent moisture content. The allowable withdrawal load p is approximately one-fifth of the ultimate load. A table of allowable withdrawal loads is given in Table A.6.2b; this eliminates the need for using Eq. 6.3 in the case of commercial woods. When subjected to axial withdrawal lag screws will develop ultimate tensile strength if given the following penetrations: 7 diameters in Group I, 8 diameters in Group II, 10 diameters in Group III, and 11 diameters in Group IV, see Table A.6.4.

Lag screws should not be loaded in withdrawal from end grain of wood. If this condition cannot be avoided, the allowable load may be taken as not more than three-fourths of that for withdrawal from side grain.

The proportional limit lateral loads for lag screws inserted in side grain and loaded parallel to the grain of a piece of seasoned wood can be determined (3) from the following equation:

$$p = K\,D^2\,r \qquad (6.4)$$

in which p is the allowable lateral load, parallel to grain, in pounds; K is a constant depending on the species specific gravity, G_{12} (K = 4,820 for Douglas Fir-Larch and Southern Pine); D is the shank diameter of the lag screw in inches; and r is a correction factor which depends on the ratio t/D, thickness of member to shank diameter of screw. For t/D = 2, 3, 4, and 5, the correction factors are 0.62, 0.93, 1.07, and 1.18, respectively. When the load is applied perpendicular to grain the allowable load is given by Eq. 6.4 multiplied by a factor v which depends on D, the diameter of the lag screw; for D = ½, ¾, and 1 in., the factors v are 0.65, 0.55, and 0.50, respectively. For other angles of loading the allowable loads may be computed using Hankinson's formula or Scholten's nomograph.

When lag screws are used with metal plates, the allowable lateral loads parallel to grain may be increased 25 percent, but no increase may be made in the allowable loads when the applied load is perpendicular to grain. Again, lag screws should not be driven in end grain, because splitting may develop under lateral load. However, if so used, the allowable loads are two-thirds those for lateral resistance when lag screws are inserted into side grain and the loads act perpendicular to the grain.

The preceding formulas and recommendations for lag bolts or lag screws are contained in Table A.6.2 for the allowable withdrawal loads for normal duration of loading, and Tables A.6.3 and A.6.4 for the allowable lateral loads for wood side pieces and metal side pieces, respectively. In addition, Table A.6.5 classifies the various species of wood in four groups and lists the specific gravity of each species based on weight and volume when oven-dry.

Regarding geometrical considerations, i.e., spacings, end distances, and net section for these fasteners, NDS recommends that they be the same as for joints with bolts of a diameter equal to the shank diameter of the lag screw used. Also, NDS provisions for various conditions of lumber at fabrication and service conditions are similar to those for bolts, see Section 6.2.2.2. In addition, a modification factor, K, affecting the capacity of multiple lag-screw joints is to be used according to the design provisions of Section 6.2.2.4.

6.3.1 Illustrative Example

The connection shown in Fig. 6.14 is used to attach a vertical load to a California Redwood 6 × 12 in. surfaced-four-sides beam. Wood is seasoned to 15% moisture content, service conditions are those of covered structures, i.e., the wood remains continuously dry and at about 15% M.C. Determine the maximum load that can be safely attached to the beam.

The connection makes use of two ½-in. lag bolts attached to the bottom of the beam. The load is transmitted to the wood by the lag bolts through direct withdrawal stresses. From Table A.6.4, California Redwood close grain is classified as Group III, with specific gravity G = 0.42. Table A.6.2b, for G = 0.42 and D = ½ in., yields an allowable withdrawal load of 291 pounds

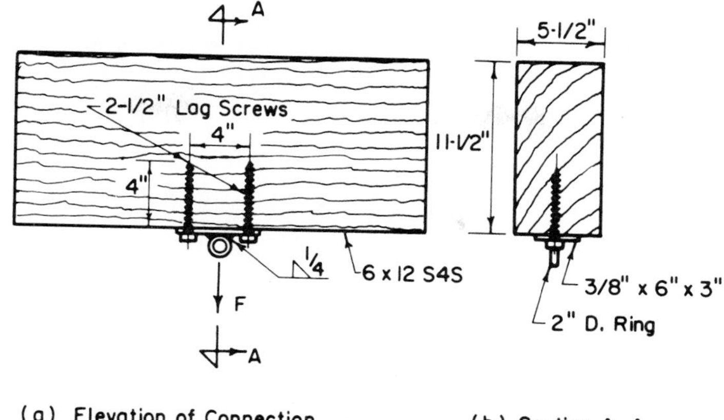

Figure 6.14. Connection of a vertical load to the soffit of a wood beam using lag screws in direct withdrawal.

per inch of penetration. Thus, F = 2 × 291 × 4 = 2,330 lb. is the maximum load that can be attached to the beam by the connection shown in Fig. 6.10.

Although this is an acceptable connection for small loads, it is advisable to suspend heavier loads from the top of the beam in direct bearing across the grain.

6.4 TIMBER CONNECTORS

An efficient method of connecting wood members together consists of using timber connectors in combination with bolts. Three general types of connectors, see Fig. 6.15, are available commercially, namely: 1.) Split-Rings, which fit into precut grooves in the wood; 2.) Toothed-Rings, which are forced into the wood as the members are pressed or clamped together; and 3.) Shear-Plates, which fit into pre-bored recesses. The latter may be used singly in making wood to metal connections, or in pairs for wood to wood connections of a demountable type.

According to Scholten (6), who researched connector joints intensively in the 1940's, principal advantages of these types are:

1. High joint efficiency.
2. Simple and practical application.
3. Small number of pieces to handle.
4. Adaptability to prefabrication for subsequent field assembly.
5. Improved appearance of joint with less exposed metal.
6. Greater fire resistance, because embedment of connections in wood reduces amount of metal exposed to fire temperatures.

Special tools are needed for the attachment of connectors to wood members. Split-rings and shear-plates require a power tool to fabricate the groove and recess, toothed-rings require a special bolt and wrench to force the ring into the wood.

Connectors are used primarily for transmitting direct axial loads, i.e., tension or compression, and shear. They are particularly suited for joints in framed structures where several members

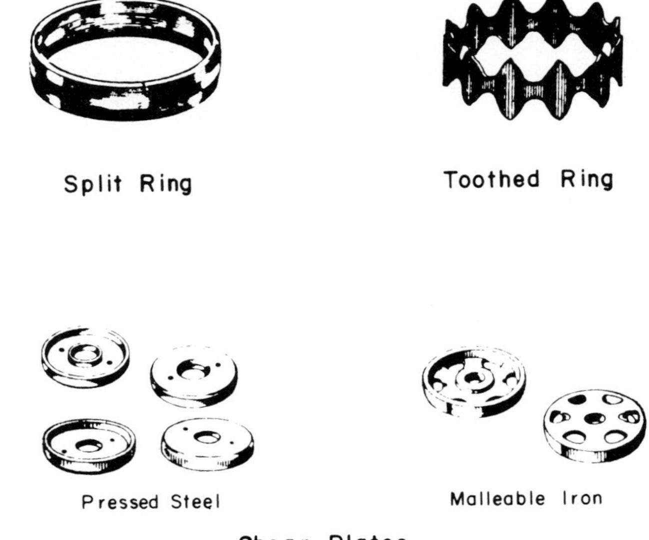

Figure 6.15. Various types of timber connectors.

meet at a common panel point, or where members must be spliced. Also, as shown later, connectors may be used for rigid connections to transmit bending moments in addition to direct axial force and shear.

For a long time connectors have been widely used in major wood construction. All types of structures are built using connectors; among these: long-span or heavy roof trusses, bowstring trusses, three-hinged arch bridges, rigid frames, and long-span roof systems for arenas or assembly halls. The versatility of connectors makes them the most widely used method for mechanical connection of wood structures.

6.4.1 Behavior and Design of Connectored Joints

Consider the simple connection shown in Fig. 6.16. It consists of a central member and two side pieces tied together by two split-rings and a ¾-or ⅞-in. bolt. The connection tranmits a direct tensile force from the central member to the side pieces.

The portions of the central member subjected to shear, compression, and tension are shown in Figs. 6.16a through d. The shaded areas indicate the corresponding parts of wood subjected to the various stresses; their value can be obtained as follows:

Shear area, see (a)

$$2\left(\frac{\pi d_1^2}{4}\right) \text{ within core}$$

$$2\ [d_2 e - \frac{1}{2}\left(\frac{\pi d_2^2}{4}\right) + 2\,\frac{ae}{2}] \text{ beyond core}$$

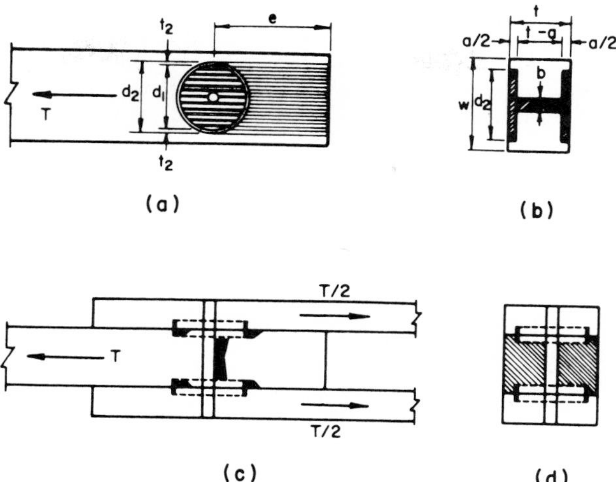

Figure 6.16. Portions of central member of a connectored joint subject to various stresses.

Compression area, see (b) and (c)

$$2\left(\frac{ad_2}{2}\right) + b\,(t_1 - a)$$

Tension area, see (d)

$$t_1\,w - ad_2 + b(t_1 - a)$$

where: d_1 and d_2 are inside and outside diameter of connector, respectively; e is end distance from center of connector to end of member; a is depth of connector; b is diameter of bolt; t_1 and w are thickness and width of central member, respectively; and t_2 is thickness of connector metal.

Past attempts to use theoretical elastic analysis to obtain the stress distribution of even the simplest joint shown in Fig. 6.16 were upset by the anisotropic structure of wood and by the presence of strength-reducing characteristics such as knots and slope of grain. Modern methods of numerical analysis, using finite element techniques, may eventually provide more insight in the mechanism of load transmission and failure of connections.

To the present day, design data for connectored joints is based on the experimental work done by Scholten (6). Two types of tests were generally performed. In one, the load was applied parallel to the grain of wood; in the other, it was applied in a direction perpendicular to the grain. All three types of connectors shown in Fig. 6.15 were tested. However, for the sake of brevity, only the behavior of split-ring connections is given here.

For load parallel to grain a typical load-slip relation obtained by Scholten on a three-member joint of clear, seasoned, and check-free wood, using 4-in. split-rings and ¾-in. bolts, is shown in Fig. 6.17. An initial slip takes place at the beginning of loading that should not be associated with the linear relation between load and slip that follows. The initial slip occurs when the ring is coming into full bearing; for the case of Fig. 6.17 the initial slip is approximately 0.019 in. The

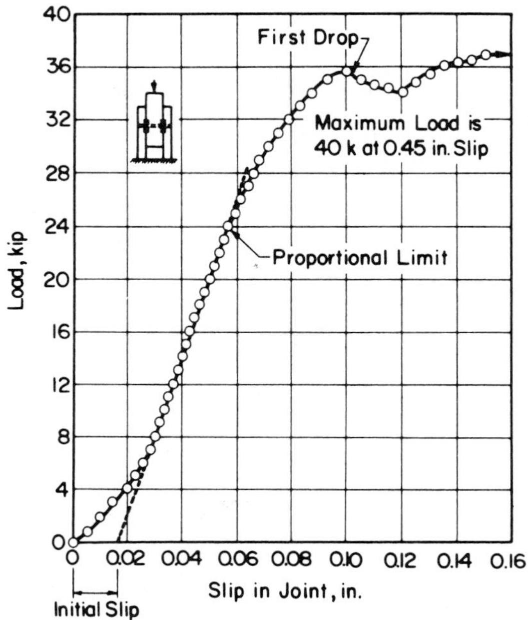

Figure 6.17. Load-slip relationship of a split-ring connectored joint, in bearing parallel to grain, for an individual test. From Ref. 6.

end of the linear relation between load and slip defines the proportional limit. The load at this point, 24 k at a slip of 0.06 in., varies between approximately one-half to two-thirds of the maximum load for various species of wood and sizes of connectors. Beyond the proportional limit the load increases with slip and follows a non-linear relation up to 36 k at a slip of 0.10 in. where the first drop occurs. This appears to be associated with shear of the wood core within the connector. The maximum load, 40 k, is obtained with much additional deformation, at 0.45-in. slip. Great ductility of the joint is evidenced by the long, almost flat portion of the load-slip relationship between first drop and ultimate, see Fig. 6.17.

Tests similar to that described above were performed for various species, for which the relations of maximum load, P, and proportional-limit load, P′, with the specific gravity, G, of the given wood, are given in Fig. 6.18. The linear relations, P = 76,000 G and P′ = 49,000 G, were obtained statistically by Scholten using experimental data. It was found that, in general, the relationships are expressed by:

$$P = K\,G \tag{6.5}$$

where: P = the load, in pounds, for two connectors and one bolt, obtained in a test of short duration; K = a constant derived by test; and G = specific gravity of the wood, oven-dry, based on volume at test.

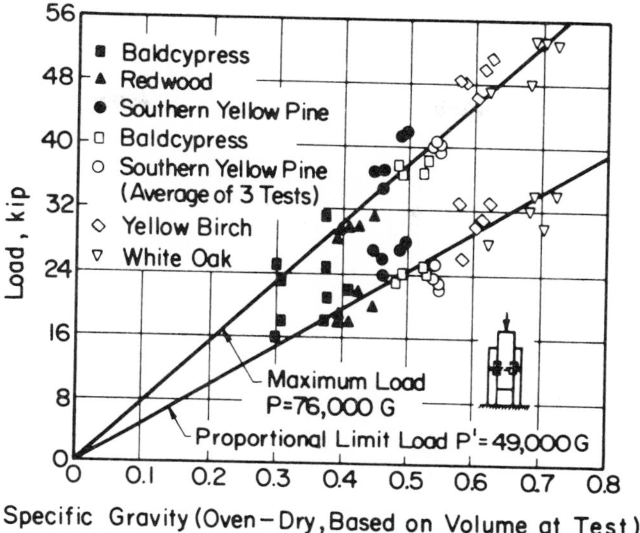

Figure 6.18. Variation of proportional limit and maximum loads in bearing parallel to grain with specific gravity for split-ring joints consisting of two 4 in. split-rings and a ¾ in. bolt.

This equation influences present design specifications. Part V of NDS (1), on timber connector joints, classifies all commercial woods into four groups according to their specific gravities, see Table A.6.5. Allowable loads for any given connector depend on the group classification of the wood. Dense Southern pine and Douglas fir-larch are classified in group A; any other type of Southern pine and Douglas fir-larch are classified in group B. Equation 6.5 is independent of the size of split-rings; various sizes were tested with similar results, see Fig. 6.19. Linear relations between proportional-limit loads and specific gravity are given in the figure.

In addition to the size of the connector and the type of wood used, other factors affect the strength of a given joint. Some of these are: 1.) the direction of the applied force with reference to the grain of wood, i.e., the angle of load to grain; 2.) the thickness, and 3.) the width of the wood member; 4.) edge and 5.) end distances; 6.) spacing between connectors; 7.) duration of loading; 8.) moisture condition of the wood; and 9.) net section of member.

Tests with connectored joints loaded in bearing perpendicular to grain (6) reflected a similar relation to specific gravity. However, the values were appreciably lower than those for which the load was parallel to grain. The slope of the load-slip curve is less; initial slip is less, but slip at proportional limit is larger. For design purposes NDS (1) gives Table A.6.6 for split-rings, A.6.7 for toothed-rings, and A.6.8 for shear-plates. Capacities for load parallel to grain, P, and load perpendicular to grain, Q, are given directly by these tables. For angles of load to grain between 0° and 90° a uniform reduction in load was observed in the tests; Hankinson's formula or Scholten's nomograph, see Chart A.2.3, may be used to determine N, the capacity of the connector at a given angle of load to grain. NDS (1) recognizes this variation; the necessary values of P and

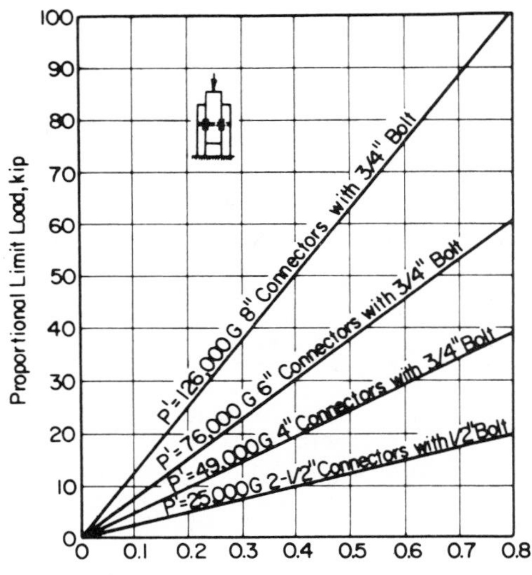

Figure 6.19. Variation of proportional-limit load in bearing parallel to grain with specific gravity for joints using split-rings of different diameters.

Q may be obtained from the above referred tables in the Appendix. In lieu of this, the value of N can be obtained directly using load charts A.6.10a through A.6.14a. Detailed information on the use of the load charts is given in the Appendix, see A.6.9.

The effect on capacity of connection of the thickness of the central member is shown in Fig. 6.20. It can be seen that the maximum load increases with an increase in thickness of member to approximately 3 in. for the 4-in. connectors (2-in. for the 2½ in. connectors); for greater thicknesses the maximum load remains fairly constant. The proportional-limit load also increases with an increase in thickness of the member but reaches a constant value at a relatively smaller thickness. Again provisions for thickness are incorporated in Tables A.6.6, 7, and 8. Minimum net thicknesses are specified; loads for connectors installed in lumber of net thickness other than those shown in the charts may be obtained by straight-line interpolation for thicknesses intermediate to those given. It can be seen from the variation shown in Fig. 6.20 that the NDS-indicated straight-line interpolation can be easily justified. The number of faces of the piece with connectors on the same bolt affects net thickness. For small thicknesses the presence of two connectors on the same bolt, a condition known as two-face loading, results in a reduced capacity. This is taken into consideration by the above referred design tables and charts, in which a distinction is made between one and two-face loading.

The minimum recommended width of member for use with the 2½ and 4-in. connectors is nominal 4-in. and 6-in. material, respectively. Tests were performed by Scholten with widths varying from a small width equal to the outside diameter of the connectors to a width of 6½ in. In the case of the smallest width, the maximum load was only 80 to 85 percent of those obtained with the minimum recommended width. The results showed only small increases in maximum

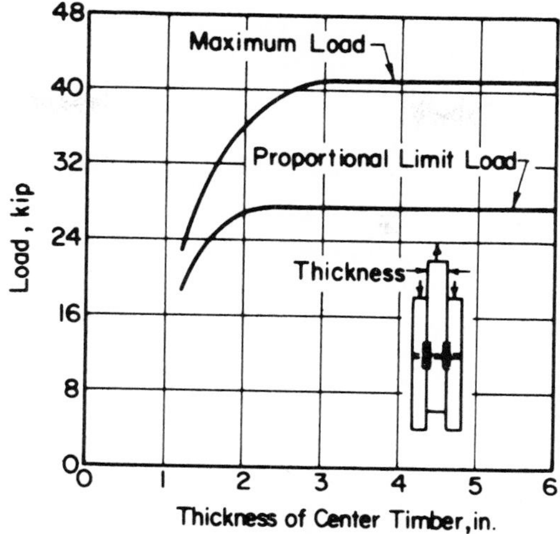

Figure 6.20. Load vs. thickness of central member for 4-in. split-ring joints in bearing parallel to grain.

load with increases in width over the minimum recommended width. Thus, only a minimum edge distance (defined by the distance from the center of the connector nearest the edge to the edge of the member) is necessary. NDS recommends, for bearing parallel to grain, minimum edge distances of 1¾ in. for 2½-in. size split-rings and 2⅝-in. size toothed rings and shear-plates, and 2¾ in. for the 4 in. size. In the case of bearing perpendicular to grain, the distance to the edge against which the load is acting, i.e., the loaded-edge distance, has significant influence on the strength of the joint. Minimum values, at a reduced capacity, are those given above for bearing parallel to grain. Full capacity is developed only if the loaded-edge distances are increased to 2¾ in. and 3¾ in. for the small and large size connectors, respectively. For design purposes the graphical information on edge distances contained in Charts A.6.10b through A.6.14b can be used.

The end distance of a connector is measured between the center of the bolt hole and the end of the member; it has great influence on the behavior of the joint when the connectors are bearing parallel to the grain of the wood. Results of tests of connectored joints with various end distances are shown in Fig. 6.21. The maximum load increases with end distance from the smallest end distance which incorporates the entire connector in the member, i.e., 2-in. for a 4-in. split-ring, to an optimum 7 in. (6-in. for the 2½ S.R.). For larger end distances the maximum load remains fairly constant. When the end distance is less than optimum, tests show that the load is reduced uniformly to 62.5 percent at half the optimum end distance; hence, the load may be obtained by direct linear interpolation, see Appendix charts. In addition, when the end surface of a member is not at right angles to the length, see Fig. 6.22, as occurs in some roof trusses, the end distance, measured parallel to the center line of the piece from any point in the center half of the connector diameter perpendicular to the center line of the piece, must not be less than the end distance required for a square-cut member. The perpendicular distance, from center of the connector to sloping end cut of a member, must not be less than the required edge distance.

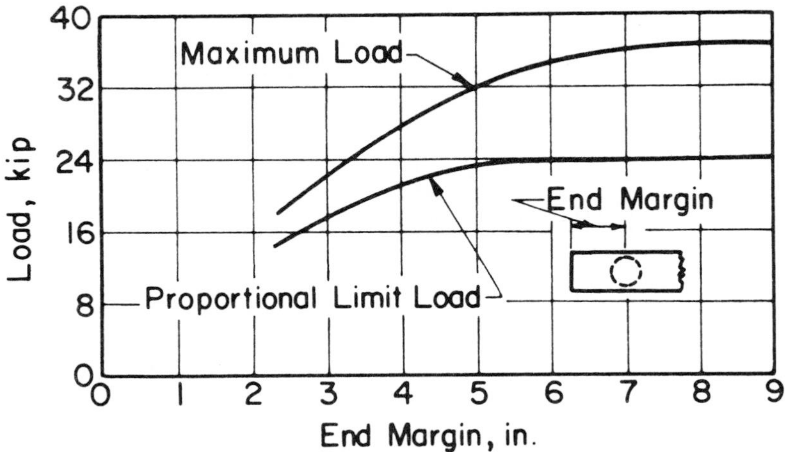

Figure 6.21. Load vs. end distance for a 4-in. split-ring joint in bearing parallel to grain.

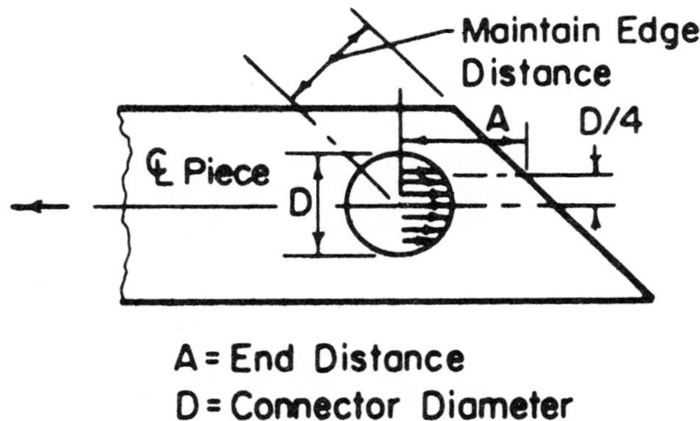

Figure 6.22. End distance for member with sloping end cut.

In joints with multiple connectors it is important to know the minimum spacing between connectors for which full capacity can be obtained. Results of tests indicate that, with an increase of spacing, the maximum load increases from the smallest spacing to a constant value at approximately 6½ in. for 2½ in. connectors; it requires slightly more than 9-in. spacing for the 4-in. connectors, see Fig. 6.23. It was also found that the proportional-limit load increases with an increase in spacing of connectors but reaches a constant value at a smaller spacing than does the maximum load. The provisions of NDS (1) are based on these results. For members loaded at an angle of grain other than 0° or 90°, with connector axes at various angles with grain, the spacing is determined in accordance with the spacing charts given in the Appendix, see A.6.10b through A.6.14b and detailed instructions in A.6.9.a.

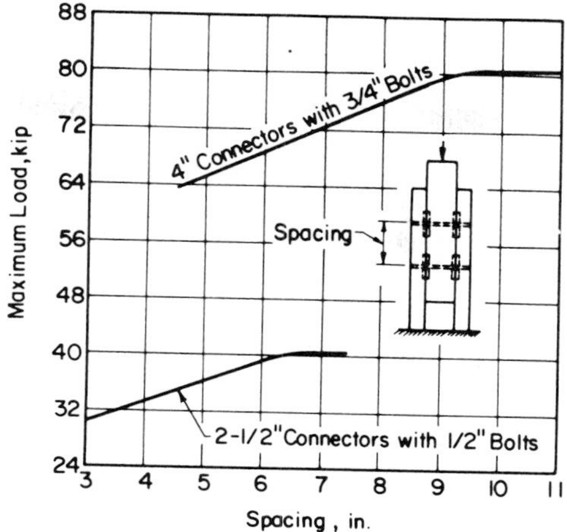

Figure 6.23. Variation of maximum load with longitudinal spacing for a split-ring joint consisting of 4 connectors and 2 bolts in bearing parallel to grain. From Ref. 6.

Even when the recommended spacing is maintained between connectors in multiple-fastener joints ($N > 2$), recent research (19, 20) indicates that a modification factor, K, is necessary to determine the actual capacity of the joints. NDS(1) has included timber connectors among the various fasteners for which a determination of K is necessary; see the design provisions of Section 6.2.2.4.

The allowable connector loads given in the Appendix charts are applicable directly for normal loading conditions, $\bar{\alpha} = 1.0$, see Section 2.13. For all other recognized conditions of loading, i.e., permanent, snow, 7-day duration, wind, or earthquake and impact, use conventional load-duration factors $\bar{\alpha} = 0.9$, 1.15, 1.25, 1.33, and 2.0, respectively.

Moisture conditions of lumber affect the strength properties of wood, which in turn affect the capacity of connectored joints. The design provisions assume first, that wood is seasoned to 15 percent moisture content to a depth of ¾ in. from the surface before fabrication and second, that it will remain dry in service. If connectors are used in lumber which is fabricated before it is seasoned and which later is seasoned, either before erection or while in the structure, 80 percent of the tabulated connector loads apply. Intermediate connector values may be used for lumber partially seasoned when fabricated. Independently of the moisture condition at fabrication, if wood is to remain wet in service, only 67 percent of the given capacities apply for all connectors. To account for the strength-reducing effect which occurs after pressure-impregnation of lumber with fire-retardant chemicals, it is recommended only 90 percent of tabulated connector capacities be taken if the lumber is kiln-dried after treatment. If the lumber is treated with the fire-retardant chemicals but not kiln-dried after treatment, only 80 percent of tabulated connector capacities apply.

Consideration must be given, in addition to the strength of the joint for bearing parallel to the grain, to the area of the net section at the joint. The latter is defined as the net area remaining after reduction for bolt holes and connectors at that section of the member, taken at right angles to the direction of the load, which is subjected to maximum stress. The net area may be determined by subtracting the projected area of the connector grooves and the bolt hole from the full cross-sectional area of the member. The total projected area for connectors and bolt holes is given in Table A.6.15. NDS also provides that where connectors are staggered, adjacent connectors, with parallel-to-grain spacing equal to or less than one connector diameter, shall be considered as occurring at the same critical section. The required net area in tension and compression members is determined by dividing the total load transferred through the critical section by an allowable stress F_t for tension members, and F_c for compression members; for the species and grade of lumber used, see Table A.2.5 for F_t or F_c.

Following are three illustrative examples of analysis and design of connectored joints. Other examples of use of connectors may be found in Sections 6.5, 8.7, 10.2, and 10.3.

6.4.1.1 Illustrative Example. Staggered Connectors

The free-body diagram of a tension member at a joint is shown in Fig. 6.24. Connectors are 2½ in. split-rings with ½-in. bolts. Wood is Southern pine Select Structural KD, surfaced-four-sides and used at 15% moisture content. The loads N, acting in the near face of the member, are due to the simultaneous action of permanent loads and wind; the angle of load to grain is 15° and the angle of axis to grain is 30°. Determine the allowable load N that can be transmitted through the near-face split-rings, see Fig. 6.20b.

The geometry of the connection is examined first. The following information is obtained from Charts A.6.10a and b. Since the member is under tension, the necessary end distance for 100 percent capacity is 5½ in.; the necessary loaded-edge distance for a 15° angle of load to grain is 2⅛ in. These dimensions are satisfied in Fig. 6.24a; in addition, the unloaded-edge distance is larger than necessary, 2⅛ in. instead of 1¾ in., for the sake of symmetry of the connection about the center line of the member. The spacing of connectors, R, for full capacity, is determined from the spacing chart at the intersection of the ellipse corresponding to 15° angle of load to grain with the 30° radial line representing the angle of axis to grain. Using a point compass, or similar device, and the scale given in the coordinate axes, this reads $R = 5.1$ in.; vertical and horizontal projections are $v = 2.5$ in. and $h = 4.4$ in. These minimum values are satisfied by the dimensions shown in Fig. 6.24a. Thus, as far as geometry is concerned, the connectors may develop full capacity.

Southern pine Select Structural lumber is classified under Group B, see Table A.6.5. For a member with two faces loaded and more than 2 in. thick Table A.6.6 renders $P = 2{,}730$ lb. and $Q = 1{,}620$ lb. for 1¾ in. loaded-edge distance and $Q = 1{,}940$ lb. for 2¾ in. or more loaded-edge distance. Since the actual loaded-edge distance is 2⅛ in., see Fig. 6.20, Q can be obtained by linear interpolation as follows: $Q = 1{,}620 + (1{,}940 - 1{,}620)\left(\dfrac{2.13 - 1.75}{2.75 - 1.75}\right) = 1{,}740$ lb. Thus, for $P = 2{,}730$ lb., $Q = 1{,}740$ lb., and $\theta = 15°$, Hankinson's formula yields $N = 2{,}630$ lb. Also, directly from Chart A.6.10a, $N = 2{,}630$ lb. Since wind-loading conditions prevail, capacity may be increased to $N = 1.33 \times 2{,}630 = 3{,}500$ lb. No reduction for moisture content condition is required since the wood is seasoned when fabricated and remains seasoned afterwards.

In some cases development of the full capacity of the connectors is not necessary. Assume tht only 85 percent of the capacity is needed; the dimensions shown on Fig. 6.24a may be reduced accordingly. Thus, from the end and edge distance charts, see A.6.10b, is obtained 4⅜-in. end

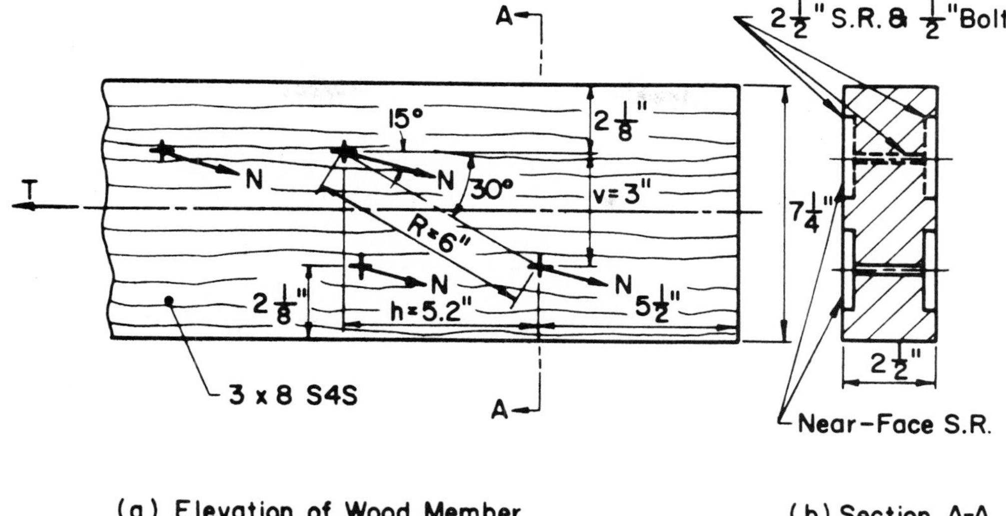

Figure 6.24. Details of connection.

distance and 1¾-in. edge distance. The spacing, R, for 50 percent capacity is 3.5 in., independently of the angle of load to grain. Hence, for 85 percent capacity, using linear interpolation, R = 3.5 + (5.1 − 3.5) (85 − 50/(100 − 50) = 4.6 in. Therefore, the space required for the connection can be reduced at the expense of capacity.

6.4.1.2 Illustrative Example. Truss Joint

This example consists of two parts. Part 1 deals with the joint shown in Fig. 6.25 corresponding to the bottom chord of a roof truss; Part 2 is a similar joint for a monochord truss, as shown in Fig. 6.26 dense wood, Group A in Table A.6.5, surfaced and used at 15% moisture content. The loads acting on the members are due to the simultaneous action of dead load and snow.

Part 1. The connection is made up of four 4-in. split-rings and ¾-in. bolts, as shown in Figs. 6.25a and b. Determine whether or not the joint is adequate.

This example is typical of the design process of connectored joints. First, the designer, based on experience or intuition, conceives a solution to transmit the various loads at the joint. (Note that all member axes intersect at a common point to prevent eccentricities which cause bending and shear stresses in the wood.) Following this initial step, analysis of the solution determines if it is acceptable; if it is not, analysis usually points out the direction for necessary corrections to the original design.

The initial step in the analysis of the given joint is the checking of the equilibrium of forces; this is shown in Fig. 6.25a. Free-body diagrams of the members of the connection may then be drawn, see Fig. 6.25c. In the case of the left diagonal, which consists of two members, equilibrium is obtained with a force of 5 kips provided by the split-ring which connects it to the adjacent chord member. The connector load is shown in dashed lines because it occurs in the far face of the figure.

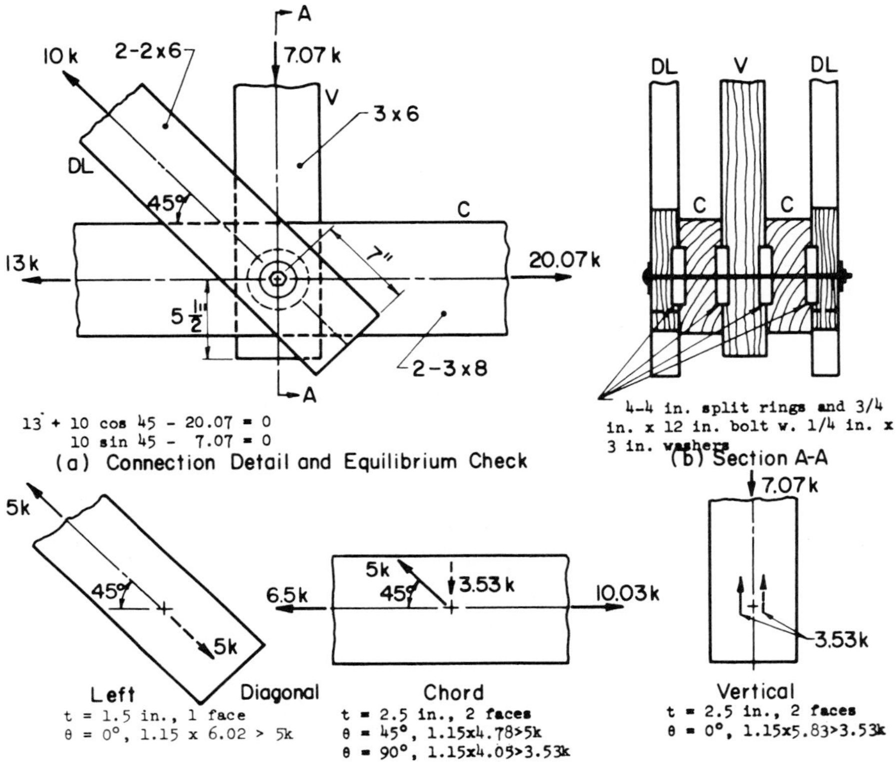

(c) Free Body Diagrams and Capacity Check

Figure 6.25. Analysis of connectored joint in a truss under snow-loading conditions. Group A wood.

The capacity of the connector in this case is determined by: 1.) the thickness of the member, 1.5 in.; 2.) the number of faces loaded, one; 3.) the angle of load to grain, $\theta = 0°$; and, 4.) the nature of the loading conditions, $\bar{\alpha} = 1.15$. From Table A.6.6, $N = P = 6.02 \times 1.15 = 6.92 \tau 5$ kips. The same procedure is used for the analysis of free-body diagrams of the chord and the vertical; the results are given in Fig. 6.25c. Note that each connectored face is checked for adequacy to transfer the applied force. The preceding analysis shows the fundamental process of transferring forces at the joint: as far as strength is concerned the connection is acceptable.

End distances for the diagonal, 7-in., and the vertical, 5½ in., guarantee development of full capacity of the corresponding connectors. Edge distance is $5.5/2 = 2.75$ in. for the diagonal and vertical, also acceptable for full capacity, according to Chart A.6.11b. The edge distance in the chord is $\frac{7.25}{2}$ = 3.62 in., good for full capacity as a loaded edge. The net section of the members is determined using the geometric properties of the 4-in. split-ring, as given in Chart A.6.11.a, and assuming 13/16 in. for the bolt hole. The following is obtained: diagonals, $8.25 - [2.24 + 0.81\ (1.5–0.5)] = 8.25 - 3.05 = 5.20$ in.^{2}; chord, $18.13 - [2.24 \times 2 + 0.81\ (2.5–1)] = 18.13 - 5.69 = 12.44$ in.^{2}; and vertical, $13.75 - 5.69 = 8.06$ in.^{2}. To simplify the preceding calculations Table A.6.15 gives the combined area of connectors and bolts directly. For the given wood: $F_t = 1.30$ ksi and $F_c = 1.70$ ksi. Thus, the capacity of one member of the diagonal is $1.15 \times 1.30 \times 5.20 = 7.77\ k > 5\ k$; that of one member of the chord is $1.15 \times 1.30 \times 12.44 = 18.60\ k > 10.03\ k$; and finally, the capacity of the vertical is $1.15 \times 1.70 \times 8.06 = 15.75\ k > 7.07\ k$.

The joint is adequate.

Part 2. Trusses are built frequently with all component members in the same vertical plane in what is known as monochord design. Joints in these trusses can be accomplished in a number of ways. Light trusses may use light-gage steel plates, such as shown in Fig. 6.34, which are driven into the wood by means of pressure, or glued and nailed plywood plates. Heavier trusses use thicker steel plates and shear-plate connectors such as shown in Fig. 6.26. The joint is similar to that of Part 1 of the example problem. Loads are identical; however, the diagonal and chord are slightly larger in size to account for the fact that only single members are used. Shear-plates are used, instead of split-rings, to connect the wood to the steel plates. All members have the same actual thickness; 2½ in. in this case. The adequacy of the joint is to be ascertained.

The process is the same as before. Free-body diagrams, see Fig. 6.26c, are drawn to determine the capacity of the connection. For this kind of connection the angle of load to grain, in all cases, is zero. The connectors are found adequate to transmit the load. The figure also shows the steel plate in equilibrium under the system of forces contributed by the wood members. It is shown that a thickness of ¼ in. is adequate for the steel plate. (Note that if a thinner plate is used buckling of the vertical leg should be checked). Since end and edge distances of the various shear-plates, to their respective wood members, are similar to those of Part 1, there is no need to check these again. Only the net sections of members remain to be checked; it may be done as a simple exercise. The connection as a whole is adequate.

In a monochord truss shop-prefabrication of members and careful layout of connecting steel plates for the least waste result in economical structures.

6.4.1.3 Illustrative Example. Multiple-Member Truss Joint

The joint shown in Fig. 6.27a corresponds to the bottom chord of a truss with redundant members. Analysis of the statically indeterminate structure, under normal loading conditions, renders the member loads shown in the figure. The connection is made up of six 2½ in. split-rings and one ½-in. bolt, see Fig. 6.27b. Wood is Douglas fir Select Structural surfaced dry and used at 19% moisture content; allowable stresses are: $F_c = 1.4$ ksi, $F_t = 1.2$ ksi, $E = 1{,}800$ ksi. Determine whether or not the joint is adequate.

There is a difference between the analysis of this joint and that of Part 1 of the preceding example. For this joint the internal reactions at the various interfaces are not readily known but must be determined, see Fig. 6.27c. For instance, consider the free-body diagram of a chord

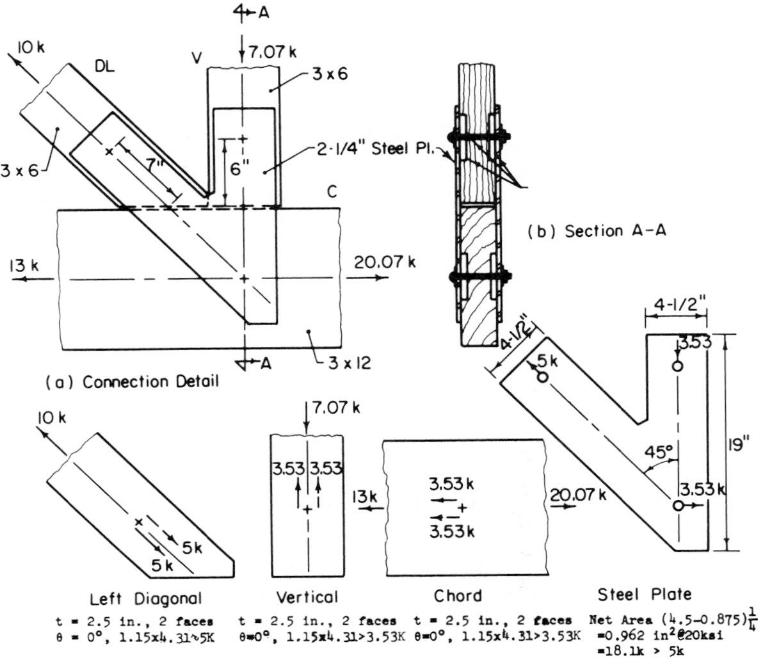

(c) Free Body Diagrams And Capacity Check

Figure 6.26. Analysis of connectored joint in a monochord truss under snow-loading conditions.

member. Reaction T_1 represents the action of the right diagonal, DR, on the connector face between it and the chord member. Reaction T_1 and its angle of load to grain θ_1, are determined using the equilibrium equations for the chord member. The analysis is continued with the right diagonal, where the force T_1 is known and T_2, θ_2 are the unknowns. Finally, at the vertical both reactions must be identical due to symmetry. The whole process, including the checking of force against capacity for each connector, is given in Fig. 6.27c. The mechanics of the process should be followed until satisfied of the adequacy of the connection. As an interesting variation in the connection, study the analysis of a similar joint obtained by making an exchange of vertical planes between left and right diagonals, see Problem 7.

6.4.2 Connectors in End Grain

The preceding discussion considered the frequent case of connectors installed in surfaces that are parallel to the general direction of the grain of the member. Some unusual applications, however, may require connections at the end of square-cut members, or at the sloping surface of a member cut at an angle to its axis.

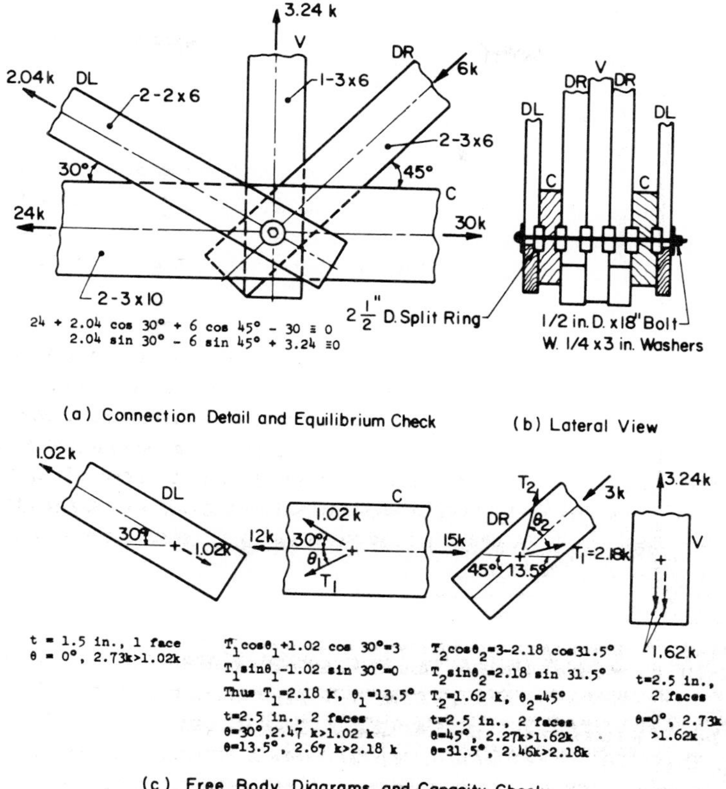

Figure 6.27. Analysis of connectored joint with indeterminate internal reactions in a truss under normal loading conditions; Group B wood.

Design of these connections is possible through successive use of Hankinson's formula (Eq. 2.1). Consider that P and Q, the design values for a connector unit in a side-grain surface, are known. The design values P_α and Q_α for the same connector unit in a sloping surface may be obtained using Hankinson's formula once α, the least angle between the sloping surface and the general direction of the wood fibers, is determined. If the applied load acts at an angle θ with the axis of cut of the sloping surface, measured in the plane of the sloping surface, then, an additional application of Hankinson's formula using P_α, Q_α and θ will render N_α. The latter is the design value for a connector unit in a sloping surface, when the direction of load is at an angle θ from the axis of the cut.

Actual design provisions, formulas and illustrations for connectors in end grain may be found in Table A.6.16.

6.5 RIGID CONNECTIONS

Timber connectors are used to provide rigid connections between prefabricated portions of glued-laminated structures. Long-span rigid frames and arches are usually prefabricated in several pieces to facilitate transportation from the laminating plant and subsequent erection at the job site. Since these structures may be subjected to the simultaneous action of thrust N, bending moment M and transverse shear Q, the connections that join the prefabricated parts together, and the structure to the foundation, are designed to resist these effects. Various types of rigid connections have been used by designers (7, 8); because of their importance two of these, the rigid-plate and the multiple-plate connections, see Figs. 6.28 and 6.29 respectively, are studied in detail.

6.5.1 Rigid-Plate Connections

The connection shown in Fig. 6.28 consists of two steel plates, one at each side, tied together by means of shear-plates and connecting bolts. The distance between rows of connectors is the gage, g; that between columns of connectors is the pitch, p. The dimensions e and c correspond to the distances of end connectors to the wood and the steel plate, respectively.

The free-body diagram of Fig. 6.28a shows two sets of generalized forces, N_1, M_1, Q_1 and N_2, M_2, Q_2, acting on the member at sections 1 and 2 situated at opposite sides of the connection. These forces are determined using an elastic analysis which assumes full continuity at the connection. Let N, M and Q be the set of forces that would act on Section C, in the case of a continuous structure. For a structure erected in parts these forces are assumed transmitted by the connection, from one side to the other, through the two side steel plates. The distribution of these forces to the individual connectors is a highly indeterminate problem. However, a simplified analysis is possible if it is assumed that: 1.) the deformation of the steel plates is negligible, i.e., the steel plates are rigid, and 2.) the effects of the various forces may be superposed linearly.

Consider first the action of the moment M on the set of connectors to the left of Section C. A kinematically admissible set of connectors forces develops only if the rotational displacement of the connected member occurs about the centroid of the group of connectors. This can be shown as follows. Assume that an infinitesimal rotation, $\delta\phi$, of the member occurs about a given point 0, see Fig. 6.30. The displacement of an individual connector i is given by $d_i\delta\phi$, where d_i is the distance of connector i to point 0. The forces F_i developed at the various connectors are proportional to the corresponding displacements $d_i\delta\phi$ as follows.

$$\frac{F_1}{d_1\delta\phi} = \frac{F_2}{d_2\delta\phi} = \dots \frac{F_i}{d_i\delta\phi} = \dots \frac{F_n}{d_n\delta\phi} \qquad (6.6)$$

The various forces may be expressed in terms of the force at a given connector; for convenience F_n, the force at connector n, is selected. The preceding relations give:

$$F_1 = F_n\, d_1/d_n,\ F_2 = F_n\, d_2/d_n,\ F_i = F_n\, d_i/d_n,\ \dots\ F_n \qquad (6.7)$$

The longitudinal X_i and transverse Y_i components of the force at a connector may be determined as $F_i \sin\phi_i$ and $F_i \cos\phi_i$ respectively, see Fig. 6.30, where ϕ_i is the angle of the vector d_i with axis x. Substituting $\sin\phi_i = y_i/d_i$, $\cos\phi_i = x_i/d_i$ where x_i and y_i are the coordinates of the connector using 0 as origin, yields:

$$X_i = \frac{F_i y_i}{d_i}, \qquad Y_i = \frac{F_i x_i}{d_i} \qquad (6.8)$$

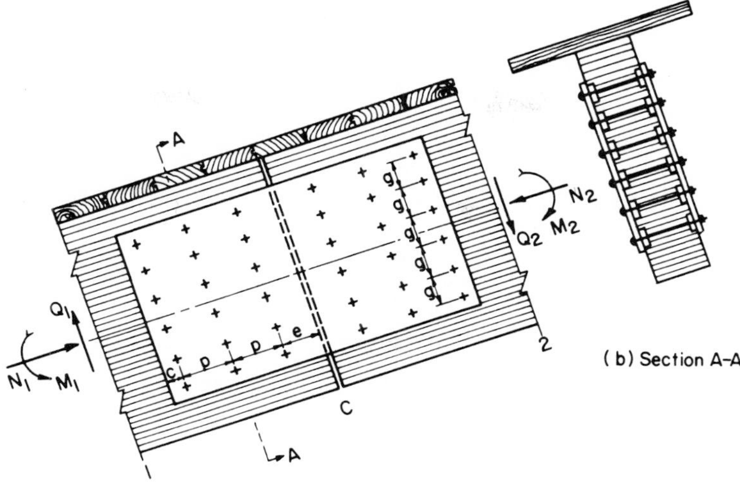

Figure 6.28. Typical rigid-plate connection between prefabricated glued-laminated members.

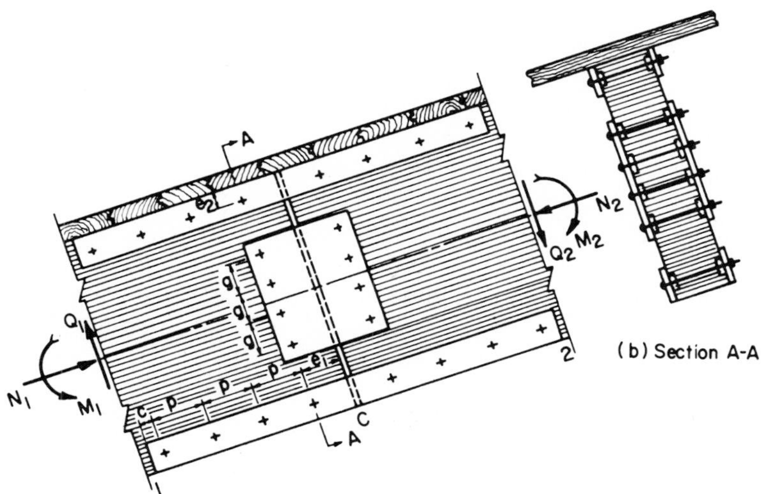

Figure 6.29. Typical multiple-plate connection between prefabricated glued-laminated members.

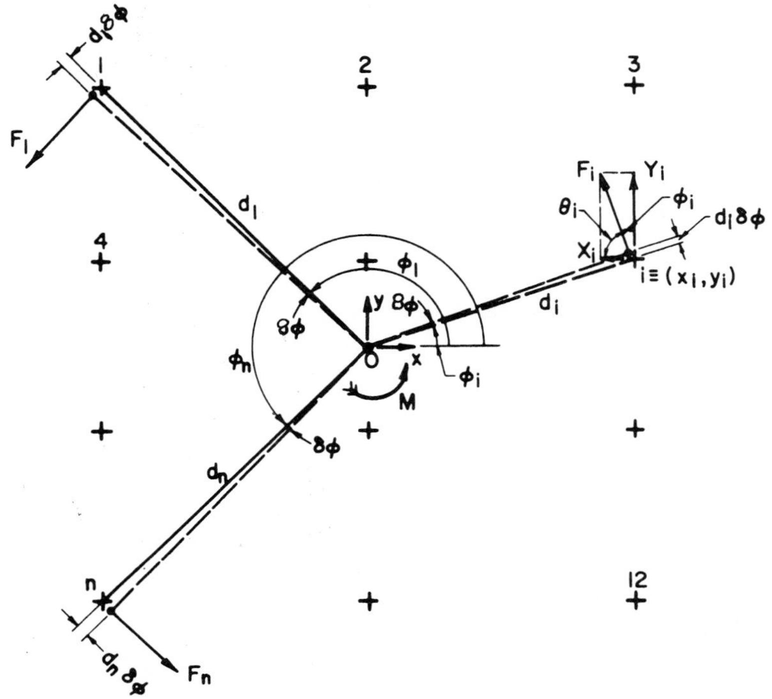

Figure 6.30. Connector forces induced by moment acting on rigid-plate connection.

Substituting F_i in terms of F_n as given by Eq. 6.5, yields:

$$X_i = F_n \frac{y_i}{d_n} \quad , \quad Y_i = F_n \frac{x_i}{d_n} \tag{6.9}$$

The sum of all X_i and Y_i gives the total longitudinal and transverse components of the resultant force. The latter is equal to zero since only the action of M is being considered. Thus,

$$\sum_1^n X_i = \frac{F_n}{d_n} \sum_1^n y_i = 0$$
$$\sum_1^n Y_i = \frac{F_n}{d_n} \sum_1^n x_i = 0 \tag{6.10}$$

Since F_n and d_n are zero only in the trivial cases where M = 0 or where the nth connector coincides with point O, it follows that:

$$\sum_1^n y_i = 0$$
$$\sum_1^n x_i = 0 \tag{6.11}$$

meaning that point O, origin of coordinates and the point about which rotation takes place, coincides with the centroid of the connectors. The total moment of the various forces F_i about point O.

i.e., $\sum_1^n F_i d_i$ is equal to the moment M. Substituting the value of F_i given by Eq. 6.7, yields:

$$M = \frac{F_n}{d_n} \sum_1^n d_i^2 \tag{6.12}$$

where the term $\sum_1^n d_i^2$ is the polar moment of inertia of the set of connectors about the centroid; Eq. 6.12 yields the value of F_n. Any other force F_i can be also determined from Eq. 6.12 using the corresponding d_i value instead of d_n.

It may be convenient to determine the term $\sum_1^n d_i^2$ as the sum of the terms $\sum_1^n x_i^2$ and $\sum_1^n y_i^2$. For the sake of convenience the longitudinal, X_i, and transverse, Y_i, components of a given force, F_i, may be determined from Eqs. 6.9 and 6.12 as follows:

$$X_i = \frac{M y_i}{\sum_1^n x_i^2 + \sum_1^n y_i^2} \qquad Y_i = \frac{M x_i}{\sum_1^n x_i^2 + \sum_1^n y_i^2} \tag{6.13}$$

The effect of the axial force N on the connectors is to add algebraically the amount N/n to the longitudinal component of a given connector, where n is the number of connectors. Similarly, the effect of the transverse force Q is to add algebraically the amount Q/n to the transverse component of a given connector. Thus, the final values of X_i and Y_i are

$$X_i = \frac{M y_i}{\sum_1^n x_i^2 + \sum_1^n y_i^2} + \frac{N}{n} \qquad Y_i = \frac{M x_i}{\sum_1^n x_i^2 + \sum_1^n y_i^2} + \frac{Q}{n} \tag{6.14}$$

The final value of the force F_i and its angle of load to grain θ_i acting at a given connector i, see Fig. 6.30, may be obtained from the following equations:

$$F_i = \sqrt{X_i^2 + Y_i^2} \qquad \theta_i = \tan^{-1} \frac{Y_i}{X_i} \tag{6.15}$$

The connection is satisfactory when each F_i is equal to, or less than, the corresponding capacity at θ_i for the given type connector and loading conditions. To clarify these concepts an example of rigid-plate connection is given as follows.

6.5.1.1 Illustrative Example. Rigid-Plate Connection

The connection shown in Fig. 6.28 between two glued-laminated portions of a rigid frame is subjected to N = 15 k, M = 1,800 k-in., and Q = 10 k. Wood for the laminations is dense Southern pine and connectors are pairs of 4-in. shear-plates with ¾-in. bolts. The values of gage and pitch are g = 6 in. and p = 9 in., respectively; also e = 7 in., c = 2 in. The loads are due to the simultaneous action of permanent loads and snow, i.e., $\bar{\alpha} = 1.15$. Since the frames are covered, dry conditions of use can be expected at all times. The object of the analysis is the determination of the adequacy of the connection.

The force F_i and angle θ_i acting at each connector can be determined from Eq. 6.15. Due to symmetry, the centroid of the group of connectors at each side of the connection is at the intersection of the member axis with the middle column of connectors, see Fig. 6.31. The polar moment of inertia of the group can be determined as follows:

$$\sum_1^{18} d_i^2 = \sum_1^{18} x_i^2 + \sum_1^{18} y_i^2$$

Considering both sides of the connection and taking advantage of symmetry it is easy to see that:

$$\sum_1^{18} x_i^2 = (2)\ (6)\ (2)\ (9^2) = 1{,}944 \text{ conn.} \times \text{in.}^2$$

$$\sum_1^{18} y_i^2 = (2)\ (2)\ (3)\ (15^2 + 9^2 + 3^2) = 3{,}780 \text{ conn.} \times \text{in.}^2$$

Thus, $\sum_1^{18} d_i^2 = 5{,}724$ conn. × in.^{2}. The longitudinal and transverse components of force at connection A3 (x = 9 in., y = 15 in.), for instance, may now be determined using Eqs. 6.14 as follows:

$$X_{A3} = \frac{1{,}800 \times 15}{5{,}724} + \frac{15}{2 \times 18} = 4.71 + 0.42 = 5.13 \text{ k}$$

$$Y_{A3} = \frac{1{,}800 \times 9}{5{,}724} - \frac{10}{2 \times 18} = 2.83 - 0.28 = 2.55 \text{ k}$$

The total force F_{A3} and angle of load to grain θ_{A3} may be determined from Eq. 6.15:

$$F_{A3} = \sqrt{(5.13)^2 + (2.55)^2} = 5.73 \text{ k}$$

$$\theta_{A3} = \tan^{-1} \frac{2.55}{5.13} = 26.4°$$

All other forces may be determined using the same procedure; for convenience the process should be tabulated for hand computations.

The results are shown in Fig. 6.31. It is easy to show that the system of connector forces is induced by the external forces N, M, Q. Equilibrium requires that

$$N = \sum_1^{18} X_i = (5.13 + 3.25 + 1.36 - 0.53 - 2.41 - 4.30)\ 3 \times 2$$

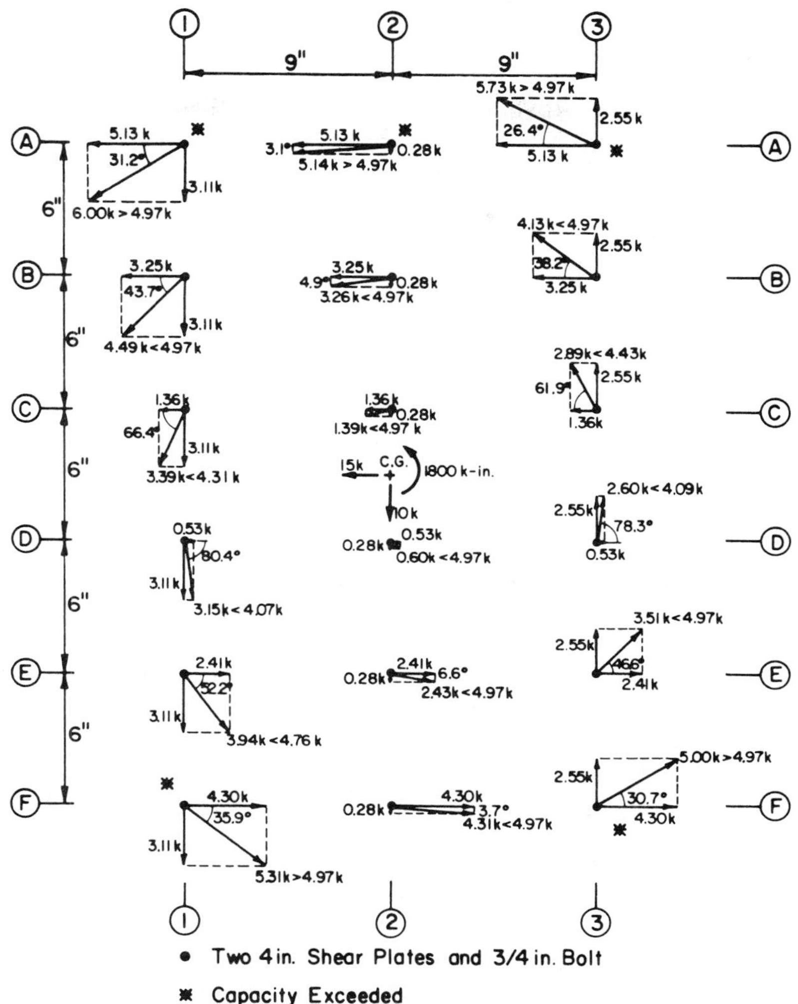

Figure 6.31. Connector forces created at a rigid-plate connection by a given system of forces and moment.

$$Q = \sum_{1}^{18} Y_i = (3.11 + 0.28 - 2.55)\, 6 \times 2$$

$$M = \sum_{1}^{18} (X_i y_i + Y_i x_i) = [(5.13 + 4.30)\, 3 \times 15 + (3.25 + 2.41)\, 3 \times 9 + (1.36 + 0.53)\, 3 \times 3 + (2.55 + 3.11)\, 6 \times 9]2$$

from which N = 15 k, Q = 10 k, M = 1,800 k-in. as given.

For each connector the force and its angle of load to grain are shown. Each force is compared to the capacity of the connector; since for any given wood the latter depends on the angle load of to grain, the capacities shown in Fig. 6.31 are not constant. For example, the capacity of the connectors A3 may be determined as follows. From NDS (1), for a 4-in. shear-plate and ¾-in. bolt, Group A, 2 faces loaded and thickness more than 3½ in., P = 5.03 k, Q = 3.50 k, with a maximum value of 4.97 k for all loadings except wind. For the calculations of allowable loads an eighteen percent increase in P only is allowed, see Footnotes 2 and 3, Table A.6.8, due to the presence of a metal side plate; therefore, $P = 1.18 \times 5.03 = 5.935$ k. Thus, for connector A3 at $\theta = 26.4°$ ($\sin\theta = 0.445$, $\sin^2\theta = 0.198$), $P/Q = 1.695$, Hankinson's formula (Eq. 2.2), yields:

$$N_{A3} = \frac{P}{1 + \left(\frac{P}{Q} - 1\right)\sin^2\theta} = \frac{5.935}{1 + (1.695 - 1)(0.198)} = 5.21 \text{ kip}$$

Since $\bar{\alpha} = 1.15$, because of snow-loading conditions, $N_{A3} = 1.15 \times 5.21 = 5.99$ k. However, this exceeds the 4.97 k limitation set by NDS. Therefore, the capacity of this particular connector is limited to 4.97 k. In the case of connector A3 the force imposed by the connection, 5.73 k, exceeds the capacity. Observation of the results shown in Fig. 6.31 discloses the fact that this is also true for connectors A1, A2, F1, and F3, all of which have been marked with an asterisk; obviously, this should not occur in a well-designed connection. It is necessary to strengthen the connection in such a way that the individual capacity of any connector is not exceeded. Various solutions are possible, namely: 1) more connectors at the same g and p values, 2) the same number of connectors with increased values of g and p, and 3) a combination of the preceding two. For the given problem all three solutions were considered in a number of trials. This was possible using a computer program developed by M. Fordham (9). A solution involving the same number of connectors as before was obtained by increasing g from 6 to 8 in. and p from 9 to 10 in. The results are shown in Table 6.3, the actual computer output. The connectors are identified by coordinates, the origin of which is the lower corner of the connnection. The following information is given: X and Y components of each force, the total force, its angle of load to grain, and the corresponding connector capacity. The last column of Table 6.3 gives the ratio of force to capacity for each connector.

Notice that all ratios are less than unity; thus, the connection is acceptable since the capacity of the individual connectors is exceeded nowhere. It is also noted that the concept of modification factor for multiple fasteners has not been applied to this connection. The research on which the modification-factor concept is based, as shown in Section 6.2.2.4, was done on connections for which the loads applied to the fasteners have all a unique direction. This contrasts with the results of the analysis given in Table 6.3 which indicates a different direction and intensity of loading for each connector. Again, in view of the fact that all connectors are subjected to loads that are smaller, by reasonable margins, than the allowable loads for the respective directions, the design of the connection may be considered safe.

The steel plates usually have a minimum thickness. Maximum stresses occur at the section defined by the first column of connectors. Let the thickness of the plate be ¼ in. and its depth 44 in. The properties of the net section, assuming bolt holes 13/16 in. in diameter, are $A = 9.78$ in.^{2}

TABLE 6.3. ANALYSIS OF RIGID-PLATE CONNECTION.

Units = Kips, Inches, Degrees
Positive Values = Sagging Moments, Tensile Axial Force, Counterclockwise Shears
Origin of Coordinates at Lower Left Corner of Connection
Number of Connectors per Bolt = 2 Conditions of Loading, Factor = 1.150
Hankinson's Formula, P = 5,935k Parallel and Q = 3,500k Normal to Grain
Moment = 1,800.00 Shear = −10.00 Axial Force = −15.00
Columns = 3 Rows = 6 Pitch = 10.000 Gage = 8.000
Number of Bolts = 18
Position of Centroid = (10.00, 20.00)
Metal Plate, Factor = 1.180
NDS Maximum Connector Capacity = 4.970

Connector x	Coordinates y	X−Component Moment	X−Component Total	Y−Component Moment	Y−Component Total	Angle to Grain	Total Force	Allowable Force	Tot. Force / Allow. Force
0.0	0.0	3.947	3.531	−1.974	−2.251	32.5	4.187	4.970	0.843
0.0	8.000	2.368	1.952	−1.974	−2.251	49.1	2.980	4.885	0.610
0.0	16.000	0.789	0.373	−1.974	−2.251	80.6	2.282	4.070	0.561
0.0	24.000	−0.789	−1.206	−1.974	−2.251	61.8	2.554	4.430	0.577
0.0	32.000	−2.368	−2.785	−1.974	−2.251	39.0	3.581	4.970	0.721
0.0	40.000	−3.947	−4.364	−1.974	−2.251	27.3	4.911	4.970	0.988
10.000	0.0	3.947	3.531	0.0	−0.278	4.5	3.542	4.970	0.713
10.000	8.000	2.368	1.952	0.0	−0.278	8.1	1.971	4.970	0.397
10.000	16.000	0.789	0.373	0.0	−0.278	36.7	0.465	4.970	0.094
10.000	24.000	−0.789	−1.206	0.0	−0.278	13.0	1.238	4.970	0.249
10.000	32.000	−2.368	−2.785	0.0	−0.278	5.7	2.799	4.970	0.563
10.000	40.000	−3.947	−4.364	0.0	−0.278	3.6	4.373	4.970	0.880
20.000	0.0	3.947	3.531	1.974	1.696	25.7	3.917	4.970	0.788
20.000	8.000	2.368	1.952	1.974	1.696	41.0	2.586	4.970	0.520
20.000	16.000	0.789	0.373	1.974	1.696	77.6	1.736	4.103	0.423
20.000	24.000	−0.789	−1.206	1.974	1.696	54.6	2,081	4.668	0.446
20.000	32.000	−2.368	−2.785	1.974	1.696	31.3	3.261	4.970	0.656
20.000	40.000	−3.947	−4.364	1.974	1.696	21.2	4.682	4.970	0.942

Sum X−Force Moment = 0.0 Sum X−Force Total = − 15.0000 Sum Y−Force Moment = 0.0000 Sum Y−Force Total = − 9.9999

and I = 1,547 in.[4] Under the action of M = (½)1,800 = 900 k-in. and N = (½) 5 = 7.5 k the stresses in the extreme fibers of the plate are, respectively:

$$f = \frac{M^{(d/2)}}{I} \pm \frac{N}{A} = \frac{900 \times 22}{1547} \pm \frac{7.5}{9.78} = \begin{matrix} 13.6 \text{ ksi, compression} \\ 12.0 \text{ ksi, tension} \end{matrix}$$

both of which are less than 20 ksi allowed. The maximum shearing stress is (3/2)(10/2)/9.78 = 0.77 ksi < 12.5 ksi allowed. Maximum bearing stress on the plate cocurs at the connector with coordinates x = 0, y = 40.0 in., see Table 6.3 For this case F = 4.91 k and the bearing stress of the bolt on the plate is 4.91/(¼ × ¾) = 26.2 ksi < 1.5 F_u, where F_u = 58 ksi is minimum for A36 steel. The dimensions and thickness of the plates are acceptable. Thus, 2 steel plates ¼″ × 3′-8″ × 5′-0″ may be used for the connection.

Designers are advised that rigid-plate connections prevent free vertical shrinkage of the wood members and induce tensile stresses across the grain that may cause splitting of the wood at the connection. Use of rigid-plate connections should not be objectionable in seasoned wood having a moisture content during erection lower than that to be attained under service conditions.

6.5.2 Multiple-Plate Connections

It may be more convenient, especially under the action of large bending moments, to use a rigid connection made up of multiple plates as shown in Fig. 6.29. A larger lever arm for the forces resisting the moment is obtained when flange plates are used. It can be shown that the contribution that the web plate makes to resisting moment is quite small in comparison, and may be easily neglected. The reverse is true regarding support of the transverse force Q. For this case the main contribution is that of the connectors at the web plate; the connectors at the flange plates contribute very little. The phenomenon is similar to that of the distribution of flexural and shear stresses in an I-section, i.e., the flanges provided most of the flexural strength, while the central portion contributes the largest shear force. The longitudinal load N is distributed among all connectors in equal shares. A variation of this type of connection consists of flanges made up of steel cover plates above and below the laminated member. For this case, the I-section behavior is even more true. The cover plates are tied to the laminated member through shear-plates and lag screws, see Section 6.3. The minimum penetration of lag screws for the development of full capacity varies between 7 and 11 diameters depending on the type of wood; it is 10 diameters for Southern pine and Douglas fir, see Section 501-D-1 of Ref. 1.

6.5.2.1 Illustrative Example. Multiple-Plate Connection.

The problem is the same as that discussed in Section 6.5.1.1. The solution in this case is a multiple-plate rigid connection similar to that shown in Fig. 6.29, except that six sets, instead of four sets of 4 in. shear plates and ¾ in. bolts are used for the flange plates and only two sets, instead of four sets, are used for the web plates. General dimensions of one half of the connection are, using the notation shown in Fig. 6.29, g = 6 in., p = 9 in., e_1 = 7 in., e_2 = 2 in., c = 2 in. The section of the rigid frame at the connection is 8.75 in. wide by 48 in. deep and is subjected to M = 1800 k-in., N = 15 k, and Q = 10 k which are due to the simultaneous action of permanent loads and snow, i.e., α = 1.15. Again, the object of the problem is the determination of the adequacy of the connection.

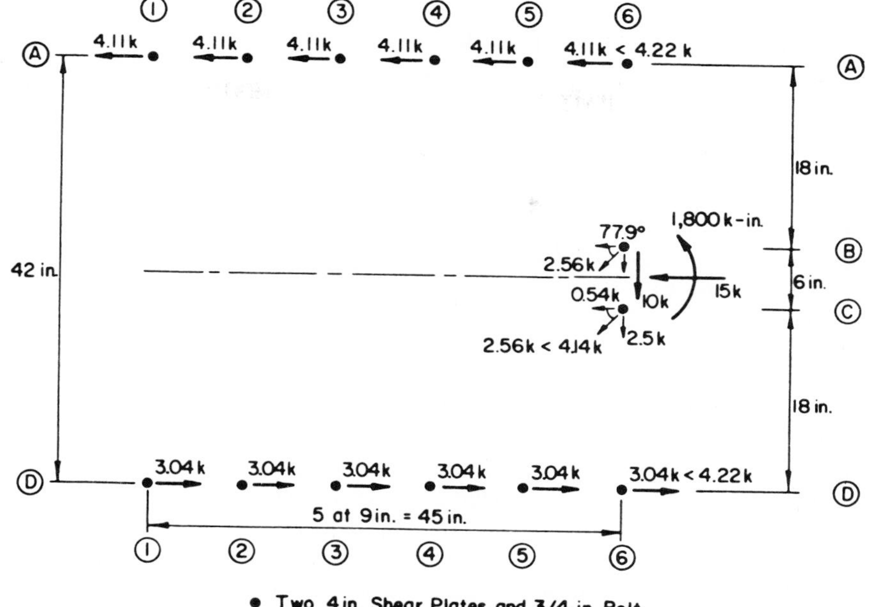

Figure 6.32. Connector forces created at multiple-plate connection by a given system of forces and moment.

The results are shown in Fig. 6.32. The force at the various connectors may be determined as follows:

Top flange, row A, numbers 1 through 6:

$$f = \frac{\frac{1800}{42}}{2 \times 6} + \frac{15}{2\,(6 + 2 + 6)} = 3.571 + 0.536 = 4.11 \text{ k}$$

Bottom flange, row D, numbers 1 through 6:

$$F = 3.571 - 0.536 = 3.04 \text{ k}$$

Web, B6 and C6:

$$X = \frac{15}{2\,(6 + 2 + 6)} = 0.536 \text{ k}$$

$$Y = \frac{10}{2 \times 2} = 2.50 \text{ k}$$

from which $F = 2.56$ k at $\theta = 77.9°$.

The capacity of a single 4 in. shear plate and ¾ in. bolt, for all loadings except wind loading, is limited by NDS (1), see Note 3, Table A.6.8, to 4.97 kips. The capacity of the multiple connectors in rows A and D is the modified capacity of the single connector times the number of connectors. The modification factor, K, can be determined using Table A.6.26 as follows. For the determination of the cross-sectional area of the main wood member, A_1, consider its width equal to 8.75 in. and depth equal to 6 in. The latter is a conservative assumption since it only takes into account the top and bottom 6 in. of the section. Thus, $A_1 = 8.75 \times 6 = 52.5$ in.^{2}. Assume two ¼ × 6 in. steel plates for the top and bottom flanges of the connection; thus, $A_2 = (2)(¼)(6) = 3$ in.^{2} and $A_1/A_2 = 52.5/3 = 17.5$. Table A.6.26 gives K = 0.85 for N = 6 and the preceding values of A_1/A_2 and A_1. The capacity of a connector in rows A and D is modified to 0.85 × 4.97 = 4.22 k. This is greater than 4.11 k and 3.04 k, the loads imposed at rows A and D, respectively.

The capacity of a 4 in. shear plate and ¾ in. bolt in rows B and C can be determined as follows. Table A.6.14a yields 3.60 k for Group A, two faces loaded, $t > 3½$ in., $\theta = 77.9°$. The modification factor for multiple fasteners, from Table A.6.26 for N = 2, is K = 1.00. Thus, the capacity for the given loading conditions, $\alpha = 1.15$, is 1.15 × 1.00 × 3.60 = 4.14 k, which is greater than the 2.56 k exerted.

Since capacity is not exceeded at any connector, the connection shown in Fig. 6.32 is adequate.

The steel plates may be designed as before. Assume ¼ × 6 in. plates for the top and bottom flanges. The net section of the plate at the end connector is 6 × ¼ − 13/16 × ¼ = 1.30 in.^{2}; the compression force is 4.11 × 6 = 24.66 k. Thus, the stress equals 24.66/1.30 = 18.97 ksi $<$ 20 ksi allowable. The web plate is subjected to 5 k shear and 2 × 0.54 = 1.08 k compression. Assuming a ¼ × 10 in. plate with 2.10 in.^{2} of net section, stresses are: $v = (3/2)(5/2.1) = 3.57 < 12.5$ ksi and $f = 1.08/2.1 = 0.51$ ksi < 20 ksi. Thus, the necessary connection plates are four ¼″ × 6″ × 7′-6″ flange plates and two ¼″ × 10″ × 1′-6″ web plates.

As a final check it can be shown that the equilibrium equations applied to the set of connector forces shown in Fig. 6.28 render the given values of N, M and Q acting at the connection.

The multiple-plate connection uses a total number of 56–4 in. shear plates and 28–¾ in. bolts as compared to a total number of 72–4 in. shear plates and 36–¾ in. bolts for the rigid plate of the Illustrative Example in section 6.5.1.1. Thus, it seems that for the given case, the solution shown in Fig. 6.32 is more economical. This is due, in no small respect, to the relatively large value of M compared to N and Q. For cases where the latter are the predominant effects, a rigid-plate connection may render the most economical solution.

6.5.3 Final Considerations

It is obvious that the design procedures used for rigid connections are conservative. Particuarly in the case of the rigid-plate of Example 1, see Table 6.3, a number of connectors are subjected to forces which are well below the allowable. The factor of safety of the connection has been equated to that of the connector subjected to the largest force. However, it is likely that if the connection is tested to ultimate loading, the factor of safety will be larger. This can be explained as follows. The load-slip characteristics of a connector, as shown in Fig. 6.17, indicate that very large loads can be sustained, and even increased, through large deformations. This ductile behavior would permit all connectors, especially those that are close to the centroid, to develop large forces at ultimate; these forces are certainly larger than those indicated by an elastic analysis. The result would be the sustainment of ultimate loads larger than those than can be predicted using a linear analysis.

A research program of actual testing and analysis of behavior of rigid connections is necessary for improved design methods.

6.6 NAILS AND SPIKES

According to Stern (10) a nail is a "straight, slender fastener, usually pointed and headed; designed to be driven; to hold two or more pieces together or to act as support." Fasteners of this kind with a length six inches or less are generally called nails; spikes include heavy nails and nails that are longer than six inches. There are 4 classes of this type of fasteners, e.g., box nails, common and threaded, hardened-steel nails and spikes. The latter are made of high-carbon steel wire, headed, pointed, annularly or helically threaded, and tempered to provide greater strength than common wire nails of corresponding size.

The types of nail heads, nail points, coatings, metal treatments, etc., are so varied and extensive that it is beyond the scope of this work. Refer to: 1) ASTM Standard F547–77 (11) for standard definitions of terms relating to nails and figures depicting various types of nail heads and points and 2) Stern's work (10, 12) for definitions and sizes of nails and for a complete discussion of behavior and design of nails and spikes in Southern pine wood.

Resistance of nailed connections can be developed by direct withdrawal from side grain of a wood member, or laterally, when driven into side or end grain of a wood member.

The resistance of a nail to direct withdrawal from a piece of wood was determined from tests at the U.S. Forest Products Laboratory (3). The maximum withdrawal load from side grain per lineal inch of penetration, p_u, in the member holding the point is given by:

$$p_u = 6{,}900\ G^{2.5}\ D \tag{6.16}$$

where G is the specific gravity of the wood based on oven-dry weight and volume*; and D is the diameter of the nails in inches. The allowable withdrawal load from side grain, p, is given by an equation similar to Eq. 6.16 except the coefficient 1,380 for a safety factor equal to five, is used instead of 6,900. In spite of this evidence of strength, NDS (1) recommends that nails and spikes not be loaded in direct withdrawal. However, for cases where this cannot be avoided, it gives Table A.6.18 for allowable withdrawal loads for nails or spikes driven perpendicular to the fiber of seasoned wood, or unseasoned wood which will remain wet. Structural design based on withdrawal of nails or spikes from end grain of wood rather than side grain, is to be avoided completely.

A great loss in withdrawal strengh of a common nail occurs when it is driven into green wood that is allowed to season, or into seasoned wood that is subjected to cycles of wetting and drying, or into wood which has been pressure-impregnated with fire-retardant chemicals. Only one-fourth of the load value given in Table A.6.18 applies; however, for threaded, hardened-steel nails, the full tabulated load can be used in all cases. For the frequently used case of toe-nailing (which consists of slant driving a nail or group of nails through the end or edge of an attached member into a main member, as in stud-to-sill assemblies), tests have shown that the allowable withdrawal load, for all conditions of seasoning, is equivalent to two-thirds of the tabulated values.

Loads at joint slips of 0.015 in. for a common nail or spike driven into the side grain (perpendicular to the wood fibers), according to tests at Forest Products Laboratory (3), is given by

$$p = K\ D^{1.5} \tag{6.17}$$

*If G_{12}, the specific gravity of wood based on oven dry weight and volume at 12 percent moisture content were used, Eq. 6.16 would change to $p_u = 7850\ G_{12}^{2.5}D$.

in which p represents the lateral loads in pounds at a 0.015 in. slip of the joint; K is a constant depending on the species, K = 2200 for Southern Pine and Douglas Fir Coast type. Other values of K are given in Ref. 3; and D, the diameter of nail or spike in inches. It is estimated that the ultimate lateral nail loads for softwoods may approach 3.5 times the loads expressed by Eq. 6.17, and for hardwoods they may be 7 times as great. Also, the joint slip at maximum load is over 20 times 0.015 inch.

The allowable loads listed in Table A.6.19 have even greater factors of safety against ultimate loads. However, minimum depths of penetration are necessary for the development of the allowable loads. This information is given in Table A.6.19 in combination with the allowable lateral loads for common nails, threaded nails, and common spikes. As in the case of lag screws, the wood species are classified in four comprehensive groups, labeled I through IV, in Table A.6.5. For penetration less than specified, the allowable load may be determined by linear interpolation between zero and the tabulated load. A minimum penetration of one-third of that specified is required. Refer to other provisions, as given in NDS (1), regarding: 1) nails or spikes in double shear and various thicknesses of side members (see illustrative example, Section 6.6.1); 2) effects of moisture content of wood at fabrication of joint; 3) effects of pressure-impregnation of wood with fire-retardant chemicals; and 4) use of metal side plates.

Allowable load in lateral resistance for nails and spikes driven in end grain (parallel to fibers) is two-thirds of the corresponding value in Table A.6.19. In the case of a toe-nailed joint allowable lateral loads are only five-sixths of the corresponding tabulated values.

As a guide to the designer, the current practice of nailing of wood-frame structures using common nails is given in Table 6.4.

An illustrative example of design of a tension splice using the lateral strength developed by nails driven in side grain is given in Section 6.6.1.

6.6.1 Illustrative Example

The bottom chord of a simply-supported light truss is made up of a 2 × 8 S4S wood, surfaced at 15% moisture content. The length of the truss is such that a splice is necessary at midspan. The force in the bottom cord at midspan, due to normal loading conditions, is 3,000 lb. in tension; bending is negligible. If a ¾-in. plywood splice plate is used at each side of the connection, as shown in Fig. 6.33 determine the required number of nails.

Assume that the given wood belongs to Group II, see Table A.6.4. Also, assume that the nailing characteristics of the plywood splice plates are not greatly different from those of solid wood (3), thus allowing the use of design tables without corrections. The total thickness of the splice is 2 × ¾ + 1.5 = 3-in. Thus 10d, 3-in. long, threaded, hardened-steel nails may be used. The nails fully penetrate all members in the three-member joint; thus, they are subjected to double shear. The allowable load, given in Table A.6.19, is P = 94 lb./ nail. This may be increased by one-third for double shear, see provisions of NDS (1), when each side member is not less than one-third the thickness of the center member, and by two-thirds when each side member is equal in thickness to the center member. In this case, the splice plates are each ½ the size of the center member. Using linear interpolation, the allowable increment due to double-shear action becomes.

$$\triangle P = 94\left[\frac{1}{3} + \frac{2/3 - 1/3}{1 - 1/3}\,(1/2 - 1/3)\right] = 39 \text{ lb.}$$

TABLE 6.4.* RECOMMENDED NAILING SCHEDULE.
Using Common Nails

Joint	Nailing
Joint to sill or girder, toe nail	3-8d
Bridging to joint, toe nail each end	2-8d
Ledger strip	3-16d at each joist
1″ x 6″ subfloor or less to each joist, face nail	2-8d
Wider than 1″ x 6″ subfloor to each joist, face nail	3-8d
2″ subfloor to joist or girder, blind and face nail	2-16d
Sole plate to joist or blocking, face nail	16d at 16″ o.c.
Top plate to stud, end nail	2-16d
Stud to sole plate, toe nail	4-8d
Doubled studs, face nail	16d at 24″ o.c.
Doubled top plates, face nail	16d at 16″ o.c.
Top plates, laps and intersections, face nail	2-16d
Continuous header, two pieces	16d at 16″ o.c. along each edge
Ceiling joists to plate, toe nail	3-8d
Continuous header to stud, toe nail	4-8d
Ceiling joists, laps over partitions, face nail	3-16d
Ceiling joists to parallel rafters, face nail	3-16d
Rafter to plate, toe nail	3-8d
1-inch brace to each stud and plate, face nail	2-8d
1″ x 8″ sheathing or less to each bearing, face nail	2-8d
Wider than 1″ x 8″ sheathing to each bearing, face nail	3-8d
Built-up corner studs	16d at 24″ o.c.
Built-up girder and beams	20d at 32″ o.c. at top and bottom and staggered; 2-20d at ends and at each splice
2-inch planks	2-16d at each bearing

*Adapted from Ref. 13.

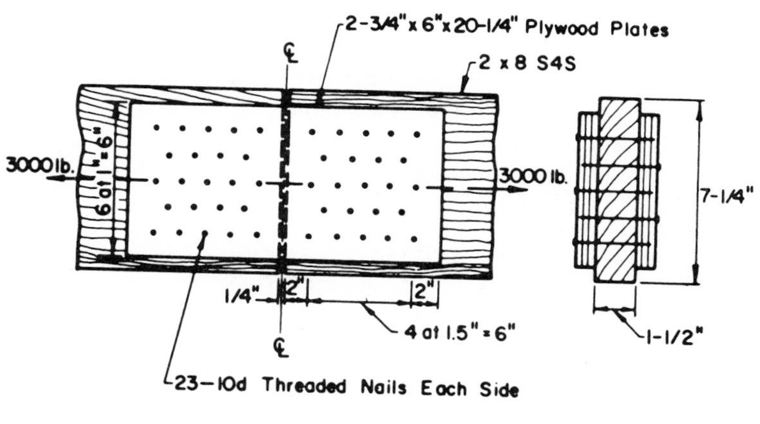

Figure 6.33. Spliced connection of a tension member using plywood side plates and threaded nails.

Thus, $P = 94 + 39 = 133$ lb./nail. The required number of nails is $3{,}000/133 = 22.6$; use 23 at each side of the connection. The spacing of the nails and other details of the connection are shown in Fig. 6.33.

6.7 WOOD SCREWS

Wood screws are used for the same purposes as nails, but their withdrawal capacity is larger. Thus, common use of wood screws is in attaching fixtures and hangers to wood, and in light structural connections. The ultimate withdrawal resistance of wood screws, p, in pounds per inch of penetration of threaded part into side grain of the member holding the point, is given by:

$$p = 15{,}700\ G^2D \tag{6.18}$$

where G is the specific gravity of the wood based on oven-dry weight and volume at 12 percent moisture content and D is the diameter of the shank. As is the case with lag screws, loading of wood screws in withdrawal is not recommended; however, when unavoidable, Table A.6.20 gives the evaluation of Eq. 6.18 for several sizes of wood screws and specific gravities of wood; the factor of safety against ultimate of the tabulated values is somewhat larger than five.

The allowable loads for lateral resistance, at any angle of load to grain, when the wood screw is inserted in the side grain of the main member and a wood side piece is used are given by the following equation:

$$p = KD^2 \tag{6.19}$$

where K is a constant which depends on the wood species and D is the diameter of the shank. The group classification of species given by Table A.6.4 holds for this case also. For groups I, II, III, and IV, the values of K are 4,800, 3,960, 3,240, and 2,530, respectively. The latter are based on proportional-limit loads divided by 1.6 and are expected to allow a maximum slip of 0.01 in. Based on these values of K and Eq. 6.19, allowable lateral loads for wood screws are given in Table A.6.21. Where metal side plates are used the allowable lateral loads may be increased 25 percent.

In order to prevent damage to the threads of the screw, it must be inserted in its lead hole by turning with a screw driver, not by driving with a hammer. To facilitate insertion, soap or other lubricant may be used on the screws, or in the lead holes. Provisions regarding condition of lumber at fabrication and service conditions are similar to those affecting bolts and lag screws, see Section 6.2.

6.8 CONNECTOR PLATES

Connector plates are made of light gage galvanized steel plates in which teeth or plugs have been punched out. The plates are embedded in the wood for the purpose of transferring load. The allowable load for a given plate connector was determined through tests (18) on seasoned lumber as the smallest of: 1) the load at wood-to-wood slip of 0.03 in. divided by 1.6, and 2) by dividing the ultimate test load by 3.0. Design loads so determined are subject to adjustment for load duration, see Section 2.13. If the lumber is unseasoned the allowable loads are reduced 20 percent. Also, if the metal plate connectors are installed in lumber pressure-impregnated with fire-retardant chemicals, the allowable loads are reduced 10 per cent if the wood is kiln-dried after treatment and 20 per cent if it is not.

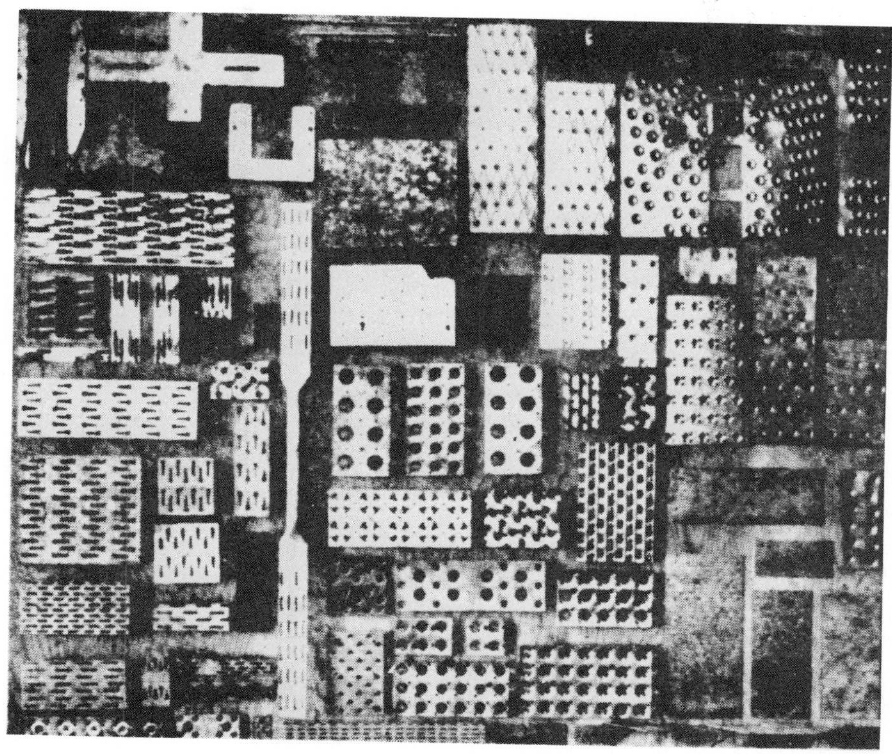

Figure 6.34. Steel plate connectors of various types: pre-drilled flat and deformed plates (right), pronged plates (center bottom), and toothed-plates (left). From Ref. 12.

The Truss Plate Institute (14) specifies toothed-metal connector plates for shop-fabricated wood trusses. The plates require supplemental nails when the teeth are embedded in a roller operation but not when the teeth are preset in a press. Final embedment in the latter operation may be by roller press. Connector plates are made of 20-gage (0.036 in. thickness) galvanized-steel plates. Various designs are possible and strength as well as design load must be determined by test. The minimum quality of steel is specified at 33 ksi yield point stress and 45 ksi ultimate strength; a minimum elongation of 20 percent in 2 inches is also specified.

As an example of design of connector plates consider the spliced connection shown in Figure 6.33. Assume that 20-gage galvanized steel connector plates with ⅜ in. long teeth spaced 1 in. on centers in rows 1⅜ in. apart are to be used. The plates are also punched for supplemental nail holes, ⅛ in. in diameter at 1 in. by 2¼ in. spacings. Standard tests (18) performed on these plates yielded the following design stresses, 180 and 170 lb. per square inch of plate area depending on whether or not supplemental nails are used. Thus, if non-nailed connector plates were used the area required at each side of the spliced connection shown in Fig. 6.33 would be $3{,}000/170 = 17.6$ in.^{2}. Since two plates must be used, one on each face, there need be an area of $17.6/2 = 8.8$ in.^{2} at each side of the connection. Therefore, each connector plate should have a minimum total area of $8.8 \times 2 = 17.6$ in.^{2}. Consider using two 4 in. wide by 6 in. long 20-gage connector plates.

The capacity of the proposed steel plates to take the tensile force could be determined if their net width were known. Assume that the punched holes occupy 1 in. of width at the critical cross-section; the net cross-sectional area of a plate would be $(4-1)(0.036) = 0.108$ in.^{2}. At an allowable tensile stress of $0.6\ F_y = 0.6 \times 33{,}000 = 19{,}800$ lb./in.^{2}, this renders $0.108 \times 19{,}800 = 2{,}140$ lb per plate, or 4,280 lb for the pair, which exceeds the design load (3000 lb) by a substantial margin. Thus, use two 4×6 in. connector plates.

6.9 MISCELLANEOUS CONNECTORS

Some additional connecting devices are available for various uses. Among these are sheet-metal anchors, explosive-driven pins and studs, drift bolts, spike-grids, and clamping plates.

Sheet-metal anchors are used mainly for light-weight buildings. A number of designs are available (15); in particular, joist and beam hangers, see Fig. 6.35, and framing anchors, see Fig. 6.36. The hangers provide a neat method of supporting joists when headers are necessary, as for example, in the case of floor openings. Also, hangers may be used to reduce the total depth of the floor system by connecting joists to supporting girder at the same level. Other types of hangers for heavier loads are available, see the Timber Construction Manual (7). Frame-anchors can be

TECO-U-GRIP joist and beam hangers

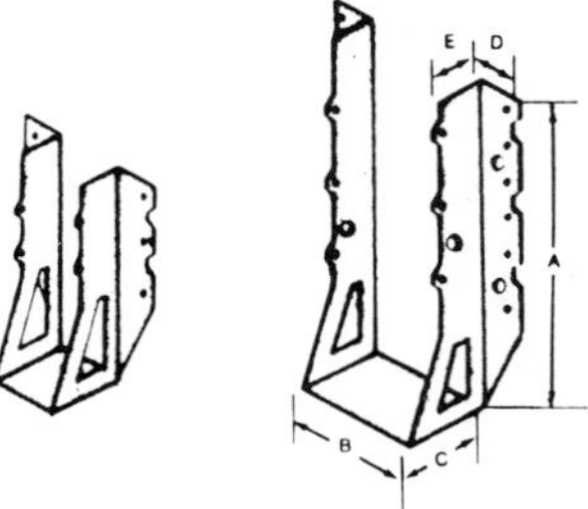

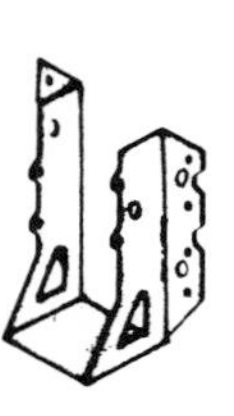

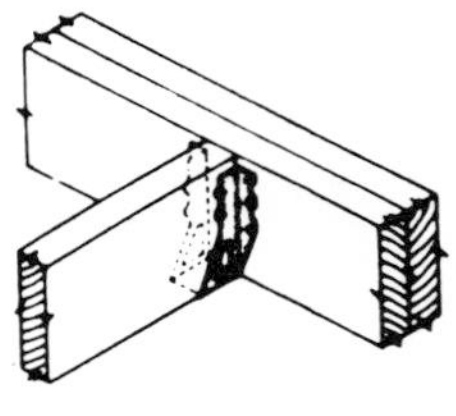

type	recommended joist or beam size	steel gauge	dimensions A	B	C	D	E	nails (packed in each carton) header	joist	nail dia. equiv. to	wire gauge	length	recommended safe working values (¼ ultimate)
24	2x4	18	3¼″	1 9/16″	2″	⅞″	⅞″	4	2	8d	11	1¼″	400 lbs.
A28	2x6 to 2x10	18	5″	1 9/16″	2″	1″	1″	6	4	10d	9	1½″	900 lbs.
B28	2x10 to 2x14	18	8½″	1 9/16″	2″	1″	1″	10	6	10d	9	1½″	1200 lbs.
A36	3x6 to 3x10	16	5¼″	2 9/16″	2¾″	1¼″	1½″	8	4	20d	6	2⅛″	1700 lbs.
B36	3x10 to 3x14	16	8½″	2 9/16″	2¾″	1¼″	1½″	12	6	20d	6	2⅛″	2800 lbs.
A46	4x6 to 4x10	16	5¼″	3 9/16″	2¾″	1¼″	1½″	8	4	20d	6	2⅛″	1700 lbs.
B46	4x10 to 4x14	16	8½″	3 9/16″	2¾″	1¼″	1½″	12	6	20d	6	2⅛″	2800 lbs.
AD6	2-2x6 to 2-2x10	16	5¼″	3¼″	2¾″	1¼″	1½″	8	4	20d	6	2⅛″	1700 lbs.
BD6	2-2x10 to 2-2x14	16	8½″	3¼″	2¾″	1¼″	1½″	12	6	20d	6	2⅛″	2800 lbs

If desired, most types of TECO-U-GRIPS are available in heavier gauge steel, although no increase in safe working values will result.

Recommended safe working values may be increased ⅓ (or as provided by local practice) for wind or earthquake loading.

Values for all types except Type 24, A28(26) and B28(26) can be increased 350 lbs. if two ⅜″x2½″ lag bolts are used (one each per header flange).

Figure 6.35. Joist and beam hangers. Courtesy of Timber Engineering Company.

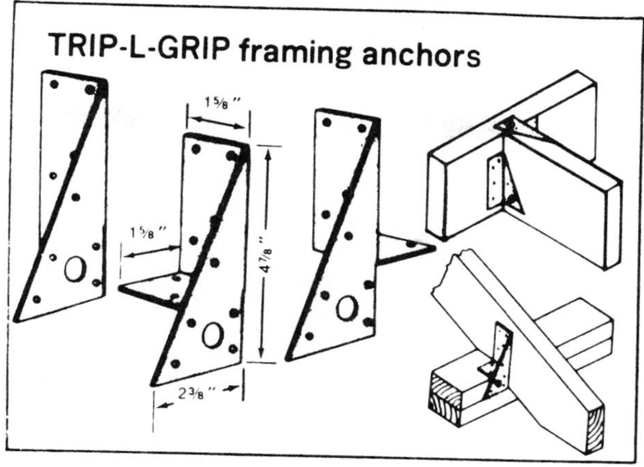

recommended safe working values

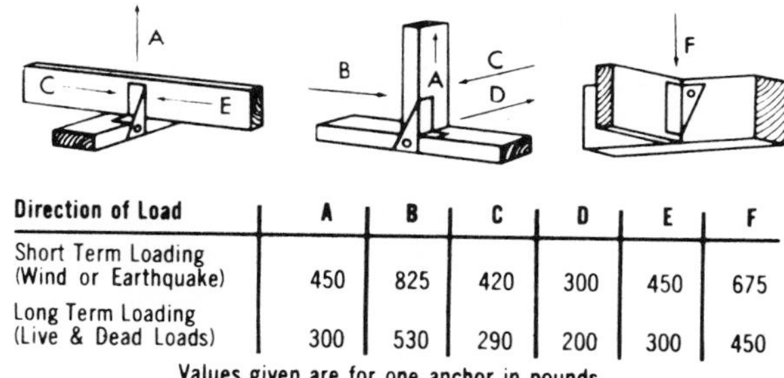

Direction of Load	A	B	C	D	E	F
Short Term Loading (Wind or Earthquake)	450	825	420	300	450	675
Long Term Loading (Live & Dead Loads)	300	530	290	200	300	450

Values given are for one anchor in pounds.

Figure 6.36. Framing anchors. Courtesy of Timber Engineering Company.

used to anchor rafters to supporting studs, and studs to overhead and foundation sills. Thus, wind-induced uplift and lateral forces can be more adequately resisted. Other applications of sheet-metal anchors and recommended safe working values are shown in Fig. 6.36.

Anchorage of wood to steel or concrete, especially when one side is inaccessible, is possible through the use of explosive-driven pins and studs. Some typical uses are the attachment of heavy pins and timber decks to steel or concrete beams, wood nailing strips to steel beams, and wood sills to concrete foundations. The savings in installation time compensate the higher cost of these fasteners as compared to studded bolts. The use of washers of appropriate thickness and diameter is recommended to develop the full effectiveness of these fasteners. Information on the load-carrying capacity of these fasteners can be obtained from the individual manufacturers (16).

Drift bolts are used to fasten heavy timbers together as in the case of pile caps to supporting piles. Washer and nut are not used at the end of the bolt which remains embedded in the wood.

Ultimate withdrawal load p, per inch of penetration, of a drift bolt is given by

$$p = 6{,}600\ G^2D \tag{6.20}$$

where G is the specific gravity based on oven-dry weight and volume at 12 percent moisture content of the wood and D is the diameter of the drift bolt in inches. Eq. 6.20 divided by a factor of safety not less than five against ultimate will give allowable withdrawal load, per inch of penetration, if the drift bolts are driven into prebored holes having a diameter ⅛ in. less than that of the bolt diameter (1,3). Regarding lateral resistance of a drift bolt some vagueness exists. It is agreed that the lateral capacity shall not exceed, and ordinarily be taken as less than, that for a common bolt of the same diameter, see Section 6.2. To compensate for the lack of washer and nut, it is considered good practice to provide for additional penetration of drift bolts into members.

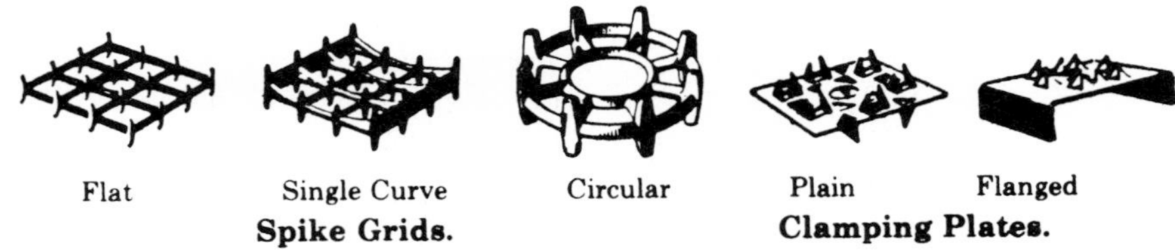

Figure 6.37. Spike-grids and clamping plates.

Spike-grids and clamping plates, see Fig. 6.37, are used in heavy timber construction. The former may be used also for cases where the contact faces are either flat or curved, as for example, at the connection between diagonal bracing and round piles, see Fig.9.3. Clamping plates are used for connecting timbers lapped at right angles, such as for fastening guard timbers to ties on open deck trestles or bridges.

A new fastener consisting of a steel angle with holes for nails and anchor bolt has been developed recently (17) to connect timber piles directly to pile caps, of either wood or concrete. An example of application is given in Fig. 9.3. Design and load data for spike-grids, clamping plates, and pile-cap connectors are given in Table A.6.22, 23, and 24, respectively.

6.10 STANDARD METHODS OF TESTING

For the sake of comparison of test data on the strength and performance of fasteners, standard methods of testing are used. ASTM Standard tests for nail and screw withdrawal, lateral nail or screw resistance, bolted or timber-connectored joints and tensile tests of plate-type connector joints are given in Ref. 18. The use of these methods allows comparison with existing data and eliminates variables in test results that would otherwise occur.

PROBLEMS

1. Design the connection of Fig. 6.12, see Section 6.2.2.4, if T = 13.4 kips, using splice plates of A-36 steel, f_y = 36 ksi, instead of the 2 × 6 wood side members shown. Use ¾-in. bolts.
2. Design the connection of Fig. 6.13, see Section 6.2.2.5, if the tension load in the vertical ties is increased from 5 k to 7.5 k. Designs using: (a) ¾-in. bolts; and, (b) ¾ in. bolts and 2⅝-in. shear-plates. For the latter case assume wood is Group A. Compare both solutions.

3. Consider the connection shown in Fig. 6.24, see Section 6.4.1.1. Assume that the angle of load to grain of loads N, acting on the near-face split-rings shown, is not 15° as in Fig. 6.24, but 45°. Determine the allowable loads N that can be transmitted by the connection.
4. The connection shown in Fig. 6.25, see Section 6.4.1.2, of the text is modified as follows: 1) the angle between the left diagonal and the chord becomes 60 degrees instead of 45 degrees; 2) the force in the diagonal is 8.18 k instead of 10 k and; 3) the left force in the new chord is 15.98 k instead of 13 k. Determine the adequacy of the connection, assuming that any other information is as in Part 1 of Section 6.4.1.2.
5. The connection shown in Fig. 6.26, see Section 6.4.1.2, is modified as described in the preceding problem. Determine the adequacy of the connection, assuming that any other information is as in Part 2 of Section 6.4.1.2.
6. The connection shown in Fig. 6.38 is subjected to the action of forces due to snow-loading conditions. Wood is Group A. The structure is under cover, thus, dry conditions of use apply. Determine the adequacy of the connection. Show free-body diagrams of each member indicating forces acting on the respective faces of the connectors. Also, check the geometry of the connection, i.e., the end and edge distances, and the spacing of the connectors.

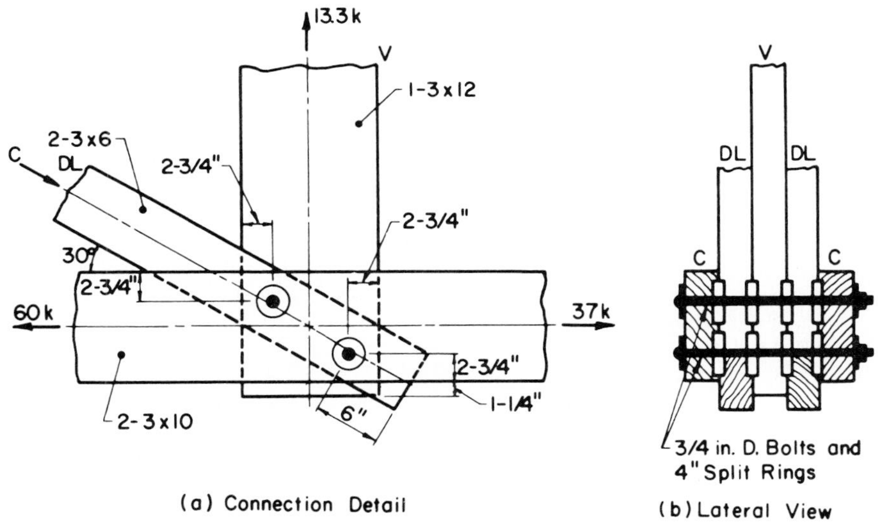

Figure 6.38. The connection in problem 6.

7. The connection shown in Fig. 6.27, see Section 6.4.1.3, is modified by exchanging the vertical planes of the right diagonal and the chords, see Fig. 6.39. Analyze the new joint; determine the forces acting at the various connectors. Draw free-body diagrams of each member and check individual capacities.

   Ans. Left diagonal: 1.02 k at connector, $\theta = 0°$, fourth quadrant. Right diagonal: 1.02 k at front connector, $\theta = 75°$, second quadrant; 3.41 k at back connector, $\theta = 16.8°$, first quadrant. Chord: 3.41 k at front connector, $\theta = 28.2°$, third quadrant; 1.62 k at back connector, $\theta = 90°$, first quadrant. Vertical: 1.62 K each at front and back connector, $\theta = 0°$, third quadrant.

   Notes: 1) angle θ is angle of load-to-grain, measured locally at each member; 2) forces at connector are assumed as acting away from center; 3) quadrants are taken in conventional order.

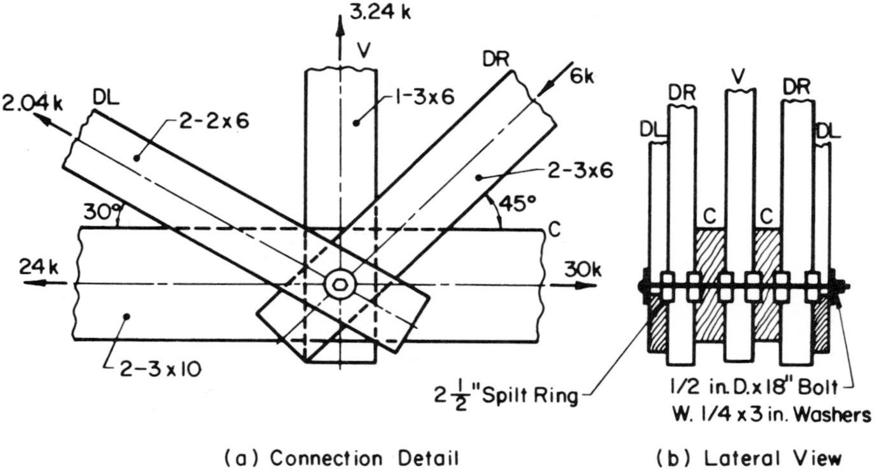

Figure 6.39. The connection in problem 7.

8. Analyze the connection shown in Fig. 6.40. Determine the force T in the verticle for necessary equilibrium. Make free-body diagrams of each member, showing the forces at the face of the respective connectors. Determine the adequacy of the connection if wood is Group A, surfaced-four-sides, and used under covered conditions. Loads correspond to wind-loading conditions.

   Ans. T = 5.64 k. Free-body diagrams as follows. Left diagonal = 4 k at connector, second quadrant, $\theta = 0°$, t = 1.5 in., 1 face loaded, allowable load = 6.02 × 1.33 k. Front chord = 4 k at front connector, fourth quadrant, $\theta = 45°$, t = 2.5 in., 2 faces loaded, allowable load = 4.78 × 1.33 k; 4 k at back connector, second quadrant $\theta = 45°$, rest is similar to front connector. Vertical = 4 k at front connector, fourth quadrant, $\theta = 45°$, t = 2.5 in., 2 faces loaded, allowable load = 4.78 × 1.33 k; 4 k at back connector, third quadrant, $\theta = 45°$, rest is similar to front face. Back chord: 4 k at front connector, first quadrant, $\theta = 45°$; 4 k at back connector, third quadrant, $\theta = 45°$, same capacity as front chord. Right diagonal: 4 k at front connector, first quadrant, $\theta = 45°$, capacity same as left diagonal. The connection is adequate.

9. Consider the connection shown in Fig. 6.28 and 6.31, see Section 6.5.1.1. Assume that g = 8 in. and p = 10 in., instead of 6 in. and 9 in., respectively, as shown in Fig. 6.31. If N = 15 k, M = 1,800 k-in., and Q = 10 k as before, determine, for each connector: (a) X and Y components of force; (b) the total force; (c) the angle of loads to grain; and, (d) connector capacity.

   Ans. See Table 6.3 in body of text.

10. Design the anchorage detail of a glued-laminated beam that is subjected to a reaction consisting of a conventional vertical component equal to 10 k, and a horizontal component equal to 2 k, both due to normal loading conditions. The vertical component is transformed to a 5 k uplift under wind-loading conditions. The beam is 30 ft. long with a section 5⅛ × 18 in. actual dimensions. Consider glued-laminated wood to have the following allowable stresses: F_b = 2.2 ksi, F_t = 1.6 ksi, F_c = 1.5 ksi, $F_{c\perp}$ = 0.45 ksi, F_v = 0.2 ksi, and E = 1,800 ksi. The beam rests on a concrete wall for which the allowable bearing stress is 1.2 ksi. Design a connection such as shown in Fig. 6.41 specifying the dimensions of required A-36 steel plates and bolts, f_y = 36 ksi.

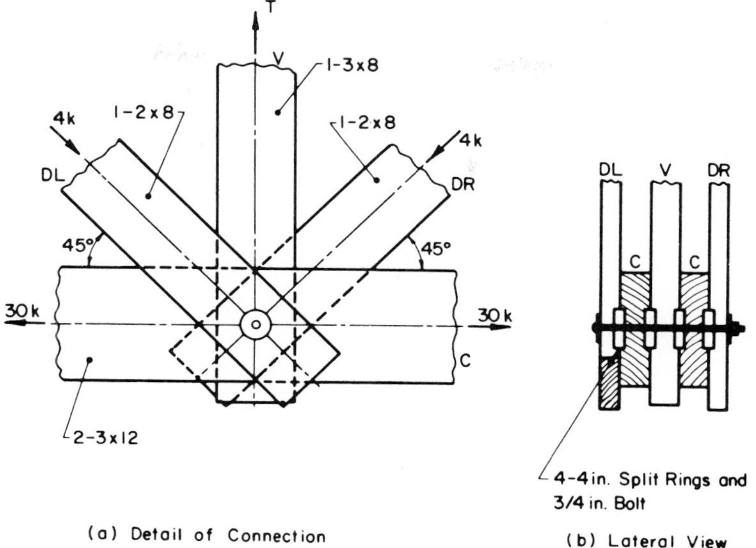

Figure 6.40. The connection in problem 8.

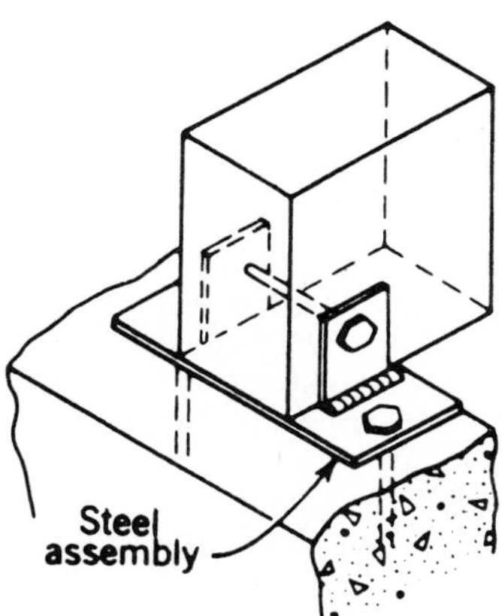

Figure 6.41.

11. For the beam of the preceding problem, consider that the other support is not subjected to horizontal forces since it is free to move. Design this support, as shown in Fig. 6.42, to resist only the 10 k vertical reaction and the 5 k uplift, using the information given in the preceding problem.

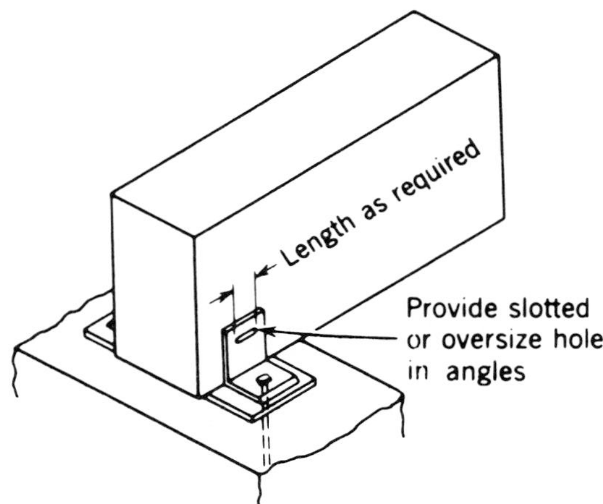

Figure 6.42. Beam anchorage—typical slip joint. Slotted or oversize holes at one or both ends of beam permit horizontal movement under lateral deflection or deformation. From Ref. 8.

12. Typical connections of beams or purlins, to other beams or girders, are shown in Fig. 6.43. Consider the case of a glued-laminated purlin, combination symbol 20F, having the following allowable stresses: $F_b = 2.0$ ksi, $F_t = 1.6$ ksi, $F_c = 1.5$ ksi, $F_{c\perp} = 0.45$ ksi, $F_v = 0.2$ ksi, and $E = 1{,}700$ ksi, with $3\frac{1}{8} \times 12$-in. actual dimensions of section. The end reaction of the purlin is 2.5 k, due to normal loading conditions. Since the structure is covered, design the connection assuming dry conditions of use. Consider both the saddle, and the partially concealed types of connection shown in Fig. 6.43. Compare both solutions.

13. For convenience purposes, long glued-laminated beams are fabricated in parts, to be connected upon erection, in layouts such as shown in Fig. 7.29. Consider a suspended beam, $6\frac{3}{4} \times 24$-in. in section, with a vertical reaction of 12 kips that is to be transmitted to the supporting beam through a saddle-connection such as shown in Fig. 6.44. Consider glued-laminated wood having the following allowable stresses: $F_b = 2.4$ ksi, $F_t = 1.6$ ksi, $F_c = 1.5$ ksi, $F_{c\perp} = 0.45$ ksi, $F_v = 0.2$ ksi, $E = 1{,}800$ ksi. Determine the free-body force diagram and equilibrium couples. Design the connection specifying the dimensions of required A-36 steel plates and lag bolts. Assume the steel has a yield point stress, $f_y = 36$ ksi.

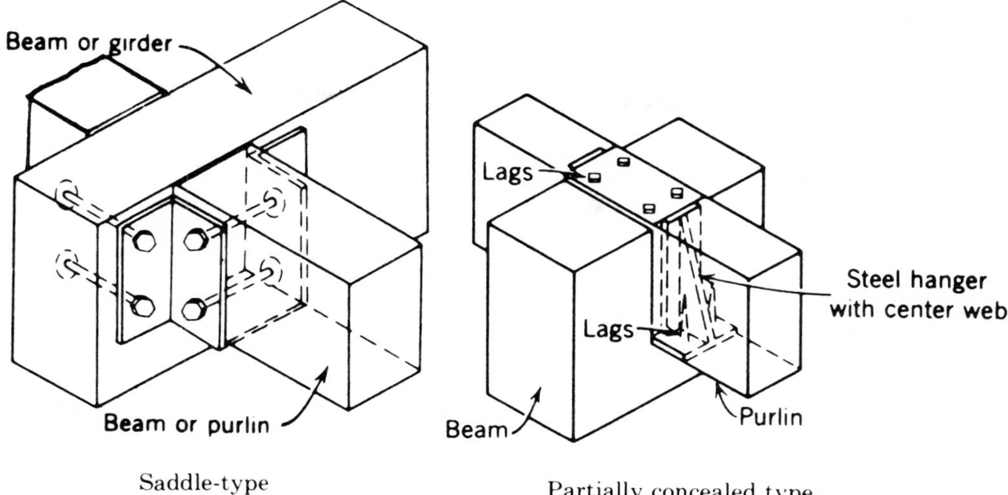

Figure 6.43. Connections of beams or purlins to other beams or girders. From Ref. 8.

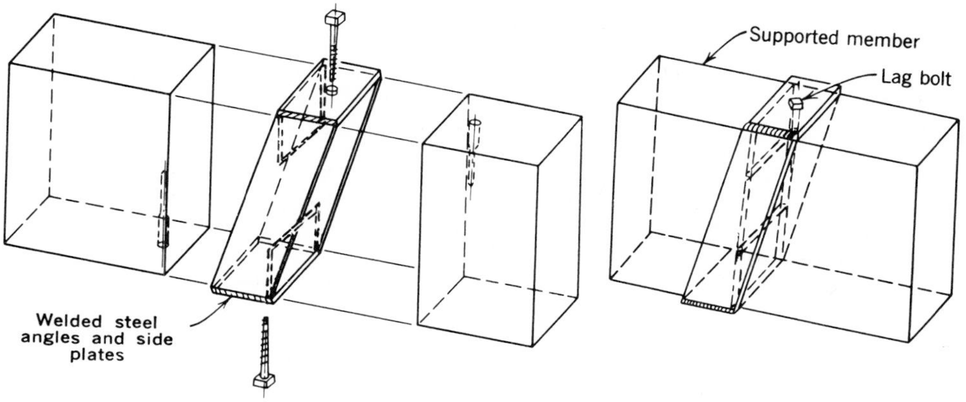

Figure 6.44. Simple saddle-type connection. Used for light loads. Ends of members are cut square. The couple created by the vertical eccentric reaction is resisted by a couple of horizontal forces developed by lag bolts through the saddle-bearing plate. No dapping is required. From Ref. 8.

REFERENCES

1. "National Design Specification for Stress-Grade Lumber and Its Fastenings," 1977 Edition. National Forest Products Association, Washington, D.C.
2. Trayer, G. W. "The Bearing Strength of Wood Under Bolts," Tech. Bul. No. 332, U.S. Dept. of Agriculture, Washington, D.C., Oct. 1932.
3. Forest Products Laboratory, "Wood Handbook," U.S. Department of Agriculture Handbook No. 72, U.S. Government Printing Office, Washington, D.C., 1955, revised Aug. 1974. Chapter on Timber Fastenings.
4. Engineering News-Record, 17 November 1927, p. 805.
5. McKaig, T. H., "Building Failures," McGraw-Hill Book Company, New York, 1962.
6. Scholten, J. A., "Timber-Connector Joints. Their Strength and Design," Technical Bulletin No. 865, United States Department of Agriculture, Washington, D.C., March 1944.
7. American Institute of Timber Construction, "Timber Construction Manual," John Wiley and Sons, New York, Second Edition 1974.
8. American Institute of Timber Construction, "Timber Construction Standards," AITC 100–65, Fourth Edition 1969.
9. Fordham, M., "Digital Computer Program for Rigid Connection of Wood Beams," unpublished report for CE 369, Civil Engineering Department, University of Illinois at Urbana-Champaign, May 1971.
10. Stern, E. G., "Mechanical Fastening of Southern Pine—A Review," Virginia Polytechnic Institute Research Division No. 87, December 1969.
11. "Standard Definitions of Terms Relating to Nails for Use with Wood and Wood-Base Materials," Designation F547–77, American Society for Testing and Materials, Book of ASTM Standards, Part 4, 1978.
12. Stern, E. G., "Nails-Definitions and Sizes, a Handbook for Nail Users," Bulletin No. 61, Wood Research Laboratory, Virginia Polytechnic Institute, 1967.
13. National Forest Products Association, "National Standard for Wood-Frame Construction," Technical Services Division, October 1966.
14. "Design Specifications for Metal Plate-Connected Wood Trusses," TPI-78, Truss-plate Institute, 100 W. Church St. Fredrick, MD 21701; 1978.
15. Timber Engineering Company, "Structural Wood Fasteners," TECO Publication No. 101, January 1970.
16. Sweet's Catalog Service, "Architectural Catalog File," Vol. 2, Structural Systems, accessories. F. W. Dodge Corporation, New York, 1971.
17. Timber Engineering Company, "TECO TEN-CON Connectors, Product Design and Specification Sheet," Publication No. 138, 1969.
18. "Standard Methods of Testing Metal Fasteners in Wood," Designation D1761–77, American Society for Testing and Materials, Book of ASTM Standards, Part 22, 1978.
19. Cramer, C. O., "Load Distribution in Multiple-Bolt Tension Joints," Journal of the Structural Division, ASCE, Vol. 94, No. ST5, Proc. Paper 5939, May, 1968, pp. 1101–1117.
20. Kunesh, R. H., and Johnson, J. W., "Strength of Multiple-Bolt Joints," Forest Research Laboratory Report T-24, Oregon State University, March 1968.

Chapter 7
Behavior and Design of Wood Beams

7.1 INTRODUCTION

The behavior of wood beams has been the object of numerous investigations and theories for many years. In this country, the U.S. Forest Products Laboratory at Madison, Wisconsin, has contributed significantly to the understanding of behavior through the results of experimental investigations and theories. These have influenced to a great extent present design practice.

It was realized early that various phenomena, peculiar to wood beams, refuse explanation through classic elastic analysis. Evidently, the origin of these phenomena stems from the anisotropic nature of the material and the presence of strength-reducing characteristics typical of wood, such as knots, slope of grain, checks and others. In this chapter, experimental data are analyzed, and various assumptions and theories are given to explain such phenomena as the influence of the depth and shape of section on strength, and the behavior in shear of checked beams. In addition, lateral stability, initial and time-induced deflections and end bearing of straight beams are fully discussed. Throughout the chapter, the relation between actual behavior and present design practice of wood beams is particularly stressed, Simple Illustrative Examples are given in this chapter. Additional examples of design of wood beams for actual structures are given in Chapters 9, 10 and 11.

The behavior of curved, glued-laminated wood members, commonly used is studied in detail. The shortcomings of the conventional theory for stress analysis of straight members are shown; the existing experimental data are used for the derivation of a correction factor to be used in the design of curved members. Also discussed are the important problems of radial stresses in curved members subjected to bending, and tapered beams.

7.2 ELASTIC THEORY OF BEHAVIOR IN BENDING

Present design practice of wood beams (1)(2)(3) is based on the premise that wood behaves as a linearly elastic material. The following assumptions are made to study the behavior of a beam subjected to pure bending:

Plane sections, normal to the centroidal axis of the beam, remain plane after bending. In other words, strains are linearly distributed in a section as a result of bending.

Stresses and strains in a section are linearly related by the modulus of elasticity, E, of the material.

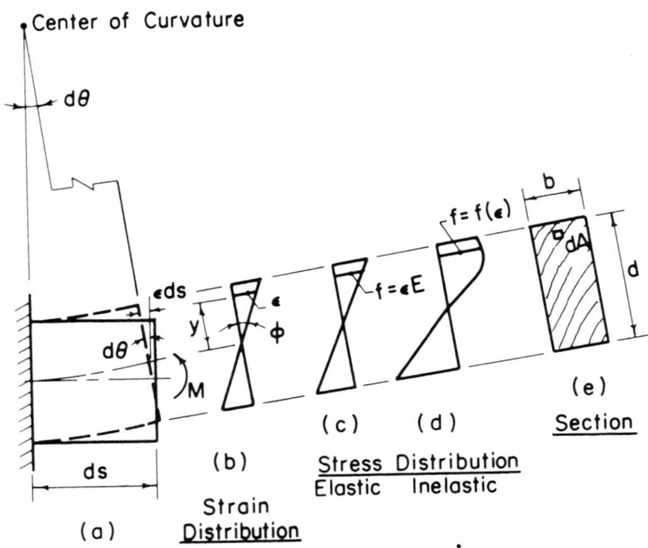

Figure 7.1. Bending of wood member of infinitesimal length.

Bending of a member of infinitesimal length, ds, is shown in Fig. 7.1a. The preceding assumptions lead to the strain and stress distributions shown in Figs. 7.1b and c, respectively. Integration of infinitesimal moments fydA leads to the classical equation (4):

$$f = \frac{My}{I} \tag{7.1}$$

where: f = stress at a given fiber
M = bending moment in the section
y = distance from the neutral axis to the given fiber
I = moment of inertia of the section about the neutral axis.

A given bending moment is positive, by convention, if it subjects the top fibers of the beam to compression strains and the bottom fibers to tensile strains. A negative moment induces opposite effects. For any given bending moment the fibers at a certain level in the section remain unstrained and by definition, that level is the neutral axis of the section. It can be shown that for a linearly elastic material the neutral axis also contains the geometric centroid of the section.

The curvature, ϕ, induced in a section by a given bending moment, M, is defined as the angle that the faces of an element with unit length, which were originally parallel to each other, make after deformation. From Fig. 7.1a, by definition:

$$\phi = \frac{d\theta}{ds} \tag{7.2}$$

But $d\theta = \epsilon ds/y$. Hence,

$$\phi = \frac{\epsilon}{y} \tag{7.3}$$

where, ϵ is the strain at any given fiber situated at a distance, y, from the neutral axis. For a linearly elastic material, $\epsilon = f/E$, where f is given by Eq. 7.1. Substituting $(My/I)/E$ for ϵ in Eq. 7.3 yields the classical expression

$$\phi = \frac{M}{EI} \tag{7.4}$$

where EI is the stiffness of the elastic section. The relation between moments and curvatures is linear for a linearly elastic material.

A design satisfies a given bending moment if the stress given by Eq. 7.1 at $y = a$, where a is the distance from the centroid of the section to the extreme fiber, is equal to, or lower than, F_b, the allowable stress in bending.

7.2.1 Inconsistencies of Actual Behavior

Experimental research in specimens of clear wood has shown the following departures from simple elastic behavior:

1. The neutral axis does not contain the geometric centroid of the section. From initial loading to failure, the neutral axis remains below the centroid of a section subjected to positive bending moment.
2. The relation between moment and curvature is approximately linear in the initial stages of loading. At a point known as the proportional limit, the ratio M/ϕ beings to decrease with increased plastification of the wood fibers subjected to compression. Beyond the proportional limit, the variation of M with ϕ is no longer linear.
3. The maximum bending stress given by Eq. 7.1 at the proportional limit, is larger than the corresponding stress at proportional limit of wood under direct compression.
4. The modulus of rupture, R, defined as the maximum stress given by Equation 7.1 when failure in bending occurs, is larger than the strength of wood under direct compression.
5. The modulus of rupture depends on the depth and shape of the section. For any shape of section, as depth increases, the modulus of rupture decreases.

The linear elastic theory does not explain any of these results. In order to do so, several theories have been put forward, namely: the inelastic theory (5)(6), the theory of supporting action (7), and the statistical strength theory (8). A discussion of these theories, their relevance to present design practice and Illustrative Examples of application are given in the following Sections.

7.3 INELASTIC THEORY OF BEHAVIOR IN BENDING

The principal assumptions made for the inelastic analysis of bending of wood beams are the following:

1. Strains are linearly distributed in a section as a result of bending.
2. The relation between stresses and strains in a section subjected to bending is the same that exists between stresses and strains in direct compression and tension.
3. The stress-strain diagrams in direct compression and direct tension for clear wood, at a given moisture content, are available.

4. Failure occurs when the ultimate compression strain is attained in the extreme fiber of the section.
5. The member is free of any strength-reducing characteristics of wood, such as knots and cross grain, that may affect the validity of any of the preceding assumptions.

The initial assumption is the same as in the elastic theory. The second permits the determination of the stress diagram, see Fig. 7.1d, for any given strain distribution; provided that the stress-strain diagrams are given, as stated in the third assumption. Because of the fact that tensile strength in most species of wood is greater than compressive strength, failure in bending is assumed to occur initially in compression. The fourth assumption expresses the fact that an inelastic stress distribution in the section is developed, and maximum capacity achieved, only after considerable straining of the extreme compression fibers. The initial compression failure is followed by a splitting failure of the fibers under tension. The classical failure in bending of clear wood specimens follows a compression-tension sequence.

The stress-strain diagrams in direct compression and tension, to which assumption 3 refers, are obtained in specimens of clear wood. Any defects in the wood of a given member subjected to bending will affect the results; hence assumption 5. Obviously, the inelastic theory of bending applies exactly to specimens of clear wood and approximately to high quality wood members. Materials of low grade, in which there is appreciable cross grain, knots and other defects, will not meet the preceding assumptions and their ultimate behavior under load may not be predicted exactly. The compression-tension failure sequence of clear wood may be reversed in the presence of defects which reduce the tensile strength of wood. Failure may even occur before plastification of the compression fibers takes place. For such cases, Eq. 7.1 may very well represent the actual conditions in the section at ultimate.

7.3.1 Stress-Strain Diagram of Clear Wood

Let Fig. 7.2 represent the stress-strain diagrams in direct compression and direct tension, for a given specimen of clear wood, at a certain moisture content. It can be noticed that the initial portions of both diagrams may be represented by straight lines, the slopes of which are slightly different. This is due to the fact that E_c, the modulus of elasticity in direct compression, is smaller than E_t, the modulus of elasticity in direct tension, by only a few percent. The linear variation ends at the proportional limit, which is attained in direct compression at a stress f_c^p and strain ϵ_c^p, as shown in Fig. 7.2. In direct tension, the proportional-limit stress, f_t^p, is attained at a strain ϵ_t^p. For most species of wood, $f_t^p > f_c^p$.

The stress-strain diagram in direct compression shows a distinct flat end portion that terminates at the ultimate strain, ϵ_c^u. The maximum stress, f_c^u, remains constant in the end portion. Between the proportional limit (which marks the end of the straight line variation) and the flat end portion, the diagram shows a smooth curve of diminishing slopes. If the initial straight line is extended to where it intersects the backward extension of the flat portion, the intersection occurs at a point with strain $\alpha\epsilon_c^u$ and stress f_c^u. The parameter, α, relates this point to the ultimate strain ϵ_c^u; the range of α is between 0.5 and 1.0. The trapezoidal stress-strain diagram thus created may replace the actual diagram in practical applications, presumably with small error. The stress-strain diagram in direct tension may be considered a straight line between the origin and an ultimate point defined by a strain ϵ_t^u and stress f_t^u. In most species of wood $f_t^u > f_c^u$.

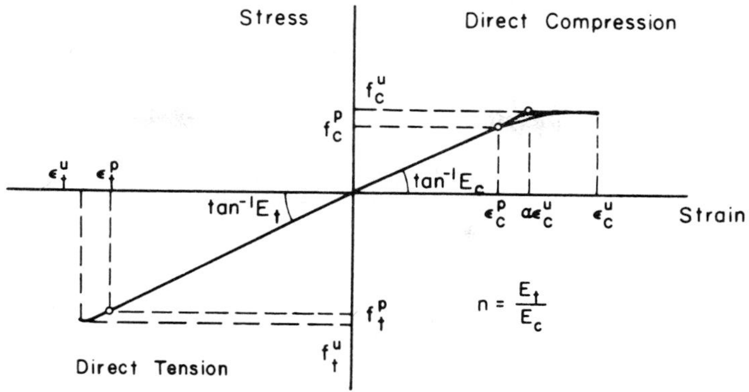

Figure 7.2. Stress-strain diagram of clear wood in direct compression and direct tension.

7.3.2 Loading Stages

The behavior of a clear-wood beam may be obtained by increasing the strain in the extreme fiber of the critical section until the ultimate strain in compression is reached. Two distinct stages of loading will be considered, namely; elastic, for which the strain at the extreme fiber ϵ_c is in the range $0 < \epsilon_c \leqslant \alpha\epsilon_c^u$; and inelastic, for which the range in ϵ_c is given by $\alpha\epsilon_u^c < \epsilon_c \leqslant \epsilon_c^u$. Ultimate occurs when $\epsilon_c = \epsilon_c^u$.

7.3.3 Elastic Stage

A rectangular section under a small bending moment M is shown in Fig. 7.3. The strain distribution is determined by a compressive strain ϵ_c, where $\epsilon_c \leqslant \epsilon_c^p$, and a tensile strain ϵ_t which occur at the top and bottom fibers of the section, respectively. The stress distribution is bilinear with a vertex at the neutral axis. The latter is a consequence of the difference between moduli of elasticity in direct compression and direct tension of wood. Internal equilibrium of forces in the section yields the following equations:

$$\frac{1}{2} f_c ab = \frac{1}{2} f_t (d\text{-}a)b$$
$$M = \frac{1}{3} f_c a^2b + \frac{1}{3} f_t(d\text{-}a)^2b \qquad (7.5)$$

where a is the distance from the neutral axis to the top fiber.

The linear strain distribution gives the relation between the strains ϵ_t and ϵ_c as $\epsilon_t = \epsilon_c (d\text{-}a)/a$. Substituting $\epsilon_t = f_t/E_t$ and $\epsilon_c = f_c/E_c$ in the preceding expression, yields the following relation between the stresses:

$$f_t = nf_c \frac{d\text{-}a}{a} \qquad (7.6)$$

where $n = E_t/E_c$, may be defined as the modular ratio. The range of n is limited to: $1 < n \leqslant 1.05$. Thus, no substantial errors occur in practice when n is taken equal to one. Substituting Eq. 7.6 into the first of Eq. 7.5 yields:

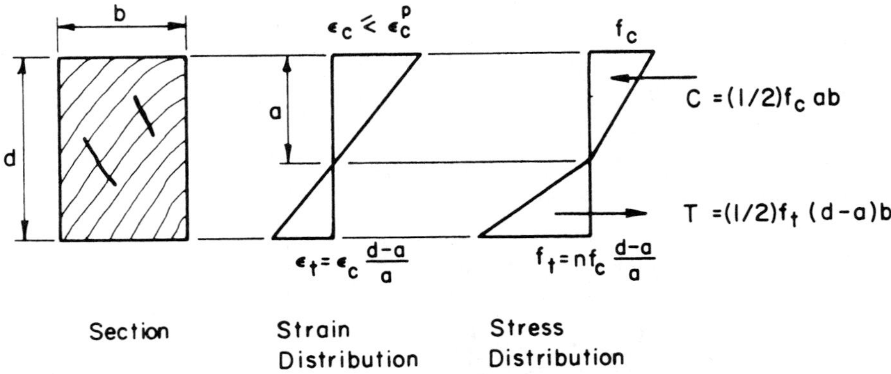

Figure 7.3. Strain and stress distributions in a rectangular section subjected to a small bending moment.

$$a = d \frac{\sqrt{n}}{1 + \sqrt{n}} \tag{7.7}$$

For values of n > 1, Eq. 7.7 gives a > d/2. Since the centroid of a rectangular section lies at a distance d/2 from the top, the preceding inequality shows that the neutral axis is below the center of the section. Substituting Eq. 7.6 and Eq. 7.7 into the second of Eq. 7.5 yields:

$$M = f_c \frac{bd^2}{3} \frac{\sqrt{n}}{1 + \sqrt{n}} \tag{7.8}$$

The preceding relation may be used to determine the moment at proportional limit for bending, $M_{P.L.}$, which occurs for $f_c = f_c^p$. Let f_b^p, the bending stress at proportional limit, be defined by $M_{P.L.}/(bd^2/6)$. It can be shown that the relation between f_c^p, see Fig. 7.2, and f_b^p is given by:

$$f_b^p = f_c^p \frac{2\sqrt{n}}{1 + \sqrt{n}} \tag{7.9}$$

which yields $f_b^p > f_c^p$ for n > 1. This justifies experimental results at proportional limit, for which the bending stress determined using Eq. 7.1 was found larger than the axial compressive stress at proportional limit.

It can be shown that the preceding equations can generate the linearly elastic case, if n is made equal to one.

7.3.4 Inelastic Stage

For a given wood member, this stage begins when $\epsilon_c > \epsilon_c^p$, see Fig. 7.2 However, for the idealized trapezoidal stress-strain diagram that is used in this discussion because of its simplicity, the inelastic stage begins when $\epsilon_c > \alpha\epsilon_c^u$. It ends with failure of the beam, at $\epsilon_c = \epsilon_c^u$. The strength of the beam when ultimate occurs can be obtained from Fig. 7.4. Using Fig. 7.2 and assumption 2, a trapezoidal stress distribution in compression and a triangular one in tension can be obtained.

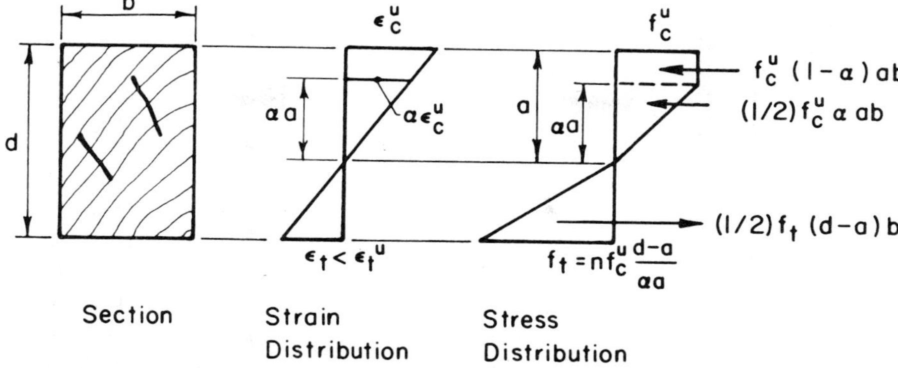

Figure 7.4. Strain and stress distributions in a rectangular section subjected to ultimate moment.

The trapezoidal distribution simplifies the determination of the compressive force and may be considered as an acceptable approximation to the actual stresses. Internal equilibrium of the section yields the following equations:

$$f_c^u (1 - \alpha)\, ab + \frac{1}{2} f_c^u \alpha ab = \frac{1}{2} f_t (d - a) b$$

$$M_u = \frac{1}{2} f_c^u a^2 b (1 - \alpha^2) + \frac{1}{3} f_c^u a^2 b \alpha^2 + \frac{1}{3} f_t b (d - a)^2 \tag{7.10}$$

where f_t, the tensile stress in the bottom fiber is assumed smaller than f_t^u, the ultimate stress in direct tension. The distance from the top fiber to the neutral axis, a, can be obtained by substituting the relation between f_t and f_c^u shown in Fig. 7.4 in the first of Eqs. 7.10. Thus,

$$a = d \frac{m}{1 + m} \tag{7.11}$$

where:

$$m = \sqrt{\frac{n}{\alpha (2 - \alpha)}} \tag{7.12}$$

Substituting Eqs. 7.11 and 7.12 in the second of Eqs. 7.10 gives:

$$M_u = f_c^u bd^2 \left(\frac{m}{1 + m}\right)^2 \left(\frac{3 - \alpha^2}{6} + \frac{2 - \alpha}{3m}\right) \tag{7.13}$$

It can be shown that, for $n > 1$ and $\alpha < 1$, Eq. 7.12 gives $m > 1$ which, when substituted in Eq. 7.11 yields $a > d/2$. Therefore, at ultimate, the neutral axis is also below the mid-depth of the section, as it was for the initial stages of loading.

The modulus of rupture, R, is the maximum stress at which failure occurs assuming that the section behaves elastically to the end. It is determined using Eq. 7.1, which in the case of a rectangular section gives

$$R = \frac{M_u}{bd^2/6} \tag{7.14}$$

R is not the real value of maximum stress at ultimate. However, its evaluation in many tests leads to the impression that a section in bending can sustain a higher compressive stress than can be sustained by the same section under axial compression. This results from ignoring the actual distribution of stresses in the section subjected to bending. The inelastic theory can be used to verify that the modulus of rupture, determined from Eq. 7.14, is always greater than the maximum stress in compression, f_c^u. Comparing Eqs. 7.13 and 7.14 yields:

$$\frac{R}{f_c^u} = \left(\frac{m}{1 + m}\right)^2 \left[3 - \alpha^2 + \frac{2}{m}(2 - \alpha)\right] \tag{7.15}$$

The variation of the ratio R/f_c^u can be obtained from the above relation as a function of the parameter α, as shown in Table 7.1. Only in the case of $\alpha = 1$, which represents a linear elastic variation, is the modulus of rupture equal to the ultimate direct compressive stress in the wood. For smaller values of α, the stress distribution is inelastic at ultimate and $R > f_c^u$. A close observation of Table 7.1 shows that the relation between the ratio R/f_c^u and α is almost linear and can be expressed approximately by:

$$\frac{R}{f_c^u} = 2 - \alpha \tag{7.16}$$

Substituting the above value of R in Eq. 7.14 leads to the following simple expression for ultimate moment of a wood section

$$M_u = f_c^u \frac{bd^2}{6}(2 - \alpha) \tag{7.17}$$

Equation 7.17 renders good results for values of $\alpha > 0.5$ and $n < 1.05$.

TABLE 7.1. VARIATION OF m, a/d AND R/f_c^u WITH α, FOR $n = 1$.

α	m	a/d	R/f_c^u Eq. 7.15	R/f_c^u (approx) Eq. 7.16
1.0	1.00	0.500	1.00	1.0
0.9	1.01	0.502	1.09	1.1
0.8	1.02	0.505	1.20	1.2
0.7	1.05	0.512	1.31	1.3
0.6	1.10	0.524	1.41	1.4
0.5	1.16	0.537	1.52	1.5

If used within these limits, M_u may be evaluated with errors smaller than 5 percent on the safe side.

In summary, it has been shown that the inelastic theory is able to explain the first four inconsistencies that were observed in the tests. Only the variation of modulus of rupture with depth and shape of section may not be explained using this theory.

7.3.5 ILLUSTRATIVE EXAMPLE

A 6 × 14 rough sawn beam of a certain wood species is subjected to bending in the laboratory. The stress-strain diagrams, in direct compression and direct tension, for the given 12 percent moisture content of the beam at the time of testing, are as shown in Fig. 7.5 in which $E_c = 2{,}000$ ksi, $E_t = 2{,}100$ ksi, $f_c^u = 7$ ksi, $\epsilon_c^u = 0.005$, $f_c^p = 6.4$ ksi, $\epsilon_c^p = 0.0032$, $\alpha = 0.7$, $\epsilon_t^p = 0.0048$, $f_t^p = 10$ ksi, $\epsilon_t^u = 0.0055$ and $f_t^u = 11$ ksi. Discuss bending conditions in the section at proportional limit and at failure, assuming the beam is of clear wood, in spite of the fact that this is not a likely occurrence in members of such dimensions. Determine also the moment-curvature relationship for the given section.

The end of the actual linear variation between stresses and strains in direct compression occurs at $\epsilon_c^p = 0.0032$, $f_c^p = 6.4$ ksi. The neutral axis can be determined by substituting d = 14 in. and n = 2.100/2,000 = 1.05 in Eq. 7.7. Thus,

$$a = 14 \frac{\sqrt{1.05}}{1 + \sqrt{1.05}} = 7.08 \text{ in.}$$

The bending moment at proportional limit, $M_{P.L.}$, can be obtained from Eq. 7.8 using $f_c = 6.4$ ksi and b = 6 in. as follows:

$$M_{P.L.} = (6.4) \frac{(6)\ (14)^2}{3} \frac{\sqrt{1.05}}{1 + \sqrt{1.05}} = 1.270 \text{ k-in.}$$

The proportional-limit stress for bending, given by Eq. 7.9 is $f_b^p = 6.47$ ksi. Hence, $f_b^p > f_c^p$. The curvature of the section at this stage of loading is given by Eq. 7.3. Thus, $\phi = 0.0032/7.08 = 4.52 \times 10^{-4}$ rad/in.

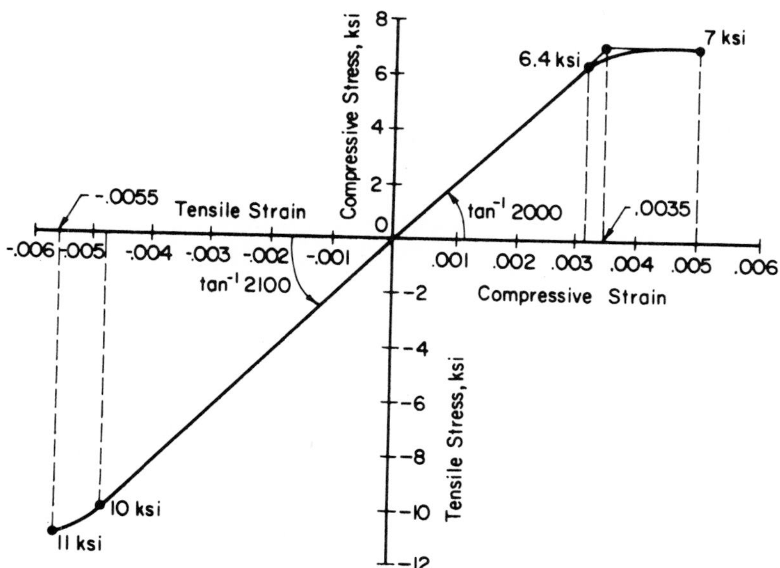

Figure 7.5. Stress-strain diagram of wood for beam of illustrative example in section 7.3.5.

Failure occurs when the maximum strain in the wood reaches the ultimate value given by the stress-strain diagram for direct compressive stresses, i.e., $\epsilon_c^u = 0.005$. When ultimate occurs, the position of the neutral axis can be determined as follows. For $\alpha = 0.7$ and $n = 1.05$, Eq. 7.12 yields:

$$m = \sqrt{\frac{1.05}{(0.7)(2 - 0.7)}} = 1.074$$

and from Eq. 7.11:

$$a = 14 \frac{1.074}{1 + 1.074} = 7.25 \text{ in.}$$

The ultimate moment in the section is given by Eq. 7.13. Thus,

$$M_u = (7)\,(6)\,(14)^2 \left(\frac{1.074}{1 + 1.074}\right)^2 \left(\frac{3 - 0.7^2}{6} + \frac{2 - 0.7}{3 \times 1.074}\right) = 1814 \text{ k-in.}$$

Hence, using Eq. 7.14, the modulus of rupture, $R = 9.26$ ksi; a value which is 32 percent larger than the ultimate stress in direct compression of the given wood. Values of $R = 9.10$ ksi and $M_u = 1{,}784$ k-in., may be obtained using approximate Eqs. 7.16 and 7.17, respectively. The approximate values are within 3 percent, on the safe side, of the actual results. The strain and stress distributions in the section at ultimate are shown in Fig. 7.6a. It is seen that the total compression and tension forces are in equilibrium; also, $M_u = 1{,}820$ k-in., $\simeq$ 1814 k-in. as above, is obtained by taking moments of these forces about the neutral axis of the section.

The curvature of the section at ultimate, from Eq. 7.3, is $\phi_u = 0.005/7.25 = 6.90 \times 10^{-4}$ rad/in. The moment-curvature relationship for the section is given in Fig. 7.7. There is an initial straight line portion to the proportional limit, P.L., beyond which, the relationship becomes non-linear all the way to ultimate, U.

In technical literature on limit design of structures and behavior of structures under dynamic loads (9), the ductility of a material, or a member, is often defined as the ratio of the deformation at ultimate to that at yield or proportional limit. The ductility of the section of a beam, K, may be defined as the ratio of curvature at ultimate, to curvature at yield or proportional limit. For this case, $K = 6.90/4.52 = 1.5$; a small amount when compared to sections of structural steel or reinforced concrete beams. Had failure occurred prematurely, due to reduced tensile strength and before inelastic compressive stresses are fully developed, the ultimate moment and curvature would be smaller. Ductility of the section in the latter case is reduced, and the behavior of the section may be considered brittle. In this respect, consider the following two illustrative hypothetical cases.

Case 1. The data are the same as that of the preceding example except that, because of the presence of a knot and the resulting local cross-grain in the bottom of the section, it is assumed that the tensile strength of the wood is reduced to $f_t^u = 5$ ksi at $\epsilon_t^u = 0.00238$; the varation between zero and ultimate in the tension diagram is linear. It is wished to determine ultimate moment and curvature.

The solution is shown in Fig. 7.6b. Because of the presence of the knot, failure occurs in a tension-compression sequence once the ultimate tensile strain is achieved. A substantial loss in moment capacity and ultimate curvature results from the presence of this defect; the beam behaves in a brittle way.

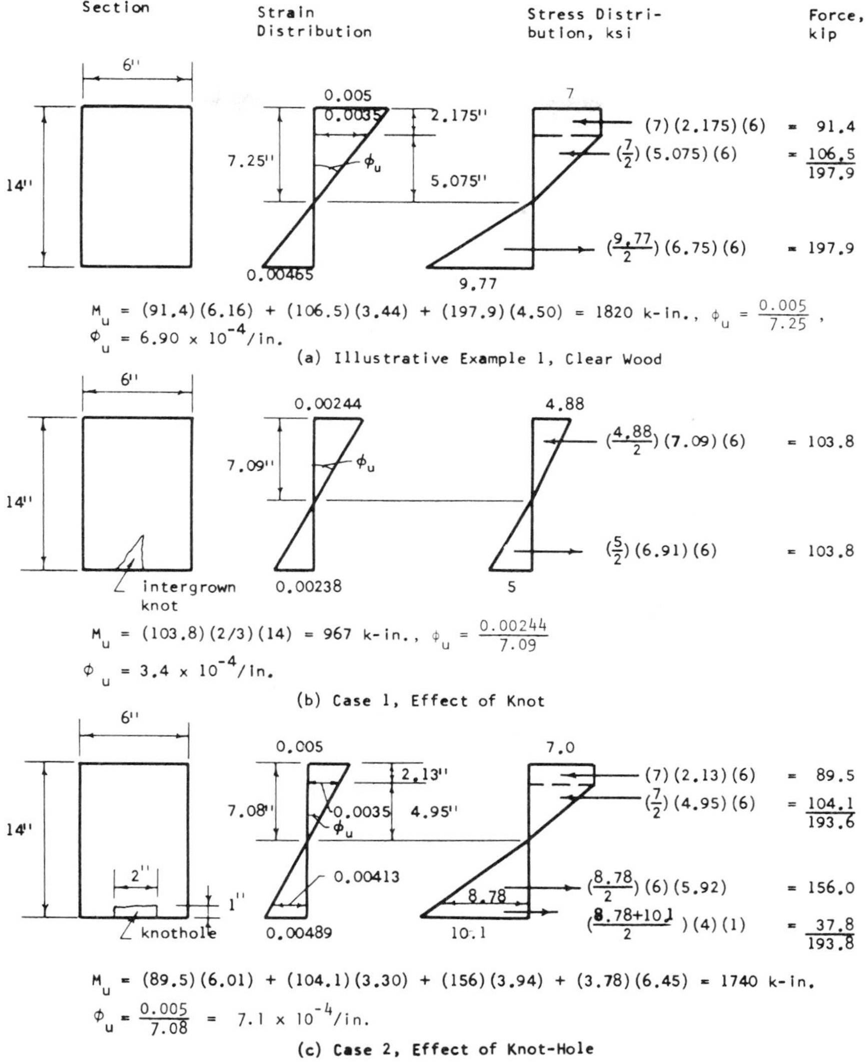

Figure 7.6. Conditions at ultimate in critical section of beams of illustrative example in section 7.3.5.

Case 2. The data are the same as that of example 1 except that a knothole, approximately 2 in. in diameter and 1 in. deep, exists in the bottom face of the beam at the critical section. Assuming that no stress concentrations occur because of the presence of the knothole, determine moment and curvature at ultimate of the reduced section.

The solution was obtained using an iteration procedure. Initially, a value for the distance from the top fiber to the neutral axis, a, is selected arbitrarily; strain and stress diagrams are determined using a, ϵ_c^u, and the given stress-strain relation of Fig. 7.5. It is likely that $C \neq T$, in which case

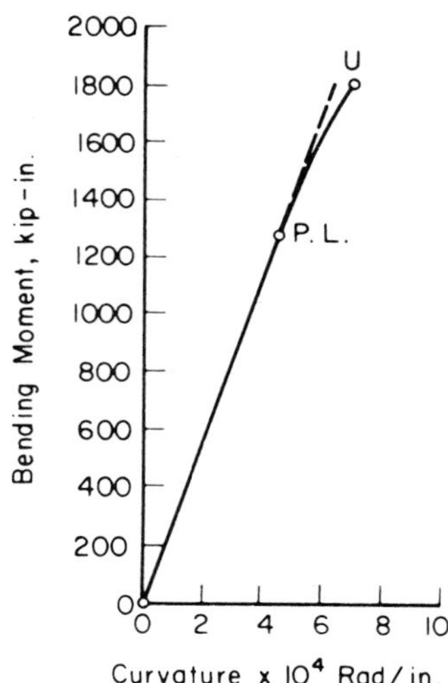

Figure 7.7. Moment-curvature relationship in section of beam.

the solution is not acceptable, since only C − T = 0 is correct. Another value of a is selected and the process repeated; say C ≠ T for this second case also. A better value of a may be obtained by linear interpolation between the pairs of values, a and C − T, that were obtained for the first two cases. This should give an acceptable value for a. However, if a greater accuracy is desired, the process may be continued iteratively using always the last two values of a and C − T for new values of a.

The final solution is shown in Fig.7.6c. Again, the presence of this defect results in loss of moment capacity when compared to the clear-wood section of Illustrative Example in spite of the fact that the compression block is fully developed and failure occurs in the compression-tension sequence. See also problems 22 and 23.

7.3.6 Load-Deflection Relationship

Consider the following example. A 20 ft-long simply-supported 6 × 14 in. wood beam is loaded at the middle-third points by concentrated loads, as shown in Fig. 7.8.a. Material properties and moment-curvature relationships are as indicated in Figs. 7.5 and 7.7, respectively. Determine the deflections at midspan, δ, for the proportional limit and ultimate stages of loading, respectively. Plot the variation of transverse load, P, vs midspan deflection, δ. Determine additional midspan deflections caused by shear strains.

Because of symmetry of beam and loads it is easy to determine the midspan deflection, δ, using Mohr's second theorem. The latter gives the vertical distance, t, from a given point to the tangent through another point in the beam. For the given case the tangent at midspan is horizontal, which makes the vertical distance to it from the support equal to the midspan deflection, δ.

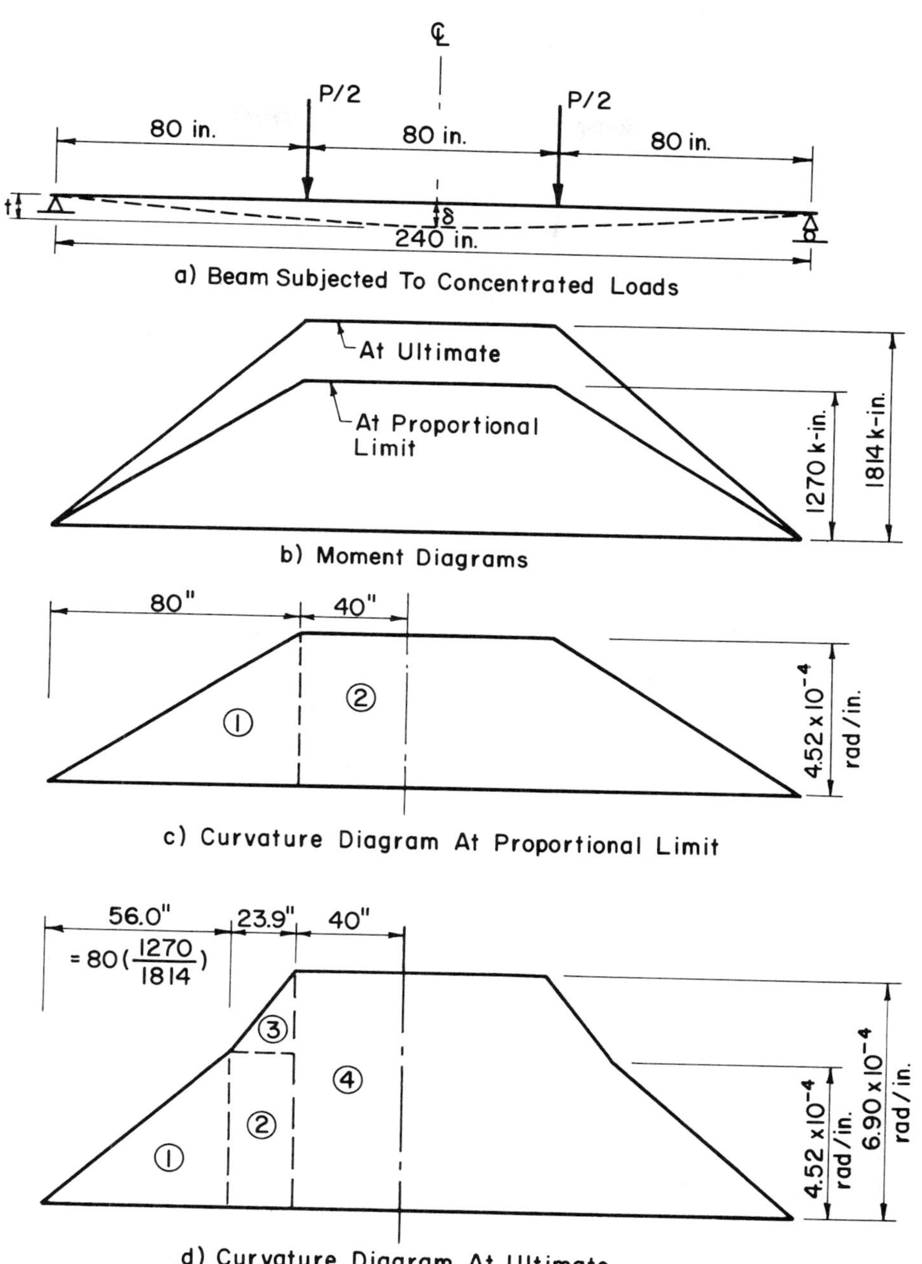

Figure 7.8. Determination of midspan deflections using curvature diagrams.

Thus, applying Mohr's theorem in its general form, i.e. using areas of actual curvature diagram, ϕ, instead of M/EI areas, one obtains:

δ = sum of moments of half-span curvature-diagram areas about support, or, symbolically:

$$\delta = \sum_{1}^{n} A_i \bar{x}_i$$

where A_i and $\bar{x}_i$ are partial areas and the corresponding distances of their centroids to the support, respectively.

The moment diagrams at proportional limit and at ultimate are given in Fig. 7.8b. The curvature diagrams at proportional limit and at ultimate are given in Figs. 7.8 c and d, respectively. The various partial areas have been numbered to facilitate identification in the calculations that follow.

a.) At Proportional limit, Fig. 7.8 c

$$\delta = \frac{1}{2}(80)(4.52 \times 10^{-4})\left(\frac{2}{3} \times 80\right) + (40)(4.52 \times 10^{-4})\left(80 + \frac{40}{2}\right)$$

$$= 0.96 + 1.81 = 2.77 \text{ in.}$$

and $$P = \frac{2M}{a} = \frac{2(1270)}{80} = 31.7 \text{ k}$$

b.) At Ultimate, Fig. 7.8 d

$$\delta = \frac{1}{2}(56.0)(4.52 \times 10^{-4})\left(\frac{2}{3} 56.0\right) + (23.9)(4.52 \times 10^{-4})\left(56.0 + \frac{23.9}{2}\right)$$

$$+ \frac{1}{2}(23.9)(6.90 \times 10^{-4} - 4.52 \times 10^{-4})(56.0 + \frac{2}{3} 23.9)$$

$$+ (40)(6.90 \times 10^{-4})\left(80 + \frac{40}{2}\right)$$

$$\delta = 0.47 + 0.74 + 0.20 + 2.76 = 4.17 \text{ in.}$$

and $$P = \frac{2\,(1814)}{80} = 45.3 \text{ k}$$

The variation of P vs δ given in Fig. 7.9 is a bilinear relationship. Note that the behavior is very close to that of a beam made of a linearly-elastic material; the deflections could have been obtained with reasonable accuracy using the first term of Eq. 2.6. Failure of the beam is abrupt, without warning signs of impending disaster; ductility and energy-absorption capacity are quite small.

It is noted that the calculations did not include the contribution of shear strains in the portion of span between the supports and the concentrated loads. If the latter were taken into account, a shear deflection, δ_s, could be determined using the second term of Eq. 2.6, as follows:

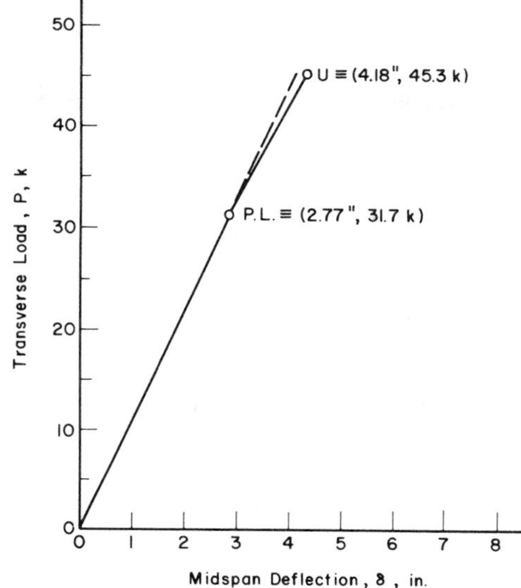

Figure 7.9. Load-deflection relationship for a given wood beam.

$$\delta_s = \frac{Pak}{2\,AG}$$

These deflections may be added to those caused by bending moments alone.

For the given beam: P = 31.7 k at P. L. and 45.3 k at ultimate, a = 80 in., G ~ E/16 = 2000/16 = 125 k/in.^{2}, κ = 1.2 and A = 6 × 14 = 84 in.^{2}. Substituting these values in the preceding equation yields δ_s = 0.14 and δ_s = 0.21 in., for proportional-limit and ultimate-loading conditions, respectively. The total deflections increase to 2.77 + 0.14 = 2.91 in. and 4.18 + 0.21 = 4.39 in., respectively.

7.4 THEORY OF SUPPORTING ACTION OF BEHAVIOR IN FLEXURE

Newlin and Trayer (7) proposed the theory of supporting action to explain the variation of the modulus of rupture with depth and shape of the section, for any given type of wood. The theory was checked experimentally by Dietz (10) and by tests at the U.S. Forest Products Laboratory, reported by Freas and Selbo (11). It constitutes the basis for present design specifications of size and form factors.

The theory assumes that wood fibers under compression act as minute columns restrained, against buckling as a whole and against individual buckling of their cell walls, by the lignin which serves to stiffen and cement them to the adjacent fibers. In a wood block subjected to axial loading, all fibers are equally strained and tend to buckle simultaneously, thereby offering one another little support. In a beam subjected to bending, the compressive strain falls from a maximum at the extreme fiber to zero at the neutral axis, beyond which it changes to tension. Hence, according to this theory, support of the extreme fiber by the other less restrained fibers is possible in bending.

The amount of support that a given fiber may render depends on the strain to which it is subjected and on its distance to the extreme fiber. Support decreases with both increasing distance and level of straining. The fibers directly under the top of the beam are very close; however, they are too highly strained themselves to be able to offer much support. Those close to the neutral axis are strained very little but are far from the top. Optimum support is offered somewhere in between. Dietz determined that maximum support is given by fibers at approximately ⅓ of the beam depth from the compression side.

The variation of modulus of rupture with depth is explained by this theory using the strain gradient of the section. For a linear strain variation, the strain gradient is the ratio of strain at a given section to its distance to the neutral axis; the definition being similar to that of curvature of the section. Consider two beams: one deep, the other shallow, and both subjected to bending. Assume that the same strain is created at the extreme fiber of the critical section of both beams. The deep section will obviously have a smaller strain gradient than the shallow section. The strains fall off slowly, and the fibers under small strains, being far distant, may not offer much support. In the shallow section, however, strains fall off rapidly, the fibers under small strains are quite close to the top fibers and render more effective support. Thus, the extreme fiber of a shallow section being more effectively supported, is able to stand greater strains than that of a deeper section. Consequently, the modulus of rupture is higher in shallow beams.

7.4.1 Size Factor for Rectangular Sections

The variation of modulus of rupture with depth of section has been determined experimentally by various investigations (7)(12) and is graphically shown in Fig. 7.10 from tests on Douglas fir beams. The ratio of modulus of rupture of beams, R_d, to modulus of rupture of 2 in. deep specimens, R_2, is known as the strength ratio. It is plotted in Fig. 7.10 against the depth of the beam, d, in inches. A definite decrease in the ratio R_d/R_2 with depth is evidence. There are some large departures of plotted points from the curve but, in general, the fit of the curve is good. The following equations have been obtained for the variation of the strength ratio with depth.

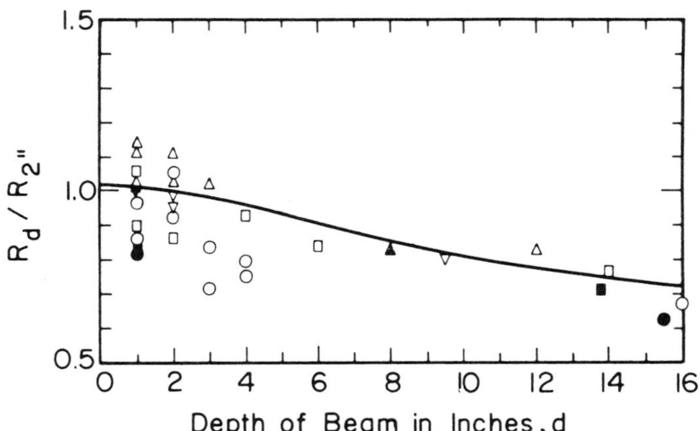

Figure 7.10. Relation between the ratio of modulus of rupture of individual beams to modulus of rupture of 2 × 2 × 30 in. control beams, and the depth of the beam. Adapted from Fig. 94A of Ref. 11.

$$\frac{R_d}{R_2} = 1.07 - 0.07\sqrt{d/2} \tag{7.18a}$$

$$\frac{R_d}{R_2} = \left(\frac{5}{8}\right)\frac{d^2 + 143}{d^2 + 88} \tag{7.18b}$$

Equation 7.18a was derived by Newlin and Trayer (7) in 1924 from data obtained from beams having 12 in. maximum depth. In 1947, Dawley and Youngquist developed Eq. 7.18b from data obtained from beams having depths up to 16 inches. Equation 7.18b was published by Freas and Selbo (11) in 1954. The size factor, C_d, stems directly from Eq. 7.18b. For convenience, C_d is defined as the ratio R_d/R_{12}, where R_{12} denotes the modulus of rupture of a 12 in. deep beam. It is admissible to express C_d as $(R_d/R_2)/(R_{12}/R_2)$. For d = 12 in., Eq. 7.18b yields $R_{12}/R_2 = (5/8)(287/232)$ which, when substituted in the preceding expression for C_d, yields:

$$C_d = 0.81\,\frac{d^2 + 143}{d^2 + 88} \tag{7.19}$$

Equation 7.19 was used in NDS prior to the 1971 edition for $d > 12$ in.; for values of d smaller than 12 in., NDS suggested the use of $C_d = 1$. The basis for present design practice on size factor is discussed in Section 7.5

7.4.2 Form Factor for I and Box Sections

Tests made with beams having I and box sections have shown that their modulus of rupture is smaller than that of a rectangular section of equal depth. The shape of the section seems to also influence the results. In the light of the supporting action theory, the phenomenon can be easily explained by the smaller support offered to extreme fibers of the compression flange. A form factor for these sections may be developed assuming that the modulus of rupture consists of two parts that can be added together. These are: 1) the compressive strength of the material and 2) the excess of modulus of rupture of a rectangular section of the same height over the compressive strength, multiplied by a support factor. This may be expressed mathematically as follows:

$$R_f = C + (R_d - C)\,C_g \tag{7.20}$$

where: R_f is the modulus of rupture of an I or box section; R_d is the modulus of rupture of a rectangular section of equal depth; C is the compressive strength, and C_g is the support factor, defined as the ratio of the support afforded the outer fibers in the I or box section to the support afforded them in a rectangular section.

The form factor, C_f, is defined as the ratio R_f/R_{12}. Dividing Eq. 7.20 by R_{12} leads to:

$$C_f = \frac{C}{R_{12}} + \left(\frac{R_d}{R_{12}} - \frac{C}{R_{12}}\right)C_g \tag{7.21}$$

where $C_d = R_d/R_{12}$ is given by Eq. 7.19, and the ratio C/R_{12} may be obtained from the experimental results shown in . Fig. 7.11. As before, there are some departures of plotted points from the curve, but again the fit may be considered good. The equation of the curve of Fig. 7.11 is:

$$\frac{R_d}{C} = \frac{d^2 + 143}{d^2 + 88} \tag{7.22}$$

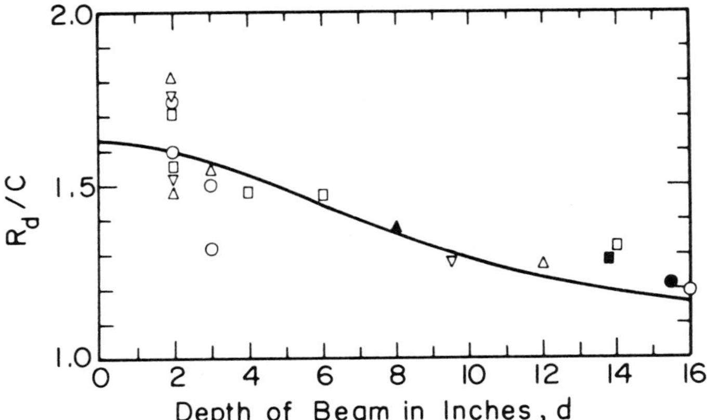

Figure 7.11. Relation between the ratio of modulus of rupture of individual beams to compressive strength parallel to grain of control specimens, and the depth of the beam. Adapted from Fig. 95A of Ref. 11.

For d = 12 in., $R_d = R_{12}$ and Eq. 7.22 yields $C/R_{12} = 232/287$. Substituting in Eq. 7.21 yields:

$$C_f = 0.81 \left[1 + \left(\frac{d^2 + 143}{d^2 + 88} - 1\right) C_g\right] \tag{7.23}$$

The preceding equation can be used, see Appendix D of NDS (1), to evaluate the form factor of a given I or box section, after the support factor C_g has been evaluated.

7.4.3. Support Factor

The supporting action that the fibers at various levels give to the extreme fiber in compression, see Fig. 7.12, has been obtained by various investigators (7) (10). A close approximation to the actual curve is given (11) by the following equation:

$$y = x\,(1 - x)^2 \tag{7.24}$$

where y is the supporting influence of a fiber of thickness dx situated a relative distance x from the extreme compression fiber. Hence, support rendered by an infinitesimal fiber is ydx and total support by all fibers is $\int_0^1 y\,dx$. The support factor for a fiber at the top of the flange is defined as the ratio between the total support given by all fibers over the thickness of the flange only, to the total support given by all the fibers over the full depth of the section. By definition:

$$\text{Support Factor} = \frac{\int_0^p y\,dx}{\int_0^1 y\,dx} = \frac{\frac{p^2}{2} - \frac{2p^2}{3} + \frac{p^4}{4}}{\frac{1}{2} - \frac{2}{3} + \frac{1}{4}} = p^2(6 - 8p + 3p^2) \tag{7.25}$$

where: p = t/d is the relative thickness of the flange, t is the thickness of the flange and d is the depth of the section.

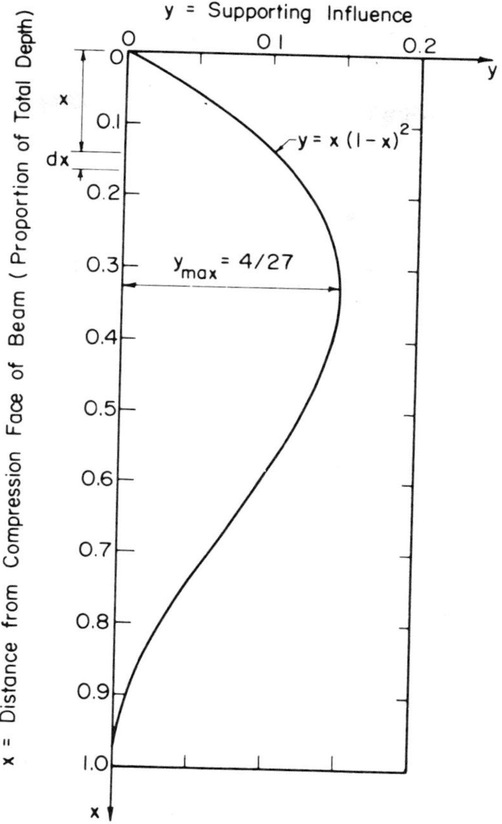

Figure 7.12. Supporting influence, on extreme fiber in compression, of fibers at various points in the depth of a rectangular beam. From Fig. 96 of Ref. 11.

In the case of an I beam, full support is given over a width equal to that of the web; partial support, see Eq. 7.25, is given over a width equal to $b - b'$, where b is the width of the flange, and b' is the width of the web.

Let the relative thickness of the web be given by q, where $q = b'/b$. The support factor for the section, C_g, is given as the sum of the support factor for the portion of flange over the web (which is equal to one by definition) times the relative width of web, q, plus the support factor of the remaining portion of the flange, given by Eq. 7.25, times the relative width of the latter. Hence, the final expression for the support factor is:

$$C_g = q + p^2(6 - 8p + 3p^2)(1 - q) \tag{7.26}$$

where all symbols have been defined before. For a rectangular section $q = 1$, $p = 1$ and Eq. 7.26 gives $C_g = 1$. Hence, $C_f = C_d$ as Eq. 7.23 degenerates into Eq. 7.19 for $C_g = 1$. For any I or box section, $0 < C_g < 1$.

If the depth of the section increases indefinitely, C_f, as determined by Eq. 7.23 becomes asymptotic to a lower bound given by $C_f = 0.81$, while the influence of C_g becomes quite small. For values of $d > 48$ in., ignoring C_g completely, i.e. $C_g = 0$, results in less than 2 percent error on the safe side of the determination of C_f. An Illustrative Example of determination of C_g and C_f for the section of a box beam is given in Section 10.3.

7.4.4 Form Factors for Beams of Circular Section and Beams of Square Section Placed with Diagonal Vertically

Wood poles and masts have circular cross sections; for architectural effects a member with a square section is sometimes placed like a perfect rhombus, with a section diagonal in a vertical plane. The form factors for these members are 1.18 for the circular section and 1.414 for the rhomboidal section (12). It can be shown that with the given form factors the strength of these sections is equal to that of corresponding square sections of the same cross-sectional area.

7.5 STATISTICAL STRENGTH THEORY OF BEHAVIOR IN BENDING

Modulus of Rupture, R, is the maximum stress at which bending failure occurs in a beam that behaves elastically to the end, see Eg. 7.14. A recent investigation by Bohannan (8) explains the variation of the modules of rupture of wood beams with depth and shape of section from a statistical point of view. The general statistical theory of strength given by Weibull (13) is used. This theory assumes that failure of a specimen will occur when the stress in the specimen is the same as the stress that would cause failure of the weakest element of volume if tested independently. Thus, the theory assumes the existence of a "weakest link" where failure, once initiated, propagates without additional loads being applied. Since final failure in bending of a wood beam is a tension failure of brittle nature, it might be assumed that a cascade-type tension failure occurs when any element of volume fails in the tension region of the beam. It follows that the tensile strength may be dependent upon the volume of the beam, rather than on its depth and shape. However, the application of this theory to observed data was not as satisfactory as expected. Considering the actual wood beam, it was concluded that the size effect on modulus of rupture depends, not on the volume of the beam but on its aspect area, i.e. depth times length, and on the method of loading. In other words, it was concluded that the effect is independent of the width of the beam.

A comparison of beams of different volumes showed that if the depth and length are both decreased by ½, the modulus of rupture increases by 8 percent. This agreed very well with existing data on Douglas fir beams that showed the average modulus of rupture of beams $1 \times 1 \times 16$ in., 7.7 percent greater than the average modulus of rupture of $2 \times 2 \times 28$ in. beams.

The influence of the method of loading was experimentally verified by comparison between beams tested with two-point loading, with 2 equal concentrated loads symmetrically placed, and beams loaded at midspan only. The average modulus of rupture of a center-loaded beam was found to be 11 percent greater than the modulus of rupture of a similar beam loaded at third-span points. The result is not surprising, when one compares the regions of maximum moment in both beams. For the beam with the two-point loading, the maximum moment covers a third of the span. For the beam loaded at midspan only, the maximum moment occurs theoretically at midspan; practically, however, it may be considered extended to a length at least equal to the depth of the beam. Still, the region of maximum moment is smaller in beams loaded at midspan only. Hence, the probability of encountering regions of low strength that will reduce the modulus of rupture is greater in the beam with the two-point loading.

The variation of the strength ratio R_d/R_2 with the depth of the beam, according to this theory is given by the following expression:

$$\frac{R_d}{R_2} = \left(\frac{2}{d}\right)^{1/9} \tag{7.27}$$

Eq. 7.27 is plotted in Fig. 7.13 for beams having constant span-depth ratio and loaded with 2 equal concentrated loads. This was done for the purpose of comparison with the results of Newlin and Trayer, given by Eq. 7.18a, and the results of Dawley and Youngquist, given by Eq. 7.18b. For beam depths between 8 in. and 130 in., the variation of strength ratio with depth given by Bohannan, Curve 3, lies above that of Dawley and Youngquist, Curve 2. For $d > 130$ in. the reverse occurs. Since the form factor specified by NDS in editions prior to 1971 was based on Curve 2, the results of Bohannan's investigation confirmed that it was conservative. For the design of laminated beams, in the 20 to 60 in. range, a 5 percent increase in the depth factor given by NDS seemed justified.

The National Design Specifications (1) and the American Institute of Timber Construction (14) have adopted Bohannan's results for the determination of the variation of the strength ratio with depth in glued-laminated beams. This variation is known as the size factor. In keeping with the 12 in. depth convention, the preceding equation can be transformed easily into:

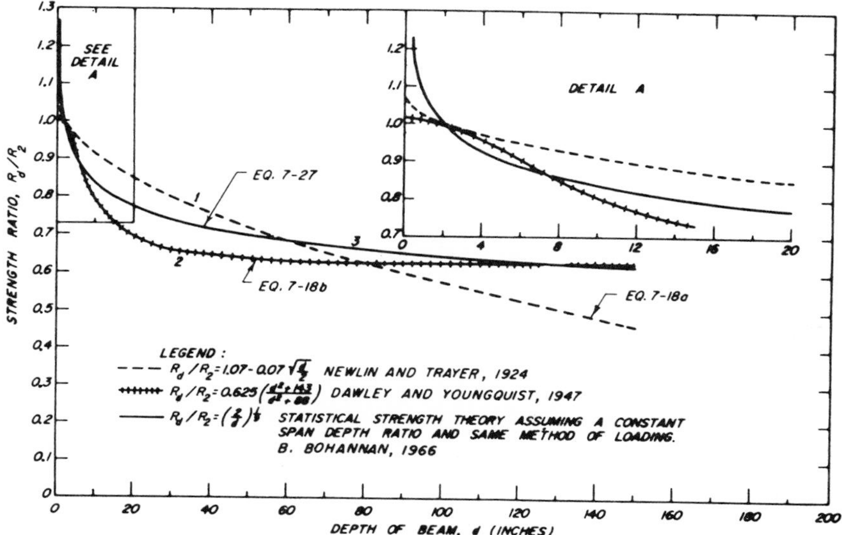

Figure 7.13. Strength radio, R_d/R_2 of beams having a depth, d, to beams having a depth of 2 inches. To compare curves one and two (based on older theories) to the statistical strength of material concept, curve three, a constant span-depth ratio and the same method of loading were assumed in calculating values in the latter theory. Adapted from Fig. 15 of Ref. 8.

$$C_F = \frac{R_d}{R_{12}} = \left(\frac{12}{d}\right)^{1/9} \tag{7.28}$$

The value of C_F given by Eq. 7.28 is considered accurate in the case of a simply-supported beam uniformly loaded and with a span-depth ratio $L/d = 21$. In all other cases, it may be considered as reasonably accurate.

Where more accuracy is desired one may calculate percentage changes in C_F using the values given in Table 7.2 for simply-supported spans. Thus, percentage changes are tabulated for three loading conditions and five span-to-depth ratios, L/d. Linear interpolation is permitted between the latter to obtain the percentage change in C_F for any given value of L/d. For continuous and cantilevered beams one may determine the size factor assuming equivalent (same maximum moment) simply-supported members with a uniformly distributed load. This should result in a slightly conservative size factor being applied to the design of span types other than simple spans (22).

Usually the change in C_F is quite small and hardly worth the effort spent beyond the use of Eq. 7.28, unless the beam has a small L/d ratio and is subjected to a single concentrated load. In the latter case an increase in C_F of almost 15 percent is possible.

7.5.1 Illustrative Example

Determine C_F for a 5⅛ × 24 in. glued-laminated girder, simply supported over a span of 28 ft, and loaded by beams at the middle-third points.

From Eq. 7.28 obtain:

$$C_F = \left(\frac{12}{d}\right)^{1/9} = \left(\frac{12}{24}\right)^{1/9} = 0.926$$

From Table 7.2, for $L/d = 28 \times 12/24 = 14$ and third-point loads obtain percentage changes of 2.3 and -3.2, respectively. Thus,

$$C_F = (0.926)\,(1 + 0.023)\,(1 - 0.032) = 0.917$$

The change is quite small and hardly worth the effort.

TABLE 7.2. VARIATION OF SIZE FACTOR, C_F.*

A. Loading Condition for Simply-Supported Beams:	Percent Change
Single Concentrated Load	7.8
Uniform Load	0
Third-Point Load	−3.2
B. Span-to-depth Ratio:	
7	6.3
14	2.3
21	0
28	−1.6
35	−2.8

*From Ref. 22

7.6 SUMMARY OF BENDING OF WOOD BEAMS AND PROVISIONS FOR DESIGN

The behavior of wood beams under flexure has been discussed in the light of various theories and experimental evidence. The linear elastic theory was found deficient in the explanation of various peculiarities of wood beams. The inelastic theory, based on the direct relation between bending stresses and strains and the stress-strain relations of clear wood in direct compression and tension, explained the discrepancies between the elastic theory and the tests. The theory was used to develop expressions for moment and curvature at various stages of loading, including failure. It is true that failure was assumed to occur upon attainment of the ultimate compressive strain; a fact that is likely to be true in specimens of clear wood but seldom in actual beams, because of characteristics that reduce tensile strength.

The ductility of clear wood beams is relatively small when compared to beams made of structural steel or reinforced concrete. In the case of commercial solid-sawn wood beams of low grade quality, that are weak in tensile strength, failure occurs before plastification of the compression zone of the beam. The moment-curvature and load-deflection relationships of such beams are linear with an abrupt end at ultimate. This type of behavior may be considered brittle. The use of Eq. 7.1 for the determination of modulus of rupture, i.e. the maximum bending stress at ultimate, is accurate for these beams.

The inelastic theory failed to explain the variation of the modulus of rupture with the depth and shape of wood beams. An explanation was given by the theory of supporting action which assumes that support of the extreme fiber in compression, against buckling and premature failure, is rendered by adjacent fibers. The depth effect was explained by means of the strain gradient of the section which, for a given ultimate strain, increases with decreasing depth of the section. The theory found application in the derivation of size and form factors presently used in design.

A more modern approach to the phenomenon is given by an application of the statistical strength theory to wood beams. Research on rectangular sections only, disclosed the fact that the effect depends not only on the depth of the beam, but also on its length and the method of loading.

7.6.1 Design Provisions

Present design practice of wood beams uses the following modified version of Eq. 7.1 to check that f, the extreme fiber stress of a given section subjected to a moment M, is equal to, or less than, $F_b\overline{\alpha}$; where F_b is the allowable extreme fiber stress in bending and $\overline{\alpha}$ is the duration of loading factor, see Section 2.13. Thus, a section is acceptable if:

$$f = \frac{M}{C_f\, S} \leqslant \overline{\alpha} F_b \tag{7.1a}$$

where S is the section modulus and C_f is the form factor for I and box sections given by Eq. 7.23. For rectangular beams $C_f = C_F$ and is given by Eq. 7.28. The variation of C_F with depth is given in Fig. 7.14, for sections with a depth smaller than 12 in., $C_F = 1$. For circular sections $C_f = 1.18$, and for square sections placed with diagonal vertically $C_f = 1.414$.

A distinction is made between engineered uses (single) and repetitive-member uses in defining the allowable unit stress, F_b, see Table A.2.5. This acknowledges the difference that exists between repetitive members such as joists, rafters, studs, planks and decking (that are spaced not more than 24 inches, are not less than 3 in number, and are joined by a floor or roof adequate to support the design load) where a load distribution capacity exists, and single members such as a beam,

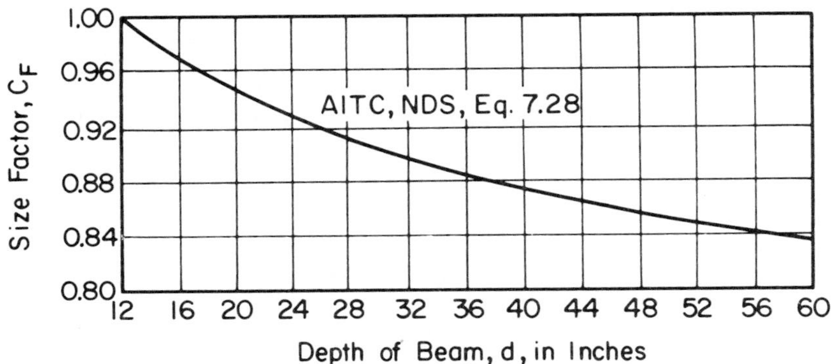

Figure 7.14. Variation of size factor with depth of beam: for rectangular sections only.

girder or post where each individual piece carries its full design load. The listed values of F_b for repetitive-member use are approximately 15 percent larger than for engineered use.

In the case of a beam subjected to concentrated loads, it is likely that lateral distribution to adjacent beams will make the loads that are actually carried smaller than the loads directly applied to the beam. Lateral distribution depends on the stiffness of the floor which is, in turn, dependent on the spacing of the supporting beams and on the kind of floor itself. Table 9.1 gives the fraction of concentrated load at midspan that is carried by a given beam. The evaluation of lateral distribution effects when the load is at one quarter of the span, usually a critical section for the determination of shear (see Section 7.9), is shown in Fig. 7.15. The latter gives the actual portion of load carried by the beam when the load is at one quarter of the span, as a function of the actual portion carried when the load is at midspan.

The design provisions of NDS (1) for bending are illustrated in the following example. It may be noted that the beam used in the example may correspond to a warehouse floor subjected to traffic of lifting and placement equipment.

7.6.2 Illustrative Example

Check the capacity in bending of a system of 4 × 10 surfaced-four-sides floor beams spaced 2 ft. apart. Each beam supports 50 lb./ft. of uniformly distributed load which includes its own weight, plus the weight of a 4-inch wood deck nailed to it. In addition, any beam is subject to a moving concentrated load of 6 kips due to normal loading conditions. The span is 10 ft. and the wood is such that allowable stresses in bending, as listed in Table A.2.5, are F_b = 1.95 ksi engineered uses and 2.25 ksi repetitive-member use. Wood is used at 15 percent maximum moisture content.

Lateral Load Distribution

Maximum moment in the beam occurs when the concentrated load is at midspan. However, not all of the load is carried by the beam directly below the load. A lateral distribution to the adjacent beams occurs due to the flexibility of the beam-floor system. The actual load on the beam may be determined using Table 9.1. For a spacing of the beams S = 2 ft., the actual load on the beam is P = (S/4)6, i.e., P = 3 kips.

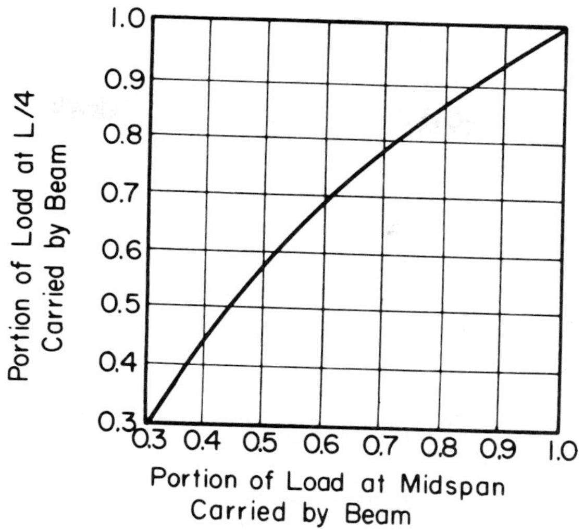

Figure 7.15. Determination of portion of concentrated load at ¼ point of span for a given portion at midspan, see Table 9.1. Adapted from Table A–1 of Ref. 1.

Maximum Moment at Midspan

For w = 0.05 k/ft. and P = 3 k at midspan, the maximum moment is given by:

$$M = \frac{1}{8}wL^2 + \frac{1}{4}PL = \frac{1}{8}(0.05)(10)^2 + \frac{1}{4}(3)(10) = 8.125 \text{ k-ft.} = 97.5 \text{ kip-in.}$$

Maximum Flexural Stress

Using Eq. 7.1a, see Section 7.6.1, with $C_F = 1(d < 12 \text{ in.})$ and sections modulus $S = (3.5)(9.25)^2/6 = 49.91 \text{ in.}^3$, yields:

$$f = \frac{M}{S} = \frac{97.5}{49.91} = 1.95 \text{ ksi}$$

The allowable stress is 1.95 ksi multiplied by the load duration factor $\overline{\alpha} = 1$, corresponding to normal loading conditions. Since the sheathing provides full lateral restraint to the beams, see Section 7.10, there is no need to modify the allowable stress as given. However, note that the allowable stress used is that specified for engineered uses, and not 2.25 ksi which is specified for repetitive-member uses. This is due to the fact that the repetitive condition of the member has already been used in the allowance for lateral distribution of the concentrated load.

The beam meets the requirements for flexure. However, the capacity to resist shear, the maximum deflection and the end support of the beam must be checked before final acceptance of the solution. This is done elsewhere in the Chapter; see examples in Section 7.9, 7.11 and 7.12.

7.7 DISTRIBUTION OF SHEAR STRESSES IN UNCHECKED BEAMS

The stress distribution in a section of an elastic beam subjected to pure flexure is linear. If bending is produced by transverse loading the theory of elasticity shows that, in the presence of shear stresses, the linearity of the stress distribution is only approximately true. However, this is ignored in the derivation (4) of the classical distribution of shearing stresses in a given section. The following relation is obtained

$$v = \frac{VQ}{Ib} \tag{7.29}$$

where: v = shear stress at a given level of the section situated at a distance y from the centroid.
V = total shear force at the section.
Q = first moment of area above the given level, about the centroid of the section.
I = moment of inertia of the section about its centroid.
b = width of the section at the given level.

In view of the fact that the flexural stress distribution in wood beams is almost linear up to the proportional-limit stage of loading, it seems acceptable to determine the distribution of shear stresses by means of Eq. 7.29. For a rectangular section of width b and depth d, $I = bd^3/12$ and the first moment of area above a level situated at a distance y from the centroid of the section is

$$Q = \frac{b}{2}\left(\frac{d^2}{4} - y^2\right) \tag{7.30}$$

Substituting the preceding values in Eq. 7.29 yields

$$v = \frac{V}{bd}\left(\frac{3}{2} - 6\frac{y^2}{d^2}\right) \tag{7.31}$$

The shear distribution given by Eq. 7.31 is parabolic as shown in Fig. 7.16a with a maximum value

$$v = \frac{3}{2}\,\frac{V}{bd} \tag{7.32}$$

at the neutral axis. Hence, the maximum shear stress in a rectangular section is 50 percent larger than the average stress. The shear-stress distribution for a box or I section with n webs of thickness t_w is shown in Fig. 7.16b. The maximum stress is given by Eq. 7.29 where $b = nt_w$ and Q is determined for $y = 0$. Usually, in box or I sections with plywood webs, the maximum shear stress is only slightly larger than a fictitious average shear stress in the webs given by V/nt_wd. For a circular cross-section the maximum shear stress is 33⅓ percent larger than the average stress.

The preceding theory is applicable to wood beams that are free of checks, shakes or splits. Glued-laminated beams and beams of box or I sections with plywood webs are usually free of these strength-reducing characteristics, and for them the classical theory applies. However, solid-sawn wood beams are subjected to creation of checks during seasoning, the presence of which violates the assumptions made in the derivation of Eq. 7.29. Hence, the actual distribution of shearing stresses in checked beams may not be obtained by the classical method.

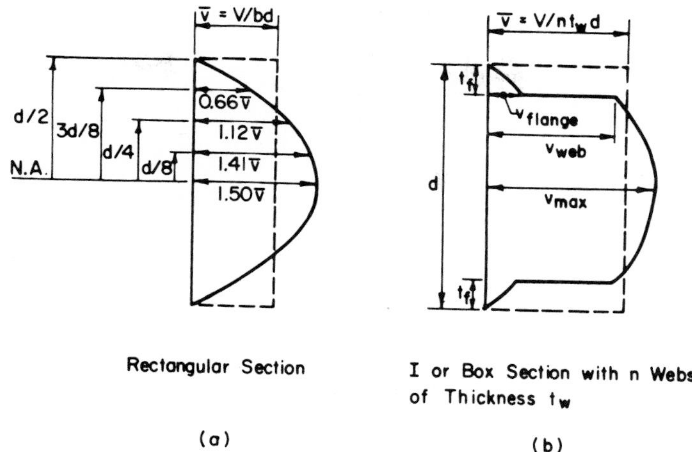

Figure 7.16. Distribution of shear stresses in a section of a beam subjected to transverse loading.

7.8 DISTRIBUTION OF SHEAR STRESSES IN CHECKED BEAMS

The random presence of checks of various lengths and penetrations in solid-sawn wood beams makes the determination of shear distribution a highly indeterminate process. In 1934, Newlin et al (15) of the U.S. Forest Products Laboratory at Madison, Wisconsin, presented a rational method of analysis of wood beams containing checks in the vertical faces. The results of this investigation have influenced design specifications to date, as will be shown later.

The use of the method given in the preceding Section for the computation of horizontal shearing stresses at the neutral plane of checked wood beams during testing, gave the FPL investigators stresses 2 to 3 times the ultimate shearing stress of the material. This obvious impossibility only pointed out the discrepancy between the actual maximum shearing stress and the one given by Eq. 7.32. There was no doubt that the actual maximum shearing stress was smaller.

The analysis of a checked beam was accomplished using the equations of the Theory of Elasticity. This required the creation of a model. Fig. 7.17a shows a section of the simply-supported beam used in the analysis. Longitudinal checks of uniform depth in the middle of the lateral (vertical) faces run all along the span of the model beam. The latter was considered subjected to a concentrated load P at a distance x from one of the supports. The solution of the equations of plane stress used in the analysis led to the following relation:

$$R = B + \frac{A}{x^2} \tag{7.33}$$

where: R = reaction at a given support a distance x from the concentrated load P
B = (2/3) bdv
A = $(Ebd^2/6)u$
v = mean shear-stress over full width of beam
u = mean longitudinal displacement.

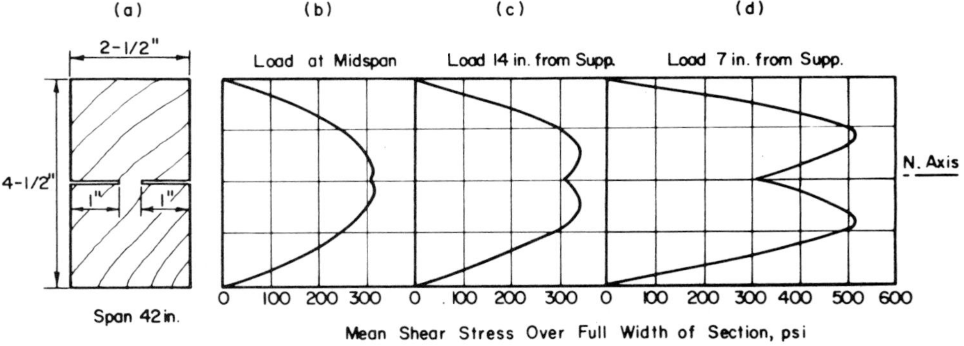

Figure 7.17. Theoretical variation of horizontal shear stress, with distance from neutral axis of checked beams, for various positions of a single concentrated load. From Fig. 2 of Ref. 15.

Equation 7.33 was one of the significant results of the analysis, and expresses the reaction at the nearer support as the sum of two portions, B, which is associated with shearing stress in the neutral plane in the usual way, and A/x^2, which is not associated with such a stress. The latter portion becomes of rapidly increasing importance as the load approaches the support.

The theoretical variation of the mean shearing stress over the full width of the beam, with depth, is shown in Fig. 7.17 for three positions of the concentrated load P. It should be noted that the stress as plotted is considered to be uniform across the width of the section. The actual average stress at the neutral plane is equal to that plotted in Fig. 7.17 times the ratio of the total width of the section to the width of the throat at the neutral plane. From the curves it is clear that the shear stress is not a maximum at the neutral plane but at points somewhere above and below this plane. The shape of these curves prompted Newlin et al (15) to give the following interpretation to the two terms of Eq. 7.33. The first term, B, is the reaction associated in conventional theory of beams with the mean shearing stress, v, in the neutral plane. The second term, A/x^2, may be attributed to the action of the upper and lower halves of the beams as two independent beams. Hence, the two parts B and A/x^2 of the reaction are referred to as the "single-beam" and "two-beam" portions of the reaction, respectively. It is the presence of the two-beam portion of the reaction, which increases rapidly as the point of loading approaches the support that accounts for the fact, found in all the tests, that the point of application of the minimum load for failure by shear is at a considerable distance from the support.

The variables A and B of Eq. 7.33 were determined by means of tests of carefully matched beams loaded to failure. The following practical relation between the single-beam and the two-beam portions of the reaction was obtained:

$$\frac{B}{\left(\frac{A}{x^2}\right)} = \frac{(x/d)^2}{2} \tag{7.34}$$

where x is the distance of the concentrated load P to the support. From statics, the reaction R at the support is:

$$R = P\,\frac{L - x}{L} \tag{7.35}$$

where L is the length of the beam. Substituting the values of A/x^2 from Eq. 7.34 and R from Eq. 7.35 in Eq. 7.33 yields:

$$B = \frac{P(L - x)\,(x/d)^2}{L[2 + (x/d)^2]} \tag{7.36}$$

For the case of a simple beam uniformly loaded throughout the entire span with a total load W, a very close approximation to the actual expression for B is given by

$$B = \frac{9}{10}\,\frac{W}{2}\left(1 - 2\frac{d}{L}\right) \tag{7.37}$$

In view of the fact that there was approximately 10 percent of "two-beam reaction" in the tests made at Forest Products Laboratory from which safe stresses were derived, Newlin et al (15) recommended the inclusion of a factor 10/9 in Eqs. 7.36 and 7.37. Using this provision, and changing B to V′, to conform with NDS (1) notation, the preceding equations can be written as follows:

$$V' = \frac{10}{9}\,\frac{P\,(L - x)\,(x/d)^2}{L\,[2 + (x/d)^2]}, \text{ for concentrated loads.} \tag{7.36a}$$

and

$$V' = \frac{W}{2}\left(1 - \frac{2d}{L}\right), \text{ for uniform loads.} \tag{7.37a}$$

Equation 7.36a may be represented in dimensionless form, V'/P versus x/L, as shown in Fig. 7.18 for various values of L/d between 6 and 20. The maximum value of V'/P for $L/d = 6$ occurs at $x/L = 0.36$ as $V'/P = 0.50$ and for $L/d = 20$ at $x/L = 0.19$ as $V\ /P = 0.79$. The maximum value of V'/P for a given span ratio L/d may be obtained by setting $dV'/d(x/L)$ to zero. The following relation is obtained:

$$\left(\frac{\bar{x}}{L}\right)\left[6 + \left(\frac{L}{d}\right)^2\left(\frac{\bar{x}}{L}\right)^2\right] = 4 \tag{7.38}$$

where $\bar{x}$ is the distance of the concentrated load P to the support which makes V′ a maximum. If $X = \bar{x}/d$, Eq. 7.38 may be transformed into the more convenient expression

$$\frac{L}{d} = \frac{X}{4}\,(6 + X^2) \tag{7.38a}$$

the solution of which is shown in Fig. 7.19 with L/d and $X = \bar{x}/d$ as abscissa and ordinate, respectively. Hence, for any given L/d, the value of $\bar{x} = Xd$ can be obtained graphically from Fig. 7.19. In addition, two approximate rules used by NDS, namely: 1) the load at 3d from the support, and 2) the load at L/4 from the support, are also plotted in Fig. 7.19. The average of rules 1 and 2, which is also given in Fig. 7.19 is the best approximation to the exact location of the concentrated load in checked beams.

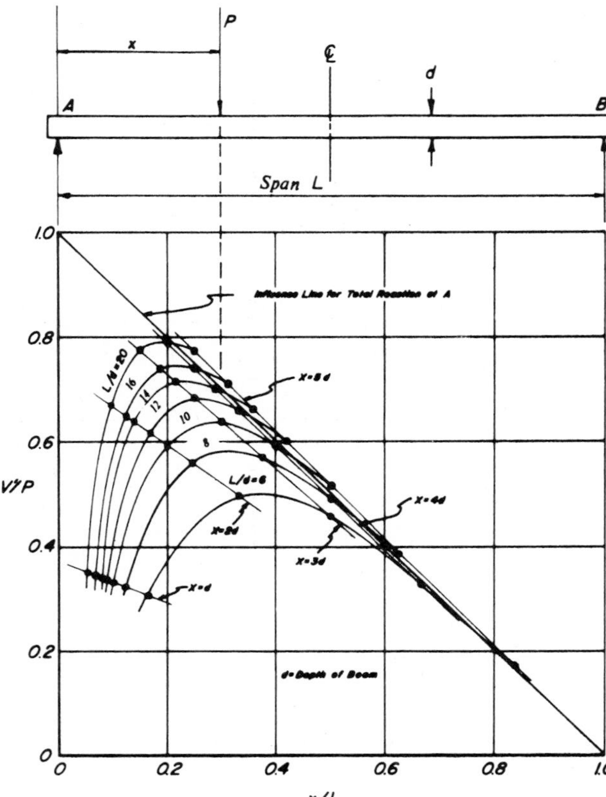

Figure 7.18. Accurate determination of shear force V' due to "single beam" action in a checked solid-wood beam subjected to a concentrated load P.

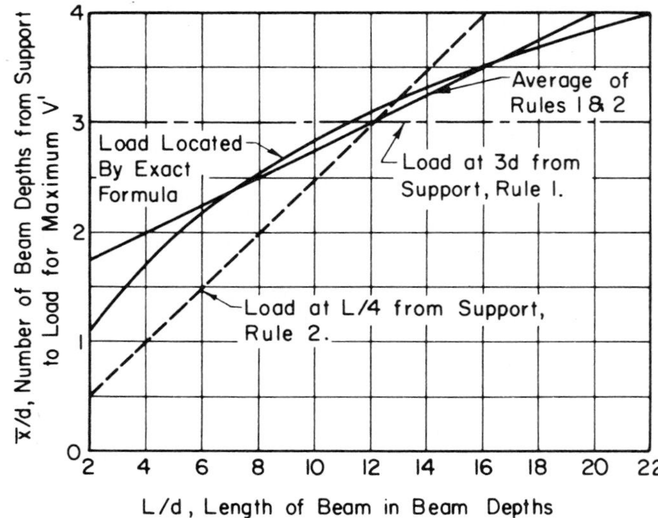

Figure 7.19. Determination of distance of concentrated load from support, for maximum shear in checked beams.

7.9 SUMMARY OF SHEAR IN WOOD BEAMS AND PROVISIONS FOR DESIGN

It is reasonable to assume that the distribution of shear stresses in a section of a wood beam subjected to bending follows the elastic theory, if the member is free of checks. This is the case in glued-laminated beams and box or I beams made of plywood webs and seasoned unchecked lumber or laminated wood flanges. For these cases, Eq. 7.29 leads to shear-stress distributions as shown in Fig. 7.16 with a maximum stress at the neutral plane of the section.

Solid wood beams containing checks have a shear-stress distribution influenced by a "two-beam" action which may reduce the mean shear stress over the full width at the neutral plane considerably, see Fig. 7.17. Investigations at Forest Products Laboratory (15) showed that the reaction at a support in checked beams is the sum of a conventional "single-beam" reaction and a "two-beam" reaction. A relation between the two portions was found, see Eq. 7.34, and from it a final expression, see Eqs. 7.36 and 7.37 for the actual portion of reaction transmitted through conventional simple-beam action.

7.9.1 Design Provisions

Present design practice of wood beams uses Eq. 7.29 to check that the maximum horizontal shear stress, v, of a given section subjected to a shear force, V, is equal to, or less than, an allowable stress, F_v, affected by the duration of loading factor, $\bar{\alpha}$, see Section 2.13. Thus, a section is acceptable if

$$v = \frac{VQ}{Ib} \leqslant \bar{\alpha}F_v \qquad (7.29a)$$

For rectangular sections Eq. 7.29a becomes:

$$v = \frac{3}{2}\,\frac{V}{bd} \leqslant \bar{\alpha}F_v \qquad (7.32a)$$

The value of F_v is given in Tables A.2.5 and A.3.1, for solid-sawn and glued-laminated beams, respectively.

When calculating the shear V, a designer may take into account any relief due to lateral distribution to adjacent members, see Table 9.1 and Fig. 7.15, and consider only the actual portion of loading carried by the members. Also, all loads within a distance from either support equal to the depth of the beam may be neglected. A single moving load, or one moving load that is considerably larger than any of the others, should be placed at a distance from the support equal to the depth of the beam, keeping the others in their normal relation. Two or more moving loads of about equal weight and in proximity, should be placed in the position that produces the highest shear; again neglecting any load within a distance from the support equal to the depth of the beam.

Most wood beams having a span to depth ratio, $L/d > 10$, for which design is governed by flexure and deflection requirements, will meet the provisions for shear easily. However, this may not be true for solid-sawn beams of small span-depth ratios and subjected to the action of heavy concentrated loads. For these special cases, which account for a small percentage of beams designed in actual practice, the results of the investigation by Newlin et al (15) are used. Thus, for solid-sawn beams only, which do not satisfy Eq. 7.32a, the shear V may be determined as before; except that if there is a single moving load or one moving load that is considerably larger than any of the others, it should be placed at a distance from the support equal to 3 times the depth of the beam, or at the quarter point, whichever is closer. All other loads may be considered in the usual manner. The allowable horizontal shear stress F_v, against which v is checked, is increased to the values given in Table 7.3. This is reasonable, since very large adjustment factors were used in the determination of allowable shear stresses, see Table 2.4, to compensate uncertainty in the calculation of shear stresses in checked beams.

TABLE 7.3. ALLOWABLE HORIZONTAL SHEAR STRESS F_v FOR CHECKED SOLID-SAWN BEAMS.†

	Maximum Moisture Content		
	Unseasoned	19 percent	15 percent
Aspen	85	90	95
Balsam Fir	85	95	95
Black Cottonwood	70	75	80
California Redwood	115	120	130
Coast Sitka Spruce	90	95	100
Coast Species	90	95	100
Douglas Fir—Larch	130	140	145
Douglas Fir—South	130	140	145
Eastern Hemlock—Tamarack	120	130	135
Eastern Spruce	95	105	110
Eastern White Pine	90	95	100
Eastern Woods	85	95	95
Engelmann Spruce/Alpine Fir	95	105	110
Hem—Fir	105	110	115
Idaho White Pine	95	100	105
Lodgepole Pine	95	105	110
Mountain Hemlock	130	140	150
Northern Aspen	90	95	100
Northern Pine	100	105	110
Northern Species	90	95	100
Northern White Cedar	85	95	100
Ponderosa Pine—Sugar Pine	100	105	110
Red Pine	100	110	115
Sitka Spruce	105	115	120
Southern Pine	125	135	145
Spruce—Pine—Fir	95	105	110
Western Cedar	100	105	110
Western Hemlock	125	135	145
Western White Pine	90	100	105
White Woods (Western Woods, West Coast Woods, Mixed Species)	95	100	105

† From National Design Specification Art. 3.4.4, See Ref. 1.

The preceding method will take care of most solid-sawn beams that do not meet the initial requirements. However, if in spite of the reduction in value of V and and the increase in allowable stress F_v, the section still does not meet the shear requirement, the more accurate value of V = V′ may be used as given by Eqs. 7.36a and 7.37a. The shear stress v is determined as before, and checked against the values of F_v given in Table 7.3.

The provisions of the American Association of State Highway and Transportation Officials (16), for the determination of the critical shear force acting on the section of a wood beam or stringer, are given in the Illustrative Example of Section 9.3. The design provisions for shear are illustrated in the following example.

7.9.2 Illustrative Example

Check the capacity against horizontal shear stress of the beam in the example of Section 7.6, using the provisions of NDS. Assume that the allowable shear stress is 0.095 ksi.

Rough Determination of Maximum Shear Stress

Load placed at a distance from the support equal to the depth of the beam, i.e., x = 9.25 in. Due to the closeness of the moving load to the support, no lateral distribution factor may be assumed.

$$V = \frac{P}{L}(L - x) + \frac{1}{2}w(L - 2x), \text{ where } P = 6\text{k}, L = 10 \text{ ft and } x = 9.25/12 \text{ ft}$$

$$V = \frac{6}{10}\left(10 - \frac{9.25}{12}\right) + \frac{1}{2}(0.05)\left(10 - 2 \times \frac{9.25}{12}\right) = 5.54 + 0.21 = 5.75\text{k}$$

$$v = \frac{3}{2}\frac{V}{A} = \frac{3}{2}\frac{5.75}{3.5 \times 9.25} = 0.266 \text{ ksi} > 0.095 \text{ ksi, N.G.}$$

However, the design provisions covered in Section 7.9.1 allow a closer determination of maximum shear stress and an increased allowable stress, see Table 7.3 for solid-sawn beams which do not qualify as above. For the given beam assume that the increased allowable shear stress is 0.145 ksi.

Closer Determination of Maximum Shear Stress

In view of the fact that this is a solid-sawn beam which did not qualify under the preceding provision, the end shear V may be determined by placing the load at a distance from the support equal to 3d, i.e. 27.25 in., or L/4, i.e. 30 in., whichever is closer. The position of the load close to the quarter of the span justifies lateral distribution to the adjacent beams, as indicated in Fig. 7.15. The portion of the concentrated load at midspan carried by the beam was 0.5 of the load. Hence, when the load is applied at the quarter of the span, Fig. 7.15 indicates that the beam carries 0.58 of the load. Thus,

$$V = \frac{0.58\,P}{L}(L - 3d) + V_D$$

$$V = \frac{0.58 \times 6}{10}\left(10 - \frac{27.25}{12}\right) + 0.21 = 2.69 + 0.21 = 2.90\text{k}$$

$$v = \frac{3}{2}\frac{V}{A} = \frac{3}{2}\frac{2.90}{3.5 \times 9.25} = 0.134 \text{ ksi} < 0.145 \text{ ksi}$$

Accurate Determination of Maximum Shear Stress

There is no need to calculate this, as the section has met the requirements for shear. However, for the sake of illustration, it is convenient to proceed with the final recourse of solid-sawn beams to meet the required shear capacity.

For $L/d = 120/9.25 = 13$, Fig. 7.19 yields $\bar{x}/d = 3.2$. Therefore, $\bar{x} = 3.2 \times 9.25 = 29.6$ in. Substituting in Eq. 7.36a and adding the dead-load component of shear, $V_D = 0.21$ k, yields

$$V' = \frac{10}{9} \frac{0.58 \times 6}{120} \frac{(120 - 29.6)(3.2)^2}{[2 + (3.2)^2]} + 0.21 = 2.43 + 0.21 = 2.64\text{k}$$

or directly from Fig. 7.18, by interpolation for $L/d = 13$, $V'/P = 0.7$. Hence, $V' = 2.42$k due to the concentrated load only. Thus,

$$v = \frac{3}{2} \frac{2.64}{3.5 \times 9.25} = 0.122 \text{ ksi} < 0.145 \text{ ksi}$$

The beam meets the requirements for shear.

7.10 LATERAL STABILITY OF WOOD BEAMS

A beam subjected to bending may fail prematurely by lateral-torsional buckling (31). Consider the simply-supported laterally-unrestrained beam shown in Fig. 7.20 which is subjected to bending caused by end couples M_x. In addition to bending about the strong axis of its cross-section the beam simultaneously twists and bends laterally, see Section 1.1 and plan view.

As a result, a lateral bending moment, $M_x\beta$, and a twist moment, $M_x\theta$, are induced. Three independent equations of equilibrium can be established between the moments M_x, $M_y = M_x\beta$ and $M_z = M_x\theta$ and displacements v, u and β, respectively. The solution of these equations yields an expression (17) for the critical value of end moments, M_{cr}, that induce lateral torsional buckling, as follows:

$$M_{cr} = \frac{\pi}{L} \sqrt{\frac{E\,I_y\,G\,J}{1 - \frac{I_y}{I_x}}} \qquad (7.39)$$

where:
- $E\,I_y$ = bending stiffness about weak axis of section
- $G\,J$ = torsional stiffness of section
- I_y/I_x = ratio of moments of inertia of section about weak and strong axis, respectively
- L = length of beam

For a rectangular section of width b and depth d, $I_y/I_x = b^2/d^2$ and $J = (db^3/3)(1 - 0.63\,b/d)$. Substituting these values in Eq. 7.39 yields:

$$M_{cr} = \frac{\pi}{L} \frac{E\,d\,b^3}{6} \sqrt{\frac{G}{E}} \sqrt{\frac{1 - 0.63\,(b/d)}{1 - (b/d)^2}} \qquad (7.40)$$

as given by Hooley and Madsen (18).

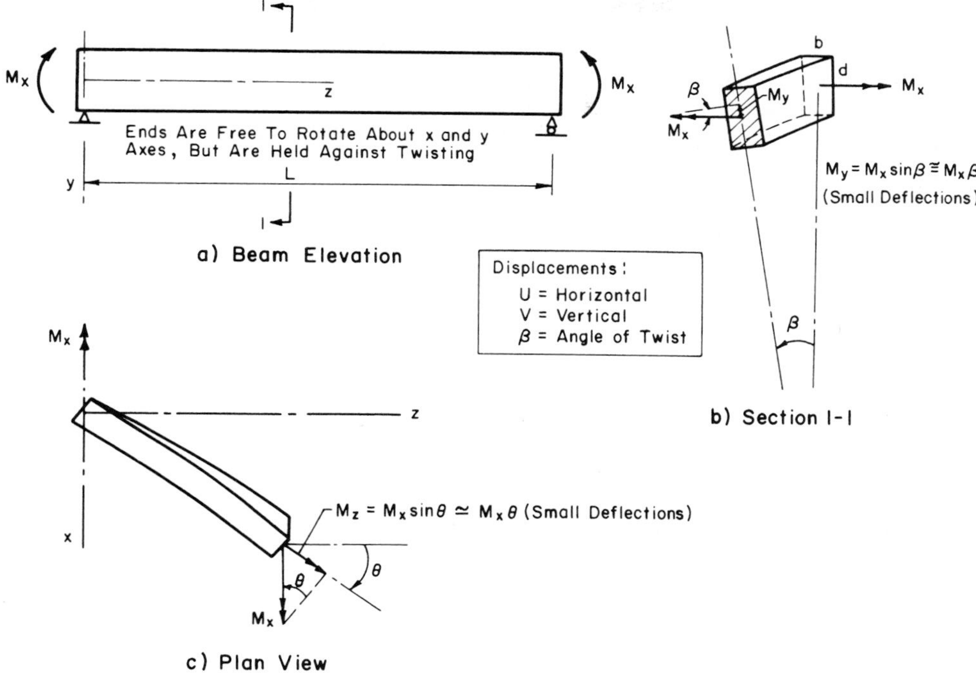

Figure 7.20. Lateral-torsional buckling of a wood beam subjected to end moments, M_x. Moments M_y and M_z induce lateral bending and torsion, respectively.

Now, call R the right hand radical of the equation. It can be shown, by solving dR/d (b/d) = 0, that the minimum value of R is equal to 0.943, when b/d = 0.354. Also, let G/E = 0.064, a low value for wood (12). If the critical stress, f_{cr}, is defined by $f_{cr} = M_{cr}/(bd^2/6)$ substitution of the preceding terms in Eq. 7.40 yields:

$$f_{cr} = \frac{0.75\ E}{Ld/b^2} \tag{7.41}$$

Expressions for f_{cr} are given by Hooley and Madsen (18) for other support conditions and types of loads; their results were verified by extensive laboratory testing. Also, the relative position of the load with respect to the longitudinal axis of the beam affects the results. As might be expected, loads applied on top of a beam result in the smallest values of critical stress, f_{cr}, and cause lateral instability sooner than loads applied closer to or below the longitudinal axis of the beam. A table is given in the paper (18) for evaluation of this effect.

The preceding equations are true, only if lateral-torsional buckling occurs in the Eulerian range, at a stress smaller than the proportional-limit stress of the given wood. Some inelastic behavior occurs if the critical stress is above this limit; for very short beams, or long beams with continuous lateral support, full development of strength may be expected without lateral torsional buckling.

7.10.1 Design Provisions

Present design provisions originate from the investigation by Hooley and Madsen (18). The following parameters are defined:

$$C_s = \sqrt{\frac{d\ell_e}{b^2}}$$

$$C_k = \sqrt{\frac{3E}{5F_b}} \tag{7.42}$$

where: C_s = slenderness factor; may not exceed 50
C_k = beginning of Eulerian range; the stress is equal to 2/3 of proportional-limit stress
ℓe = effective length of beam, a function of ℓ_u, see Table 7.4
ℓ_u = unsupported length of beam, in.
F_b = allowable stress in bending, Table A.2.5.

TABLE 7.4. LATERAL STABILITY OF RECTANGULAR BEAMS DETERMINATION OF EFFECTIVE LENGTH OF COMPRESSION FLANGE.†

Type of Span	Type of Load	ℓ_e/ℓ_u
Single	Conc. at midspan	1.61
Single	Unif. distributed	1.92
Single	Equal end Moments	1.84
Cantilever	Conc. at end	1.69
Cantilever	Unif. distributed	1.06
Single or Cant.	Any	1.92, Safe.

†From National Design Specifications, See Ref. 1.

Three possible ranges in behavior are defined as a function of the slenderness factor, C_s. For $C_s \leqslant 10$, i.e. for short beams, the full allowable unit stress in bending, F_b, may be used. In the intermediate beam rage $10 < C_s \leqslant C_k$, the allowable unit stress in bending, $F_{b'}$, shall be determined from the following formula

$$F_{b'} = F_b\left[1 - \frac{1}{3}\left(\frac{C_s}{C_k}\right)^4\right] \tag{7.43}$$

and for the range $C_k < C_s \leqslant 50$, i.e. for long beams, the allowable unit stress in bending, $F_{b'}$, shall be determined by the following formula

$$F_{b'} = \frac{0.4\,E}{C_s^2} \tag{7.44}$$

The variation of the ratio $F_{b'}/F_b$ with C_s is shown in Fig. 7.21 for a given case. The three possible ranges in behavior can be easily identified. It can be seen that Eq. 7.43 plots as a transition curve between long and short-beam behavior. $C_s = C_k$ defines the beginning of the Eulerian range, or the long-beam behavior; $C_s = 10$, the end of the short-beam range. A small vertical step is necessary for continuity at $C_s = 10$, caused by the fact that the transition curve gives $F_{b'}/F_b <$ 1 for any value of C_s other than zero.

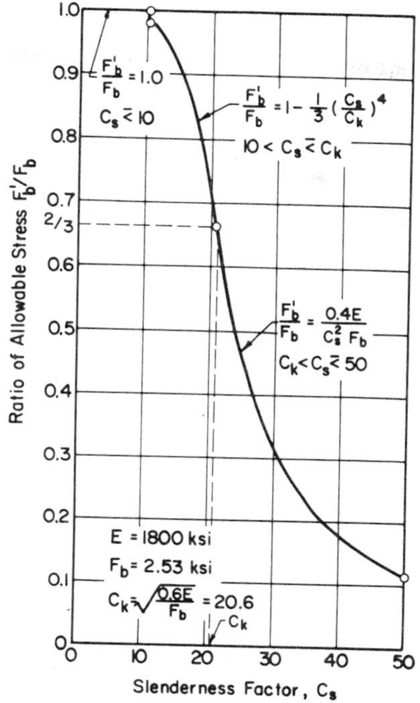

Figure 7.21. Variation of allowable stress in bending, F_b, with slenderness factor, $C_s = \sqrt{\frac{d\ell_e}{b^2}}$.

The design value of extreme fiber in bending, $F_{b'}$, determined in accordance with these formulas need not be modified by the size factor, C_f, except for the short-beam range, $C_s < 10$. In no case shall $F_{b'}$ exceed the full design value at extreme fiber in bending, F_b, modified by size factor, C_f.

Experience shows that simply-supported beams, laterally-restrained all along the length by decking, are not subject to lateral-torsional buckling; their allowable bending stress is $F_{b'} = F_b$ as if $C_s = 0$. Girders supporting a system of joists, that restrains them laterally at a reasonable spacing, are seldom affected and usually $F_{b'} = F_b$, although checking of Eqs. 7.42 through 44 is required.

In cantilevered and continuous construction, or cantilever-suspended systems (Section 11.8), compression stresses are generated in the bottom portion of the beams. These regions may be of substantial lengths and are usually laterally unrestrained. It is here that lateral instability may cause the value of the design bending stress, $F_{b'}$, to be seriously reduced below F_b. This will result in larger beam cross-sections unless lateral bracing is implemented to the bottom portions of the beam which are subjected to compression stresses.

The lateral stability of beams with build-up cross sections, like box or I beams with plywood webs and wood flanges, is complex and involved. Tests at the Forest Products Laboratory (12) have indicated that box and I beams of usual proportions present no undue problems of lateral stability if the spans are not excessive, and the ratio of moment of inertia about the neutral axis to moment of inertia about an axis perpendicular to the neutral axis does not exceed 30. Table 7.5 contains conservative rules (12) that may be used to check a design for lateral stability. The design specifications (19) for box and I beams using plywood webs are contained in Table 7.5.

TABLE 7.5. LATERAL STABILITY OF I AND BOX BEAMS.†

Range of I_x/I_y:	Lateral Support Required:
Less than 5	None
5 to 10	Compression flange at supports
10 to 20	Tension and compression flanges at supports
20 to 30	One edge either top or bottom, held in line
30 to 40	Bridging or bracing at S < 8 ft.
Greater than 40	Continuous restraint of compression flange

†From Wood Handbook, see Ref. 12, or Plywood Design Specification, Supplement No. 2, see Ref. 19

A design aid is given in Appendix A.7.1. Also, see Section 10.2.3.5 for an example.

7.10.2 Modification of $F_{b'}$ for Duration of Loading

Duration of loading, other than normal, affects the design stress in bending but not the modulus of elasticity of wood. Consider a loading condition such that its modification factor is $\overline{\alpha}$, see Section 2.13. It follows that the design bending stress, F_b, may be taken as $\overline{\alpha}\ F_b$ but E should remain the same. As a result, the formula for determination of $F_{b'}$ for long beams, i.e., $F_{b'} = 0.4\ E/C_s^2$, remains the same as before. However, C_k, the lower limit of slenderness factor at which long beam behavior begins, is changed. Instead of determining C_k by means of Eq. 7.42 it is necessary to use

$$C_k = \sqrt{\frac{3E}{5\ \overline{\alpha}\ F_b}} \tag{7.42a}$$

This changes the lower limit for long beam behavior making it smaller as $\overline{\alpha}$ increases. Only for the case of permanent loading conditions, were $\overline{\alpha} = 0.9$, would the value of C_k increase.

The intermediate beam range remains the same, i.e. between a slenderness factor of 10 and the modified value of C_k. Determination of $F_{b'}$ in this range requires modification of Eq. 7.43 as follows:

$$F_{b'} = \overline{\alpha}\ F_b\left[1 - \frac{1}{3}\left(\frac{C_s}{C_k}\right)^4\right] \tag{7.43a}$$

where C_k is given by Eq. 7.42a.

Finally, in the short beam range, $C_s < 11$, instead of $F_{b'} = F_b$ used $F_{b'} = \overline{\alpha}\ F_b$.

The complete variation of $F_{b'}$ with C_s for a given wood is shown for $\overline{\alpha} = 1.15$ (wind) in Fig. 7.22. For the purposes of comparison the variation for $\overline{\alpha} = 1.0$ is also shown. It is clearly noted that the main effect of an increase in $\overline{\alpha}$ is in the short range, for which the full increase of $F_{b'}$ to $\overline{\alpha}\ F_{b'}$ is available. The increase is gradually reduced in the intermediate range and tapers off to zero at a slenderness factor C_k. The graph also shows that the long beam range is greater for the case of $\overline{\alpha} > 1$ at the expense of the intermediate beam range which has become smaller.

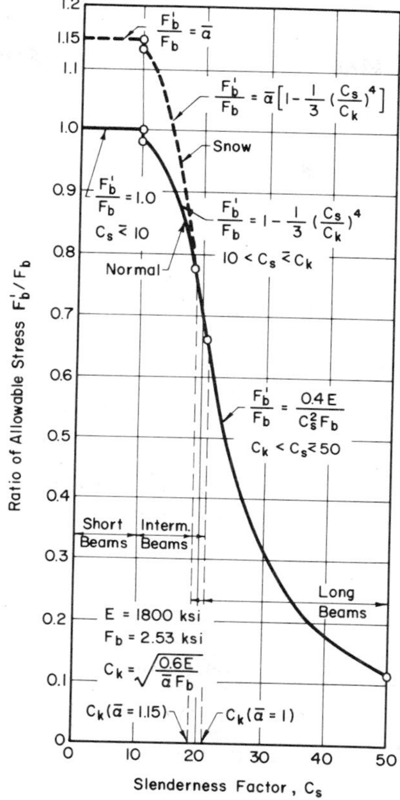

Figure 7.22. Effect of load-duration factor $\bar{\alpha}$ on allowable stress in bending, $F_{b'}$.

As in the case of normal loading conditions the design value at extreme fiber in bending, $F_{b'}$, determined in accordance with these formulas for values of $\bar{\alpha}$ other than 1.0 need not be modified by the size factor, C_f, except for the short beam range, $C_s < 10$. In no case, again, shall $F_{b'}$ exceed the full design value at extreme fiber in bending, F_b, modified by size factor, C_f, and load duration factor, $\bar{\alpha}$.

7.11 DEFLECTION OF WOOD BEAMS

Under service load conditions wood beams behave like linear elastic members. The classic methods (20) for the determination of displacements in elastic structures due to bending can be used to find maximum deflections in wood beams. A collection of useful formulas is given in references 21 and 22 and those for a few selected cases are shown in Fig. 7.23.

In addition to deflections due to bending strains, wood beams are subjected to shear deflections due to the integrated effects of shear strains. The second term of the formulas in Fig. 7.23 account for the shear deflections. There, the coefficient κ is 6/5 for rectangular sections and 10/9 for

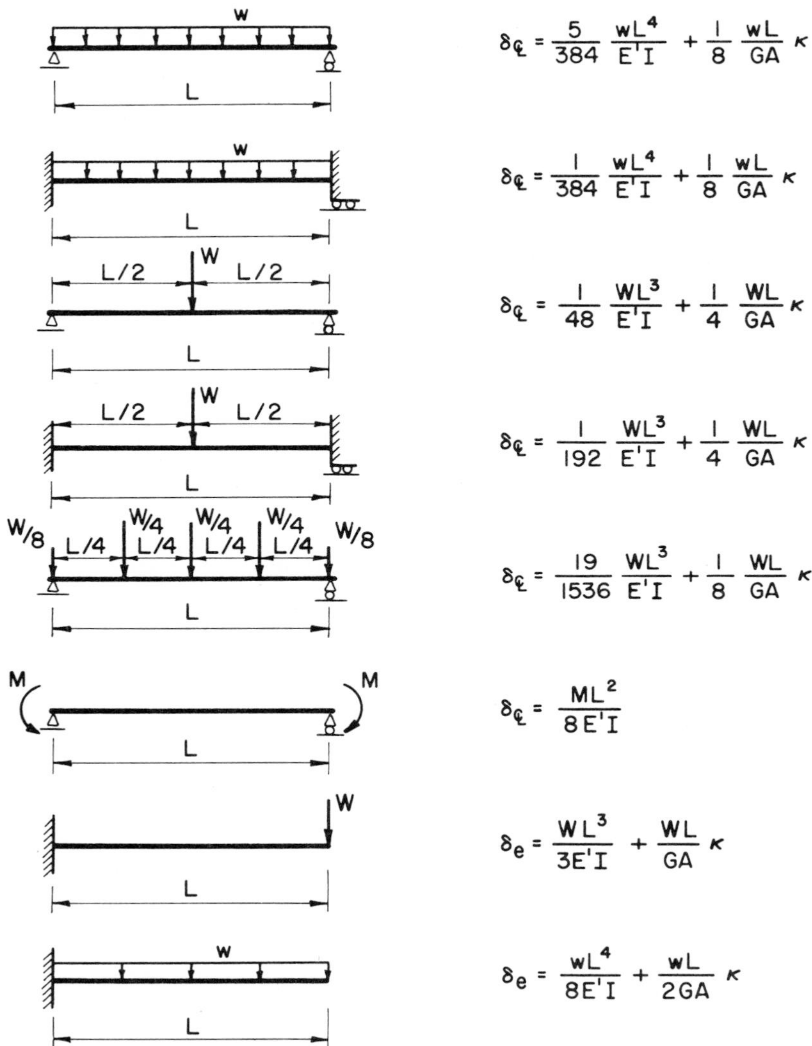

Figure 7.23. Deflections for various beam loadings.

circular sections; A is the cross-sectional area of the beam; and G is the shear modulus or modulus of rigidity which may be taken approximately as E′/16, where E′ is the true modulus of elasticity.

Generally, shear deflections may be as much as 10 percent of the total deflection of a beam. However, designers usually don't calculate shear deflections; they are included in the deflections due to bending stresses. This is possible because the listed modulus of elasticity, E, for any given species of wood, is an apparent value that is about 10 percent smaller than the true modulus of elasticity, E′. When using this approximate method substitute for E for E′ in the first term of the formulas in Fig. 7.23 and neglect the second term.

Only in some cases, such as box beams and I beams with plywood webs, or short beams subjected to large concentrated loads, may the approximate method render inaccurate results. For these cases an actual independent calculation of the total deflection, including shear deflections, is in order; see Section 10.3.2 or Reference 19. Additional examples of calculation of shear deflection are given in Section 7.11.1, which follows, and Sections 7.15.7 and 7.15.9.

There is a creep effect in wood beams that may not be ignored. Initial deflections increase with time. This is specially true in green timbers, which may sag appreciably, if allowed to season under load; however, partially seasoned material may also sag to some extent. In thoroughly seasoned wood members, there are small changes in moisture content but little permanent increase with time (12). It may be estimated that the initial deflection of unseasoned wood due to permanent loading may double with the passage of time. For glued-laminated timber and seasoned sawn lumber a factor of 1.5 affecting the deflection due to permanent loading is usually enough.

Recommended deflection limitations for solid-sawn and glued-laminated beams (23) are given in Table 7.6. In addition, recommended values of the ratio minimum camber to dead-load deflection for glulam beams are also given. Camber is important in long-span beams to eliminate unpleasant sagging and to give instead a graceful flat-arch like appearance. In addition, for the case of long horizontal roof beams, camber may be necessary to avoid potential ponding of rain water.

The ponding phenomenon, as studied by Sawyer (24), Chinn et al. (25) and others, may cause increasing deflection in the beam which in turn will allow more accumulation of water, thereby increasing the load on the beam. The deflection of a simply-supported uniformly-loaded beam under ponding water can be estimated closely (12) by multiplying deflection under design loading (without ponding) by a magnification factor, γ, as follows

$$\gamma = \frac{1}{1 - \dfrac{W' L^3}{\pi^4 EI}}$$

where; W' is total load of 1 in. depth of water on roof area, A, supported by the beam. In other words, $W' = 5.2\,A$ where, if A is expressed in sq. ft., W' is obtained in pounds; L is beam span (in.); E is modulus of elasticity (lb/in.^{2}) and I is the moment of inertia (in.^{4}). Instability occurs as the term $W'L^3/\pi^4 EI$ approaches unity, in which case the magnification factor becomes very large; this should be considered an indication of certain collapse in the beam caused by ponding.

It has been estimated (26) that a positive slope, or camber, equivalent to ¼ in. per foot of horizontal distance between the level of the drain and the high point of the roof, in addition to the recommended minimum camber given in Table 7.5, may be enough to avoid the ponding of water.

Determination of ultimate deflections under increasing transverse load, for research in the behavior of wood beams, may be accomplished as indicated in Section 7.3.6.

7.11.1 Illustrative Example

Determine the maximum deflection of the beam in the example of Section 7.6.2.

The 4 × 10 S4S beam is simply supported over a 10 ft span and subjected to 50 lb/ft uniformly-distributed load and a 3 kip concentrated load at midspan. The maximum elastic deflection, under total load, is given by:

$$\delta = \frac{5}{384}\frac{wL^4}{EI} + \frac{WL^3}{48\,EI}$$

where w = 50 lb/ft = 0.00417 k/in., W = 3 k, L = 120 in., E = 1900 ksi, and I = (1/12) (3.5) $(9.25)^3$ = 230.8 in.^{4}. Substituting these values in the preceding equation yields:

$$\delta = \left(\frac{5}{384}\right) \frac{(0.00417)(120)^4}{(1900)(230.8)} + \frac{(3)(120)^3}{(48)(1900)(230.8)} = 0.026 + 0.246 = 0.272 \text{ in}$$

This value of midspan deflection includes the effects of shear strains, which were obtained indirectly when E = 1900 ksi was used instead of the true average value of the modulus of elasticity, E′ = 1.1 × 1900 = 2,090 ksi.

For the purpose of comparison, consider the determination of the actual value of midspan deflection using the exact formulas given in Fig. 7.23. Thus,

$$\delta = \frac{5}{384} \frac{wL^4}{E'I} + \frac{1}{8} \frac{wL}{GA} \kappa + \frac{1}{48} \frac{WL^3}{E'I} + \frac{1}{4} \frac{WL}{GA} \kappa$$

Substituting the same values as before (except E′ = 2090 ksi, G = E′/16 = 2090/16 = 131 ksi, A = 3.5 × 9.25 = 32.4 in.^{2}, and κ = 1.2 for rectangular section) in the preceding equation one obtains:

$$\delta = \frac{5}{384} \frac{(0.00417)(120)^4}{(2090)(230.8)} + \frac{1}{8} \frac{(0.00417)(120)(1.2)}{(131)(32.4)} + \frac{1}{48} \frac{3\,(120)^3}{(2090)(230.8)}$$

$$+ \frac{1}{4} \frac{(3)(120)(1.2)}{(131)(32.4)} = 0.023 + 0.00002 + 0.224 + 0.0025.$$

$$= 0.272 \text{ in.}$$

The result matches that obtained using the simpler method recommended for designers.

The above deflection occurs immediately after application of the live load. Due to the effects of passage of time the dead-load deflection will double and the total deflection will increase to $\delta = 2\,\delta_D + \delta_L = 2 \times 0.026 + 0.246 = 0.298$ in. ~ 0.3 in.

The deflection to span ratio for total load is 0.3/120 = 1/400 < 1/240, and for applied load only is 0.246/120 = 1/488 < 1/360, both of which are acceptable under the provisions of Table 7.5. No camber is possible in a solid-sawn beam; however, the dead load deflection is so small that no camber appears necessary.

7.11.2 Illustrative Example

For the beam of the preceding example determine the effects that variability in modulus of elasticity of the wood may have on the actual deflection.

Since V = 0.25 is the variability of modulus of elasticity, E, of visually-graded sawn lumber,* a 5-percent exclusion value of E for the latter may be obtained, see Section 2.3, as follows

$$E_{5\%} = E - (0.25 \times 1.645)\,E = 0.589\,E$$

Considering that E = 1900 ksi one obtains $E_{5\%}$ = 0.589 × 1900 = 1119 ksi which is a value that should be exceeded by 95 percent of all beams of the given grade. Thus, in the worst of cases, the total deflection could increase to (0.3) (1900/1119) = 0.51 in.

*For machine stress-rated sawn lumber, V = 0.11; and for glued laminated timber with six or more laminations, V = 0.10.

Such considerations are not usually made by designers when dealing with conventional structures like this one, but are reserved for applications where deflections may be critical.

7.12 END BEARING OF WOOD BEAMS

Wood beams may be supported at their ends by direct bearing across the grain, or through metal connections; the latter are studied in detail in various examples of Chapters 6 and 10.

TABLE 7.6. RECOMMENDED DEFLECTION LIMITATIONS FOR BEAMS.[1]

	Applied load only[2]	Total load	Minimum Ratio Camber D.L. Deflections[5]
Roof beams:			
Industrial[3]	L/180	L/120	1.5
Commercial and Institutional[4]			
a) without plaster ceiling	L/240	L/180	1.5
b) with plaster ceiling	L/360	L/240	1.5
Floor beams:	L/360	L/240	1.5
Bridge Stringers:			
a) Highway	L/200 to L/300		2
b) Railway	L/300 to L/400		2

1. From AITC Standard 102–65, see Ref. 23. For special uses, such as beams supporting vibrating machines, beams over glass windows and doors, etc., more severe limitations may be required.

2. Applied load is live load, wind load, snow load, etc., or any combination thereof.

3. For glued-laminated members ample camber is to be applied to offset deflection. For sawn members, this classification applies to construction for which appearance is not of prime importance and for which adequate drainage is provided to avoid ponding.

4. Applies to churches, schools, residences and other buildings for which appearance, absence of visible deflection, and minimizing of plaster cracking are of prime importance.

5. Parabolic or circular and built into a glued-laminated member by introducing curvature opposite to that due to bending. If drainage is to be provided by camber, add to the above listed values as required.

The behavior of wood in compression across the grain depends on the proportion of the loaded area to the total area under bearing. As shown in Section 2.2 the supporting action of the wood fibers adjacent to the loaded area causes an increase in strength over that which would have resulted if the entire area had been covered. This phenomenon is acknowledged in design (1) by admitting an increase in the value of $F_{c\perp}$, the allowable unit stress in compression perpendicular to grain. For bearing of less than 6 inches in length and not nearer than 3 inches to the end of a member, the maximum allowable stress, $F'_{c\perp}$, is given by the following expression:

$$F'_{c\perp} = F_{c\perp}\left(1 + \frac{0.375}{\ell_b}\right) \tag{7.46}$$

where ℓ_b is the length of bearing, in inches, measured along the grain of the wood. The preceding equation is based on experimental results at the U.S. Forest Products Laboratory (12). For the cases where the bearing area is closer than 3 inches to the end of a member, and for all bearings

6 inches or more in length at any other location, the maximum allowable unit stress in compression across the grain should be taken as $F_{c\perp}$. No allowance need be made for greater stresses at the inner edge of the bearing area of wood beams.

7.12.1 Illustrative Example

The beams in the example of Section 7.6 are supported in end bearing by a 6 × 24 in. glued-laminated girder, to which they are toe-nailed. Determine the suitability of the support.

The actual width of the girder is 5⅛ in. In view of the fact that the support is less than 6 inches long, the beams may protrude 3 inches beyond the width of the supporting girder to obtain maximum allowable bearing stress. Assume that from Table A.2.5 the allowable stress in compression across the grain for the given wood, used at 15 percent moisture content, is $F_{c\perp} = 0.405$ ksi. Using $_b = 5.12$ in., Eq. 7.46 yields

$$F'_{c\perp} = F_{c\perp}\left(1 + \frac{0.375}{\ell_b}\right) = (0.405)\left(1 + \frac{0.375}{5.12}\right) = 0.435 \text{ ksi}$$

Thus, the capacity of the support is (0.435)(5.12)(3.5) = 7.8 k. The maximum reaction of a beam occurs when the concentrated load is directly above the support. For this case, no lateral distribution to adjacent beams can take place, and $R = P + \frac{wL}{2} = 6 + (0.050)\left(\frac{10}{2}\right) = 6.25$ k < 7.8 k; therefore, the beam has adequate support.

The need for the protruding length of the beams may be questioned at this point. The minimum length of bearing may be deternined directly by: 6.25/(0.405 × 3.5) = 4.4 inches. Evidently, the beams need only bear this amount on the supporting girder to satisfy the specifications. A cautious designer may not wish to use less than 4 in. bearing in any case, unless proper fastenings are used, to prevent complete loss of support due to the action of lateral earthquake or wind induced forces.

The design of the given floor system may cost less if a larger section of beam and wood of lower grade are selected.

7.13 BENDING OF CURVED WOOD MEMBERS

Glued-laminated curved members are used in frames and arches. Seldom is the ration d/R, member depth to initial radius of curvature, great enough to cause concern over the accuracy of the linear stress analysis used for straight beams, see Section 7.2. However, at the knees of rigid laminated frames, the ratio d/R may be as large as 0.3 or more, thereby requiring the use of a more realistic stress analysis.

It can be shown that in a curved linearly elastic member of rectangular section, bd, subjected to bending moments, M, the strain and stress reach their greatest values at the extreme fiber closest to the center of curvature. The reason becomes obvious from observation of Fig. 7.24a. The elongations of the top and bottom fibers, due to bending, are not too different. However, the respective lengths of the fibers may differ substantially as the ratio d/R increases. The strain in the shortest fiber, that closest to the center of curvature, is evidently the greatest see Fig. 7.24 (a). The maximum stress is given by (4):

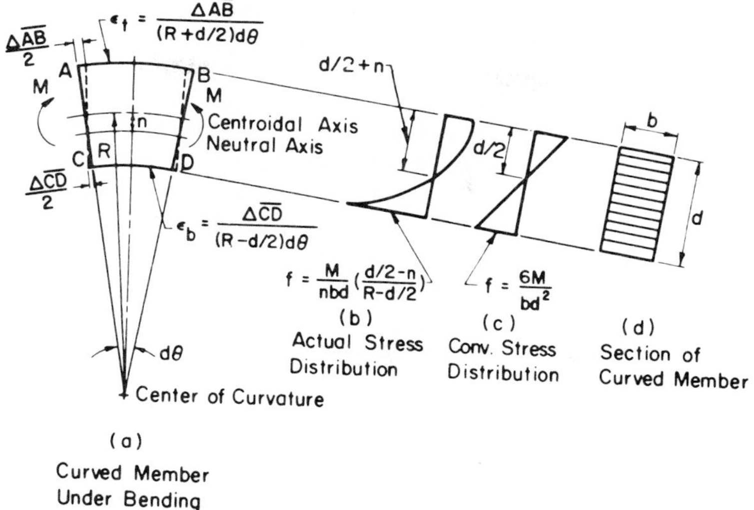

Figure 7.24. Behavior of curved elastic member under bending.

$$f = \frac{M}{nbd} \quad \frac{\left(\frac{d}{2} - n\right)}{\left(R - \frac{d}{2}\right)} \tag{7.47}$$

where n is the distance of the neutral axis to the centroidal axis of the beam measured towards the center of curvature, and R is the radius of curvature of the centroidal axis. For a rectangular section, n is given by the solution of:

$$\log_e \frac{1 + \frac{d}{2R}}{1 - \frac{d}{2R}} = \frac{\frac{d}{R}}{1 - \frac{n}{R}} \tag{7.48}$$

For $d/R < 0.3$, accurate values of n/d may be obtained from the following simple expression:

$$\frac{n}{d} = \frac{d}{12R} \tag{7.48a}$$

The ratio between the maximum stress given by the actual theory of deep curved members to the maximum stress given by the straight beam theory, see Section 7.2, may be used as a correction factor, C_r, to the latter. It is given by the following expression

$$C_r = \frac{\frac{1}{2} - \frac{n}{d}}{6\left(\frac{n}{d}\right)\left(\frac{R}{d} - \frac{1}{2}\right)} \tag{7.49}$$

where the ratio d/R is given, and the ratio n/d can be found from Eq. 7.48a for values of d/R $\leqslant$ 0.3 and from Eq. 7.48 for d/R > 0.3. For example, if d/R = 0.3, Eq. 7.48a and 7.49 yield n/d = 0.025 and C_r = 1.12, respectively. Thus, f = 1.12 M/(bd²/6), i.e. the maximum stress is 12 percent larger than the conventional result.

The experimental determination of the variation of strength with curvature was given by Wilson (27), from end-thrust tests of arches built of Southern pine, Sitka spruce and Douglas fir. The results are shown in Fig. 7.25, for moment at proportional limit and maximum moment, expressed as percentages of corresponding values for the straight member. The variation is with respect to the ratio t/R, where t is the thickness of the lamination and R is the radius of curvature of the center line of the member. Both curves show a reduced capacity with increasing values of t/R.

It was observed by Freas and Selbo (11) that a portion of the increased deficiency of strength with increased curvature came from the use of stress equations applicable to straight members; these are in error when applied to deep sharply curved members. It was concluded, however, that this error is relatively small and that the actual reason for the loss of strength shown in Fig. 7.25, is the creation of stresses induced in bending the laminations to form.

The provisions of NDS(1) for curved portions of members use Wilson's results in the determination of the allowable unit stress in bending. A curvature factor, C_c, given by

$$C_c = 1 - 2{,}000\left(\frac{t}{R}\right)^2 \tag{7.50}$$

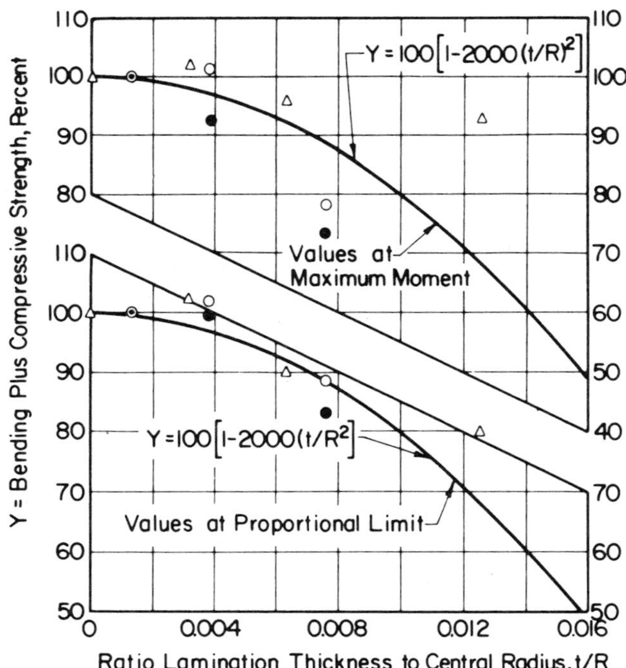

Figure 7.25. Strength of laminated members as related to curvature. From Fig. 99 of Ref. 11.

modifies, by multiplication, the allowable unit stress in bending of straight members. The ratio t/R is limited to 1/100 for hardwoods and Southern pine, and 1/125 for other softwoods.

The similarity between Eq. 7.50 and the expressions given in Fig. 7.25 indicates the origin of the curvature factor. It becomes apparent that C_c accounts also for the increase in elastic stresses in curved members. In view of this, a designer using the provisions of NDS to check the capacity of a section, need not use the correction factor C_r as given by Eq. 7.49. The value of C_c for the example given above may be found as follows. Let d = 30 in. and t = ¾ in. The value of R = 30/0.3 = 100 in. meets the limitations on t/R = 3/4/100 = 1/133 < 1/100. Thus, from Eq. 7.50, $C_c = 1 - 2{,}000\,(1/133)^2 = 0.888$.

7.14 RADIAL STRESSES IN CURVED WOOD MEMBERS

It can be shown that radial stresses are also created in curved members subjected to bending moments. Let Fig. 7.21a represent an infinitesimal portion of a curved beam of rectangular section, bd, under the action of moments, M, only. The geometry of the given portion is defined by $d\theta$, the angle at the center of curvature and by R, the radius of the centroidal axis. It has been shown, see Section 7.13, that the actual distribution of flexural stresses is nonlinear; however, the error of using a linear distribution in the determination of radial stresses for actual laminated members is quite small. Hence, for the sake of simplicity, it is assumed in the following derivations that the distribution of flexural stresses in the section is linear.

A free-body diagram of the curved member above a given level is shown in Fig. 7.26b. The distance from the bottom fiber of the free portion to the centroidal axis is y_o. The compression forces, C, created by the bending moments, are in equilibrium with a radial force, T_r, of tensile nature. Equilibrium of forces about a radial axis of symmetry requires that

$$2\,C \sin\frac{d\theta}{2} \simeq C d\theta = T_r \tag{7.51}$$

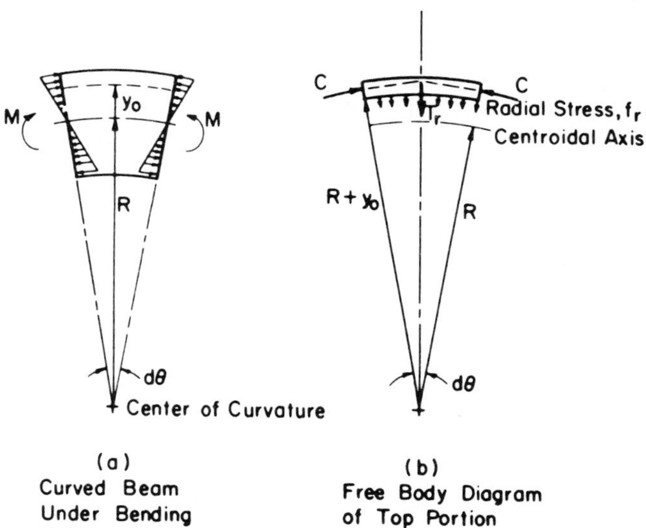

Figure 7.26. Determination of radial stresses due to bending in curved beams. Linear stress distribution assumed.

The determination of C can be accomplished by integration of the differential forces f b dy, where $f = My/I$, between the limits y_o and $d/2$ as follows:

$$C = \int_{y_o}^{d/2} \frac{My}{I} bdy = \frac{Mb}{2I}\left(\frac{d^2}{4} - y_o^2\right) \tag{7.52}$$

Assuming that the radial stress f_r is uniformly distributed over the area b $(R + y_o)\, d\theta$, the radial force T_r can be expressed as

$$T_r = f_r b(R + y_o)d\theta \tag{7.53}$$

Substituting the values of C and T_r and $I = bd^3/12$ in Eq. 7.51 yields

$$f_r = \frac{3}{2}\,\frac{M}{bd(R + y_o)}\left[1 - \left(\frac{y_o}{d/2}\right)^2\right] \tag{7.54}$$

The variation of f_r versus y_o is almost parabolic; for values of $d/R < 0.1$, the difference between the actual variation and a parabola is insignificant. At the extreme fibers, $y_o = \pm\, d/2$, and $f_r = 0$; the maximum value of f_r occurs at the centroidal axis, $y_o = 0$, for which Eq. 7.53 gives:

$$f_r = \frac{3}{2}\,\frac{M}{bdR} \tag{7.55}$$

an expression first presented in 1939 by T. R. Wilson (27). Radial stresses are of a compressive nature in the curved member shown in Fig. 7.26 if applied moments are of opposite sign. A simple rule may be used to determine the nature of the radial stresses created. Bending moments that reduce the initial curvatures of a member, see Fig. 7.26, induce tensile radial stresses; conversely, bending moments that increase initial curvatures induce compressive radial stresses.

The provisions of NDS for checking the capacity of curved members, indicate the use of Eq. 7.55 to determine the maximum value of radial stress. Allowable stresses in radial tension may not exceed ⅓ the allowable stress for horizontal shear. For Douglas fir and Larch, the preceding applies for wind or earthquake loads; for other types of load, radial tension may not exceed 15 psi. However, where mechanical reinforcement is provided to resist all radial tension stress, the foregoing limits do not apply. Compressive radial stresses shall be limited to the allowable stress in compression perpendicular to the grain.

When designing a curved glued-laminated beam of variable cross section such as a double-tapered curved beam, the radial stress, f_r, is computed by the equation:

$$f_r = K_r \frac{6M}{bd_c^2} \tag{7.55a}$$

Where: M = bending moment at midspan in inch pounds
b = width of cross section in inches
d_c = depth of cross section at the apex in inches
K_r = radial stress factor determined from the following polynominal approximation to K_r:

$$K_r = A + B\left(\frac{d_c}{R_m}\right) + C\left(\frac{d_c}{R_m}\right)^2 \tag{7.55b}$$

where: R_m = radius of curvature at the center line of the member at midspan in inches
A, B, and C = constants as given below

β (1)	A (2)	B (3)	C (4)
0.0	0.0	0.2500	0.0
2.5	0.0079	0.1747	0.1284
5.0	0.0174	0.1251	0.1939
7.5	0.0279	0.0937	0.2162
10.0	0.0391	0.0754	0.2119
15.0	0.0629	0.0619	0.1722
20.0	0.0893	0.0608	0.1393
25.0	0.1214	0.0605	0.1238
30.0	0.1649	0.0603	0.1115

and β = angle between the upper edge of the member and the horizontal in degrees. Values of K_r for intermediate values of β may be interpolated linearly.

The allowable radial stresses are the same as those given for members of constant cross section. Since the preceding relationship was developed specifically for a variable cross section glued-laminated beam, it is not recommended that this design equation be applied to arches having varying cross sections such as occurs in a Tudor arch. Since arches are normally stressed in radial compression, radial stresses seldom control design of arches.

7.15 BENDING OF TAPERED WOOD MEMBERS

It is not difficult to fabricate glued-laminated timber beams, with a tapered profile, to meet architectural requirements, see Section 3.3.4. Tapered members are used to provide pitched roofs, and frequently in arches and rigid frames. Various profiles of simply-supported beams are shown in Fig. 7.27.

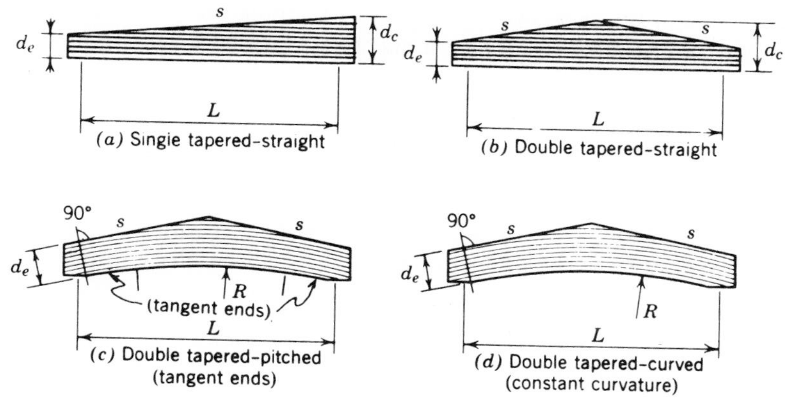

Figure 7.27. Simple-span tapered beam forms, (s designates sawn surface.).

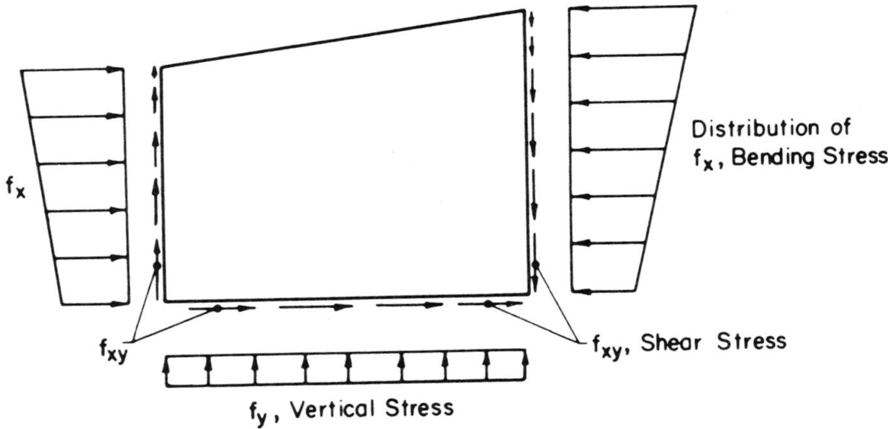

Figure 7.28. Distribution of bending, shear and vertical stresses in a differential element of a tapered beam subjected to transverse loading.

Design of tapered wood beams is based on research conducted at Forest Products Laboratory, as reported by Maki and Kuenzi, see Ref. 28. The investigation included the theoretical development of expressions for the evaluation of bending, shear and vertical stresses, see Fig. 7.28, and deflections. The validity of these expressions was verified with experimental research on tapered beams made of isotropic materials such as aluminum, and orthotropic materials such as wood. Based on the result of this investigation, design criteria for tapered wood beams were developed. In what follows, insight is given in 1) the nature of the stresses created by tranverse load; 2) the test results at Forest Products Laboratory (28); 3) an ultimate interaction formula; 4) deflections; and, 5) design criteria. Finally, an illustrative example of design of a tapered beam is fully discussed.

7.15.1 Bending Stresses

Consider a tapered beam, as shown in Fig. 7.27. Assuming that the elastic theory of bending applies for this case see Section 7.2, the bending stress, f_x, at a given section, bd, caused by a moment, M, is given by:

$$f_x = \frac{12M}{bd^3}\left(y - \frac{d}{2}\right) \tag{7.56}$$

where, y is the distance of the non-tapered surface to the point where the stress is f_x. If both M and d are functions of x, as in the case of the beams shown in Fig. 7.27 for example, it is reasonable to assume that there may be a section in the beam at which an absolute maximum value of f_x occurs at its extreme fibers. This section can be found from:

$$\frac{df_x}{dx} = 0 \tag{7.57}$$

which yields

$$\frac{6}{bd^2}\left[\frac{dM}{dx} - \frac{2M}{d}\frac{dd}{dx}\right] = 0 \tag{7.58}$$

from which

$$d = \frac{2M\left(\frac{dd}{dx}\right)}{\left(\frac{dm}{dx}\right)} \tag{7.59}$$

and the maximum value of f_x at such a section is

$$f_x = \frac{3}{2bM}\frac{\left(\frac{dM}{dx}\right)^2}{\left(\frac{dd}{dx}\right)^2} \tag{7.60}$$

where $\frac{dM}{dx}$ and $\frac{dd}{dx}$ are the shear force and slope of taper at the section, respectively. In the case of a member of constant taper, such as shown in Figs. 7.27a and b and 7.29, dd/dx = tan θ, where, θ, is the angle between the tapered and the non-tapered sides. Also, dM/dx is constant in spans subjected to concentrated loads, whether simply-supported or cantilevered.

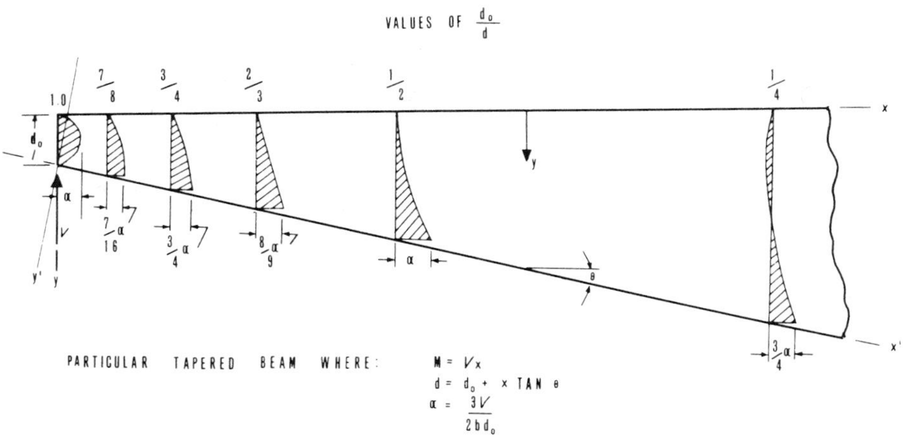

Figure 7.29. Shear-stress distributions for a tapered beam, under a concentrated load at midspan. From Ref. 28.

7.15.2 SHEARING STRESSES

The evaluation of shear stress, f_{xy}, was accomplished by Norris, see Appendix I of Ref. 28, using equilibrium considerations on a free-body diagram of a section of the tapered beam. The following expression was obtained:

$$f_{xy} = \frac{6M}{bd^2}\left[3\left(\frac{y}{d}\right)^2 - 2\left(\frac{y}{d}\right)\right]\frac{dd}{dx} + \frac{6}{bd}\left[\frac{y}{d} - \left(\frac{y}{d}\right)^2\right]\frac{dM}{dx} \tag{7.61}$$

where, f_{xy} is the shear stress at any point y (measured positive from the non-tapered face, see Fig. 7.29) in a section of width, b, depth, d, subjected to a moment M. It can be shown that at the non-tapered surface, $y/d = 0$, the shear stress is zero, as it should be in order to satisfy the boundary condition. Also, it may be shown that the maximum shear stress in a section, as given by conventional Equation 7.32 for the neutral axis, occurs in a tapered beam only at a section where $M = 0$, i.e. at the support of a simple-span beam or at the free end of a cantilevered beam. The maximum shear at other sections, as one moves away from the support, occurs below the mid-depth of the section. Beyond a relatively short distance from the support, the maximum shear stress, f_{xy}, at any given section occurs at the tapered side; thus, for $y/d = 1$, Eq. 7.61 yields the value of the maximum shear stress as follows:

$$f_{xy} = \frac{6M}{bd^2}\left(\frac{dd}{dx}\right) \tag{7.62}$$

Shear-stress distributions at various sections of a particular tapered beam are given in Fig. 7.29. Note that beyond the section where $d = (4/3)d_0$, the maximum shear stress f_{xy} occurs at the tapered side. The absolute maximum shear stress occurs at a section where $d = 2d_0$ and is equal to $(3/2)(V/bd_0)$; this makes it of equal intensity to the maximum shear stress at the support.

7.15.3. VERTICAL SRESSES

Analysis of the vertical equilibrium of an element of a tapered beam led (28) to the determination of an approximate relationship for the vertical stress existing in the beam. It was found that the vertical stress is given by:

$$\begin{aligned} f_y = {} & \frac{6M}{bd^2}\left(\frac{y}{d}\right)^2\left(4\frac{y}{d} - 3\right)\left(\frac{dd}{dx}\right)^2 + \frac{6M}{bd^2}\left(\frac{y}{d}\right)\left(1 - \frac{y}{d}\right)\frac{d^2d}{dx^2} + \\ & \frac{12}{bd}\left(\frac{y}{d}\right)^2\left(1 - \frac{y}{d}\right)\left(\frac{dd}{dx}\right)\left(\frac{dM}{dx}\right) + \frac{1}{b}\left[1 - 3\left(\frac{y}{d}\right)^2 + 2\left(\frac{y}{d}\right)^3\right]\frac{d^2M}{dx^2} \end{aligned} \tag{7.63}$$

and at $y/d = 0$:

$$f_y = \frac{1}{b}\frac{d^2M}{dx^2} \tag{7.64}$$

and at $y/d = 1$:

$$f_y = \frac{6M}{bd^2}\left(\frac{dd}{dx}\right)^2 \tag{7.65}$$

It can be shown that the vertical stress is maximum at the tapered side, i.e. at $y/d = 1$.

For the particular beam considered in Fig. 7.29 for which $\frac{d^2d}{dx^2} = 0$ and $\frac{d^2M}{dx^2} = 0$, Eq. 7.63 is reduced to:

$$f_y = \frac{6M}{bd^2}\left(\frac{y}{d}\right)^2\left(4\frac{y}{d} - 3\right)\left(\frac{dd}{dx}\right)^2 + \frac{12}{bd}\left(\frac{y}{d}\right)^2\left(1 - \frac{y}{d}\right)\frac{dd}{dx}\frac{dM}{dx} \qquad (7.66)$$

The depth of section d at which the absolute maximum value of f_y occurs is double the depth of the section at the support, i.e., $d = 2d_o$. The absolute maximum value of the stress is given by:

$$f_y = \frac{3V}{2bd_o}\tan\theta \qquad (7.67)$$

For the usual case of beams, such as shown in Fig. 7.27, vertical stresses at the tapered side are compressive in nature. Conversely, for beams such as shown in Fig. 7.29, vertical stresses at the tapered side are of a tensile nature.

7.15.4 Summary of Stresses

From the preceding discussion the general values of the bending, shear and vertical stresses existing at the taper are given by:

$$\begin{aligned} f_x &= \frac{6M}{bd^2} \\ f_{xy} &= \frac{6M}{bd^2}\left(\frac{dd}{dx}\right) \\ f_y &= \frac{6M}{bd^2}\left(\frac{dd}{dx}\right)^2, \text{ respectively.} \end{aligned} \qquad (7.68)$$

For a particular beam with uniformly varying depth, i.e. $dd/dx = \tan\theta$, these relationships can be written as follows:

$$\begin{aligned} f_x &= \frac{6M}{bd^2} \\ f_{xy} &= \frac{6M}{bd^2}\tan\theta \\ f_y &= \frac{6M}{bd^2}\tan^2\theta \end{aligned} \qquad (7.69)$$

Finally, for the particular beam shown in Fig. 7.29, the absolute maximum values of the bending, shear and vertical stresses in the beam occur at the taper of a section with depth $d = 2d_o$ and have the following values:

$$\begin{aligned} f_x &= \frac{3}{2}\frac{M}{bd_o^2} \\ f_{xy} &= f_x\tan\theta \\ f_y &= f_x\tan^2\theta \end{aligned} \qquad (7.70)$$

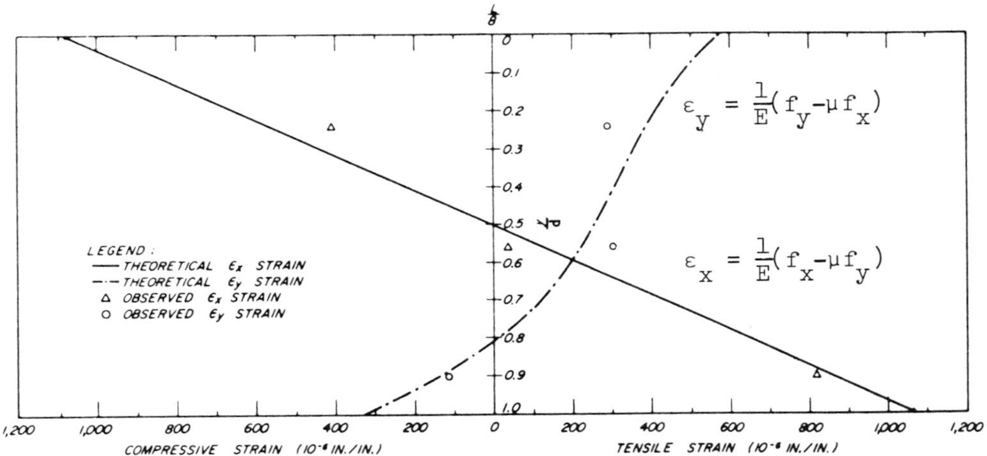

Figure 7.30. Comparison of theoretical and observed bending and vertical strains at section of beam No. 1 under a concentrated midspan load of 2V = 7,000 pounds. Adapted from Ref. 28.

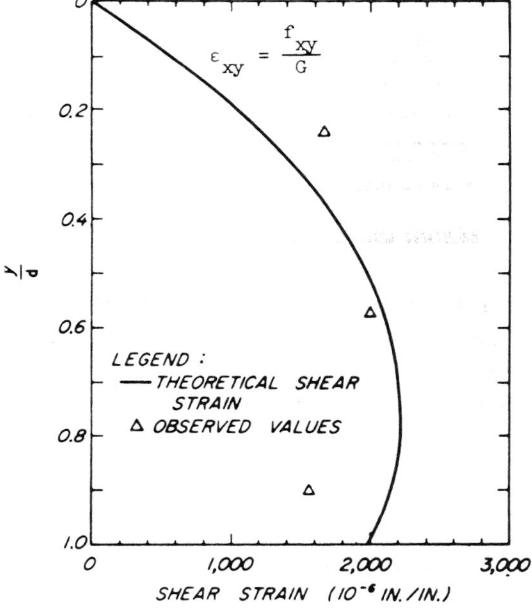

Figure 7.31. Comparison of theoretical and observed shear-strain distributions at section of beam No. 1 under a concentrated midspan load of 2V = 7,000 pounds. Adapted from Ref. 28.

7.15.5 Test Results

Tests at Forest Products Laboratory (28) on tapered beams made of elastic, isotropic materials, such as aluminum, confirmed the reliability of the elastic analysis on which the preceding relations are based. The applicability of the analysis to anisotropic wood members was tested using 3 symmetrical double-tapered beams. These were constructed from planks of Sitka spruce carefully chosen to be straight grained and free from strength-reducing characteristics, to obtain the most reliable results possible. The beams were subjected to concentrated loads at midspan, similarly to the particular case shown in Fig. 7.29.

Typical results are shown in Figs. 7.30 and 7.31 for beam No. 1, which had a span of 75.36 in., depth at the support of 4.71 in. and midspan depth of 9.42 in. Fig. 7.30 compares the theoretical and observed bending and vertical strains at a section a distance L/8 from the support. Fig. 7.31 compares the theoretical and observed shear-strain distributions at the same section. While perfect agreement was not achieved, a tendency toward agreement is present.

Based on the results of the investigation, Maki and Kuenzi (28) concluded that fundamental Eqs. 7.56, 7.61, and 7.66 very closely approximate the stress situation existing in tapered members with uniformly varying cross section.

7.15.6 Interaction Formula

The preceding discussion showed that it is possible for bending, shear and vertical stresses to be combined at one point in the beam; the combined stresses affect the ultimate strength of the wood. Norris (29) has proposed the use of an interaction equation to determine the strength of an orthotropic material subjected to combined stresses in a two-dimensional stress system. For the case of a tapered wood beam subjected to bending, vertical and shear stresses at the taper, this equation may take the following form:

$$\left(\frac{f_x^u}{F_x^u}\right)^2 + \left(\frac{f_y^u}{F_y^u}\right)^2 + \left(\frac{f_{xy}^u}{F_{xy}^u}\right)^2 = 1 \tag{7.71}$$

where f_x^u, f_y^u, and f_{xy}^u are the bending, vertical and shear stresses, respectively, existing at some point at ultimate; F_x^u, F_y^u and F_{xy}^u are the respective ultimate stresses, that is, the stress at which the member would fail were each stress acting alone.

7.15.7 Deflections

Complicated mathematical expressions are available (28) for the determination of deflections of tapered wood beams. From these formulas, and in order to facilitate calculations of maximum deflections, the graphs shown in Figs. 7.32 and 7.33 are given. Both graphs are for straight-tapered, simply-supported beams only. Fig. 7.32 considers maximum deflections, Δ_B, due to bending under uniformly distributed loading, while Fig. 7.33 deals with maximum deflections, Δ_B, due to bending under concentrated loads acting at midspan. The additional deflection due to shear, Δ_S, may be computed, with only small error, as follows:

Uniformly distributed load:

$$\Delta_S = \frac{3WL}{20\ Gbd_o} \tag{7.72}$$

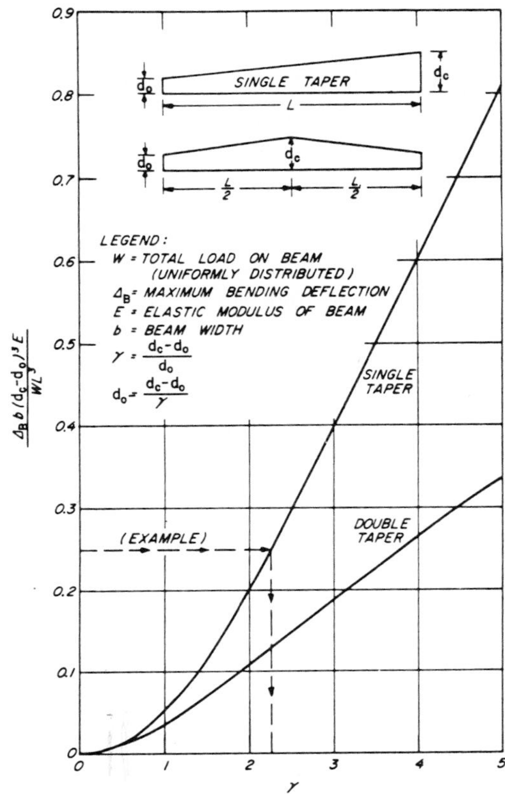

Figure 7.32. Graph for determining tapered beam deflection under uniformly distributed load. Adapted from Ref. 28.

Concentrated load at midspan:

$$\Delta_S = \frac{3PL}{10\ Gbd_o} \tag{7.73}$$

where $G \simeq E/16$, is the shear modulus of wood, see Section 2.2.6, W is the total load on beam uniformly distributed; P is the concentrated midspan load, and all other terms have been defined before. The two preceding formulas are conservative, since they give the shear deflection of a prismatic beam of uniform depth d_o, the minimum depth at the support.

In addition to maximum vertical deflections, the designer of double-tapered curved beams should provide for horizontal displacement of the supports. Considering that one end of the beam is to remain anchored, the horizontal displacement, Δ_H, of the other end is given approximately by the following expression:

$$\Delta_H = \Delta_V \frac{2h}{L} \tag{7.74}$$

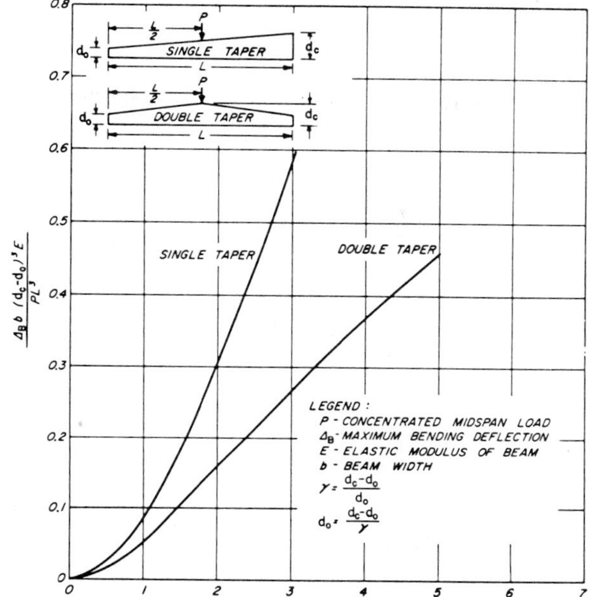

Figure 7.33. Graph for determining tapered beam deflection under concentrated mid-span load. Adapted from Ref. 28.

where: h is the rise of the beam, measured vertically from axis at end of beam to axis at centerline, L is the span length and Δ_V is the maximum vertical deflection, due to bending and shear, determined as before. The designer must allow this displacement to occur by providing elastomeric bearing plates, rollers, lubricated plates or other devices. In the case of a beam supported by flexible columns, offering small lateral resistance, a rigid attachment may be acceptable.

The determination of deflections for other types of tapered beams, including continuous beams, can always be determined using numerical methods of integration (30).

7.15.8 Design Criteria

As in the case of other wood structures, tapered beams are designed to satisfy both strength and serviceability. The interaction equation, adapted to allowable stresses, is used to check the first criterion. In this respect, Eq. 7.71 is modified as follows:

$$\left(\frac{f_x}{F_x}\right)^2 + \left(\frac{f_y}{F_y}\right)^2 + \left(\frac{f_{xy}}{F_{xy}}\right)^2 \leq 1 \qquad (7.75)$$

where: f_x, f_y and f_{xy} are the bending, vertical and shear stresses, respectively, at the critical section determined as before and:

F_x = allowable stress in bending, F_b, modified for duration of loading, but not for depth effect.

F_y = allowable stress in compression perpendicular to grain, $F_{c\perp}$, modified for duration of loading. When the taper cut is on the tension side (not a recommended practice), the allowable stress in tension perpendicular to grain, $F_{t\perp}$, should be used instead in accordance with Section 7.14.

F_{xy} = allowable horizontal shear stress, F_v, modified for duration of loading.

The determination of the critical section of tapered beams has been solved for various simple cases (28). The frequent cases of single, or symmetrically double-tapered, beams is considered as follows. It can be shown that the absolute maximum bending, shear and vertical stresses occur all at the taper of a section with depth, d, given by:

$$d = 2d_o \left(\frac{d_o + L \tan \theta}{2d_o + L \tan \theta} \right) \tag{7.76}$$

and have values given by Eqs. 7.69. The designer should also check the maximum shear at the neutral axis of the beam at the support, as given by Eq. 7.32. In addition, for curved beams with constant cross section, the maximum radial stress occurring at the center of the midspan section should be checked using Eq. 7.55.

In the case of tapered beams subjected to various load systems, and continuous beams, it may be necessary to check the interaction equation, Eq. 7.75, at various reasonably-spaced sections of the beam. This procedure may prove more expeditious than the exact determination of the critical section through analytical means.

The serviceability criterion is satisfied if the maximum deflection is below accepted standards, see Table 7.5. For this purpose, Figs. 7.32 and 7.33 may be used to determine the maximum deflection of a given beam due to bending, Δ_B; also, the graphs can be used to determine the size of a tapered beam, the deflection of which under load is below a certain value. In both cases, once the beam section is determined, the shear deflection, Δ_S, should be calculated, see Eqs. 7.72 and 7.73; if necessary, the beam size should be increased so that the total deflection, $\Delta_B + \Delta_S$ does not exceed allowable values. In all cases, however, the designer may wish to provide camber equal to at least 1.5 times the permanent deflection of the beam, or as suggested by Table 7.5. AITC (22) recommends that, when midspan camber exceeds 2 in., camber equal to ¼ of the midspan camber be sawn into the tapered or compression face.

7.15.9 Illustrative Example

Design a double-tapered straight glued-laminated roof beam, such as shown in Fig. 7.27.b to meet the following requirements: span 60 ft., spacing 16 ft., roof slope 1/12, dead load (including self-weight of the beam) 15 psf, and snow load 30 psf; maximum deflection limited to 1/180th of the span, i.e. 4 in. The roof decking is applied directly to the beams, thus providing continuous lateral support to the top, compression flange. Assume the following allowable stresses: $F_b = 2.4$ ksi, $F_v = 0.165$ ksi, $F_{c\perp} = 0.415$ ksi, $E = 1{,}800$ ksi, $G = E/16 = 112.5$ ksi. Provide a minimum camber equal to 1 and ½ times the dead load deflection. A possible solution is as follows:

1. Allowable Stresses
   For this design, the governing loading condition is total load, i.e. dead plus snow load, for which $\bar{\alpha} = 1.15$. Thus, the set of given allowable stresses becomes $F_b = 2.4 \times 1.15 = 2.76$ ksi, $F_v = 0.165 \times 1.15 = 0.190$ ksi and $F_{c\perp} = 0.415 \times 1.15 = 0.477$ ksi. E and G remain as given.

2. Minimum Dimensions
Assume that the width of the section, b, is 5⅛ in. and that the effective depth of the section, a distance d from the face of the support, is 30 in. Thus, the effective span for shear, see Section 7.9, is $60 - 2(30/12) = 55$ ft. The design load is $30 + 15 = 45$ psf; since the beams are spaced at 16 ft., the uniformly distributed load on the typical beam is $0.045 \times 16 = 0.72$ k/ft. Shear, V, at the section a distance d from the support is

$$V = 0.72 \times \frac{55}{2} = 19.8 \text{ k}$$

The minimum depth, d, that satisfies the allowable shear stress is given by:

$$d = \frac{3}{2}\,\frac{19.8}{(5.125)(0.190)} = 30.5 \text{ in.}$$

Thus, the minimum end depth, d_o, may be obtained from the given taper of the beam as follows:

$$d_o = 30.5 - \left(\frac{1}{12}\right)(2.5 \times 12) = 28 \text{ in.}$$

Use $d_o = 30$ in.

3. Midspan Depth, d_c

$$d_c = 30 + (30 \times 12)\left(\frac{1}{12}\right) = 60 \text{ in.}$$

4. Maximum Deflection
First determine the coefficient γ as follows:

$$\gamma = \frac{d_c - d_o}{d_o} = \frac{60 - 30}{30} = 1.0$$

then, using $\gamma = 1.0$ enter Fig. 7.32 and obtain

$$\frac{\Delta_B b(d_c - d_o)^3 E}{WL^3} = 0.035$$

Substituting $b = 5.125$ in., $d_c - d_o = 30$ in., $E = 1{,}800 \times 1.10 = 1{,}980$ ksi, $W = 0.72 \times 60 = 43.2$ k, $L = 60 \times 12 = 720$ in. in the preceding equation, and solving for Δ_B, yields
$\Delta_B = 2.06$ in.

The additional deflection at midspan, due to shear, may be computed from Eq. 7.72 as follows

$$\Delta_S = \frac{3WL}{20\ Gbd_o} = \frac{(3)(42.2)(720)}{(20)(112.5)(5.125)(30)} = 0.27 \text{ in.}$$

The total deflection, $\Delta_{TL} = 2.06 + 0.27 = 2.33$ in. < 4 in. Therefore, the beam meets the required serviceability.

5. Camber at Midspan

The minimum ratio of camber to dead load deflection for this type of beam, see Table 7.6, is 1.5. The dead load deflection can be determined from the total deflection as follows:

$$\Delta_{DL} = \Delta_{TL}\frac{w_{DL}}{w_{TL}} = 2.33\frac{0.24}{0.72} = 0.78 \text{ in.}$$

Thus, the midspan camber $= 1.5\ \Delta_{DL} = (1.5)(0.78) = 1.2$ in. Since this camber is less than 2 in., no tapered-face camber is required.

6. Critical Section

The critical section occurs at a section the depth of which is given by Eq. 7.76 as follows

$$d = 2d_o\left(\frac{d_o + L\tan\theta}{2d_o + L\tan\theta}\right)$$

Substituting in the preceding equation $d_o = 30$ in., $L = 720$ in., $\tan\theta = 1/12$ yields:

$$d = (2)(30)\ \frac{30 + (720)(1/12)}{(2)(30) + (720)(1/12)} = 45 \text{ in.}$$

The critical section is situated at a distance from the end, x, given by $d = d_o + x\tan\theta$. Substituting $d = 45$ in., $d_o = 30$ in. and $\tan\theta = 1/12$ gives $x = 180$ in. $= 15$ ft.

7. Maximum Stresses at Critical Section

The bending moment, M, at the critical section is given by:

$$M = w\frac{x}{2}(L - x) = \left(\frac{0.72}{12}\right)\frac{180}{2}\ (720 - 180) = 2{,}910 \text{ k-in.}$$

The bending, f_x, shear, f_{xy}, and vertical f_y, stresses at the taper side of the critical section may be computed using Eqs. 7.69 as follows:

$$f_x = \frac{6M}{bd^2} = \frac{(6)(2{,}910)}{(5.125)(45)^2} = 1.68 < 2.76 \text{ ksi}$$

$$f_{xy} = f_x\tan\theta = (1.68)(1/12) = 0.14 < 0.19 \text{ ksi}$$

$$f_y = f_x\tan^2\theta = (1.68)(1/12)^2 = 0.012 < 0.477 \text{ ksi}$$

All of these stresses are below the respective allowable stresses.

8. Combined Stresses at Critical Section

The interaction formula given by Eq. 7.75 must be satisfied. Thus, substituting $f_x = 1.68$ ksi, $F_x = 2.76$ ksi, $f_{xy} = 0.14$ ksi, $F_{xy} = 0.19$ ksi, $f_y = 0.012$ and $F_y = 0.477$ ksi in Eq. 7.75 gives:

$$\left(\frac{1.68}{2.76}\right)^2 + \left(\frac{0.14}{0.19}\right)^2 + \left(\frac{0.012}{0.477}\right)^2 = 0.915 < 1$$

Therefore, the beam also meets the required strength and may be considered as a possible solution to the problem.

PROBLEMS

1. Solve the Illustrative Example of Section 7.3.5 considering $E_t = E_c$, i.e., n = 1. Compare results and draw conclusions.

   Ans. At proportional limit: a = 7 in. M = 1,254 k-in. $\phi = 4.57 \times 10^{-4}$ rad/in. At ultimate: m = 1.05, a = 7.17 in., M = 1.800 k-in., $\phi_u = 6.97 \times 10^{-4}$ rad/in. Results of both sections are within 1 percent; thus, effects of variation of n are negligible.

2. The stress-strain diagram shown in Fig. 7.2 is modified as follows. The horizontal portion between strains $\alpha\epsilon_c^u$ and ϵ_c^u is given a certain slope which results in an ultimate stress f_c^u at a strain ϵ_c^u and a stress βf_c^u at a strain $\alpha\epsilon_c^u$; Fig. 7.2 corresponds to the case where $\beta = 1$. Ascending slopes result for all $\beta < 1$; descending slopes for $\beta > 1$. Assuming that ultimate moment occurs when the ultimate compression strain ϵ_c^u is attained, determine the following:
   1) Distance of neutral axis to extreme fiber in compression, a.
   2) Ultimate moment, M_u.
   3) Show that if $\beta = 1$, the equations resulting from items 1 and 2 degenerate into Eqs. 7.11 and 7.13, respectively.
   4) Show that if n = 1 and $\alpha = \beta$, the conventional linear-elastic equations are obtained.

   Ans.

   1) $a = d\dfrac{m}{1 + m}$

   2) $M = f_c^u \dfrac{bd^2}{6}\left(\dfrac{m}{1 + m}\right)^2[\beta(1 + \alpha) + (2 - \alpha - \alpha^2) + \dfrac{2}{m}(\beta + 1 - \alpha)]$

   where $n = \dfrac{E_t}{E_c}$ and $m = \sqrt{\dfrac{\beta n}{\alpha(\beta + 1 - \alpha)}}$

3. Using the results of problem No 2 determine the ultimate moment and curvature of the section discussed in the Illustrative Example of Section 7.3.5. The same data apply except that, instead of $\beta = 1$, as in the Illustrative Example, consider $\beta = 0.9$ for this case. Note E_c changes to 6.3/0.0035 = 1800 ksi

   Ans. m = 1.118, a = 7.39 in. M_u = 1,720 k-in., $\phi_u = 6.76 \times 10^{-4}$ rad/in.

4. The beam of Fig. 7.8 is loaded at the ends by a compression force, N = 50 kips, acting along the longitudinal axis of the beam, in addition to the transverse loads shown at the middle-third points. Determine the moment, M, and curvature, ϕ, at attainment of proportional limit and ultimate. You may consider the beam adequately braced against lateral-torsional buckling. Hints: Use an iteration method to determine a, the distance of top fiber to neutral axis, for which C−T = 50 k. For the sake of consistency refer M to the longitudinal axis of the beam.

   Ans.: At proportional limit: a = 7.79 in., M = 1152 k-in., $\phi = 4.11 \times 10^{-4}$ rad/in. At ultimate: a = 7.72 in., M = 1685 k-in., $\phi = 6.48 \times 10^{-4}$ rad/in.

5. For the beam of the preceding problem determine the midspan deflections, δ, due to bending and the transverse loads, P, at proportional limit and at ultimate. Hints: Ignore shear deflections due to P and secondary deflections due to N-δ effects. Determine P from $M = N\delta + (P/2)$ a.

   Ans.: At proportional limit: δ = 2.52 in., P = 25.7 k;
   At Ultimate: δ = 3.92 in., P = 37.2 k

6. Plot M-ϕ and P-δ curves for the preceding beam using the results obtained in Problems 4 and 5. Superpose the corresponding values obtained in Section 7.3.5 of the text for the case of N = 0 and compare results.
7. Given the moment-curvature relationship of Fig. 7.7 determine the P−δ diagram for δ at midspan and at the cantilever ends of the beam shown below. Solve for the stages of loading corresponding to attainment of proportional limit and ultimate. Plot the deflection configuration of the beam for each case.

   Ans. P_{PL} = 42.33 k, δ_{CL}^{PL} = 1.08 in., δ_{cant} = − 5.42 in.; P_U = 60.67 k, δ_{CL} = 1.56 in., δ_{cant} = − 7.95 in.

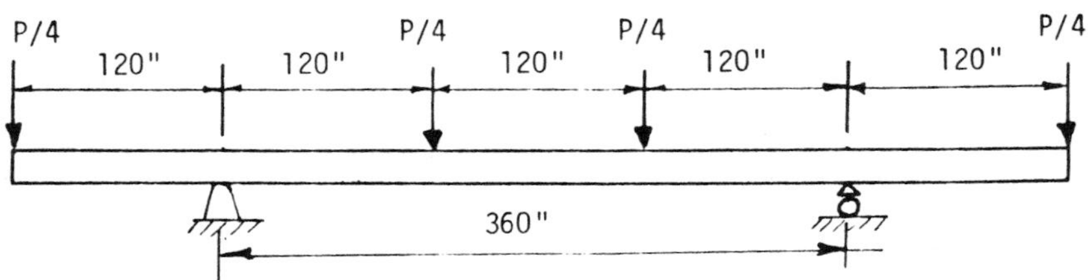

Figure 7.34. The beam of problem 7.

8. A floor is supported by 2 × 12 in. joists, at 16 in. spacing, on a 16 ft. span. The joists are of a wood species and moisture content such that allowable stresses are: F = 1.90 ksi, for repetitive-member uses; F_t = 1.10 ksi; F_v = 0.095 ksi; $F_{c\perp}$ = 0.405 ksi; F_c = 1.45 ksi; and E = 1,900 ksi. Assume normal loading conditions, continuous lateral support and full bearing on 6 in. (nominal) wide girders. The maximum deflection under the total load should not exceed 1/240th of the span nor 1 in. Determine the allowable uniform load, including self-weight, that the joists may take and specify whether it depends on bending, shear, end bearing or deflection.
9. A glued-laminated beam 6.75 in. wide × 30 in. deep spans 40 ft. simply supported. It carries the following uniformly distributed load: w_{DL} = 0.1 k/ft., which includes its own weight, and w_{LL} = 0.7 k/ft.; the beam is braced laterally at 5 ft. spacing. Assume normal loading conditions and dry conditions of use. Allowable stresses are: F_b = 2.4 ksi; F_t = 1.6 ksi; F_c = 1.5 ksi; $F_{c\perp}$ = 0.45 ksi and 0.385 ksi, for tension face and compression face, respectively; F_v = 0.165 ksi and E = 1,800 ksi. Determine the following items:
   1) Section properties.
   2) Design moment and shear, reaction at support.
   3) Deflection, under dead and total load; suggested camber.
   4) Check lateral stability.
   5) Allowable moment and shear.

   Ans.
   1) A = 202 in.^{2}, S = 1,012 in.^{3}, I = 15,187 in.^{4}, C_F = 0.903.
   2) M = 1,920 k-in., V = 14k, R = 16k.
   3) δ_{DL} = 0.21 in., δ_{TL} = 1.69 in., 3/8 in. camber at midspan.
   4) ℓ_e= 115 in., C_s = 8.7, C_k = 21.2, C_s < 10, F_b = 2.4 ksi.
   5) M = 2,190 k-in., V = 22.3k.

10. Twenty feet long solid-sawn timber beams, spaced 8 ft. apart and simply supported on masonry walls (6 in. at each end), carry a roof deck nailed directly to the beams. The loads are uniformly distributed and consist of the dead load of the deck and roofing materials, estimated at 10 psf, and the design snow load of 30 psf. Consider the beams made of wood having the following allowable stresses: F_b = 1.50 ksi, F_t = 1.00 ksi, F_v = 0.09 ksi, $F_{c\perp}$ = 0.40 ksi, F_c = 1.25 ksi, and E = 1,800 ksi. Determine the adequacy of the design, if the beams are 4 × 16's.

11. Design a simply-supported, glued-laminated roof beam to meet the following requirements: span, L = 50 ft.; spacing, s = 20 ft.; dead load of supported joists (on 20 in. centers) and nailed sheathing w_{DL} = 10 psf; snow load on roof w_{SL} = 30 psf; deflection limits, δ_{TL} = L/180; camber, c = $1.5\delta_{DL}$ Use wood with the following allowable stresses: F_b = 2.6 ksi, F_v = 0.165 ksi, $F_{c\perp}$ = 0.45 ksi and E = 1,800 ksi. Consider the top flange of the beam braced laterally by the joists. (Hint: follow step-by-step process used in Illustrative Example of Section 10.2.)

12. Savings in multiple-span structures are obtained through the use of cantilevered and suspended-span members which may reduce substantially the maximum design bending moment. A continuous beam is thus transformed if, at certain sections, hinges are placed that may transmit shear and longitudinal forces but no bending moment. Consider the case of a three-span beam as shown in Fig. 7.35. In order to determine the optimum position of the hinges it is necessary to obtain the following information:
    1) Distance X that will make the negative moment at the interior support equal to the positive moment at midspan of the central member.
    2) Same as above for distance Y, see Fig. 7.35c.
    3) Distance X that will make the negative moment of the interior support equal to the maximum positive moment in the exterior span.

    Now, assuming the particular case $L_e = L_c = L$, do the following:
    4) Determine the moments at the interior support, midspan of the central span and maximum positive moment at the exterior span of the continuous beam.
    5) Same as item above for beams of parts 1, 2 and 3. Tabulate the results of all four solutions and indicate the one leading to the smallest prismatic beam.

    Ans.

    1) $\dfrac{wL_c^2}{16} = \dfrac{w}{2}(L_c - 2X)X + \dfrac{wX^2}{2}$, $X = 0.1464L_c$

    2) $\dfrac{wL_c^2}{16} = \dfrac{w}{2}(L_e - Y)Y + \dfrac{wY^2}{2}$

    3) $\dfrac{wL_e^2}{11.65} = \dfrac{w}{2}(L_c - 2X)X + \dfrac{wX^2}{2}$

    4) and 5)

	$\dfrac{M_e^+}{wL^2}$	$\dfrac{M^-}{wL^2}$	$\dfrac{M_c^+}{wL^2}$	
Continuous Beam	0.080	− 0.100	0.025	
Case 1	0.0957	− 0.0625	0.0625	X/L = 0.1464
Case 2	0.0957	− 0.0625	0.0625	Y/L = 0.125
Case 3	0.0859	− 0.0859	0.0391	X/L = 0.220

Case 3 renders the smallest prismatic beam.

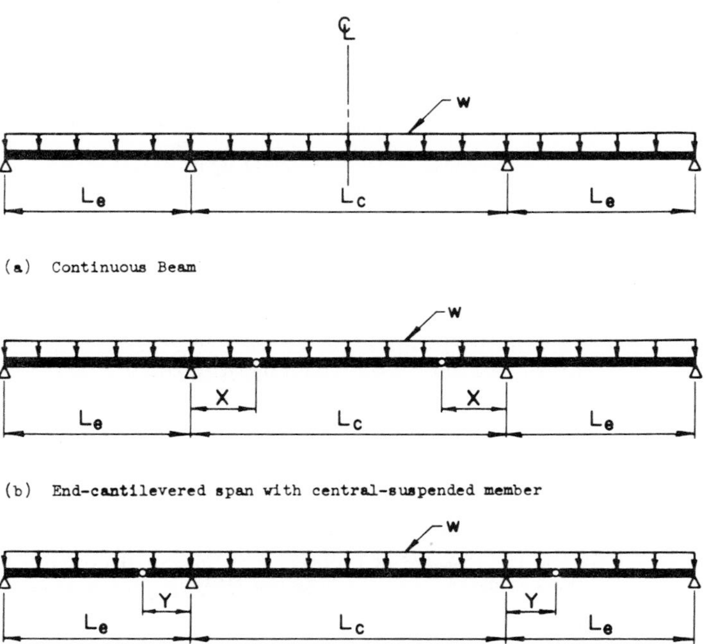

Figure 7.35. The beams in problems 12, 13, 14.

13. Consider a three-span beam such as shown in Fig. 7.35 for which $L_e = 0.8L_c$. Determine the distances X and Y as required in parts 1, 2 and 3 of the preceding problem. Analyze the continuous beam, and the beams shown in Fig. 7.35, using the values of X and Y determined before. Tabulate the results of all four solutions and indicate the one leading to the smallest prismatic beam.
14. A beam such as shown in Fig. 7.35 is built in 3 parts to be hinged in place; consider L_e = 50 ft. and L_c = 100 ft. The hinges may be assumed to transmit only shear and longitudinal loads but no bending moment. A glued-laminated beam with a uniform prismatic section is desired. The permanent load on the beam, including self-weight, is 200 lb./ft.; the live load is 300 lb./ft. uniformly distributed. Determine the following:
    a) Distance X of the hinge to the interior support that makes the maximum positive moment, at midspan of the suspended span, equal to the negative moment over the support.
    b) Shear to be transmitted by hinges; shears and reactions at the supports; maximum bending moment. Check the possibility of uplift at an exterior support when the live load is placed only in the interior span.
    c) Section of the beam. Assume continuous lateral support offered by the deck.
    d) Design the hinge connection. Study at least two types, determining size and number of fasteners, in each case. Show details.

Ans.

a) $X = 14.6$ ft.

b) $V_{hinge} = 17.7$ k, $V_{supp.} = 25$ k, $R_{supp} = 43.7$ k, $M^+ = M^- = 312$ k-ft. Uplift forces equal to 1.24 k may be developed at the exterior supports when the live load is placed only in the central span.

15. Beam tables are useful design tools for initial proportioning of possible sections. Consider the following conditions for the design of floor beams, under normal loading conditions, using standard glued-laminated members:

1) Simple spans only.
2) Beam weight must be subtracted from total load-carrying capacity.
3) Design for uniform loads of 40 psf live load and 10 psf dead load.
4) Allowable stresses are: $F_b = 2.4$ ksi (reduced by size factor), $F_v = 0.165$ ksi, $E = 1{,}800$ ksi.
5) Deflection limits: 1/360th of span for live load only.
6) Lateral stability is provided to the compression flange of the beam by the deck.
7) Use Table A.3.4 for section properties.

Write a computer program that will list the best solution for spans ranging between 20 ft. and 60 ft., in increments of 2 ft., and beam spacings of 8, 12 and 16 ft. Using hand computations check the adequacy of at least one group of solutions given in the answers below.

Ans.

The following sample answers are obtained from "Structural Glued-Laminated Timber," by AITC, publication No. 6.5/Ai, Jan. 1972.

Span, ft.	Spacing, ft.	Section
20	8	3⅛ × 16½
	12	5⅛ × 15
	16	5⅛ × 18
30	8	5⅛ × 21
	12	5⅛ × 22½
	16	5⅛ × 25½
40	12	6¾ × 28½
	16	6¾ × 31½
50	12	6¾ × 34½
	16	8¾ × 34½
60	12	8¾ × 39
	16	8¾ × 42

16. It is possible to show that the shear distribution across a certain section of a tapered beam subjected to transverse load is linear. In this respect, determine: (a) the depth, d, of that section, (b) the equation that gives shear-stress distribution at the section and (c) the point at which the maximum shear stress occurs and its intensity. (Hint: Use Eq. 7.61.)

Ans.

a) $d = 3M\,(dd/dx)/(dM/dx)$

b) $f_{xy} = 2\,(y/d)(dM/dx)/bd$

3) At $y/d = 1$, $f_{xy} = 2\,(dM/dx)/bd$

17. It is possible to show that the shear distribution across a certain section of a tapered beam subjected to transverse load is a function of $(y/d)^2$ alone. Determine:
(a) the depth d of that section, (b) the equation that gives shear-stress distribution at the section, and (c) the point at which the maximum shear-stress occurs and its intensity.

Ans.
a) $d = 2M\,(dd/dx)/(dM/dx)$
b) $f_{xy} = 3\,(y/d)^2\,(dM/dx)/bd$
c) At $y/d = 1$, $f_{xy} = 3\,(dM/dx)/bd$

18. Let f_x, f_{xy} and f_y be the bending, shear and vertical stresses at the taper, $y/d = 1$, of a given tapered beam. The stresses are referred to axes x and y, as shown in Fig. 7.29. Show that these stresses can be transformed to the x′, y′ coordinate axes, also shown in the figure, by means of the following transformation equations:

$$f_{x'} = f_x\cos^2\theta + f_y\sin^2\theta + 2f_{xy}\sin\theta\cos\theta$$
$$f_{y'} = f_x\sin^2\theta + f_y\cos^2\theta - 2f_{xy}\sin\theta\cos\theta \qquad (7.77)$$
$$f_{x'y'} = -f_x\sin\theta\cos\theta + f_y\sin\theta\cos\theta + f_{xy}(\cos^2\theta - \sin^2\theta)$$

Using Eqs. 7.77, show that $f_{y'}$, and $f_{x'y'}$ are zero, as can be expected from the boundary conditions at the taper.

19. At the tapered sides ($y/d = 1$), of test beam No. 1, Figs. 7.30 and 7.31 show the following strains: $\epsilon_x = 1{,}060 \times 10^{-6}$, $\epsilon_y = -320 \times 10^{-6}$, $\epsilon_{xy} = 2{,}000 \times 10^{-6}$. Assuming elastic isotropic behavior of the material, determine the corresponding bending, vertical and shear stresses. Use $E = 1{,}800$ ksi, $G = 110$ ksi, and $\mu = 0.45$. (Hint: Use the stress-strain relationships given in the respective Figures.)

Ans.
$f_x = 2.07$ ksi, tension
$f_y = 0.35$ ksi, tension
$f_{xy} = 0.22$ ksi

20. Consider a simply-supported tapered beam, such as shown in Figs. 7.27a and b, subjected to a uniformly distributed load, w. Show that the maximum bending stress occurs at a section where the depth, d, is given by Eq. 7.76. (Hint: Use Eq. 7.59 with $M = \frac{wLx}{2} - \frac{wx^2}{2}$, $d = d_e + x\tan\theta$.) Also show that the maximum bending stress is given by $f_x = \frac{3\,wL^2}{4\,bd_e\,(d_e + L\tan\theta)}$ (Hint: Use Eq. 7.60.)

21. The tapered glued-laminated beams designed in the Illustrative Example of Section 7.15 were spaced 16 ft. apart. This spacing requires a 4 in. thick laminated roof deck. If the beams were spaced only 12 ft. apart, a 3 in. thick deck would be sufficient, and possible savings in cost may ensue. Design the typical beam for this case, assuming that all other information given in the Illustrative Example applies here also.

22. Determine the ultimate moment and ultimate curvature for the section shown in Case 1 of Fig. 7.6, assuming that the intergrown knot shown occurs in the top of the section instead of at the bottom. Assume that because of the presence of the intergrown knot the compression strength is reduced to $f_c^u = 6$ ksi at $e_c^u = 0.003$; note that E_c remains at 2000 ksi.

    Ans. $a = 7.08$ in., $M_u = 1{,}190$ k-in, $\theta_u = 4.2 \times 10^{-4}$ rad/in.

23. Determine the ultimate moment and ultimate curvature for the section shown in Case 2 of Fig. 7.6, assuming that the knothole shown occurs in the top of the section instead of at the bottom. Hint: Iterate to determine the distance of the neutral axis to the top of the beam, until $\dfrac{C - T}{C + T} < 0.01$.

    Ans. $a = 7.40$ in., $M_u = 1{,}680$ k-in., $\theta_u = 6.7 \times 10^{-4}$ rad/in.

REFERENCES

1. "National Design Specification for Stress-Grade Lumber and its Fastenings," 1977 Edition. National Forest Products Association, Washington, D.C.
2. American Institute of Timber Construction, "Timber Construction Standards," AITC 100–69, Fifth Edition, 1969, Englewood, Colorado 80110.
3. American Plywood Association, "Plywood Design Specifications," 1966. Published by American Plywood Association, Tacoma, Washington.
4. Den Hartog, J. P., "Strength of Materials," Dover Publications, Inc., New York.
5. Bach, C., and Baumann, R., "Elastizitat und Festigkeit," 9 Auflage, Julius Springer. Berlin, 1924, pp. 300–308.
6. Ramos, Agustin N., "Stress-Strain Distribution in Douglas Fir Beams Within the Plastic Range," U.S. Forest Products Laboratory, Report No. 2231, December 1961.
7. Newlin, J. A., and Trayer, G. W., "From Factors of Beams Subjected to Transverse Loading Only," National Advisory Committee for Aeronautics Report No. 181, 1924. Reprinted as Forest Products Lab. Report No. 1310, 1941.
8. Bohannan, Billy, "Effect of Size on Bending Strength Wood Members," Forest Products Laboratory, U.S. Department of Agriculture, Research Paper FPL 56, May 1966.
9. Blume, J. A., Newmark, N. M., and Corning, L. H., "Design of Multistory Reinforced Concrete Buildings for Earthquake Motions," Portland Cement Association, Chicago, Illinois. 1961.
10. Dietz, A. G. H., "Stress-Strain Relations in Timber Beams of Douglas Fir," ASTM Bulletin No. 118, October 1942, pp. 19–27.
11. Freas, A. D., and Selbo, M. L., "Fabrication and Design of Glued-Laminated Wood Structural Members," U.S. Forest Products Laboratory Technical Bulletin No. 1069, February 1954.
12. Forest Products Laboratory, "Wood Handbook," Agriculture Handbook No. 72, U.S. Department of Agriculture, Washington, D.C. Revised August 1974.
13. Weibull, W., "A Statistical Theory of the Strength of Materials," Swedish Royal Inst. Eng. Res. Proc., Stockholm, Sweden, 1939.
14. American Institute of Timber Construction, "Standard Specifications for Structural Glued-Laminated Timber," AITC 203–70, AITC, Englewood, Colorado 80110.
15. Newlin, J. A., Heck, G. E., and March, H. W., "New Method of Calculating Longitudinal Shear in Checked Wooden Beams," American Society of Mechanical Engineers Transactions, Vol. 56, pp. 739–744, 1934.
16. American Association of State Highway and Transportation Officials, "Standard Specifications for Highway Bridges," Twelfth Edition, 1977, published by AASHTO, Washington, D.C. 20004.

17. Flint, A. R., "The Lateral Stability of Unrestrained Beams," Engineering, Vol. 173, 1952, pp. 65–67, 99–102.
18. Hooley, R. F., and Madsen, B., "Lateral Stability of Glued-Laminated Beams," Journal of the Structural Division, ASCE, Vol. 90, No. ST3, Proc. Paper 3948, June, 1964, pp. 201–218.
19. American Plywood Association, "Design of Plywood Beams," Supplement No. 2 to Plywood Design Specification, by APA, Tacoma, Washington, 1968.
20. Timoshenko, S. P., and Young, D. H., "Theory of Structures," Second Edition, McGraw-Hill Book Company, New York, 1964.
21. Roark, R. J., "Formulas for Stress and Strain," McGraw-Hill Book Co., Inc., New York, 1954.
22. American Institute of Timber Construction, "Timber Construction Manual," John Wiley and Sons, New York, Second Edition, 1974.
23. American Institute of Timber Construction, "Standards for the Design of Structural Timber Framing," AITC 102–69, Timber Construction Standards, Fifth Edition, 1969.
24. Sawyer, D. A., "Ponding of Rainwater on Flexible Roof Systems," Journal of the Structural Division, ASCE, Vol. 93, No. ST1, Proc. Paper 5094, February 1967, pp. 127–147.
25. Chinn, J., Mansouri, A. H., and Adams, S. F., "Ponding of Liquids on Flat Roofs," Journal of the Structural Divison, ASCE, Vol. 95, No. ST5, Proc. Paper 6539, May 1969, pp. 797–807.
26. Hanssler, R. W., "Roof Deflection Caused by Rainwater Pools," Civil Engineering, Vol. 32, No. 10, October 1962, pp. 58–59.
27. Wilson, T. R. C., "The Glued-Laminated Wooden Arch," Technical Bulletin No. 691, U.S. Department of Agriculture, Washington, D.C., October 1939.
28. Maki, A. C., and Kuenzi, E. W., "Deflection and Stresses of Tapered Wood Beams," U.S. Forest Service Research Paper FPL 34, 54 pp., illust., September 1965.
29. Norris, C. B., "Strength of Orthotropic Materials Subjected to Combined Stresses," Forest Products Laboratory Report No. 1816, 24 pp. illust., 1950.
30. Newmark, N. M., "Numerical Procedure for Computing Deflections, Moments, and Buckling Loads," Transactions, ASCE, V. 108, 1943, pp. 1161–1188.
31. Structural Stability Research Council, "Guide to Stability Design Criteria for Metal Structures," Third Edition ed. by B. G. Johnston, John Wiley and Sons, New York, 1976.

Chapter 8
Behavior and Design of Wood Columns

8.1 INTRODUCTION

Structural members subjected mainly to compression stresses are called columns. As such, they are used in buildings and bridges to support girders and to transmit vertical loads to the foundations. Other examples of compression members include arches in general and the top chords and some web members of simply-supported trusses.

It is unlikely for any compression member to be subjected to axial compression only. The case of a column under axial load is only an idealized case which rarely occurs in actual practice. Even in columns designed for axial loading, small eccentricities of the load with respect to the column axis are created due to construction tolerances, initial curvatures or material nonhomogeneity; these eccentricities create bending moments in the section of the column. Therefore, most columns are designed for the simultaneous action of axial loads and bending moments due to lateral loads induced by wind or earthquake and rigid connections in framed structues; in others, gravity loading alone creates bending moments. Hence, it is reasonable to design wood columns for a minimum eccentricity of the load even if analysis indicates an absence of bending moments.

A column subjected to a load P with an eccentricity measured from the axis of the column is shown in Fig. 8.1. Theoretically, any transverse section of the column is subjected to an axial

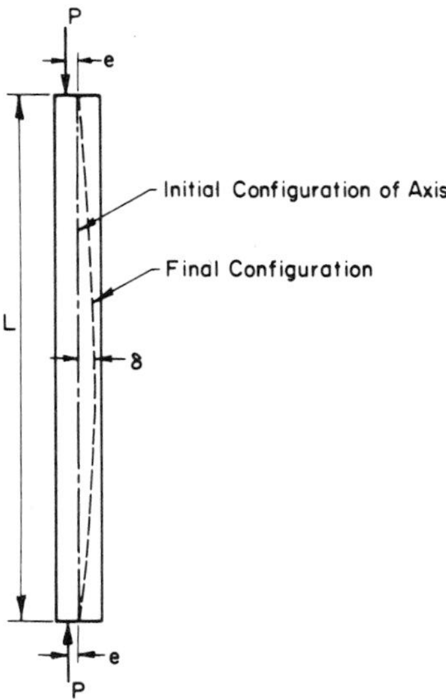

Figure 8.1. Wood column under eccentric loading.

load, P, and a moment, M = Pe. Due to the combined action of P and M, the column bends; the maximum midspan deflection δ is given by (1):

$$\delta = \frac{M_0 L^2}{8EI} \lambda(u) \tag{8.1}$$

where $\lambda(u) = 2(1 - \cos u)/u^2 \cos u$, $u = (\pi/2)\sqrt{P/P_{cr}}$ and $P_{cr} = \pi^2 EI/L^2$ is the critical load of a braced column with length L, modulus of elasticity, E, and sectional moment of inertia, I. The function $\lambda(u) > 1$ is a magnification factor that when multiplied by the deflection of the member without axial load, yields the actual deflection. The final eccentricity of the load at the midspan section is $e + \delta$. The moment at midspan is increased from $M_o = Pe$ to $M_{max} = P(e + \delta)$. In general, a good approximation of the maximum moment for values of $P/P_{cr} < 0.6$ can be found from the following equation (2):

$$M_{max} = M_o + \frac{P\delta_o}{1 - P/P_{cr}} \tag{8.2}$$

where M_o and δ_o are, respectively, the moment and deflection without regard to the added moment caused by deflection. In short columns the difference between M_{max} and M_o is negligible; not so in long columns where secondary effects may be substantial.

The relationship between the acting load and the bending moment for a given eccentricity, when the load is increased from zero to ultimate, is useful since it permits the study of behavior of the column in the entire range of load. If this relationship is developed for the critical section of the column it is called path to failure. Typical paths to failure of a short and a long column are labeled A and B, respectively, in Fig. 8.2. For short columns this relationship is nearly a straight line; for slender columns the path to failure is distinctly nonlinear. Curve A shows that failure occurs as the axial load is still increasing and the wood crushes. Curve B shows that the

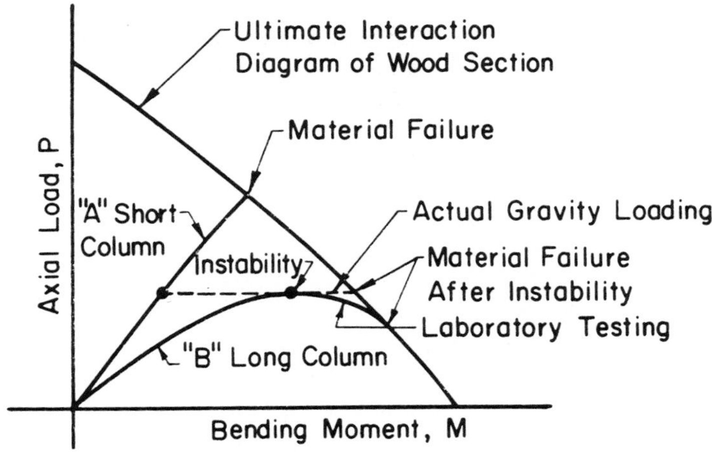

Figure 8.2. Paths to failure of wood columns subjected to eccentric loading.

slender column is likely to develop instability, defined by a negative dP/dM, at which point because of the deformation, the bending moment increases under a decreasing axial load. Thus, in a long column the largest axial load that the column is capable of supporting is reached before the actual full crushing strength of the wood is developed. For both short and long columns, material failure occurs when the path to failure intersects the envelope of all possible combinations of ultimate axial load and moment of the section for the given moisture content and duration of loading. This envelope is known as the ultimate interaction diagram, see Fig. 8.2.

In laboratory testing of a long column material failure is reached after increased deformations, at the intersection of the path to failure with the interaction diagram. For a long column under actual gravity load, material failure occurs at the intersection of the horizontal tangent through the peak with the interaction diagram. In a long column curve A represents the behavior of the end sections; the horizontal distance between curves B and A, for any given load, is the additional moment at the critical section due to secondary deformations.

The importance of the ultimate interaction diagram is thus shown. In what follows the diagram is determined and from it, the linear version used in design is derived.

8.2 DETERMINATION OF INTERACTION DIAGRAM

There are many possible combinations of axial load P and bending moment M of a given duration that can create failure conditions in a section of a wood column at a given moisture content. These ultimate loading conditions can be plotted as points of a graph in which the ordinates represent axial loads and the abscissas repesent bending moments. The curve that connects all the points in the graph is the ultimate interaction diagram, UID, of the section. Any combination of axial load and bending moment which plots inside the area bound by this curve can be resisted by the given section.

A typical interaction diagram for a rectangular wood section is shown in Fig. 8.3; its determination can be accomplished using the following assumptions:

1. Strains are linearly distributed in a section subjected to eccentric loading.
2. The relation between strains and stresses in the section due to eccentric loading is the same that exists between strains and stresses in direct compression and tension.
3. The stress-strain diagrams in direct compression and tension for clear wood are available for a given duration of loading (7 to 8 minutes) and moisture content.
4. Failure occurs when either the ultimate strain in compression or the ultimate strain in tension is attained at an extreme fiber of the section.
5. The section belongs to a member that is free of any strength-reducing characteristics of wood that may affect the validity of any of the preceding assumptions.

These assumptions are similar to those given in Section 7.3. The UID obtained using the preceding assumptions is valid only for a specimen of clear wood at a given moisture content and duration of loading; an impractical case, at best. The actual UID for a given wood section would depend on nearby strength-reducing characteristics such as knots, slope of grain, compression wood and others which affect the strength of the wood. However, the limitations in validity of a clear wood UID are compensated by the insight and understanding of the behavior of a section under eccentric loading that is obtained.

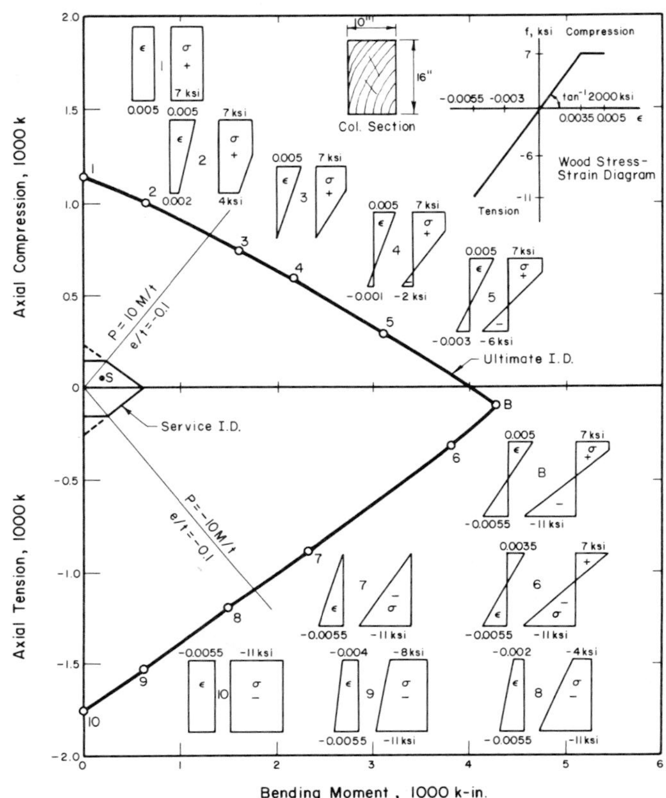

Figure 8.3. Ultimate interaction diagram axial load vs. bending moment for a given wood section.

The various combinations of P and M that create failure in a given wood section can be obtained using the relations of equilibrium for ultimate strain distributions that satisfy assumption No. 4. Let P_u and M_u be the external ultimate load and moment, respectively, on the section. In order to satisfy the equilibrium of forces, the sum of the internal forces in the section plus the external force P_u must equal zero. Also, the sum of the moments of the internal forces plus the external moment M_u must equal zero. This leads to two equations the solution of which yields a pair of values, P_u and M_u, which cause failure of the given section.

A systematic way to determine points of the interaction diagram is illustrated in Fig. 8.4. The upper portion of the diagram may be obtained as follows. One may start, see point 1 of Fig. 8.3, with a uniform compressive strain distribution over the section at an ultimate strain ϵ_c^u. Then, other ultimate strain distributions, see points 2 through 5 in Fig. 8.3, can be obtained by rotating about point C, in Fig. 8.4, until the strain distribution labeled "Balanced" is reached. The latter corresponds to simultaneous failure in tension $\epsilon_t = \epsilon_t^u$ and compression $\epsilon_c = \epsilon_c^u$ at the extreme fibers, see point B in Fig. 8.3. The bottom portion of the interaction diagram can be obtained by

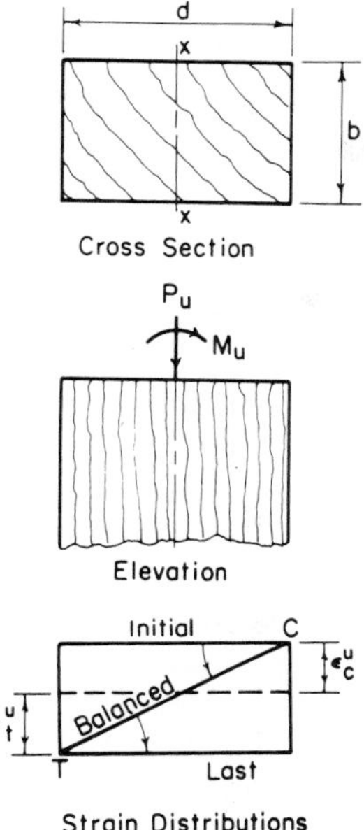

Figure 8.4. Determination of ultimate interaction diagram about axis X-X for a given wood section.

rotating the strain distributions of Fig. 8.4 about point T, see points 6 through 9 in Fig. 8.3, until a uniform strain distribution ϵ_t^u is achieved in the section; the last distribution gives point 10 on Fig. 8.3. The bottom portion of the interaction diagram is of interest only in the design of members subjected to the combined action of tensile axial loads and moments. It can be pointed out that the intersection of the interaction diagram with the horizontal axis corresponds to ultimate conditions due to pure bending in the absence of axial load. The moment for this point may be obtained exactly from Eq. 7.15 or approximately from Eq. 7.17, see Section 7.3.

The presence of strength-reducing characteristics in the wood of a given structural member can be recognized in the determination of the UID for the section. If the defect is such that the tensile strength is reduced, as for the case of slope of grain, the interaction diagram is reduced accordingly. For the case of reduced compressive strength due to some defect in the wood, as presence of compression wood for instance, the UID is also reduced.

A service interaction diagram can be obtained easily, if allowable stresses for wood at a given moisture content are assigned for design using normal loading conditions and service loads. Since allowable stresses are usually small it is likely that they will fall in the straight line portion of the corresponding stress-strain diagrams of the given wood. It can be shown that for a linear stress-strain relation, both top and bottom portions of the interaction diagram are straight lines, see Fig. 8.3. This shape changes somewhat if a minimum eccentricity ratio $e/t = 0.1$ for column design is assumed. The service interaction diagram is reduced as shown in Fig. 8.3. The intersection of the line $P = 10\ M/t$ with the diagram determines the maximum service axial load for the given section. Thus, the top portion of the diagram is truncated into a horizontal line. Similar considerations can be made for the bottom portion of the diagram; for ultimate design of wood columns, the reduced UID may also be truncated to account for minimum eccentricity.

An illustrative example of application of the preceding concepts is given as follows.

8.2.1 Illustrative Example

A wood column with a rectangular section 10 in. by 16 in., actual dimensions, is subjected to the simultaneous action of axial load P_u and moment M_u about its strong axis of bending. The stress-strain diagram of clear wood in direct compression and tension is given in Fig. 7.5 except, for simplification purposes, $E_t = E_c = 2{,}000$ ksi and $f_t^u = 11$ ksi at $\epsilon_t^u = 0.0055$. The problem consists of eight parts as follows:

Part 1. Determine the UID of a section of clear wood. The solution is given in Fig. 8.3. The UID is the curve connecting points labeled 1 through 11, including point B; the strain and stress distributions for the various points are also shown. These points were determined following the systematic procedure shown in Fig. 8.4. As an example, the specific determination of point 5 is given in Fig. 8.5. The ultimate compression strain $\epsilon_c^u = 0.005$ (in the extreme fiber to the right) and a tensile strain $\epsilon_t = 0.003$ (in the extreme fiber to the left) determine an ultimate strain distribution. The corresponding stress distribution and the internal forces acting in the section can be determined from the given strain distribution and stress-strain relations. Static equilibrium of the section yields the following equations:

$$210 + 245 - 180 - P_u = 0$$

$$210\left(8 - \frac{1}{2} \times 3\right) + 245\left(8 - 3 - \frac{1}{3} \times 7\right) + 180\left(8 - \frac{1}{3} \times 6\right) - M_u = 0$$

from which $P_u = 275$ k, $M_u = 3{,}100$ k-in. All other points shown in Fig. 8.3 were obtained using the same numerical procedure. The case of pure bending, $P_u = 0$, can be easily obtained from Eq. 7.17, for $\alpha = 0.0035/0.005 = 0.7$, as $M_u = (1/6)(10)(16)^2(7)(2 - 0.7) = 3{,}900$ k-in.

Since stress distributions for the given case are linear between points 6 and 10 it follows that the corresponding portion of the interaction diagram is a straight line.

Part 2. The column may be used for the determination of the UID of a wood section with reduced strength due to the presence of defects. Assume that because of slope of grain the tensile strength of the given wood is reduced to 6 ksi at an ultimate strain $\epsilon_t^u = 0.003$; the stress-strain diagram in direct compression is not affected. It is wished to determine the UID for this case.

Note that the upper branch of the reduced UID is identical to that shown in Fig. 8.3 from point 1 to point 5. The latter corresponds to the balanced conditions for this case, since both maximum compressive and tensile strains have been reached simultaneously. The lower branch of the UID may be obtained following the procedure indicated in Fig. 8.4, where $\epsilon_t^u = 0.003$ and $f_t^u = 6$ ksi. The results are indicated in Fig. 8.6. This shows that the presence of slope of grain in the member reduces the capacity of the section.

Failure under pure bending, $P_u = O$, is determined by the reduced tensile strength of the section; the collapse mechanism follows a tension-compression sequence in this case. Since $\epsilon_t^u < \alpha\, \epsilon_c^u$, i.e. $0.003 < 0.0035$, then $M_u = (1/6)(10)(16)^2(6) = 2{,}560$ k-in.

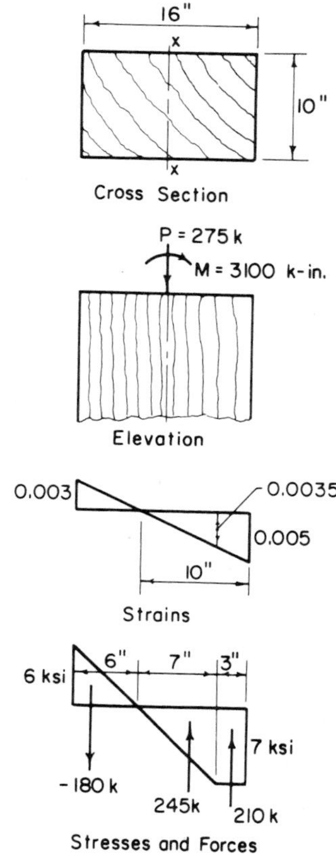

Figure 8.5. Determination of ultimate axial load and bending moment in a given column about axis X-X for a given strain distribution.

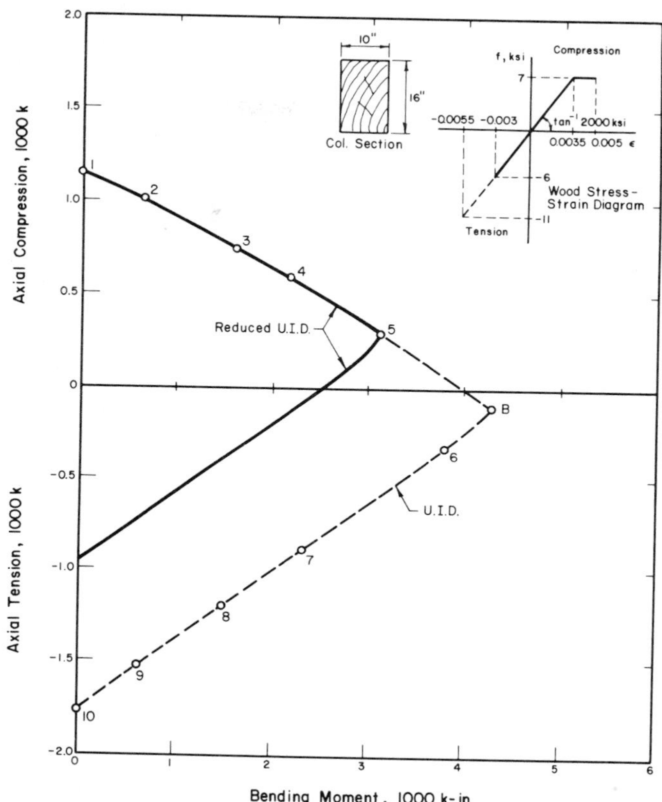

Figure 8.6. Ultimate interaction diagram reduced by a wood defect that affects its tensile strength.

Part 3. Due to the presence of hypothetical defects, the compression stress-strain diagram of wood is reduced to a straight line from zero to $\epsilon_c^u = 0.0035$, $f_c^u = 7$ ksi. Hence, the flat portion of the diagram shown in Fig. 7.5 is reduced to zero. It is wished to determine the reduced UID for this case.

There is no need for additional calculations as points 1, and 6 through 10 of Fig. 8.3 belong also to the given case. Thus, the bottom branch between points 6 and 10, is identical to that determined in the first part of the problem. Since the compression stress-strain diagram is now a straight line, the top branch of the UID is also a straight line between points 1 and 6; see Fig. 8.7. Thus, the given wood defect results in a loss of capacity of the section.

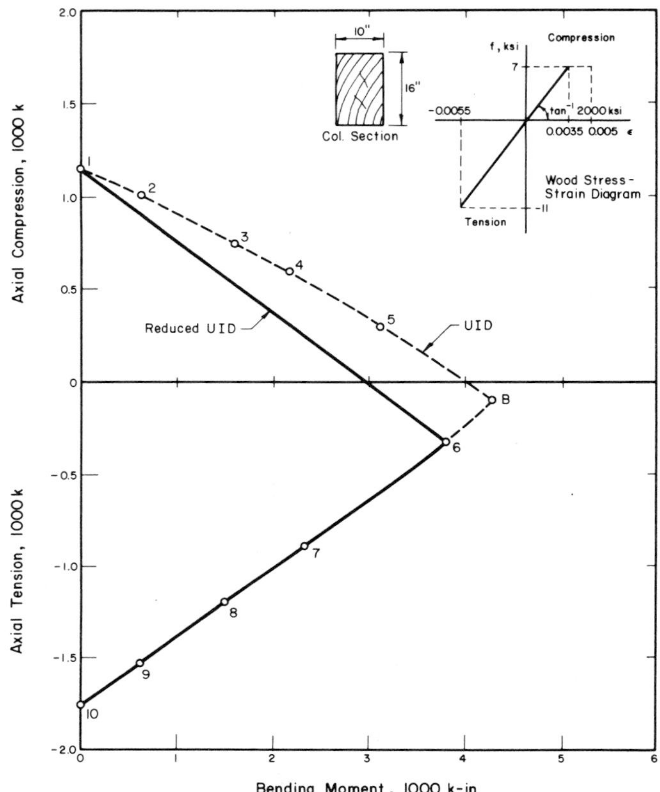

Figure 8.7. Ultimate interaction diagram reduced by a wood defect that affects its compressive strength.

If the wood defects of parts 2 and 3 occur at the same section, the reduced UID is given in Fig. 8.8 with a balanced point labeled "b" at the intersection.

Part 4. For the purpose of design using working stress procedures the column is stress-graded for normal loading conditions and 19 percent maximum moisture content as follows: allowable stress parallel to grain in direct compression F_c = 1.3 ksi, in direct tension F_t = 1.6 ksi and in bending F_b = 1.6 ksi. It is wished to determine the sevice interaction diagram for this case.

The solution is obtained with only three points, namely: 1) maximum compression load P = 1.3 (10 × 16) =208 k, M = 0, 2) maximum tension load, P = − 1.6 (10 × 16) = − 256 k, M = 0 and 3) pure bending moment P = 0, M = (1/6) (10) $(16)^2$ (1.6) = 684 k-in. The diagram is shown in Fig. 8.3. One can readily notice the substantially smaller size of the service interaction diagrams as compared to the UID of the clear wood section.

Part 5. Assume that the column given in part 4 is 14 ft long and is subjected to the combined action of longitudinal load P = 40 kips with an eccentricity of 3 in. and a distributed uniform

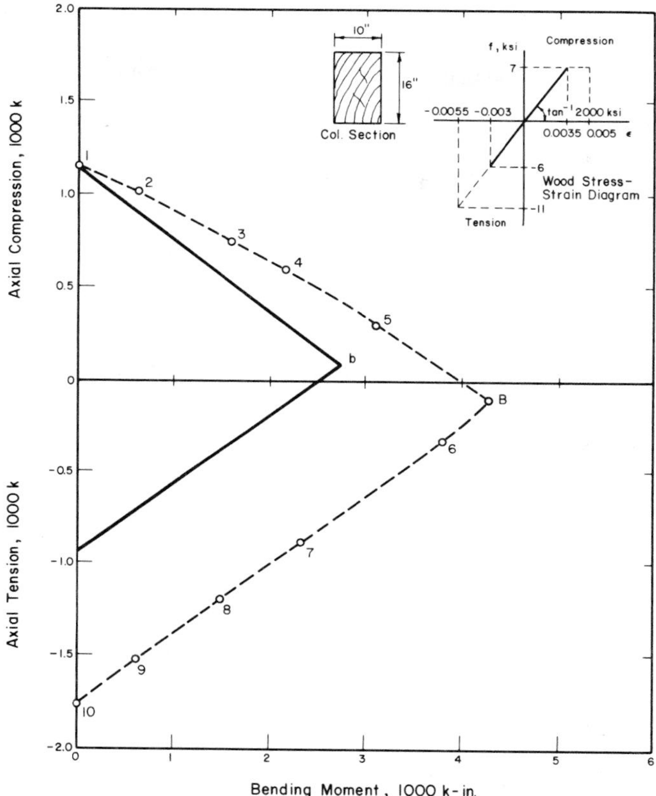

Figure 8.8. Ultimate interaction diagram reduced by wood defects that affect both its tensile and compressive strengths.

size load w = 0.245 kips per ft, all due to normal loading conditions, see Fig. 8.9. The ends of the column are hinged; relative lateral displacements of the ends, and flexure about the weak axis of the section, are prevented by bracing. Determine the adequacy of the 10 × 16 in. section of the given column to support the loads.

The conditions at the end sections of the column are P = 40 k, M = 40 × 3 = 120 k-in.; at midspan P = 40 k, M_o = 120 + (1/8) (0.245/12) (14 × 12)^2 = 192 k-in. Thus, the midspan section is the critical section. The actual moment at midspan, considering the column deformations that take place under load, is given by Eq. 8.2 in which

$$P_{cr} = \pi^2 \frac{EI}{L^2} \tag{8.3}$$

$$\delta_0 = \frac{5}{384} \frac{wL^4}{EI} + \frac{ML^2}{8EI} \tag{8.4}$$

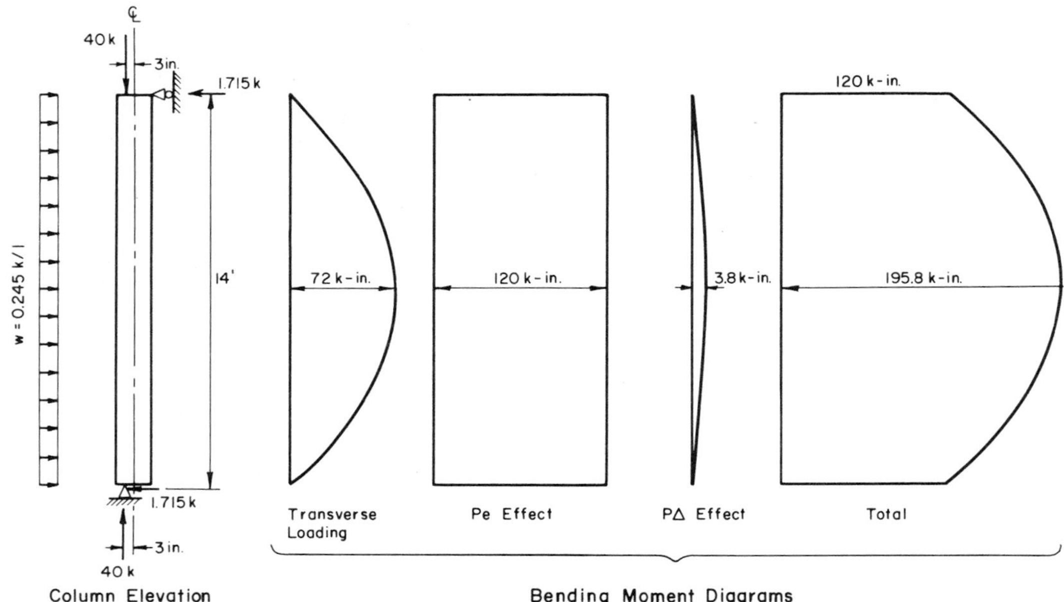

Figure 8.9. Wood column subjected to eccentric thrust and uniformly-distributed transverse load.

Substituting $w = (0.245/12)$ k/in., $M = 40 \times 3 = 120$ k-in., $L = 168$ in., $E = 2{,}000$ ksi and $I = 3{,}413$ in.^{4} in Eqs. 8.3 and 8.4 yields $P_{cr} = 2{,}380$ k and $\delta_0 = 0.093$ in. Then, from Eq. 8.2, since $P/P_{cr} < 0.6$,

$$M_{max} = 192 + \frac{(40)\ (0.093)}{1 - \dfrac{40}{2{,}380}} = 195.8 \text{ k-in.}$$

A summary of results for six sections of the column spaced 1.4 ft apart between the support and midspan is given in the following Table.

TABLE 8.1. BEHAVIOR UNDER LOAD OF COLUMN SHOWN IN FIGURE 8.9.

Section	Distance from Support, ft	Moment, k-in.	Top Strain 10^{-3} in./in.	Curvature 10^{-4} rad/in.	Deflection in.
At Support	0	120	0.266	0.176	0
2	1.4	147.2	0.298	0.216	0.0326
3	2.8	168.5	0.322	0.247	0.0591
4	4.2	183.7	0.340	0.269	0.0786
5	5.6	192.8	0.351	0.282	0.0906
At Midspan	7.0	195.8	0.354	0.287	0.0947

It follows that the midspan section is subjected to P = 40 k, M = 195.8 k-in.; point S in Fig. 8.3 corresponds to these conditions. Since point S plots inside the service interaction diagram of the given section, the column may be considered adequate to support the given loads.

Part 6. Determine the behavior of the beam-column of Part 5 under increasing thrust, while the transverse load is kept constant at w = 0.245 k/ft. Assume that there is adequate lateral bracing against bending about the weak axis of the cross section.

Research conducted by G. Ruzicka and the author, on beam-columns, generated a computer program that renders the analytical behavior of a given beam-column from initial loading to failure. The program uses Newmark's method (16) for the computation of deflected configurations, and a Newton-Raphson technique (17) to determine the actual curvatures and strains in all sections on which the beam-column is divided for analysis.

Use of this tool rendered the results shown in Fig. 8.10. Four curves showing the variation of thrust with midspan moment are given. They are labeled L/d = 5, 10.5, 20 and 40, respectively. These curves are the paths to failure for four independent beam-columns. The results sought are given by the curve labeled L/d = 10.5, which corresponds to a 14 ft column length and a 16 in. section depth.

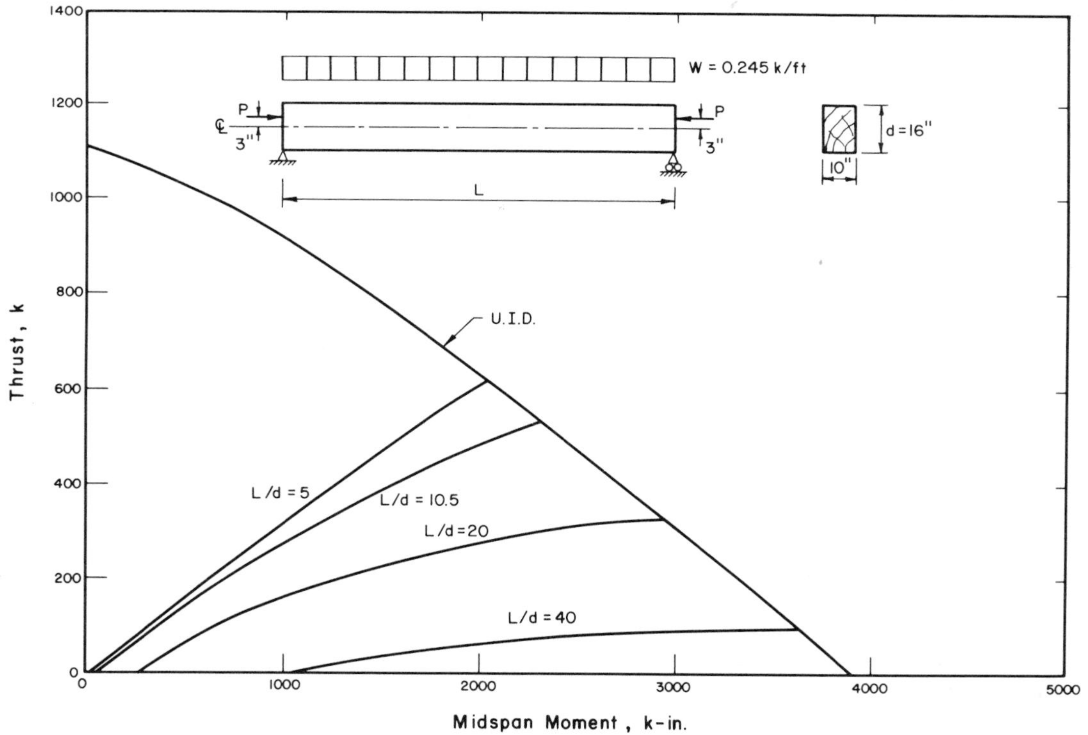

Figure 8.10. Variation of midspan moment with thrust for various slenderness ratios (constant transverse load).

Failure occurs at the intersection of the curve with the upper branch of the Ultimate Interaction Diagram for the given wood cross-section. It is noted that the deviation of the paths to failure from a straight-line variation is caused by the P-Δ effect principally, but also by material non-linearities all of which increase as the slenderness ratio, L/d, increases. Premature failure of the structure, caused by strength-reducing characteristics in the wood, could also be ascertained by using the corresponding reduced UID, as shown in the preceding sections of this problem.

Part 7. Determine the behavior of a 10 × 16 in. beam-column subjected to constant end moments, M = 1000 k-in., and increasing thrust. As before, assume that there is adequate lateral bracing against bending about the weak axis of the cross-section.

The results are shown in Fig. 8.11 for slenderness ratios L/d = 5, 20 and 40. Small increments in the midspan moment occur for the case where L/d = 5; here the P-Δ effects caused by the thrust acting on the deflected configuration are negligible. Not so as the slenderness ratio increases to 20 and 40, respectively. Particularly in the latter case the midspan moment at failure is magnified 260 percent from an initial value, M = 1000 k-in., for zero thrust to a value close to 3600 k-in. at a thrust of 105 k.

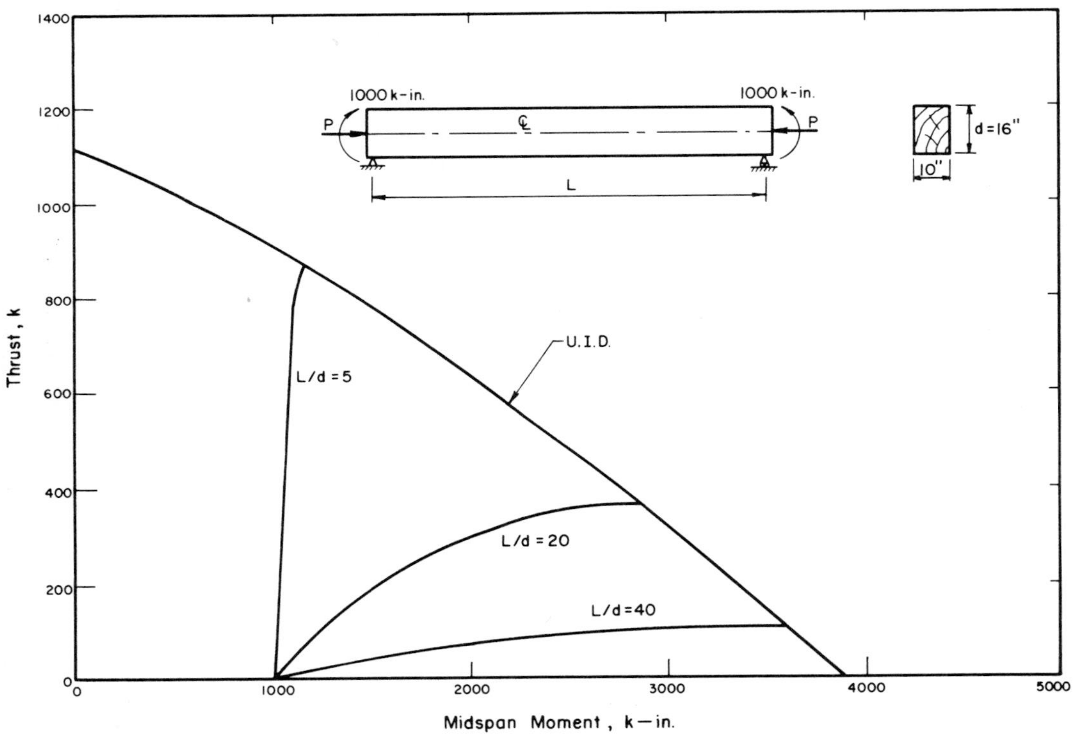

Figure 8.11. Variation of midspan moment with thrust for various slenderness ratios (constant end moments).

Part 8. Determine the behavior of a 10 × 16 in. beam-column subjected to eccentric thrusts. Consider two cases: one with small eccentricity, e = 0.2 in., and the other with a relatively large eccentricity, e = 8 in. Consider that adequate lateral bracing against bending about the weak axis of the cross-section has been provided.

The results are shown in Figs. 8.12 for e = 0.2 in., and Fig. 8.13 for e = 8 in., for various slenderness ratios. Deviation of the various paths to failure from the corresponding straight lines is more noticeable as the slenderness ratio, L/d, increases. This, again, is caused principally by the P-Δ effect which magnifies the maximum moment at midspan. In Fig. 8.12 buckling occurs for both cases shown; note that the maximum thrust is obtained well before failure of the cross-section occurs at the intersection of the path to failure with the upper branch of the UID.

For the purpose of comparison, the reader may wish to work out the solution of this illustrative example considering flexure about the weak axis of the section.

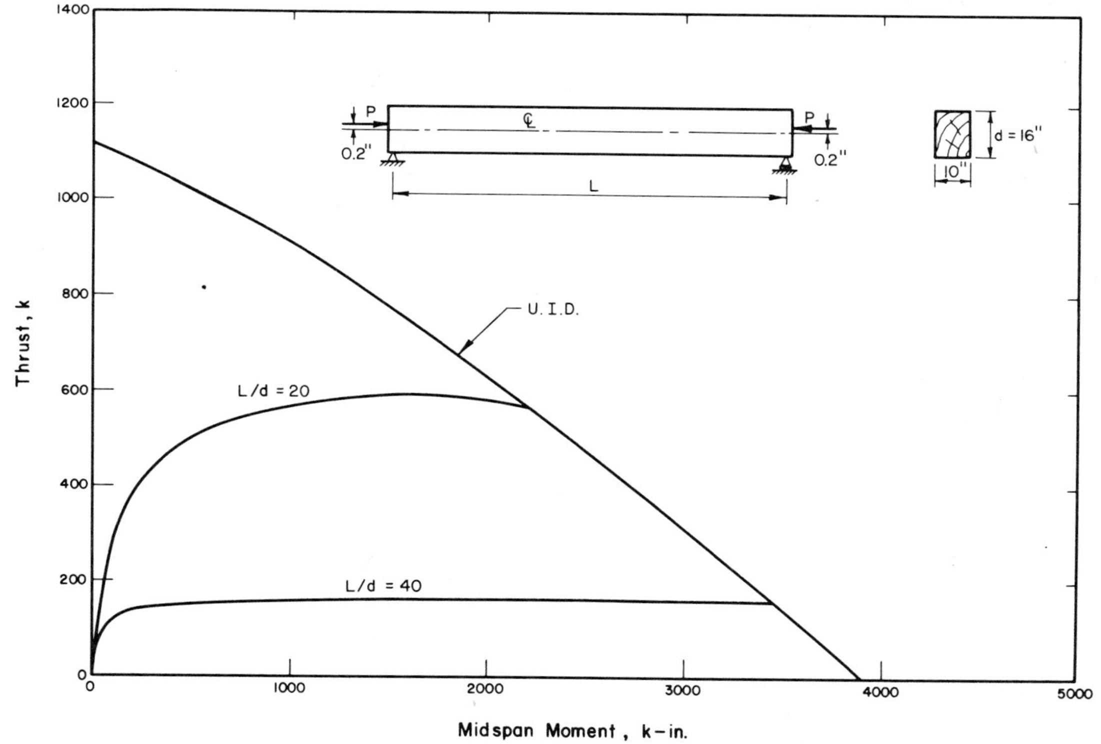

Figure 8.12. Variation of midspan moment with thrust for various slenderness ratios (constant eccentricity, e = 0.2 in.).

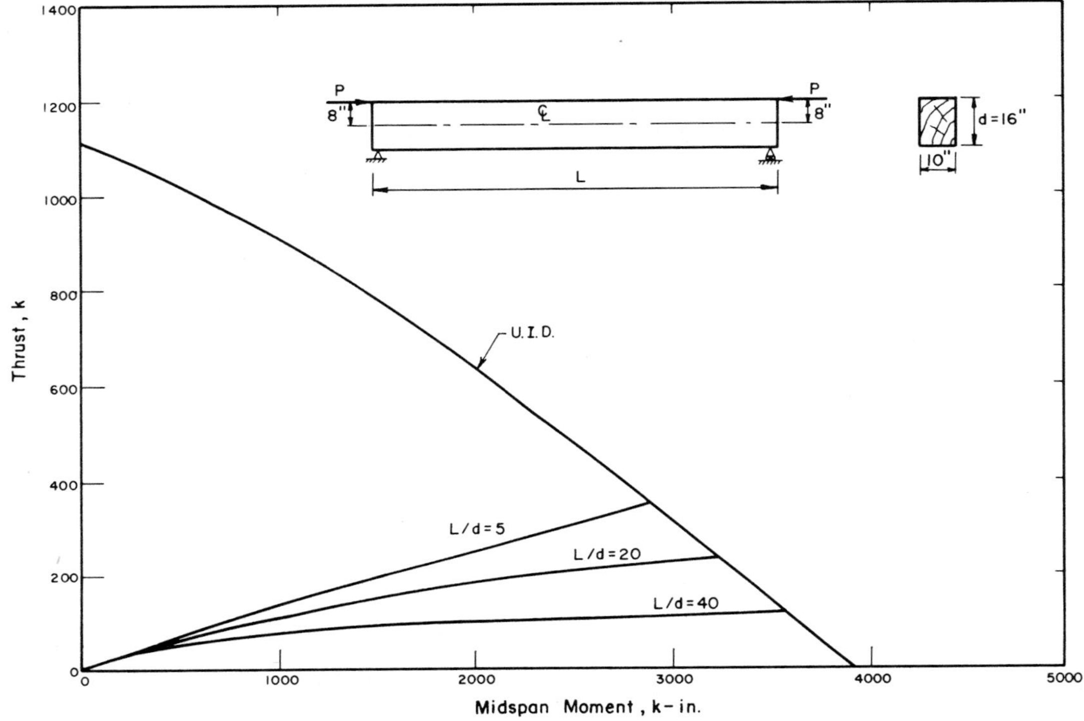

Figure 8.13. Variation of midspan moment with thrust for various slenderness ratios (constant eccentricity, e = 8 in.).

8.3 DESIGN OF AXIALLY-LOADED COLUMNS

The behavior of a wood column subjected to axial loading only depends on its slenderness ratio. Long slender columns fail at a critical load, P_{cr}, which is given by the Euler formula. For the classical case of a double-hinged, laterally-braced column, Euler's formula renders:

$$P_{cr} = \pi^2 \frac{EI}{L^2} \tag{8.3}$$

where E is the modulus of elasticity of wood, I the moment of inertia of the column cross-section and L is the length of the column. The critical load may be obtained also for various end rotational restraints, and for cases where the column is free to sway, using the concept of effective length, kL, see Fig. 8.14. Thus,

$$P_{cr} = \pi^2 \frac{EI}{(kL)^2} \tag{8.5}$$

for all cases. The coefficient of effective length, k, varies between 0.5 and 1 for braced columns and between 1 and ∞ for unbraced columns. For any given case the actual value of k may be

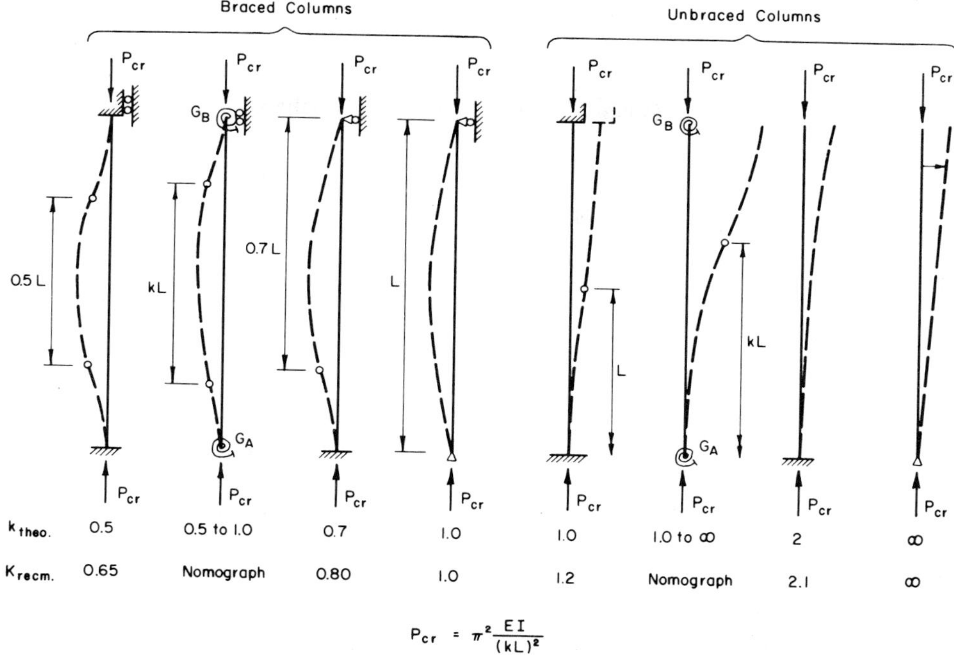

Figure 8.14. Effective length factor, k, for centrally-loaded columns with various end conditions.

determined using the alignment charts developed by Julian and Lawrence, see Fig. 8.15, or the graphs given by Gurfinkel and Robinson (Ref. 8, p. 163) for columns with uninhibited sidesway.

The critical stress, f_{cr}, may be defined as the compressive stress in the column cross-section when the critical load is attained. Thus, $f_{cr} = P_{cr}/A$. Also, $I = Ad^2/12$ for rectangular cross-sections. Substituting these values in the critical-load formula, Eq. 8.5, yields

$$f_{cr} = \frac{\pi^2}{12} \frac{E}{(k\ L/d)^2} \tag{8.6}$$

The design stress $F_{c'}$, may be determined by dividing the critical stress, f_{cr}, by a suitable reduction factor. A value of 2.74 for the latter has been selected to obtain the formula presently used by National Design Specification (3):

$$F_{c'} = 0.3 \frac{E}{(k\ L/d)^2} \tag{8.7}$$

in which the effective slenderness ratio k L/d is taken as the larger value of k L/d about either axis of bending. Usually, buckling occurs about the weakest axis of bending, with k L/d defined by the narrow side of the rectangular section. However, an exception may occur if additional bracing (in the form of girts at midheight, for example) were provided to the column.

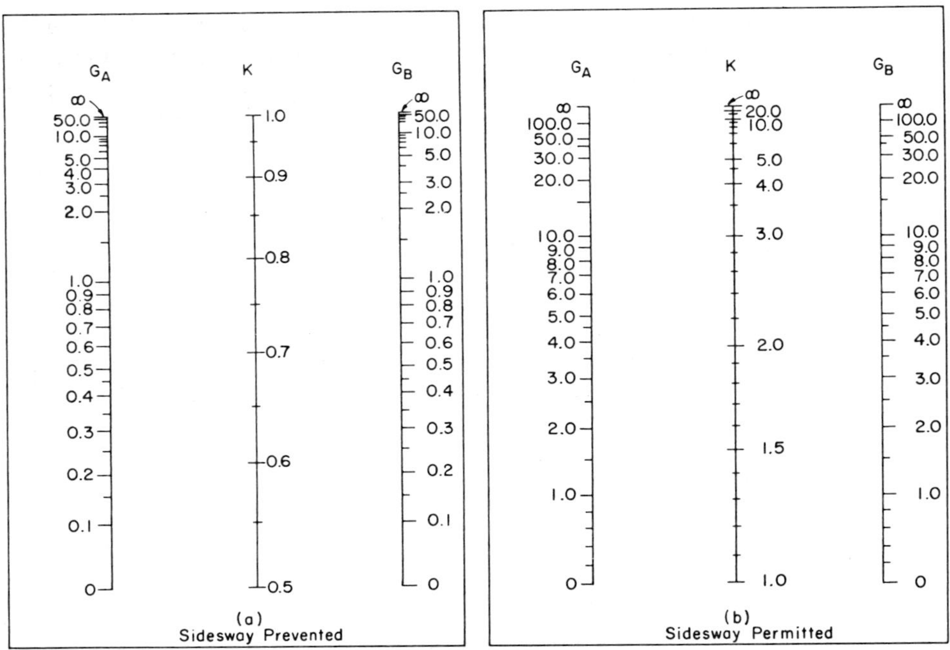

Figure 8.15. Nomograph for effective length of columns. Note that $G = \Sigma\,(EI/L)_{cols}/\Sigma\,(EI/L)_{beams}$. Taken from Ref. 2.

The design stress determined by Eq. 8.7 is valid for long columns i.e. those columns with effective slenderness ratios between K and 50. The latter is an upper value set arbitrarily to limit the maximum effective slenderness of wood columns; the lower value K is given by the effective slenderness at which $F_{c'} = (2/3)\,F_c$. Thus, substituting these values in Eq. 8.7 renders

$$\frac{2}{3}\,F_c = 0.3\,\frac{E}{K^2}$$

from which

$$K = 0.671\sqrt{\frac{E}{F_c}} \tag{8.8}$$

There is a range in effective slenderness for which columns would fail by crushing in compression parallel to grain if subjected to increasing axial load. These are short columns for which the design stress is taken as F_c and the range in effective slenderness is 0 to 11.

Columns with effective slenderness in the range 11 to K are considered intermediate columns. Failure of these members when subjected to axial compression occurs by inelastic buckling with critical stresses exceeding the proportional limit of wood in compression parallel to grain. The critical stress varies with the effective slenderness ratio of the given column between the Euler stresses for long-column behavior and the crushing stress of the wood in compression parallel to grain. An empirical formula for this variation, already reduced to design stresses, is:

$$F_{c'} = F_c\left[1 - \frac{1}{3}\left(\frac{kL/d}{K}\right)^4\right] \tag{8.9}$$

It can be shown that for an effective slenderness ratio, $kL/d = K$, both Eqs. 8.7 and 8.9 render $F_{c'} = (2/3)\ F_c$; also, both have the same rate of change of $F_{c'}$ with kL/d, i.e. $d\ F_{c'}/d\ (kL/d) = -(4/3)\ (F_c/K)$. Thus, there is a smooth transition between intermediate and long-column calculation of design stresses. However, between short and intermediate columns, at $kL/d = 11$, Eq. 8.9 gives a value somewhat smaller than F_c, thereby creating a minor discontinuity at that point. The whole variation of design stress, $F_{c'}$, with effective slenderness ratio, kL/d, is shown in Fig. 8.16 for a given wood for which $F_c = 1.3$ ksi and $E = 1{,}800$ ksi.

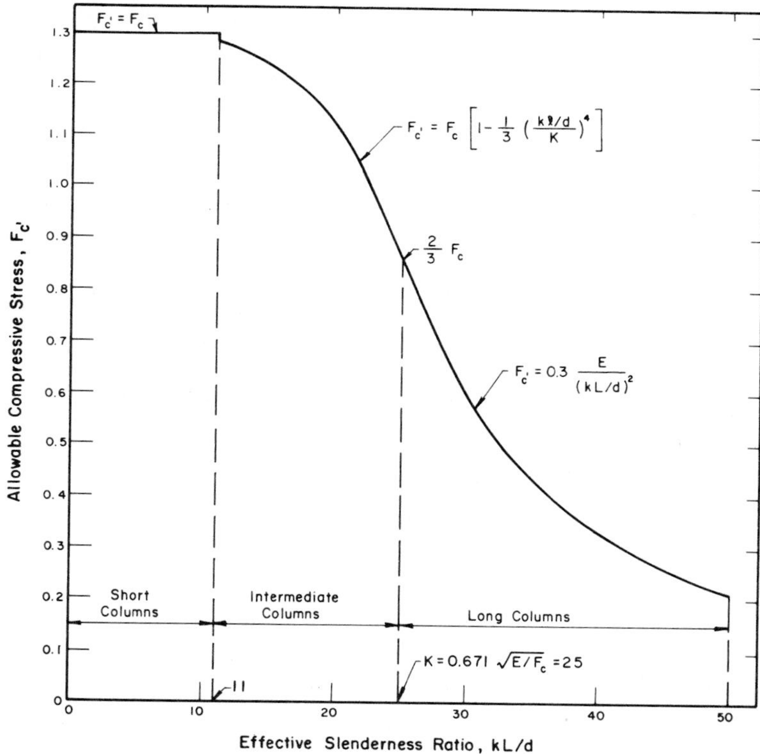

Figure 8.16. Design compressive stress vs. effective slenderness ratio for normal loading conditions. Variation drawn for $F_c = 1.3$ ksi, $E = 1800$ ksi, $\overline{\alpha} = 1.0$.

8.3.1 Actual Factor of Safety of Long Columns

The actual factor of safety of long columns depends on the coefficient of variation in modulus of elasticity for visually-graded sawn lumber (0.25), machine stress-rated sawn lumber (0.11), and glued-laminated wood (~0.10). Thus, a 5-percent lower exclusion value on pure bending modulus of elasticity (the latter elasticity is 1.03 times the average modulus-of-elasticity values in Table A2.5) would render $E_{5\%} = E\,(1.03)\,(1 - 1.645 \times 0.25) = 0.606\,E$ and a critical load $P_{cr} = (\pi^2/12)\,(0.606\,E)/\,(kL/d)^2 = 0.498\,E/(kL/d)^2$. Since the allowable load is $P = 0.3\,E/(kL/d)^2$ it follows that the true factor of safety may be as low as $P_{cr}/P = 1.66$.

For glued-laminated timber and machine stress-rated lumber, the 2.74 combined reduction factor represents less than 0.01-percent lower exclusion value on pure bending modulus of elasticity $[E_{0.01\%} = E\,(1.03)\,(1 - 3.74 \times 0.11) = 0.606\,E]$ with a 1.66 factor of safety. In other words, 99.99 percent of all long glulam and machine stress-rated wood columns have a factor of safety greater than 1.66. It can be shown also that 95 percent of these columns have a factor of safety larger than 2.3.

It appears that long columns made of visually-graded sawn lumber, because of their larger variability in E, render smaller factors of safety than long columns made of glued-laminated wood or machine stress-rated lumber. For the sake of consistency, it may be justified to design long columns made out of visually-graded sawn lumber using a modified version of Eq. 8.7 with a 0.22 coefficient instead of 0.3.

No such reduction is presently recommended by National Design Specification, although it suggests (Appendix G, Ref. 3) that, where unusual hazard exists, a larger reduction factor may be appropriate for sawn-lumber column designs. Cautious designers may wish to consider long columns made of visually-graded sawn-lumber as one such case.

8.3.2 Modification of $F_{c'}$ for Duration of Loading and Other Causes

Duration of loading other than normal affects the design stress in compression parallel to grain but not the modulus of elasticity of wood. Consider a loading condition such that its modification factor is $\bar{\alpha}$, see Section 2.13. It follows that the design compressive stress, F_c, may be taken as a $\bar{\alpha}\,F_c$ but E should remain the same. As a result, the formula for determination of $F_{c'}$ for long columns, i.e. $F_{c'} = 0.3\,E/(kL/d)^2$, remains the same as before. However, K, the lower limit of effective slenderness ratio at which long column behavior begins, is changed. Instead of determining K by means of Eq. 8.8, it is necessary to use

$$K = 0.671\sqrt{\frac{E}{\bar{\alpha}F_c}} \tag{8.10}$$

This changes the lower limit for long-column behavior making it smaller as $\bar{\alpha}$ increases. Only for the case of permanent loading conditions, where $\bar{\alpha} = 0.9$, would the value of K increase.

The intermediate-column range remains the same, i.e. between an effective slenderness ratio of 11 and the modified value of K. Determination of $F_{c'}$ in this range requires modification of Eq. 8.9 as follows

$$F_{c'} = \bar{\alpha}\,F_c\left[1 - \frac{1}{3}\left(\frac{kL/d}{K}\right)^4\right] \tag{8.11}$$

where K is given by Eq. 8.10.

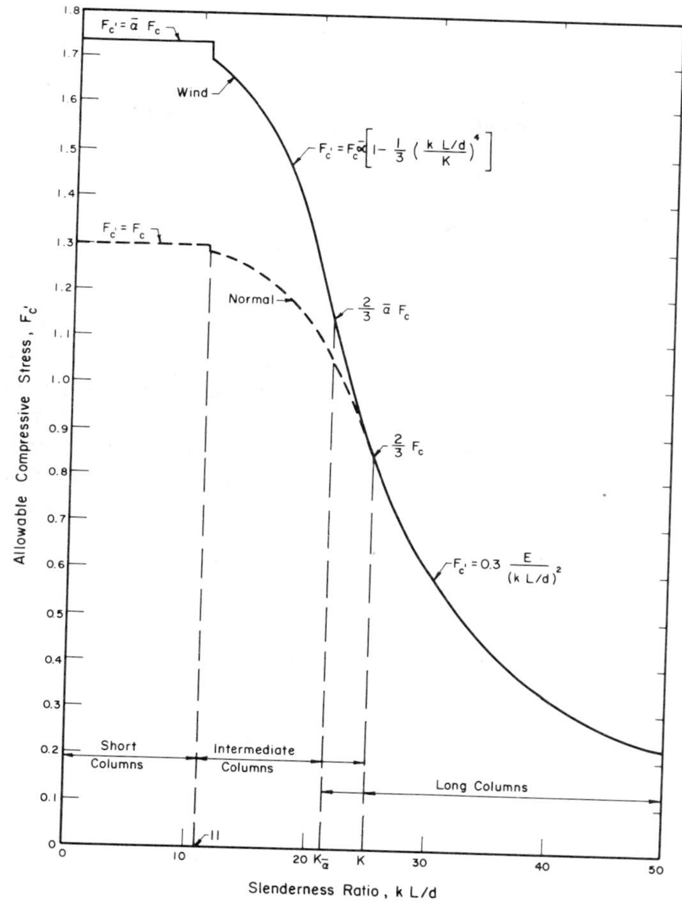

Figure 8.17. Design compressive stress vs. slenderness ratio for normal ($\bar{\alpha} = 1$) and wind ($\bar{\alpha} = 1.33$) loading conditions. Variation drawn for $F_c = 1.3$ ksi and $E = 1{,}800$ ksi.

Finally, in the short column range (for which $kL/d < 11$) instead of $F_{c'} = F_c$, use $F_{c'} = \bar{\alpha}\ F_c$.

The complete variation of $F_{c'}$ with kL/d for a given wood is shown for $\bar{\alpha} = 1.33$ in Fig. 8.17. For the purpose of comparison the variation for $\bar{\alpha} = 1$ is also shown. It is clearly noted that the main effect of an increase in $\bar{\alpha}$ is in the short column range, for which the full increase of $F_{c'}$ to $\bar{\alpha}\ F_c$ is available. The increase is gradually reduced in the intermediate range and tapers off to zero at an effective slenderness ratio K. The graph also shows that the long-column range is greater for the case of $\bar{\alpha} > 1$ at the expense of the intermediate-column range which has become smaller.

8.3.3 Illustrative Example

Determine an acceptable section surfaced-four-sides (S4S), for a laterally-braced double-hinged column 14 ft long, subjected to an axial load P = 40 kips due to snow-loading conditions ($\overline{\alpha} = 1.15$). The column is to be used in an exposed structure and may be subjected to wetting for an extended period of time. Use wood pressure-impregnated with preservatives. Consider wood for which $F_c = 1.3$ ksi, $F_b = 1.6$ ksi and E = 1800 ksi. (Note that these design stresses indicate no ordinary wood but a rather high-quality grade of posts and timbers; see Table A2.5 for comparison with actual grades).

Trial 1. Consider an 8 × 8 S4S section (7.5 × 7.5 in. actual dimensions) for which A = 56.2 in.^{2}, and $f_c = P/A = 40/56.2 = 0.71$ ksi. For a double-hinged laterally-braced column k = 1 and k L/d = 14 × 12/7.5 = 22.40. Also $K = 0.671\sqrt{E/F_c}$ where F_c is the design value in compression parallel to grain. The latter is affected by the duration of load factor, $\overline{\alpha} = 1.15$, and by a 0.91 reduction factor (see footnotes of Table A2.5) because the exposed condition of the column may cause the moisture content of the wood to exceed 19 percent in service (note that if the wood used were Southern Pine no reduction factor would be required). The design value of modulus of elasticity, E, is not affected. Thus, $F_c = 1.15 \times 0.91 \times 1.3 = 1.36$ ksi, E = 1800 ksi, and $K = 0.671\sqrt{1800/1.36} = 24.41$. It follows that $11 < k\,L/d < K$ which makes the 8 × 8 S4S timber an intermediate column.

The design compression stress, $F_{c'}$ is

$$F_{c'} = F_c\left[1 - \frac{1}{3}\left(\frac{kL/d}{K}\right)^4\right] = 1.36\left[1 - \frac{1}{3}\left(\frac{22.4}{24.41}\right)^4\right]$$

$$= 1.04 \text{ ksi}$$

and the ratio of actual to design compression stresses is $f_c/F_{c'} = 0.71/1.04 = 0.68 < 1$. Thus, the 8 × 8 S4S section is acceptable although, at first view, perhaps somewhat overdesigned.

Trial 2. Consider a 6 × 8 S4S section (5.5 × 7.5 actual dimensions) for which A = 41.2 in.^{2} and $f_c = 40/41.2 = 0.971$ ksi. Also K = 24.41, k = 1 and k L/d = 14 × 12/5.5 = 30.55. Since $k\,L/d > K$ the column is long and the design stress need be calculated using $F_{c'} = 0.3\,E/(k\,L/d)^2 = 0.3 \times 1800/(30.55)^2 = 0.579$ ksi. Thus, $f_c/F_{c'} = 0.971/0.579 = 1.68 > 1$ which makes the 6 × 8 S4S section unacceptable.

Use an 8 × 8 S4S timber.

8.4 DESIGN OF WOOD BEAM-COLUMNS

The provisions of the "National Design Specification for Stress Grade Lumber and its Fastenings" (3) regarding design of beam-columns, i.e. members subjected to a combination of end longitudinal loads, transverse loads and eccentricity of the longitudinal loads are contained in:

$$\frac{f_c}{F_{c'}} + \frac{f_b + f_c\left(\frac{6e}{d}\right)(1 + 0.25\,J)}{F_{b'} - J\,f_c} \leq 1 \tag{8.12}$$

where: f_c = stress in compression parallel to grain induced by axial load,
f_b = flexural stress induced by transverse loads only,

$F_{c'}$ = design value in compression parallel to grain that would be permitted if axial compressive stress only existed, see Section 8.3.

$F_{b'}$ = design value for extreme fiber in bending that would be permitted if flexural stress only existed, see Chapter 7.

$1 + 0.25\ J$ = magnification factor which varies between 1 and 1.25

J = a parameter which varies between 0, for effective slenderness ratios ≤ 11, and 1 for effective slenderness ratios $\geq K$, see Fig. 8.18.

K = value of effective slenderness ratio of which $F_{c'} = (2/3)\ F_c$,

e = maximum end eccentricity,

L = length of column,

d_1 = side of rectangular column measured in the direction perpendicular to the transverse loads

d_2 = side of rectangular column measured in the direction parallel to the transverse loads.

For the purposes of using Eq. 8.12 the following information is required.

$$J = \frac{kL/d_2 - 11}{k - 11} \tag{8.13}$$

and

$$K = 0.671 \sqrt{\frac{E}{\bar{\alpha}\ F_c}} \tag{8.14}$$

8.4.1 Derivation of Design Interaction Equation

The origin of Eq. 8.12 may be traced back to Newlin's research (4) in the late 1930's. Its derivation which was reviewed and reaffirmed by Wood (5) in 1961 is based on the following simple assumptions:

1. Wood behaves as a homogeneous isotropic linearly-elastic material under service loads.
2. The service interaction diagram axial load vs bending moment is linear.
3. The column is prismatic, has a rectangular section, and is hinged at both ends.
4. The column is prevented from relative lateral displacements at the ends and is bent in single curvature by the applied loading.
5. The column is subjected to an eccentric longitudinal load and to transverse loads symmetrical about the midspan.

At the critical section of the column the bending moment is equal to M, the moment produced by the transverse loads, plus Pe, the moment produced by the eccentric longitudinal load times a magnification factor, $1 + 0.25\ J$. The latter is due to the so-called P-Δ effect in which additional moments are created by the action of the longitudinal load on the configuration of the column. Using the classical double hinged column with an initial sinusoidal distribution of curvatures caused by transverse loads and the secant formula (see Ref. 1, pp. 31–32 and pp. 13–14, respectively) Newlin obtained a value for the magnification factor equal to $\pi^2/8 \simeq 1.25$. Since the magnification factor in the design equation is identified as $1 + 0.25\ J$, this leaves J with a maximum value of one for long columns. For short columns where, obviously, no magnification

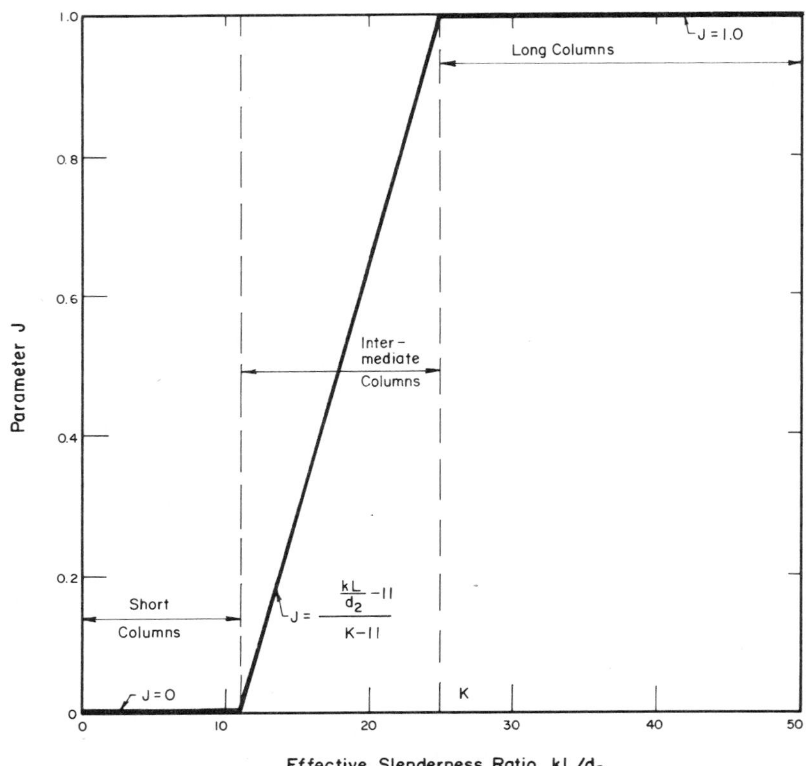

Figure 8.18. Variation of parameter J with the effective slenderness ratio of the beam-column in the direction of bending.

of moments exists, $J = 0$. For the intermediate-column range, $11 < k\,L/d_2 < K$, for which $0 < J < 1$ a linear variation for J between 0 and 1 has been taken, as follows

$$J = \frac{k\,L/d_2 - 11}{k - 11} \tag{8.13}$$

The whole variation of J with the effective slenderness ratio is given in Fig. 8.18. In general, for design purposes the maximum moment is $M + Pe\,(1 + 0.25\,J)$ as compared to the more accurate value rendered by Eq. 8.2.

The maximum bending stress $f_{b,max}$ is obtained by dividing the maximum moment by the value of the section modulus, S. Thus, $f_{b,max} = M/S + (Pe/S)\,(1 + 0.25\,J)$. In the case of a rectangular section in which $S = Ad/6$, the preceding equation can be written as $f_{b,max} = M/S + (P/A)\,(6e/d)\,(1 + 0.25\,J)$. If now $f_b = M/S$ and $f_c = P/A$ one obtains, by substituting in the previous expression, $f_{b,max} = f_b + f_c\,(6e/d)\,(1 + 0.25\,J)$. The terms f_b and f_c are the flexural stress induced in the cross-section by the transverse loads only and the compressive stress induced by the axial load, respectively.

The assumption of a linear interaction diagram, see Section 8.2, adapted to design stresses leads to

$$\frac{f_c}{F_{c'}} + \frac{f_b + f_c\,(6e/d)\,(1 + 0.25\,J)}{F_{b'}} \leq 1 \qquad (8.15)$$

in which $F_{c'}$ and $F_{b'}$ are the design stresses in compression parallel to grain and bending, respectively, for the given beam-column. Determination of $F_{c'}$ and $F_{b'}$ is discussed in Section 8.3 and Chapter 7, respectively.

It is noted that $F_{b'}$ is the design value for extreme fiber in bending that would be permitted if flexural stress only existed; in other words, it is to be determined by considering the lateral-torsional behavior of the member about an axis perpendicular to the axis of bending. Experience in the design of wood beam columns indicates that usually, $F_{b'} = F_b$, where F_b is the design stress in bending. Only in the case of columns with rectangular sections such that the product of aspect ratio (d_2/d_1) and column-slenderness ratio (L/d_1) exceeds 60 would $F_{b'} < F_b$; this situation may occur only in columns that are both thin and tall. The depth factor, indicated by the coefficient C_F, see Chapter 7, reflects the variation of design bending stress with depth for rectangular sections. For cases where $d > 12$ in. $F_{b'}$ will be given by the term $C_F F_b$. To summarize, $F_{b'}$ is to be taken as the lesser value of $F_{b'}$ (based on lateral stability of the member acting as a beam subjected to flexure only) and $C_F F_b$ (based on strength affected by depth of the cross-section).

Using a conservative approach Newlin suggested that the denominator of the bending term in the interaction equation be the difference between the design bending stress, $F_{b'}$, and J times the axial stress in the section, f_c; thus, a net design bending stress, $F_{b'} - Jf_c$, is used. For the case of short columns ($J = 0$) this term becomes just $F_{b'}$ while for long columns ($J = 1$) it simply becomes $F_{b'} - f_c$. Again, it is only for the range of intermediate columns that J would need to be evaluated to obtain the net design bending stress.

Substitution of $F_{b'} - Jf_c$ for $F_{b'}$ in Eq. 8.15 yields

$$\frac{f_c}{F_{c'}} + \frac{f_b + f_c\,(6e/d)\,(1 + 0.25\,J)}{F_{b'} - Jf_c} \leq 1 \qquad (8.16)$$

which is the design interaction equation required by National Design Specification (3).

8.4.2 Discussion of Design Equation

Interaction equation 8.16 was derived for the simple case of a prismatic wood column of rectangular cross-section with both ends hinged and prevented from relative lateral displacement. Use of the interaction equation has been extended to all other columns, namely, to those with rotational end restraints and to those with uninhibited sidesway by means of the effective-length concept. The latter is the length between theoretical inflection points, at the instant of buckling, of a column subjected to axial loading only, see Fig. 8.14.

The effective length, kL, replaces the actual length of the column, L, in the determination of the axial load that the column can sustain and thus, affects the term $F_{c'}$, the design axial stress in the interaction equation. For columns with sidesway prevented k varies between 0.5 (both ends fixed) and 1.0 (both ends hinged). Use of $k = 1$ is always conservative for columns with ends prevented from lateral displacements. However, for columns with uninhibited sidesway, where k ranges between 1.0 (both ends fixed) and infinity (both ends hinged) a realistic evaluation of kL is important.

The effect of k appears indirectly also in the second term of the interaction equation. There, it is used for evaluation of the parameter J, which in turn magnifies the moment due to eccentricity of the longitudinal load, P, in intermediate and long columns. The term (1 + 0.25 J) is used to account for the second-order moments that develop in these columns as a result of additional eccentricity caused by the deflection configuration. In addition, J, is used to reduce the design bending stress, F_b, by the quantity Jf_c in the denominator of the second term of the interaction equation.

The maximum value of the term (1 + 0.25 J) is 1.25 since $J \leq 1$. The true magnification factor, however, may increase somewhat beyond that value because of the fact that the denominator of the second term of the interaction equation is $F_{b'} - Jf_c$. The nature of the deflection configuration, i.e. whether the column is bent in single curvature (larger magnification) or double curvature (smaller magnification), does not enter directly in the calculations affecting determination of moment magnification. Thus, in spite of the use of the effective-length concept, the magnification of bending moment in the interaction equation seems somewhat contrived when compared to the approach taken for modern design of steel and reinforced-concrete beam-columns. There, a more rational determination is made using the term $C_m/(1 - P/P_e)$* as magnifying factor for the bending moment. This sets no arbitrary limitation for the magnification factor and accounts well for the effects on the latter caused by bending of the column in either single or double curvature.

It seems also advisable to design wood columns for a certain eccentricity of the thrust. The concept of the axially-loaded column, subjected to zero bending is an idealization that seldom occurs in actual practice. Wood columns ought to be designed for eccentricity, e, corresponding to the maximum moment which can accompany the thrust but not less than a specified minimum eccentricity. The latter may be taken as 1 in. or 0.1 d_1 or 0.1 d_2 about either principal axis, respectively. This recommendation, which is similar to that followed in the design of reinforced-concrete columns in the 1971 Building Code of American Concrete Institute (9), results in ultimate and service interaction diagrams truncated at the intersections with the lines $e/t = \pm 0.1$, see Fig. 8.3. It is noted that the 1977 Code (18) substituted this requirement with an equivalent requirement calling for truncation of the UID at 80 percent of the axial strength of the column.

The effect of sustained loads is theoretically considered in the value of $\overline{\alpha}$ which directly affects the design stresses used. However, the effect of sustained loading in creating additional deformations and bending stresses has not been considered in the derivation of the interaction equation. Experimental determination of strength, and theoretical development of interaction diagrams for ultimate axial load and bending moment, are presently needed for improved methods of design. It appears that additional basic research is necessary in the behavior of wood columns.

8.4.3 Illustrative Example

Determine an acceptable section, surfaced-four-sides (S4S), for a column 14 ft long subjected to the combined action of longitudinal load P = 40 kips with an eccentricity of 3 in. at each end and a uniformly distributed transverse load of 0.245 kips per ft, all due to normal loading conditions ($\overline{\alpha} = 1.0$). Sidesway of the column is prevented by bracing at the end supports but rotation of the

*Where $C_m = 0.6 - 0.4\, M_1/M_2$ is a reduction factor which depends on the magnitude and sign of the bending moments M_1 and M_2 at the ends of the column; P is the axial load; and $P_e = \pi^2 EI/(kL)^2$ is the elastic critical load in the plane of the applied moments.

latter is unrestrained, see Fig. 8.9. The column is to be used in a covered structure and will remain dry thereby requiring no downward adjustments in the allowable stresses due to increasing moisture content. Use wood for which $F_c = 1.3$ ksi, $F_b = 1.6$ ksi and $E = 1{,}800$ ksi.

Note that this column is the same as that of Part 5 of the example in Section 8.2.1.

Trial 1. Assume a 12 × 16 (S4S) section (11.5 × 15.5 in., actual dimensions) for which $A = 178.2$ in.², $S = 460.5$ in.³, and $C_f = (12/15.5)^{1/9} = 0.972$. The long face of the column is placed parallel to the transverse loads for maximum effectiveness of the rectangular section to resist bending; thus, $d_1 = 11.5$ in. and $d_2 = 15.5$ in. For a double-hinged laterally-braced column, $k = 1$, and maximum slenderness (about the weak axis) $kL/d_1 = (14 \times 12)/11.5 = 14.61$. Also, $K = 0.671\sqrt{E/F_c} = 0.671\sqrt{1800/1.3} = 24.97 \sim 25$. It follows that $11 < kL/d_1 < 25$ which makes the 12 × 16 in. timber an intermediate column.

The design compression stress, $F_{c'}$, is

$$F_{c'} = F_c\left[1 - \frac{1}{3}\left(\frac{kL/d}{K}\right)^4\right] = 1.3\left[1 - \frac{1}{3}\left(\frac{14.61}{25}\right)^4\right] = 1.25 \text{ ksi}$$

The design bending stress, $F_{b'}$, requires determination of the design value that would be permitted if flexural stress only existed. Thus, using the formulas and Table in Section 7.10, one obtains: 1) the effective length of compression flange $\ell_e = 1.92 \times 14 = 26.88$ ft (for a single type of span subjected to uniformly distributed load) 2) the slenderness factor $C_s = \sqrt{\ell_e d/b^2} = \sqrt{(26.88 \times 12)(15.5)/(11.5)^2} = 6.15$ and 3) $C_k = \sqrt{3E/5F_b} = \sqrt{(3)(1800)/(5)(1.6)} = 25.98 \sim 26$. Because $C_s < 10$, no correction due to lateral stability is required and the design bending stress may be taken as $F_{b'} = C_F F_b = 0.972 \times 1.6 = 1.56$ ksi.

The stresses f_c and f_b may be calculated as follows: $f_c = P/A = 40/178.2 = 0.224$ ksi and $f_b = M/S = (1/8)(0.245/12)(14 \times 12)^2/460.5 = 72.0/460.5 = 0.156$ ksi.

The variable $J = (kL/d_2 - 11)/(K-11) \leq 1$ depends on the effective slenderness ratio about the actual axis of bending, for which $kL/d_2 = (14 \times 12)/15.5 = 10.84$. Since $L/d_2 < 11$ take $J = 0$.

Using interaction equation 8.16 one obtains:

$$\frac{f_c}{F_{c'}} + \frac{f_b + f_c\left(\frac{6e}{d}\right)(1 + 0.25\,J)}{F_{b'} - Jf_c} = \frac{0.224}{1.25}$$

$$+ \frac{0.156 + 0.224\left(\frac{6 \times 3}{15.5}\right)(1 + 0)}{1.56 - 0}$$

$$= 0.179 + 0.267 = 0.446 < 1, \text{ OK}$$

The 12 × 16 S4S section is acceptable, although it may be considered somewhat big. A closer section may be obtained using $A_2 = (0.446)(178.2) = 79.5$ in.² and $S_2 = (0.446)(460.5) = 205.4$ in.³. However, the designer is advised, in view of the fact that a smaller section with larger slenderness will result from this process, to use values somewhat larger than A_2 and S_2 in his selection in order to compensate for the smaller design stresses. Various S4S sections such as 8 × 14, 10 × 12, 8 × 16 and others may be selected.

Trial 2. Check the smallest of the possible solutions, i.e. 8 × 14 S4S (7.5 in. × 13.5 in. actual dimensions) for which A = 101.2 in.^{2}, S = 227.8 in.^{3}, and $C_f = (12/13.5)^{1/9} = 0.987$. As before, the long side is placed parallel to the transverse loads. Thus $kL/d = (14)(12)/7.5 = 22.4$ and, since K = 25, it follows that $11 < kL/d < K$ and the column is again of intermediate length. It follows that $F_{c'} = 1.3\ [1 - (1/3)\ (22.4/25)^4] = 1.02$ ksi. Also, since $\ell_e = 26.88$ as before and $C_s = \sqrt{\ell_e\, d/b^2} = \sqrt{(26.88 \times 12)\ (13.5)/(7.5)^2} = 8.80 < 10$ the design bending stress need not be corrected by lateral stability and may be taken directly as $F_{b'} = C_f\, F_b = (0.987)\ (1.6) = 1.58$ ksi.

The stresses f_c and f_b are $P/A = 40/101.2 = 0.395$ ksi and $M/S = 72/227.8 = 0.316$ ksi, respectively. The value of J is $[(14 \times 12/13.5) - 11]/(25 - 11) = 0.10$. Substitution in the interaction equation renders:

$$\frac{0.395}{1.02} + \frac{0.316 + 0.395\left(\dfrac{6 \times 3}{13.5}\right)(1 + 0.25 \times 0.1)}{1.58 - 0.10 \times 0.395}$$

$$= 0.387 + 0.556 = 0.943 < 1, \text{0K}$$

and the section is acceptable.

Use an 8 × 14 S4S timber.

Additional possible solutions for this example have been obtained using a computer program. The solutions are shown in Table 8.2 for rough-sawn sections and surfaced-four-sides sections of solid wood. The most economical sections in both categories, i.e. those with the least cross-sectional area, have been identified. In the rough-sawn category a 6 × 18 in. section (108 in.^{2}) is the most economical with an 8 × 14 in. (112 in.^{2}) section running a close second. Among the S4S sections the most economical is the 8 × 14 in. (101.25 in.^{2} actual) section; this checks the solution obtained previously using hand calculations.

8.4.4 Illustrative Example

Consider the same design problem as before, i.e., a double-hinged braced column subjected to P = 40 k, e = 3 in., w = 0.245 k/ft, and made out of wood for which $F_c = 1.3$ ksi, $F_b = 1.6$ ksi and E = 1800 ksi. Consider, however, that wind-loading conditions apply ($\bar{\alpha} = 1.33$) and the column is braced at midheight by wood girts attached to the wide face of the section.

Trial 1. Assume a 6 × 12 S4S section (5.5 × 11.5 in. actual dimensions) for which A = 63.2 in.^{2}, S = 121.2 in.^{3}, and $C_f = 1$. For a double-hinged column braced also at midheight about its weak axis of bending, k = ½, and the slenderness ratio about the weak axis is $kL/d_1 = (½)(14 \times 12)/5.5 = 15.27$. About the strong axis of bending, however, the slenderness ratio would be $kL/d = (14 \times 12)/11.5 = 14.61 < 15.27$. Thus, the maximum slenderness ratio is 15.27. Also, $K = 0.671\ \sqrt{E/\bar{\alpha}\, F_c} = 0.671\ \sqrt{1800/(1.33 \times 1.3)} = 21.65$. (Note that in the previous calculation the design compression stress F_c has been taken as $\bar{\alpha}\, F_c$ but E has not been affected by $\bar{\alpha}$, as required by NDS). It follows that $11 < kL/d < K$ which makes the 6 × 12 S4S section an intermediate column.

TABLE 8.2. DESIGN OF A WOOD COLUMN SUBJECTED TO AXIAL LOAD AND BENDING.*

Various Possible Solutions

Rough Sawn Sections

Section	Area in.^{2}	Interaction Equation
6 × 18†	108	0.925
6 × 20	120	0.802
6 × 22	132	0.705
6 × 24	144	0.628
8 × 14	112	0.810
8 × 16	128	0.652
8 × 18	144	0.547
10 × 12	120	0.816
10 × 14	140	0.619
12 × 12	144	0.668
12 × 14	168	0.508

Surfaced-Four-Sides Sections

Section	Area in.^{2}	Interaction Equation
6 × 22	118.25	0.887
6 × 24	129.25	0.790
8 × 14†	101.25	0.944
8 × 16	116.25	0.749
8 × 18	131.25	0.626
8 × 20	146.25	0.535
10 × 12	109.25	0.941
10 × 14	128.25	0.699
10 × 16	147.25	0.553
12 × 12	132.25	0.760
12 × 14	155.25	0.566

*P = 40 k, e = 3 in., M = 72 k-in., Normal loading conditions, L = 168 in., Double-hinged and braced at the ends. F_c = 1.3 ksi, F_b = 1.6 ksi, E = 1800 ksi.

†Most economical in corresponding category.

The design compression stress, $F_{c'}$ is

$$F_{c'} = \bar{\alpha}\, F_c \left[1 - \frac{1}{3}\left(\frac{kL/d}{K}\right)^4\right] = (1.33 \times 1.3)\left[1 - \frac{1}{3}\left(\frac{15.27}{21.65}\right)^4\right] = 1.59 \text{ ksi}$$

The design bending stress, $F_{b'}$, requires determination of the design value that would be permitted if flexural stress only existed. Since the unsupported length of beam $\ell_u = 14/2 = 7$ ft, the effective length of compression flange, see Section 7.10, is $\ell_e = 1.92 \times 7 = 13.44$ ft. The slenderness factor $C_s = \sqrt{\ell_e\, d/b^2} = \sqrt{(13.44 \times 12)\,(11.5)/(5.5)^2} = 7.83 < 10$, thus no correction due to lateral stability is required and the design bending stress may be taken as $F_{b'} = C_f\, \bar{\alpha}\, F_b = 1.0 \times 1.33 \times 1.6 = 2.13$ ksi.

The stresses f_c and f_b are $P/A = 40/63.2 = 0.633$ ksi and $M/S = 72/121.2 = 0.594$ ksi, respectively. The magnification factor is $J = (L/d - 11)/(K\text{-}11) = (14.61 - 11)/(21.65 - 11) = 0.34$. Using interaction equation 8.16 one obtains:

$$\frac{f_c}{F_{c'}} + \frac{f_b + f_c\left(\frac{6e}{d}\right)(1 + 0.25\,J)}{F_b - Jf_c} = \frac{0.633}{1.59} + \frac{0.594 + 0.633\left(\frac{6 \times 3}{11.5}\right)(1 + 0.25 \times 0.34)}{2.13 - 0.34 \times 0.633}$$

$$= 0.398 + 0.872 = 1.27 > 1, \text{ N.G.}$$

The 6 × 12 S4S section is not acceptable.

Trial 2. Consider using a section having $A = 1.27 \times 63.2 = 80.3$ in.^{2} and $S = 1.27 \times 121.2 = 153.9$ in.^{3}. Take a 6 × 14 S4S ($A = 74.2$ in.^{2} and $S = 167.1$ in.^{3}) for which, although the cross-sectional area is somewhat smaller than indicated the section modulus is larger and may compensate for this apparent deficiency. It follows that: $C_f = 0.987$, $kL/d_2 = (\frac{1}{2})(14 \times 12)/5.5 = 15.27$, $kL/d_1 = 14 \times 12/13.5 = 12.44$, $K = 21.65$, $11 < kL/d_1 < 21.65$ and $F_{c'} = (1.33 \times 1.3)[1 - (\frac{1}{3})(15.27/21.65)^4] = 1.59$ ksi. Also, $\ell_e = 1.92 \times 7 = 13.44$ ft, $C_s = \sqrt{(13.44 \times 12)(13.5)/(5.5)^2} = 8.48 < 10$ and $F_{b'} = (0.987)(1.33)(1.6) = 2.10$ ksi. Compression and flexural stresses are $f_c = 40/74.2 = 0.539$ ksi and $f_b = 72/167.1 = 0.431$ ksi, respectively; the magnification factor $J = (12.44 - 11)/(21.65 - 11) = 0.14$.

Use of the interaction equation renders

$$\frac{0.539}{1.59} + \frac{0.431 + 0.539\left(\frac{6 \times 3}{13.5}\right)(1 + 0.25 \times 0.14)}{2.10 - 0.14 \times 0.539}$$

$$= 0.339 + 0.580 = 0.919 < 1, \text{ OK}$$

and the section is acceptable.

Use a 6 × 14 S4S section for the column.

8.4.5 Illustrative Example

The problem is the same as before, except that the load has no initial end eccentricity. Solutions are to be found, using laminated wood only, for various conditions of end restraint in both braced and unbraced columns; the purpose being to study the effect of these variables on design. In view of the quality and control that exist in the manufacture of laminated sections, higher design stresses $F_c = 2.0$ ksi, $F_b = 2.6$ ksi, and $E = 1{,}800$ ksi are selected for the design.

Figure 8.19 shows the solutions obtained using a computer program developed by W. Korkosz. Each section is plotted using its cross-sectional area as abscissa and the corresponding value on interaction equation 8.16 as ordinate. Sections that are acceptable for the conventionally supported wood column (double-hinged and braced at the ends) occupy the middle portion of the figure.

Braced columns with their ends rotationally restrained, are satisfied with lighter sections, and their solutions are plotted to the left of those corresponding to the double-hinged braced column. Figure 8.18 shows the solutions, in order of increasing stiffness, for colunns with one end fixed and the other hinged, and for columns with both ends fixed. The acceptable sections become lighter as the stiffness of the end restraints is increased.

Unbraced columns, on the other hand, require heavier sections, as shown in Fig. 8.19. Solutions satisfying the conditions of unbraced columns appear to the right of those obtained for the double-hinged braced column. The same two cases of rotational end restraint were considered for unbraced columns as for braced columns and again the lighter sections occurred for the stiffer end restraints. In spite of the presence of rotational end restraints, unbraced columns are subjected to sway, have larger effective slenderness and require heavier sections than the double-hinged braced column. Small rotational end restraints in unbraced columns will require very heavy sections and may even lead to instability. Unbraced columns should be avoided, both for economical and strength considerations; however, if unavoidable, stiff rotational end restraints should be provided.

In line with preceding analyses, the effective length, kL, was taken instead of the unsupported length, L, to determine the actual slenderness ratio, kL/d_2 to be used in the interaction equation. (For this purpose, it was assumed that the end conditions shown for the column in the plane of the figure are the same for the plane perpendicular to it. However, this may not be necessarily

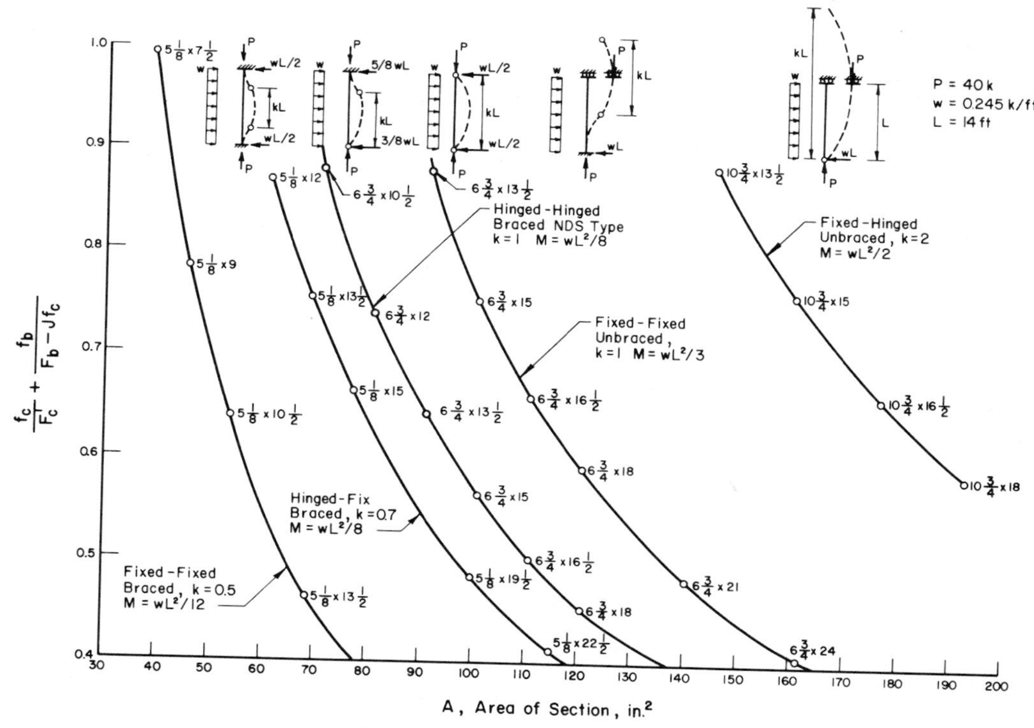

Figure 8.19. Solution of design problem of section 8.4.5. Note that braced columns require lighter sections than unbraced ones. Also, within each category, stiffer end-rotational restraints lead to lighter sections.

true in all cases, as a column may be strongly restrained rotationally by a large girder in one plane and only slightly so in a perpendicular plane by light joists). For braced columns $k \leq 1$ and may be as low as 0.5 if both ends are fully restrained. While the latter may be a theoretical condition, end fixities of appreciable amount can be easily obtained with moment-resistant connections such as those suggested by the American Institute of Timber Construction (11, 12). It is apparent that using $k = 1$ leads to conservative designs of braced columns. However, for unbraced columns, where $k > 1$ and may reach much higher values with weak rotational restraints, use of the effective slenderness ratio is considered essential for safe design. Since the stability of an unbraced column depends entirely on the rotational restraints provided by its end connections, a careful evaluation of the latter is necessary for a safe determination of the effective slenderness. Certainly this is a case where the use of the interaction design equation with L/d_2 instead of kL/d_2 may lead to an unsatisfactory and even dangerous section.

An additional example of column design is given in Section 10.3.3 for an interior column of a building.

8.4.6 Lateral Bracing

The solution of the preceding design problems was based on the assumption that the columns would be prevented from sidesway by lateral bracing. It is the designer's responsibility to make sure that lateral bracing is available in the structure. Usually, there would be provided vertical X-bracing, or walls, either made of wood studs and plywood panels, masonry, reinforced concrete or steel panels, that would carry the wind or seismic forces acting on the structure. Two such systems, orthogonal to each other, are required to brace the structure in all directions. Design of these systems may be governed by either wind or seismic forces in conventional structures. The capacity of these systems to provide lateral bracing to the columns is usually adequate. To make sure, however, a designer would do well to check if each system may sustain, independently of the other and without exceeding the design stresses, a lateral load applied at the top of each column equal to 2 percent of the gravity load.

Concern about lateral bracing of columns applies even more during the erection stage of the structure, as it is then that vulnerability to failure may be greatest. It behooves a designer to alert erectors of the need to provide temporary lateral bracing until such time as the structure is completed.

8.5 DESIGN OF SPACED COLUMNS

A spaced column is formed of two or more individual members with their longitudinal axes parallel, separated at the ends and middle points of their length by blocking and joined at the ends by timber connectors capable of developing the required shear resistance. Fig. 8.20 shows an elevation and transverse section of a typical spaced column. The over-all unsupported length, L, is measured from center to center of lateral supports of continuous spaced columns, as in multistory building columns and compression chords or web members of trusses, and from end to end of simple spaced columns as in single story building columns.

The individual leaves of the spaced column are tied together by end and spacer blocks. A spacer block is necessary at midlength; in long columns two spacer blocks at the third points may be required. Spacer blocks must be of equal thickness and at least as thick as the individual leaves;

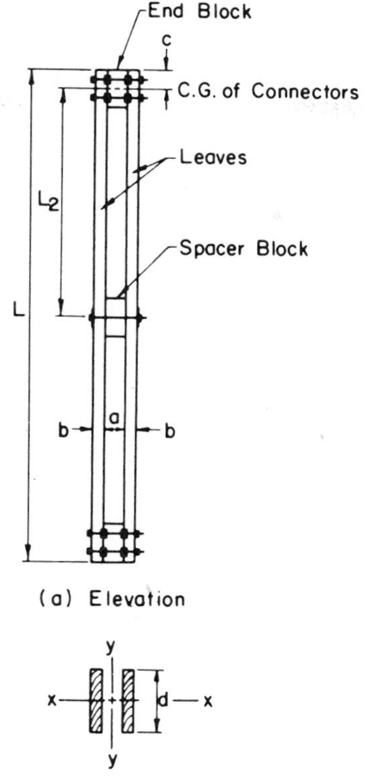

Figure 8.20. Typical spaced column.

their direction of grain should be parallel to the length of the column. The minimum length of end spacer blocks is determined by the end distances required by the connectors; their width should be equal to that of the individual leaves. Bolting or spiking is generally adequate for middle spacer blocks.

The position of the centroid of connectors at each end block is measured by the distance c. Since the capacity of a spaced column depends on the distance c, the latter is an important variable. Two conditions are recognized in the specifications, namely: "a," for which $c \leq L/20$ and "b," for which $L/20 \leq c \leq L/10$. For a spaced column with only one intermediate spacer at midspan, the unbraced length for a given leaf is the distance between the centroids of the connectors at the end block and the midlength block. Thus, the unbraced length for conditions "a" and "b" can be conservatively estimated, for $c = 0$ and $c = L/20$ as $L/2$ and $0.9\ L/2$, respectively. End condition "b" leads to smaller effective slenderness for the leaves of a given spaced column. The term F_c, allowable stress in compression parallel to grain, for a leaf of a given spaced column is a function of its effective slenderness as shown before and may be obtained as follows:

$$F_{c'} \leq \frac{0.3\ E}{\left(\frac{L/2}{b}\right)^2}\ \text{, condition "a"}$$

$$F_{c'} \leq \frac{0.3\ E}{\left(\frac{0.9\ L/2}{b}\right)^2}\ \text{, condition "b"} \qquad (8.18)$$

For design purposes the preceding expressions for $F_{c'}$ have been conservatively reduced (3, 6), to Eqs. 8.19.

$$F_{c'} = \frac{0.75\ E}{(L/b)^2}\text{, condition "a"}$$

$$F_{c'} = \frac{0.90\ E}{(L/b)^2}\text{, condition "b"} \qquad (8.19)$$

These formulas are valid in the long-column range for which $K < L/b \leq 80$, where:

$$K = 1.06\sqrt{\frac{E}{F_c}}\text{, condition "a"}$$

$$K = 1.16\sqrt{\frac{E}{F_c}}\text{, condition "b"} \qquad (8.20)$$

The meaning of K is identical to that used for solid columns, i.e. it represents the slenderness ratio at which $F_{c'} = (2/3)\ F_c$.

Spaced columns with individual members having an L/b ratio in the intermediate-column range $11 < L/b < K$, are designed using

$$F_{c'} = F_c\left[1 - \frac{1}{3}\left(\frac{L/b}{K}\right)^4\right] \qquad (8.21)$$

which, as in the case of solid columns, provides the best transition between the Euler-type behavior of long columns (strength based on elastic stability) and the behavior of short columns which depend solely on the compressive strength of wood parallel to grain. Thus, for short columns having an L/b ratio of 11 or less

$$F_{c'} = F_c \qquad (8.22)$$

The variation of the ratio $F_{c'}/F_c$ with slenderness for given values of E and F_c is shown in Fig. 8.21 for both end conditions when $\overline{\alpha} = 1$. The slenderness ratios L/b and L_2/b, see Fig. 8.21 are limited to maximum values of 80 and 40, respectively, for individual leaves of a spaced column. The ratio L/d is limited to 50, as in the case of solid columns.

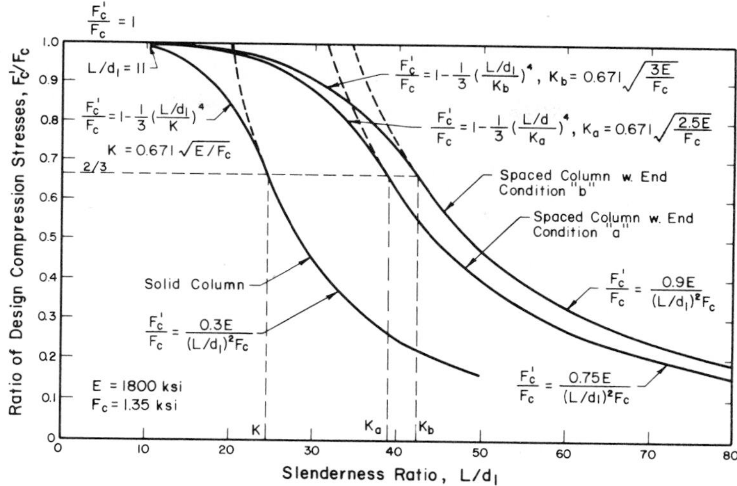

Figure 8.21. Ratio of design compressive stresses vs. slenderness ratio for spaced columns. The variation for a solid column is also shown for the purpose of comparison.

The load capacity of the spaced column is given by the smallest of:

1.) The sum of the products of $F_{c'}$ as given by Eqs. 8.19 through 8.22, times the cross-sectional area of each individual leaf.
2.) The capacity of a solid column bending about axis x-x, see Fig. 8.20, as given by Eq. 8.16.
3.) The capacity of a built-up column, bending about axis y-y, the geometrical properties of which are those of the built-up section.

The latter determination is required in Great Britain (13). The built-up column is considered as capable of bending about axis y-y; its effective length is multiplied by the modification factor K_{20} given in Table 8.3. This reflects the influence of the load-slip characteristics of the method of connection of the individual leaves on the stiffness of the column; the fact that K_{20} diminishes with increasing stiffness of the connection, for any given value of the ratio a/b, see Fig. 8.20, implies that glued end blocks and spacers render the stiffest spaced columns. Table 8.3 is traced by Booth and Reece (14) to the investigations of Pleskov in Russia in the 1930's and Niskannen in Finland in 1961.

TABLE 8.3. MODIFICATION FACTOR K_{20} FOR THE EFFECTIVE LENGTH OF SPACED COLUMNS ACTING AS BUILT-UP COLUMNS†.

Value of ratio a/b*:	0	1	2	3
Method of Connection:				
Nailed	1.8	2.6	3.1	3.5
Screwed or Bolted	1.7	2.4	2.8	3.1
Connectored	1.4	1.8	2.2	2.4
Glued	1.1	1.1	1.3	1.4

†From British Standard Code of Practice CP 112, Art. 314 g, Table 20.
*See Fig. 8.20.

The only spaced-column action that is recognized by NDS is that created by connectored end blocks and bolted spacers. It can be shown using Table 8.3 that, for the minimum design requirements set forth by NDS, a/b = 1 in Fig. 8.20, the capacity of the built-up column is greater than that given by the sum of the capacities of the individual leaves; thus, for NDS spaced-columns there is no need to calculate the capacity of the built-up column. Also it can be shown that, in the absence of bending moments of the capacity of the long spaced column is the same for both axes of bending when the ratio d/b equals $\sqrt{2.5}$ and $\sqrt{3}$ for conditions "a" and "b," respectively. In view of the small amount of research available on spaced columns under combined action of axial load and bending moment about axis y-y, design for this condition is not encouraged.

To satisfy NDS requirements, the size and number of connectors in each mutually contacting surface of end block and individual leaf at each end of a spaced column must be such as to provide a load capacity in pounds equal to the required cross-sectional area in square inches of one of the individual leaves times the appropriate end-spacer block constant given in Table 8.4. Once the section of a spaced column has been proportioned, the number and size of the required connectors are determined using Table 8.4 for the given species of wood.

An Illustrative Example of design of a spaced column is given in Section 10.2.4, using glued-laminated wood, for an interior column of a building.

TABLE 8.4. END SPACER BLOCK CONSTANT, PSI, FOR CONNECTOR—JOINED SPACED COLUMNS†

L/b*	Group A	Group B	Group C	Group D
0 to 11	0	0	0	0
15	38	33	27	21
20	86	73	61	48
25	134	114	94	75
30	181	155	128	101
35	229	195	162	128
40	277	236	195	154
45	325	277	229	181
50	372	318	263	208
55	420	358	296	234
60 to 80	468	399	330	261

†From NDS (3), Section 3.8, Table 3A.
*Constants for intermediate L/b ratios may be obtained by linear interpolation.

8.6 DESIGN OF BUILT-UP COLUMNS

In the absence of large timbers or glued-laminated members, built-up columns have been used consisting of planks or smaller square timbers nailed, spiked or bolted together. Two types are used (6), namely: a) parallel planks and cover plates and b) planks boxed around a solid core, see Fig. 8.22.

It has been found (15) that no arrangement of pieces with any kind of mechanical fastenings will make a built-up column fully equal in strength to a one piece column of comparable material

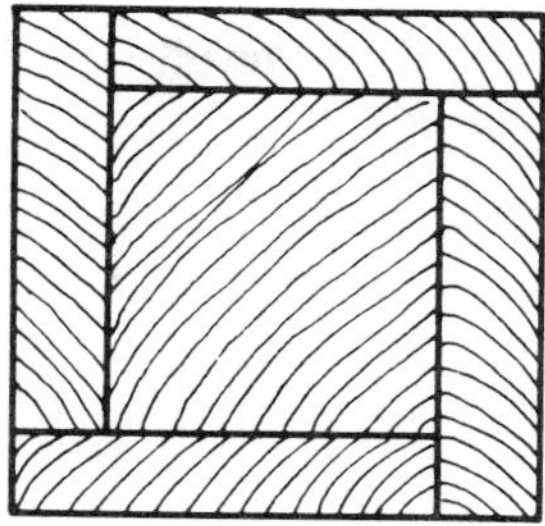

(a) Planks and Cover Plates (b) Boxed Solid Core

Figure 8.22. Typical sections of built-up-columns with mechanical fastenings.

and dimensions. The strength of a built-up column can be expressed as a percentage of the strength of a one piece column of the same dimensions and quality. Table 8.5 gives the percentage of strength in relation to the slenderness ratio of the built-up column. For slenderness ratios of 10 or greater the strength of the column is independent of whether the individual pieces are full length or not. However, for $L/b < 10$ pieces butted end to end fail at 75 to 80 percent of the crushing strength of full-length pieces.

TABLE 8.5. BUILT-UP COLUMNS WITH MECHANICAL CONNECTORS.†

L/b	Percentage of Strength of One Piece Column
6	82
10	77
14	71
18	65
22	74
26	82

†From Wood Handbook, see Ref. 6.

8.7 CONNECTIONS OF WOOD COLUMNS

Various design details for connections of columns to supporting foundations are shown in Fig. 8.23, 10.5 and 11.9. Anchorage details for columns subjected to substantial transverse shears, in addition to axial load, are shown in Fig. 11.6. In all cases, the column bearing elevation should be at least three inches above grade or finished floor if a grade-type floor construction is used. In locations where column base anchorages are subject to damage by moving vehicles, protection of the columns from such damage should be considered.

Examples of connections of columns to beams or girders, are shown in Figs. 10.5, 10.10 and 11.14. Other design details may be found in Ref. 11.

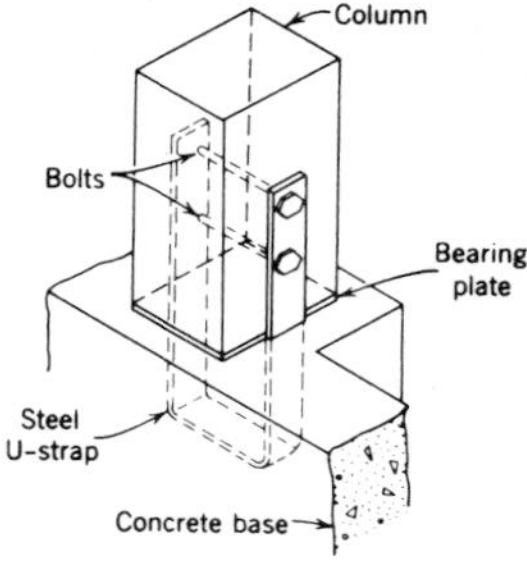

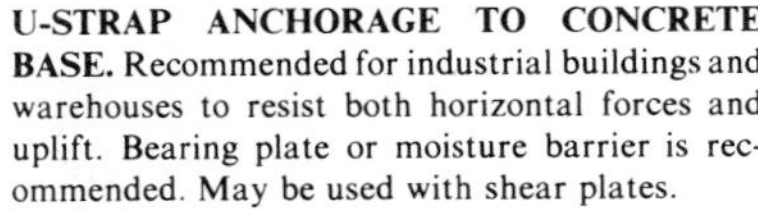
U-STRAP ANCHORAGE TO CONCRETE BASE. Recommended for industrial buildings and warehouses to resist both horizontal forces and uplift. Bearing plate or moisture barrier is recommended. May be used with shear plates.

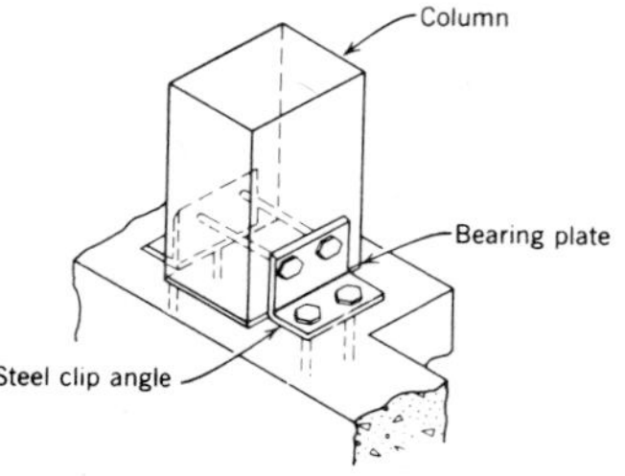

CLIP ANGLE ANCHORAGE TO CONCRETE BASE. Recommended for industrial buildings and warehouses to resist both horizontal forces and uplift. Bearing plate or moisture barrier is recommended.

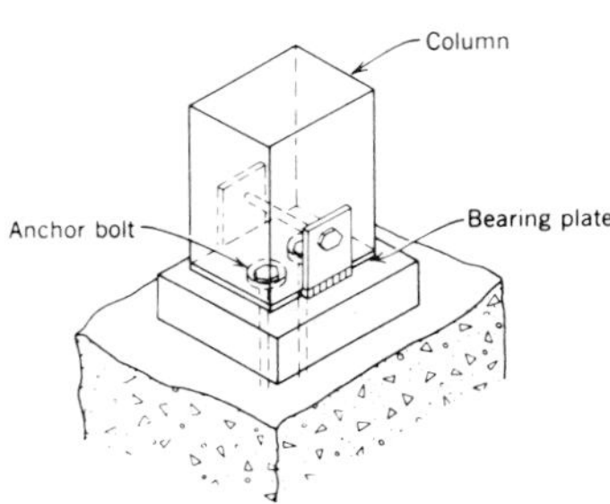

RAISED FOOTING COLUMN ANCHORAGE. For use where pedestal is limited in size. Resists some horizontal forces and uplift.

Figure 8.23. Various design details for column anchorages. From Ref. 11.

8.7.1 Connections Through Bearing on End Grain

A number of connection details for column support allow transfer of thrust through direct bearing on the end grain of wood. Design stresses for bearing parallel to grain are listed in Table A8.1 for commercial species available in the United States. Modification factors caused by duration of load, fire-retardant treatments, moisture service conditions and others, affect the design stresses listed.

Comparison of the design stresses given in Table A8.1 with the corresponding design stresses in compression parallel to grain (Table A2.5) indicates that, generally, the former are larger than the latter. Thus, it would be unusual to have design of a column governed by end bearing details to comply with the specifications, however, stresses in end-grain bearing should be calculated (using the net area in bearing) and checked against the design stresses. If the stress in end-grain bearing exceeded 75 percent of the corresponding value for the given wood, as listed in Table A8.1, a steel plate, (inserted with a snug fit between abuting ends) would be required.

The design values for end grain in bearing apply also to end-to-end bearing of compression members provided there is adequate lateral support and the end cuts are accurately squared and parallel. When the thrust is transferred to a member at an angle of load to grain the maximum bearing value may be determined by the Hankinson formula (see Section 2.2) using the design value for end grain in bearing parallel to grain from Table A8.1 and the corresponding design value for compression perpendicular to grain as listed in Table A2.1.

PROBLEMS

1. Parts 1 through 5 of the illustrative example of Section 8.2 were solved in the text assuming that bending of the column takes place about the strong axis of the section. Solve assuming that bending is about the weak axis of the section. Compare both solutions.

   Ans.

   Part 1. Points 1 and 10 are the same as in Fig. 8.3; Point B: P = 115 k tension, M = 2,640 k-in.; Point 5: P = 275 k compression, M = 1,936 k-in.; Point 6: P = 320 k tension, M = 2,400 k-in.

   Part 2. Points 1 and 5 are the same as in Part I; also for: $\epsilon_c = 0.0035$, $\epsilon_t = -0.003$, P = 81 k compression, M = 1,730 k-in. Max. Tens. force $\epsilon_c = \epsilon_t = -0.003$, P = 960 k, tension.

   Part 3. Points 1, 6 and 10 are the same as in Part I; the U.I.D. consists of two straight lines.

   Part 4. Service Interaction Diagram: P = 208 k compression, M = 0; P = 256 k, tension, M = 0; P = 0, M = 427 k-in.

   Part 5. $P_{cr} = 930$ k, $\delta_o = 0.24$ in.; Point given by $M_{max} = 202$ k-in., P = 40 k plots inside service interaction diagram. Column is adequate.

2. Determine the UID for a circular section, D = 16 in., assuming the same stress-strain diagram used in the solution of the illustrative example of Section 8.2.

   Ans. (compression is positive)

Point	ϵ_t	ϵ_2	P, kips	M, k-in.
1	0.0050	0.0050	1,409	0
2	0.0029	0.0050	1,386	143
3	0.0008	0.0050	1,087	1,243
4	−0.0013	0.0050	700	2,255
5	−0.0034	0.0050	292	3,190
B	−0.0055	0.0050	−122	4,089
7	−0.0055	0.0029	−523	3,394
8	−0.0055	0.0008	−946	2,545
9	−0.0055	−0.0013	−1,369	1,697
10	−0.0055	−0.0034	−1,792	848
11	−0.0055	−0.0055	−2,214	0

3. Determine three acceptable S4S sections, for a column 19 ft. long subjected to the combined action of longitudinal load P = 40 kips with an eccentricity of 3 in., and a midspan moment M = 72 k-in. due to transverse loading. Assume $\overline{\alpha}$ = 1.0 and such wood that F_c = 1.3 ksi, F_b = 1.6 ksi and E = 1,800 ksi. The column is to be used in a covered structure, hence, no adjustments in the allowable stresses are needed.

   Ans. In ascending order of areas the following sections are acceptable:
   10 × 14, α = 0.825; 8 × 18, α = 0.866; 12 × 12, α = 0.859;

4. Solve problem 3 assuming that the column is only 10 feet long.

   Ans. 6 × 18, α = 0.843; 8 × 14, α = 0.856; 10 × 12, α = 0.861.

5. The interior column of the building discussed in Section 10.2 was designed as a spaced column, see Section 10.2.4. For the given conditions, design a solid wood column instead using solid-sawn wood or glued-laminated timber.

6. The interior column of the building discussed in Section 10.3 was designed as a double-hinged column braced against relative lateral displacements of its ends, see Section 10.3.3. Design the column assuming that it is hinged to the foundation pedestal, but fully fixed to the supported beam. Consider two cases: (a) full lateral bracing, and (b) no lateral bracing.

7. In the preceding problem, design the column assuming that it is rotationally fixed to both the foundation pedestal and the supported girder. Design for (a) full lateral bracing and (b) no lateral bracing.

8. Determine the moment-curvature diagram of a 6 in. × 14 in. (actual dimensions) wood beam-column 20 ft long subjected to a 50 k constant thrust. Consider the stress-strain diagram for wood as given in Fig. 7.5 of this text. Calculate M and ϕ for two stages of loading, namely: 1) Attainment of proportional limit stress in compression, σ = 6.4 ksi, and 2) at failure.

   Ans. At proportional limit, a = 7.79 in., M = 1151 k-in., $\phi = 4.11 \times 10^{-4}$ rad/in. At ultimate: a = 7.72 in., M = 1685 k-in., $\phi = 6.48 \times 10^{-4}$ rad/in.

9. Consider the beam-column of the preceding problem subjected to two transverse loads, each P/2, situated at the middle third points, in addition to the 50 k thrust applied at the ends. Using the results of the preceding problem, determine the variation of transverse load, P, vs midspan deflection, δ, for the member.

   Ans. At proportional limit, P = 25.3 k, δ = 2.8 in.
   At ultimate: P = 36.6 k, δ = 4.4 in.

REFERENCES

1. Timoshenko, S. P. and Gere, J. M., "Theory of Elastic Stability," McGraw-Hill Book Co., Inc., New York, 1961.
2. "The Column Research Council Guide to Design Criteria for Metal Compression Members," 2nd Ed., Johnston, B. C., ed., John Wiley and Sons, Inc., New York, N.Y., 1966.
3. "National Design Specification for Wood Construction," National Forest Products Association, Washington, D.C., 1977 Ed.
4. Newlin, J. A., "Formulas for Columns with Side Loads and Eccentricity," Building Standards Monthly, Dec., 1940.
5. Wood, L. W., "Formulas for Columns with Side Loads and Eccentricity," Forest Products Laboratory, U.S. Dept. of Agriculture, Report No. 1782, Oct., 1961.

6. Wood Handbook, Forest Products Laboratory, U.S. Dept. of Agriculture Handbook No. 72, pp. 160–161. First Edition, 1955.
7. "Commentary on the Specification for the Design, Fabrication and Erection of Structural Steel for Buildings," American Institute of Steel Construction, Manual of Steel Construction.
8. Gurfinkel, G., and Robinson, A. R., "Buckling of Elastically Restrained Columns," Journal of the Structural Division, ASCE, Vol. 91, No. ST6, Proc. Paper 4574, December, 1965, pp. 159–183.
9. ACI Committee 318, "Proposed Revision of ACI 318–63: Building Code Requirements for Reinforced Concrete," Journal of the American Concrete Institute, Vol. 67, No. 2, Feb., 1970, pp. 77–186.
10. Gurfinkel, G., "Design of Wood Columns for Flexural and Axial Loading," Journal of the Structural Division, ASCE, Vol. 96, No. ST11, Proc. Paper 7701, November, 1970, pp. 2389–2408.
11. AITC 104–69, "Typical Construction Details," Timber Construction Standards, AITC 100–69, American Institute of Timber Construction, 5th Ed., 1969.
12. Timber Construction Manual, American Institute of Timber Construction, John Wiley & Sons, Inc., New York, 2nd, Ed., 1974.
13. "British Code of Practice CP 112:1967," British Standards Institution, 2 Park St., London, W. 1, Great Britain.
14. Booth, L. G., and Reece, P. O., "The Structural Use of Timber. A Commentary on the British Standard Code of Practice CP 112," E. & F. N. Spon Ltd., London, 1967.
15. Scholten, J. A., "Built-up Wood Columns Conserve Lumber," Engineering News Record, Vol. 107, No. 9, 1931.
16. Newmark, N. M., "Numerical Procedure for Computing Deflections, Moments, and Buckling Loads," Transactions, ASCE, Vol. 108, pp. 1161–1234, 1943.
17. Gurfinkel, G. R., and Robinson, A. R., "Determination of Strain Distribution and Curvature in a Reinforced Concrete Section Subjected to Bending Moment and Longitudinal Load," ACI JOURNAL, Proceedings V. 64, No. 7, July 1967, pp. 398–403.
18. ACI Standard 318–77, "Building Code Requirements for Reinforced Concrete," by American Concrete Institute, Box 19150, Redford Station, Detroit, Michigan, 48219.
19. "Wood Handbook: Wood as an Engineering Material," by Forest Products Laboratory, U.S. Department of Agriculture, Agriculture Handbook No. 72, Revised August 1974.

Chapter 9
Design of Wood Bridges

9.1 INTRODUCTION

Historically, wood bridges were among the first to be used by man. In the eastern part of the United States early settlers built many covered wood bridges which still remain. For many years, railroads have been using timber trestles so extensively that it has been estimated by the American Railway Engineering Association that over eighteen hundred miles of timber bridges and trestles, made up of untreated and preservative-treated wood, are in service on the major railroads of United States and Canada (12). Presently, only preservative-treated wood is used in new trestles because of improved durability and reduced maintenance.

The simplest type of wood bridge used consists of simply supported stringers on which a wood deck is nailed, spiked, or doweled in place. A wearing surface is used on top of the wood deck to protect it against abrasion from wheel loads. Coverings of asphalt plank, bituminous or asphaltic mats, and oiled gravel ballast have been used as wearing surfaces. The bridge stringers usually bear on a heavy timber beam, or pile cap, that ties together the top of the supporting pile-columns. The Illustrative Example in Section 9.3 discusses the design of a bridge of this type.

Composite wood-concrete bridges use the necessary deck to advantage by making it an integral part of the structure of the bridge. Composite action between the wood sub-deck and the concrete slab is developed by means of shear connections placed at the interface between the two materials. This type of bridge is not limited to simple spans. Continuity of composite action may be achieved for negative moments, with placement of reinforcing steel at the top of the concrete deck. Two types of sections have been used, namely: the composite wood subdeck-concrete slab, and the composite wood stringer-concrete slab. Both types are fully discussed in this chapter, see Section 9.4 and 9.6. In addition, Illustrative Examples of design of a continuous composite slab-bridge of the first type is given in Section 9.5 and of a simply-supported long-span composite beam-bridge of the second type in Section 9.7.

Efforts to improve the nail-laminated wood deck, as spacing between stringers increased with use of glued-laminated timbers, have led to development of the glulam deck, a relatively new product in wood-bridge engineering. As its name indicates this deck is laminated together in the factory, and not in the field, of nominal 2 in. dimension lumber using glue. Panels 4 ft wide, and as long as the bridge is wide, are then attached to the stringers by means of vertical lag bolts, and to adjacent panels through horizontal steel dowels. The advantages of this product over the nail-laminated deck are numerous and are fully discussed in Section 9.8. An Illustrative Example is given in Section 9.9 of design of a simply-supported 60 ft long bridge using a glulam deck, on glued-laminated stringers spaced as much as 7 ft. 3 in. apart.

The twelfth edition (1977) of the Standard Specifications for Highway Bridges of the American Association of State Highway and Transportation Officials (1), and Interim Specifications Bridges 1978 and 1979, have been used extensively in the design examples.

9.2 PROVISIONS OF AASHTO SPECIFICATIONS AFFECTING DESIGN OF WOOD BRIDGES

Two systems of highway live loading are provided, the H loading and the HS loadings, the dimensions and weights of both of which are shown in Fig. 9.1. The H loading consists of a two-axle truck with 80 percent of its weight concentrated in its rear axle and only 20 percent in the front axle. Three classes of H loading exist, namely H-20-44, H-15-44 and H-10-44; the first number indicating the gross weight of the standard truck in tons. The use of a second number, indicating the year in which the loading specification was approved, was instituted with the publication of the 1944 edition of AASHTO. It will remain unchanged until such time as the loading specification is revised.

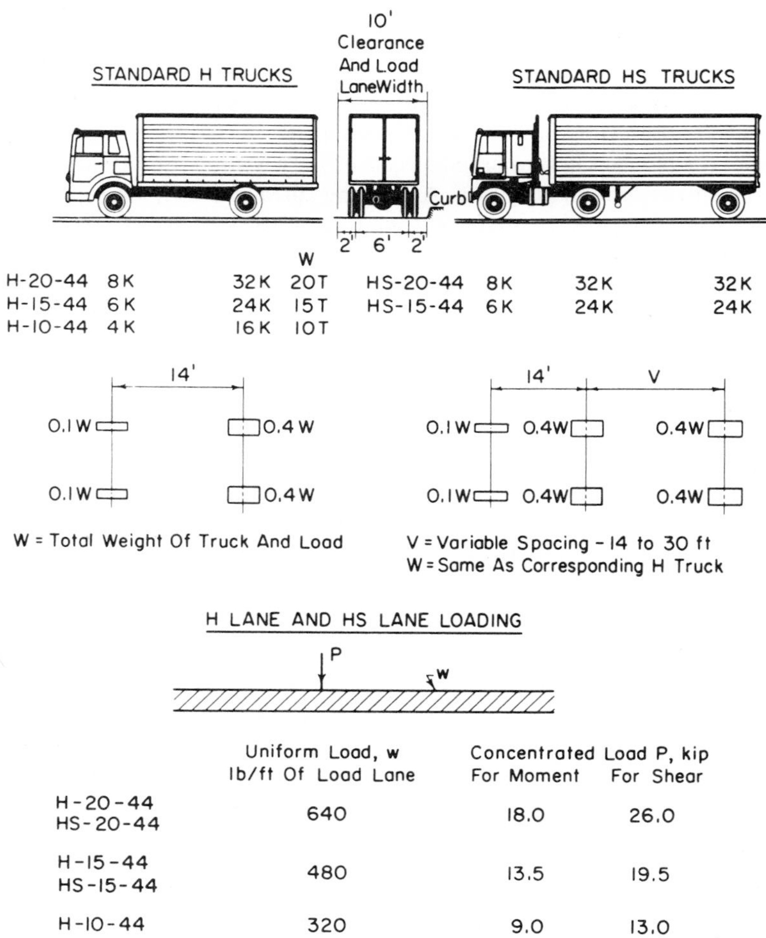

Figure 9.1. Standard H trucks, HS trucks and lane loadings according to AASHTO specifications for highway bridges.

The HS loading consists of a tractor truck with semi-trailer. It is designated by the letters HS followed by a first number indicating the gross weight in tons of the tractor truck and by a second number indicating year as in the H loading. The variable axle spacing, 14 to 30 ft., has been introduced to approximate more closely the tractor trailers now in use. The variable spacing also provides more satisfactory loading for continuous spans, in that heavy axle loads may be so placed on adjoining spans as to produce maximum negative moments. Two classes of HS loading exist, namely HS-20-44 and HS-15-44. Loading HS-15 is 75 percent of loading HS-20. The latter consists of a regular H–20 loading followed, at variable spacing, by a rear axle weighing 16 tons.

AASHTO requires that for truck highways, or for other highways which carry, or which may carry, heavy truck traffic, the minimum live load be HS-15-44 loading. Also, it requires that bridges supporting Interstate Highways be designed for HS-20-44 loading or an Alternate Military Loading of two axles four feet apart with each axle weighing 24,000 pounds, whichever produces the greatest stress.

A system of so-called lane loads, corresponding to the H and HS wheel loadings, may be used as a simpler method of calculating moments and shears. Each lane loading, see Fig. 9.1, consists of a uniform load per linear foot of traffic lane combined with a single concentrated load (or two concentrated loads, in two adjoining spans, for maximum negative moment in the case of continuous spans) so placed on the span as to produce maximum moment or shear at any given section. The concentrated load and uniform load shall be considered as uniformly distributed over a 10-foot width on a line normal to the center line of the lane. The design moments and shears for the bridge of Illustrative Example in Section 9.7 are determined by lane loading rather than by wheel loading because of the long span of the bridge. For the case of smaller spans, as in Illustrative Examples in Sections 9.3 and 9.5, the wheel loading governs design.

The position of the standard truck that creates maximum bending moment in a simply-supported span is given in Fig. 9.2. It can be shown that the midspan section of the bridge bisects the horizontal distance between the heavy axle and the center of gravity of the truck. For small bridges (spans smaller than approximately 26.6 ft.), maximum moment is obtained when the heavier axle is placed at midspan. The design moment is the sum of the maximum live-load moment and the dead-load moment at the critical section. It is conservative to use the dead-load moment at midspan in the determination of the total maximum moment. In the case of continuous bridges an envelope of maximum moments at various sections for each span must be determined before design can be accomplished. For this purpose, it is convenient to draw influence lines (2) for bending moment and shear for the various sections selected in order to place the loading in the positions imposing maximum effects.

The distribution of the wheel loads on the bridge deck is necessary in design. Article 1:3.1B of AASHTO (1) provides the following empirical method for the determination of bending moments in stringers and longitudinal beams. No longitudinal distribution of the wheel loads is assumed; lateral distribution is considered by applying to the stringer the fraction of a wheel load (both front and rear) given by Table 9.1.

For the determination of end shears and end reactions in transverse floor beams and longitudinal beams and stringers, no longitudinal distribution of the wheel load is again assumed for the wheel load adjacent to the end at which shear is being determined. For maximum effect, the wheel load adjacent to the end may be assumed as placed directly above the stringer or floor beam without lateral distribution to adjacent elements. For loads in other positions on the span, the

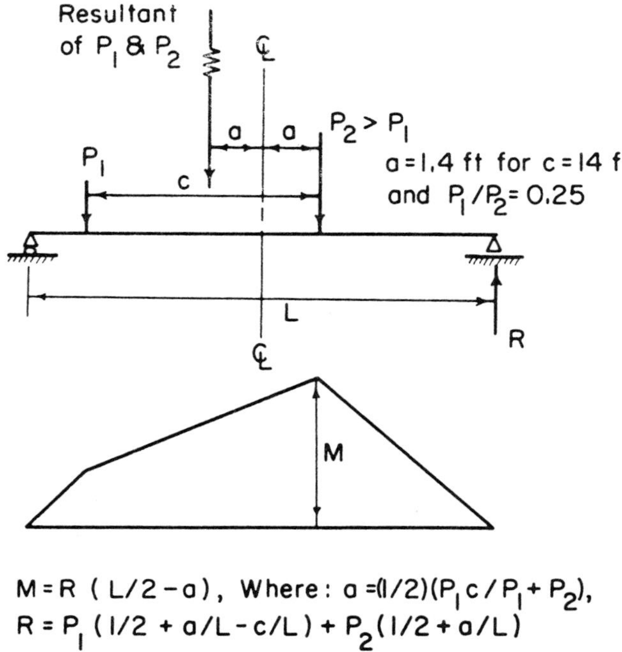

Figure 9.2. Position of truck for maximum moment in simply-supported bridge. For $L < 2c—a$ approximately, use $M = PL/4$.

distribution is determined by the method described above for bending moment; i.e. using Table 9.1. A special provision is made for the determination of horizontal shear stresses in solid timber rectangular sections only. Further refer to Section 9.3.2 of this chapter for the statement of the rule, and its application to the determination of maximum shear force in an interior stringer of Illustrative Example in Section 9.3.

The additional effects of impact due to the moving traffic may be neglected for wood bridges. This reflects the increased capacity of wood to absorb shock and loads of very small duration. For the case of composite wood-concrete-steel bridges, the effects of impact on concrete and steel should be considered. The incremental effect of impact is expressed as a fraction of the live-load stress given by the following expression:

$$I = \frac{50}{L + 125} \tag{9.1}$$

where I is the impact fraction (maximum thirty percent) and L is the length in feet of the portion of the span which is loaded to produce the maximum stress in the member.

Other AASHTO provisions are examined in the following sections as they become necessary for particular aspects of design.

TABLE 9.1. ACTUAL FRACTION OF WHEEL LOADS FOR THE DETERMINATION OF BENDING MOMENTS IN INTERIOR STRINGERS†.

	Traffic Lanes	
	One	Two or More
Timber Floor:		
Plank	S/4.0	S/3.75
Strip 4″ thick or multiple layer floors over 5″ thick	S/4.5	S/4.0
Strip 6″ or more thick	S/5.0 If S > 5′ use footnote 2	S/4.25 If S > 6.5′ use footnote 2
Concrete Floor:		
On timber stringers	S/6.0 If S > 6′ use footnote 2	S/5.0 If S > 10′ use footnote 2

†After Table in Article 1.3.1.B of AASHTO Specifications (1).

S = average stringer spacing in feet.

1. No longitudinal distribution of wheel loads shall be assumed.
2. In this case the load on each stringer shall be the reaction of the wheel loads, assuming the flooring between the stringers to act as a simple beam.

9.3 ILLUSTRATIVE EXAMPLE

Design a two-lane highway trestle suitable for H-20-44 traffic with bents at 17 ft. spacing and a minimum 24 ft. width of roadway. For the laminated decking use rough sawn wood; any unevenness to be covered by the asphaltic wearing surface. The stringers are to be designed using surfaced-four-sides, S4S, timbers.

The section shown in Fig. 9.3 is analyzed for adequacy and to select the required grade of wood. Various other possible solutions for the trestle are suggested at the end of the example. A step-wise annotated process for the design of the bridge follows.

9.3.1 Design of Deck

Article 1.3.4 of AASHTO (1) specifies the distribution of wheel loads for the calculation of bending moments; see Figs. 9.4 (a) and (b). For the case of a laminated transverse flooring it specifies $b_d = 15$ inches normal to the direction of the span and $b_t = 20$ inches in the direction of the span due to the H-20 truck loading. Consider a 4 in. thick deck, spike-laminated out of 2 × 4 rough-sawn strips (see Fig. 9.4 (b)) and spanning across the supporting stringers; for the purpose of analysis, the structural unit of deck would be 4 in. thick and 15 in. wide. The span of this unit, L, is the clear distance between stringers plus one-half the width of the stringer, $S - b_s/2$ in Fig. 9.4 (a), but shall not exceed the clear span plus the deck thickness $(S - b_s + t)$. Hence for the deck of the bridge shown in Fig.9.3 for which S = 25 in., $b_s = 7.5$ in. and t = 4 in. one obtains $L = 25 - 7.5/2 = 21.25$ in. not to exceed $L = 25 - 7.5 + 4 = 21.5$ in. Thus, L = 21.25 in.

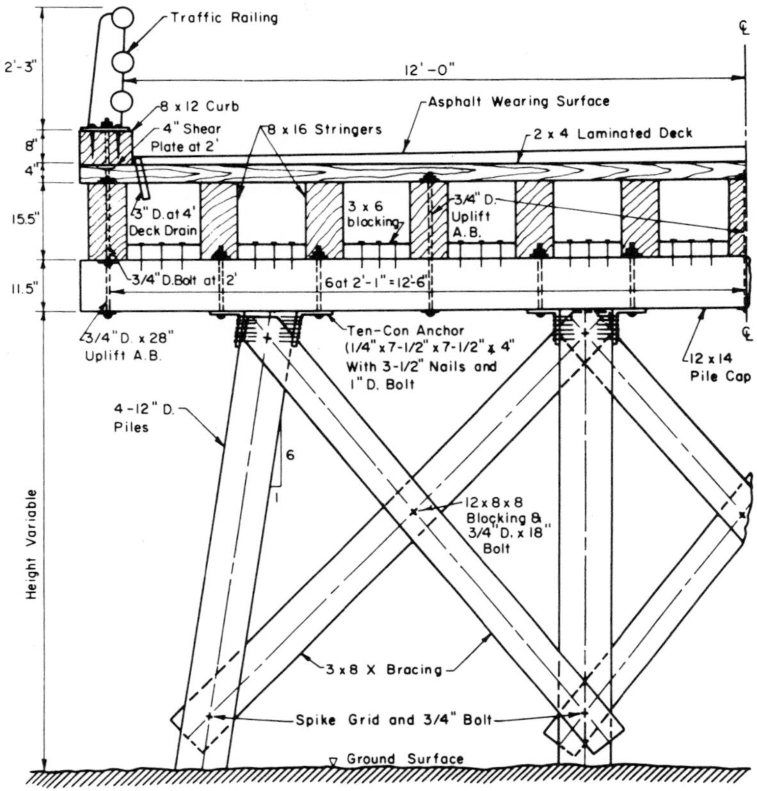

Figure 9.3. Transverse section of trestle.

Loads on Structural Unit Deck

Dead Load:	Self-weight: (4/12) (50)	=	16.7 psf
	Asphaltic Concrete: (3/12) (150)	=	37.5
	Total	=	54.2 psf

Hence, for the 15 in.-wide structural unit deck w_D = (15/12) (54.2) = 67.8 lb/ft = 0.0056 k/in.

Live Load: Rear tire of HS-20-44 truck weighs 16 kip, uniformly distributed over b_t = 20 in.
Hence w_L = 16/20 = 0.8 k/in.

Maximum Bending Moment

Article 1.3.4.c of AASHTO allows the use of 80 percent of the bending moment obtained for a simple span if the deck is continuous over more than two spans. This being the case for the given bridge it is necessary to calculate first the simple-span moment, M_i. The latter may be calculated using Fig. 9.4.c for the position of loads shown. Thus,

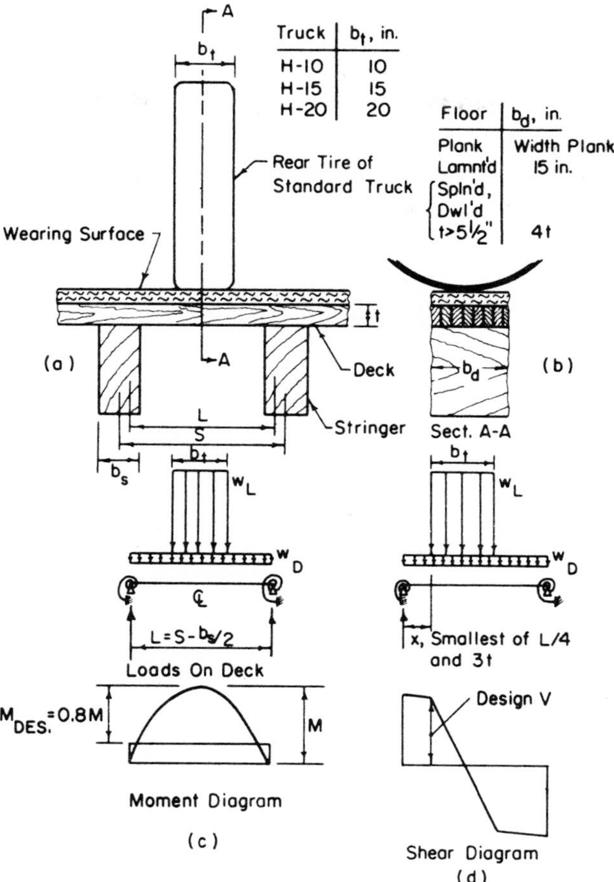

Figure 9.4. Positions of rear tire of standard truck for maximum moment and shear on bridge deck.

$$M_i = \frac{1}{8} w_D L^2 + \frac{wL\, b_t}{4}\left(L - \frac{b_t}{2}\right) \tag{9.2}$$

Substituting $w_D = 0.0056$ k/in., $w_L = 0.8$ k/in., $b_t = 20$ in. and $L = 21.25$ in. yields

$$\begin{aligned} M_i &= \frac{1}{8}(0.0056)(21.25)^2 + \frac{(0.8)(20)}{4}\left(21.25 - \frac{20}{2}\right) \\ &= 0.32 + 45.0 = 45.32 \text{ k-in.} \end{aligned}$$

from which

$$M = 0.8\, M_i = (0.8)(45.32) = 36.3 \text{ k-in.}$$

Note that the moment contribution of the self-weight of the wood deck and wearing surface (0.32 k-in.) is quite small and could have been ignored without major consequences.

Maximum Shear

The position of loading for maximum shear in the deck is shown in Fig.9.4 (d). The section where the maximum shear stress occurs is situated at a distance X from the support by the smallest of 3t = 3 × 4 = 12 in. or L/4 = 21.25/4 = 5.31 in. Hence, X = 5.31 in. For the given deck the quantity b_t + X = 20 + 5.31 = 25.31 in. exceeds L = 21.25 in. Thus, part of the live load will be outside the deck span for maximum shear, V. The latter quantity may be calculated as follows

$$V = w_D\left(\frac{L}{2} - X\right) + w_L\,\frac{(L-X)^2}{2L}, \text{ for } b_t + X > L\ * \tag{9.3}$$

Substituting w_D = 0.0056 k/in., w_L = 0.8 k/in., X = 5.31 in. and L = 21.25 in. in the preceding expression yields:

$$V = 0.0056\left(\frac{21.25}{2} - 5.31\right) + 0.8\,\frac{(21.25 - 5.31)^2}{2 \times 21.25}$$

$$= 0.03 + 4.78 = 4.81 \text{ k}$$

Again, the shear contribution of the self weight of the deck and the wearing surface is quite small when compared to that of the wheel load.

Capacity of Deck

For the 4 × 15 in. unit deck, having an actual thickness of 4 in. and a width of 15 in., A = 4 × 15 = 60 in.^{2}, S = (1/6) (15) (4^2) = 40 in.^{3}. Then

$$f = \frac{M}{S} = \frac{36.3}{40} = 0.908 \text{ ksi}$$

$$v = \frac{3}{2}\,\frac{V}{A} = \frac{3}{2}\left(\frac{4.81}{60}\right) = 0.12 \text{ ksi.}$$

Use 2 × 4 wood for which F_b = 2.3 ksi >> 0.908 ksi and F_v = 0.125 ksi > 0.120 ksi. It may be noted that the governing criterion in the selection of wood for the deck is its capacity to resist horizontal shear stresses. This is typical of flexural members with a small ratio of span to thickness and which are subjected to large transverse loads. If wood of lesser quality (say F_v = 0.080 ksi) were specified, it would be necessary to use a 6 in.-thick deck instead. Note that it is not necessary to check the end span of the deck, in spite of a possible larger moment, because of the controlling effect of shear.

The deck should be nailed together with nails of sufficient length to penetrate two strips and half of a third. For the given deck, 40 pennyweight nails (0.225 in. D., 5 in. long) nailed at 18 in. may be considered adequate; also, each deck strip should be fastened to the stringers by toe-nailing with 20 pennyweight nails (0.192 in. D., 4 in. long) through alternate laminations.

Deflections of the 4 in. thick nail-laminated deck over a span of 21 in. are negligible and need not be calculated.

*For the case when $b_t + X \leq L$ the maximum shear may be calculated using

$$V = w_D\left(\frac{L}{2} - X\right) + \frac{w_L\, b_t}{L}\left(L - X - \frac{b_t}{2}\right)$$

9.3.2 Design of Interior Stringer

8 × 16 S4S (7.5 × 15.5 in. actual dimensions) spaced 25 in. apart.

Dead Load:

Weight of Deck	$= w_D S$	$= (54.2)\ (25/12)$	= 113 lb/ft
Weight of Stringer	$= bd\gamma$	$= \dfrac{(7.5)\ (15.5)\ (50)}{144}$	= 40
		w_D	= 153 lb/ft

Live Load: Rear wheel of H-20-44 truck weighs 16 k and is assumed riding directly over typical interior stringer. Due to lateral distribution to adjacent stringers (see Table 9.1), the actual load on the stringer, P_L, is given for the case of a timber floor 4 in. thick strip and two traffic lanes on the bridge. Thus, $P_L = (S/4)\ 16$, where S = 25 in. = 2.08 ft., is the spacing between stringers. Hence,

$$P_L = \left(\frac{2.08}{4}\right)(16) = 8.32 \text{ k.}$$

The front wheel may be treated similarly. However, it is not considered in this design because the bridge span is too small to require the presence of both front and rear wheels for the conditions of maximum moment, shear and deflection.

Maximum Moment

Maximum moment for this bridge stringer occurs with the rear wheel at midspan. It can be determined using

$$M = w_D \frac{L^2}{8} + P_L \frac{L}{4} \tag{9.4}$$

Substituting $w_D = 0.153$ k/ft, $P_L = 8.32$ k and, L = 17 ft in the preceding formula renders

$$M = (0.153)\ \frac{(17)^2}{8} + (8.32)\left(\frac{17}{4}\right) = 5.53 + 35.36$$

$$= 40.9 \text{ k-ft} = 491 \text{ k-in.}$$

Maximum Shear

Article 1.10.2.A of AASHTO Specifications (1) defines the section of maximum shear as occurring at a distance X from the support equal to three times the depth of the beam, or at the quarter point, whichever is closer to the support. For the given case, X is the smaller of 3 (15.5/12) = 3.87 ft or 17/4 = 4.25 ft. Therefore, take X = 3.87 ft. The shear due to dead load at the critical section is

$$V_D = \frac{w_D}{2}\ (L - 2\ X) = \frac{0.153}{2}\ (17 - 2 \times 3.87) = 0.71 \text{ k}$$

In order to account for lateral distribution to adjacent stringers (and for the phenomenon of reduced shearing stresses due to checking, see Section 7.8) AASHTO's Article 1.10.2A provides that V_L, the shear due to live load be given by

$$V_L = \frac{1}{2}(0.6\, V_{LU} + V_{LD}) \tag{9.5}$$

where V_{LU} and V_{LD} are, respectively, the shear due to undistributed wheel loads, and the shear due to wheel loads distributed laterally as specified for moment. For this case

$$\begin{aligned} V_{LU} &= (0.4\, W)\left(\frac{L - X}{L}\right) \\ V_{LD} &= \left(\frac{S}{4}\right) V_{LU} \end{aligned} \tag{9.6}$$

where W = 20 Ton = 40 k for an H–20–44 truck, L = 17 ft, X = 3.87 ft, and S = 2.08 ft. Substituting these values in the preceding formulas yields

$$V_{LU} = (0.4 \times 40)\left(\frac{17 - 3.87}{17}\right) = 12.36 \text{ k}$$

$$V_{LD} = \left(\frac{2.08}{4}\right)(12.36) = 6.43 \text{ k}$$

Therefore:

$$V_L = \frac{1}{2}(0.6 \times 12.36 + 6.43) = 6.92 \text{ k}$$

and the total shear at the critical section is given by

$$V_T = V_D + V_L = 0.71 + 6.92 = 7.63 \text{ k}$$

Capacity of Stringer

For the surfaced-four-sides 8 × 16 in. stringer (7.5 × 15.5 in. actual dimensions) A = 116 in.^{2}, S = 300 in.^{3}, C_F = 0.97, I = 2,327 in.^{4}. The maximum bending and shear stresses f and v, respectively, may be determined as follows

$$f = \frac{M}{C_F\, S} = \frac{491}{0.97 \times 300} = 1.69 \text{ ksi}$$

$$v = \frac{3}{2}\,\frac{7.63}{116} = 0.099 \text{ ksi}$$

Select a grade and species of wood with the following design stresses: F_b = 1.70 ksi, F_v = 0.105 ksi, and E = 1,600 ksi. Note that the maximum bending and shear stresses are smaller than the corresponding design stresses, F_b and F_v, respectively, for the grade selected. Hence, the grade of wood specified can be considered acceptable.

Maximum Deflection at Midspan

Deflection due to uniformly-distributed dead load is given by

$$\delta_D = \frac{5}{384}\,\frac{wL^4}{EI} \tag{9.7}$$

where w = 153 lb/ft = 0.013 k/in., L = 17 × 12 = 204 in., E = 1600 k/in.^{2} and I = 2327 in.^{4}. Substituting in the preceding formula renders

$$\delta_D = \frac{5}{384}\,\frac{(0.013)\,(204)^4}{(1600)\,(2327)} = 0.08 \text{ in.}$$

A very small deflection, even considering the effects of creep which may double its value to 0.16 in. with time.

Additional deflection occurs under live load. For a concentrated load P at midspan, the deflection is given by:

$$\delta_L = \frac{1}{48}\,\frac{P_L{}^3}{EI} = \frac{1}{48}\,\frac{(8.32)\,(204)^3}{(1600)\,(2327)} = 0.40 \text{ in.}$$

where P = 8.32 k for the rear wheel of the H-20-44 riding directly over a stringer and E, I, L have the same values used before. The deflection-span ratio is $\delta_L/L = 0.40/204 \sim 1/510$, which is less than 1/300, the minimum value for highways stringers required by AITC Standard 102–69 (3) as set forth in Table 7.5.

9.3.3 Miscellaneous

A transverse section of the bridge and its supporting foundation is shown in Fig. 9.3. The pile cap which supports the stringers remains to be designed and is left to the reader as an exercise in beam design. It is usually a 12 × 14 in. timber, providing more than the minimum 6 in. bearing for the stringers, great stiffness and small deflections. Uplift-preventing anchor bolts are used to connect selected stringers to the pile cap. Blocking is also used between stringers to allow transmission to the foundation of horizontal forces due to wind or traffic. The piles are designed to act as columns and transmit the vertical reactions to the ground. In addition, x bracing connecting all four piles is used for lateral stability against the action of horizontal loads normal to the axis of the bridge. Along the same line of thinking, the two outside piles are battered outwards to improve resistance to lateral loads.

Construction of this type of trestle is very simple, requiring only light construction equipment for its erection. Various other solutions for this bridge are possible. Fewer, but heavier, stringers could be used by increasing the thickness of the deck. The type of deck itself may be changed by using planks or a doweled floor not less than 5½ in. thick. Only a comparison of various solutions can render the most economical solution for a given trestle.

9.4 COMPOSITE WOOD-CONCRETE SLAB

This type of deck may be used not only in bridges but also for floors subjected to heavy live loads as in piers and warehouses. It consists of a concrete slab cast on a continuous laminated deck made up of small-dimensioned timbers placed on edge and spiked together laterally as shown in Fig. 9.5. The wood sub-base acts as formwork for the concrete slab. It must support its own

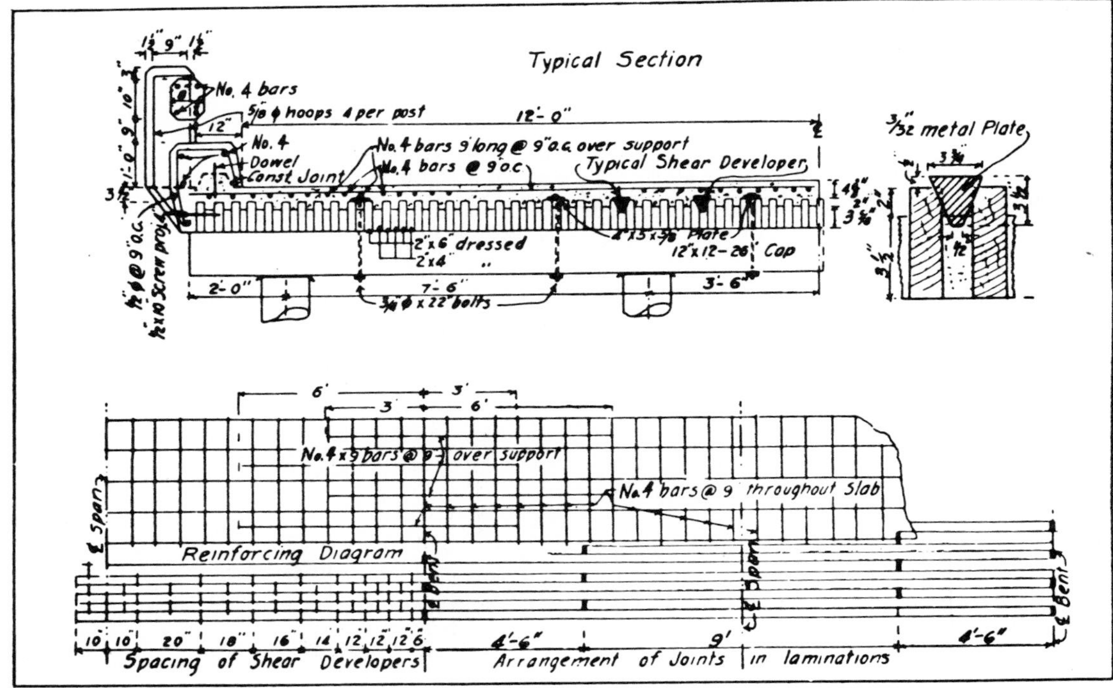

Figure 9.5. Plan and transverse semi-section of composite slab bridge.

weight and that of the concrete slab, plus a nominal construction live load, until composite action is obtained. The latter action is used to support the live load on the bridge. On simply-supported spans, and at midspan of continuous bridges, the wood deck will be subjected to compressive stress in the top fibers and tensile stresses in the bottom fibers under permanent loads; see Fig. 9.6a. Composite action of the deck under live loads will increase the wood tensile stresses substantially but will only increase the wood compressive stresses slightly. The stress distribution under total load is also shown in Fig. 9.6a. It is likely that the tensile stress in the bottom fiber will govern the design of the wood deck for the given section. The compressive stress at the top of the composite deck should be equal to, or less than, the allowable stress for concrete.

Similar considerations can be made for a section of a continuous bridge subjected to negative moment. Fig. 9.6b shows the stress distributions corresponding to a section over an interior support. It can be seen that the bottom fiber again controls the design of the wood deck; in this case it is subjected to compressive stress. The reinforcing steel takes most of the tensile force created by the negative moment over the support. For both cases shown in Fig. 9.6 concrete and steel stresses should be incremented by the impact coefficient. This is not necessary in wood because the impact stresses will not exceed 1.3 times the static stresses, while the allowable stresses under impact conditions may be taken as two times the normal allowable stresses.

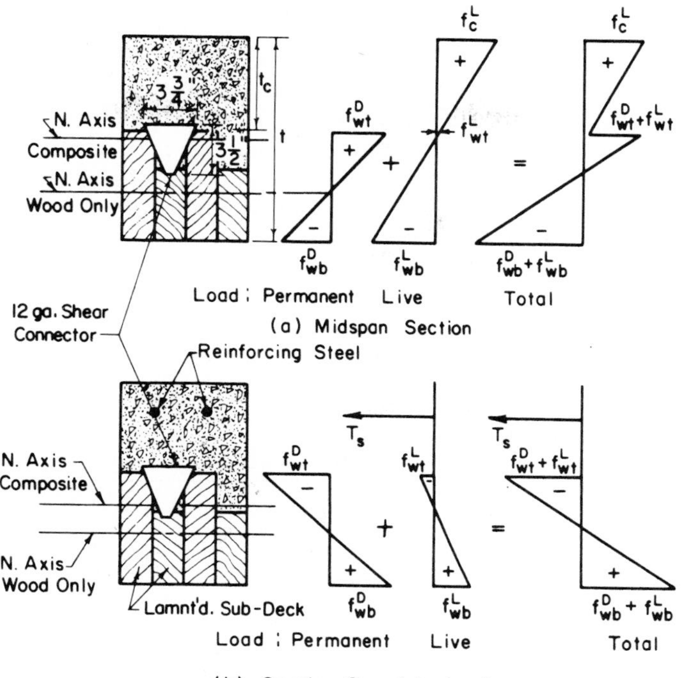

Figure 9.6. Stress distributions at various stages of loading in composite wood-concrete and wood-steel sections. Compressive stresses are positive.

The development of composite action requires provision to resist the horizontal shear stresses that exist at the junction between wood and concrete. The natural bond between the two materials is insufficient to resist much horizontal shear or diagonal tensile stresses, both of which tend to separate the component materials. Some mechanical device is needed to connect wood and concrete together. Several types of shear developers and combinations thereof have been used for this purpose, namely: triangular plates as shown in Fig. 9.6, railroad spikes and lag screws (4)(5). A solution that requires no mechanical devices and has been used in the northwest United States provides for castellated dapping in the wood laminations for transmission of horizontal shear stresses and prevention of slip, and grooves in the sides of the tall laminations to resist diagonal tensile stresses and prevent uplift separation (6).

A comprehensive study of the behavior of composite slabs with mechanical connections was performed by Richart and Williams at the University of Illinois (4). Their tests included the standard determination of behavior to ultimate loading of the composite slab 28 days after casting of the concrete. In addition, they studied possible effects of shrinkage by testing specimens after two and a half years of casting, and fatigue effects by submitting the specimens to repeated loads. From an analysis of the experimental results Richart and Williams concluded that composite action, while not perfect, was possible by means of mechanical connections. In particular, triangular plates combined with uplift-preventing spikes performed most satisfactorily in producing integral beam action, high strength and stiffness with only small slips. Of the various sizes of plates tested the 12-gage plates were most economical. Castellated dapping in combination with lag screws or spikes was considered fair but dapping alone gave poor results. Volume change due to shrinkage or temperature was measured over a 2½ year period and no effect was found on strength, as compared with specimens tested 28 days after fabrication. Fatigue effects due to a considerable number of repetitions of the design load were negligible. However, under the repeated application of a much higher load, 2.5 times the design load, permanent sets developed. In spite of these, there was satisfactory structural behavior when rapid loading to failure was resumed.

Allowable shear loads for various connections are given in Table 9.2.

TABLE 9.2. ALLOWABLE SHEAR LOADS FOR VARIOUS CONNECTIONS IN A COMPOSITE WOOD-CONCRETE DECK†*.

Connection	Angle with Vertical	Allowable Shear
6½ × ⅝ R.R. spikes	0°	1,280 lb.
6½ × ⅝ R.R. spikes	45°	1,875 lb.
6 × ⅝ lag screws	0°	1,240 lb.
8 × ⅝ lag screws	45°	2,150 lb.
Daps	—	1,070 psi**
Triangular Plates, with uplift spikes	0°	1,500 lb.

†After Richart and Williams, see Reference (4).
*Minimum factor of safety is 3.75.
**Of vertical bearing surface.

9.4.1 Analysis of Continuous Decks

Because of continuity, transverse loads acting on the deck create both positive and negative moments. Positive moments have been conventionally considered as moments which create compressive stresses in the top and tensile stresses in the bottom of the member to which they are applied. In continuous spans, positive moments occur in the central portions; negative moments in the regions over the supports. To resist the negative moments and the tensile stress induced by them, reinforcing steel bars must be used because of the negligible tensile strength of concrete. Hence, the deck is structurally a composite wood-concrete, and composite wood-steel member. Individual sections are of one type or the other, depending on whether they are subjected to positive or negative moment.

Extensive literature exists on various methods of analysis of homogeneous linearly-elastic structures. However, these methods are not directly applicable to the analysis of a structure for which the stiffness EI of any given section depends on the nature of the moment affecting it. The solution to the four cases of Fig. 9.7, was obtained by Richart and Williams (4, 5), and is given in Fig. 9.8. Values of k, and the ratio M_1/M_o, see Fig. 9.7, are plotted versus I_o/I_1, the ratio of

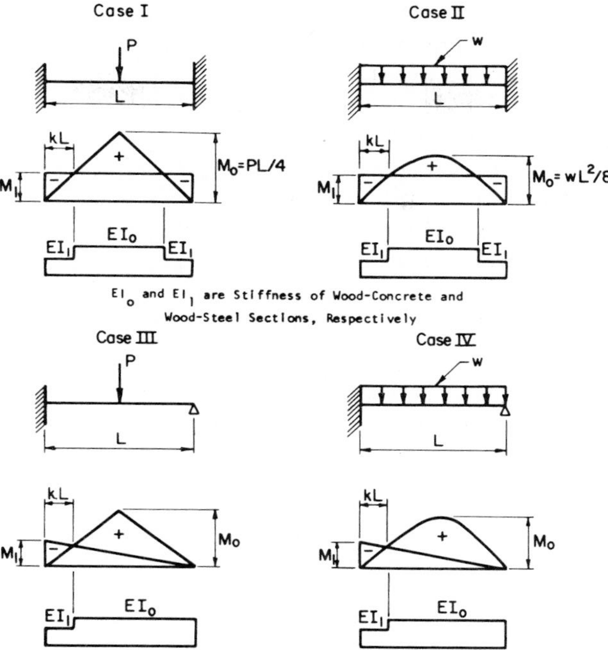

Figure 9.7. Moment and stiffness distributions in wood-concrete-steel composite beams under various loadings and support conditions.

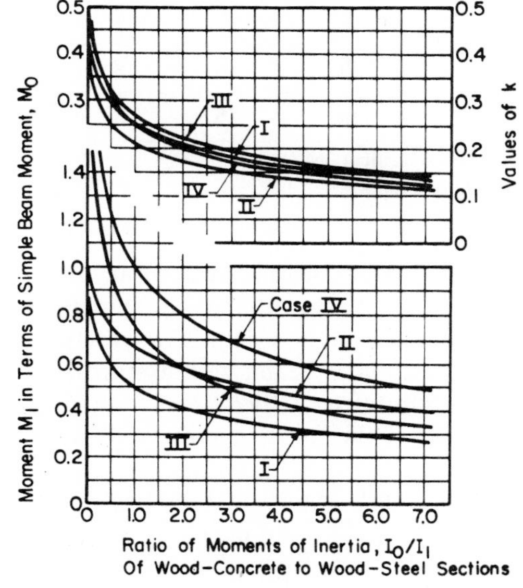

Figure 9.8. Moments and distances to points of inflection in continuous composite wood-concrete-steel beams.

moments of inertia of the composite wood-concrete and wood-steel sections. The case of a conventional linearly-elastic homogeneous material is given by $I_0/I_1 = 1$. For the type of composite deck studied, the ratio I_0/I_1 is larger than one; sometimes much larger, as in the case of composite slabs for which $E_c/E_w = 2$. All curves shown in Fig. 9.8 are asymptotic to the horizontal axis; in other words, for very large values of the ratio I_0/I_1 both k and M_1/M_0 tend to zero. This means that for large values of the ratio I_0/I_1 the member behaves as if it were virtually hinged at the ends. Conversely, small values of the ratio I_0/I_1 for cases I and II lead to $k = 0.5$ and $M_1/M_0 = 1$ and for cases III and IV $M_1/M_0 = 2$ and 4, respectively. This means that when the ratio I_0/I_1 approaches zero, the member behaves as a double cantilever hinged at midspan in cases I and II, and a cantilever of full length for cases III and IV.

The fixed-end beam closely represents an interior span of a continuous beam having several equal spans. The beams with one end fixed and one end free represents one span of a continuous beam of two equal spans. For cases of partial continuity the curves do not apply directly. AASHTO Specifications for Highway Bridges (1) provide a practical design method using the coefficients shown in Table 9.3. However, a study of the curves of Fig. 9.8 should enable the experienced designer to determine a reasonable distribution of moments for any particular case.

TABLE 9.3. DISTRIBUTION OF BENDING MOMENTS IN CONTINUOUS WOOD-CONCRETE DECKS†*.

	Uniform Dead Load				Live Load			
	Wood Subdeck		Composite Slab		Conc. Load		Unif. Load	
Span	Pos.	Neg.	Pos.	Neg.	Pos.	Neg.	Pos.	Neg.
Interior	50	50	55	45	75	25	75	55
End	70	60	70	60	85	30	85	65
Two span**	65	70	60	75	85	30	80	75

†After Table in Art, 1.3.5.B of AASHTO Specifications (1).
*Maximum bending moments in percent of simple-span moment.
**Continuous beam of two equal spans.

For a bridge having unequal spans, or where more accuracy in the determination of moments is desired, Table 9.3 need not be used. Instead, a simple numerical method of analysis should render answers after only a few cycles of iteration. The method is better illustrated with an example. Consider a three-span continuous beam, with a 40-ft central span and two equal end spans at 32 ft each. The beam is subjected to 3 k/ft uniformly-distributed load. Also, the ratio of moments of inertia between that of the wood-concrete section, I_0, and that of the wood-steel section, I_1, is $I_0/I_1 = 2$.

Determination of bending moments begins by assuming a prismatic beam with constant moment of inertia, I, throughout its length, see Fig. 9.9.1. Analysis of this beam, by any of the classical available methods (slope deflection, moment distribution, superposition, etc.) renders the bending moment diagram shown under Step 1 of the figure, where positive and negative moments

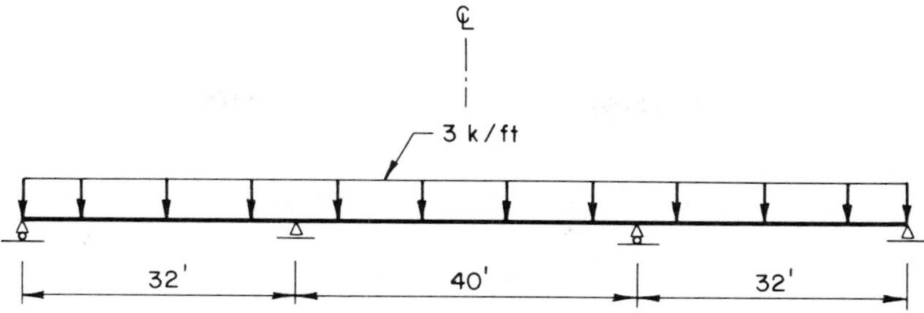

1) Initial Analysis On Prismatic Beam, Step 1

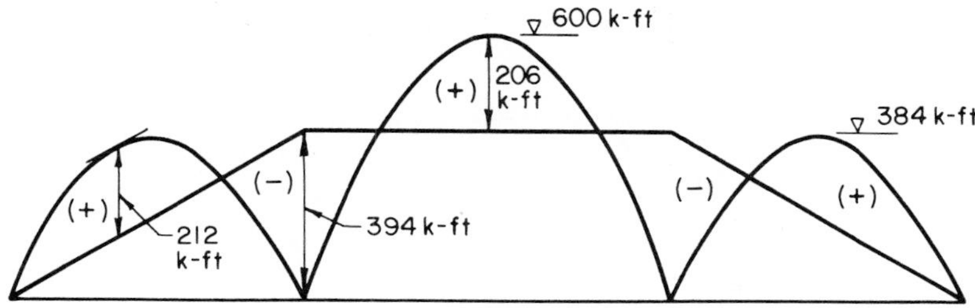

Moment Diagram for Prismatic Beam

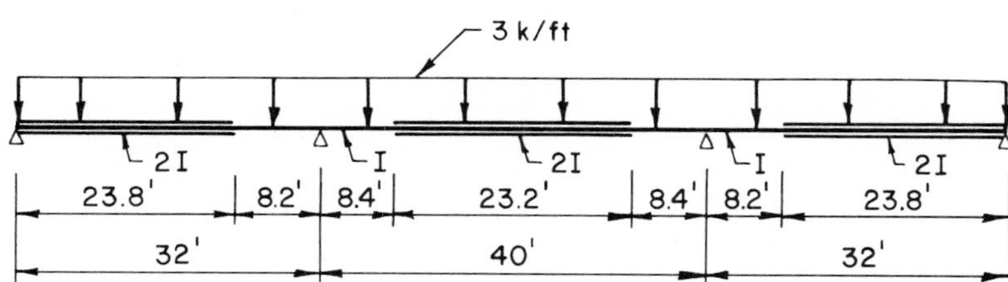

2) Nonprismatic Beam with I's Determined from Preceding Moment Diagram, Step 2

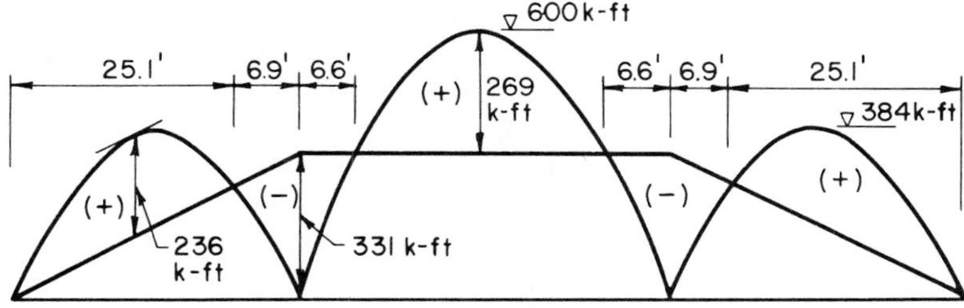

Moment Diagram for Nonprismatic Beam of Step 2

Figure 9.9. Iteration method for analysis of continuous composite beam.

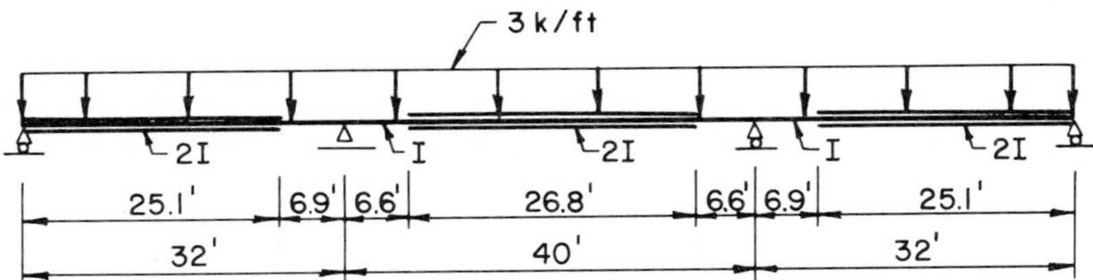

3) Nonprismatic Beam with I's Determined from Preceding Moment Diagram, Step 3

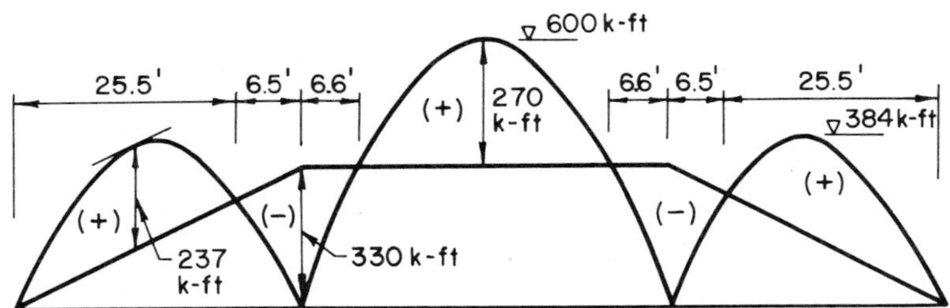

Moment Diagram for Nonprismatic Beam of Step 3

Figure 9.9. continued.

are identified. The specific portions of the continuous beam where these moments occur can be easily determined by simple statics from the moment diagram.

The beam shown in step 2 has been chosen to have the moment of inertia of any given cross section correspond with the sign of the bending moment affecting it. Thus, the end portion of the side spans, for a length of 23.8 ft from the end support, is subjected to positive bending moment and, therefore, the moment of inertia of its cross-section is taken as $I_0 = 2I$; for identical reasons, a 23.2 ft central portion of the interior span is equally treated. A portion of the span over each interior support is subjected to negative moment and, hence, its moment of inertia is taken as $I_1 = I$.

Analysis of the nonprismatic beam of step 2, under the 3 k/ft loading, renders another moment diagram. Comparison between the two moment diagrams indicates that the position of the inflection points and the maximum moments have changed appreciably. Another cycle of iteration is required using now the beam shown in part 3 of Fig. 9.9, for which the distribution of I's corresponds with the moment diagram obtained for the previous cycle. The resulting moment diagram indicates a negligible change in maximum moments and position of inflection points. For all practical purposes and within 3 significant figures, convergence has been achieved. Further iteration is not necessary.

Comparison of the moment diagrams shown in steps 1 and 3 of Fig. 9.9 shows the difference between the prismatic beam of uniform moment of inertia, and the composite beam with variable moment of inertia. Note that all maximum positive moments have increased (212 to 237, and 206 to 270 k-ft) and, as a consequence, the maximum negative moment has decreased (394 to

330 k-ft). This result could have been expected, as the ratio of positive to negative moments of inertia, I_o/I_1, exceeds unity, thereby making the positive regions of the beam more attractive to bending moment.

Solution of this problem using Figs. 9.7 and 9.8 should render only an approximate solution. Thus, consider the end span first, Case IV in Fig. 9.7. For $I_o/I_1 = 2$ one obtains $M_1/M_o = 0.8$ in Fig. 9.8. Thus, the fixed-end moment $M_1 = 0.8\ (wL^2/8)$, where $w = 3$ k/ft and $L = 32$ ft. Substitution of these values in the preceding formula renders $M_1 = 307$ k-ft. For the central span now, consider case II in Fig. 9.7 with $I_o/I_1 = 2$ for which Fig. 9.8 renders $M_1/M_o = 0.58$. Thus, $M_1 = 0.58\ (1/8)\ (3)\ (40)^2 \geqq 348$ k-ft. It follows that the true value of the negative moment over the interior supports lies between 307 and 348 k-ft. If one simply would take the average of these two values then a negative moment equal to 328 k-ft would be obtained. If one assumed prismatic-member behavior and made a simple moment distribution over the support (only one cycle is required, considering symmetry for the central span and a hinged end at the far end of the side span) one would obtain a negative moment equal to 334 k-ft. Either one of these results is quite close to 330 k-ft, the true answer, and their average, $(328 + 334)/2 = 331$ k-ft, is only 0.3 percent off. Therefore, in this particular example, an excellent approximation was obtained using Richart's results; this situation however, need not be true for all cases.

9.4.2 Design Considerations

Permanent loads on the bridge or floor deck, such as the weight of the wood sub-deck and the fresh concrete plus a small live load, are to be taken by the wood sub-deck alone. Any other loads that occur after the concrete has hardened are taken by composite action. This is true for any live load on floor decks and for vehicular traffic on bridges. The wood sub-deck ought to be sufficiently stiff to avoid unsightly sagging. The maximum deflection should comply with the provision of Table 7.5.

According to AASHTO, wheel loads on this type of bridge are distributed over a transverse width of 5 feet for bending moment and a width of 4 feet for shear. The maximum bending moment created by the wheel loads on a simply-supported span may be determined as shown before in Fig. 9.2. For continuous decks, however, both positive and negative moments shall be distributed in accordance with Table 9.3. It can be seen that the values given in the table for live load effects reflect Richart and Williams' theoretical analysis of fixed-end and continuous composite spans. The small ratio of maximum negative to positive moments under both concentrated and uniform loads, is only because of the large ratio I_o/I_1 (see Fig. 9.8) typical of wood-concrete composite slabs.

In long continuous bridges, it may be necessary to introduce a discontinuity at the support at intervals in order to provide expansion joints in the concrete. Spans adjacent to expansion joints should be treated as end spans. To provide for the increased positive moment in the end span, greater thickness of slab may be necessary. The large positive moment may be avoided altogether by a slight length reduction in the end span.

For the determination of the properties of the transformed wood-concrete and wood-steel sections used in the determination of stresses, it is necessary to know the modular ratios E_c/E_w and E_s/E_w; where E_c, E_s and E_w are the moduli of elasticity of concrete, steel, and wood, respectively. Article 1.3.5.C of AASHTO Specifications (1) provides the following relations: $\frac{E_c}{E_w} = 1$, for $t_c < \frac{t}{2}$; $\frac{E_c}{E_w} = 2$, for $t_c \geqslant \frac{t}{2}$ and $\frac{E_s}{E_w} = 18.75$. The net concrete thickness and the overall thickness of the composite slab are denoted by t_c and t, respectively. For the determination of the moment of inertia of the wood-steel transformed section at the supports, only two-thirds of the wood section should be taken into account since every third lamination is usually spliced at the support.

The capacity of the deck to take horizontal shear stresses can be checked using the following relation:

$$v = \frac{VQ}{Ib} \tag{9.8}$$

where v is the horizontal shear stress at a distance y from the neutral axis; V is the shear force in a section situated at a distance of three times the thickness of the deck from the face of the support; Q is the first moment of that area of the section above the level determined by y, about the neutral axis; I is the moment of inertia of the section; and b is the width of the section, taken as 5 ft. in the case of highway bridges. The maximum shear is determined by adding individual shear stress distributions in the deck, before and after composite action sets in. It is likely that the maximum shear stress will not occur at the same horizontal section for both distributions; however, an exact determination is laborious and not warranted. Instead, an upper bound for the maximum horizontal shear stress may be found by adding the individual maximum stresses. For this type of bridge, shear is not a determining factor in the selection of thickness of deck.

The spacing s, in inches, of the shear connectors that guarantee development of composite action may be found by means of the following equation:

$$s = 4\frac{St}{V_L g} \tag{9.9}$$

where S = shear capacity of a given connector, see Table 9.2, lb.
V_L = shear force due to live load plus impact only, per foot of critical section, 1b.
t = thickness of the deck, in.
g = width of groove, i.e. width of lamination, in.

The minimum spacing determined above may be increased gradually towards the center of the span as V_L decreases. A maximum spacing of 24 in. between triangular plates is usually observed.

9.5 ILLUSTRATIVE EXAMPLE

A composite wood-concrete slab is required for a continuous highway bridge with 16 ft. approach spans and 18 ft. interior spans. The bridge is to be designed for AASHTO's H-15-44 traffic with two lanes and a minimum 24 ft. width between inside faces of curb.

For the laminated wood sub-deck use a grade and species of surfaced-four-sides wood. Allowable stresses are: F_b = 1.65 ksi (repetitive member uses), F_v = 0.090 ksi and E = 1,600 ksi, see Table A.2.5. Concrete with a standard 28-day strength f_c' = 3 ksi and allowable compressive stress in bending f_c = 1.2 ksi; and reinforcing steel with yield point stress f = 40 ksi and allowable tensile stress f_s = 20 ksi are to be used for the top deck. Unit weights of wood and concrete are 50 and 150 pcf, respectively.

Design of this type of deck consists of checking a given section for adequacy. Tables A.9.1 and 2, may be used for the initial proportioning of the section. The process that follows herein is then used to check maximum stresses and deflection as indirect measures of strength and serviceability, respectively. In the case of an overdesigned section, corrections may be applied to decrease the amount or quality of materials used. Conversely, if the section is underdesigned, additional material may be added and the new section is analyzed again. The detailed analysis of a given section for the proposed bridge follows.

9.5.1 Loads

Dead load, see Fig. 9.10a

Wood Sub-deck	(4.5/12)(50)	= 19 lb./ft.^{2}
Concrete	(5.5/12)(150)	= 69
	w_D	= 88 lb./ft.^{2}

Live load, Rear wheel of H-15-44 weighs 12k

For moment: 12/5 = 2.4 k/ft.
For shear: 12/4 = 3.0 k/ft.

9.5.2 Bending Moments

Simple-span moments, k-ft.

	Approach Span	Interior Span
Dead Load	$(0.088/8)\,(16)^2 = 2.82$	$(0.088/8)\,(18)^2 = 3.56$
Live Load	(2.4/4) (16) = 9.60	(2.4/4) (18) = 10.80

Design moments, k-ft.

Section	Dead Load	Live Load
Midspan Approach	(0.7) (2.82) = 1.97	(0.85) (9.60) = 8.15
1st Int. Support	(0.6) (2.82) = 1.69	(0.30) (9.60) = 2.88
Midspan Int. Span	(0.5) (3.56) = 1.78	(0.75) (10.80) = 8.10
Interior Support	(0.5) (3.56) = 1.78	(0.25) (10.80) = 2.70

9.5.3 Section Properties

Wood sub-deck; see Fig. 9.10b

	Area in.^{2}	y, in.	Ay in.^{3}	$y-\bar{y}$ in.	$A(y-\bar{y}^2)$ in.^{4}	I in.^{4}
6 in. of 2 × 6	33.0	2.75	90.75	0.39	5.0	83.2
6 in. of 2 × 4	21.0	1.75	36.75	−0.61	7.8	21.4
	54.0		127.50		128.8	104.6

Hence, $\bar{y} = \dfrac{127.50}{54.0} = 2.36$ in.

and I = 12.8 + 104.6 = 117.4 in.^{4}
or $I = (\frac{1}{3})\,(6)[(5.5)^3 + (3.5)^3] - (54.0)\,(2.36)^2 = 117.4$ in.^{4}

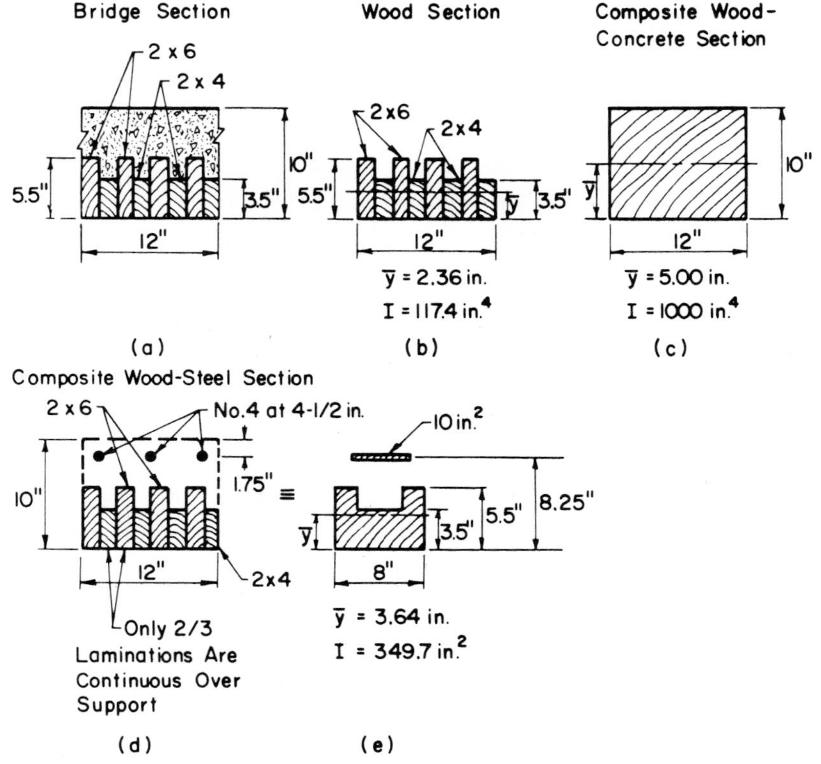

Figure 9.10. Section properties of composite slab bridge.

Composite wood-concrete deck; see Fig. 9.10c

Since the net concrete thickness, 4.5 in., is less than half the overall depth of the composite section, AASHTO species $E_c = E_w$. The section can then be considered as homogeneous and I may be determined simply as $(1/12)(12)(10)^3 = 1{,}000$ in.^{4}

Composite wood-steel deck; see Figs. 9.10d and e

The steel reinforcement is transformed as follows:
0.2 (in.^{2}/No. 4) $\times$ (12/4.5) 18.75 = 10.0 in.^{2}

	Area in.^{2}	y in.	Ay in.^{3}	$y-\bar{y}$ in.	$A(y-\bar{y})^2$ in.^{4}	I in.^{4}
No. 4 @ 4½ in.	10.0	8.25	82.5	4.61	212.5	—
4 in. of 2 $\times$ 6	22.0	2.75	60.5	0.89	17.4	55.5
4 in. of 2 $\times$ 4	14.0	1.75	24.5	1.89	50.0	14.3
	46.0		167.5		279.9	69.8

Hence, $\bar{y} = \frac{167.5}{46.0} = 3.64$ in.

and $I = 279.9 + 69.8 = 349.7$ in.^{4}
or $I = (1/3)(4)[(5.5)^3 + (3.5)^3] + (10)(8.25)^2 - (46)(3.64)^2 = 349.7$ in.^{4}

9.5.4 STRESSES

Dead load stresses; wood sub-deck only

	M, k-in.	I, in.^{4}	My/I, ksi*
Midspan Approach	23.7	117.4	0.476, (T)
1st Int.Support †	20.3	78.4	0.611, (C)
Midspan Int. Span	21.4	117.4	0.430, (T)
Interior Support†	21.4	78.4	0.645, (C)

Notes: *All stresses computed at bottom fiber, y = 2.36 in.
†Moment of Inertia taken as two-thirds of solid section.
(C) compression, (T) tension.

Live load and combined stresses in composite section

Section	Material	M k-in.	y in.	I in.^{4}	Live Load stress, ksi	Combined stress, ksi
Midspan-Approach	wood	97.8	5.00	1000	0.489(T)	0.965(T)
	concrete	127.1*	5.00		0.636(C)†	0.636(C)
1st Int.-Support	wood	34.6	3.64	350	0.360(C)	0.971(C)
	steel	45.0*	4.61		11.11(T)††	11.11(T)
Midspan-Int. Span	wood	97.2	5.00	1000	0.486(T)	0.916(T)
	concrete	126.5*	5.00		0.633(C)†	0.633(C)
Interior-Support	wood	32.4	3.64	350	0.337(C)	0.982(C)
	steel	42.1*	4.61		10.40 (T)††	10.40 (T)

Notes: *Stress multiplied by 1.3 to account for impact effects.
†Stress in concrete is $(E_c/E_w)(M_y/I)$, where $E_c/E_w = 1$ and M_y/I is stress in equivalent wood.
††Stress in steel is $(E_s/E_w)(M_y/I)$, where $E_s/E_w = 18.75$ and M_y/I is stress in equivalent wood.
(C) compression, (T) tension.

All stresses are below the allowable stresses for the various materials. In addition, it can be shown that the bridge meets the overload provision of Art. 1.2.4 of AASHTO for infrequent heavy loads.

9.5.5 SHEAR

Maximum horizontal shear stress

Dead Load Only:

$$V_D = (1/2)(0.088)\left(18 - 2 \times 3 \times \frac{10}{12}\right) = 0.572 \text{ k}$$

$$v_D = \frac{(0.572)(12 \times 2.36 \times 2.36/2)}{(117.4)(12)} = 0.014 \text{ ksi}$$

Live Load:

$$V_L = (3)\left(18 - 3 \times \frac{10}{12}\right)/18 = 2.59 \text{ k}$$

$$v_L = \frac{3}{2} \times \frac{2.59}{12 \times 10} = 0.032 \text{ ksi}$$

Hence, $v_T = 0.046$ ksi < 0.090 ksi.

Spacing of shear connectors

The spacing of 12-ga. triangular plates used as shear connectors may be determined from Eq. 9.9 with S = 1,500 lb., see Table 9.2; $V_L = 2{,}590 \times 1.3 = 3{,}370$ lb., 30% increase for impact; t = 10 in. and g = 1.5 in., as follows:

$$s = 4 \times \frac{1{,}500 \times 10}{3{,}370 \times 1.5} = 11.9 \text{ in.}$$

The spacing may then be specified as follows: Use 12 in. spacing from support through one fourth of the span increased gradually to 24 in. at midspan and staggered in adjacent rows. In addition, use 60d nails inclined 45° and protruding 2½ inches at 12 in. spacing in adjacent rows to prevent uplift.

9.5.6 Deflection

The deflection of the wood sub-deck under permanent loads, before composite action, may be determined approximately using elastic assumptions. The approach span, because of lack of continuity at its end support and in spite of being shorter than the interior spans, is likely to experience the maximum deflection in the bridge. Assuming full continuity at the interior support, the maximum deflections of a propped cantilever under uniform load is given by

$$\delta = \frac{1}{185} \frac{wL^4}{EI} \tag{9.10}$$

In this case, E = 1,600 ksi, w = (0.088/12) k/in., L = 192 in., I = 117.4 in.^{4}, which when substituted in Eq. 9.6 yields $\delta = 0.28$ in., an acceptable value under any standard.

The section just analyzed meets all the requirements. It is an acceptable solution for the bridge. Figure 9.5 gives a transverse section and plan of the bridge as shown in actual construction sheets.

9.6 COMPOSITE WOOD-CONCRETE BEAM

In contrast with the composite wood-concrete slab, this type of deck is not limited to short spans and may be used effectively for modern Interstate Highway bridges. Composite wood-concrete beam bridges consist of longitudinal wood stringers, preferably of the laminated type, over which a concrete slab is cast in place. The stringers support their own weight and that of the concrete slab and forms, acting as conventional beams. Adequate stiffness and stability must be provided in the stringers to take care of this non-composite temporary state, as only after the concrete has hardened will composite action develop. Traffic loads on the bridge are resisted by the composite section. For composite action to resist all loads, including the dead load of the bridge, shores have to be placed under the stringers until concrete has hardened. This procedure reduces the necessary section and cost of the stringers. A designer may consider using it for cases of limited total depth of bridge section. However, the combined cost of shoring and additional shear connections needed at the wood-concrete interface may make its use uneconomical for most other cases.

Development of composite action depends on adequate connections between wood and concrete to transmit the shear stresses without slip. Various solutions have been used (8), namely: castellated dapping only, ½ to ¾ in. deep, all cut into the top of the wood stringer; castellated dapping with nails or spikes partially driven into the top to prevent vertical separation of the concrete and the wood; lag bolts at 45° inclination to the horizontal to take the shear stresses and prevent separation; and recently, in the laboratory (9) (10), epoxies to glue wood stringers and precast concrete slabs together.

The negligible extensibility of the wood stringers restrains the free expansion and contraction of the concrete due to changes in temperature and the relative length of these bridges. The shear force developed between the concrete slab and the wood stringers depends on α, the coefficient of expansion of concrete, ΔT the change in temperature, E_c the modulus of elasticity of concrete, A_c the area of the concrete slab and L the length of the bridge. For a theoretical case of infinite restraint, the shear force between wood and concrete at each side of midspan would be $\alpha(\Delta T)E_cA_c/(L/2)$. Since the wood stringers have a finite stiffness that allows deformations of the slab, it is considered that the shear force developed is much smaller and can be safely taken as: $(1/3)\alpha(\Delta T)E_cA_c/L$, see Chapter 4 of Ref. 11. The horizontal shear stresses created by this force are considered uniformly distributed at both sides of midspan and are added to the stresses due to the live load at the interface. Shear connectors are designed to take the sum of the two effects.

An example of design of a simply-supported bridge follows to illustrate these concepts. A given section is analyzed for strength and serviceability. Two procedures are followed to check strength. First, maximum bending and shearing stresses under service loads are determined and compared against corresponding allowable stresses. This follows the conventional working-stress method of design. Secondly, the ultimate moment capacity of the composite section is determined. The ratio of ultimate moment capacity to maximum moment due to service loads, may be considered a measure of the factor of safety of the bridge. Serviceability of the solution is checked by the determination of maximum deflection and the provision of shop-fabricated camber to the laminated stringers. In view of the importance of shear connections at the wood-concrete interface, great emphasis is given in the example to this aspect of design, and various other solutions are indicated. Finally, various design details such as diaphragms, bearing plates and anchorage to the supports are briefly discussed.

9.7 ILLUSTRATIVE EXAMPLE

Design a two-lane simply-supported, composite wood-concrete bridge, having a 60-ft span, 24 ft width of roadway, and a 2-ft wide pedestrian sidewalk at each side, to carry H-20-44 truck traffic. Use pressure-treated glued-laminated wood stringers with the following design stresses: $E = 1{,}800$ ksi, $F_b = 2.4$ ksi, $F_t = 1.6$ ksi, $F_c = 1.5$ ksi, $F_v = 0.2$ ksi, and $F_{c\perp} = 0.45$ ksi. Also, for the composite deck use concrete with a standard 28-day strength $f'_c = 3$ ksi and design stress in bending $f_c = 0.4\, f'_c = 0.4 \times 3 = 1.2$ ksi, and reinforcing steel ASTM A615 Gr. 40 with yield point stress $f_y = 40$ ksi and design tensile stress $f_s = 20$ ksi. The bridge is to be designed for erection without shoring.

Following is a step by step annotated process for the design of a typical interior stringer of the bridge.

9.7.1 Loads

Dead Load, see Fig. 9.11:

Consider a 6 in. thick continuous concrete slab supported by typical glulam beams, 8¾ in. wide and 36 in. deep, spaced 4 ft apart. Each beam would carry, in addition to its own weight at 50 lb/ft³, the weight of a 4 ft-wide 6 in. thick concrete slab at 150 lb/ft³. Thus, for a typical beam, the dead load per unit length is:

Concrete slab (4) (6/12) (150)	=	300 lb/ft
Glulam beam (8.75 × 36/144) 50	=	110
w_D	=	410 lb/ft = 0.41 k/ft

Live Load: H-20-44

9.7.2 Bending Moments

Due to Dead Load:

$$M_D = \frac{1}{8} w_D L^2 = \left(\frac{1}{8}\right) (0.41) (60)^2 = 184.5 \text{ k ft}$$

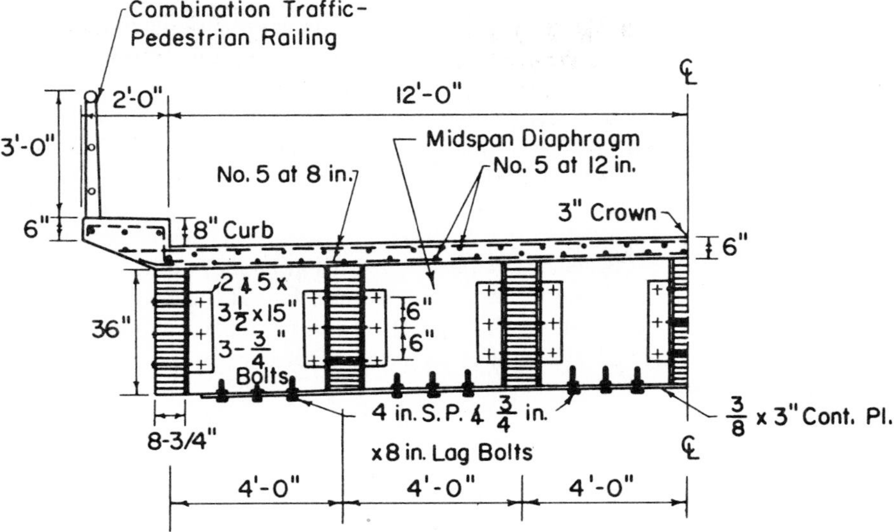

Figure 9.11. Transverse semi-section of composite wood-concrete beam bridge.

Due to Live Load:

It is necessary to check whether the standard-truck loading or the lane loading governs design.

Consider first the determination of maximum bending moment per lane, due to the standard-truck loading. From Fig. 9.2, for $P_1 = 4$ ton $= 8$ k, $P_2 = 16$ ton $= 32$ k, c = 14 ft, a = 1.4 ft, and L = 60 ft, obtain:

$$R = P_1\left(\frac{1}{2} + \frac{a}{L} - \frac{c}{L}\right) + P_2\left(\frac{1}{2} + \frac{a}{L}\right)$$
$$= 8\left(\frac{1}{2} + \frac{1.4}{60} - \frac{14}{60}\right) + 32\left(\frac{1}{2} + \frac{1.4}{60}\right)$$
$$= 2.32 + 16.75 = 19.07 \text{ k}$$
$$M = R\left(\frac{L}{2} - a\right) = (19.07)\left(\frac{60}{2} - 1.4\right) = 545.4 \text{ k-ft/lane}$$

To determine M caused by the standard lane loading it is necessary to place a uniformly distributed load $w_L = 0.64$ k/ft throughout the span plus a concentrated load P = 18 K at midspan. Thus,

$$M = \frac{1}{8} w L^2 + \frac{PL}{4}$$
$$= \frac{1}{8}(0.64)(60)^2 + \frac{(18)(60)}{4} \tag{9.11}$$
$$= 288 + 270 = 558 \text{ k-ft/lane}$$

This moment is larger than that generated by the standard-truck. Hence, M = 558 k-ft/lane is the design moment.*

Only one set of truck wheels (front and rear) may bear on the typical interior beam at one time; this makes the maximum moment per beam only one half that sustained by the whole lane. Also, because of lateral distribution to adjacent beams a smaller moment need be considered for design. From Table 9.1, in the case of a concrete floor on timber stringers and two or more traffic lanes, the actual fraction of wheel load to be used for the determination of bending moments in interior stringers is S/5, where S is the average spacing between stringers. For the given bridge S = 4 ft, and therefore:

$$M_L = \left(\frac{4}{5}\right)\left(\frac{558}{2}\right) = 223.2 \text{ k-ft}$$

(Another example of determination of M_L, for a bridge with stringers spaced 7.25 ft apart, is given in Section 9.9.)

* Also directly available from Appendix A of AASHTO Specifications (1).

9.7.3 SECTION PROPERTIES

Non-Composite Section (Wood beam only)

Area = $8.75 \times 36 = 315$ in.^{2}
Section Modulus = $(1/6)\,(8.75)\,(36)^2 = 1890$ in.^{3}
Moment of Inertia = $(1/12)\,(8.75)\,(36)^3 = 34020$ in.^{4}
Depth factor, C_F = $(12/36)^{1/9} = 0.885$

Composite Section (Wood beam plus concrete flange)

Effective flange width: 1/4 span = (1/4) 60 = 15 ft
beam spacing = 4 ft
12 times slab thickness = (12) (6/12) = 6 ft
} select 4 ft

Assume modular ratio $E_w/E_c = 1$

Properties of Composite Section

	Section in.	Area in.^{2}	y in.	Ay in.^{3}	$y - \bar{y}$ in.	$A\,(y - \bar{y})^2$ in.^{3}	I in.^{4}
Concrete-Slab	48 × 6	288	3	864	−10.97	34,658	864
Wood-Beam	8.75 × 36	315	24	7,560	10.03	31,689	34,020
	Sum:	603		8,424		66,347	34,884

$$\bar{y} = \frac{\Sigma A_i y_i}{\Sigma A_i} = \frac{8{,}424}{603} = 13.97 \text{ in.}$$

Hence: $I = 66{,}347 + 34{,}884 = 101{,}231$ in.^{4} $\sim$ 101,230 in.^{4}

Section moduli for the bottom (wood) and top (concrete) fibers of the section, S_w and S_c, respectively, are obtained as follows:

$$S_w = \frac{I}{y_w} = \frac{101{,}230}{36 + 6 - 13.97} = 3{,}611 \text{ in.}^3$$

$$S_c = \frac{I}{y_c} = \frac{101{,}230}{13.97} = 7{,}246 \text{ in.}^3$$

9.7.4 STRESSES

Dead load stresses at midspan (wood stringer only)

$$f_w = \frac{M_D}{S} = \frac{184.5 \times 12}{1890} = 1.17 \text{ ksi, tension}$$

Live load stresses at midspan (composite section)

$$\Delta f_w = \frac{M_L}{S_w} = \frac{223.2 \times 12}{3{,}611} = 0.742 \text{ ksi, tension}$$

$$f_c = \frac{M_L}{S_c} = \frac{223.2 \times 12}{7{,}246} = 0.370 \text{ ksi, compression}$$

The additional stress Δf_w in the wood beam is for the bottom fiber only. The stress computed for concrete must be multiplied by the factor $(1 + I)$ where I is the impact coefficient given by Eq. 9.1 as follows

$$I = \frac{50}{L + 125} = \frac{50}{60 + 125} = 0.270$$

Thus, the actual stress in the concrete, f_c, is equal to $0.370\ (1 + 0.270) = 0.470$ ksi, compression.

Combined stresses at midspan

Bottom of wood stringer	=	$1.17 + 0.74 = 1.91 \text{ ksi} < 0.885 \times 2.4 = 2.12$ ksi
Top of concrete slab	=	$0.47 \text{ ksi} < 1.2$ ksi

Stresses are below design stresses for wood and concrete.

9.7.5 SHEAR

Maximum Shear Force

It occurs in a section at a distance equal to the depth of the wood beam from the center of the support. Thus,

$$V_D = \frac{w_D}{2}(L - 2d)$$

$$V_L = \frac{P_L}{2}\left(\frac{L-d}{L}\right) + \left(\frac{1}{2}\right)\left(\frac{w_L}{2}\right)\frac{(L-d)^2}{L} \tag{9.12}$$

where $w_D = 0.41$ k/ft is the dead load on the typical interior stringer including selfweight; $P_L = 26$ k is the concentrated load per lane for shear-force determination, using standard-lane loading, see Fig. 9.1; and, $w_L = 0.64$ k/ft is the uniformly distributed load for standard lane loading. Substituting these values in the preceding formulas yields:

$$V_D = \frac{0.41}{2}(60 - 2 \times 3) = 11.07 \text{ k}$$

$$V_L = \frac{26}{2}\left(\frac{60 - 3}{60}\right) + \left(\frac{1}{2}\right)\left(\frac{0.64}{2}\right)\frac{(60 - 3)^2}{60}$$

$$= 12.35 + 8.66 = 21.01 \text{ k}$$

No lateral distribution to adjacent beams has been considered in the determination of maximum shear force due to live load. This is due to the close distance (3 ft) of the concentrated load to the support and the negligible deflection of the stringer at the section. To determine V_L for any section situated at a distance X from the support use X instead of d in the preceding formula for V_L and multiply by the lateral distribution factor (Use linear interpolation between values given by Table 9.1 for midspan and Fig. 7.13 for $X = L/4$.)

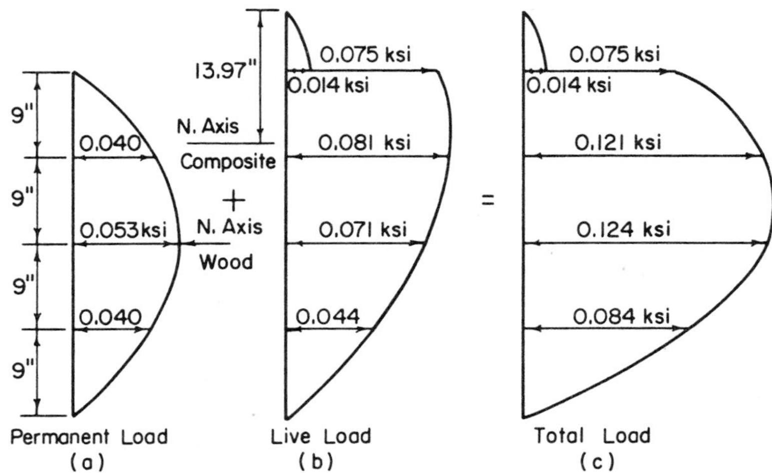

Figure 9.12. Shear stress distributions at the critical section at various stages of loading.

Shear Stresses due to Dead Load

The distribution of horizontal shear stresses in the wood beam due to dead load can be determined by means of Eq. 9.4 as shown in Chapter 7. The result is the well known quadratic parabola shown in Fig. 9.12 (a) with maximum stress occurring at mid depth of the wood beam. The latter value is given also by $v = (3/2)\ (V/A) = (3/2)\ (11.07/8.75 \times 36) = 0.053$ ksi.

Shear Stresses due to Live Load and Combined Load

For this case V = 21.0 k, I = 101,230 in.[4] and b = 48 in. for $0 < y \leq 6$ and b = 8.75 in. for $6 \leq y \leq 44$ in., where y is the vertical distance from the top fiber of the concrete slab to the given section. Hence, application of Eq. 9.8 to the composite wood-concrete section results in the following expressions

$$v = \frac{VQ}{Ib} = \frac{(21.0)\ [(48\ y)\ (13.97 - \frac{y}{2})]}{(101{,}230)\ (48)}, \quad 0 \leq y \leq 6 \text{ in.}$$

and

$$v = \frac{VQ}{Ib} = \frac{(21.0)\ [3159 + 8.75\ (y - 6)\ (10.97 - \frac{y}{2})]}{(101{,}230)\ (8.75)}, \quad 6 \leq y \leq 42 \text{ in.}$$

The resulting shear-stress distribution is shown in Fig. 9.12 (b). A sharp discontinuity occurs at the interface of wood and concrete because of the abrupt change in width.

The total horizontal shear stress distribution is shown in Fig. 9.12 (c), which is the sum of distributions (a) and (b). The maximum stress is approximately 0.125 ksi, which is below the 0.2 ksi design shear stress, and occurs at a horizontal section between the neutral axes of the non-composite section and the composite section. Note that the maximum shear stress is smaller than a stress given by the sum of the individual maximum stresses of Figs. 9.12 (a) and (b). Hence, if the latter shear stress (which is easy to determine and may be considered an upper bound for the actual maximum horizontal shear stress) is less than the design shear stress for the given wood no further investigation would be required.

Spacing of Shear Connectors

Shear connections are required at the concrete-wood interface to guarantee composite action by transferring horizontal shear stresses and preventing separation between concrete slab and wood beam. The minimum spacing of the shear connectors will occur at the critical section of the beam where the horizontal shear stresses due to live load are maximum. In addition, the effects of shrinkage and temperature-induced expansion or contraction of the concrete slab create additional shearing stresses, the transfer of which also takes place through the shear connectors. From Table 9.2, assuming two 8 × 5/8 in. D. lag bolts at 45° inclination, p = 2 × 2.15 = 4.3 k per pair. Spacing of these connectors may be determined using Table 9.4 first and then Fig. 9.13.

An alternate solution with fewer connectors is possible using 4 in. shear plates shop-attached to the wood beam by 3/4 in. D lag bolts through a continuous 1/4 × 4 in. steel plate. To the latter plate there would be attached 3/4 in. D. × 3 in. long headed studs each with a capacity to transfer a shear force of 11.5 k, as used in standard steel-concrete composite construction; preferably, these studs could be shop-welded to the steel plate prior to attachment to the wood beams. To reduce the cost of formwork for the cast-in-place concrete slab use corrugated steel sheets laid over the wood beams directly. Formwork and shear connections may be eliminated by using precast concrete slabs laid on the wood stringers over a layer of epoxy (9); however, the

TABLE 9.4. DETERMINATION OF HORIZONTAL SHEAR-STRESS DISTRIBUTION OF WOOD-CONCRETE INTERFACE AND REQUIRED CONNECTOR SPACING.

Section	Max. Shear Force,[1] k	Lat. Distrib. Factor[2]	Actual Shear Force,[3] V_L, k	Live Load Shear Stress,[4] v_Lb, k/in.	Temperature Shear Stress,[5] $v_\alpha b$, k/in.	Total Shear Stress,[6] v_tb, k/in.	Connector Spacing,[7] in.
3 ft. from support	21.00	1.0	21.0	0.83	0.13	0.96	4.5
(1/4) L	15.15	0.9	13.6	0.54	0.13	0.67	6.4
Midspan	8.90	0.8	7.1	0.28	0.13	0.41	10.5

1. Using formulas in Section 9.7.4 for Maximum Shear Force Caused by Standard Lane Loading (w_L = 0.64 k/ft., P_L = 26 k)
2. Relative amount of shear taken by beam after lateral distribution to adjacent beams. Factors obtained by linear interpolation between 0.8 at midspan and 1.0 at support.
3. Product of preceding two columns.
4. From $v_Lb = 1.27\ V_L\ Q/I$ at interface concrete-wood, where $Q = 6 \times 48 \times (13.97 - 6/2) = 3160$ in.^{3}, $I = 101{,}230$ in.^{4} and 1.27 accounts for impact.
5. From $v_\alpha b = (1/3)\ A_c\alpha\ (\Delta T)\ (E_c/L)$ where $A_c = 6 \times 48 = 288$ in.^{2}, $\alpha = 5.5 \times 10^{-6}$/degree F, $\Delta T = 60°F$, $E_c = 3000$ ksi, and $L = 720$ in.
6. Sum of v_Lb and $v_\alpha b$.
7. Spacing of two 5/8 × 8 in. lag bolts at 45° inclination at an allowable shear of 2.15 k each is given by $4.30/v_tb$.

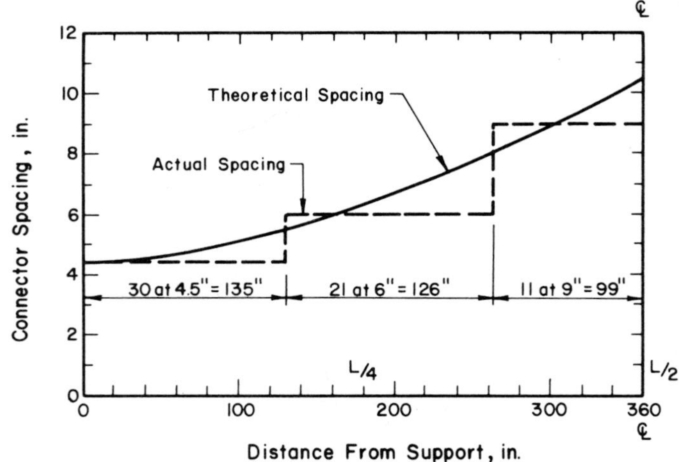

Figure 9.13. Determination of spacing of shear connectors for composite wood concrete beam.

designer should be aware that the use of epoxies in composite wood-concrete construction is still in an experimental stage with little information available as yet on effect of repeated loading, impact and long-term behavior.

9.7.6 DEFLECTION

The deflection of the bridge under permanent loads, before composite action, may be determined using elastic analysis. The maximum deflection occurs at midspan and is given by Eq. 9.2 in which $w_D = (0.41/12)$ k/in., $L = (60 \times 12) = 720$ in., $E = 1800$ ksi, and $I = 34{,}020$ in.^{4}. Thus,

$$\delta = \frac{5}{384}\frac{w_D L^4}{EI} = \frac{5}{384}\left(\frac{0.41}{12}\right)\frac{(720)^4}{(1800)(34020)} = 1.95 \text{ in.}$$

Considering the variability in modulus of elasticity of glued-laminated timber at 10 percent, and a 5-percent exclusion value, a lower bound for E would be given by $1800\ (1 - 0.10 \times 1.645) = 1{,}504$ ksi. This may increase δ to $1.95\ (1800/1504) = 2.33$ in. It does not seem reasonable to apply to this deflection an additional 1.5 factor for creep caused by long-term loading because of the fact that most of it will occur after the much stiffer composite section is established. All facts considered, it appears that $\delta = 2.5$ in. may be a reasonable upper bound value for midspan deflection caused by dead load. This deflection may be compensated by laminating camber into the stringers. In order to prevent unsightly sagging, and to compensate for long-time effects, it is reasonable to provide at least double the estimated dead-load deflection. Hence, a 5 in. midspan camber should be specified. Also, a parabolic camber configuration would call for variable camber along the span as follows: 2.19, 3.75 and 4.68 in. at distances 1/8, 1/4 and 3/8 of the span length from each end, respectively.

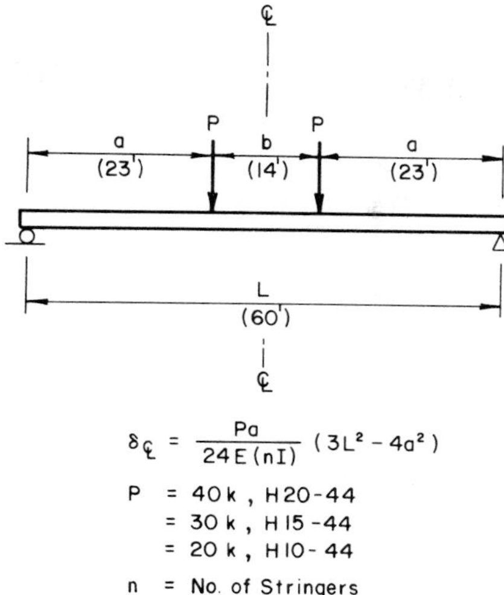

Figure 9.14. Determination of midspan deflection due to live load in a two-lane stringer bridge.

The deflection at midspan due to live load may be determined from a position of the wheels as shown in Fig. 9.14. The latter depicts two H-20-44 trucks in opposite directions (each concentrated load, P, representing one rear axle of a truck, 32 k, plus one front axle of the adjacent truck, 8k). The stiffness of the bridge is taken as the sum of that of all stringers; the presence of the midspan diaphragm may be assumed to force uniform displacement of the bridge at midspan. For $n = 7$ stringers, $I = 101{,}230 \text{ in.}^4$ (the composite moment of inertia of one stringer and corresponding concrete slab), δ may be calculated as follows:

$$\delta = \frac{Pa}{24\, En\, I}(3L^2 - 4a^2) \tag{9.13}$$

$$= \frac{(40)\,(23 \times 12)}{(24)\,(1800)\,(7 \times 101{,}230)}(3 \times 60^2 - 4 \times 23^2)\,(12)^2$$

$$= 0.45 \text{ in.}$$

This gives a deflection-span ratio of $0.45/(60 \times 12) = 1/1600$, which is well below the range of 1/200 to 1/300 allowed for highway bridges. Because of the large stiffness attained by wood stringers and concrete slab acting compositely it is unlikely that this type of bridge will ever fail to meet the live-load deflection criterion.

9.7.7 Ultimate Strength in Bending

The ultimate moment capacity and curvature of the composite wood-concrete section may be theoretically determined using fundamental principles of statics and strain compatibility. Ultimate is assumed to occur upon attainment of the crushing strain, $\epsilon_u = 0.003$, in the concrete. A conservative estimate of the total compression force developed by the concrete is $0.7\, f_c^1\, A_c$ where

A_c is the area of the slab. The stresses in wood may be expressed in terms of a, the unknown distance of the top fiber of the composite section to the neutral axis. Substituting the expressions for stresses in the equilibrium equation of horizontal forces yields the value of a. The ultimate moment is determined by taking moments of the various forces about any convenient point, in this case about the neutral axis. The ultimate curvature, ϕ_u, of the section is a measure of its energy absorption capacity and may be used to compare ductility of various solutions. The determination of ϕ_u is made using fundamental Eq. 7.3. The whole process is shown in Fig. 9.15 for the composite section corresponding to an interior stringer of the bridge.

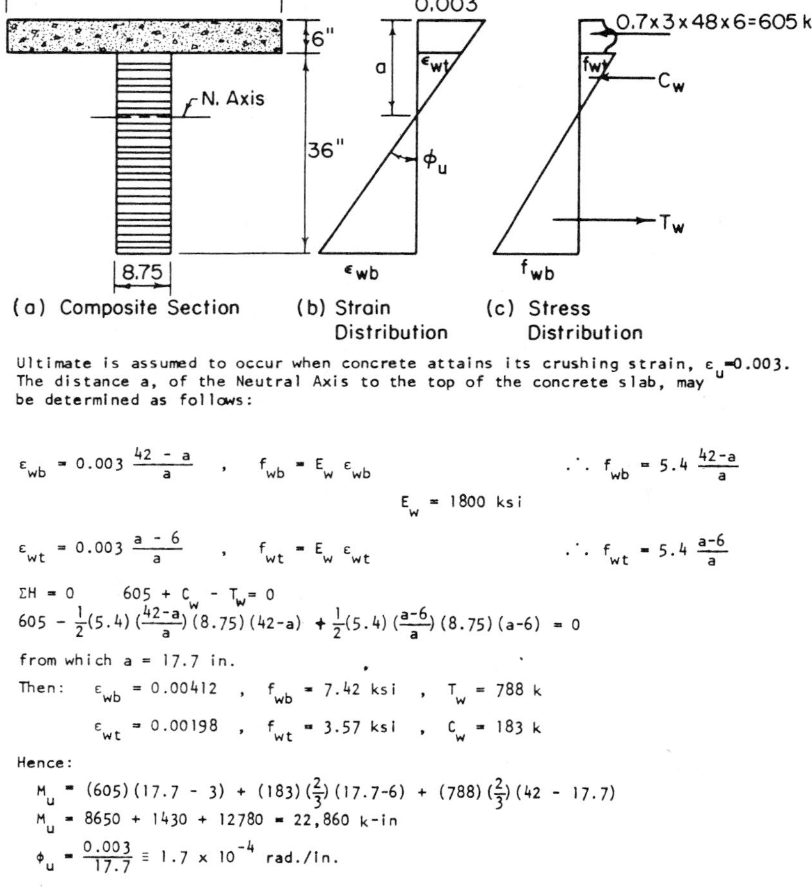

Figure 9.15. Determination of ultimate moment and curvature of composite wood-concrete section.

9.7.8 Miscellaneous Details

Transverse and longitudinal sections of the bridge are shown in Figs. 9.11 and 9.16, respectively. Each stringer is supported at one end by an elastomeric bearing plate to allow for longitudinal displacements, and at the other end it is held in place by an anchor bolt and a steel bearing plate.

At both ends of the bridge, transverse diaphragms are placed to tie all stringers together and to assist in transferring to the abutments wind and other horizontal forces perpendicular to the longitudinal axis of the bridge. A diaphragm has also been placed at midspan to improve both the lateral distribution of traffic loads and the over-all stiffness of the bridge. Continuity of the midspan diaphragm has been obtained at the bottom through a continuous steel plate connected to the stringers by shear plates, and at the top by composite action with the reinforced concrete slab, see Fig. 9.11. Instead of solid diaphragms it is possible to use X-type bracing (consisting of steel angles or solid timber members) attached to the stringers as required.

During construction of this type of bridge it is necessary to provide sufficient bracing to the stringers to avoid lateral buckling and collapse. For this purpose, good practice would indicate placement of the midspan diaphragm soon after erection of the first pair of stringers.

Other solutions for this type of bridge are possible by varying the spacing of the stringers. A smaller number of heavier stringers is possible if the thickness and reinforcement of the concrete slab are increased. However, it should be noted that this may only compound the difficulties of transmission of horizontal shear stresses at the interface by requiring a greater amount of shear connectors. Before deciding on the final section of the bridge, a designer would do well for the sake of cost comparison to try at least another solution.

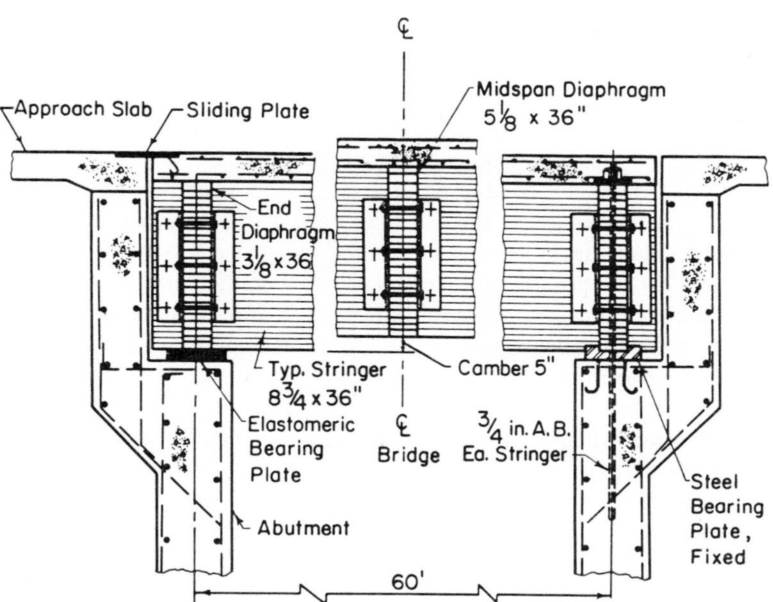

Figure 9.16. Longitudinal section of composite wood-concrete beam bridge.

9.7.9 Continuous Spans

Extension of this type of bridge to continuous spans is possible. In the regions over the supports, negative moments due to live loads may be taken in composite action. For this purpose, steel reinforcement should be placed in the top portion of the concrete slab to furnish the required tensile force, while the bottom portion of the wood stringer would provide compression strength.

The composite wood-steel section may be analyzed using the transformed section in the same way as shown for the composite slab. Hence there would be both a negative and a positive moment of inertia. Analysis of this type of continuous bridge under live load using linear elastic behavior is possible if the difference between positive and negative moments of inertia were not large; otherwise, a numerical analysis such as indicated in section 9.4.1 would be indicated.

Lateral stability of the bottom portion of the wood stringers should be checked over the length of span where negative moments occur. Diaphragms should be placed between the stringers at all supports over which the bridge is continuous and, if necessary also at, or close to, points of inflection along the span. Early erection of the diaphragms, soon after placement of the bridge stringers has begun, would provide also lateral stability to the latter, at a stage in the construction process where it is most valuable.

9.8 GLULAM DECKS

Section 9.3 discussed the design of a conventional timber highway trestle consisting of a wood deck made of 2-in. wide lumber, nail-laminated together, and toe nailed to the supporting wood stringers. The latter, which are spaced only 25 in. apart, make deflections of the deck negligible and keep it tight and securely attached. These factors guarantee adequate performance and durability of nail-laminated decks.

However, for increased spacing of stringers, nail-laminated decks are not acceptable. Substantial deflections, and the swelling and shrinkage due to cyclic wetting and drying (in spite of protective wearing surfaces), cause a gradual loosening of the nails and eventually of the entire deck. As a result, water from rain or snow may penetrate into the deck and stringers causing the entire structure to become unserviceable in a short time.

In the early 1970's, research conducted at U.S. Forest Products Laboratory, see Refs. 13 through 16, on glued-laminated decks showed that remarkable potential existed for this product in highway bridges. Glulam decks consist of panels, generally 4 ft wide and the length of the roadway width, which are shop-manufactured by gluing together pieces of nominal 2-in. (1.5 in. actual) dimension lumber using a waterproof adhesive; 32 laminations are required to make a 4 ft-wide panel. Thickness depends on nominal size of dimension lumber which ranges between 4 and 16 in.; because of resurfacing on both sides after manufacture, the finished thickness of the deck will be slightly less than the actual depth of the lumber.

The same grade of wood is used throughout the panel, which looks like a conventional glued-laminated beam except that the load is applied to it parallel to the wide face of the laminations. This causes the panels to act as vertically-laminated beams where all laminations, and not just the end ones, are fully stressed by the applied load.

Various comparisons of behavior under load of glulam panels and nail-laminated deck showed the obvious superiority of the former. Under concentrated load the longitudinal deflection configuration of glulam panel is a smooth bell-shaped curve, indicating effective load distribution, while the deflection configuration of the nailed-laminated deck is greatly truncated, with most of the deflection occurring directly under the load and hardly any elsewhere.

Theoretical studies of the glulam deck behaving as an orthotropic plate were made to determine the necessary design parameters. These are the primary moment and shear, M_x and R_x, respectively, the secondary moment and shear, M_y and R_y and the deflection δ. This work was checked experimentally as well, with excellent agreement reported (13).

To prevent relative displacements and rotations of adjacent panel edges it is necessary to provide connectors between panels, see Fig. 9.17. Tests conducted at Forest Products Laboratory (13) showed that the best possible connectors were steel dowels. Design of the dowels required to transfer the secondary moments and shears across adjacent panels, is also based on analytical and experimental work at Forest Products Laboratory. Use of the theory of beams on elastic foundations to analyze the behavior of the dowels required determination of the foundation modulus, k_0, of wood. This called for experimental work on bearing of steel dowels on wood across the grain, see Ref. 13; the value of k_0, measured in pounds per square inch per inch of deformation, was obtained as the stress at proportional limit in compression across the grain divided by the slip (deformation) at that level. The average foundation modulus, k_0, for 18 specimens was 180,000 pounds/in.^{3}. In the theory of beams on elastic foundations the fourth root of k_0 is used; hence slight variations of k_0 do not materially affect the results.

A number of charts for evaluation of all design parameters were published (13, 14, 15, 17) as a result of the investigation conducted at Forest Products Laboratory. From these, a series of simple formulas were derived that have been adopted by AASHTO (1) for design of glulam decks.

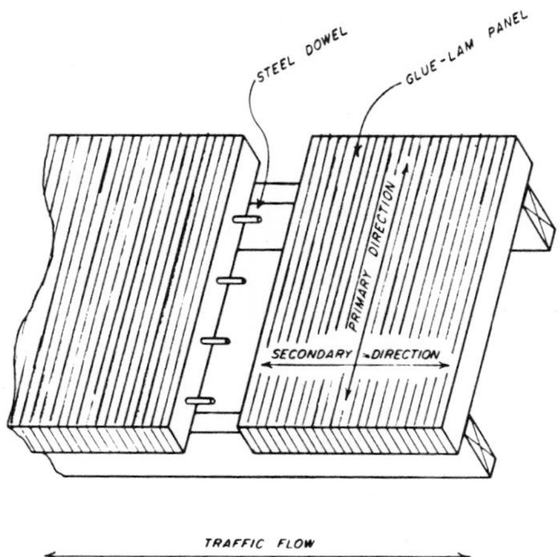

Figure 9.17. Glulam deck on wood stringers. Taken from Ref. 14.

9.8.1 Design Formulas

The deck thickness, t, is determined as the greater of:

$$t = \sqrt{\frac{6\ M_x}{F_b}} \tag{9.13}$$

and

$$t = \frac{3\ R_x}{2\ F_v} \tag{9.14}$$

where: M_x = primary bending moment in lb/in.
R_x = primary shear (lb/in.)
x = denotes direction perpendicular to longitudinal stringers
F_b = design bending stress, psi, based on load applied parallel to the wide face of the laminations
F_v = design shear stress, psi, based on load applied parallel to the wide face of the laminations.

The quantities M_x and R_x are calculated using the following formulas:

$$M_x = P\ (0.51 \log_{10} s - K) \tag{9.15}$$

and

$$R_x = 0.034\ P \tag{9.16}$$

where: P = design wheel load (lb)
s = effective deck span (the lesser of the clear distance between stringers plus one half the width of one stringer, and the clear span plus the floor thickness)
K = design constant depending on design load. Thus K = 0.44, 0.47 and 0.51 for H-10, H-15 and H-20, respectively.

Determination of the minimum size and spacing of the steel dowels required to transfer the load between panels is based on the following interaction equation:

$$n = \frac{1000}{\sigma_{PL}}\left(\frac{\overline{R}_y}{R_D} + \frac{\overline{M}_y}{M_D}\right) \tag{9.17}$$

where: n = number of steel dowels of a given diameter and length required for the given span, s
σ_{PL} = proportional limit stress in compression perpendicular to grain. For Douglas Fir or Southern Pine, use 1000 psi (for other woods at 12 percent moisture content use 1.5 times the value listed in Table A.2.2 under "compression perpendicular to grain, fiber stress at proportional limit").
$\overline{R}_y$ = total secondary shear transferred, lb.
$\overline{M}_y$ = total secondary moment transferred, in.-lb.
R_D = shear capacity of individual dowel, lb, see Table 9.5.
M_D = moment capacity of individual dowel, in.-lb, see Table 9.5.

Determination of $\overline{R}_y$ and $\overline{M}_y$ depends on the effective deck span, s. Thus, for $s \leq 50$ in.

$$\overline{R}_y = 6\,\frac{Ps}{1000}$$

$$\overline{M}_y = \frac{Ps}{1600}\,(s - 10) \tag{9.18}$$

and for s > 50 in.

$$\overline{R}_y = \frac{P}{2s}\,(s - 20)$$

$$\overline{M}_y = \frac{Ps}{20}\left(\frac{s - 30}{s - 10}\right) \tag{9.19}$$

The dowels shall be checked to ensure that the maximum stress, σ, caused by the interaction of $\overline{R}_y$ and $\overline{M}_y$ does not exceed the elastic limit of the steel given conservatively by 0.75 F_y where F_y is the yield-point stress of the steel. For the determination of σ the following interaction equation is used:

$$\sigma = \frac{1}{n}\,(C_R\,\overline{R}_y + C_M\overline{M}_y) \tag{9.20}$$

where: C_R and C_M are coefficients directly related to the inverse of the cross-sectional area and section modulus, respectively, of the steel dowel, see Table 9.5. The quantities n, $\overline{R}_y$ and $\overline{M}_y$ have been defined before.

TABLE 9.5. DOWEL DESIGN DATA.

Diameter, in.	Shear Capacity, R_D, lb	Moment Capacity, M_D, lb-in.	Section Coefficients, C_R, in.$^{-2}$	Section Coefficients, C_M, in.$^{-3}$	Length Required in.
0.5	600	850	36.9	81.5	8.5
0.625	800	1340	22.3	41.7	10.0
0.75	1020	1960	14.8	24.1	11.5
0.875	1260	2720	10.5	15.2	13.0
1.0	1520	3630	7.75	10.2	14.5
1.125	1790	4680	5.94	7.15	15.5
1.25	2100	5950	4.69	5.22	17.0
1.375	2420	7360	3.78	3.92	18.0
1.5	2770	8990	3.11	3.02	19.5

No formula is given for the determination of panel deflection, δ; instead Fig. 9.18 is offered. Deflections may seldom govern design of glulam decks.

Finally, Table 9.6 is given to allow designers an immediate estimate of the approximate size of glued-laminated stringers and the total number of steel dowels required for the glulam panels. Note that the table is limited to a 30 ft roadway width and to glued-laminated stringers made of 24 F combination symbol Douglas Fir and Larch or Southern Pine under dry conditions of use. Naturally, all wood is considered treated with preservatives.

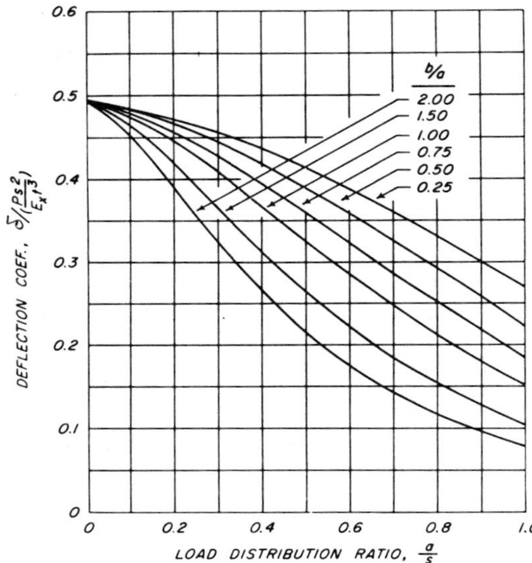

Figure 9.18. Deflection coefficients for glulam deck. Taken from Ref. (14).

TABLE 9.6. TYPICAL DESIGNS FOR 30 FT. WIDE ROADWAY BRIDGES USING GLULAM STRINGERS AND GLULAM DECKS.*

Span, ft	HS 15-44[1] Stringers Size in. × in.	No.	Dowels† Total No.	HS 20-44[2] Stringers Size in. × in.	No.	Dowels† Total No.
20	6¾ × 27	5	104	6¾ × 29½	5	136
24	6¾ × 30	5	130	6¾ × 30	5	170
28	6¾ × 31½	6	162	6¾ × 31½	5	204
32	8¾ × 34½	5	182	8¾ × 36	5	238
36	8¾ × 39	5	208	8¾ × 40½	5	272
40	8¾ × 43½	5	234	10¾ × 40½	5	306
44	8¾ × 42	6	220	10¾ × 43½	5	300
48	10¾ × 45	5	286	10¾ × 48	5	330
52	10¾ × 48	5	312	10¾ × 51	5	360
56	10¾ × 49½	5	338	12¼ × 49½	5	390
60	10¾ × 52½	5	364	12¼ × 52½	5	420
64	12¼ × 52½	5	390	12¼ × 55 ½	5	450
68	12¼ × 54	5	416	12¼ × 58½	5	480
72	12¼ × 57	5	442	12¼ × 61½	5	510
76	12¼ × 58½	5	468	12¼ × 64½	5	540
80	12¼ × 61½	5	494	12¼ × 66	5	570

*Adapted from Sheet GL-2 of "Glulam Bridge Systems. Plans and Details" by American Institute of Timber Construction. Wood is combination symbol 24F Douglas Fir and Larch or Southern Pine under dry conditions of use.

†1½ in. D × 19½ long dowels.

1. Deck thickness 5⅛ in.

2. Deck thickness 6¾ in.

9.8.2 Miscellaneous Details

Data for specifying size of dowels, and the number required per panel, to transfer secondary moments and shears between the adjacent panels is given in Fig. 9.19. The information may be used for HS-15 and HS-20 truck loading and for bridge stringers spaced between 2 ft 8 in. and 6 ft 8 in. apart.

Details for attachment of glulam panels to the supporting stringers are given in Figs. 9.20 and 9.21.

9.9 ILLUSTRATIVE EXAMPLE

Design a two-lane simply-supported bridge having a sixty-foot span and a 30 ft roadway width without sidewalks. The bridge consists of a glulam deck on glued-laminated stringers designed to carry H-20-44 truck traffic. Use pressure-treated wood stringers with the following design stresses: E = 1800 ksi, F_b = 2.4 ksi, F_t = 1.6 ksi, F_c = 1.5 ksi, F_v = 0.2 ksi, and $F_{c\perp}$ = 0.45 ksi. A layer of asphaltic concrete, 3 in. thick, will cover the wood deck to protect it and to provide the required wearing surface of the bridge.

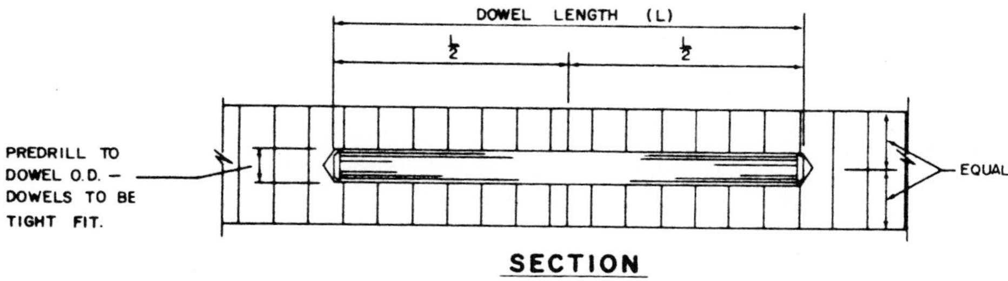

TABLE A – DECK DOWEL REQUIREMENTS

SPACING BETWEEN GIRDERS	HS – 15		HS — 20	
	DOWEL SIZE	NUMBER OF DOWELS PER BAY (N)	DOWEL SIZE	NUMBER OF DOWELS PER BAY (N)
2'-8"	1" Φ x 14.5"	4	1" Φ x 14.5"	4
3'-4"	1" Φ x 14.5"	5	1 1/4" Φ x 17"	4
4'-0"	1 1/4" Φ x 17"	4	1 1/2" Φ x 19.5"	4
4'-8"	1 1/2" Φ x 19.5"	4	1 1/2" Φ x 19.5"	5
5'-4"	1 1/2" Φ x 19.5"	5	1 1/2" Φ x 19.5"	6
6'-0"	1 1/2" Φ x 19.5"	5	1 1/2" Φ x 19.5"	7
6'-8"	1 1/2" Φ x 19.5"	6	1 1/2" Φ x 19.5"	8

DOWEL SPACING

Figure 9.19. Dowel design data. Taken from Ref. (18).

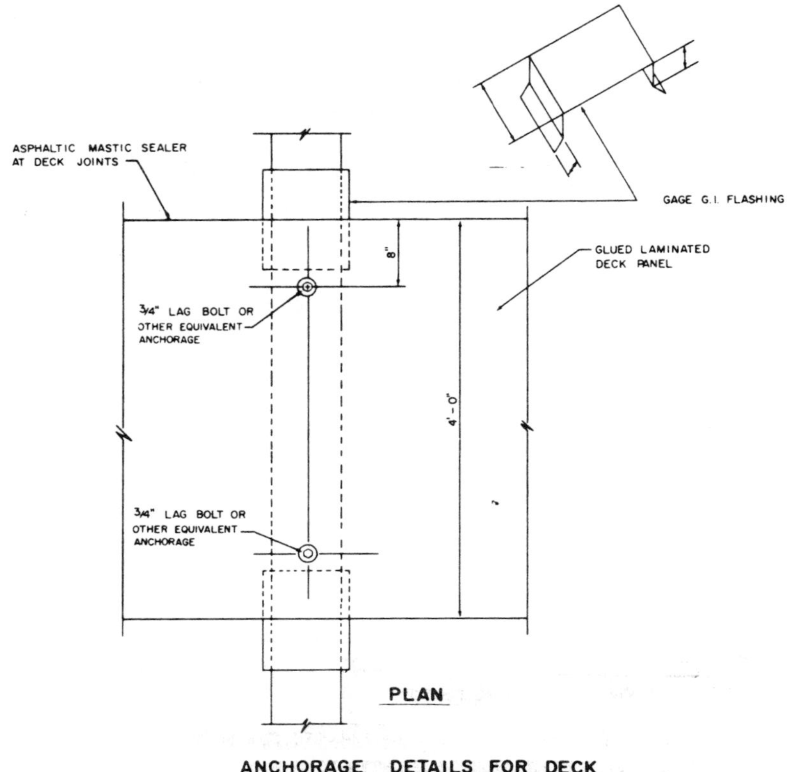

Figure 9.20. Plan view of deck attachment to stringers. Taken from Ref. (18).

9.9.1 Design of Glulam Deck

Consider a 6¾ in. thick glulam deck and a 3 in. thick asphaltic-concrete wearing surface. The glulam deck is continuous over five stringers, spaced 7 ft 3 in. apart, and each assumed 12¼ in. wide.

Loads on Structural Unit Deck

Dead Load:

Self weight	(6.75/12) (50)		= 28.1 psf
Asphaltic concrete	(3/12) (150)		= 37.5
		Total	= 65.6 psf

The dead load, considering a 1 in. unit width of deck, is 65.6/12 = 5.5 lb/ft.

Live Load:

Rear tire of H-20-44 weighs 16 k. Thus P = 16 k.

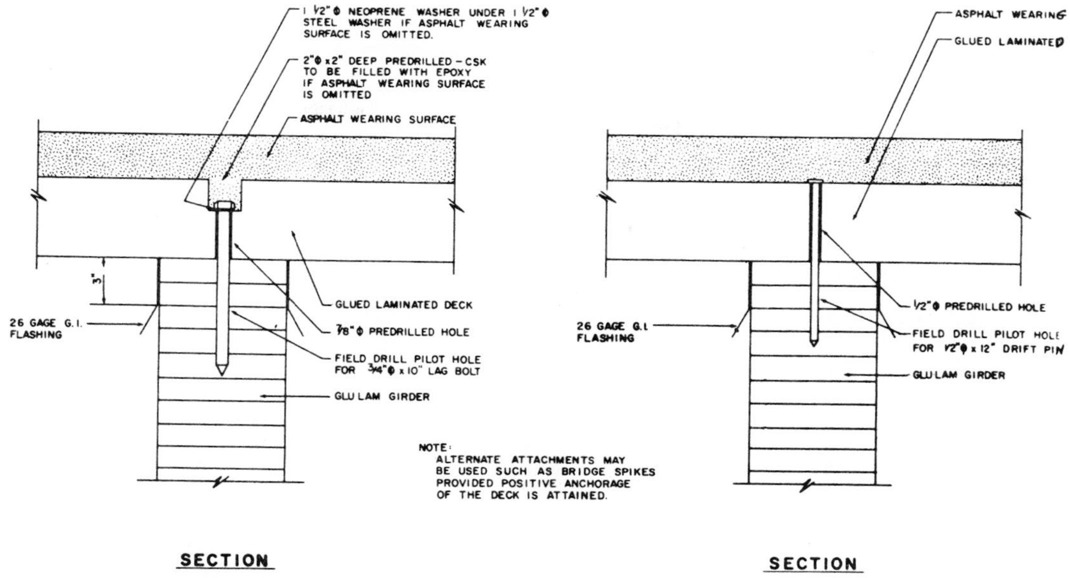

Figure 9.21. Attachment of glulam deck to supporting stringers. Taken from Ref. (18).

Effective Span

The effective span, s, is the clear distance between stringers plus one half the width of the stringer, $S - b/2$, but shall not exceed the clear span plus the deck thickness, $S - b + t$. Hence, for $S = 7$ ft 3 in. $= 87$ in., $b = 12.25$ in. and $t = 6.75$ in., the above conditions render $s = 87 - 12.25/2 = 80.9$ in. and $s = 87 - 12.25 + 6.75 = 81.5$ in. Thus, $s = 80.9$ in.

Maximum Bending Moment

It is given by the sum of the moments due to dead load, M_D, and live load, M_L. Thus,

$$M = M_D + M_L$$

where: $M_D = 0.8\,(1/8)\,w_D\,s^2$ and $M_L = P\,[0.51 \log s - K]$. For the given deck $w_D = 5.5$ lb/ft, $s = 80.9$ in., $P = 16{,}000$ lb, and $K = 0.51$ for H-20 trucks. Substituting these values in the preceding formulas yields

$$\begin{aligned} M &= (0.80)\,(1/8)\,(5.5)\,(80.9/12)^2\,12 + (16{,}000)\,[0.51 \log 80.9 - 0.51] \\ &= 300 + 7409 = 7709 \text{ lb–in./in.} \end{aligned}$$

Note that the moment contribution of the weight of the glulam deck and wearing surface is only 4 percent of the total bending moment.

Maximum Shear

As in the previous case the maximum shear is the sum of the shear due to dead load, V_D, and the shear due to live load, V_L. Thus

$$V = V_D + V_L$$

where $V_D = w_D\,(s/2 - x)$, and $V_L = R_x = 0.034$ P. For the given deck $w_D = 5.5$ lb/ft, s = 80.9 in. and x is the critical section for shear stress given as the smallest of 3 t = 3 × 6.25 = 18.75 in. or s/4 = 80.9/4 = 20.23 in. Hence, x = 20.2 in. Also, P = 16,000 lb. Substituting these values in the preceding formulas yields:

$$V = \left(\frac{5.5}{12}\right)\left(\frac{80.9}{2} - 20.2\right) + (0.034)\,(16000)$$
$$= 9 + 544 = 553 \text{ lb}$$

Note that the shear contribution of the weight of the glulam deck and wearing surface is only 1.6 percent of the total shear.

Required Thickness

The thickness of the glulam deck is determined from strength requirements to resist the maximum moment and shear imposed by the combined effects of dead load and traffic. Thus,

$$t \geq \sqrt{\frac{6M}{F_b}} \tag{9.13}$$

and

$$t \geq \frac{3V}{2\,F_V} \tag{9.14}$$

to satisfy bending moment and shear, respectively. For the given deck M = 7709 lb-in./in. and V = 553 lb. Consider using Southern Pine No. 2 wood for which $F_b = 1{,}550$ psi and $F_v = 165$ psi (see Table 7.2, for 15 percent maximum moisture). Substituting these values in the preceding formulas gives:

$$t \geq \sqrt{\frac{(6)\,(7709)}{1{,}550}} = 5.46 \text{ in.}$$

$$t \geq \frac{3}{2}\,\frac{553}{165} = 5.03 \text{ in.}$$

For the given wood it is necessary to select a 6¾ in.-thick glulam deck. Only if better quality material were used, for which $F_b = 1800$ psi or higher, would a 5⅛ in.-thick glulam deck be acceptable.

Dowel Design

Determination of the minimum size and spacing of the steel dowels required to transfer the loads between glulam panels is based on the following formula

$$n = \left(\frac{1000}{\sigma_{PL}}\right)\left(\frac{\overline{R}_y}{R_D} + \frac{\overline{M}_y}{M_D}\right) \tag{9.17}$$

where: n is the number of steel dowels required for the given span, s; σ_{PL} is the proportional limit stress in compression perpendicular to grain; R_y and M_y are the total secondary shear and moment, respectively, transferred to an adjacent panel; and R_D and M_D are the shear and moment capacities of the selected individual dowel, see Table 9.5. For the given deck:

$$s = 80.9 \text{ in.} > 50 \text{ in.}$$

$$P = 16{,}000 \text{ lb}$$

$$\sigma_{PL} = 1000 \text{ lb/in.}^2 \text{ (Southern Pine)}$$

$$\overline{R}_y = \frac{P}{2s}(s - 20) = \frac{16{,}000}{(2)\,(80.9)}(80.9 - 20) = 6020 \text{ lb}$$

$$\overline{M}_y = \left(\frac{Ps}{20}\right)\left(\frac{s - 30}{s - 10}\right) = \frac{(16{,}000)\,(80.9)}{20}\left(\frac{80.9 - 30}{80.9 - 10}\right) = 46{,}460 \text{ lb–in.}$$

Also, from Table 9.5, for 1½ in. D × 19.5 in. long dowels, $R_D = 2770$ lb and $M_D = 8990$ lb–in. Substituting these values in the formula for n yields

$$n = \frac{1000}{1000}\left(\frac{6020}{2770} + \frac{46460}{8990}\right) = 2.17 + 5.17 = 7.34 \sim 8$$

Thus, use eight 1½ in. D × 19.5 in. long dowels. Spacing of the dowels may be obtained by dividing the center-to-center distance between stringers, 87 in., by the required number of dowels, 8. Thus, the spacing between dowels is $\frac{87}{8} = 10\frac{7}{8}$ in. with end dowels in each span positioned one-half of a space (5 7/16 in.) off the stringer centerlines. Dowels may then be equally spaced over the full length of the panel and this uniform pattern should reduce fabrication errors.

Dowel Strength

It is necessary to check that the individual dowels are not over-stressed by the transfer to shear and moment to the adjacent panels. For this purpose, the following interaction equation is used

$$\sigma = \frac{1}{n}\,(C_R\,\overline{R}_y + C_M\,\overline{M}_y) \tag{9.20}$$

For the given deck n = 8, R_y = 6020 lb, M_y = 46,460 lb–in. and, from Table 9.5 for 1½ in. D. dowels, $C_R = 3.11 \text{ in.}^{-2}$ and $C_M = 3.02 \text{ in.}^{-3}$. Substituting these values in the preceding formula yields:

$$\sigma = \frac{1}{8}\,(3.11 \times 6020 + 3.02 \times 46{,}460) = 19{,}880 \text{ lb/in.}^2$$

This must be compared to 0.75 F_y which gives a conservative value for elastic limit stress of steel dowles. Most cold-finished bar stock has a yield-point stress, F_y, exceeding 60,000 psi, and quenched and tempered bars can be obtained with yield points up to 130,000 psi. If the former were specified then 0.75 F_y = (0.75) (60,000) = 45,000 psi > 19,880 psi, which is far above the maximum stress generated in the steel dowels.

Deflection of Glulam Deck

Although deflection criteria is not generally used for design of bridge decks it may be of interest to determine the deflection of the glulam deck, δ_L, under the H-20-truck wheel load.

The value of δ_L is a function of the load distribution ratio, a/s, and the ratio a/b, where a is the tire width in the direction normal to the span and b is the tire width in the direction of span. For an H-20 truck and a laminated floor, AASHTO's Art. 1.3.4 specifies a = 15 in. and b = 20 in. Hence, b/a = (20/15) = 1.33 and, since s = 80.9 in., it follows that a/s = (15/80.9) = 0.18. From Fig. 9.18, the deflection coefficient is obtained as

$$\frac{\delta_L}{\left(\frac{Ps^2}{Et^3}\right)} = 0.46$$

Substituting P = 16,000 lb, s = 80.9 in., E = 1.7×10^6 psi and t = 6.75 in. in this formula yields δ_L = 0.092 in. This amounts to a deflection-span ratio δ_L/s = 0.092/80.9 = 1/879, which is quite satisfactory by any standard of comparison.

It can be shown that the midspan deflection of the glulam deck due to dead load is 0.002 in., which is a negligible value.

9.9.2 Design Typical Interior Stringer

Five stringers spaced 7 ft 3 in. apart were assumed for the design of the glulam deck. Using Table 9.6 consider that the stringers are 12¼ × 52½ in. glulam members to obtain an initial estimate of their weight.

Loads:

Weights (dead load)

Wearing Surface	(3/12) (150) (7.25)	= 272 lb/ft
Glulam deck	(6.75/12) (50) (7.25)	= 204
Stringer (est.)	$\frac{(12.25 \times 52.5)}{144}$ (50)	= 223
	w_D	= 699 lb/ft ~ 700 lb/ft

Live load: H-20-44

Maximum Bending Moment:

$$\text{Dead Load: } M_D = \frac{1}{8} w_D L^2 = \left(\frac{1}{8}\right)(0.700)(60)^2 = 315 \text{ k-ft}$$

Live Load: 558 k-ft/lane (see Appendix A of AASHTO Specifications (1) or Section 9.7.2). Determination of the fraction of this moment taken by an interior stringer is shown in Fig. 9.22; this follows the AASHTO's provisions set forth in Table 9.1. It follows that

$$M_L = (558)(0.81) = 452 \text{ k-ft}$$

Total Moment:

$$M_T = 315 + 452 = 767 \text{ k-ft}$$

Required Bending Strength

$$C_F\,S = \frac{M_T}{F_b} = \frac{767 \times 12}{(2.4)} = 3{,}835 \text{ in.}^3$$

Consider a section 12.25 in. wide by 48 in. deep for which $C_F = (12/48)^{1/9} = 0.857$, $S = (1/6)(12.25)(48)^2 = 4704$ in.^{3}, and $C_F\,S = (0.857)(4704) = 4032$ in.$^3 > 3835$ in.^{3}. Thus a 12.25 in. $\times$ 48 in. glulam section would meet the moment requirement. Stability of the stringers against lateral buckling should be of no concern once the glulam deck has been attached using a detail such as shown in Figs. 9.20 and 9.21.

Maximum Shear Force

The critical section for shear is at a distance equal to the depth of the stringer, d, from the center of the support. Assuming d = 4 ft it is possible to determine the maximum shear force on the critical section by placing on the bridge the rear wheels of two H-20-44 trucks as shown in Fig. 9.22. This causes the front wheels of the truck to be placed at a distance of 14 ft from the critical section. As a result, the actual load carried by the interior stringer, see Fig. 9.22, is 25.93 k due to the rear wheels, and (4/16) (25.93) = 6.48 k due to the front wheels.

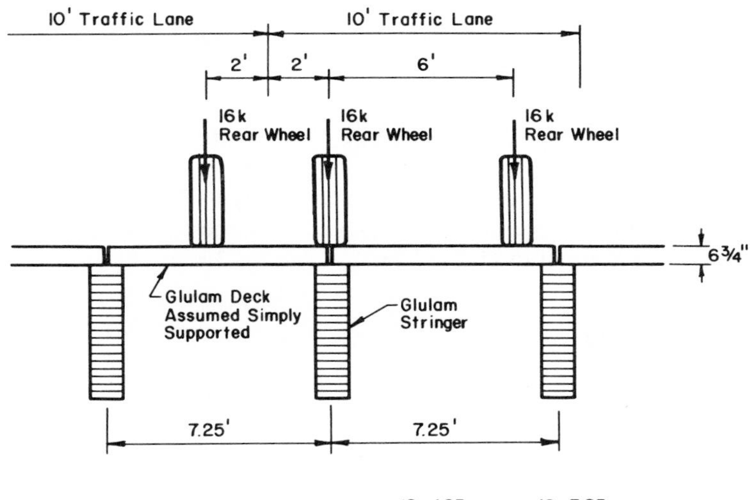

Figure 9.22. Determination of fraction of lane maximum moment taken by an interior stringer.

The maximum shear force due to live load, V_L, may be calculated for the 60 ft span, as follows:

$$V_L = 25.93\left(\frac{60 - 4}{60}\right) + 6.48\left(\frac{60 - 4 - 14}{60}\right) = 28.74 \text{ k}$$

The shear force due to dead load, V_D, may be determined at the critical section by simple statics

$$V_D = w_D\left(\frac{L}{2} - d\right) = 0.7\left(\frac{60}{2} - 4\right) = 18.20 \text{ k}$$

The total shear force $V_T = V_D + V_L = 18.20 + 28.74 = 46.94$ k. The maximum shear stress, V_{max}, for the 12.25 in. × 48 in. stringer section is given by:

$$V_{max} = \frac{3}{2}\frac{V_T}{A} = \frac{3}{2}\frac{46.94}{12.25 \times 48} = 0.120 \text{ ksi}$$

which, being less than the design shear stress, $F_v = 0.2$ ksi, makes the 12.25 in. × 48 in. section acceptable.

Deflection

Due to dead load, δ_D:

Using Eq. 9.7, where $w_D = 0.700$ k/ft $= 0.0583$ k/in., L = 60 ft = 720 in., = 1,800 ksi, and $I = (1/12)(12.25)(48)^3 = 112{,}900$ in.^{4}, yields:

$$\delta_D = \frac{5}{384}\frac{w_D L^4}{EI} = \left(\frac{5}{384}\right)\frac{(0.0583)(720)^4}{(1800)(112900)} = 1.00 \text{ in.}$$

Considering such factors as variability of modulus of elasticity (1.2), creep (1.5) and the aesthetic wish for a flat-arch type appearance for the bridge (2) it is necessary to specify midspan camber for the beams in an amount equal to (1.2) (1.5) (2) (1.0) = 3.6 in.

Due to Live Load:

The value of δ_L at midspan may be obtained using the position of H-20-44 trucks shown in Fig. 9.14. For P = 40 k, n = 5, I = 112,900 in.^{4}, E = 1800 ksi, L = 60 × 12 = 720 in. and a = 23 × 12 = 276 in., the formula gives:

$$\delta_L = \frac{(40)(276)}{(24)(1800)(5)(112{,}900)}(3 \times 720^2 - 4 \times 276^2)$$

$$\delta_L = 0.57 \text{ in.}$$

The deflection-span ratio is 0.57/720 = 1/1260 which is well below the range of 1/200 to 1/300 for highway bridges. This makes the 12.25 × 48 in. section for the bridge stringers acceptable by having met all bending, shear and deflection criteria.

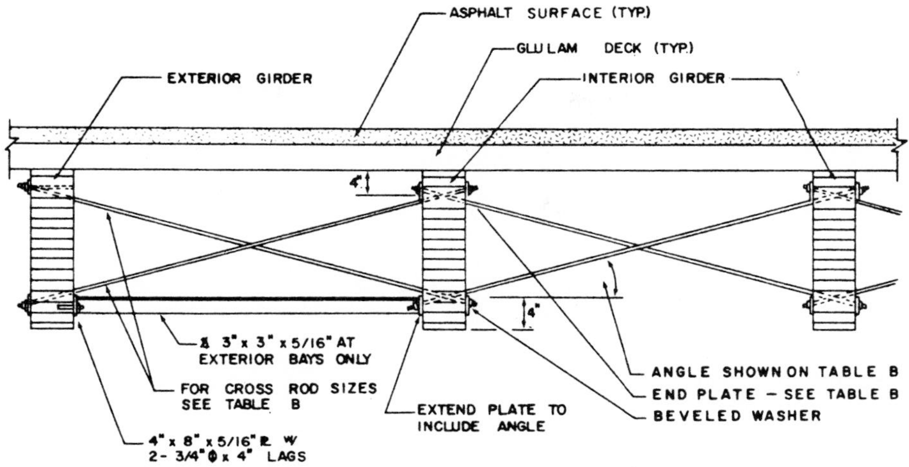

TABLE B

	HS 15		HS 20	
ANGLE	ROD SIZE	PLATE SIZE	ROD SIZE	PLATE SIZE
20°	1 1/8"φ	5" x 5" x 1/2"	1 1/4"φ	5" x 5" x 5/8"
25°	1 1/8"φ	5" x 5" x 1/2"	1 1/4"φ	5" x 5" x 5/8"
30°	1"φ	4" x 4" x 1/2"	1 1/8"φ	5" x 5" x 5/8"
35°	7/8"φ	4" x 4" x 1/2"	1 1/8"φ	5" x 5" x 1/2"
40°	7/8"φ	4" x 4" x 1/2"	1"φ	5" x 5" x 1/2"
45°	7/8"φ	4" x 4" x 1/2"	1"φ	4" x 4" x 1/2"
50°	7/8"φ	4" x 4" x 1/2"	1"φ	4" x 4" x 1/2"

Figure 9.23. Transverse cross-section of bridge. Taken from Ref. 18.

9.9.3 Miscellaneous Details

A transverse cross-section of the bridge (partial), is given in Fig. 9.23. Cross-bracing between the stringers, consisting of diagonal bars is indicated. For the given bridge, using the Table shown in the figure and $\alpha = \tan^{-1}(48 - 2 \times 4)/(87 - 12.25) = 28.2°$, the required cross-bracing would call for 1¼ in. D. rods and 5 in. × 5 in. plates ⅝ in. thick. Placement of the cross frames is required only at midspan, and at the ends of the bridge if required to transfer lateral forces to the abutments. Early attachment of the cross frames to the stringers, during the erection stage of the bridge, is recommended. The lateral stability of the stringers should be assured before placement of the glulam deck is begun.

Details of support of the glulam stringers by the contrete abutment is shown in Fig. 9.24. A cross-section of the exterior stringer showing the attachment of the required railing is given in Fig. 9.25.

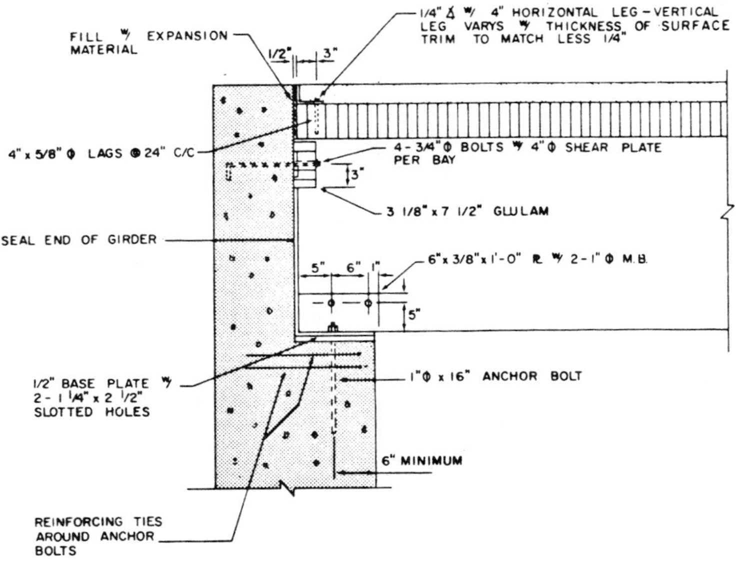

Figure 9.24. Bridge support at concrete abutment. Taken from Ref. 18.

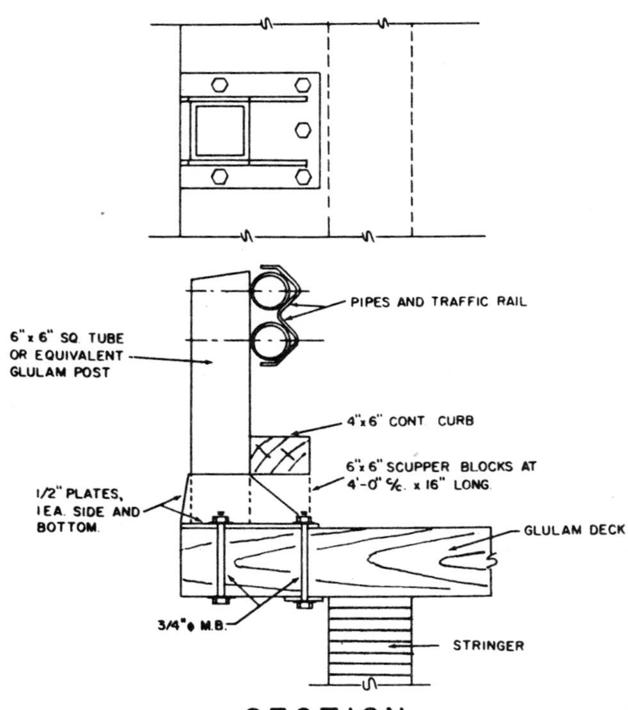

Figure 9.25. Details of railing support. Taken from Ref. 18.

PROBLEMS

1. Design a two-lane highway overpass suitable for HS-20-44 traffic, with a minimum 24 ft. width of roadway and simply-supported on abutments spaced 50 ft. apart. Design the laminated decking using rough-sawn wood with the following allowable stresses: $F_b = 2.3$ ksi, $F_t = 1.55$ ksi, $F_v = 0.125$ ksi, $F_{c\perp} = 0.475$ ksi, $F_c = 1.70$ ksi and $E = 1{,}900$ ksi. The stringers should be glued-laminated members, made of wood with the following allowable stresses: $F_b = 2.4$ ksi, $F_t = 1.6$ ksi, $F_c = 1.5$ ksi, $F_{c\perp} = 0.45$ ksi, and 0.385 ksi (for tension face and compression face, respectively), $F_v = 0.165$ ksi and $E = 1{,}800$ ksi. Any unevenness of the rough-sawn decking is covered by an asphaltic-concrete wearing-surface having an average 3 in. thickness. Hint: Follow step by step process of Section 9.3.
2. Design a wood-concrete composite slab for a two-lane, simply supported, highway bridge with only one span, 25 ft. long. A minimum 24 ft. width between inside faces of curb is desired. The live load is given by AASHTO HS-20-44 trucks. Use wood having the following allowable stresses: $F_b = 1{,}750$ psi, repetitive member use; $F_t = 1{,}000$ psi; $F_v = 95$ psi; $F_{c\perp} = 385$ psi; $F_c = 1{,}250$ psi; $E = 1{,}800{,}000$ psi, and concrete $f_c' = 3000$ psi at 28 days. Assume equal moduli of elasticity for wood and concrete. Show that the section selected meets the strength and serviceability requirements.
3. Determine the fixed-end moment, M_1, and distance from the fixed end to the point of inflection, kL, of cases I through IV of Figure 9.7. The figure represents a wood-concrete-steel composite beam subjected to various conditions of end restraint and loading.

   Ans. I) $\dfrac{I_o}{I_1} = \left(\dfrac{1 - 2k}{2k}\right)^2 \qquad \dfrac{M_1}{M_o} = 2k$

   II) $\dfrac{I_o}{I_1} = \dfrac{(1 - 2k)^3}{2k^2(3 - 4k)} \qquad \dfrac{M_1}{M_o} = 4k(1 - k)$

   III) $\dfrac{I_o}{I_1} = \left(\dfrac{1 - 2k}{2k}\right)\left(\dfrac{3 - 5k + 2k^2}{6k - 2k^2}\right) \qquad \dfrac{M_1}{M_o} = \dfrac{2k}{1 - k}$

   IV) $\dfrac{I_o}{I_1} = \dfrac{(1 - k)^4}{k^2(k^2 - 4k + 6)} \qquad \dfrac{M_1}{M_o} = 4k$

   The solution of k as a function of I_o/I_1 is difficult, except for case I. It is simpler to assign values to k and determine corresponding values of I_o/I_1. Thus, the variations of k, and M_1/M_o, with I_o/I_1 for all four cases are given in Fig. 9.8.
4. Design a wood-concrete composite slab for a warehouse floor that is continuous over several interior 20 ft. spans with end spans 16 ft. long. The live load on the floor is 200 psf. Use wood with the following allowable stresses: $F_b = 1.90$ ksi (repetitive member use), $F_t = 1.1$ ksi, $F_v = 0.095$ ksi, $F_{c\perp} = 0.405$ ksi, $F_c = 1.45$ ksi and $E = 1{,}900$ ksi. Use concrete with a strength, $f_c' = 3.0$ ksi, and allowable stress, $f_c = 0.45 f_c'$; reinforcing steel with a yield strength, $f_y = 40$ ksi, and allowable stress, $f_s = 20$ ksi. Assume $E_c/E_w = 1$ and $E_s/E_w = 18.75$. The composite floor is supported by 8¾ in.-wide glued-laminated girders. Assume unshored construction, i.e., before composite action takes place, the wood subdeck must support its own weight and that of the fluid concrete without developing deflections larger than 1/240th of the span.

5. Design the composite floor of the preceding problem, assuming full shoring of the wood subdeck. Shores are to remain until the concrete has reached a strength of 2 ksi, or a minimum of seven days after initial casting.
6. It is possible, depending on the relation between the cost of shoring a composite slab and the actual labor and material costs of the slab, that a fully-shored floor construction might be more economical than the unshored type. For the purpose of comparison, determine the volumes of wood, concrete and steel, per square foot of floor, required by the respective solutions to problems No. 4 and 5. Assign current costs for material and labor to each, and to the cost of shoring per square foot of floor, and determine the most economical solution.
7. The composite wood-concrete slab designed in Section 9.5 assumed a modular ratio $E_c/E_w = 1$. Designers may wish to know the effects that a variation in the modular ratio may have on the results. For this purpose, determine the changes in section properties and stresses in the given slab, if the modular ratio is taken as $E_c/E_w = 2$. Compare the results with those obtained in Section 9.5.
   Hint: Determine the properties of the transformed wood section by doubling the width of the concrete portion.
8. The composite wood-concrete beam bridge of Illustrative Example 3, see Section 9.7, was designed using seven 8.75 × 36 in. stringers at a spacing of 4 ft. It is wished to determine the suitability of a design that employs more stringers at a smaller spacing. For this purpose, assume eight 8.75 × 30 in. stringers at a spacing of 3.43 ft. instead. Follow the same format in your calculations as that of the Illustrative Example. Use wood for which $F_b = 2.6$ ksi. Compare the two designs.
   Ans. A. Dead load 348 lb./ft.
   B. $M_D = 156.5$ k-ft., $M_L = 191.3$ k-ft.
   C. Section Properties. Non composite Section: $A = 263$ in.^{2}, $S = 1{,}313$ in.^{3}, $I = 19{,}688$ in.^{4}, $C_F = 0.903$. Composite Section: width of flange 41.2 in., $A = 510$ in.^{2}, $\bar{y} = 12.30$ in., $I = 61{,}700$ in.^{4}, $S_w = 2{,}600$ in.^{3}, $S_c = 5{,}020$ in.^{3}.
   D. Stresses: Dead Load $f_w = 1.43$ ksi, Live Load $\Delta f_w = 0.88$ ksi, $f_c = 0.58$ ksi. Combined Load: wood stringer 2.31 ksi < 2.34 ksi; concrete deck 0.58 ksi < 1.2 ksi.
   E. Deflection: $\delta = 2.9$ in.; provide 6 in. camber at midspan.
   The proposed design is also acceptable.
9. The composite wood-concrete beam bridge of Illustrative Example 9.7, see Section 9.7, was designed using seven 8.75 × 36 in. stringers at a spacing of 4 ft. It is wished to determine the suitability of a design that employs less stringers at a greater spacing. For this purpose assume six 8.75 × 42 in. stringers at a spacing of 4.8 ft. instead. Follow the same format in your calculations as that of the Illustrative Example. Compare the two designs and that of the preceding problem.

10. The determination of ultimate strength of the composite wood-concrete section of the bridge of Section 9.7, as shown in Fig. 9.13, assumes that ultimate occurs when the concrete attains a crushing strain $\epsilon_u = 0.003$. This requires that the tensile strength in the wood be greater than the maximum tensile stress $f_{wb} = 7.42$ ksi developed at ultimate. Assuming that the tensile strength of wood is only 6 ksi, determine the ultimate moment and curvature of the composite section. Consider: 1) that the tensile stress-strain diagram of wood is linear from zero to ultimate; and, 2) the stress-strain diagram of concrete in compression is elastic from zero to a point with coordinates: $\epsilon = 0.0015$, $f = 2.8$ ksi, and plastic from there on to $\epsilon_u = 0.003$. Hint: Use an iterative method similar to that of Section 7.3.5.

11. Design a highway cantilever-suspended beam bridge of the type shown schematically in Fig. 7.29b; consider $L_e = 40$ ft., $L_c = 120$ ft. and $X = 40$ ft. The central-suspended member is prismatic; the end-cantilevered members may be tapered. The bridge is subjected to HS-20-44 traffic. A minimum 24 ft. width between inside faces of curbs is desired. Design the bridge using a laminated wood deck and glued-laminated beams. Consider that the beams are spaced transversely 3 ft. apart.
Hints: 1) Assume initially that: (a) the beams in the suspended span have a uniform section 12¼ in. × 30 in.; and (b) the end-cantilevered beams are symmetrical about the interior support and have 12¼ in. × 30 in. end sections and a central section, at the interior support, equal to 12¼ in. × 72 in. These dimensions are given only as initial estimates to allow determination of self-weight. The designer may vary these as necessary for strength and serviceability.

    2) The designer may wish to consider using a 6 in. concrete deck in the exterior spans to reduce the possible uplift at the end abutments.

    3) For the design of the end-cantilevered beams consider these members divided into 9 sections 10 ft. apart. Determine maximum moment and shear at each section due to live load. Add these to the corresponding values due to permanent loads. Draw envelopes of maximum moments and shears. Check the adequacy of the design at each section through the use of Eq. 7.75. You may consider bracing the bottom flange of the end-cantilevered members to improve their lateral stability and to secure a larger allowable bending stress.

    4) The central-suspended beams may be designed as it is done in conventional simply-supported bridges.

    5) The deflection configuration of the bridge, under permanent loads, must be determined. An adequate amount of camber should be provided for the 120 ft. span. In addition, the camber configuration should be a continuous, smooth curve with no kinks at the hinged connection of the suspended span.

    6) Design the connection between the end-cantilevered and the central-suspended beams using the hanger rod connection shown in Fig. 9.26.

    7) Draw an elevation and two transverse sections of the final solution.

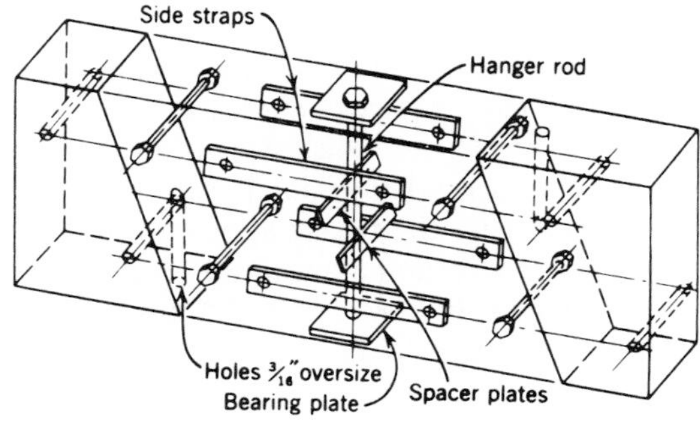

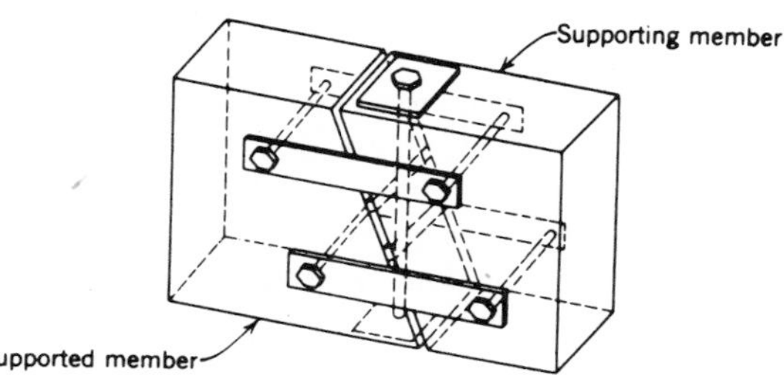

Figure 9.26. Hanger rod type—exploded view and assembly. For light, moderate, and heavy loads. In the hanger rod type of cantilever beam connection, the vertical reaction of the supported member is carried in tension by the hanger rod. The hanger rod must be sized to carry this load in tension. The hanger rod bearing plates must be sized to transfer the vertical reaction in bearing perpendicular to grain in both the supported and the supporting members. Spacer plates are used and located as close as possible to the neutral axis to provide clearance between the beam ends and to allow for end rotation in the members without crushing the beam ends at their top and bottom edges. These spacer plates are steel flat bars attached to the beams with nails through predrilled holes in the spacer plates. The side straps are used to hold the beam ends in alignment and to serve as tension ties where an axial tension tie between the beams is required. From Ref. 3.

REFERENCES

1. American Association of State Highway and Transportation Officials, "Standard Specifications for Highway Bridges," Twelfth Edition, 1977, published by AASHTO, 444 North Capital Street, N.W., Suite 225, Washington, D.C. 20004.
2. Timoshenko, S. P., and Young, D. H., "Theory of Structures," Second Edition, McGraw Hill Book Company, New York, 1964.
3. American Institute of Timber Construction, "Timber Construction Standards AITC 100–69," 5th Edition, 1969, Englewood, Colorado 80110.
4. Richart, Frank E., and Williams, Clarence B., Jr., "Tests of Composite Timber and Concrete Beams," University of Illinois Engineering Experiment Station Bulletin No. 343, May 1943.
5. Richart, Frank E., and Williams, Clarence B., Jr., "Tests of Composite Timber and Concrete Beams," Journal of the American Concrete Institute, Vol. 14, No. 4, February 1943, Proceedings, Vol. 39, pp. 253–276.
6. May, T. K., "Composite Timber-Concrete Construction," Report of Committee 7—Wood Bridges and Trestles, American Railway Engineering Association Bulletin, Vol. 56, No. 520, January 1955.
7. National Forest Products Association, "National Design Specification for Wood Construction," 1977 Edition.
8. McCullough, C. B., "Oregon Tests on Composite (Timber-Concrete) Beams," Journal of the American Concrete Institute Vol. 14, No. 5, April 1943, Proceedings, Vol. 39, pp. 429–440.
9. Pincus, George, "Bonded Wood-Concrete T Beams," Journal of the Structural Division, ASCE, Vol. 95, No. ST10, Proc. Paper 6842, Oct. 1969, pp. 2265–2279.
10. Pincus, George, "Behavior of Wood-Concrete Composite Beams," Journal of the Structural Division, ASCE, Vol. 96, No. ST10, Proc. Paper 7592, Oct. 1970, pp. 2009–2019.
11. American Institute of Timber Construction, "Timber Construction Manual," Second Edition, John Wiley and Sons, Inc., New York, 1974.
12. Magee, G. M., "Wood Research in the Railroad Industry," Journal of the Structural Division, ASCE, Vol. 93, No. ST2, Proc. Paper 5178, April 1967, pp. 105–120.
13. McCutcheon, William J., and Tuomi, Roger L., "Procedure for Design of Glued-laminated Orthotropic Bridge Decks," F.P.L. 210, U.S. Forest Products Laboratory, Madison, Wis., 1973.
14. McCutcheon, William J., and Tuomi, Roger L., "Design Procedure for Glued-Laminated Bridge Decks," Forest Products Journal, Vol. 23, No. 6, pp. 36–42, June 1973.
15. Tuomi, Roger L., "Advancements in Timber Bridges Through Research and Engineering," Proceedings of 13th Annual Colorado State University Bridge Engineering Conference, 1972.
16. Bohannan, Billy, "FPL Timber Bridge Deck Research," Journal of the Structural Division, ASCE, Vol. 98, No. ST3, Proc. Paper 8779, March 1972, pp. 729–740.
17. "Modern Timber Highway Bridges. A State of the Art Report," by American Institute of Timber Construction, July 1973.
18. "Glulam Bridge Systems. Plans and Details," Recommended by American Institute of Timber Construction, 1974.

Chapter 10
Design of Wood Buildings and Houses

10.1 INTRODUCTION

In recent years there have been erected many outstanding wood structures such as: domes, buttressed arches and rigid frames of glued-laminated wood used in large assembly halls, gymnasiums and sports arenas; and trusses, bowstring arches and conventional post and beam structures of glued-laminated or solid-sawn wood used for large warehouses and industrial buildings. "A" frames have been designed for warehouses used for grain, sugar and/or storage as well as for churches and other buildings. With the advent of plywood, surface structures such as prismatic plates and hyperbolic paraboloids are used to provide large, column-free spaces. In combination with solid-sawn lumber or laminated wood, plywood has been used to create stressed-skin panels for roofing systems and I or box sections for long-span beams.

The versatility of wood is thus shown. Designing with wood is no longer limited; it is open to imaginative ideas and new concepts. In this chapter, modern design of wood buildings is illustrated by the design of a larger Exhibition Hall in which some of the elements described above are used in two designs, each using different wood elements. Basic knowledge of the various concepts, equations, and tables contained in the preceding chapters of the text, particularly No. 6, 7 and 8, is assumed.

The first design uses glued-laminated wood solely. The roof decking, beams, girders and columns of the building are designed using various sizes and combinations available in glued-laminated construction. The cost of the structure is reduced by studying various possible layout and erection procedures for the girders.

The second design of the building uses a combination of plywood and stress-graded lumber to provide stressed-skin panels and box-girders for the roof system. Solid wood sections are used for the columns. As in the first design, the erection and construction sequence of the girders is studied in order to reduce the cost of the structure.

Because of the fact that wood houses constitute the major source of residential dwellings in the United States it is important that fundamental aspects of their design and construction be well understood. For this purpose, the last section of Chapter 10 is devoted to a discussion of various materials and arrangements used for foundations, walls, framing and sheathing of wood houses.

Other structural elements used in wood buildings and houses are discussed in Chapter 11. A collection of selected works on various aspects of design, construction and cost of wood buildings and houses can be found in the list of references at the end of this chapter.

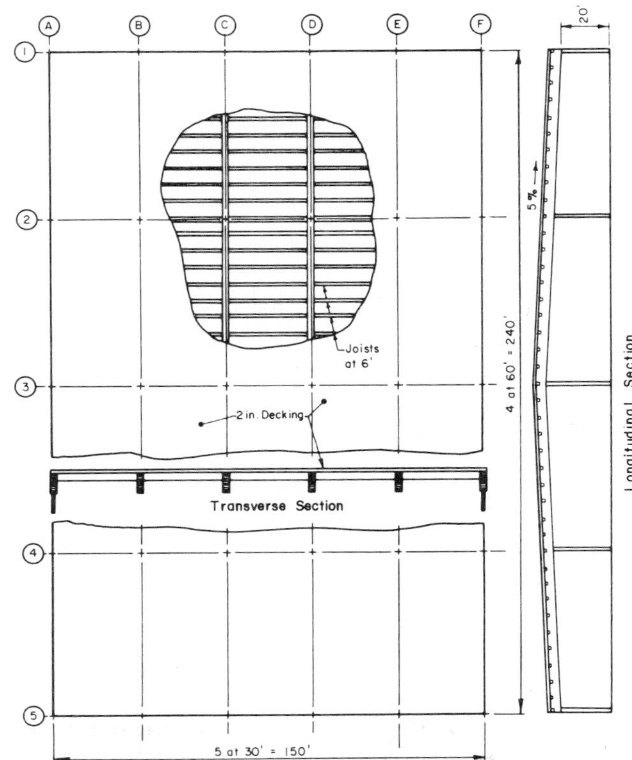

Figure 10.1. Glued-laminated wood structure of exhibition hall.

10.2 DESIGN OF A BUILDING USING GLUED-LAMINATED WOOD

A plan view of a building for an Exhibition Hall is shown in Fig. 10.1. The roof structure of the building consists of a 2-in. deck on beams spaced 6 ft. on centers. The roof beams rest on girders spaced 30 feet on centers which run along axes A to F and are in turn supported by columns spaced 60 ft. apart, see transverse and longitudinal sections of Fig. 10.1. The building is to be located in a midwestern city for which the design snow load is 30 psf. The roof of the building is provided with insulation material and built-up roofing consisting of 5-ply felt and gravel for waterproofing purposes. The object of this example is the design of the deck, typical beam, girder and columns together with their necessary connections. The provisions of NDS (1) and AITC (2) are used.

10.2.1 Design of Deck

The deck spans 6 ft. between beams, see plan detail of Fig. 10.1, and may be laid continuous over several beams. Material for the deck consists of wire-brushed 2 × 6 in. tongue-and-groove wood planks 12 ft. long, surfaced and used at 15 percent moisture content and having the following allowable stresses: F_b = 1.15 ksi (single member), $1.15F_b$ = 1.3 ksi (repetitive member), F_v =

0.080 ksi, and E = 1,500 ksi. The planks are staggered in such a way that continuous planks alternate with terminating planks over each beam, as shown in Fig. 11.19c. For other possible span and end joint arrangements of decking see Ref. 3, or Section 11.7.

The dead weight of the roof can be estimated as follows:

2 in. wood deck	4 psf
insulation	2
vapor barrier and built-up roof	6
	w_D = 12 psf

Assuming a live load, w_L of 30 psf for snow, the total load on the deck, w_T, is 42 psf. The maximum moment and shear in the deck occur over the supporting beams and are given by $M = w_T L^2/10$ and $V = (½)\ w_T\ L$, respectively. Substituting w_T = 42 psf × 1 ft. = 42 lb./ft. and L = 6 ft. yield: M = 151 lb.-ft./ft. = 1.814 k-in./ft. and V = 0.126 k/ft. The preceding values are created by total loading on two 6 in. wide planks. The section resisting shear is that of both 6 in. planks. However, the section resisting the moment is that of the continuous plank only, for which $S = (1/6)\ (6)\ (1.5)^2 = 2.25$ in.³. Hence, Eq. 7.la and 7.32a yield:

$$f = \frac{M}{S} = \frac{1.814}{2.25} = 0.807 \text{ ksi} < 1.15 \times 1.3 \text{ ksi}$$

$$v = \frac{3}{2}\frac{V}{A} = \frac{3}{2}\frac{0.126}{2 \times 1.5 \times 6} = 0.011 \text{ ksi} < 1.15 \times 0.080 \text{ ksi}$$

where $\bar{\alpha} = 1.15$ for snow loading and $F_b = 1.3$ ksi is the stress corresponding to repetitive-member use of the given wood. It is obvious that the 2 in. decking (1.5 in. actual thickness) meets the necessary strength requirements. Shear-stress is always low for decking subjected to uniform loading; therefore, its computation is usually disregarded. It is noted that the same result for the stress f can be obtained using Eq. 11.27 with S = 4.5 in³. and Table 11.5 from which $C_2 = 0.1$ and $C_3 = 0.5$, see Section 11.7.

The serviceability of the deck may be measured by its maximum deflection. The latter may be estimated, see Eq. 11.26 and Table 11.5, as follows:

$$\delta = 0.009174\ \frac{wL^4}{EI} \tag{10.1}$$

where, for the given case, w = 42 lb./ft. = 0.0035 k/in., L = 6 ft. = 72 in., E = 1,500 ksi, $I = (1/12)(2 \times 6)(1.5)^3 = 3.37$ in.⁴. Substituting these values in Eq. 10.1 yields $\delta = 0.17$ in. This represents a deflection—span ratio of approximately 1/420, which is acceptable under the usual deflection requirements. Attach deck planks to supporting beams with two 16d (3.5 in. long) nails at each end.

The 2 in. decking meets strength and serviceability requirements; thus, it can be used safely. The capacity of the deck is far from exceeded, a fact that indicates that larger spans are possible. However, this requires greater spacing between supporting beams, as a result of which fewer but heavier beams may be necessary. Consideration of this variation may be used as an exercise. The design proceeds in the assumption that the 6 ft. spacing between beams is maintained.

10.2.2 Design of Beam

Beams can be considered as simply-supported members spanning between girders and carrying the weight of the deck and the design snow load. The substantial number of beams in the building requires a careful selection of this element if an economical structure is to be obtained. The design of a typical beam can be accomplished using the following procedure.

10.2.2.1 Determination of Loads

An estimate of the size of the beam is necessary, in order to add its own weight to the loads to be supported. An accurate determination is not needed, a rough estimate is enough. For a span of 30 feet and a span-depth ratio of 20, the depth of the laminated beam may be taken as 18 in. If the aspect ratio (depth to width) of the section is taken as about 3, the width of the beam may be selected as 5⅛ in. actual. Assuming a unit weight of 40 lb./ft.^{3}, the estimated weight of the beam is given by $(5.12 \times 18/144)(40) = 26$ plf. Since the beams are spaced 6 ft. apart, the uniformly distributed load on the typical beam is:

$$\begin{aligned} \text{Dead Load: } 12 \times 6 + 26 &= 98 \text{ plf} \\ \text{Live Load: } 30 \times 6 &= \underline{180\quad} \\ w_T &= 278 \text{ plf} \end{aligned}$$

Note that the dead load is a very minimum. The designer should be careful to consider such additional loads as air conditioning and heating ducts, sprinkler systems, and other possible loads hung from the overhead structural system which are usually included in these occupancies. Refer to Problem 2 of this chapter.

10.2.2.2 Design Criteria

The governing loading condition for the typical beam is "total loading," with allowable stresses increased fifteen percent because of the load-duration factor for snow. This is true because for total loading the ratio load-to-allowable-stress is larger than the corresponding ratio for the permanent loading condition; i.e. $278/1.15 > 98/0.9$. For a combination type 24 F, under dry conditions of use, the following may be considered as allowable stresses* for design: $F_b = 2.4$ ksi, $F_t = 1.6$ ksi, $F_c = 1.5$ ksi, $F_v = 0.2$ ksi, $F_{c\perp} = 0.385$ ksi, and $E = 1{,}800$ ksi. These stresses, multiplied by the load-duration factor 1.15, may not be exceeded in service. Hence, for snow loading conditions: $F_b = 2.76$ ksi, $F_t = 1.84$ ksi, $F_c = 1.72$ ksi, $F_v = 0.23$ ksi and $F_{c\perp} = 0.443$ ksi. The modulus of elasticity $E = 1{,}800$ ksi is not affected. Finally, for serviceability, the elastic deflection of the beam may not exceed the limits indicated in Table 7.5.

10.2.2.3 Required Value of C_FS

The maximum bending moment in the beam occurs at midspan and is equal to $\frac{1}{8} wL^2$, where $w = 0.278$ k/ft. and L is the clean span of the beam. Assuming that the width of the supporting girder is 8¾ in., $L = 30 - 0.73 = 29.3$ ft. and $M = (\frac{1}{8})(0.278)(29.3)^2 = 29.9$ k-ft. The required value of the product C_FS can be found from Eq. 7.1a as follows:

$$C_F\, S = \frac{M}{\alpha F_b} = \frac{29.9 \times 12}{1.15 \times 2.4} = 130 \text{ in}^3$$

*See Table A.3.1.

Bending is satisfied by any section with a C_FS equal to, or larger than, 136 in.^{3}

10.2.2.4 Required Area of Section

The maximum shear in the beam affecting determination of horizontal shear stresses, occurs at a distance, d, from the face of the supports. In this case, assuming d = 18 in., the shear force may be determined as $V = w\left(\frac{L}{2} - d\right) = (0.278)\left(\frac{29.3}{2} - 1.5\right) = 3.66$ kip. The cross-sectional area, $A = bd$, is given by Eq. 7.32a as follows:

$$A \geqslant \frac{3}{2}\frac{V}{\overline{\alpha}F_v} = \frac{3}{2}\frac{3.66}{1.15 \times 0.2} = 24 \text{ in}^2$$

Shear is satisfied by any section that has an area equal to, or larger than, 24 in.^{2} Note at this point that the small value of A indicates shear is not a governing factor in the selection of the final section of the beam.

10.2.2.5 Required Moment of Inertia

The maximum deflection of the beam occurs at midspan. According to Table 7.5, for roof beams of commercial buildings without plaster ceiling, the deflection due to applied load only, δ_{LL}, should not exceed L/240; the maximum deflection due to tatal load, δ_{TL}, should not exceed L/180, where L is the clear span of the beam. Thus, for L = 29.3 ft. = 351 in., $\delta_{LL} \leq 1.46$ in. and $\delta_{TL} \leq 1.95$ in. The moment of inertia, I, of the section is governed by the loading that makes the ratio W/δ a maximum. This occurs for total loading since: $278/1.95 > 180/1.46$. Hence, I, can be determined from Eq. 7.45 for P = 0 as follows:

$$I \geq \frac{5}{384}\frac{wL^4}{E\delta} \tag{10.2}$$

where: $w = (0.278/12)$ k/in., L = 351 in., E = 1,800 ksi, $\delta = 1.95$ in. Substituting these values in the above expression yields $I \geq 1{,}300$ in.^{4}.

10.2.2.6 Selection of Final Section

The section selected for the typical beam must meet the conditions: $A \geq 24$ in.^{2}, $S \geq 136$ in.^{3}, $I \geq 1{,}300$ in.^{4}. The following sections, listed in increasing order of cross-sectional area, are all acceptable solutions:

Dimensions	Properties				
Actual, in.	C_F	A, in.^{2}	S, in.^{3}	C_FS, in.^{3}	I, in.^{4}
3⅛ × 17¼	0.960	54	155	149	1,335
5⅛ × 15	0.975	77	192	187	1,440
6¾ × 13½	0.987	91	205	202	1,380

It is likely that the section with smaller cost has also the smallest cross-sectional area. Thus, select 3⅛ × 17¼ in. beams. Because of the relatively large, 5.5, aspect ratio of the section, the lateral stability of the beams would be questionable were it not for the fact that continuous lateral restraint is provided by the wood decking, to which they are nailed.

Minimum camber for the beams equal to 1.5 times the dead load deflection may be determined as: $1.5\ [(5/384)(w_D L^4/EI)] = (1.5)(5/384)(0.098/12)(351)^4/(1800 \times 1335) = 1.00$ in. Additional camber is necessary to provide for drainage toward the girders. Assuming an equivalent slope of ¼ in./ft., the additional camber is (¼)(15) = 3.75 in. Therefore, specify 4.75 in. for the final camber at midspan. A parabolic configuration of camber gives 2, 3½ and 4½ in. for the sections situated at ⅛, ¼ and ⅜ of the span, respectively, from the end.

10.2.2.7 Connection of Beam to Girder

The reaction transmitted by the beam to a supporting girder is R = 4.07 kips. Using ¾ in. bolts with L = 6.75 in., L/d = 6.75/3/4 = 9, and angle of load to grain, $\theta = 90°$, Table A.6.1 for Group 3 wood yields Q = 1.85 k. Since beams are connected at both sides of the girder, the number of bolts required is $2R/\overline{\alpha}\,Q = 2 \times 4.07/(1.85 \times 1.15) = 3.83$; hence, use 4 bolts, two at each side of the beam. Because only two bolts per row are used, Table A.6.26 gives a modification factor for multiple connectors, K = 1.00, and therefore the capacity of the connection is not reduced. The connection is shown in Fig. 10.2; a saddle welded to the connecting angles is provided to facilitate erection of the beams. The saddle is also used to transmit the total reaction of the beam to the supporting girder; a single ¾ in. bolt connecting the beam to the supporting angles provides capacity against uplift and longitudinal forces.

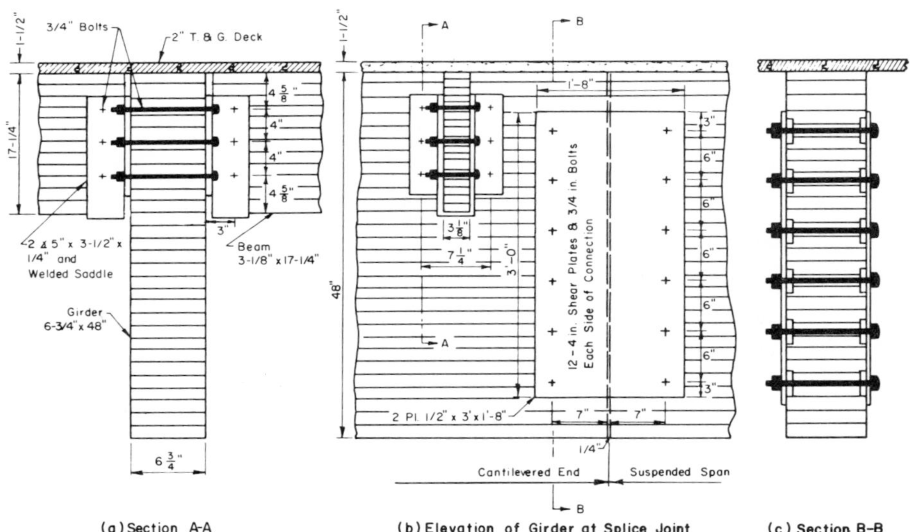

Figure 10.2. Connection details of typical beam to girder and between girders.

10.2.3 Design of Girder

The first solution that may occur to a designer for this particular structure is that of four simply-supported spans. However, upon preliminary examination it becomes apparent that this leads to the largest girders and possibly greatest cost. Let the load be assumed as uniformly distributed on the girder; an acceptable simplification in view of the close spacing of the beams. The moment at midspan, for the case of simply-supported individual girders is: $(1/8)w(60^2) = 450$ w, where w is the uniformly distributed load. For the case of a fully continuous girder, two equal spans (exterior wall to center line of building), the maximum moment is of the same magnitude, except it is negative and occurs over the interior support. Hence, a fully continuous girder would be stiffer, but not lighter, than two simply supported spans.

Optimum design is likely using the following solution. Figs. 10.3 a, b and c show a semi-elevation of the proposed girder with its corresponding moment and shear diagrams due to load. It is possible to determine the position of an inflection point that makes the maximum negative and positive moments equal in absolute value. In actual practice, a shear-type connection may be placed at that section to act as a virtual hinge and hence, to transmit no bending moment. The distance a, see Fig. 10.3b, may be determined by the following equation

$$\frac{wa^2}{2} + \frac{w(L - a)}{2}a = \frac{w(L - a)^2}{8} \qquad (10.3)$$

from which a = 0.172 L. For this case, L = 60 ft., a = 10.32 ft., and $M^- = M^+ = 309$ w.

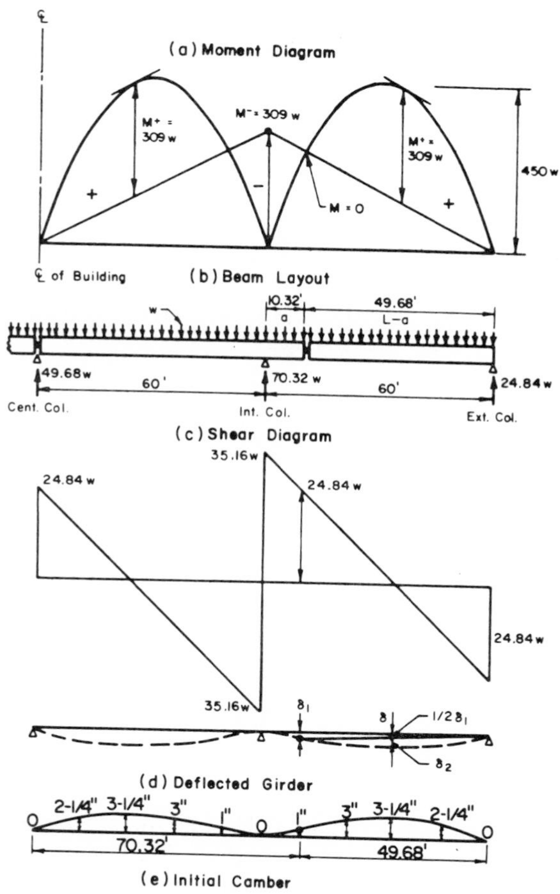

Figure 10.3. Behavior of girder under service load.

Because of symmetry, the maximum positive moment in the interior span is also equal to 309 w. This solution minimizes the maximum design moment, and should render the lightest possible girders. Other possible locations of hinges for optimization of moment diagrams are given in the literature (3). See also Chapter 7.

10.2.3.1 Determination of Loads

The dead load on the girder may be determined by assuming that the width of the girder is 8¾ in. and its depth is approximately one twentieth of the span, i.e. 36 in. The estimated self-weight is (8.75 × 36/144)(40) = 88 lb./ft. The combined weight of beams and roof deck was determined as 98 pounds per foot of beams. The reaction of beams at both sides of the girder, due to dead load, is 98 × 30 = 2,940 lb. which leads to 2,940/6 = 490 lb./ft. of equivalent uniformly distributed load on the girder. The total dead load is then: 490 + 88 = 578 lb./ft., or about 0.58 k/ft. The live load is 180 × 30/6 = 900 lb./ft. = 0.90 k/ft.; the total load on the girder is 0.90 + 0.58 = 1.48 k/ft.

10.2.3.2 Required Value of C_FS

The same criteria used in the design of the beam, can be applied in this case also. Let M = 309 w = (309)(1.48) = 457 k-ft. From Eq. 7.1a

$$C_FS = \frac{M}{\bar{\alpha}F_b} = \frac{457 \times 12}{1.15 \times 2.4} = 1{,}987 \text{ in}^3$$

Bending is satisfied by any section that has a C_FS value equal to, or larger than, 1,987 in.^{3}.

10.2.3.3 Required Area of Section

The maximum shear in the girder at a distance d = 3 ft. from the face of the interior support, see Fig. 10.3c, is V = (35.16 − 3) 1.48 = 47.6 k. The cross-sectional area required, A, is:

$$A \geq \left(\frac{3}{2}\right)\left(\frac{47.6}{0.23}\right) = 310 \text{ in.}^2$$

Shear is satisfied by any section that has an area equal to, or larger than, 310 in^2.

10.2.3.4 Required Moment of Inertia

The maximum deflection of the girder occurs approximately at midspan of the suspended span. It can be determined from Fig. 10.3d as

$$\delta = \delta_2 + \frac{1}{2}\,\delta_1 \tag{10.4}$$

where: $$\delta_2 = \frac{5}{384}\,\frac{w(L-a)^4}{EI} \tag{10.5}$$

$$\delta_1 = \frac{w(L-a)\,a^3}{6\,EI} + \frac{w\,a^4}{8\,EI}$$

and a = 10.32 ft. = 124 in., L − a = 49.68 ft. = 596 in., E = 1,800 ksi, w = 1.48 k/ft. = 0.123 k/in. Let δ, for appearance purposes, be less than (1/240)(60 × 12) = 3 in. The required moment of inertia, I = 39,900 in.^{4}, is obtained by substituting the preceding values in Eq. 10.4. Hence, any section with moment of inertia larger than, or equal to, 39,900 in.^{4} will meet the deflection criterion.

10.2.3.5 Selection of Final Section

The final section selected for the girder must meet the conditions: $A \geq 310$ in.^{2}, $C_FS \gg 1987$ in.^{3}, $I \geq 39{,}900$ in.^{4}. The following sections, listed in increasing order of cross-sectional area, are all acceptable solutions:

Dimensions Actual, in.	Properties				
	C_F	A, in.^{2}	S, in.^{3}	C_FS, in.^{3}	I, in.^{4}
6¾ × 48	0.857	324	2,590	2,220	62,210
8¾ × 43½	0.867	381	2,760	2,393	60,020
10¾ × 39	0.877	420	2,730	2,394	53,140

The first section has the smallest area and meets all the previous requirements. However, because of the relatively large aspect ratio of the section, 7.1, the lateral stability of the girder needs careful checking, see Section 7.10. In view of the fact that the girder is subjected to negative moment, due to its cantilevered portion, the maximum unsupported length, ℓ_u, occurs for the bottom flange and is equal to 10.32 ft. This is the distance between the column support and an inflection point, at both sides of the column, for which the bottom flange is subjected to compression stresses. This distance governs as it is more than the unsupported length of the top flange, which is equal to the spacing of the joists, i.e. 6 ft. The effective length of beam, ℓ_e, required for the determination of the factor C_s is $1.69\ell_u$ which corresponds to a cantilever beam with a concentrated load at the unsupported end. Then, $\ell_e = (10.32)(12)(1.69) = 209$ in. and from Eq. 7.42:

$$C_s = \sqrt{\frac{\ell_e d}{b^2}} = \sqrt{\frac{209 \times 48}{(6.75)^2}} = 14.8$$

$$C_k = \sqrt{\frac{3E}{5\bar{\alpha}F_b}} = \sqrt{\frac{3 \times 1800}{5 \times 1.15 \times 2.4}} = 19.8$$

Since $10 < C_s < C_k$ the allowable unit stress in bending, F_b, may be determined from Eq. 7.43 as follows:*

$$F_{b'} = \bar{\alpha}F_b\left[1 - \frac{1}{3}\left(\frac{C_s}{C_k}\right)^4\right] = 1.15(2.4)\left[1 - \frac{1}{3}\left(\frac{14.8}{19.8}\right)^4\right] = 2.47 \text{ ksi}$$

The allowable stress is reduced from 2.76 ksi to 2.47 ksi. Hence, the required section modulus for $d = 48$ in. is: $S = (457 \times 12)(0.857 \times 2.47 = 2591$ in$^3 \simeq 2590$ in^3, which makes this section acceptable. Should this not have been the case, however, several options are open to the designer, namely, he may: 1) require bracing of the bottom flange of the girder to the joists, 2) increase the depth of the section using the same width or 3) use a wider and perhaps shallower section.

A final note about the self-weight of the girder. The section adopted weighs (6.75 × 48/144) 40 = 90 lb./ft., which is 2 lb./ft. more than the weight initially estimated. The total load is 1.482 k/ft., instead of 1.48 k/ft., or an increase of 0.1 percent which hardly requires further considerations.

Camber for the beam is shown in Fig. 10.3e, with the vertical scale exaggerated for illustration purposes. It is easy to see that the parabolic camber configuration follows approximately the

*See also design aid in Appendix A.7.1.

mirror image of the deflection configuration of the girder under total load. For the given amount of camber the designer can be reasonably sure of 1) no visible sagging of the girder when fully loaded and 2) a pleasant flat-arch effect when the live load is absent or small.

10.2.3.6 Connection between Suspended and Cantilevered Girders

The shear force transmitted through the connection is given in Fig. 10.3c as V = 24.84 w. For the given girder w = 1.48 k/ft. and hence, V = 36.8 k. This relatively large shear may be transmitted by steel side plates connected by 4 in. shear-plates with ¾ in. bolts to the laminated girders. The final connection is shown in Figs. 10.2b and c. Each side of the connection will be subjected to a shear force of 36.8 kips and a moment of 36.8 × 7.12 = 262 k-in. due to the eccentricity of the shear force. The forces in the connectors may be obtained by vectorial summation of the vertical component due to the shear force and the horizontal component due to the moment.

The vertical component Y_i is the same for all connectors; it is given by 36.8/12 = 3.07 kip. The horizontal component, X_i, varies for each connector with the corresponding distance to the neutral axis, y_i; it is given by Eq. 6.14 as follows: $X_i = My_i/(\Sigma y_i^2 + \Sigma x_i^2)$ where M = 262 k-in.; y_i = 3, 9, 15 in. for connectors labeled, 1, 2, 3, respectively, in Fig. 10.4; and $\Sigma y_i^2 = (2)(2)(3^2 + 9^2 + 15^2) = 1{,}260$ in.². Since all $x_i = 0$, it follows that $\Sigma x_i^2 = 0$. Hence, the horizontal components for connectors 1, 2 and 3 are 0.62, 1.87 and 3.12 kip, respectively.

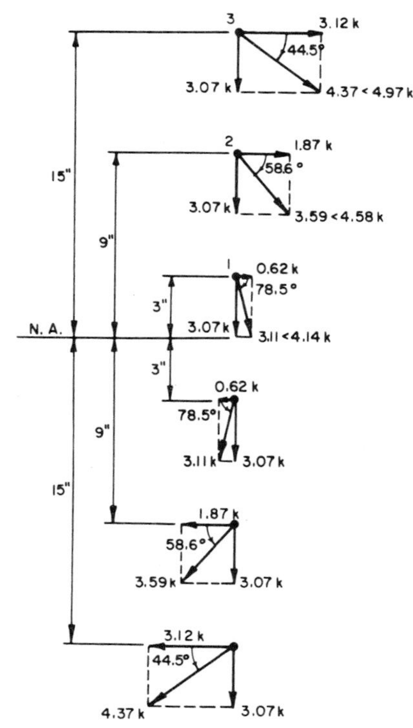

Figure 10.4. Behavior of connectors at splice joint between girders under service load.

Connector 3 is subjected to the worst combination of the group. The resultant of the component forces $F = \sqrt{(3.12)^2 + (3.07)^2} = 4.37$ k, at an angle of load to grain $\theta = \tan^{-1}\dfrac{3.07}{3.12} =$ 44.5°. The allowable load for the 4 in. connector can be determined as follows. For a thickness larger than 3⅝ in., two faces loaded, $\theta = 44.5°$ and Group A wood, Table A.6.13a gives 4.45 kip capacity for a 4 in. shear-plate with ¾ in. bolt. In addition, the load duration factor increases the capacity to the smallest of $4.47 \times 1.15 = 5.14$ k or 4.97 k (see Table A.6.8, Note 3). Connectors 1 and 2 are also subjected to forces smaller than allowable, as shown in Fig. 10.4. Hence, the connection is acceptable.* The reader is referred to Fig. 9.15 for an alternate solution, see problem 3 of this chapter.

10.2.4 Design of an Interior Column

Columns in the building are considered braced against relative lateral displacement of the ends. This is true because the laminated decking of the roof system acts as a stiff translational restraint, in the form of a continuous diaphragm, for the transfer of wind forces and inertial forces generated by seismic motions to the rigid walls in the perimeter of the building. Ignoring the effects of possible rotational end restraints (a conservative approach) allows design of the column as a braced double-hinged member for which the effective length is the actual length. Consider using glued laminated wood for which $F_c = 2.2$ ksi, $F_b = 2.2$ ksi for load parallel to wide face of laminations, $F_b = 2.6$ ksi for load perpendicular to wide face of laminations and $E = 2000$ ksi.

10.2.4.1 Design for Spaced Column Action

The interior column of the building supports the longitudinal girder at its continuous end and is subject to a reaction of 70.32 w, see Fig. 10.3.b, where $w = 1.48$ k/ft. Thus $P = 70.32 \times 1.48 = 104$ kip. The length of the column is 22 ft 7½ in. from finished floor to bottom of girder. Because the connection of the girder takes place along the depth of the girder, see Fig. 10.5, the transmission of the load to the column is at an elevation higher than that of the bottom of the girder. Assuming that it takes place at the elevation of the center line of the 4 ft deep girder, the total length of the column will be $22.8 + 4/2 = 24.8$ ft $= 297$ in.

Two laminated leaves may be used for the spaced column. The minimum width of the leaves, b, is given by the maximum slenderness ratio, $L/b = 80$. Thus, $b = 297/80 = 3.71$ in. Since use of a standard width is desirable to avoid increased manufacturing costs, the practical minimum width for the laminated leaves is 5⅛ in. However, in view of the appreciable length of the column, and to avoid penalization in allowable stress due to large slenderness, it is convenient to use at least a 6¾ in. width. Then $L/b = 297/6.75 = 44$. Assuming condition "a" prevails one may determine K from Eq. 8.20 as $K = 1.06\sqrt{E/\bar{\alpha} F_c} = 1.06\sqrt{2000/1.15 \times 2.2} = 29.8$. It follows that $K < L/b < 80$ and the member is in the long-column range.

The allowable compressive stress, $F_{c'}$, as given by Eq. 8.20, may be determined as follows:

$$F_{c'} = \frac{0.75\ E}{(L/b)^2} = \frac{0.75\ (2000)}{(44)^2} = 0.775 \text{ ksi}$$

*It is noted that the concept of modification factor for multiple fasteners has not been applied to this connection. The research on which the modification-factor concept is based, as shown in Section 6.2.2.4, was done on connections for which the loads applied to the fasteners have all a unique direction. This contrasts with the results of the analysis shown in Fig. 10.4, which indicates various intensities and directions of loading for the connectors.

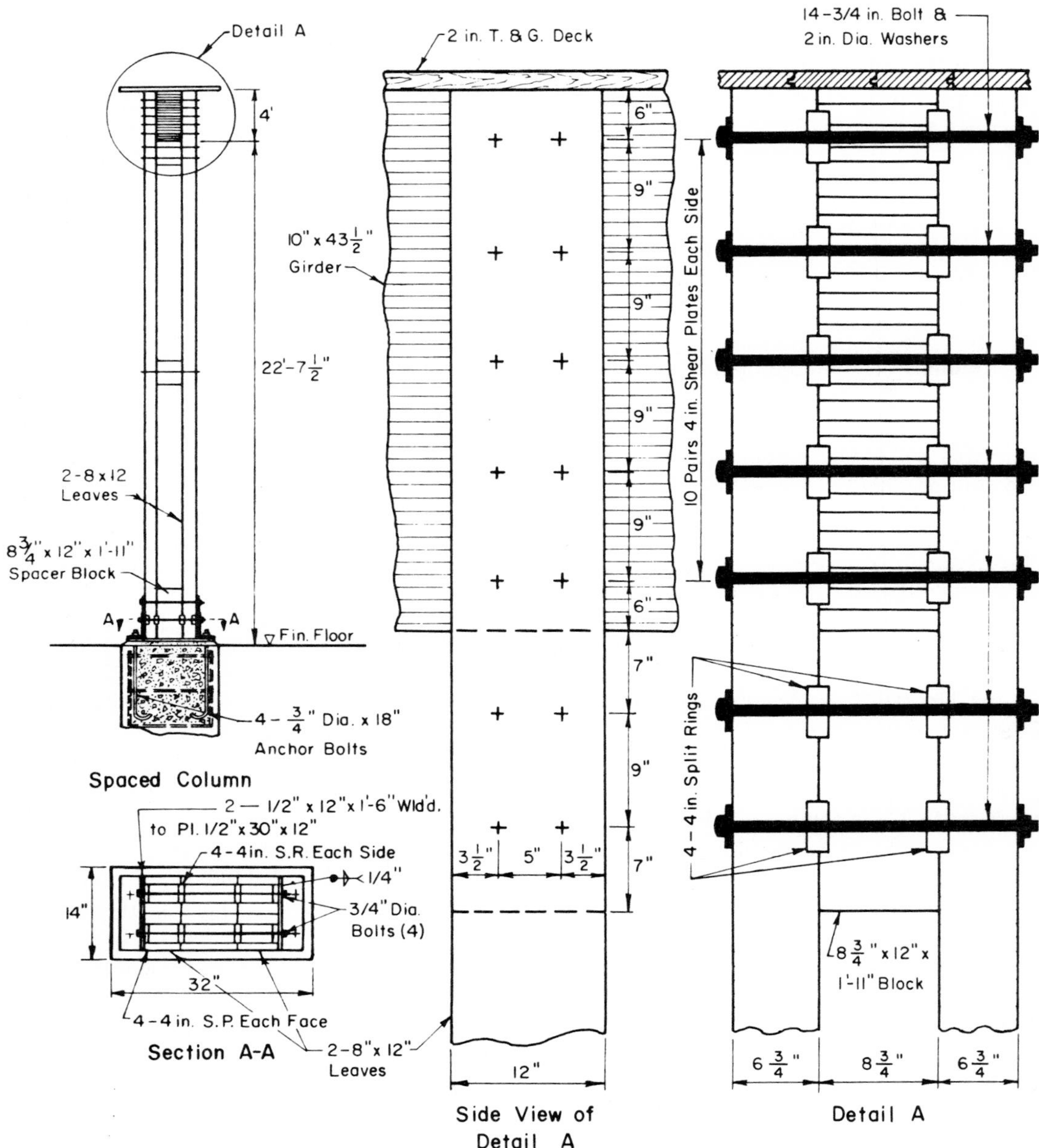

Figure 10.5. Elevation of interior column and details of connections to girder and foundation pedestal.

Hence, the required cross-sectional area is given by:

$$A = \frac{P}{F_{c'}} = \frac{104}{0.775} = 134 \text{ in.}^2$$

for which two 6.75 × 12 in. leaves with an actual cross-sectional area of 162 in.^{2} would be satisfactory.

10.2.4.2 Design for Solid Column Action

It is necessary to check the design for its capacity to resist the action of combined axial load and bending moment about the strong axis. Because of the type of rigid connection used between girder and column, see Fig. 10.5, some bending moment may be transmitted to the column. The moment, however, may not be large because: 1) the column stiffness is much smaller than that of the girder, 2) the actual stiffness of the connection is relatively small and 3) under full loading, symmetry requires zero rotation of the joint thereby effectively making the column moment equal to zero. Only for snow loading placed asymmetrically, in one span and not the other, would there be a rotation at the joint and hence, a moment generated in the column. In lieu of an exact analysis the load may be considered as having a minimum eccentricity of 1 in. or 0.1 d, whichever is greatest, see Section 8.4. Using leaves as deep as 12 in. the minimum eccentricity is 1.2 in.

The section properties of two 6.75 × 12 in. leaves are: $A = 2 \times 6.75 \times 12 = 162$ in.^{2}, $S = 2 \times (6.75 \times 12^2/6) = 324$ in.^{3} and $I = 2\,(6.75 \times 12^3/12) = 1944$ in.^{4}. Also $kL/d = 297/12 = 24.8$ and $K = 0.671\sqrt{E/\bar{\alpha}\,F_c} = 0.671\sqrt{2000/1.15 \times 2.2} = 18.9$. Since $K < kL/d < 50$ the section renders long-column behavior and $J = 1$. Also, from Eq. 8.7 the design compressive stress, $F_{c'}$, is given by:

$$F_{c'} = 0.3\,E/(kL/d)^2 = 0.3 \times 2000/(24.8)^2 = 0.976 \text{ ksi.}$$

The design bending stress, $F_{b'}$, see Section 7.15, is found by using $\ell_e = 1.84\,\ell$ where ℓ is one half the unsupported length because of the midspan spacer between column leaves. Thus $\ell_e = 1.84 \times (297/2) = 273$ in. It follows that $C_s = \sqrt{\ell_e\,d/b^2} = \sqrt{(273)(12)/(6.75)^2} = 8.5$. Since $C_s < 10$ and $C_F = 1.0$ the design bending stress, $F_{b'} = \bar{\alpha}\,F_b = 1.15\,(2.6) = 2.99$ ksi.

For eccentric load only the interaction formula given by Eq. 8.12 is reduced to:

$$\frac{\dfrac{P}{A}}{F_{c'}} + \frac{\left(\dfrac{P}{A}\right)\left(\dfrac{6e}{d}\right)(1 + 0.25\,J)}{F_{b'} - J\dfrac{P}{A}} \leq 1.0$$

Thus,

$$\frac{\dfrac{104}{162}}{0.976} + \frac{\left(\dfrac{104}{162}\right)\left(\dfrac{6 \times 1.2}{12}\right)(1 + 0.25)}{2.99 - \dfrac{104}{162}} = 0.863 < 1.0$$

which shows that the spaced column with two 6.75 × 12 in. leaves is acceptable.

There remains only the determination of the number of connectors required for the end block to develop spaced-column action. From Table 8.4, with L/d = 44 and Group A wood, the end spacer block constant is 315 psi. The required shear capacity is (315)(6.75)(12) = 25.7 kips. The number of 4 in. split-rings with ¾ in. bolt needed is 25.7/(6.14 × 1.15) = 3.62; say four. Section AA of Fig. 10.5 shows a detail of the end block of the column.

Column Connection to Girder

The vertical reaction of the girder on the interior column is R = 104 kip. It is transmitted to the column by the connection shown in detail A of Fig. 10.5. It is seen that the reaction is transferred from the girder to the column in two ways, namely through two sets of six 4 in. shear-plate connectors (one set at each side of the connection), the capacity of which are determined by compression perpendicular to the grain of the girder, and by direct bearing on the central block of the column. The added capacity of the two systems must be equal to, or larger than, the reaction to be transmitted.

The capacity of the two sets of six 4 in. shear-plate connectors, which connect the girder to the leaves of the spaced-column, can be determined as follows. The two sets of six connectors are laid out in four rows of three connectors each. Thus, the capacity, as given by Eq. 6.2, is 4 Nf K P, where P is the capacity of a single 4 in. shear-plate and K is the modification factor for multiple, $N_f = 3$, fasteners. The value of P can be obtained from Table A.6.8 for Group A wood, $\theta = 90°$, $t > 3½$ in., 2 faces loaded, as P = 3.500 kip; since $\alpha = 1.15$ it follows that $P = 1.15 \times 3.5 = 4.025$ kip. The value of K can be obtained from Table A.6.25, for $A_1 = 5 \times 12 = 60$ in.², $A_2 = 2 \times 5 \times 12 = 120$ in.², $A_1/A_2 = 0.5$ and $N_f = 3$, as K = 1.00. Therefore, the capacity of this system is (12) (1.00) (4.025) = 48.30 kip.

The capacity of the two sets of six split-ring connectors, which connect the central block to the two leaves of the column, can be determined also as 4 N_f K P, where P, in this case, is the capacity of a single 4 in. split-ring and K is the modification factor for multiple, $N_f = 3$, fasteners. The value of P can be obtained from Table A.6.6 for Group A wood, $\theta = 0°$, $t > 3$ in., 2 faces loaded, as P = 6.14 kip; since $\alpha = 1.15$, it follows that $P = 1.15 \times 6.14 = 7.06$ kip. The value of K is the same as before, i.e., K = 1.00. Therefore, the capacity of this system is (12) (1.00) (7.06) = 84.72 k.

The total capacity is the sum of the individual capacities of each system, 48.30 + 84.72 = 133 k > 104 k. This is more than necessary and may justify a reduction in the number of 4 in. shear-plates that connect the girder and the column directly. However, it is wise to keep the extra capacity in the connection. The reasons for this are: the prime importance of the connection and the relatively small additional cost involved.

The connection is suited for simple erection procedures. The girder may be erected, aligned and placed on the bearing block of the supporting columns by a crane in one simple step. For stability purposes, the column connections should be finished before erection of the suspended girders.

A useful exercise would be to complete this problem with the design of: 1) exterior and central columns; 2) connection between the suspended girder and the exterior column and between the girder and central column; and 3) the special beam connection at the column-girder joint. The reader is also referred to the problems at the end of the chapter for further design considerations.

10.3 DESIGN OF A BUILDING USING PLYWOOD AND STRESS-GRADED LUMBER

A building for an Exhibition Hall was designed in the preceding section using glued-laminated wood exclusively. The same building may be designed using plywood and stress-graded lumber as shown in Fig. 10.6. The roof structure consists of staggered 4 × 32 ft. stressed-skin panels, made up of plywood panels and sandwiched lumber stringers, with a total thickness of 6¼ in., see Fig. 4.7. The panels rest on 75 ft. long prefabricated box-beams, see Fig. 4.6, made up of 4 ft. wide plywood webs and lumber flanges, spanning 50 ft. between columns and cantilevering 25 ft. towards the center of the building. A combined moment and shear connection at the center ties the abutting members together restoring continuity to the box-girder. In the following sections the typical roof structural unit, girder and interior column are designed. Roofing materials and snow loads are considered equal to the values given in Section 10.2. The provisions of NDS (1) and the American Plywood Association (4) (5) and (6) are used in the design.

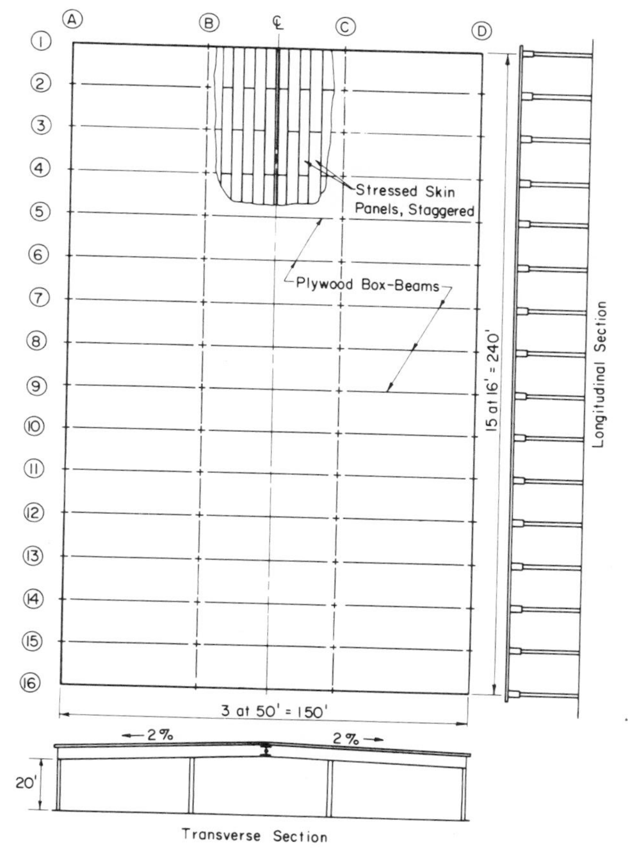

Figure 10.6. Structure of exhibition hall using plywood and stress-graded wood.

10.3.1 Design of Flat Plywood Stressed-skin Panel

This type of decking is obtained by gluing together plywood panels and lumber stringers, see Fig. 10.7. The resulting product is usually four feet wide by the desired length; the maximum length is determined by the length of lumber which is available. Structurally, the panel acts as a series of built-up wide flange beams; the plywood panels, which provide sheathing, take the flexural stresses and the lumber stringers the shear stresses. In this fashion, the sheathing, which in conventional design is only used to carry the load across to the supporting stringers or joists, is also used in stressed-skin panels as an integral part of the stringers. This results in smaller stringers and less over-all weight for the roof or floor system.

The initial work of proportioning a stressed-skin panel for a given loading and span can be facilitated by load-span tables (7). In lieu of the tables, a reasonable initial section may be obtained by taking a span-depth ratio not larger than 20 for simply-supported panels and 30 for continuous spans, and by assuming that the flanges take all flexural stresses and the stringers all shear stresses. Once a section has been selected, it must be proved acceptable by checking its strength and serviceability. As is usual in wood design, strength requirements are met when the maximum stresses in bending, horizontal shear and rolling shear are smaller than, or equal to, the corresponding allowable stresses. Serviceability is adequate when the maximum longitudinal deflection of the stressed-skin panel and the local transverse deflection of the plywood panels are smaller than, or equal to, the corresponding allowable deflections.

The determination of maximum stresses requires an elastic analysis of the stressed-skin panel under loading, in order to find the maximum bending moment and shear. The properties of the section are determined by assuming perfect composite action (no-slip) between the plywood panels and the wood stringers. In view of the fact that moduli of elasticity of plywood panels and wood stringers may not be equal, a transformed section is needed for exact calculations. However, the difference between results obtained using a transformed section and results obtained using a homogeneous section with an average modulus of elasticity is small. This is due to the fact that the moduli of elasticity of plywood and lumber are fairly similar. Thus, the refinement of the

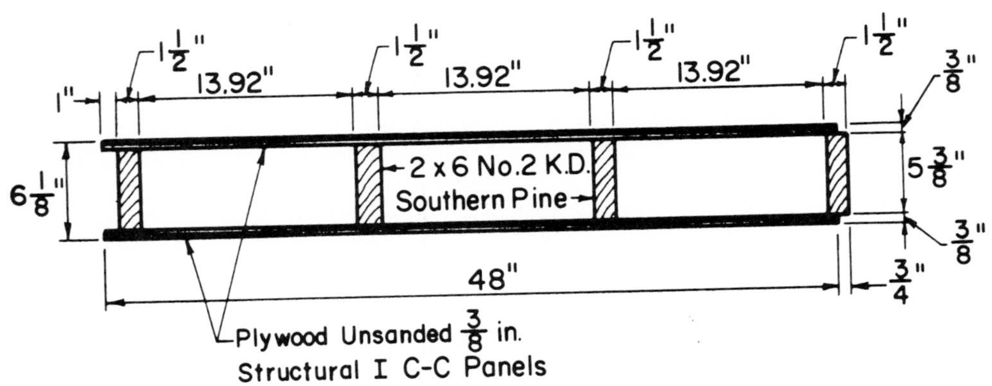

Figure 10.7. Typical stressed-skin panel for roof structure. Note that an additional ⅛ in. has been deducted from the depth of the stringers to allow for resurfacing. Face grain of plywood panel runs parallel to 2 × 6 stringers.

transformed section is deemed necessary in only a few special cases. Once the maximum moment, shear and the section properties are determined, the stresses can be obtained from Eqs. 7.1a and 7.29.

There is a limitation on the width of flange that can be considered effective in resisting longitudinal bending of the panel. This limitation reflects shear lag in the flange and the tendency of plywood panels to deflect inwards when stringers are widely spaced. A basic spacing of stringers, w_e, see Table 10.1, is used to take this tendency into account. The distance w_e represents the amount of plywood panel which may be considered to act compositely with the stringer, at each side, for bending-stress calculations. Thus, for an interior stringer of width b, the effective width of flange may not exceed $b + w_e$. In spite of this, the full 4-ft. width of plywood may be used for the determination of deflections, for the design of splice plates and the analysis of shear (5).

TABLE 10.1. STRESSED-SKIN PLYWOOD PANELS BASIC SPACING, w_e, IN INCHES FOR VARIOUS PLYWOOD THICKNESSES†.

Thickness	Face Grain	
	Parallel to Stringers	Normal to Stringers
¼ in. Sanded	10.3	11.6
5/16 in. Unsanded	11.9	16.8
⅜ in. Unsanded (3-ply)	14.2	20.1
⅜ in. Sanded (3-ply)	16.4	16.4
⅜ in. Sanded (5-ply)	18.1	20.2
½ in. Unsanded and Sanded*	23.2	28.5
⅝ in. Unsanded and Sanded*	29.1	35.6
¾ in. Unsanded and Sanded**	38.2	38.2
⅞ in. Unsanded and Sanded*	41.6	48.1
1″ Unsanded	45.5	58.9
1″ Sanded	54.5	47.9
1⅛″ 2–4–1	53.4	57.3

† From Section 2.2.1 of Reference 5
*Use only 5-ply panels
**5-ply and 7-ply

The resultant moment of inertia of the full section is called the gross moment of inertia. In those cases where the spacing of the stringers exceeds $b + w_e$, the gross moment of inertia will be larger than the moment of inertia used in the calculations of bending stresses. The maximum deflection determined using the modulus of elasticity E given in Ref. 1 is likely to be larger than the actual maximum. This is true because the listed values of E are 10 percent smaller than the actual values to approximate the effects of shear deformations. For the case of stressed-skin panels, this can be too conservative. An accurate determination of the maximum deflection may be accomplished as the sum of the deflection due to bending, using the actual modulus of elasticity, and the deflection due to shear (5). However, the additional complication may not be warranted in the majority of cases.

The allowable stresses, against which the maximum stresses are checked, are given in Table A.4.10 for the plywood panels, and Table A.2.5 for the lumber stringers. The allowable stresses are modified by the load duration factors as usual. In addition, for the plywood panels, a reduction factor is applied to provide against buckling of the skins due to bending. The reduction factor, R.F., is a function of the ratio c/w_e, where c is the clear distance between stringers and w_e is the basic spacing of the stringers given in Table 10.1. For values of $c/w_e \leq 0.5$, R.F. = 1. For $c/w_e \geq 1.0$, R.F. = ⅔. For the range $0.5 < c/w_e < 1.0$ the reduction factor varies linearly between 1 and ⅔ and can be expressed algebraically as:

$$\text{R.F.} = 1 - (1/3)(2\,c/w_e - 1) \qquad (10.6)$$

The design of a stressed-skin panel for the building shown in Fig. 10.6 illustrates the preceding concepts. The length of the panel is 32 ft., making it able to span across three supporting beams. The panel being continuous over a central support reduces the number of panels by almost one half, as compared to simple spans, thus cutting handling and erection costs without increasing the size of the required section or the cost of fabrication. Also, the two-span length of the panel allows a staggered layout as shown in Fig. 10.6, which enhances the diaphragm action of the roof against lateral loads. In the following, a panel section is proportioned and its adequacy is examined.

10.3.1.1 Determination of Loads

The dead weight of the roof can be determined as follows:

stressed-skin panel	5 psf
insulation	2 psf
vapor barrier and built-up-roof	6 psf
w_D =	13 psf

The live load, w_L = 30 psf; hence, the total load w_T = 43 psf. For the full 4 ft. width of panel, the loads are: w_D = 52 lb./ft. w_L = 120 lb./ft. and w_T = 172 lb./ft.

10.3.1.2 Determination of maximum moment and shear

Both maximum moment and shear occur at the section over the continuous support and are given by:

$$M = (1/8)(172)(16)^2 = 5{,}504 \text{ lb.-ft.} = 66.05 \text{ k-in.}$$
$$V = (5/8)(172)(16) = 1{,}720 \text{ lb.} = 1.72 \text{ k}$$

10.3.1.3 Initial proportioning of the section

The minimum depth of the stressed-skin panel may be determined by selecting a span-depth ratio equal to thirty. Then, d = 16 × 12/30 = 6.4 in. Assuming 2 × 6 stringers plus top and bottom ⅜ in. plywood panels gives 6.25 in. for the total depth of section. For Southern Pine Structural I C-C exterior plywood panels (species group 1), Table A.4.10 yields F_c = 1.65 ksi. Using the largest stress reduction factor, R.F. = ⅔, see Section 10.3.1, and $\overline{\alpha}$ = 1.15, the load duration factor for snow loading yields F_c = (1.65)(⅔)(1.15) = 1.26 ksi. The force component of the resisting couple is 1.26 A_f and its lever arm is 6.250 − 0.375 = 5.875 in. Hence, moment equilibrium requires that

$$A_f\,(1.26)(5.875) = 66.05 \text{ k-in.} \qquad (10.7)$$

where $A_f = 8.92$ in.^{2} is the cross-sectional area of the flanges. From Table A.4.9b, a 3/8 in. unsanded panel provides $(2.40)(4) = 9.60$ in.^{2}. The necessary number of 2×6 stringers may be determined from shear considerations. For $F_v = 0.080 \times 1.15 = 0.092$ ksi, $A = 1.5 \times 5.5 = 8.25$ in.^{2} and $V = 1.72$k, the number of 2×6 stringers is

$$n = \frac{3}{2}\,\frac{1.72}{(0.092)(8.25)} = 3.4$$

Four stringers is the minimum number used to prevent excessive transverse deflection of the top plywood panel under load. The section shown in Fig. 10.7 is a reasonable initial trial for the solution.

10.3.1.4 Section properties

The following calculations determine the area A, moment of inertia I, section modulus S, and first moment of area above the neutral axis Q, respectively, of the given section.

$$A = (4)(2.4)(2) + (4)(1.5)(5.375) = 19.2 + 32.2 = 51.4 \text{ in.}^2$$

$$I = (2)(9.6)(3.063 - 0.188)^2 + (4)(0.048)(2) + (4)(1/12)(1.5)(5.375)^3$$
$$= 158.8 + 0.4 + 77.7 = 236.9 \text{ in.}^4$$

$$S = \frac{236.9}{3.063} = 77.4 \text{ in.}^3$$

$$Q = (4)(2.4)(3.063 - 0.188) + (4)\left(1.5 \times \frac{5.375}{2}\right)\left(\frac{5.375}{4}\right)$$
$$= 27.6 + 21.7 = 49.3 \text{ in.}^3$$

10.3.1.5 Allowable stresses

From Table A.4.10, the allowable stress in compression parallel to grain for Structural I C-C panel, species group 1, is 1.65 ksi. The allowable stress in tension is 2 ksi. Therefore, the panel in compression, i.e. the bottom panel at the interior support, controls the design in bending. The reduction factor is a function of c/w_e, where $c = 13.92$ in., see Fig. 10.7, and $w_e = 14.2$ in. from Table 10.1. Thus, $c/w_e = 0.98$, which when substituted in Eq. 10.6 yields R.F. $= 0.68$. Also, $\bar{\alpha} = 1.15$. Hence, the allowable stress in bending is $(0.68)(1.15)(1.65) = 1.29$ ksi.

The allowable stress in shear through the thickness of the plywood panel for normal loading conditions is 0.25 ksi, see Table A.4.10; for snow loading the allowable stress becomes 0.287 ksi. For rolling shear, the same table gives 0.075 ksi; for snow loading it becomes 0.086 ksi.

10.3.1.6 Maximum stresses

The maximum stress in bending may be determined using Eq. 7.1a for $M = 66.05$ k-in. and $S = 77.4$ in.^{3} as follows:

$$f = \frac{66.05}{77.4} = 0.854 \text{ ksi} < 1.29 \text{ ksi}$$

The maximum horizontal shearing stress may be determined using Eq. 7.29, for $V = 1.72$ k, $Q = 49.3$ in.^{3}, $I = 236.9$ in.^{4} and $b = 4 \times 1.5 = 6$ in., as follows

$$v = \frac{(1.72)(49.3)}{(236.9)(6)} = 0.057 \text{ ksi} < 0.092 \text{ ksi}$$

10.3.1.7 Rolling Shear

Maximum shear stress v in the plane of the plies occurs at the interface between the face ply and the first interior ply. The area of parallel plies above this level, A, and its first moment about the neutral axis, Q, may be determined from Table A.4.12. For a ⅜ in. unsanded Structural I plywood panel with face grain parallel to stringers A = 4.8 in.^{2}, $y' = 0.05$ in., where y' is the distance between the outside surface of the panel and the centroid of the area A. Hence, $y - y' = 3.063 - 0.05 = 3.01$ in. and $Q = (4.8)(3.01) = 14.45$ in.^{3}. The product vb, where v is the rolling shear and b is the sum of the glueline widths of stringers in contact with the plywood panel, may be found from Eq. 7.29 as follows:

$$vb = \frac{(1.72)(14.45)}{(236.9)} = 0.101 \text{ k/in.}$$

The allowable value of vb may be found as the sum of the individual products of width of stringers times allowable stress. For interior stringers, the allowable stress in rolling shear is 0.086 ksi; for exterior stringers, due to stress concentrations, the allowable stress is taken as $(\frac{1}{2})(0.086) = 0.043$ ksi. Hence, the allowable value of vb, see Fig. 10.7, is given by $(2)(1.5)(0.086) + (1.5 + 0.75)(0.043) = 0.355$ k/in. > 0.101 k/in., i.e., the rolling shear stress is within acceptable limits.

10.3.1.8 Transverse deflection of top plywood panel between stringers

A stressed-skin panel must be checked for deflection of its top skin under the loading. It is reasonable to assume that in a unit, such as that shown in Fig. 10.7, a large amount of rotational restraint exists for the panel spanning across the stringers. Assuming 80 percent of full fixity,* the maximum deflection may be determined approximately from:

$$\Delta = \frac{wL^4}{300\ EI} \tag{10.8}$$

where w, the weight of the top panel only plus superimposed loading is given by $w = 1 + 2 + 6 + 30 = 39$ psf $= 0.0033$ k/in., $L = 13.92$ in., $E = 1{,}800$ ksi, see Table A.4.10, and $I = 0.003$ in.^{4}, see Col. 10 of Table A.4.9b. Substituting these values in Eq. 10.8 yields $\Delta = 0.075$ in. The deflection-span ratio is 1/185, which is admissible.

10.3.1.9 Longitudinal deflection of unit between supports

Maximum deflection for a two-span continuous beam under uniform loading occurs when one of the spans is loaded and the other unloaded. For this case, it can be shown (8) that the maximum deflection is given by:

$$\delta = \frac{wL^4}{110\ EI} \tag{10.9}$$

where $w = 43 \times 4 = 172$ lb./ft. $= 0.0143$ k/in., $L = 192$ in., $E = 1{,}800$ ksi, $I = 236.9$ in.^{4}. Substituting these values in Eq. 10.9 yields $\delta = 0.41$ in. The deflection-span ratio is $1/470 << 1/240$, which is quite small and hence, acceptable. In view of the stiffness of the stressed-skin panel, provisions for camber appear unnecessary.

*A conservative estimate, since APA (5) allows 100 percent of full fixity.

A more accurate determination of deflection, resulting in smaller values, is possible (5) because the value of E used before is low by 10 percent. Normally, the use of a reduced modulus of elasticity accounts approximately for deflections due to shear. However, the latter are quite small in stressed-skin panels thereby making the approximate deflection somewhat big. An exact deflection configuration may be obtained as the sum of the deflections due to bending, determined with modulus of elasticity 1.1E, and the deflections due to shear using $G = E/16$. These additional calculations do not seem warranted in most cases.

10.3.2 Design of Plywood Box Beam

A frequent application of plywood to buildings is obtained when it is used in combination with lumber or laminated wood to make long-span beams of I or box sections. The beam may be any depth; however, the full 48 in. size of the plywood panel is often used. In this type of construction, the plywood panels which form the web of the beam are glued to the solid or glued-laminated wood used for the flanges; extra webs and plywood fillers may be added as necessary to carry large shears (6).

The initial design of this type of beams is similar to that of steel plate girders. The flanges are designed to carry the flexural stresses created by bending moment and the web to take the shearing stresses. Generally, equal flanges are used, notwithstanding the fact that larger allowable tensile stress in glued-laminated wood leads to smaller tensile flanges. Once the initial determination of the section is done, a full elastic analysis follows, after which corrections to the initial design may be introduced as required.

The wood components of the given section may have different moduli of elasticity. This requires the use of a transformed section, as shown in Section 9.5.3 for composite wood-concrete section, or the determination of the equivalent EI of the section (9). However, for the usual case, the difference between moduli in wood is small; also, using a homogeneous material with the lower modulus of elasticity gives results which are easier to obtain and which are conservative in general. Thus, a homogeneous section can be assumed for the determination of section properties.

Parallel to grain material of the plywood web is taken as being continuous for the determination of the section properties, only if the panels are spliced by means of a scarf joint; or on both sides, full depth, by splice plates.* The plywood panels may be butt jointed to transmit shear only; for this case, the splice-plate need only be as deep as the clear distance between flanges. If such joints in the webs are staggered 24 in. or more, then only one web is counted in computing net I for bending stress calculations; if the joints are closer than 24 in., the contribution of the webs is neglected.

It is important to emphasize again a phenomenon peculiar to plywood construction, namely rolling shear. The allowable stress at the glue lines between flanges and plywood webs is limited to the strength of the plywood in shear acting in the interface between the face ply and the first interior ply. Allowable stresses for rolling shear are given in Table A.4.10 and are affected by a 50 percent reduction for stress concentration.

The design of I or box-beams is completed when web stiffeners are provided. Stiffeners are necessary to improve the elastic stability of the web and to distribute concentrated loads and reactions at the supports. Stiffeners are glued both to the web panels and to the lumber flanges.

*All parallel-grain material may be included, regardless of butt joints, for shear and deflection calculations (6).

They are usually made of 2 in. dimension lumber, and are equal in width to the lumber flange between webs, allowing for splice-plates, if any. Under concentrated loads, the web stiffeners must be designed so that the unit stress of the flanges, in compression across the grain, is not exceeded. Finally, the lateral stability of the compression flange must be checked as suggested in Section 7.10.

In the following sections, the design of a box-beam is given for the structure shown in Fig. 10.6. The beams are prefabricated with a length of 75 ft. that allows them to be erected over the exterior and interior columns and span all the way to the center of the building. In the initial erection stage, in which the beam is supporting its own weight only, the interior span acts as a double cantilever. Before any additional loads are placed on the structure, the midspan moment and shear-connection between beams is completed. Erection of the roof system proceeds thereafter; hence, the beam acts as a continuous three-span member to support all additional loads. The design of the beam takes into consideration the effects of both, the erection and final stages of loading.

10.3.2.1 Determination of loads

Assume a box section 4 ft. deep made up of two 1 in. plywood webs and lumber flanges consisting of six 2 × 6 glued stringers. The weight of the beam can be estimated as follows:

$$
\begin{aligned}
&\text{2 webs at 3 psf} \times \text{4 ft.} && = 24 \\
&\text{2 flanges at 6 } (1.5 \times 5.5/144)(40) && = \underline{28} \\
&\qquad\qquad w_1 && = 52 \text{ lb./ft.} = 0.052 \text{ k/ft.}
\end{aligned}
$$

The roof system, consisting of stressed-skin panels, insulation, and built-up roofing, weighs 13 psf and the design snow load is 30 psf for a total load of 43 psf, see Section 10.3.1. The beams are spaced 16 ft. apart; hence, the total roof load on any given beam is $w_2 = 43 \times 16 = 688$ lb./ft. $= 0.688$ k/ft.

10.3.2.2 Determination of maximum effects on the beam

The bending moment and shear diagrams for the erection and final stages of the beam are given in Figs. 10.8 a and c, respectively. The reactions at the supports are also indicated on Fig. 10.8b. Maximum moment occurs at the section over the interior column and can be determined, see Fig. 10.8a, as follows:

$$
\begin{aligned}
M &= (312.5)(0.052) + (250)(0.688) = 16 + 172 = 188 \text{ k-ft.} \\
&= 2{,}256 \text{ k-in.}
\end{aligned}
$$

Maximum shear force occurs in the inside portion of the exterior span, see Fig. 10.8c. Maximum shear stresses are developed at the face of the support, assumed 12 inches wide; hence, the maximum shear that affects the section occurs at a distance of 0.5 ft. from the interior support and can be determined as follows:

$$V = (31.25 - 0.5)(0.052) + (30 - 0.5)(0.688) = 1.6 + 20.3 = 21.9 \text{ k}$$

Reactions R_I and R_E, at the interior and exterior supports respectively, are given by:

$$
\begin{aligned}
R_I &= (56.25)(0.052) + (55)(0.688) = 2.93 + 37.84 = 40.77 \text{ k} \\
R_E &= (18.75)(0.052) + (20)(0.688) = 0.98 + 13.76 = 14.74 \text{ k}
\end{aligned}
$$

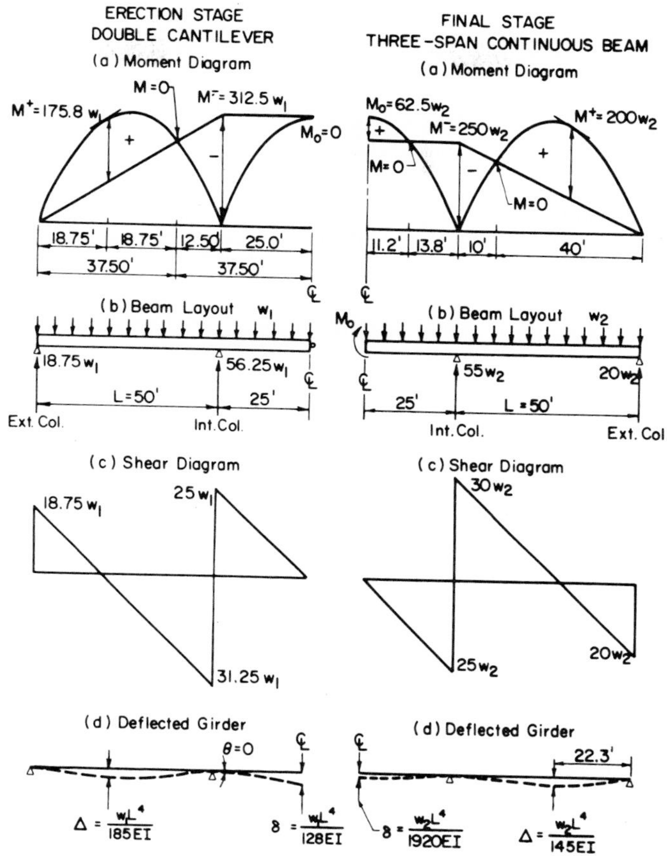

Figure 10.8. Behavior of prefabricated box girder during erection and under service load.

10.3.2.3 Initial proportioning of the section

Assume the following materials for the box-beam: Wood having allowable stresses of F_b = 1.65 ksi, F_c = 1.45 ksi, F_t = 1.10 ksi and $F_{C\perp}$ = 0.405 ksi is used for the flanges. Structural I C-C unsanded, exterior glue only, plywood panels for the webs. The area of the flanges A_f may be determined using the conventional assumption that the webs contribute nothing to resist the bending moment. Since the lumber stringers that constitute the flanges are spliced together with scarf joints not steeper than 1 in 12, the joint factor* is 0.85; hence the allowable stress in tension (see Section 5.7.2.1 of Ref. 4) is 0.85 F_t. Therefore, the allowable stress in tension is 0.85 × 1.10 = 0.935 ksi. The lowest possible value of the form factor of the section, i.e. C_f = 0.81, see Section 7.4, results in an allowable bending stress 0.81 F_b, where F_b = 1.65 ksi. Hence, the allowable stress in bending is 0.81 × 1.65 = 1.33 ksi. Thus, the governing condition for the allowable stress is given by tension parallel to grain. Considering the load-duration factor, $\overline{\alpha}$ = 1.15 for snow, the allowable stress is (1.15)(0.935) = 1.075 ksi. The lever arm of the resisting couple for 6 in.

*Other joint factors are 0.8 and 0.75, for scarf slopes of 1 in 10 and 1 in 8, respectively.

nominal dimension lumber is 48 − 5.5 = 42.5 in. Hence, the area of the flange given by

$$A_f\ (1.075)(42.5) = 2{,}256 \text{ k-in.} \tag{10.10}$$

results in $A_f = 49.4$ in.² Six 2 × 6 S4S stringers, $A_f = 49.5$ in.², may be used. Even a small deficit in area is acceptable in view of the fact that the actual contribution of the web was neglected.

The thickness of the plywood webs, t_w, may be determined by assuming that the flanges contribute nothing to resist shear. For Structural I C-C plywood panels, the allowable shear-stress in a plane perpendicular to the plies is 0.25 ksi; when affected by $\overline{\alpha} = 1.15$, it becomes $F_v = 0.287$ ksi. The full section of the panels may be used if the panels are spliced together by scarf joints, with slope of 1 in 8 or flatter. The necessary thickness t_w of the panels can be determined from:

$$(2t_w)(48)(0.287) = (1.2)(21.9) \tag{10.11}$$

as 0.96 in. In the preceding determination, the shear force was increased 20 percent to account for the fact that the maximum shear stress, see Fig. 7.13, is somewhat larger than the average shear stress. To satisfy the required area of web, two 1 in. thick plywood panels with 1.037 in. effective thickness, see Col. 1 of Table A.4.9b, are used. The section of the box-beam is shown in Fig. 10.9b.

10.3.2.4 Section properties

The equivalent thickness t_{we} of the web, for continuous parallel-to-grain material, must be determined first. From Table A.4.11, the veneer layout of a 1 in., unsanded panel, shows two 0.1 in. face plies, two 0.123 in. center bands and three 0.185 in. cross bands. Hence, $t_{we} = 2(0.100) + 2(0.123) = 0.446$ in. The same result could have been obtained from Table A.4.9b directly.

The following calculations determine the area A, moment of inertia I_x about horizontal axis of bending, moment of inertia I_y about vertical axis of bending, section modulus S, first moment of area above the neutral axis Q, support factor C_g and form factor C_f, respectively, of the given section.*

$$A = (2)(0.446 \times 48) + (2)(6 \times 1.5 \times 5.5) = 42.8 + 99.0 = 141.8 \text{ in.}^2$$

$$I_x = (2)(1/12)(0.446)(48)^3 + 2(1/12)(9)(5.5)^3 + (2)(9 \times 5.5)(24 - 2.75)^2 = 8{,}221 + 250 + 44{,}705 = 53{,}176 \text{ in.}^4$$

$$I_y = 2(1/12)(5.5)(9)^3 + 2(5.354 \times 4)(4.5 + 0.5)^2 + 2(0.574 \times 4) = 668 + 1{,}071 + 5 = 1{,}744 \text{ in.}^4$$

$$S = 53{,}176/24 = 2{,}216 \text{ in.}^3$$

$$Q = (2)(0.446)(24)(12) + (6 \times 1.5 \times 5.5)(24 - 2.75) = 257 + 1{,}052 = 1{,}309 \text{ in.}^3$$

$$C_g = (5.5/48)^2[6 - 8(5.5/48) + 3(5.5/48)^2](1 - 2 \times 0.446/9.892) + (2 \times 0.446/9.892) = 0.15, \text{ from Eq. 7.26}$$

$$C_f = (0.81)\left[1 + \left(\frac{48^2 + 143}{48^2 + 88} - 1\right)0.15\right] = 0.813, \text{ from Eq. 7.23}$$

*If resurfacing of the lumber flanges becomes necessary, calculation of section properties should be based on the final section of the beam.

In the case of deep sections, as pointed out in Section 7.4.3, ignoring C_g completely, i.e. $C_g = 0$ results in small error in the determination of C_g; for the given case C_f would have been 0.81 instead of 0.813, as above.

The exact determination of the form factor confirms the initial assumption that the allowable stress for design in bending, in this case, is tension parallel to grain; i.e. $F_b = 1.075$ ksi.

10.3.2.5 Maximum stresses

In bending, from Eq. 7.1a:

$$f = \frac{2{,}256}{2{,}216} = 1.017 < 1.075 \text{ ksi.}$$

In shear, from Eq. 7.29a:

$$v = \frac{(21.9)(1{,}309)}{(53{,}176)(2 \times 1.037)} = 0.260 < 0.287 \text{ ksi.}$$

10.3.2.6 Rolling shear

The shear-stresses between the flanges and the plywood webs must be determined and checked against the capacity of plywood to resist rolling shear-stress. This effect occurs in the first interior ply, for which the direction of grain is normal to the direction of horizontal shearing-stresses. The determination of the rolling shear-stress can be done by means of Eq. 7.29a where $V = 21.9$ k, $I = 53{,}176$ in.^{4}, $b = 5.5$ in. and Q is the first moment of area of one flange only about the neutral axis, i.e. $Q = (6 \times 1.5 \times 5.5)(24 - 5.5/2) = 1{,}052$ in.^{3}. Since both lateral faces of the flange are in contact with the plywood webs, the rolling shear-stress v is given by:

$$v = \left(\frac{1}{2}\right)\frac{(21.9)(1{,}052)}{(53{,}176)(5.5)} = 0.0394 \text{ ksi}$$

From Table A.4.10 the allowable rolling shear in the plane of the plies is 0.075 ksi for Structural I panels. However, a 50 percent reduction in allowable stress is recommended for stress concentration. Hence, the allowable stress for the given duration of loading, is $(0.075)(0.5)(1.15) = 0.0431$ ksi. Evidently, the panel has sufficient capacity to resist rolling shear.

10.3.2.7 Maximum deflections

Deflections at two sections are determined, namely, δ at midspan of the interior span and Δ, close to midspan of the exterior span; see Fig. 10.8d. It is necessary to determine the deflections as the sum of the individual deflections for the temporary stage of erection and the final stage of the beam. Thus,

$$\delta = \frac{w_1 L^4}{128 \text{ EI}} + \frac{w_2 L^4}{1{,}920 \text{ EI}} \qquad (10.12)$$

$$\Delta = \frac{w_1 L^4}{185 \text{ EI}} + \frac{w_2 L^4}{145 \text{ EI}}$$

Substituting $w_1 = 52$ lb./ft. $= 0.0043$ k/in. $w_2 = 688$ lb./ft. $= 0.0574$ k/in., $E = 1{,}800$ ksi, $I = 53{,}178$ in.^{4} and $L = 600$ in., yields:

$$\delta = 0.05 + 0.04 = 0.09 \text{ in.}$$

$$\Delta = 0.04 + 0.53 = 0.57 \text{ in.}$$

Larger deflections will occur if the live load is placed in specific patterns. Thus, the maximum value of δ occurs when only the interior span is loaded; conversely, the maximum value of Δ occurs when only the exterior spans are loaded. The time effects on deflections may be estimated by doubling the initial deflection due to permanent loading, see Section 7.11*. The permanent loading on the beam is $52 + 13 \times 16 = 260$ lb./ft. Since time deflections follow the final configuration of the structure, i.e., the three-span continuous beam, see right hand side of Fig. 10.8, the additional deflections at the given sections may be close to the following values: $(260/688)(0.04) \simeq 0.02$ in., and $(260/688)(0.53) = 0.20$ in. respectively. Therefore, the final deflections, after some time, are $\delta = 0.09 + 0.02 = 0.11$ in. and $\Delta = 0.57 + 0.20 = 0.77$ in.

It is necessary to indicate that Eqs. 10.12 give the deflection due to bending stresses only. The effect of shear can be determined exactly and added to the preceding values (6). However, it may be accounted for, approximately, by means of shear-deflection factors (9). These factors depend on L/d, the span-depth ratio. For values of L/d equal to 10, 15 and 20, the shear-deflection factors are 1.5, 1.2 and 1.0, respectively. For the given beam $L/d = 50/4 = 12.5$; hence, the shear-deflection factor, by interpolation, is 1.35. Thus, the maximum deflections become:

$$\delta = (1.35)(0.11) = 0.15 \text{ in.}$$
$$\Delta = (1.35)(0.77) = 1.04 \text{ in.}$$

The deflection-span ratio for $\Delta = 1.04$ in. and $L = 600$ in. is 1/577, well within acceptable limits. In view of the small deflections of the beam, camber is considered unnecessary.

10.3.2.8 Connection at midspan

Upon erection of the prefabricated box-beams, the interior span behaves as a double cantilever. A connection has to be designed between the two beams to obtain a three-span continuous beam. This requires capacity in the connection to transmit bending moment and shear force. From Fig. 10.8, the bending moment M_o and shear force V_o due to the action of $w_2 = 0.688$ k/ft are:

$$M_o = (62.5)(0.688) = 43.0 \text{ k-ft.} = 516 \text{ k-in.}$$
$$V_o = 0 \qquad (10.13)$$

It is necessary to design the connection to resist the maximum possible moment and shear. In this regard, the moment at the connection is a maximum when the live load is placed exclusively on the interior span. For this case it can be shown that the moment due to live load is $(0.075)\ w_3L^2$, where $w_3 = 16 \times 0.03 = 0.48$ k/ft. is the live load only. For maximum shear at the connection, the live load should be placed in the interior span at one side of the connection, and also in the exterior span at the other side of the connection. It can be shown that for this condition, V_o is given by $(19/96)\ w_3L$. Thus, the design values of M_o and V_o for the given connection can be determined as follows:

$$M_o = (62.5)(0.688 - 0.48) + (0.075)\,(0.48)\,(50)^2$$
$$= 13 + 90 = 103 \text{ k-ft.} = 1{,}236 \text{ k-in.} \qquad (10.14)$$

$$V_o = \left(\frac{19}{96}\right)(0.48)(50) = 4.75 \text{ k}$$

*This is conservative since little creep may occur in box-beams that are made of seasoned lumber and plywood.

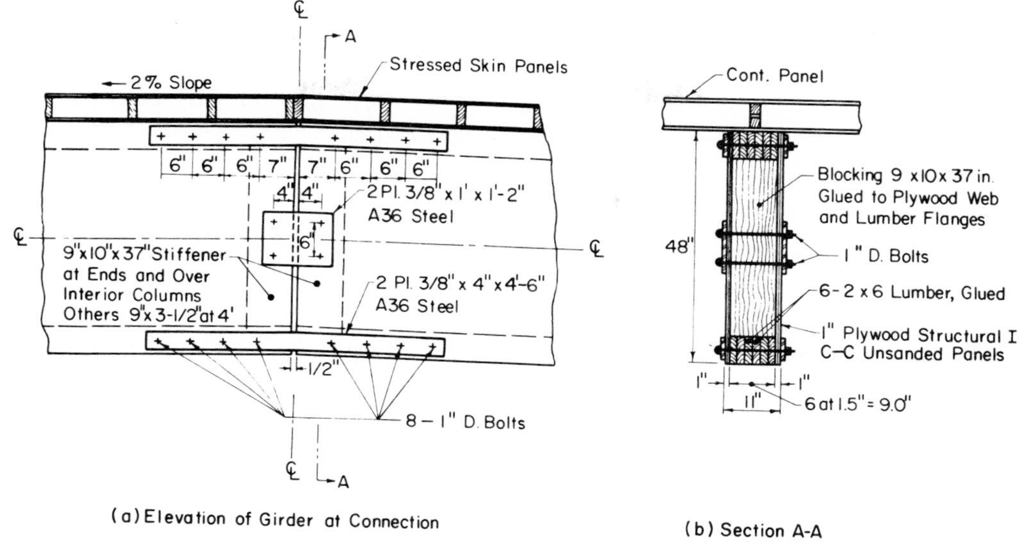

(a) Elevation of Girder at Connection

(b) Section A-A

Figure 10.9. Combined moment and shear connection between prefabricated box girders.

The connection between the beams is shown in Fig. 10.9. The top plates provide the moment capacity; the web plates provide the shear capacity. The force in the flange plates is $M_o/(\text{lever arm}) = 1236/(48 - 5.5) = 29.1$ k, and the force in the web plates is $V_o = 4.75$ k. The number of bolts necessary to transmit these forces may be determined as follows. For 1 in. dia. bolts with $L/D = 11$ and Group 3 wood, Table A.6.1 allows $P = 5.25$ k and $Q = 2.81$ k. There is a 25 percent increase allowed in P only due to metal plates, and a 15 percent increase in both P and Q, due to the snow loading condition. Hence, $P = 1.15 \times 1.25 \times 5.25 = 7.55$ k and $Q = 1.15 \times 2.81 = 3.23$ k. Let N_f and N_w be the required number of bolts at each side of the connection in the flange and in the web, respectively. Then, $N_f = 29.1/7.55 = 3.85$, use 4 bolts; and $N_w = 4.75/3.23 = 1.47$, use 2 bolts.

The above values neglected to consider the modification factor for multiple fasteners, K, that reduces the capacity of the single connector when more than two are used in a row. Therefore, it is reasonable to keep $N_w = 2$ and to increase N_f from 4 to 5, see Fig. 10.9. If this is done, the capacity of the 5 bolts in the flange plate can be determined, using Eq. 6.2, as (5) (K) (7.55), where K is the modification factor corresponding to the conditions shown in Fig. 10.9. The value of K can be obtained from Table A.6.26, as follows. For the determination of the cross-sectional area of the main wood member, A_1, consider it made up of $6 - 2 \times 6$ plus the corresponding portions of plywood web. Thus $A_1 = 6 \times 1.5 \times 5.5 + 2 \times 0.446 \times 5.5 = 50.0$ in.^{2}, consider two steel plates $3/8 \times 4$ in. for the top and bottom flanges of the connection; thus, $A_2 = 2 \times 3/8 \times 4 = 3$ in.^{2} and $A_1/A_2 = 50/3 = 16.7$. For the preceding values of A_1/A_2, A_1, and $N_f = 5$, Table A.6.26 gives $K = 0.90$. Therefore, the capacity of the 5 bolts in the flange plate is $5 \times 0.90 \times 7.55 = 33.9$ kip which is greater than the 29.1 kip imposed by the loads. Figure 10.9 shows details of the connection, including such provisions as spacing of the bolts and their end and edge distances.

10.3.2.9 Web stiffeners

The beam is subjected to uniformly distributed load all along its length except at the supports, where it is subjected to concentrated reactions. The latter are 40.77 k and 14.74 k, at the interior and exterior columns, respectively, as determined in the section on maximum effects on the beam. For the given wood of the flanges, the allowable stress in compression across the grain is $F_{c\perp} = (0.405)(1.15) = 0.466$ ksi. Thus, the necessary areas for the bearing stiffeners over the interior and exterior columns are 87.6 in.^{2} and 31.6 in.^{2}, respectively. Using the 9 in. width between plywood webs, the minimum length in the direction of the span for the bearing stiffeners is 10 in. for the interior support and 3.5 in. for the exterior support.

The stiffener at the interior end of the prefabricated beam, see Fig. 10.9, is of the same size as the one over the interior column. The reason for this is to facilitate the required shear connection between beams. Other stiffeners are used, 9 × 3½ in. in section and spaced 4 ft. apart, at the splice sections of the plywood webs.

10.3.2.10 Lateral stability of compression flange

The top flange is the compression flange of the beam for most of its length, see Fig. 10.8a. However, there is a portion of the span, at each side of the interior column where negative moment causes the bottom flange of the beam to be subjected to compression stresses. Since the top flange is fully restrained by the stressed-skin panels, which are nailed to it, only the portion of the bottom flange under compression needs some consideration.

The ratio I_x/I_y between the moments of inertia of the section with respect to the horizontal and vertical axes of bending, is a measure of the lateral stability of the compression flange, see Section 7.10. For the given beams, $I_x/I_y = 53{,}176/1{,}744 = 30.5$. For I_x/I_y between 20 and 30, Table 7.4 requires that one edge of the beam, either top or bottom, be held in line. This is true for both edges of the top flange. Since the actual ratio I_x/I_y of the beam is barely above 30, bridging or bracing of the bottom flange may be considered unnecessary.

10.3.3 Design of an Interior Column

The building columns, as those of Section 10.2.4, may be considered braced against relative lateral displacements of the ends. The staggered stressed-skin panels of the roof system provide a continuous diaphragm action that transmits any lateral forces on the structure to the rigid perimetric walls of the building. The columns are solid timber sections, S4S wood having allowable stresses as follows: $F_b = 1.65$ ksi, $F_c = 1.05$ ksi, and $E = 1{,}600$ ksi.

The interior column of the building supports the box-girder at its continuous end. The column is subjected to a load equal to $P = 41.22$ k, which is transmitted to it by direct compression between column and box-beam. The length of the column is $20 + 0.02 \times 50 = 21$ ft. from finished floor elevation to bottom of box-girder. Consider a section 10 × 12, 9.5 × 11.5 in. actual, as a possible solution. The dimensions of the column are quite small compared to those of the box-beam; thus, the stiffness ratio of column to beam at the joint, is quite small. This implies that, even in the case of a rigid connection, the moment transmitted to the column is small. A thorough analysis of the continuous structure is an unnecessary refinement at this stage.

It is conservative to design the column assuming that it is hinged at both ends about both axes, i.e. $k_1 = k_2 = 1$, and that it is subjected to end moments created by an eccentricity of the load given by the maximum of 1 in. or 0.1 times the side of the column (10), see section 8.4.

Hence, for the 10 × 12 in. section, bending is caused by an eccentricity of 0.1 × 11.5 = 1.15 in. about the strong axis and 1 in. about the weak axis. Consider the latter first. The section properties are A = 109 in.^{2}, I = 822 in.^{4}, S = 173 in.^{3}. Also, $k_2L/b = 21 \times 12/9.5 = 26.5$ and $K = 0.671\sqrt{E/\bar{\alpha}F_c} = 0.671\sqrt{1600/1.15 \times 1.05} = 24.4 < 26.5$. Since $k_2 \ell/b$ is greater than K, the section renders long-column behavior and $F_{c'} = 0.3\,E/(L/b)^2 = (0.3)(1600)/(26.5)^2 = 1.684$ ksi and J = 1.0.

The design bending stress, $F_{b'}$, is found by using $\ell_e = 1.84\ell = 1.84\,(21 \times 12) = 464$ in., see Section 7.15. Thus, $C_s = \sqrt{\ell_e d/b^2} = \sqrt{(464)\,(9.5)/11.5^2} = 5.77$. Since C_s is less than 10, and the size factor $C_F = 1.0$, the design bending stress is given by $F_{b'} = \bar{\alpha}\,F_b = 1.15\,(1.65) = 1.90$ ksi. For eccentric end load only the interaction formula given by Eq. 8.12 is reduced to:

$$\frac{\dfrac{P}{A}}{F_{c'}} + \frac{\left(\dfrac{P}{A}\right)\left(\dfrac{6e}{d}\right)(1 + 0.25\,J)}{F_{b'} - J\dfrac{P}{A}} \leq 1.0$$

Thus,

$$\frac{\dfrac{41.22}{109}}{0.684} + \frac{\left(\dfrac{41.22}{109}\right)\left(6 \times \dfrac{1.0}{9.5}\right)(1 + 0.25)}{1.90 - \left(\dfrac{41.22}{109}\right)} = 0.749 < 1$$

For bending about the strong axis, the effective slenderness if given by $k_1L/d = 21 \times 12/11.5 = 21.9 < 24.4$. Since k_1L/d is less than K but larger than 11, it follows that $F_{c'} = \bar{\alpha}\,F_c\,[1 - (1/3)\,(kL/d/K)^4] = 1.15\,(1.05)\,[1 - (1/3)\,(21.9/24.4)^4] = 0.946$. Also, $C_s = \sqrt{(464)\,(11.5)/(9.5^2)} = 7.69 < 10$ $F_{b'} = 1.15\,(1.65) = 1.90$ ksi and $J = (21.9 - 11.0)/(24.4 - 11.0) = 0.813$. Thus,

$$\frac{\dfrac{41.22}{109}}{0.946} + \frac{\left(\dfrac{41.22}{109}\right)\left(6 \times \dfrac{1.15}{11.5}\right)(1 + 0.25 \times 0.813)}{1.90 - 0.813\left(\dfrac{41.22}{109}\right)} = 0.571 < 1$$

The section is acceptable. Other solutions are shown in Table 10.2, including a 10 × 10 S4S section which would be less costly. This, however, may be compensated by the fact that, from a construction point of view, it is advantageous to use a 12 in. dimension for the column to facilitate the connection between it and the box-girder, see Fig. 10.10. There is no need for a base plate at the connection, since the compression across the grain for the lumber flanges of the box-girder, equal to 41.22/9.5×11.0=0.395 ksi, is less than the allowable 0.405 × 1.15 = 0.466 ksi for the given stress-graded wood. The two side plates shown in the figure connect the two members by means of 4 in. shear plates and ¾ in. bolts. The connection of the column to its foundation pedestal may be similar to that shown in Fig. 10.5. Both connections provide small stiffness for the transmission of bending moments to the column; thus, for this case, the initial design assumption (the column is considered as a double hinged member) is good.

An additional exercise would be to complete this problem by designing the typical exterior column (considering the local lateral action of wind) and its connection to the box-girder.

TABLE 10.2. TYPICAL COLUMN OF DESIGN EXAMPLE SUMMARY OF ACCEPTABLE SECTIONS.

Input Data:

$P = 41.22$ k, Minimum Eccentricity: 1 in. or 0.1 d, $L = 252$ in.

$E = 1600$ ksi, $F_c = 1.05$ ksi, $F_b = 1.65$ ksi,

Snow Loading Conditions ($\overline{\alpha} = 1.15$), Double-hinged column, braced against lateral displacements.

Solid-Sawn Surfaced Four Sides Sections

Nominal Size	Actual Size, in.	Area in.^{2}	Col. Design Equation
8 × 18	7.5 × 17.5	131	0.937
8 × 20	7.5 × 19.5	146	0.837
8 × 22	7.5 × 21.5	161	0.757
8 × 24	7.5 × 23.5	176	0.691
10 × 10	9.5 × 9.5	90	0.920
10 × 12	9.5 × 11.5	109	0.749
10 × 14	9.5 × 13.5	128	0.632
10 × 16	9.5 × 15.5	147	0.547

Rough Sections

Nominal Size	Actual Size, in.	Area in.^{2}	Col. Design Equation
8 × 14	8 × 14	112	0.986
8 × 16	8 × 16	128	0.857
8 × 18	8 × 18	144	0.758
8 × 20	8 × 20	160	0.680
8 × 22	8 × 22	176	0.616
8 × 24	8 × 24	192	0.563
10 × 10	10 × 10	100	0.753
10 × 12	10 × 12	120	0.620
10 × 14	10 × 14	140	0.527

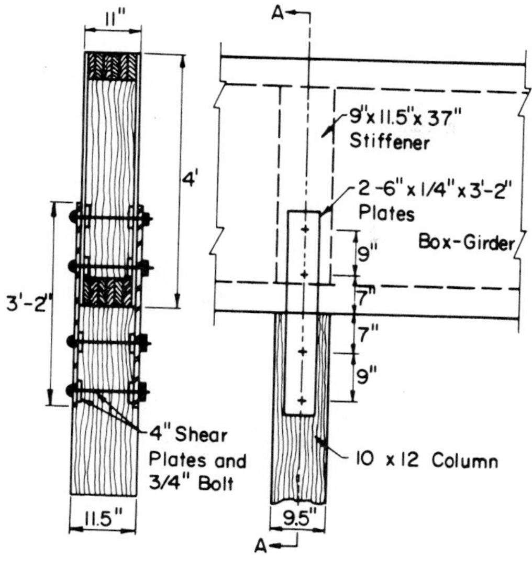

Figure 10.10. Connection details between box-girder and interior column.

10.4 DESIGN OF WOOD HOUSES

Residential construction in the United States is a major user of wood products in all shapes and forms. Wood-framed houses constitute by far the most numerous group of dwellings in this country and their overall soundness and durability have been thoroughly established.

A review of the literature in construction of wood houses shows two outstanding contributions by Anderson (19) and Dietz (20) to which the reader is referred for thorough coverage of the subject matter. In this section only the major aspects of home framing that relate to structural design are considered. Thus, the following important components of a wood-house structure are discussed: 1) foundation walls, 2) floor framing 3) wall framing and sheathing, and 4) ceiling and roof framing including sheathing.

10.4.1 Foundation Walls

Basements and crawl spaces in wood houses require perimeter walls that are usually built of masonry or concrete. Treated wood, otherwise referred to as "All-Weather Wood Foundation", has also been found acceptable for this task, see Section 11.4.

It is beyond the scope of this work to give details of concrete and masonry walls. However, it is emphasized that steel reinforcement may be necessary to prevent unsightly cracking, inwards bulging, and even failure of the walls if the presence of lateral pressure (generated for example by the combined action of saturated soil and a hydraulic head) could exceed the strength of the unreinforced wall. The concept of "equivalent fluid pressure" for the subsoil discussed in Section 11.4 can be applied to examine this condition.

Design may consider the walls laterally restrained at the top by the floor structure and subjected to vertical loads caused by the weight of floors, walls, roof and live loads. Backfilling against the walls should not be permitted until the floor structure is in place.

For added protection when wet soil conditions exist, a waterproof membrane of roofing felt or other material should be applied; hot tar or hot asphalt is often used over the membrane. Placement of a good drainage system around the perimeter of the outside walls down to the foundation should help prevent built-up of hydraulic pressures against the walls, damp basements, and wet floors. Clay or concrete draintile 4 in. in diameter and 12 in. long may be placed on top of a 2 in. gravel bed at foundation level draining toward a ditch or into a sump where the water can be pumped to a storm sewer (19).

To anchor the floor system to the foundation walls it is necessary to provide a perimetral sill plate. In highwind and storm areas where tornadic winds may cause so much havoc, well anchored sill plates are very important. Anchor bolts (½ in. D. minimum) should be embedded 18 inches or more in concrete walls and in block walls and spaced not farther than 4 ft. apart. In block walls the core of the blocks should be filled with concrete with a large plate washer used at the head end of the anchor bolts. A sill sealer is often used under the sill plate to smooth-out any irregularities of the top surface that may have occurred during casting of the concrete walls.

The foundation wall must include a supporting ledge about 5 in. wide if a masonry-veneer face is required for the outside walls. See Fig. 10.11. Note such important construction details as the base flashing to be lapped with sheathing paper, weep holes to provide drainage of any water-vapor condensation occurring in the 1 in. space between the masonry and the sheathing, and corrosion-resistant metal ties (spaced 32 in. apart horizontally and 16 in. apart vertically) to tie the brick veneer to the framework.

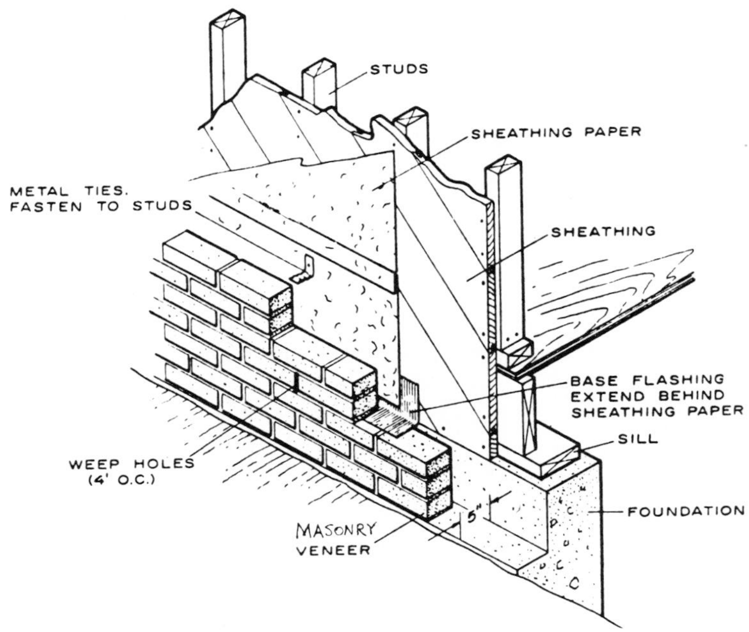

Figure 10.11. Masonry veneer face for wood-frame wall F. from Ref. 19.

10.4.2 Floor Framing

The floor framing for a typical wood house is shown in Fig. 10.12. It consists of subfloor, supporting joists and bridging, girders, and masonry piers or steel and wood posts.

The subfloor forms a working platform and a base for the finished flooring. It could be made of either 8 in. wide × ¾ in. thick square-edge or tongue-and-groove boards or ½ in. to ¾ in. plywood panels. Boards may be applied either diagonally or at right angles to the joist. Diagonal subflooring permits finish flooring to be laid either parallel, or at right angles to the joists; however, when subflooring is placed at right angles to the joists the finish flooring should be placed at right angles to the subflooring. Subflooring is generally attached to each joist using two 8 d nails for board widths smaller than 8 in. and three 8 d nails for 8 in. wide boards. Joist spacing is normally 16 in. on center, although 24 in. spacing may be used.

Plywood suitable for subflooring is standard sheathing, Structural I and II and C-C exterior grades panels. When Douglas Fir and Southern Pine plywood is used only ½ in. thick material is required assuming that the joists are spaced 16 in. apart. For such species as Western Hemlock, Western White Pine and Ponderosa Pine, ⅝ in. thick plywood panels are required. Installation should place the face grain of plywood panels at right angles to the supporting joists, staggering of adjacent panels is required for continuity by allowing end joists to occur over different joists. Nailing of plywood panels should take place at each supporting joist by means of 8 d common or 7 d threaded nails spaced 6 in. apart along the panel edges and 10 in. apart elsewhere.

Wood joists are generally 2 × 8's, 2 × 10's, or 2 × 12's (nominal dimensions) spaced 16 in. apart. Size depends on loading, length of span and the species and grade of lumbers used. Usually, the grades for joists are "Standard" for Douglas Fir, "No. 2 or No. 2 KD" for Southern

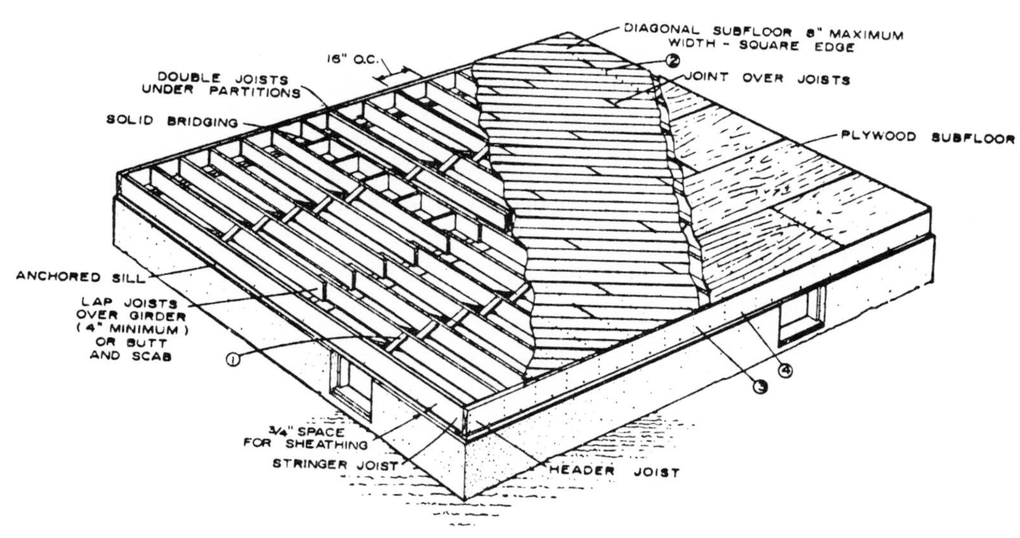

Figure 10.12. Typical floor framing. Note nailing of 1) bridging to joists, 2) board subfloor to joists, 3) header to joists and 4) toenailing header to sill. From Ref. 19.

Pine and comparable grades for other species (19). Strength alone is not sufficient for design, as adequate stiffness to control deflections and limit vibrations under moving loads is very important. A line of cross-bridging between joists for every span is required for lateral stability of the joists before the subflooring is placed, and for lateral distribution of vertical loads to adjacent joists under service conditions.

Wood girders or standard I or W steel beams are generally used to support floor joists. Wood girders usually are of the built-up type consisting of two or more members of 2-inch dimension lumber spiked together. For a two-piece girder Anderson (19) suggests 10 d nails spaced 16 in. apart, driven stagger fashion with two additional nails at each end. Solid wood girders are usually less dry and, thus, are less dimensionally stable. Commercially available glue-laminated beams in one piece are made using dry wood and couid be very desirable for their appearance where exposed. Not less than 4 in. of bearing on the masonry walls or pilasters should be provided for the ends of wood girders. A niche allowing at least ½ in. of air space separating the wood girder from the masonry is required; an end bearing steel or neoprene plate should be provided to separate the wood beam from the masonry. The bearing plate should be separated from the edge of the wall at least one inch to avoid spalling of the masonry.

To support the floor girders in house basements round steel pipes and 6 × 6 in. wood posts have been frequently used. Masonry pier have been employed in houses with crawl-spaces. Proper connection to both the overhead girder and the supporting foundation, by means of steel angles and bolts, is required for steel pipes and wood posts. Under no circumstances should these columns be left loose and unattached.

10.4.3 Wall Framing and Sheathing

Wall Framing includes primarily vertical studs and horizontal members such as soleplates, top plates and window and door headers. Usually, wall studs are 2 × 4's spaced 16 in. apart, top plates and soleplates are also 2 × 4's while window and door headers consist of pairs of 2 × 6's or deeper members depending on the span of the opening. Wood use for wall framing should be at about 15 percent moisture content and, certainly, at not more than 19 percent.

There are two general ways of wall construction, namely, platform and balloon, see Figs. 10.13 and 14 respectively. The main difference between the two is at the floor line. The wall frame in platform construction is erected above the subfloor "platform" which extends to all edges of the building. The balloon wall studs, on the contrary, extend from the sill of the first floor to the top plate or end rafter of the second floor.

Platform construction allows use of "tilt-up" of wall sections. After erection, plumbing and bracing of the wall studs and sole plates the remaining nailing is completed. Sole plates are nailed to the floor joists and headers or stringers (through the subfloor), floor and window headers are nailed to adjoining studs and plates, and corner studs are nailed together. To improve the uplift anchorage of this type of construction in hurricane areas, or areas with high winds, it is necessary to fasten the wall studs and floor framing to the anchored foundation sill, particularly when wall sheathing itself is not nailed to the sill. For this purpose, light gage steel straps are used. One end of the strap is placed under the anchored sill and the other end is attached, through three 8 d nails to a wall stud. Because of its simplicity platform construction is more often used than balloon type construction (19).

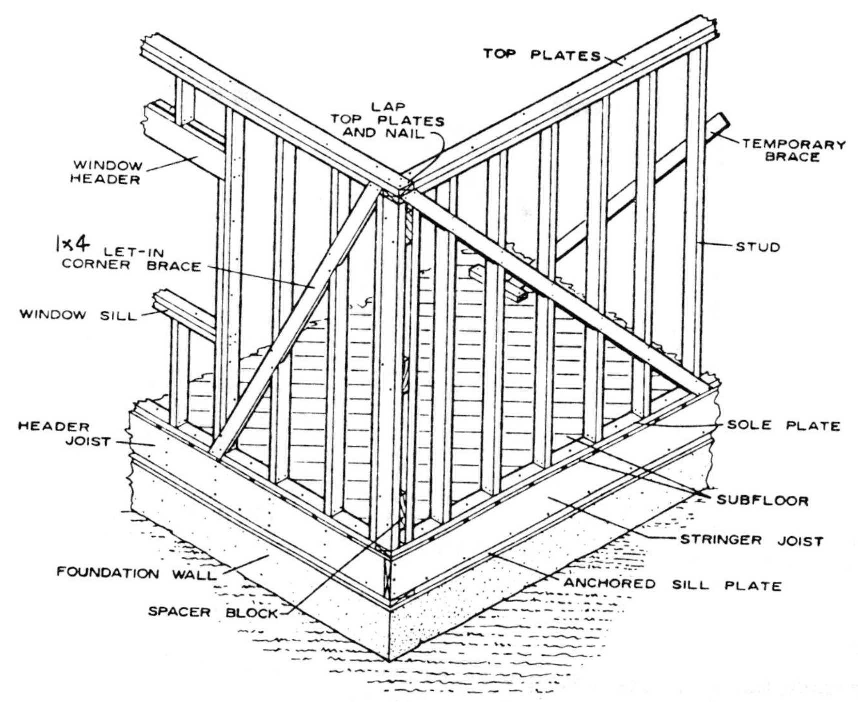

Figure 10.13. Platform construction. Lower wall studs bear on joist and subfloor platform and are not continuous with upper wall studs. From Ref. 19.

In balloon type construction the wall studs and floor joists are nailed together using 3 d nails. Both studs and joists rest on the anchored sill to which they are toenailed by means of 8 d nails. Most building codes require firestops in the form of 2 × 4 in. blocking placed between the studs to prevent the spread of fire through the open wall passages. The ends of the second-floor joists bear on a 1 × 4 in. ribbon, see Fig. 10.14, that has been let into the studs and are attached to the latter with 4 d nails at these connections. Balloon framing is preferred over platform construction in full-two story brick or stone veneer houses because there is less potential shrinkage of the exterior walls.

Wall sheathing is the outside cover of the house frame. The most common types of sheathing are boards, plywood, structural insulating board, and gypsum sheathing. Boards 6, 8 and 10 in. wide of nominal 1 in. thickness (and moisture content at or below 15 percent) may be applied either horizontally or diagonally. In the latter case if 45° diagonal sheathing were carried down to the anchored sill plate and nailed to it, greater strength and stiffness would result. Also used extensively for wall sheathing are 4 × 8 ft. plywood panels, ⅜ in. and thicker, applied vertically. Insulating boards are structural fiberboards coated or impregnated with asphalt or given other treatment to make them a water resistant product; availability is in three density-related categories namely: regular density, intermediate-density and nail-base. The first type is manufactured in ½

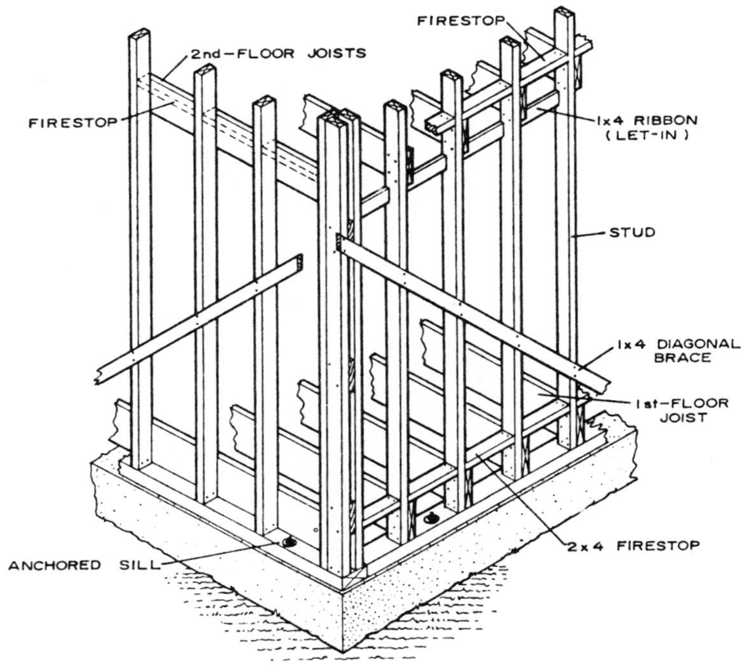

Figure 10.14. Balloon construction. Lower wall studs bear directly on anchored sill and are continuous with upper wall studs. From Ref. 19.

in. and 25/32 in. thickness, the last two only in ½ in. thickness; panel sizes are 4 × 8 ft. and 4 × 9 ft. Gypsum sheathing is composed of treated gypsum filler, faced on two sides with water-resistant paper, and with matched groove and V edges that make application easier.

Wall sheathing may be used to provide strength and stiffness to the house against racking induced by wind or seismic motions. Types of sheathing that provide adequate bracing and that need not have the corner bracing shown in Figs. 10. 13 and 14 are: 1) diagonal wood sheathing attached to each wall stud using three 8 d nails per 8 in. wide board; 2) 4 × 8 ft. plywood panels applied vertically and attached to the wall studs with 8 d nails or staples spaced 6 in. apart on all edges and not more than 12 in. elsewhere; 3) 4 × 8 ft. structural fiberboards panels (25/32 in. thick if regular grade, ½ in. thick intermediate-density or nail-base grade) applied with long edges vertical and attached to the wall studs with 1¾ in. roofing nails or staples spread 3 in. apart along all edges and 6 in. apart elsewhere; and 4) a combination of ½ in. thick 4 × 8 ft. plywood panels at each side of each outside corner of the house and ½ in. regular-density fiberboards elsewhere, all attached to the wall studs as previously described.

10.4.4 Ceiling and Roof Framing Including Sheathing

There are two basic types of house roofs, see Fig. 10.15, namely: 1) flat or slightly pitched roof in which roof and ceiling supports are furnished by one set of joists and 2) pitched roofs where both ceiling joists and rafters or trusses are required.

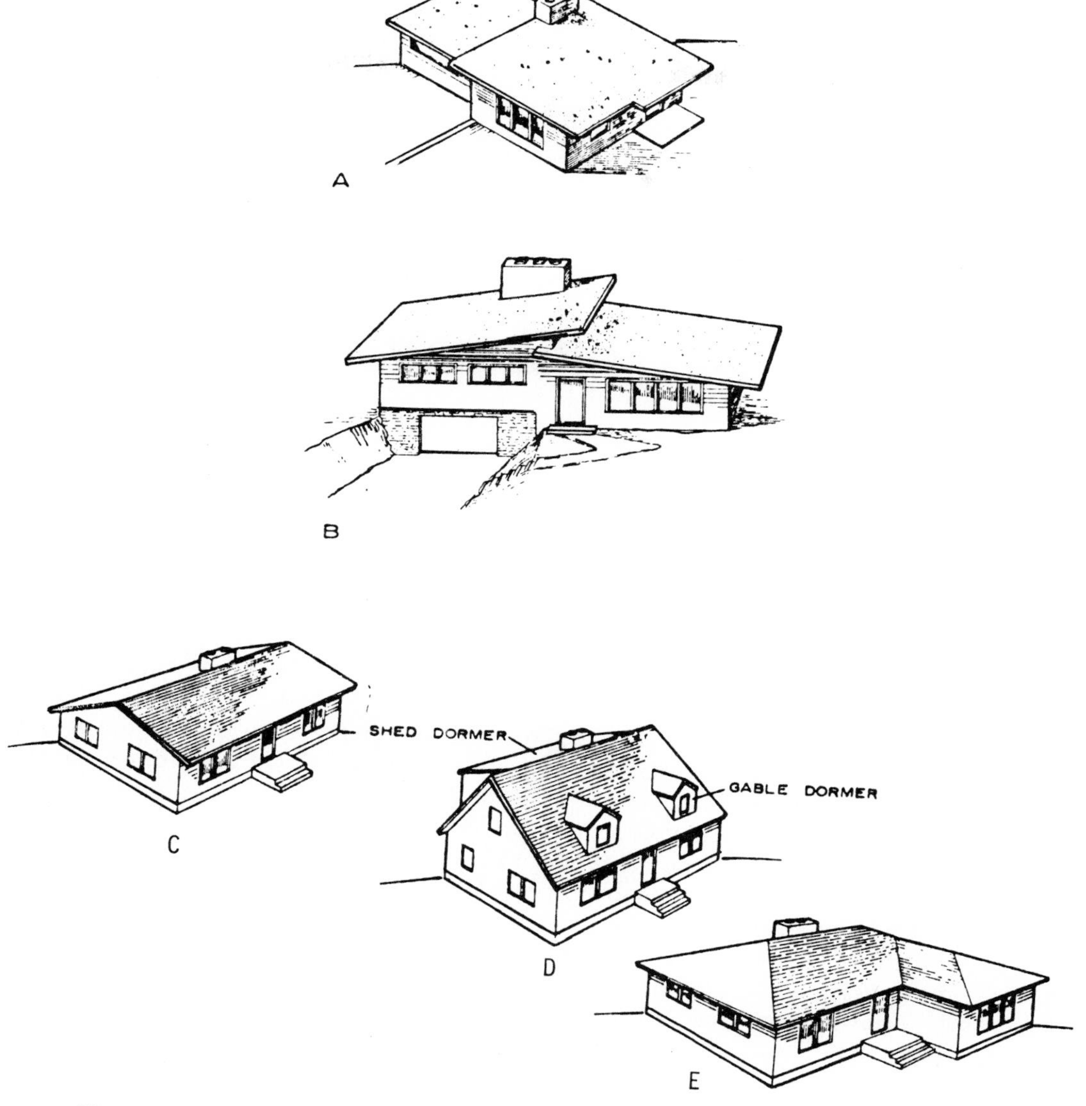

Figure 10.15. Various types of roofs: A, flat; B, low-pitched; C, Gable; D, Gable with Dormers; E, hip. The first two types need only roof joists while the others need also ceiling joists to support attic space or second floor. Adapted from Ref. 19.

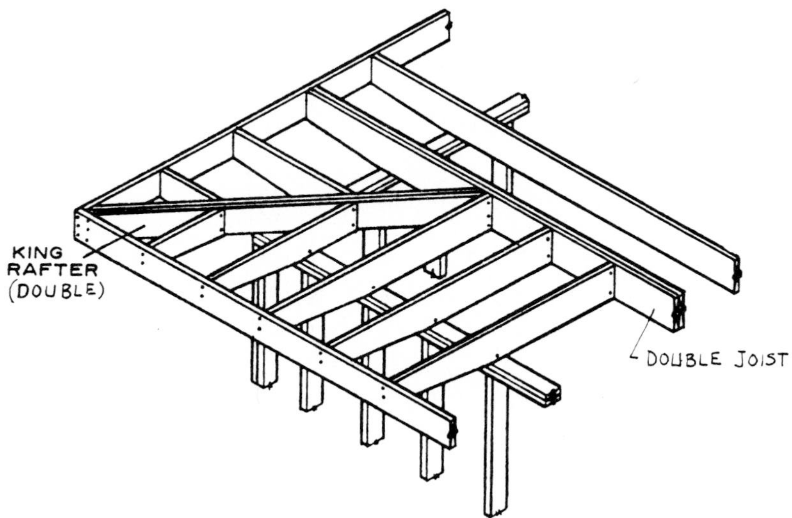

Figure 10.16. Roof rafters for flat or low-pitched roof with side and end overhangs. From Ref. 19.

Roof joists for flat roofs are commonly laid level or with a slight pitch. They are used to support the roof sheathing and roofing materials on top, and a hung ceiling underneath. When the house design involves cantilevering the roof beyond the exterior walls, lookout rafters are ordinarily used, see Fig. 10.16.

A typical design detail for framing of pitched roofs is shown in Fig. 10.17. Rafters are usually precut to length with proper angle cut at the ridge and eave, and with notches provided for the top plates; they are erected in pairs and placed against the ridge board only after the ceiling joists have been securely nailed to the plates of the interior and exterior wall frames. The reason for this sequence is, of course, to avoid having the thrust of the rafters push the exterior walls out.

Ceiling joists are principally used to support ceiling finishes. However, they may act as floor joists for second floors or attics and, as mentioned before, they tie together opposite exterior walls and interior partitions. Details of attachment of ceiling joists are shown in Fig. 10.18. Note the presence of metal straps attaching wall studs to ceiling joists. Since the latter are nailed to the roof rafters this detail indirectly serves to anchor down the roof system and increases the capacity of the roof against uplift forces caused by high winds.

Trusses are also frequently used as principal roof-supporting elements. Trusses are made of individual members arranged in triangular cells and connected together by various means which include plywood gussets (nailed, glued or bolted in place) or steel gusset plates of the type described in Section 6.5. Long-span trusses may be assembled using split-ring connectors and bolts, see Sections 6.4 and 11.2. For residential application trusses are usually spaced 2 ft apart and are designed to span from one exterior wall to the other with lengths varying from 24 to 40 ft. Since no interior bearing walls are required, greater architectural flexibility is possible for interior planning and partitions can be placed without regard to structural considerations.

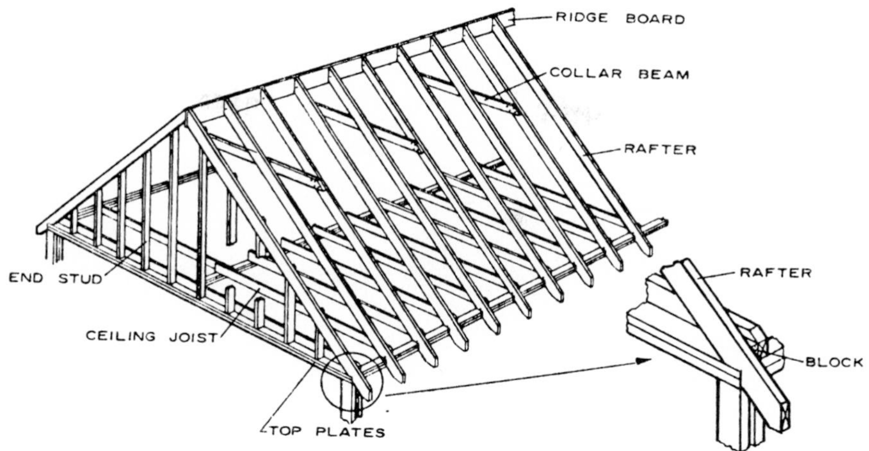

Figure 10.17. Ceiling and roof framing for pitched roofs. From Ref. 19.

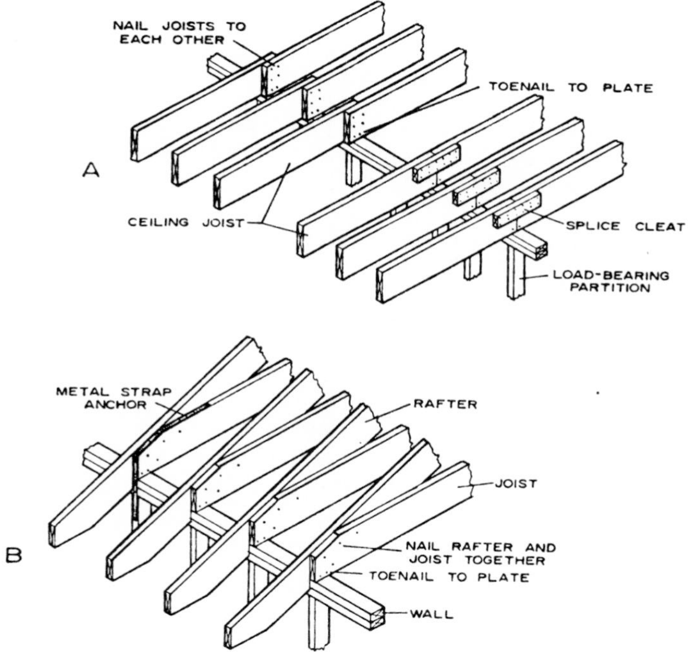

Figure 10.18. Ceiling joist connections. A, at interior partition with joists lapped or butted; B, at outside wall with joists nailed to rafters. From Ref. 19.

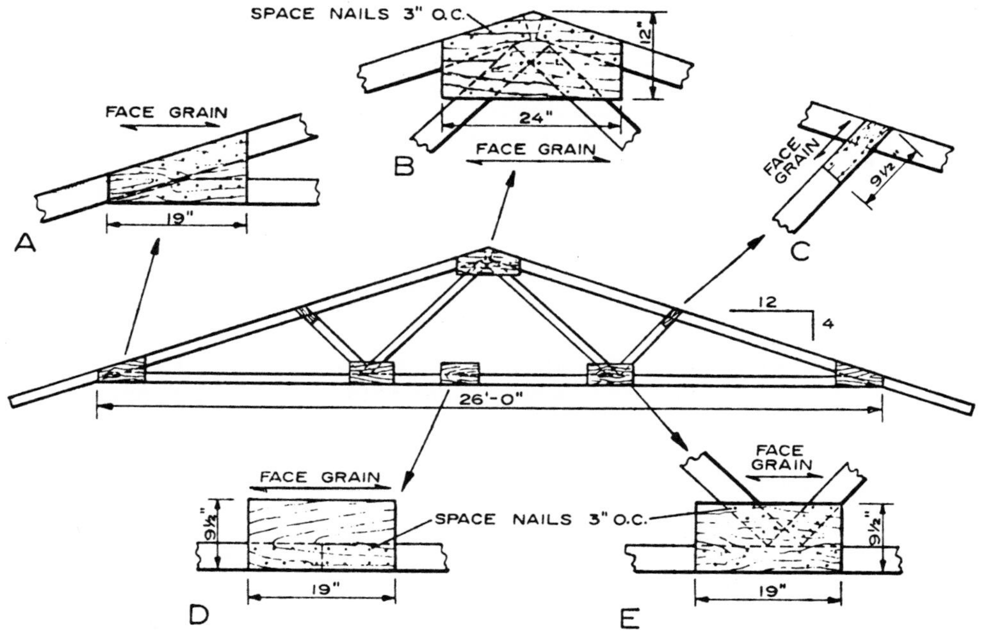

Figure 10.19. Design details for a 26-ft. span wood truss with nailed-glued plywood gusset connections. A, bevel heel gusset; B, peak gusset; C, upper chord gusset; D, splice of lower chord; E, lower-chord gusset. From Ref. 19.

An example of a roof truss often used in house construction is given in Fig. 10.19. Plywood gussets made of ½ in. thick standard plywood with exterior glue line or exterior sheathing grade plywood are nailed with 6 d nails and glued to both sides of the connections. Resorcinol glue is recommended where high relative humidity in the attic is likely, as in the southern and southeastern states; in dry and arid areas a casein or similar glue may be acceptable.

It is important, whether the roof structure is to be made out of rafters or trusses, that it be designed to sustain not less than 20 pounds per square foot of live load, in areas where no snow accumulates, and at least 30 pounds per square foot in the snow-belt states. More may be required in particular regions of the country and under certain conditions of use; the reader is referred to ANSI's Building Code Requirements for Minimum Design Loads in Buildings and Other Structures" (21) for more details.

Plywood sheathing is commonly used in house construction to span over the roof rafters or trusses. Standard sheathing-grade plywood laid with face grain perpendicular to the rafters or trusses is generally employed. To achieve racking stiffness and diaphragm action it is necessary that the plywood panels be staggered; end joints made over the center of rafters or trusses should be staggered at least 16 or 24 inches, and preferably more. To provide better penetration for nails used to attach roof shingles, plus better racking resistance and a smoother roof appearance, Anderson (19) advises using a minimum ⅜ in. thickness for plywood panels. For slate and similar heavy roofing materials ½ in. thick plywood is considered minimum for 16 in. rafter spacing.

Other wood materials are used for roof sheathing. Thus, decking consisting of 2 in. and thicker tongue and grooved planks spanning between 6 and 12 ft is often used for flat or low-pitched roof in post and beam construction, see Section 10.2 and 11.7. Fiberboard roof decking is also used except the supports are much closer together. Thus, 2 × 8 ft planks 1½, 2, and 3 in. thick should be supported by joists spaced not farther apart than 24, 32, and 48 in., respectively. Combined use of plywood and lumber, such as stressed-skin panels, see Section 10.3, is also employed for roof sheathing in wood buildings, especially when the supporting beams are spaced more than 12 ft apart.

REFERENCES

1. "National Design Specification for Stress-Grade Lumber and Its Fastenings," 1977 Edition. National Forest Products Association, Washington, D.C.
2. American Institute of Timber Construction, "Standard Specifications for Structural Glued-Laminated Timber," AITC 203–70, AITC, Englewood, Colorado 80110.
3. American Institute of Timber Construction, "Timber Construction Manual," Second Edition, John Wiley and Sons, New York, 1974.
4. American Plywood Association, "Plywood Design Specification," by APA, 1119A St., Tacoma, Washington, 98401, 1966.
5. American Plywood Association, "Plywood Design Specification. Supplement No. 3. Design of Flat Plywood Stressed-Skin Panels," by APA, 1970.
6. American Plywood Association, "Plywood Design Specification. Supplement No. 2. Design of Plywood Beams," by APA, 1968.
7. Plywood Fabricator Service, Inc., "Stressed-Skin Panels," by American Plywood Association, May 1968.
8. American Institute of Steel Construction, "Manual of Steel Construction," Seventh Edition, AISC, New York, 1970. See Beam Diagrams and Formulas.
9. Forest Products Laboratory, "Wood Handbook," U.S. Department of Agriculture Handbook No. 72, U.S. Government Printing Office, Washington, D.C. Revised August, 1974.
10. Gurfinkel, German, "Design of Wood Columns for Flexural and Axial Loading," Journal of the Structural Division, ASCE, Vol. 96, No. ST11, Nov. 1970, pp. 2389–2408.
11. Rhude, M. J., "Structural Timber Framing in U.S. Forest Products Lab," Civil Engineering, ASCE, Vol. 37, No. 8, August 1967, pp. 42–44.
12. Stadelmann, D. H., "A Structural Engineer's View of Laminated Wood Systems," Wood Preserving, Vol. 49, No. 4, April 1972, pp. 5–12.
13. Burns, Bennet W., "Glued-Laminated Timber Frame for a Crane Building," Civil Engineering, ASCE, May 1961, pp. 40–41.
14. Howard, H. S., "Structure: An Architect's Approach," McGraw-Hill Book Co., 1966, See Chapter 4.
15. Countryman, D., "Evolution of the Use of Plywood for Structures," Journal of the Structural Division, ASCE, Vol. 93, No. ST2, Proc. Paper 5166, April 1967, pp. 13–24.
16. Page, W. D., "Designing with Plywood Structural Components," Journal of the Structural Division, ASCE, Vol. 93, No. ST2, Proc. Paper 5167, April, 1967, pp. 25–31.
17. Carney, J. M., "Hurricane-resistant Plywood Construction," Civil Engineering, ASCE, Vol. 36, No. 6, June 1966, pp. 57–61.
18. Flint, T. R., "Plywood Quality Inspection and Testing," Journal of the Structural Division, ASCE, Vol. 93, No. ST2, April 1967, pp. 71–74.
19. Anderson, L. O., "Wood-Frame House Construction," Agriculture Handbook No. 73, U.S. Department of Agriculture, Washington, D.C., July 1970, 223 pages.
20. Dietz, Albert G. H. "Dwelling House Construction", The MIT Press, Cambridge, Massachusetts, Fourth Edition, 1974.
21. American National Standard "Building Code Requirements for Minimum Design Loads in Buildings and Other Structures" by ANSI A58.1–1972 by American National Standard Institute, 1430 Broadway, New York, N.Y. 10018.

Chapter 11
Miscellaneous Applications

11.1 ARCHES AND RIGID FRAMES

Glued-laminated wood provides the means to span large unobstructed areas with arches and rigid frames of wood; a number of typical shapes that are often used is shown in Fig. 11.1. In this country, the first application of glued-laminated rigid frames to building construction took place in 1935 for the service building of the Forest Products Laboratory (2); the building is still in use. Since then, design of these structures has extended to churches, community halls, garages, gymnasiums and other buildings where large column-free areas are necessary and desirable, see Figs. 3.2 and 3.4.

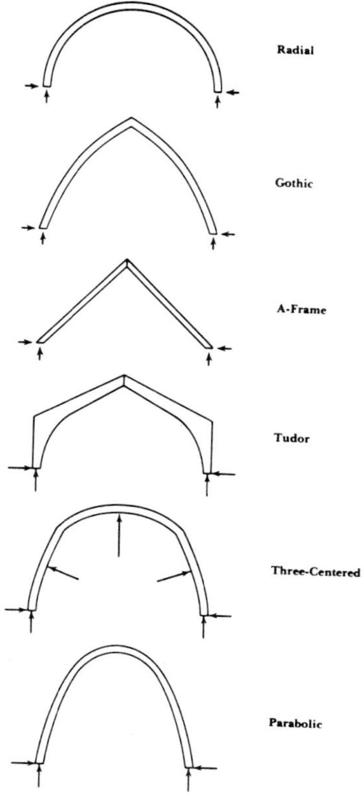

Figure 11.1. Various types of glued-laminated arches. Adapted from AITC Standard 104, see Ref. 1.

For convenience of transportation and for erection purposes many of these structures have been designed of two parts which are pin-connected (or hinged) to their supports and to each other, usually at midspan. This has a number of advantages, one of the most important being that the three-hinged structure is statically determinate and its analysis is very simple. In addition, the connections of the laminated portions to the foundations and to each other are simple because only thrust is transmitted. The foundations are also less massive than they would be if subjected to moments; the outward thrust of the legs of rigid frames, or of the ends of an arch, is taken by special steel shoes and bases anchored to a concrete base, buttress or other supporting member.

Two-hinged and fully-fixed arches or frames are less commonly used for several reasons, among which are more difficult transportation and erection procedures and cumbersome connections. Analysis of these structures using elastic theory (3) is acceptable; use of electronic computation (4) simplifies this task considerably. For those designers who are limited to conventional means of computation, various collections of formulas exist (5)(6) to simplify the work.

In the following an analysis of three-hinged structures is given with application to a design example.

11.1.1 Analysis of Three-Hinged Structures

Three-hinged arches and frames are statically determinate structures; for any given loading condition, the reactions at the supports can be determined using the equations of statics. There is no need to use the geometry of deformation of the structure.

The frame shown in Fig. 11.2 is subjected to a vertical load, w, uniformly distributed over the roof. Due to symmetry, the reactions at the supports, V and H, can be determined from $\Sigma V = 0$ and $\Sigma M = 0$ about midspan. The following is obtained:

$$V = \frac{wL}{2}$$

$$H = \frac{wL^2}{8r} \tag{11.1}$$

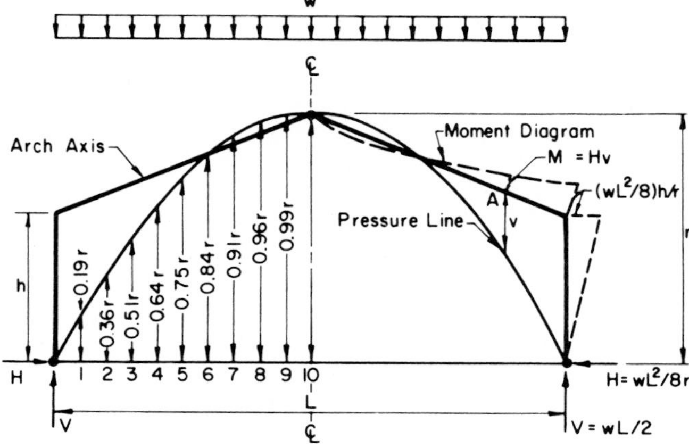

Figure 11.2. Three-hinged structure under uniform vertical load.

Once the reactions are known, the moment at any point in the axis of the frame can be easily determined. Let coordinate axes be established at the left support and let x, y be the coordinates of a given section. Taking moments of all forces to the left of the section yields:

$$M = \frac{wL}{2} \times - \frac{wL^2}{8r} y - \frac{wx^2}{2} \tag{11.2}$$

The preceding relation may be used to determine the equation of the axis of a certain arch for which no bending moments exist under uniform vertical load. Let x_p, y_p define the coordinates of the axis of such arch. Substituting x_p, y_p for x, y respectively, in the preceding equation and setting M to zero, yields:

$$y_p = 4r\left(\frac{x_p}{L}\right)\left(1 - \frac{x_p}{L}\right) \tag{11.3}$$

which is the equation of the quadratic parabola plotted in Fig. 11.2 and labeled pressure line. Physically, this parabola is the only arch configuration which, under uniformly distributed load, is not subjected to bending moment; only thrust acts on any section of this arch. Moment at any given section of an arch can be conceived of as the product of the resultant force on the section times its distance to the section; since moments are zero for the parabolic arch, it follows that the locus of the resultant force coincides with this ideal arch. The latter interpretation justifies the title "pressure line" given to this curve.

Any structure such as the rigid frame of Fig. 11.2, whose axis departs from the pressure line, is subjected to moment. The magnitude of the moment at any given section is Ff, where F is the resultant force at the given section and f is its distance to it. In the case of Fig. 11.2, it can be shown that $M = Ff = H(y - y_p)$; since H is constant for all sections, the difference in ordinates between the frame and the pressure line gives a relative measure of the distribution of moments in the frame. The points of intersection of the two configurations correspond to sections subjected to zero moment. The sections of the frame above the pressure line are subjected to negative or hogging moments (tension stresses in the outside fibers) while those below the pressure line are subjected to positive or sagging moments (tension in the inside fibers).

A similar analysis for the load shown in Fig. 11.3, which may represent, for instance, the case of snow-loading on only one half of the structure, leads to the following relations for the pressure line:

$$y_p = \frac{2r}{L} x_p \qquad \text{for } 0 \leqslant x_p \leqslant \frac{L}{2}$$

$$y_p = \frac{8r}{L^2}(L - x_p)\left(x_p - \frac{L}{4}\right) \qquad \text{for } \frac{L}{2} \leqslant x_p \leqslant L \tag{11.4}$$

where the axes of coordinates are located at the left support. Again, only an arch that has the exact configuration of the pressure line would not be subjected to bending moment under the given loading.

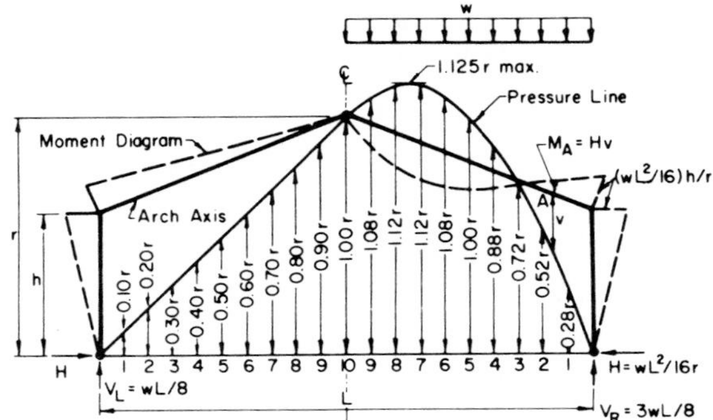

Figure 11.3. Three-hinged structure under partial uniform vertical load.

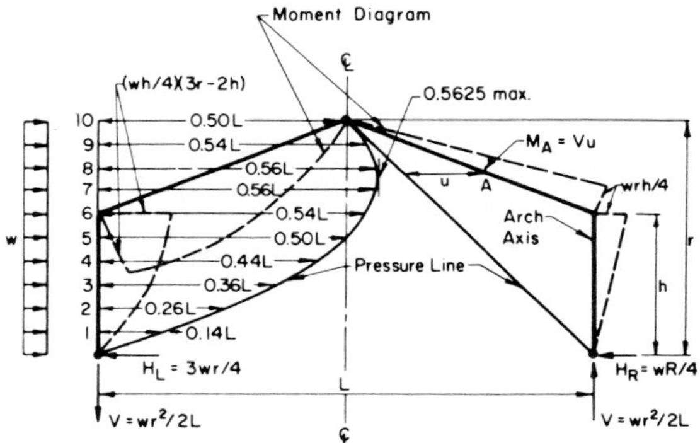

Figure 11.4. Three-hinged structure under uniform lateral load.

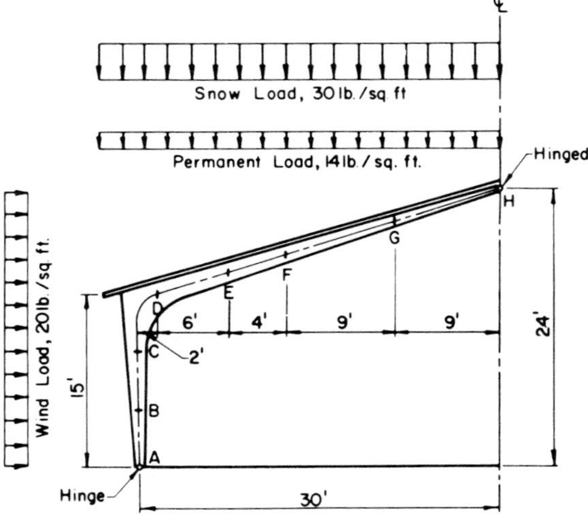

Figure 11.5. Glued-laminated three-hinged rigid frame under load.

For the case shown in Fig. 11.4, representing, for instance, the action of wind on the structure, the following relations are obtained for the pressure line:

$$x_p = \frac{L}{r^2} y_p \left(\frac{3}{2} r - y_p\right) \text{ for } 0 \leq y_p \leq r$$

$$y_p = 2r\left(1 - \frac{x_p}{L}\right) \text{ for } \frac{L}{2} \leq x_p \leq L \qquad (11.5)$$

where all notations have been defined before. The moment diagram is shown by dash lines in all three cases, see Figs. 11.2, 3 and 4.

11.1.2 Illustrative Example

Design a three-hinged frame of the characteristics and loading shown in Fig. 11.5 and the permanent loads as follows:

Weight of frame, estimated at	3.0 lb./sq. ft.
Acoustical ceiling	2.5
Roof beams	2.0
Sheathing	2.5
3-ply built-up-roof	4.0
	14.0 lb./sq. ft.

The minimum radius of the soffit of the frame is 120 in. at the knee. The frame is to be built using glued-laminated wood of variable depth. The following allowable stresses, for normal loading conditions, are to be used in the design: F_b = 2.4 ksi, F_t = 1.6 ksi, F_c = 1.5 ksi, $F_{c\perp}$ = 0.385 ksi, F_v = 0.2 ksi, and E = 1,800 ksi. The frame is under cover, thus dry conditions of use apply. Frames are spaced 14 ft apart. Loads are shown in Fig. 11.5.

The analysis of the frame follows the principles discussed previously. Three conditions of loading, as listed in Table 11.1, are considered in the analysis; the reactions for each case are given in the Table. A summary of design values at selected sections, see Fig. 11.5, is given in Table 11.2.

TABLE 11.1. HORIZONTAL AND VERTICAL REACTIONS FOR THE GIVEN FRAME.[1]

Loading Condition	V_L, kip	V_R, kip	H_L, kip	H_R, kip
Permanent plus full snow[2]	18.48	18.48	11.55	11.55
Permanent, plus snow load over left half span only	15.33	9.03	7.60	7.60
Permanent plus wind from left[3]	4.54	7.23	1.37	5.36

1. Frames are spaced at 14 ft.
2. Permanent Load = 196 lb. per ft.; Snow load = 420 lb. per ft.; Total Load = 616 lb. per ft. of horizontal projected area.
3. Wind Load = 280 lb. per ft. of vertical projected area.

TABLE 11.2. DESIGN VALUES FOR THE GIVEN FRAME.†

Section	Moment, M k-in.	Thrust, N k	Shear, Q k
A	0	18.9	10.8
B	848	18.7	10.8
C	1,694	18.6	10.8
D	1,982	20.7	0.6
E	1,344	13.4	11.7
F	826	13.0	9.3
G	313*	12.2	3.8
H	0	11.4	4.1*

†All values, except as noted otherwise, are caused by permanent loading and snow load over the full span.
*Caused by permanent loading plus snow load over half span only.

Design of these structures is a process of trial and error, which requires that the capacity of selected sections be checked for adequacy against the design values. As an example of this process consider section D for which: M = 1,982 k-in. N = 20.7 k, Q = 0.6 k; since this corresponds to the simultaneous action of permanent load and snow, the load duration factor is $\bar{\alpha} = 1.15$, see Fig. 2.14. Assume a rectangular section with the following geometric properties: b = 6.75 in., d = 28.5 in., A = 192 in.^{2}, S = 914 in.^{3}, I = 13,020 in.^{4} The size factor of the section, using Eq. 7.28 or Fig. 7.12, is $C_F = 0.908$. Since section D is in the curved section of the frame, the curvature reduction factor, C_c, see Section 7.13, applies also. For t = 0.75 in. (¾ in. laminations) and R = 120 in., Eq. 7.50 yields $C_c = 1\text{-}2{,}000\ (0.75/120)^2 = 0.922$. Thus, the allowable bending stress for the given section is

$$F_{b'} = \bar{\alpha}\ C_F\ C_c\ F_b = (1.15)\ (0.908)\ (0.922)\ (2.4) = 2.31 \text{ ksi}$$

The simultaneous action of thrust and bending moment on section D requires the use of Eq. 8.16 to check its adequacy. Since the frame is assumed continuously restrained by the sheathing it behaves as a short column and thus, J = 0 and $F_{c'} = \bar{\alpha}\ F_c = (1.15)\ (1.5) = 1.725$ ksi. Substituting the preceding values and $F_c = N/A = 20.7/192 = 0.1078$ ksi, $f_b = M/S = 1982/914 = 2.168$ ksi, e = 0 in Eq. 8.16 yields:

$$\frac{f_c}{F_{c'}} + \frac{f_b + f_c\ (6e/d)\ (1 + 0.25\ J)}{F_{b'} - J\ F_c} = \frac{0.1078}{1.725} + \frac{2.168}{2.31} = 0.0625 + 0.9385 = 1.00$$

The adequacy of the section is satisfied exactly for combined bending and thrust.

The action of the moment on the curved member produces radial stresses, the magnitude of which can be determined using Eq. 7.55. Because maximum radial stresses occur at mid-depth, the corresponding radius is R = 120 + 28.5/2 = 134.7 in. Substituting this value and M = 1,982 k-in., b = 6.75 in., and d = 28.5 in. in Eq. 7.55 yields:

$$f_r = \frac{3}{2} \frac{1{,}982}{(6.75)(28.5)(134.7)} = 0.115 < 0.385 \times 1.15 \text{ ksi} \tag{11.7}$$

The bending moment at section D is of a hogging nature, which tends to increase the initial curvature of the given member and, thus, induces compressive radial stresses. The latter are compared in the preceding equation to allowable stress in compression normal to grain times the load duration factor.

The maximum shear stress in the section may be obtained using Eq. 7.32.a. Substituting V = Q = 0.6 kip, b = 6.75 in., d = 28.5 in. yields:

$$V = \frac{3}{2} \times \frac{0.6}{6.75 \times 28.5} = 0.0047 << 0.2 \times 1.15 \tag{11.8}$$

11.1.3 Design Details

Various types of connections of light arches and rigid frames to supporting concrete foundations are shown in Fig. 11.6 which also includes a detail of support by a timber girder. Support of large, heavy arches requires true hinged connections such as the one shown in Fig. 11.7, where the bridge steel-pin that is used transmits only thrust to the foundation.

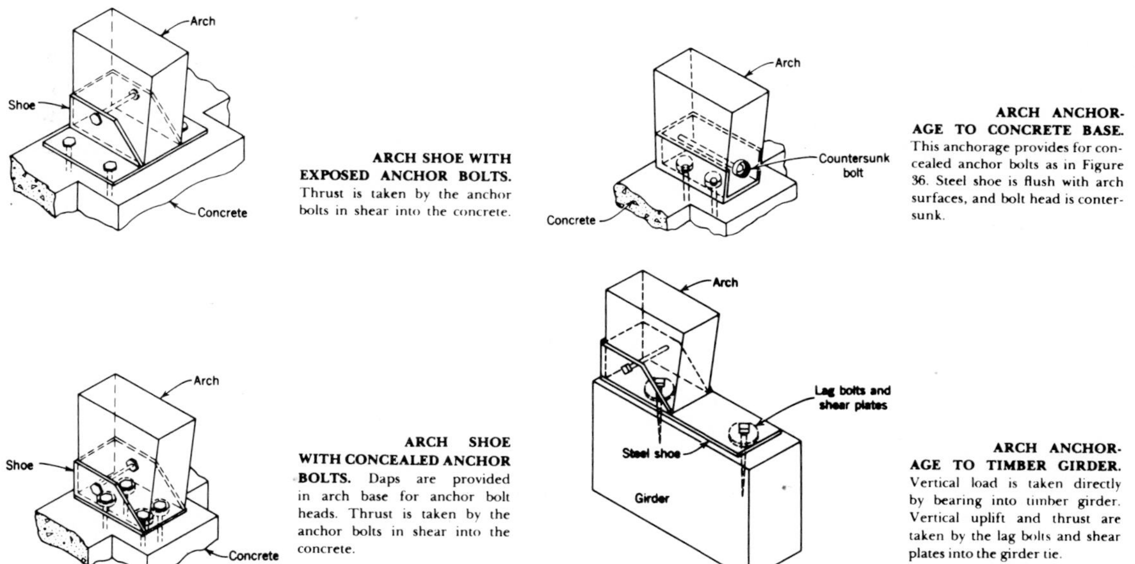

Figure 11.6. Various design details for the support of light arches. From AITC Standard 104, See Ref. 1.

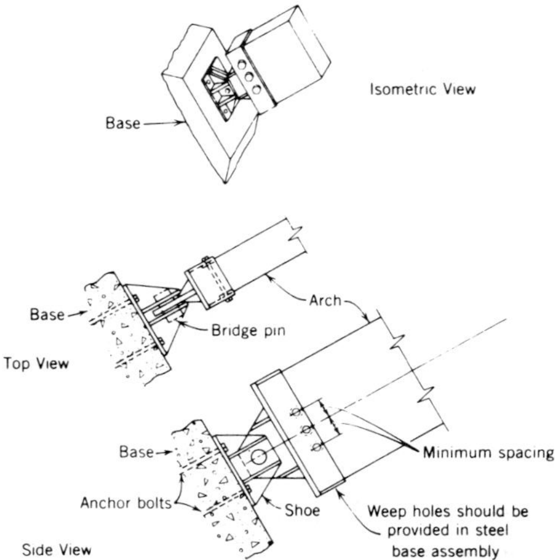

Figure 11.7. True hinge anchorage for large arches. Recommended for long-span, deep-section arches where true hinge action should be considered. Taken from AITC Standard 104, see Ref. 1.

11.2 TRUSSES

A large number of possible designs exist for wood trusses. The three fundamental types, namely, pitched, flat and bowstring trusses are shown in Fig. 11.8. Variations of these types include: scissors trusses, which are similar to pitched trusses except for the bottom chord (that is not a straight line but forms an angle at midspan) and camelback trusses which are like bowstring trusses with polygonal, rather than circular, top chords.

For design purposes, the depth-span ratio, H/L, should be as large as possible; the following ranges are recommended (1): 1/6 or deeper for pitched trusses, 1/6 to 1/8 for bowstring trusses and 1/8 to 1/10 for flat or parallel-chord trusses. When ratios lower than these are used, heavier trusses result that may deflect excessively under load.

The selection of the type of truss for any given project is influenced by functional requirements and architectural considerations. Thus, the shape of the roof, the necessary clearances, and the aesthetic appearance of the building are of importance. Almost any type of design of wood truss may be fabricated using modern fasteners, see Chapter 6. The designer may choose to use double-chord or multiple-chord trusses with a system of single-member or multiple-member webs; examples of joint connections for these types of trusses, using split-ring connectors, are given in Figs. 6.21 and 6.23. In recent times, monochord trusses have become increasingly popular. Heavy, long-span trusses use gusset steel plates with shear-plate connectors for the necessary connections of members at the joints, see Fig. 6.22, while lighter trusses and trussed-rafters are connected by means of toothed galvanized-steel plates (7), see Fig. 6.30, and glued or nailed plywood side plates (8), see Fig. 6.29. In general, the following guidelines may be used for an initial selection of the type of truss.

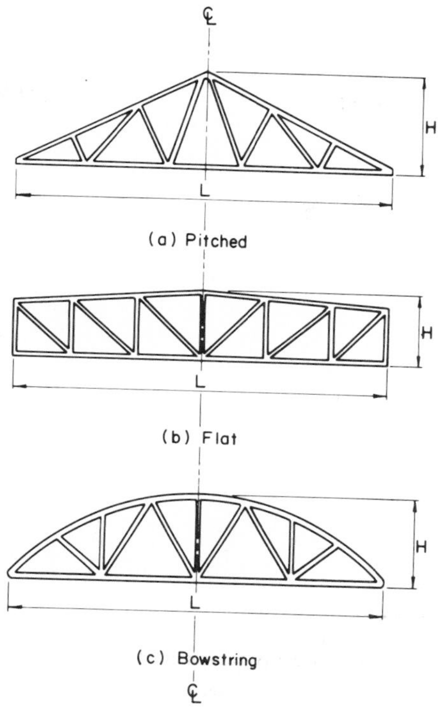

Figure 11.8. Fundamental types of wood trusses.

Pitched, triangular or A-type trusses are best suited for short to moderate spans, up to 75 ft. Major application of these trusses is in roofing of light-frame construction, as for instance, houses and small buildings. The top chord gives enough pitch in the roof to prevent large accumulations of snow and to permit easy drainage of rain-water; the flat bottom chord can be used to support the finished ceiling.

Flat or parallel-chord trusses are used frequently as support for long-span floors and for roofs of gentle and aesthetically-pleasing slopes; a minimum 3-percent slope is recommended for drainage. A flat truss can be designed in such a way as to have the supporting columns become its end members, see Fig. 11.9; this creates a rigid frame, with enough lateral stability to make knee-bracing unnecessary.

Bowstring trusses have a circular top chord with a radius approximately equal to the span of the truss; the lower chord is usually straight, except it may be cambered slightly in long spans. A circular curve is used more often than a parabolic curve for the top chord of bowstring trusses because of simpler glue-laminating procedures. A parabolic curve for the top chord would theoretically place it under thrust only and zero bending moment when subjected to uniform vertical loading; with the H/L ratios used in practice the variation between the parabola and the circular arc is very slight and not sufficient to change stresses materially.

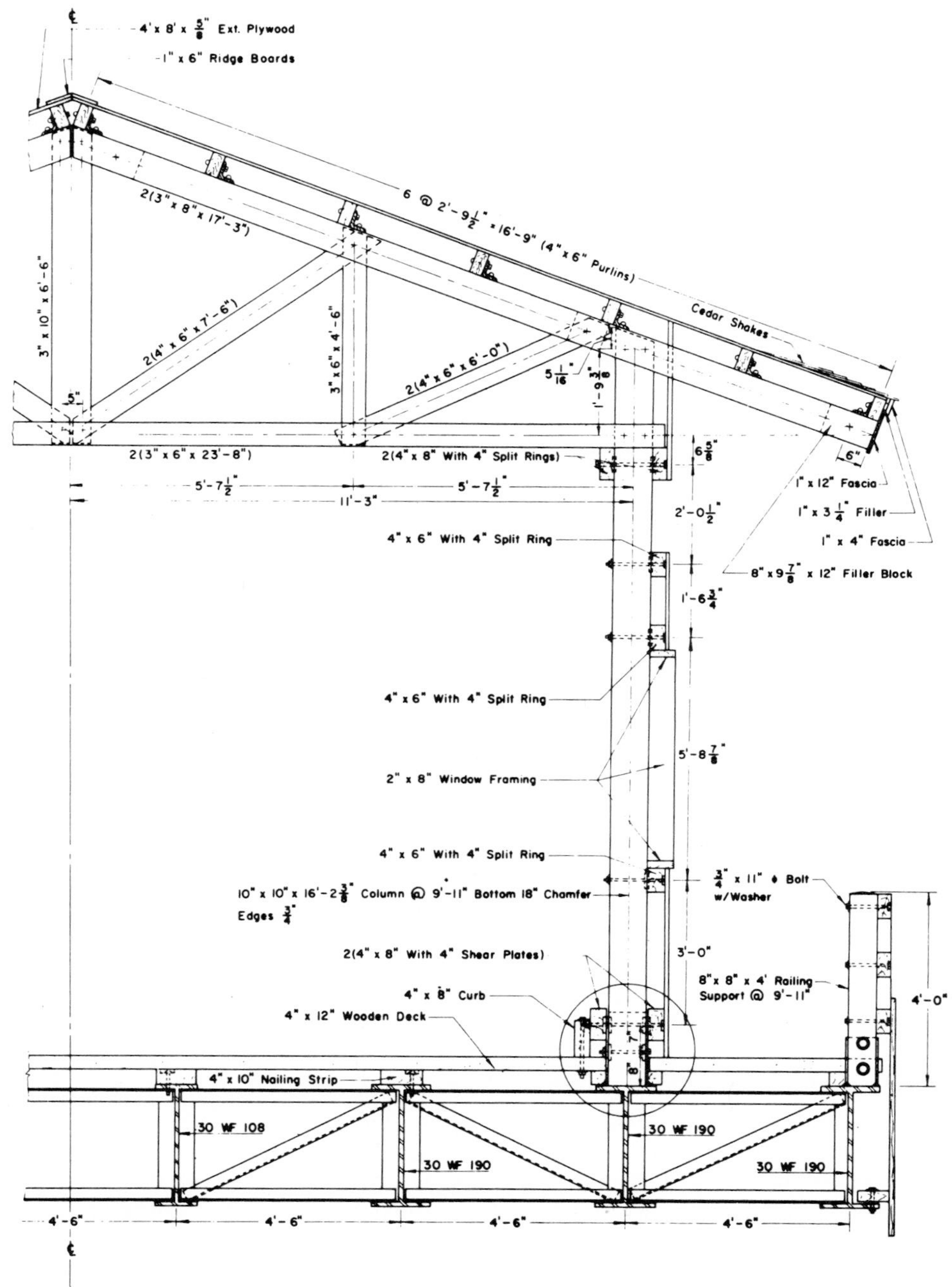

Figure 11.9. Transverse semi-section of scenic covered bridge, designed by the author, over Sangamon River in Mahomet, Illinois, see Refs. 9 and 10. The connection of the wood truss to the supporting columns allows rigid frame action to resist wind loads.

11.2.1 ANALYSIS

The initial analysis of these structures assumes simply-supported trusses with hinged connections for its members and concentrated loads applied only at the joints. Hence, simple methods of analysis may be used for the determination of member forces; these methods, both analytical and graphical, are described in the technical literature (11).

Actually, some wood trusses behave quite differently from the conditions assumed in the initial analysis. To provide stiffness in a building against wind action, some trusses may be braced to the supporting wood columns by means of knee-braces; a rigid frame action is thus created between the truss and the columns which may alter significantly the initial member forces. Bending moments may be created in truss members that are connected to each other through rigid gusset plates, as for instance in monochord trusses, or in trusses with continuous chords. Bending moments and shears are also created by external loads that are not acting directly on the joints but are otherwise distributed uniformly over the chord or concentrated at various spacings between joints.

This latter effect may be considered after the initial analysis of the truss. The chord is analyzed as a continuous beam over rigid supports; moments and shears thus determined are used for the final design of the member. The determination of effects of knee braces on trusses and columns and the secondary effects due to rigidity of the joints involves laborious hand calculations; therefore, computer analyses (4) is indicated for these cases.

11.2.2 DESIGN

Once the type, span and depth of a truss are selected it is necessary to decide on its spacing and on the desirable number of panels. The spacing of the trusses affects the size of the supported roof or floor system; as the spacing increases beyond 12 ft. the possibility of using sheathing or planks economically without beams or purlins is reduced substantially. Large spacing between trusses will require heavier beams for delivery of the loads; the result may be fewer but heavier trusses and beams. The number of panels of a truss determines the number of beams; increasing the number of panels tends to reduce bending moments in the beam-supporting chords at the expense of additional connections. Optimum design of trusses for a given project requires study of several possible solutions. This is usually facilitated by the existence of commercial solutions for a number of trusses that are available to the designer (12)(13).

An initial estimate of the weight of the truss is necessary; the following expressions (14) are considered good for this purpose:

$$\begin{aligned} &\text{Pitched trusses} &&: w = 0.64L \\ &\text{Flat trusses} &&: w = 0.043L + 1.75 \\ &\text{Bowstring trusses} &&: w = 0.038L + 0.60 \end{aligned} \tag{11.9}$$

where w is the estimated weight of the truss per square foot of horizontal surface, and L is the span in feet. Equations 11.9 are to be considered only as initial estimates; after the design is completed the actual weight of the truss should be determined and compared to w. Usually, no design corrections are necessary since the weight of the truss is a small fraction of the total load.

In addition to the weight of the truss, permanent loads include the weights of roofing, sheathing, rafters or purlins and any other permanently applied loads which will act throughout the service life of the truss. Snow and wind constitute the principal loading on roof trusses; for floor trusses, the actual design live load should be used.

The design of members of the truss is governed by axial loading; in other words, truss members are designed fundamentally to resist compression or tension loads. In the case of chords subjected to transverse bending, due to concentrated loads of beams bearing at sections other than panel points or to uniform loading of sheathing or plank systems, design is accomplished for the simultaneous action of axial load and bending; adequacy of a member is checked using Eq. 8.16. The slenderness ratio of members of the top chord is taken as the largest of ℓ/b or ℓ/d, where ℓ is the distance between panel points or beams as the case may be, b is the width, and d is the depth of the member. For continuous lateral restraint, as provided by nailed sheathing, ℓ/b is zero and ℓ/d governs. Web members of trusses are braced not only at the top chord but also at the bottom chord, since the latter is a tension member whose tendency is to remain in its original plane. Thus, it is reasonable to design web members under compression as braced columns since displacements of their ends outside the principal plane of the truss is not possible. Tension web members and the bottom chord may be designed for combined bending and axial load if the case demands it; usually, however, bending moments are small compared to tension loads and may be neglected. The adequacy of the section can be checked using Eq. 8.16 with $J = 0$ and F_t and $f_t = T/A$ used instead of $F_{c'}$, and f_c, respectively.

Finally, the design of joints in a truss should be accomplished using the principles discussed in Chapter 6. In addition, reference is made to an article by Gordon (15) describing some details of joints that have caused failure of bowstring trusses in the past.

11.2.3 Bracing

Trusses are very rigid structures in their own longitudinal plane. Outside this plane, however, trusses are slender structures subject to lateral instability unless properly braced. Bracing is necessary and quite important during erection; many collapses have occurred at this stage due to wind action and the lack of proper bracing. Good design calls for temporary bracing that can remain and be used as permanent bracing of the finished structure.

Bracing in the longitudinal direction of the building is provided for the top chord by the roof system acting as a rigid diaphragm. The bottom chord may be braced by longitudinal struts extending between trusses and by vertical X bracing, usually steel rods or cables, between trusses in alternate bays. The latter system provides for lateral load distribution in the case of overloading of any given truss. In the transverse direction of the building the effects of wind are transmitted to the supporting columns or walls through diaphragm action of the rigid roof system; the effects of sudden wind gusts on localized areas are similarly distributed.

For design purposes, it is recommended that the bracing system of the compression chord of a truss be designed to withstand a horizontal force equal to 2 percent of the compressive force in the chord; this is usually met by most roof or floor systems.

11.2.4 Camber

Deflection of trusses can be compensated, totally or partially, by initial camber. Hence, it is important to estimate deflections as accurately as possible. The determination of elastic deflections of a truss may be accomplished analytically, using virtual work, or graphically using the Williot-Mohr method (3); the results are reliable for short-time loading. However, in addition to these instantaneous deflections, inelastic and time-induced deflections occur due to creep in wood and slip of the joints, that are difficult to estimate. It is for this reason that the necessary camber, C, in inches, is often determined using the following empirical formula developed by Timber Engineering Company, see Appendix F of Ref. 16, for connector-connected trusses:

$$C = \frac{L}{H}(K_1L^2 + K_2L) \tag{11.10}$$

In the preceding formula the span L and the rise H, both expressed in feet, are defined in Fig. 11.8; K_1 and K_2 are constants. The value of K_1, independently of the type of truss, is 3.2×10^{-5}; K_2 is equal to 63×10^{-5} for bowstring trusses and 280×10^{-5} for pitched and flat trusses. For instance, the camber recommended for midspan of a bowstring truss in which L = 100 ft. and H = 12.5 ft. is $C = (100/12.5)(3.2 \times 10^{-5} \times 100^2 + 63 \times 10^{-5} \times 100) = 3.06$ in.

Both lower and upper chord are cambered in flat and parallel-chord trusses. However, for pitched and bowstring trusses only the lower chord is usually cambered; in that case conservative design demands that the effective depth of the truss, H-C, be used for the analysis of member forces.

11.3 COMBINED WOOD-STEEL TRUSSES

Wood can be used in combination with steel cables or rods to create truss-type structures known as King or Queen-trusses, see Fig. 11.10. In the past, when only solid-sawn lumber was available for design of wood structures, these trusses allowed large column-free areas. Presently, with the use of glued-laminated members, their potential has increased; as recently as 1971 King-trusses were used for the structural system of a 2000-seat hockey ring at Wesleyan University (17) with a clear span of 147 ft. Trusses consisted of a 12 × 42 in. glued-laminated member for the top chord, an 8-in. diameter inclined steel pipe as post and bridge-strand for the tension member. The adaptability of the King-truss is shown in this project for which the lower chord is not horizontal but parallels the cross-sectional outline of the sloped seating and level skating slab.

In the following, only the analysis and design procedures for the King-truss are discussed, those for the Queen-truss are somewhat similar.

11.3.1 Analysis of King-Truss

The truss shown in Fig. 11.10a is a statically indeterminate structure because the internal forces of its members cannot be determined using equilibrium equations only; consideration of the deformations of the structure is necessary. However, since the structure is simply supported, the external reactions may be found easily. Exact analysis of a King-truss can only be accomplished after the member sizes and loads are fully defined. Thus, unless the truss is proportioned by experience, a simple analysis is necessary for initial proportioning of member sizes.

Experience in the design of these trusses has shown that, for the r/L ratios used in practice, the post and cable sub-structure provide the top chord with a virtually rigid support at midspan. This assumption leads to a determination of forces, shears and moments as shown in Fig. 11.10b. The top chord of the truss behaves as a continuous beam on three rigid supports. Maximum moment is negative and occurs at the central support; since maximum shear also occurs at the

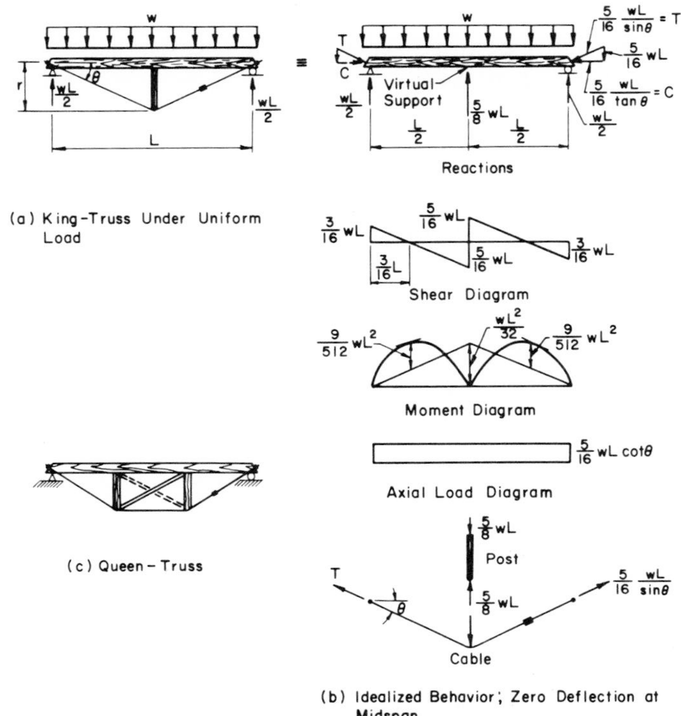

Figure 11.10. King and queen trusses and analysis of king-truss for initial design purposes.

same section, it becomes the governing section for design. In addition to moment and shear, the top chord is subjected to the action of a longitudinal load given by the horizontal projection of the tension force in the cable, see Fig. 11.10b. The initial design of the post is for column action, as a double-hinged member subjected to a maximum axial load given by (⅝) wL. The load is transmitted directly to the cable and a tensile force T is generated whose value will depend on the final angle θ that the cable makes with the beam. Since the fundamental assumption on which the initial analysis is made is that of negligible displacement at midspan, it is consistent to assume that the initial angle θ remains unchanged. The cable is designed for the force T shown in Fig. 11.10b.

Final analysis of a given King-truss can be accomplished as follows. The deflection configuration of the truss under uniform load is shown in Fig. 11.11a. The deflection Δ at midspan of the beam, see Fig. 11.11b, can be determined as the difference between δ_1, the deflection caused by the uniform load and δ_2, the upward deflection caused by the unknown reaction R of the post. Also, Δ must be equal to the sum of δ_3, the axial deformation of the post and δ_4, the vertical elongation of the cable system at midspan. Thus,

$$\Delta = \delta_1 - \delta_2 = \delta_3 + \delta_4 \tag{11.11}$$

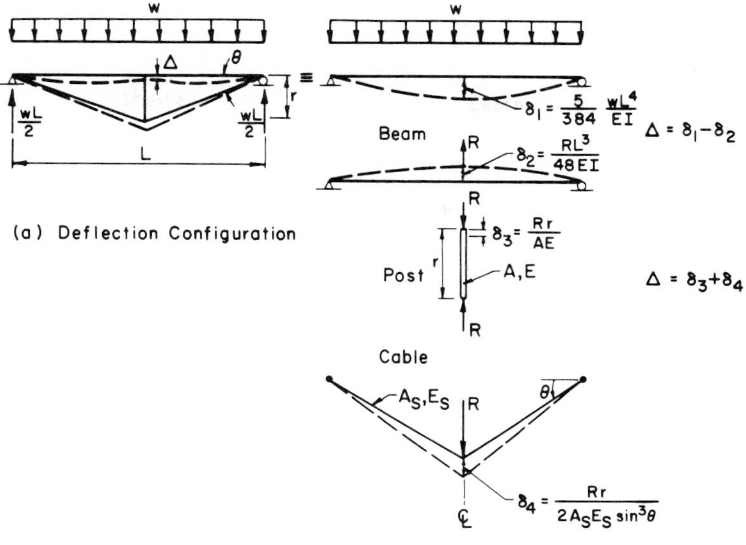

Figure 11.11. Determination of actual reaction R and midspan deflection △ for final checking of a given king-truss.

The values of δ_1, δ_2, δ_3, and δ_4 are given in Fig. 11.11b. In the determination of δ_1 and δ_2 the secondary effects of the axial load on the beam have not been considered; however, they are usually negligible. Also the additional deflections due to shear are considered compensated by the use of the reduced value of E listed for the given wood, see Section 2.2. Substituting in the preceding Equation yields.

$$\frac{5w\,L^4}{384\,E\,I_b} - \frac{R\,L^3}{48\,E\,I_b} = \frac{Rr}{A_pE} + \frac{Rr}{2\,A_s\,E_s\,\sin^3\theta} \tag{11.12}$$

Thus, the actual value of R is given by

$$R = \frac{\dfrac{5w\,L^4}{384\,E\,I_b}}{\dfrac{L^3}{48\,E\,I_b} + \dfrac{r}{A_pE} + \dfrac{r}{2\,A_s\,E_s\,\sin^3\theta}} \tag{11.13}$$

where E and E_s are the moduli of elasticity of wood and the steel cable, respectively. The latter varies between 25,000 ksi for strand cables, see Reference 18, to 29,000 ksi for solid steel bars. Once the value of R is found, the actual deflection at midspan can be obtained using Eq. 11.11. The actual value of R turns out to be less than the value given in Fig. 11.10. The same is true for both the tensile force in the cable, and the maximum moment at midspan of the beam. Thus, the initial design assumptions are conservative.

As an exercise consider if a moment larger than $wL^2/32$ can occur at midspan or at some other section. Definitely, this is theoretically possible in the case of a very shallow truss, since for a small value of R the moment may be close to that of a simply-supported beam of length L. In other words, the moment may be $wL^2/8$, or four times as much as that shown in Fig. 11.10b, if only $R = 0$. This, of course, is the upper bound for maximum moment in the top chord. However, it is easily shown that for a value of $R = 0.5$ wL, the moment at midspan is zero and maximum moment (positive) occurs at either side of midspan, in sections situated at L/4 from the supports. The value of maximum moment for this condition is equal to $wL^2/32$, just as assumed in initial design except of different sign. There is a value of R for which the maximum negative and positive moments in the beam have equal value, i.e. $M^+ = M^- = w(L/2)^2/11.65 = wL^2/46.6$ such as shown in Fig. 10.3a. For this case, it is easy to obtain $R = (wL/2) + 2(wL^2/46.6)/(L/2) = 0.586$ wL. Thus, for the range $0.5\ wL \leqslant R \leqslant 0.625\ wL$ the maximum moment in the top chord is equal to, or smaller than, the initial design value $M = wL^2/32$. The optimum case is reached when $R = 0.586$ wL for which $M = wL^2/46.6$. For the range $0 \leqslant R < 0.5$ wL, the maximum moment in the top chord is smaller than the initial design value.

The wide range in values of R, within which the initial design values are valid, justifies the initial analysis. It is quite likely that for reasonable values of the ratio r/L, i.e. for $r/L > 1/10$, the reaction R will be larger than 0.5 wL. However, it is possible to make R as large as one wants it, independently of the value of r/L, through initial prestressing of the steel cable. In that way, design for the minimum value of M requires only that a force T in the cable be provided that generates the necessary value of R. However, large prestressing is not recommended because additional shortenings of the top chord with the passage of time may dissipate these effects. Thus, it is reasonable to use $M = wL^2/32$ for design purposes; initial prestressing to provide enough camber to upset the deflection due to the actual load is advisable. As an exercise consider also the possibility of imposing end moments to the top chord, in the fashion of prestressed concrete beams (18), by connecting the ends of the cable to the top chord at sections below its neutral axis. These end moments are equal to the horizontal component of T times the eccentricity; they can be used to reduce the maximum positive moment of shallow trusses.

11.4 SHALLOW FOUNDATIONS

Conventional foundations for wood houses and other small buildings consist of masonry walls that rest on lightly-reinforced continuous footings of concrete, 18 to 24 in. wide and 8 to 12 in. thick. Often, masonry pilasters are used to reinforce the walls. In the past it has been the practice that wood framing begins at a minimum of 6 in. above ground level. This was done to avoid deterioration of untreated wood placed in contact with the soil, see Chapter 5.

In order to investigate the use of treated wood for foundations, the Forest Products Laboratory, in 1937, built an experimental 24 by 36-foot prefabricated house on a wood foundation (19). Douglas fir 6 × 6 in. vertical posts spaced at 4½ ft. on center were used to support five 2 × 8 in. nominal Southern pine planks 4.5 or 9 feet long, nailed to the outside perimeter of the posts to form a retaining wall 5 × 7.5 in. = 37.5 in. or a little over 3 feet tall. After 30 years of actual service new construction at the Laboratory required removal of the house to a new location thereby allowing an evaluation of the wood foundation. Isolated decay was found in the surface of the planks in contact with the soil and an investigation of the creosote content of the planks showed it in substandard amounts. The posts showed no decay except in two cases for which creosote content was also well below present standard amounts. It was concluded (19) that, even after 30 years and in spite of the isolated cases of decay, the wood foundation was giving excellent service.

A modern version of a wood foundation which has been accepted by the Federal Housing Administration, is shown in Fig. 11.12 for a typical basement (20). It is referred to as the "All-Weather Wood Foundation" because it adapts so readily to easy installation no matter how inclement the weather and is not dependent upon other construction trades which often are delayed by weather. The stud wall is designed to carry the transverse load created by soil pressure in addition to the axial loads due to the weight of the structure. The maximum moment, which occurs at a section situated at a distance $c = h_1 + h\sqrt{h/3H}$ from the bottom of the floor joist, see Fig. 11.12, is given by the following expression:

$$M = \frac{wh^3}{6H}\left(h_1 + \frac{2h}{3}\sqrt{\frac{h}{3H}}\right) \qquad (11.14)$$

where w is the "equivalent fluid pressure" for the soil (21). Hence, the wall is designed to withstand the pressure of a liquid that is assumed to exert the same pressure against the wall as the real backfill. A minimum value of 30 pounds per cubic foot is recommended; in the case of clayey soils of low permeability or with poorly drained soils, where actual water pressure may be exerted

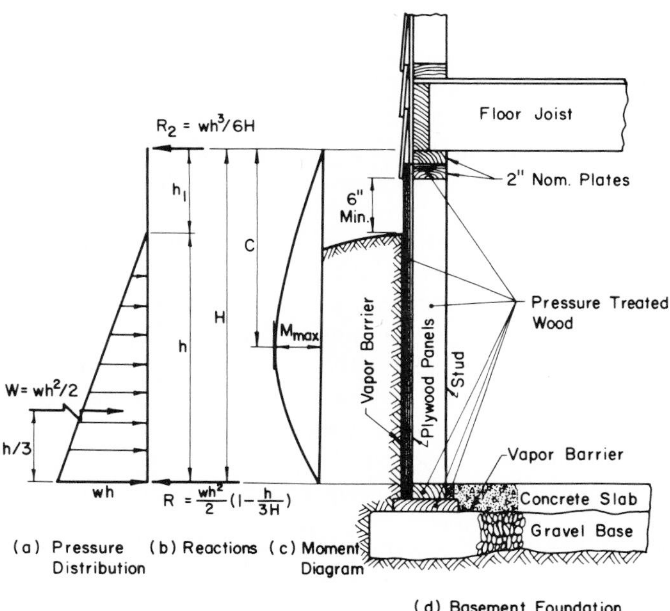

Figure 11.12. Modern design of all-weather wood foundation. Adapted from Ref. 20.

against the wall, a higher value of w should be used. Design of the stud walls must consider the combined action of axial load and moment; Eq. 8.4 should be used with the studs considered as double-hinged members whose ends are prevented from lateral displacement. For usual conditions, 2 × 4 or 2 × 6 studs at 12 or 16 in. spacing will suffice. The studs are nailed to the 2 in. nominal plates using 16d corrosion-resistant nails; the plywood panels, ½″ or thicker, are attached to the studs using 8d corrosion-resistant nails at 6 in. spacing or less on center along panel edges, and 12 in. or less on center at intermediate supports. Plywood panels are used with face grain vertical or parallel to the studs; vertical joints between panels should be located over studs.

In the case of walls which run parallel to joists it is recommended that solid bridging at 4-ft. spacing be installed between the first five rows of joists nearest the wall to transfer the lateral load from the top of the foundation studs to the rigid floor system.

11.5 DEEP FOUNDATIONS

Wood piles have been used as foundation for bridges and buildings for many centuries; the art of pile-driving was known during the Roman empire. Engineers realized early that shallow foundations on weak subsoils could fail by lack of bearing capacity or settle so much as to render the structure unserviceable. The use of piles to take the load to deeper, more resisting strata, is indicated in these cases. The many aspects and details of pile foundations is beyond the scope of this work; a number of references exist (21 through 25) in which may be found information necessary for design.

Timber piles are classified according to the manner in which their load-carrying capacity is developed. Two classes exist, namely, friction piles and end-bearing piles. The former class assumes that the load on the pile is transmitted along its length to the supporting soil by virtue of the friction developed between the soil and the pile. End-bearing piles, as the title implies, are those for which load capacity is determined by the end-bearing capacity of the soil at the pile tip. A friction pile derives some support also from the tip, while an end-bearing pile (unless supported by a very stiff stratum, or rock) also draws some support along its length.

Wood piles are naturally tapered; the widest end is the butt, the narrowest end is the tip. Because of this variation, the determination of the capacity of an end-bearing wood pile depends on the cross-sectional area of the tip and the allowable bearing stress of the supporting stratum. Long-column action, where capacity is determined by considerations of elastic stability, is not usually the case for piles fully driven in the ground; however, for piles that are only partially driven into the soil, as for instance in the case of marine piers (26) or trestles, see Fig. 9.3, long-column action may take place and should be investigated. If lateral loads are present, shears and bending moments are also induced. The simultaneous action of axial load and bending moment should be investigated using Eq. 8.16; because of the natural taper of wood piles, it is advisable to check several sections about the section of maximum moment to be sure of the adequacy of the pile.

To specify the diameter of the tip, with the corresponding minimum butt circumference, Table A.11.1 provides the necessary information; piles may be specified which range in length between 20 and 120 ft., with tip diameters between 5 and 12 inches. A similar Table A.11.2 is given for the design of friction piles; for this case, butt diameters and corresponding minimum tip circumferences are specified. It is important for the sake of strength and reasonable taper that the wood pile meet the minimum butt or tip circumferences, as the case may be; circumference measurements are taken under the bark. Some specifications (28) define allowable loads of friction piles by the diameter of the section situated at 3 feet from the butt; to prevent the use of a pile with a sharp change in diameter between the butt and the section at 3 ft. from it, ASTM (27) requires that the circumference of the butt exceed that of the governing section by not more than 8 in. Criteria for acceptance of wood piles affected by general or local crookedness, natural defects (twist of grain, knots, checks, shakes, splits), holes and scars and other characteristics are given in Ref. 25.

The design load for a given pile depends not only on its size and type of wood but also on the conditions of the subsoil and its capacity to support the load delivered by the pile. The final decision is based on studies of soil borings of the site and on test results of its various physical characteristics. In addition, test piles are driven in large jobs to determine with a greater degree of accuracy the length necessary to develop a given load. An estimate of the resistance of the pile may be obtained as a function of the number of blows of the driving hammer per inch of penetration; dynamic expressions which equate driving energy to work performed by the resisting force plus energy losses are used for this purpose. However, the correct interpretation of the results given by various formulas is the object of considerable discussion (21). On very large pile jobs full-load testing of a few test piles is recommended; the information obtained regarding load-settlement characteristics is of great value. As a guide to the designer the following values are maximum loads allowed by AASHTO Specifications (28) for the design of highway bridges. For sections 10, 12, 14 and 16 inches in diameter, at a distance of 3 ft. from the butt, the maximum loads are: 20, 24, 28, and 32 tons, respectively.

A final note is necessary on durability of timber piles. Buildings erected several centuries ago in Venice and Amsterdam, both cities famous for their canals and the closeness of the water table to the surface of the ground, are still supported by their timber-pile foundations; a remarkable example of durability of untreated wood under the water table and verification that decay-producing fungi cannot prosper in the absence of oxygen. Modern preservative treatment of wood with not less than 12 pounds of creosote per cubic foot, see Chapter 5, guarantee an indefinite durability of the pile against decay when driven into the ground, independently of the location or variations in elevation of the water table. Treatments which combine coal tar-creosote with pentachlorophenol have been found effective for wood piles used in saline or brackish waters, even against attack of marine borers. Untreated surfaces are exposed at the cut-offs that are necessary to bring the piles to the desired elevation. These should also be given protection by thoroughly brushing the cut surfaces with hot coal-tar creosote. A coat of pitch, asphalt, or similar material may then be applied over the creosote and a protective sheet material (metal, felt or saturated fabric) fitted over the pile head (29).

11.6 POLE-TYPE STRUCTURES

Man has used buildings erected on poles for housing for many centuries. The most primitive of these buildings consisted of four poles over which animal skins were stretched to shelter him from the sun and rain. In this country, pole buildings were used for farm buildings in the nineteenth century; however, they didn't last long because the poles had no preservative treatment and were easily destroyed by decay and insects (30).

Use of modern preservative treatments for wood poles provides protection against major agencies of wood destruction, see Chapter 5, and has brought an increased use of pole-type structures of all kinds in this country. Pole-buildings are back on farms for animal shelters and storage purposes; they are used also in office and residential construction. Major use of pole-type buildings occurs in beaches and low-coastal areas where the combined ravages of hurricane-induced floods, rain and wind cannot be adequately resisted using conventional construction.* Evidence of successful behavior of pole-buildings under these conditions was given by the passage of hurricane Camille in August 1969 over the coastal regions of Louisiana and Mississippi (32). A report on the behavior of wood structures in general, under the same hurricane, is given in Ref. 33.

There is nothing unconventional in the design of pole-buildings as shown by the various details of framing and spacing of poles that are given in Figs. 11.13 and 14. Thus, no special discussion of general design is given. However, because of its importance, a full discussion follows on determination of required depth of embedment and structural adequacy of wood piles.

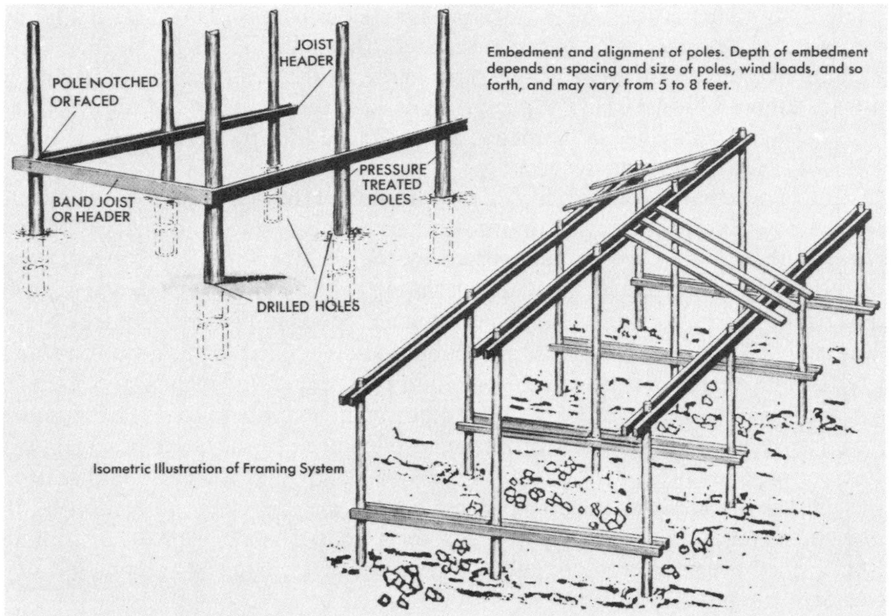

Figure 11.13. Various framing systems of pole-supported buildings. From Ref. 32.

* Use of supplemental reinforcing by special anchoring or metal strapping is recommended by Anderson (31) to improve the performance of conventional wood-frame construction against hurricane effects; full anchorage of all components, from foundation to roof rafters, is considered essential.

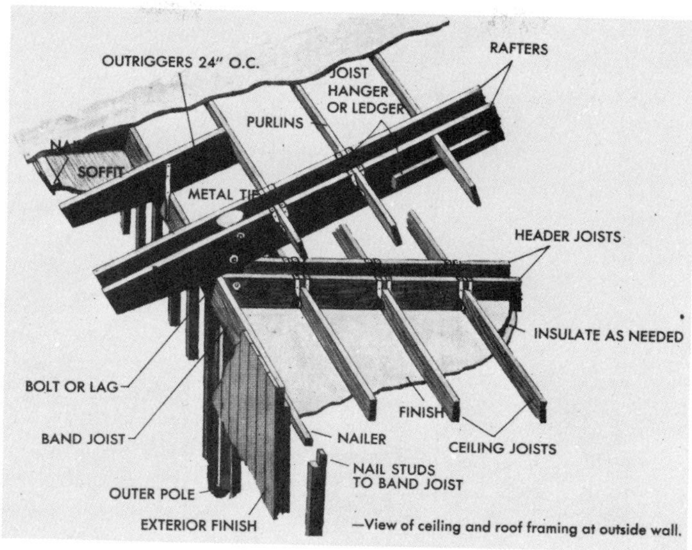

Figure 11.14. Framing details of poles to ceiling joists and roof rafters. From Ref. 32.

11.6.1 Determination of Required Depth of Embedment

Design of a pole-supported structure requires determination of the depth of embedment of the poles as required by the loads and the conditions of the subsoil; a typical case is shown in Fig. 11.15. The pole is assumed to carry a horizontal load Q due to wind or earthquake and a total vertical load P + W, where P is the weight of the structure and the live load acting on it and W is the weight of the pile. The end moment M is due to the rigid connection of the pole to the supported structure. The vertical load is resisted by the sum of P_1, the end reaction at the butt, and P_2 the lateral capacity developed by shear and friction at the interface soil-pole; see Fig. 11.15. Usually, the support of vertical loads is not the governing condition for the determination of depth of embedment; cast-in-place concrete jackets or pads, if necessary, can provide bearing support. It is the lateral load that determines the depth of embedment of poles.

The distribution of the passive pressure against the embedded portion of the pole shows two well-defined portions; one directly below the ground surface in the shape of a parabola with an average soil pressure σ_1 and a depth equal to 0.68D, and the other, below the point of rotation, see Fig. 11.15, for which the average soil pressure is σ_2, and the depth 0.32D. Obviously,

$$Q = Q_1 - Q_2 \tag{11.15}$$

where $Q_1 = \sigma_1 B(0.68D)$ and $Q_2 = \sigma_2 B(0.32D)$ are the resultant passive forces. The distribution of depths and positions of the resultants shown in the Figure were obtained by P. C. Rutledge in tests conducted at Purdue University.

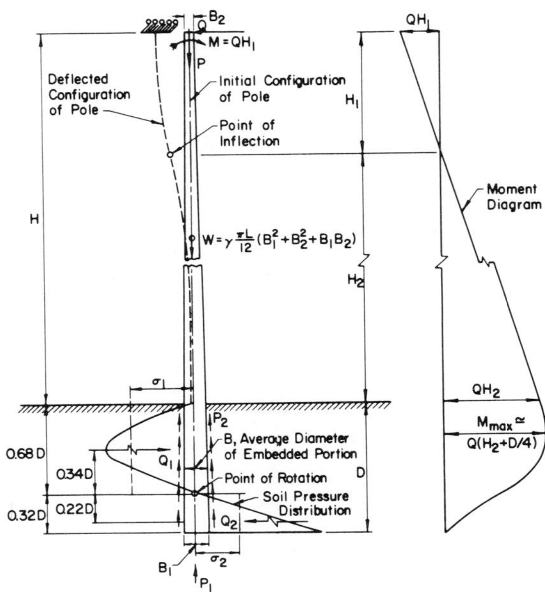

Figure 11.15. Behavior of a pole under the action of lateral loading.

The deflection configuration of the pole is also shown in the Figure. At the point of inflection, the bending moment is zero by definition; only P and Q are applied there. Hence, taking moments about Q, the following is obtained

$$Q(H_2 + 0.34D) = Q_2(0.22D + 0.34D) \tag{11.16}$$

from which

$$Q_2 = Q \frac{H_2 + 0.34D}{0.56D} \tag{11.17}$$

Substituting the preceding value of Q_2 and $Q_1 = \sigma_1 B(0.68D)$ in Eq. 11.15 yields:

$$Q = \sigma_1 B(0.68D) - Q \frac{H_2 + 0.34D}{0.56D} \tag{11.18}$$

and finally:

$$\frac{Q}{\sigma_1} = \frac{BD^2}{2.37D + 2.64\ H_2} \tag{11.19}$$

The preceding Equation can be used to determine D once the values of Q, σ_1, B and H_2 are given.

The value of Q is obtained using the design intensity of wind or earthquake for the given region in connection with the tributary area or mass, respectively, of the building. Analysis is necessary to determine the distribution of the total horizontal force among the various poles that may form the typical frame of a building. The design value of σ_1 depends on the subsoil immediately below the ground level; testing and theoretical procedures of Soil Mechanics (21) are indicated here. However, in lieu of this investigation, Table 11.3 gives allowable values for three groups of soils. As in the case of other Tables giving design values for soils that are lumped into categories, caution is indicated in the use of Table 11.3. The values of B, the average diameter of the embedded portion of the pole can be obtained from Table 11.4 for the various classes of poles that are available; it is conservative to take B as $(C/12)/\pi$, where C is the minimum circumference in inches, at 6 ft. from the butt, see Table 11.4. Finally, analyses of various pole-type structures have shown that H_2, the distance of the point of inflection of the pole to the surface of the ground, may be taken as 2H/3 with confidence.*

TABLE 11.3.† ALLOWABLE LATERAL PRESSURE FOR VARIOUS TYPES OF SOIL.

Class of Material		Allowable Value,* psf	Maximum Value, psf
Good:	Compact well-graded sand and gravel. Hard clay, Well-graded fine and coarse sand (All drained so water will not stand)	400D**	8,000
Average:	Compact fine sand. Medium clay. Compact sandy loam. Loose coarse sand and gravel (All drained so water will not stand)	200D**	2,500
Poor:	Soft clay. Clay loam. Poorly compacted sand. Clays containing large amounts of silt. (Water stands during wet season)	100D**	1,500

† From International Conference of Building Officials.

* Isolated poles, such as flagpoles or signs, may be designed using lateral bearing values equal to two times the tabulated values.

** D, embedment depth below natural grade in ft.

Substitution of the preceding values of Q, σ_1, B and H_2 in Eq., 11.19 leads to a cubic equation for D, which can be easily solved by trial and error. Only a quadratic equation is obtained if σ_1 is defined by a given value, as proposed by Patterson (34), and not as a function of D.

For the case where lateral restraint such as that of a rigid floor or a concrete pavement is provided at the ground surface, the depth of embedment may be determined from the following expression (34):

$$D^2 = 4.25 \frac{QH_2}{\sigma_1 B} \tag{11.20}$$

where all symbols have been defined before.

* For the case of supported structures that are not rigidly connected or knee-braced to the poles, the full value of H should be taken for H_2.

TABLE 11.4. DIMENSIONS OF PREFERRED STANDARD SIZES OF DOUGLAS FIR AND SOUTHERN PINE POLES.†

Class		1	2	3	4	5	6	7	9	10
Minimum Circumference at Top, in.		27	25	23	21	19	17	15	15	12
Length of Pole ft.	Ground-line to Butt, ft. *	Minimum Circumference at 6 Feet from Butt, in.								
20	4						21.0	19.5	17.5	14.0
25	5						23.0	21.5	19.5	15.0
30	5.5		34.0	32.0	29.5	27.5	25.0	23.5	20.5	
35	6		36.5	34.0	31.5	29.0	27.0	25.0		
40	6	41.0	38.5	36.0	33.5	31.0	28.5	26.5		
45	6.5	43.0	40.5	37.5	35.0	32.5				
50	7	45.0	42.0	39.0						
60	8	48.0	45.0	42.0						
70	9	51.0	48.0	45.0						
80	10	54.0	50.5							
90	11	56.0	53.0							
100	11	58.5	55.0							
110	12	60.5	57.0							
120	12	62.5	59.0							

† Adapted from Table VI of Ref. 34.

* Intended for use only when a definition of groundline is necessary in order to apply requirements relating to scars, straightness, etc.

11.6.1.1 Illustrative Example

Determine the depth of embedment, D, of a class 2 pole in compact, well-graded sand in order to support a lateral load Q = 2,000 lb. due to wind applied at H = 21 ft. from the ground surface. The pole is rotationally restrained at the top by the supported structure, as shown in Fig. 11.15, but is otherwise free to move laterally. In addition, the pole is subjected to a vertical load P = 3,000 lb. applied at the top and to its own weight W.

The determination of D depends on the action of the lateral force. Since the pole is rotationally restrained at the top, bending is in the configuration shown in the Figure, where H_2 may be estimated as $(2/3)(21) = 14$ ft. Class 2 poles in 30-ft. lengths have 34 in. of minimum circumference at 6 feet from the butt; the diameter of this section can be used conservatively, up to an embedded depth of 12 ft., as the average diameter of the embedded portion of the pole. Hence, $B = 34/\pi = 10.82$ in. $= 0.902$ ft. According to Table 11.3 a compact, well-graded sand is a good soil, for which the allowable value of average passive pressure is $\sigma_1 = 400$ D. Substituting the preceding values of Q, H_2, B and σ_1 in Eq. 11.19 leads to the following expression:

$$\frac{2{,}000}{400D} = \frac{0.902D^2}{2.37D + 2.64 \times 14}$$

the solution of which, by trial and error, yields D = 6.63 ft. Using a 30-ft. pole with this embedment leaves approximately 2.4 ft. for attachment of the top structure to the pole.

11.6.2 Structural Design of Poles

The structural suitability of a pole against simultaneous action of axial load and bending moment can be checked assuming a linear interaction diagram of the type shown in Fig. 8.3. This leads to the following expression:

$$\frac{P/A}{F'_c} + \frac{M/S}{F_b C_f} \leqslant \overline{\alpha} \tag{11.21}$$

where

P = axial load at critical section
M = bending moment at critical section
A = $\pi B^2/4$, area of critical section
S = $\pi B^3/32$, section modulus of critical section
F'_c = allowable axial stress
F_b = allowable bending stress, see Table A.11.3
C_f = form factor, 1.18 for round sections
$\overline{\alpha}$ = load duration factor, see Fig. 2.14

Critical sections of the pole are assumed to occur, see Fig. 11.15, at the connection with the supported structure and at a distance D/4 below the surface of the ground. The respective moments at these sections are QH_1, where H_1 may be taken as H/3, and $Q(H_2 + D/4)$, where $H_2 = 2H/3$ for conditions such as shown in Fig. 11.15. In the case of very slender poles, a designer would do well to magnify these moments by the so-called PΔ effect, which is a result of additional bending due to the action of P on the deflected configuration of the pole, see Chapter 8. The value of P should be increased by W, the weight of the pole, when checking the critical section at the bottom; usually γ = 50 lb/cu. ft. for pressure-treated poles.

The allowable axial stress is a function of the critical load of the pole, with the maximum value F_c dictated by the wood species, see Table A.11.3. In view of the difficulty of determining the exact critical load of a tapered column with various end restraints (numerical methods are suggested), NDS (16) recommends the use of the following expression for the allowable stress in compression:

$$F'_c = \frac{P_{cr}/2.74}{A} = \frac{3.62\ E}{(\ell/r)^2} \leqslant F_c \tag{11.22}$$

Eq. 11.22 is based on the elastic behavior of a double-hinged prismatic column with ends prevented from lateral displacements, or a double-fixed prismatic column with ends free to move laterally; the last condition approaching that of the pole shown in Fig. 11.15. A factor of safety equal to 2.74 affects the preceding Equation. The prismatic column that is equivalent to the actual tapered pole is assumed to be that determined by a section situated at a distance of one-third the length of the pole from the tip; for the case of a circular section, the value of r in Eq. 11.22 is given by one-fourth of the diameter of the equivalent section.

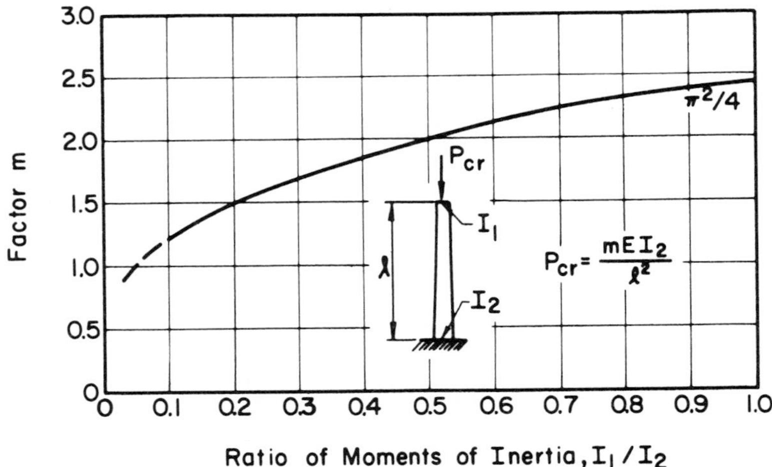

Figure 11.16. Determination of critical load in a cantilevered pole. Adapted from data given in Ref. 35.

Poles behave as vertical cantilevers when not rigidly attached to the structures they support. For such cases, Eq. 11.22 renders values that are not conservative and possibly too high. Instead, the following expression (35) should be used for the critical load:

$$P_{cr} = \frac{mEI_2}{\ell^2} \tag{11.23}$$

where I_2 is the moment of inertia of the section of the pole at ground level, and m is a coefficient given in Fig. 11.16. Hence, the allowable compression stress for a given section of a cantilevered pole is

$$F_c' = \frac{P_{cr}/2.74}{A} \tag{11.24}$$

where 2.74 is the same factor of safety used in Eq. 11.22.

The lateral force Q produces maximum shearing-stresses in the pole which are given by:

$$v = \frac{4}{3}\,\frac{Q}{A_1} \tag{11.25}$$

where A_1 is the circular cross-sectional area at the top. The value of v must be less than, or equal to, the allowable horizontal shear stress given in Table A.11.3.

11.6.2.1 Illustrative Example

Determine the structural adequacy of the class 2 pole used in the preceding example. Assume D = 6.6 ft. Consider F_c = 1.2 ksi, F_b = 2.15 ksi, F_v = 0.130 ksi, $F_{c\perp}$ = 0.26 ksi, E = 1,600 ksi, or determine actual values for a given wood species from Table A.11.3

The critical sections are the top section, for which Q = 2,000 lb., P = 3,000 lb. and M = 2,000(⅓)(21) = 14,000 lb.-ft. and the section situated at a distance 6.6/4 = 1.65 ft. from the ground surface, see Fig. 11.15. For the latter Q = 2,000 lb., P = 3,000 + W and M = 2,000 $\left(\frac{2}{3} \times 21 + 1.65\right)$ = 31,300 lb.-ft. The value of W, the weight of the pile above the critical section, may be estimated as follows. From Table 11.4, $B_2 = 25/\pi = 7.96$ in. = 0.664 ft., $B_1 \simeq 34/\pi = 10.82$ in. = 0.902 ft.; also L ≃ 25 ft. Substituting these values in the formula shown in Fig. 11.15 gives:

$$W = 50\pi\,(25)(0.664^2 + 0.902^2 + 0.664 \times 0.902)/12 = 605 \text{ lb.}$$

Thus, P + W = 3,600 lb. These loads correspond to wind loading conditions, for which $\overline{\alpha}$ = 1.33.

The value of F_c' can be determined from Eq. 11.22 using the dimensions of the section at one-third the length of the pole from the tip, i.e. 10 ft. The circumference of the section can be computed from data of Table 11.4 as follows: 25 + (34 − 25)(10/24) = 28.75 in. Hence, the radius of gyration r = (28.75/π)/4 = 2.29 in. Substituting E = 1,600 ksi, ℓ = 21 ft. = 252 in., and r = 2.29 in. in Eq. 11.22 yields:

$$F_c' = \frac{3.62 \times 1{,}600}{(252/2.29)^2} = 0.479 \text{ ksi} < 1.2 \text{ ksi}$$

Therefore, use F_c' = 0.479 ksi.

Consider the lower critical section. Substituting P = 3.6 k, M = 31.3 × 12 = 375 k-in., A = (π/4)(10.82)² = 92 in.², S = (π/32)(10.82)³ = 125 in.³, F_c' = 0.479 ksi, F_b = 2.15 ksi, C_f = 1.18 and $\overline{\alpha}$ = 1.33 in Eq. 11.21 yields:

$$\frac{3.60/92}{0.479} + \frac{375/125}{2.15 \times 1.18} = 0.08 + 1.18 = 1.26 < 1.33$$

which means the lower critical section is acceptable for simultaneous action of axial load and bending moment.

For the top critical section, using linear interpolation, B = [25 + (34 − 25)2.4/30]/π = 8.2 in. Also, P = 3 k, M = 14 × 12 = 168 k-in., A = (π/4)(8.2)² = 52.9 in.², S = (π/32)(8.2)³ = 54.2 in.³; as before F_c' = 0.479 ksi, F_b = 2.15 ksi, C_f = 1.18 and $\overline{\alpha}$ = 1.33. Substituting the preceding values in Eq. 11.21 yields:

$$\frac{3/52.9}{0.479} + \frac{168/54.2}{2.15 \times 1.18} = 0.12 + 1.22 = 1.34 \simeq 1.33$$

The section may be considered acceptable since it only exceeds the allowable by a very small margin. The capacity against shear may be checked by substituting Q = 2.0 k, A = 52.7 in.² in Eq. 11.25 as follows:

$$v = \frac{4}{3} \times \frac{2.0}{52.9} = 0.051 \text{ ksi} < 0.130 \times 1.33 \text{ ksi}$$

Therefore, the selected pole is acceptable.

11.7 DECKING

Coverage of floor and roof areas of wood buildings may be accomplished in a number of ways, namely, through wood decking, see Section 10.2.1, plywood sheathing, see Fig. 4.1 and Section 4.2, and stressed-skin panels, see Fig. 4.7 and Section 10.3.1. Because of the coverage given in the above-referred sections to the latter two types, this section is devoted to the discussion of wood decking only.

Decking may be either solid, see Fig. 11.17, or glued-laminated, see Fig. 11.18; sizes, weights and coverage factors of decking are given in Table A.11.4.

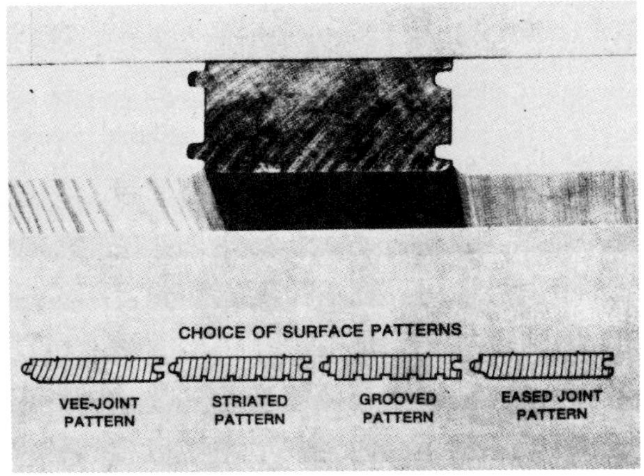

Figure 11.17. Solid decking. From Ref. 37.

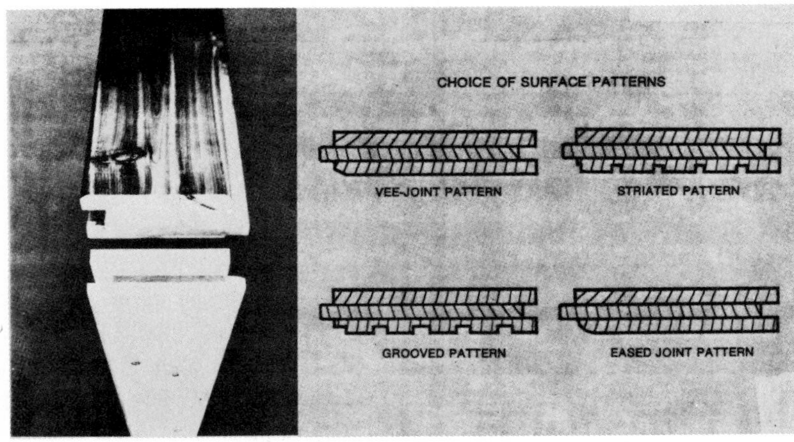

Figure 11.18. Glued-laminated decking. From Ref. 37.

Solid heavy timber decking is double tongue-and-groove, kiln-dried material of nominal 3 × 6 or 4 × 6 in. size available in two grades, namely, select and commercial. Select grade is recommended for construction for which good strength and fine appearance are desired; knots and other natural characteristics which add to the decorative character of the piece are permitted. When appearance and strength requirements are less critical commercial grade is recommended, and customarily used for the purposes served by select grade. Specific requirements for both select and commercial grades are given by AITC's Standard No. 112, see Ref. 1.

Glued-laminated decking is manufactured into single decking members from three or more kiln-dried laminations which form tongue-and-groove or double tongue-and-groove joints as shown in Fig. 11.18. Laminated decking is available in 3, 4 and 5 inch nominal thickness, and different appearance grades. Placement of laminated decking is with pattern faces down and exposed on the underside; each piece should be nailed to the supports with two 20d common nails for 3 in., 30d for 4 in., and 50d for 5 in. nominal thickness of deck. In addition, each course should be slant-nailed to the tongue of the adjacent course using 8d common nails for 3 in. and 16d for 4 and 5 in. nominal thickness of deck; nails should be spaced 30 in. apart with one nail not over 12 in. from the end of each piece. In adjacent rows nails should be staggered 15 in.

11.7.1 Layout Patterns

Five distinct layout patterns for wood decking are used in practice (37), see Fig. 11.19. Briefly, these patterns may be described as follows.

Pattern a. Simple span. All pieces are simply supported; end joints occur at all supports.

Pattern b. Two-span continuous. All pieces bear on three supports and are continuous over the interior one; all joints occur on every other support.

Pattern c. Combination simple and two-span continuous. Alternate pieces in end spans are simply supported, adjacent pieces are two-span continuous. End joints are staggered in adjacent courses and occur over supports.

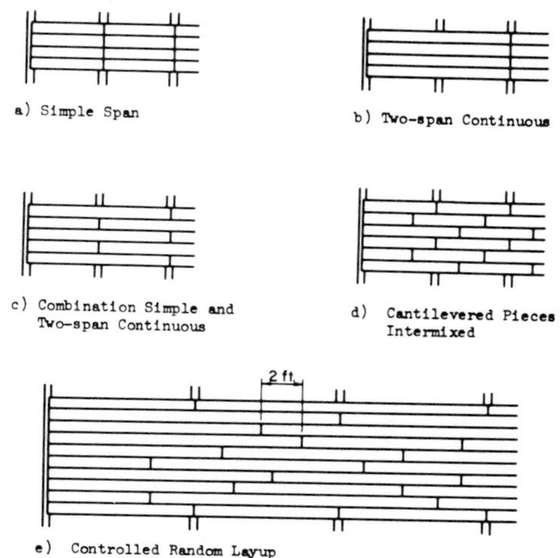

Figure 11.19. Standard layout patterns of timber decking.

Pattern d. Cantilever pieces intermixed. The starter course and every third course are simply supported. Pieces in other courses are cantilevered over the supports with end joints at alternate quarter-points or third-points of the spans, with each piece resting on at least one support. A tie between supports is provided by the simple span courses of the pattern.

Pattern e. Controlled random layup. This pattern is continuous for two or more spans. The distance between end joints in adjacent courses is at least 2 ft. Joints in the same general line (within 6 in. of being in line each way) shall be separated by at least two intervening courses. Pieces must rest on at least one support, with not more than one joint between supports in each course. To provide a continuous tie for lateral restraint for the supporting member, the pieces in at least the first and second courses, and repeating at least after each group of seven intervening courses, must bear on at least two supports with end joints in these two courses occurring in alternate spans or on alternate supports. Some other provision, such as plywood overlayment, may be made to provide continuity for the lateral bracing of the supporting member. The minimum length of pieces is 75 percent of the span length, i.e. for a 12 ft. spacing of supports the minimum length is 9 ft.

11.7.2 Design

Decking is usually designed to support uniformly distributed loads. In this respect, design for strength and serviceability demand the determination of maximum deflection, δ, of the decking and largest stress, f, generated at the critical section. The following general formulas evaluate δ and f.

$$\delta = C_1 \frac{wL^4}{EI} \tag{11.26}$$

and

$$f = C_2 \frac{wL^2}{(C_3 S)} \tag{11.27}$$

where:

w = uniformly distributed load
L = span of the deck between supports
E = modulus of elasticity of wood
I = moment of inertia of section of decking
S = section modulus of section of decking
C_1, C_2, C_3 = coefficients which depend on the layout pattern of the deck.

The coefficients C_1, C_2 and C_3 are given in Table 11.5, which is based on the specific layout patterns of the decking shown in Fig. 11.19.

TABLE 11.5. DESIGN COEFFICIENTS FOR WOOD DECKING.*

Layout Pattern	C_1	C_2	C_3
a	0.013021	0.125	1.0
b	0.005405	0.125	1.0
c	0.009174	0.100	0.5
d	0.009524	0.100	0.6667
e	0.010000	0.100	0.6667

* From data in Ref. 37.

The values of I and S for the selected type of decking must take into consideration its surface pattern, see Figs. 11.17 and 11.18; the designer should obtain specific information from the manufacturer on section properties of the decking. The values of δ and f obtained from Eqs. 11.26 and 11.27, respectively, should meet the serviceability and strength criteria established for the given structure.

Design is often simplified by means of tables which list the maximum uniformly distributed load that can be taken by decking of various sizes and species spanning a range of distances. Two layout patterns, a and e, are considered in Tables A.11.5 and A.11.6 for solid decking and laminated decking, respectively. The allowable total loads for spans ranging between 8 and 20 ft. are listed for two deflection criteria, namely 1/180th and 1/240th of the span. The tables are based on 2 months duration of loading, i.e. $\overline{\alpha} = 1.15$, which is typical of the design of roofs subjected to snow loads.

11.7.3 Illustrative Examples

Determine the net load that may be placed on a Douglas fir/Larch 3 × 6 select solid decking, placed in a controlled random layup, continuous over 3 or more equal 14 ft. spans, using a deflection limitation of 1/240th of the span. For the given conditions, Table A.11.5 gives a total load of 34 psf. The weight of the decking, from Table A.11.4, is 6.6 psf; thus, the net load that may be placed on the given decking is $34 - 6.6 = 27.4$ psf.

Determine the necessary section of Southern pine laminated decking laying continuously (pattern type e) over 15 ft. spans and subjected to a snow load of 50 psf. Maximum deflection should not exceed 1/240th of the span. Assume the weight of roof insulation and shingles as 8 psf. Consider 5 × 6 in. nominal size decking, weighing 10.5 psf, see Table A.11.4. From Table A.11.6, for the given conditions, $w = 97$ psf; the live load that the decking may take is $97 - (8 + 10.5) = 78.5 \text{ psf} > 50$ psf. Thus, the 5 × 6 in. nominal decking is acceptable. A smaller section, such as 4 × 6, or a simple span layout of the 5 × 6 in. sections, would not have rendered the necessary capacity.

An additional example of design of wood decking, using layout pattern c, is given in Section 10.2.1.

11.7.4 Inclined Decking

Equations 11.26 and 11.27, and Tables A.11.5 and A.11.6 may be used for the design of flat decks and inclined roofs with a maximum slope of 3/12. For greater slopes, see Fig. 11.20, the normal component of the load, w_N, causes bending and deflection of the decking while the tangential component of the load, w_T, is parallel to the slope and is taken by diaphragm action. Therefore, before using Eqs. 11.26 and 11.27, or entering Tables A.11.5 and A.11.6, the normal component, w_N, must be determined.

It can be shown, using simple statics, that the values of w_N, and w_T are given by Eqs. 11.28, as follows:

$$w_N = (w_L \cos\theta + w_D)\cos\theta$$
$$w_T = (w_L \cos\theta + w_D)\sin\theta \qquad (11.28)$$

where: w_L = vertical load, in pounds per square foot of horizontal projection of roof
w_D = dead load, in pounds per square foot of roof surface area
θ = angle made by plane of roof with the horizontal.

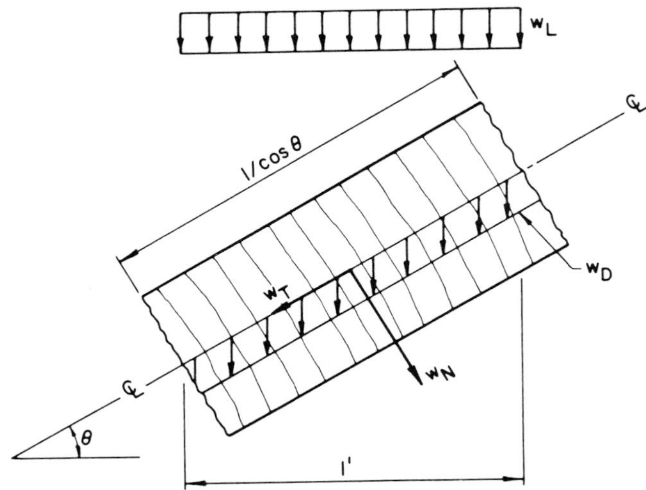

Figure 11.20. Determination of normal, w_N, and tangential, w_T, components of load on inclined decking.

As an example of use of Eqs. 11.28, consider $w_L = 60$ psf, $w_D = 10$ psf and a roof with a slope of 10 in 12. The angle θ is given by $\tan^{-1} 10/12$, from which $\theta = 39.8°$. Substitution of the preceding values in Eqs. 11.28 yields $w_N = 43$ psf and $w_T = 36$ psf. The same results are obtained by means of the nomograph given in Table A.11.7.

11.8 CANTILEVER-SUSPENDED BEAM CONSTRUCTION

In multiple-span beam design it is possible to obtain savings in cost and member size by using a measure of continuity over the supports. This can be easily obtained in wood structures using a cantilever-suspended system, where the lengths of the cantilever and suspended spans are carefully selected to minimize the design bending moment.

Consider, for example, the beams shown in Fig. 11.21. Placement of hinges, as indicated, reduces by a substantial amount the design moment. Compared to a series of simply-supported spans, or to a fully continuous beam, these solutions require smaller-size sections. This fact usually more than compensates for the cost of the hinged connection.

Continuous wood beams are difficult to attain principally because of their length. If made in one piece they would be awkward to manufacture and transport and if fabricated in two or more pieces, the rigid connections required to guarantee full continuity, see Section 6.5, may be expensive and difficult to accomplish. Connections such as required by cantilever-suspended systems, see Fig. 9.25, where no bending moment is transferred across, are much simpler and less expensive.

Beams over multiple spans can be reduced to combinations of cantilever-suspended systems by careful selection of the position of hinges. Two basic types are possible as follows.

Where continuity exists at both sides of the span under consideration, the basic type of beam is that shown to the left of Fig. 11.21. Two hinges are possible and the design moment is simply

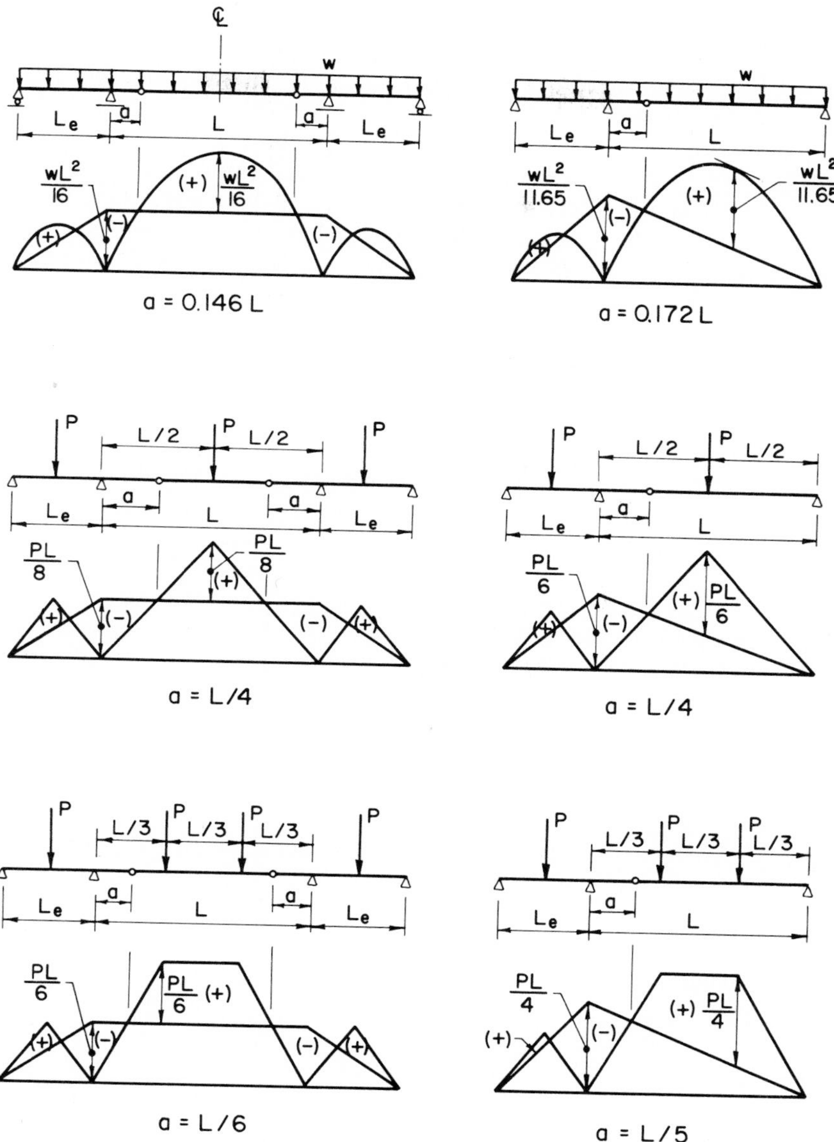

Figure 11.21. Various cantilever-suspended beams with hinges placed in optimum positions to minimize maximum moments.

one half the isostatic moment. (The latter term is the maximum moment in a simply-supported beam of identical span and loading). Determination of the position of the hinges, even in the case of asymmetrical loading, can be easily accomplished using simple free-body diagrams of the suspended span and cantilevers.

The other basic type of cantilever-suspended span, for the case where continuity exists only at one side of the span, is that shown by various examples at the right of Fig. 11.21. Only one hinge may be provided and the design moment is somewhat larger than one half the isostatic moment. Determination of the position of the hinge and the design bending moment is not as simple for this type of cantilever-suspended beam as it is for the one where two hinges are possible. To aid the designer in this task a general solution for the case of uniformly-distributed load has been obtained by the writer which is given in the following section. In addition, a numerical iteration method, suited for all kinds of loads and combinations thereof, is also explained by an illustrative example.

11.8.1 Determination of Hinge Location and Design Moment

Consider first the case of a beam subjected to uniformly distributed loads, Fig. 11.22. The beam is continuous at its right support and is conventionally cantilevered past its left support. The cantilever moment, M_c, is considered known, as it is statically determinate.

The minimum design moment for the beam is obtained by making equal the negative moment at the support, M^-, and the maximum positive moment in the span, M^+. This is possible by placing a hinge at a distance, a, from the right support. All that is required is the determination of the distance, a, for which $M^- = M^+ = M_o$.

From Fig. 11.22, and using simple statics, the value of maximum positive moment, M^+, is given by:

$$M^+ = \frac{V^2}{2w} - M_c \tag{11.29}$$

Where V, the shear force at the interior face of the left support is obtained from:

$$V = \frac{wL}{2} - \frac{M_o - M_c}{L} \tag{11.30}$$

Substituting the value of V given by Eq. 11.30 in Eq. 11.29 leads to:

$$M^+ = \frac{\left(\frac{wL}{2} - \frac{M_o - M_c}{L}\right)^2}{2w} - M_c \tag{11.31}$$

Using the condition $M^+ = M_o$, and solving Eq. 11.31 for M_o yields the following equation for determination of the design moment, M_o.

$$\frac{M_o}{\frac{wL^2}{8}} = \left[12 + \frac{M_c}{\frac{wL^2}{8}}\right] - 4\sqrt{2}\sqrt{4 + \frac{M_c}{\frac{wL^2}{8}}} \tag{11.32}$$

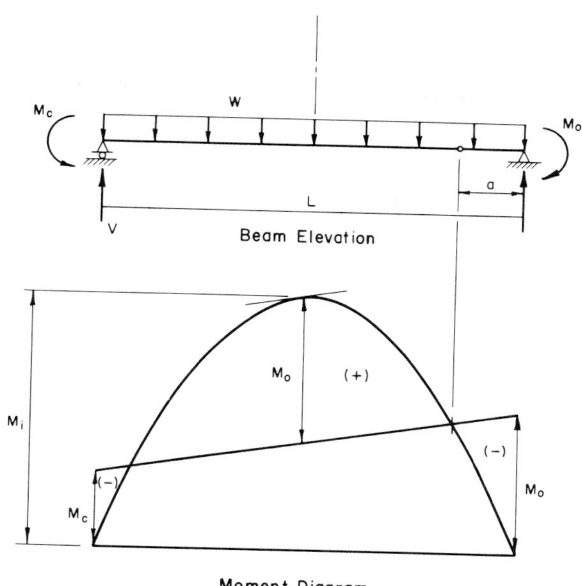

Figure 11.22. Determination of hinge location, a, to minimize maximum moment, M_o, in a continuous beam subjected to a known cantilever moment, M_c.

Determination of the distance, a, may be accomplished by establishing the condition M = 0 at the hinge. Using a free-body diagram of the portion of the beam to the right of the hinge, and taking moments about the latter, results in the following equation:

$$\left(\frac{wL}{2} + \frac{M_o - M_c}{L}\right) a - \frac{wa^2}{2} - M_o = 0 \tag{11.33}$$

from which, after rearranging some terms, one obtains:

$$\left(\frac{a}{L}\right)^2 - \left(1 + \frac{1}{4} \; \frac{M_o - M_c}{\frac{wL^2}{8}}\right)\left(\frac{a}{L}\right) + \frac{1}{4} \; \frac{M_o}{\frac{wL^2}{8}} = 0 \tag{11.34}$$

The numerical solution of quadratic Eqs. 11.33 and 11.34 is simple and can be easily done for any particular problem. However, for the purpose of illustration and to serve as design aids, consider the charts shown in Figs. 11.23 and 24. For any given values of M_c, w, and L it is possible to obtain M_o and a from these charts. Note that M_o varies between a maximum value of 0.6863 $(wL^2/8) = wL^2/11.65$ when the cantilever moment is zero, to a minimum value of $(0.5)(wL^2/8) = wL^2/16$, when the cantilever moment is also equal to the latter value. For the same two extreme cases the distance to the hinge, a, varies between 0.1716L and 0.1464L, respectively.

The beam of Fig. 10.3, see Section 10.2.3, corresponds to the case where the cantilever moment is zero. Since L = 60 ft. it follows that: a = 0.1716L = 10.30 ft., and $M = 0.6863(wL^2/8) = 309w$, as obtained in section 10.2.3.

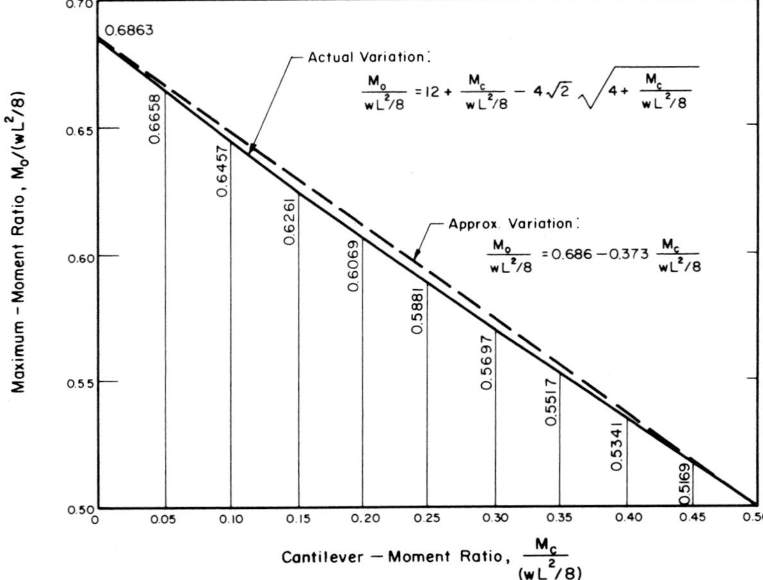

Figure 11.23. Determination of maximum moment, M_o, as a function of cantilever moment, M_c, for uniformly-distributed load, w.

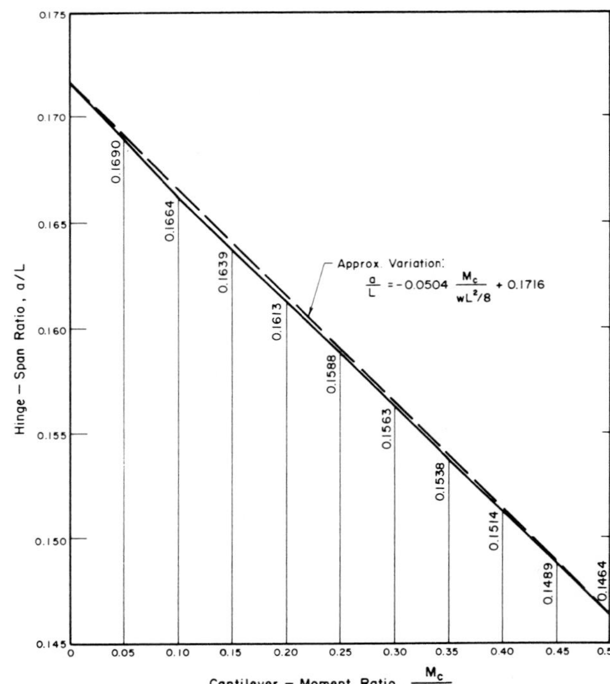

Figure 11.24. Determination of hinge position, a, as a function of cantilever moment, M_c, for uniformly-distributed load, w.

11.8.2 Illustrative Example

A glued-laminated girder spans 170 ft over three supports as shown in Fig. 11.25a. The beam carries a dead load equal to 640 lb/ft, which includes its own weight, and a live load of 600 lb/ft. Design requires placement of a hinge somewhere in the 100 ft. span to minimize the maximum design moment.

Consider the hinge situated at a distance a from the interior support. The dead and live load are uniformly distributed throughout the span. However, to obtain maximum effects on the interior span it is necessary not to place the live load on the cantilever, see Fig. 11.25a.

The following results are obtained using the notation of Fig. 11.22 and the problem data:

$$w = 0.60 + 0.64 = 1.24 \text{ k/ft},$$

$$w_D = 0.64 \text{ k/ft},$$

$$L = 100 \text{ ft}, \ L_c = 20 \text{ ft},$$

$$M_c = \frac{w_D L_c^2}{2} = \frac{(0.64)(20)^2}{2} = 128 \text{ k-ft.},$$

$$\frac{wL^2}{8} = \frac{(1.24)(100)^2}{8} = 1550 \text{ k-ft., and}$$

$$\frac{M_c}{\frac{wL^2}{8}} = \frac{128}{1550} = 0.08258$$

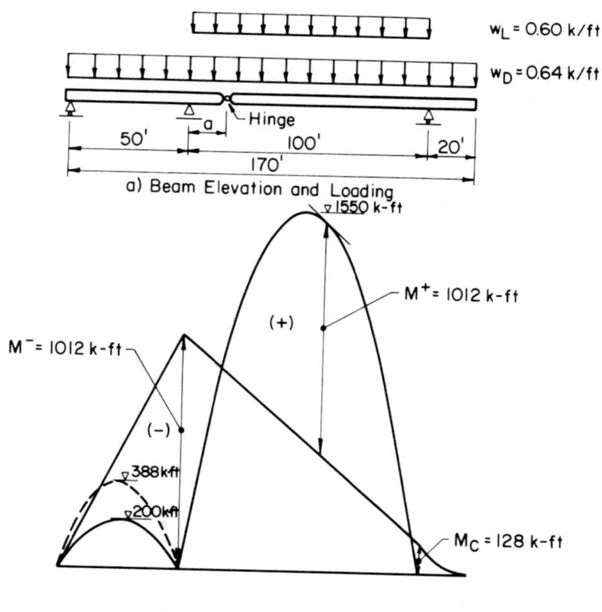

Figure 11.25. Determination of hinge location and maximum moment for a given cantilever-suspended beam.

from which, using Figs. 11.23 and 24 one obtains:

$$\frac{M_o}{\frac{wL^2}{8}} = 0.653 \text{ and } \frac{a}{L} = 0.167$$

Thus, M_o = (0.653)(1550) = 1012 k-ft. and a = (0.167)(100) = 16.7 ft. The final bending moment diagram is shown in Fig. 11.25b.

Had live load been present on the cantilever span the results would have been somewhat different. It is easy to show that the following would have been obtained: M_o = (0.622)(1550) = 964 k-ft. and a = (0.163)(100) = 16.3 ft. It is obvious that the smaller value of M_o obtained by placing live load on the cantilever should not be used for design. The designer has no guarantee that there will be live load on the 20 ft. cantilever every time there is live load on the 100 ft. interior span. Placement of live load on the 50 ft. exterior span does not change the design, as shown by the dotted bending-moment curve in Fig. 11.25b.

11.8.3 Numerical Method to Determine Hinge Location and Design Moment

Consider solving the previous problem directly, by means of a numerical iteration procedure. The general method is described in Fig. 11.26.

The cantilever moment, $M_c = 0.64\left(\frac{20^2}{2}\right) = 128$ k-ft.

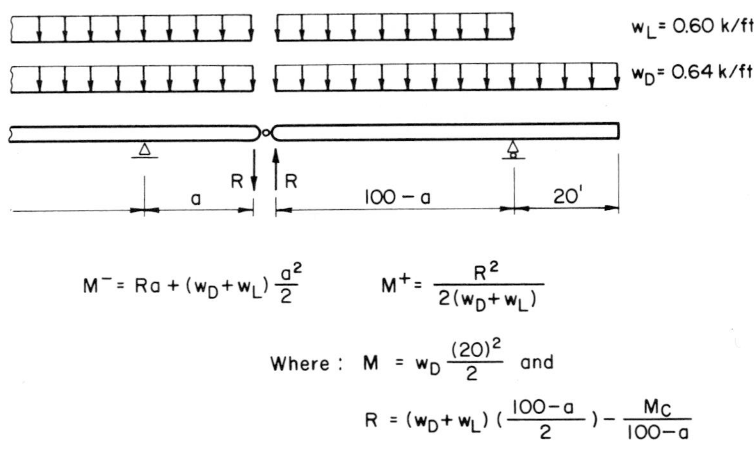

Figure 11.26. Iteration procedure to determine optimum position of hinge location, a.

Trial 1, a = 10 ft.

$$R = (0.64 + 0.60)\frac{(100 - 10)}{2} - \frac{128}{100 - 10} = 54.38 \text{ k}$$

$$M^+ = \frac{(54.38)^2}{2(0.640 + 0.60)} = 1192 \text{ k-ft.}$$

$$M^- = (54.38)(10) + (0.64 + 0.60)\left(\frac{10^2}{2}\right) = 606 \text{ k-ft.}$$

$$M^+ - M^- = 1192 - 606 = 586 \text{ k-ft.} > 5 \text{ k-ft., N. G.}$$

$$\frac{M^+ - M^-}{M^+ + M^-} = \frac{1192 - 606}{1192 + 606} = 0.33 > 0.01, \text{ N. G.}$$

Trial 2, a = 20 ft.

$$R = (0.64 + 0.60)\left(\frac{100 - 20}{2}\right) - \frac{128}{100 - 20} = 48.0 \text{ k}$$

$$M^+ = \frac{(48.0)^2}{2(0.64 + 0.60)} = 929 \text{ k-ft.}$$

$$M^- = (48.0)(20) + (0.64 + 0.60)\left(\frac{20^2}{2}\right) = 1208 \text{ k-ft.}$$

$$M^+ - M^- = 929 - 1208 = -279, \quad 279 > 5 \text{ k-ft., N. G.}$$

$$\frac{M^+ - M^-}{M^+ + M^-} = \frac{929 - 1208}{929 + 1208} = -0.13, \quad 0.13 > 0.01, \text{ N. G.}$$

An educated guess for the next trial may be obtained using linear interpolation between pairs of values ($M^+ - M^-$) and a, the results for the preceding two trials, as shown in Fig. 11.27. Thus, $a_3 = 16.8$ ft. The results for trial 3 are shown in Table 11.6. Note that at the end of trial 3 the difference between M^+ and M^- is only 6 k-ft or 0.3 percent of the sum of the two. This almost meets the absolute and relative tolerances (5 k-ft. and 1 percent, respectively) and may be good enough for beam design. The latter could take place for the average of the two moments, $M^+ = M^- = (1016 + 1010)/2 = 1013$ k-ft. If desired, a final trial (No. 4) could be carried out by first finding the value of a using the results of the last two trials and then following the established procedure. For this purpose it is found that a = 16.73 ft. The results for the final iteration cycle, see Table 11.6 are accurate to four significant figures and show $M^+ = M^- =$ 1012 k-ft. which coincides with the values obtained using the charts.

TABLE 11.6. ITERATION METHOD.

Summary of Results						
Trial	a, ft.	R, k	M^+ k-ft.	M^- k-ft.	$M^+ - M$ k-ft.	$\frac{M^+ - M^-}{M^+ + M^-}$ k-ft.
1	10	54.38	1192	606	586	0.33
2	20	48.00	929	1208	−279	−0.13
3	16.8	50.05	1010	1016	−6	0.003
4	16.73	50.09	1012	1012	0	0

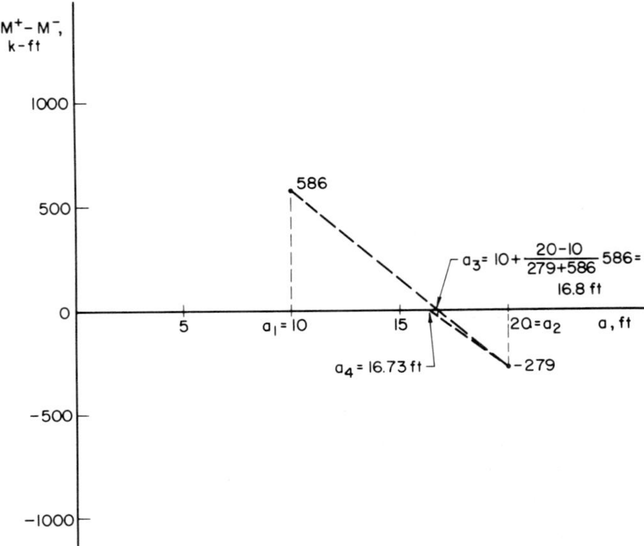

Figure 11.27. Interpolating to determine a better value for hinge location, a, to be used in iteration procedure.

11.8.4 Erection and Bracing of Cantilever-Suspended Beams

Erection of these beams is simple and straight forward. Cantilever spans are erected first. They should be attached to the supporting columns and braced laterally before the suspended beams are placed. Bracing of the beams can be accomplished by means of the joists and decking.

Special considerations of lateral stability should be given to the bottom flange of the cantilevered beams in the region where they are subjected to compression stresses (tip of cantilever to inflection point inside). Unless the supporting columns are laterally braced, i.e. prevented from displacements in a plane perpendicular to the plane of the beam, the total length of the bottom compression flange should be considered for the determination of unbraced length in the calculations of lateral stability, see Section 7.9. If the columns are laterally braced only the individual segments of beam (i.e. between the cantilever tip and the column, and between the latter and the point of inflection) need be considered. Lateral bracing of the bottom flange of the beam, by diagonals extending into the floor joists in vertical planes, or by lines of horizontal struts abutting against the bottom of the beams, is always possible.

The erection sequence and provisions for lateral bracing should preferably be specified by the designer of these systems. If these were left open to the contractor and erector of the works, for the purpose of improving competitive bidding, a provision in the contract should require submission to and approval by the designer before implementation. Close coordination of such details between designer and builder should prevent failure of the system during its erection stage.

11.8.5 Application to Floor-Joist Systems

The cantilever-suspended floor joists shown in Fig. 11.28 are structurally more efficient than conventional simple-span joists, such as those shown in Fig. 10.2. Thus, under the same loads, and for the same size and spacing of joists, the former are subjected to smaller bending stresses and deformations than the latter. It follows that cantilever-suspended joists are stronger and stiffer than simple-span joists.

Savings are possible by using joists one size smaller than would be required in the conventional system. Also, the usual overlapping of simple spans over the supports is eliminated, the joists are "in line" rather than offset, plywood floor panels are easier to place and material waste in subfloor and joists is minimized. In addition, the suspended spans make use of shorter lumber lengths, although, on the other hand, the cantilever spans require longer lumber lengths.

Information contained in Table A11.8 allows direct determination of the joist size required for building widths between 22 and 34 ft, and using standard commercial woods such as Douglas Fir, Southern Pine and Hem-fir. Selection is possible for 16 in. and 24 in. spacings between joists. The latter are designed to sustain 50 psf total load (of which 40 psf are considered live load) while deflecting not more than 1/360 of the clear span under live load. The actual layouts of cantilever and suspended joists for the various building widths is given in Table A.11.9.

Research by Baker (38) indicated the suitability of plywood splice plates to transfer the shear across the connection between the cantilever and suspended spans. The patterns shown in Fig. 11.29 call for a ½ in. thick plywood splice at each side of the joist. Fasteners are 10 d common nails driven from one side and clinched on the other in the direction of the splice face—grain by turning over the point of the nail so that it will hold fast. The nails act in double shear. No glue is required.

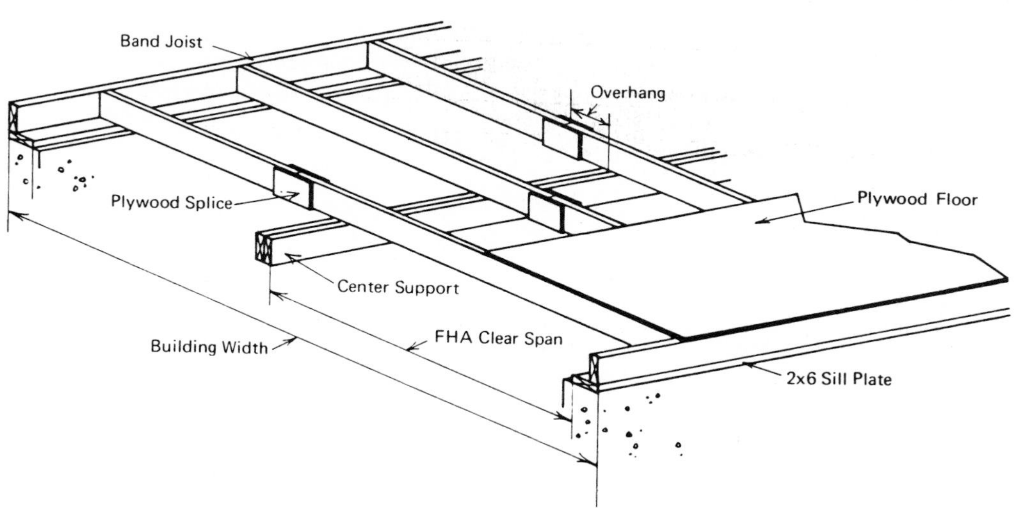

Figure 11.28. Cantilever-suspended joists used for building-floor system. Note cantilever and suspended spans alternating about center support to equalize strength and stiffness of both spans. From Ref. 39.

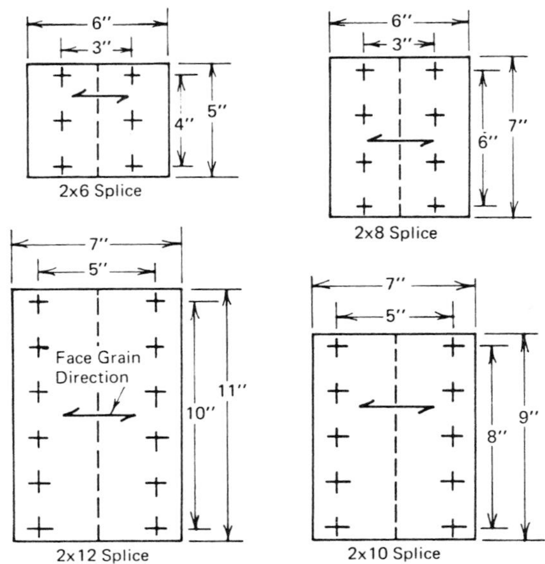

Figure 11.29. Recommended plywood splices for cantilever-suspended floor joists. From Ref. 39.

11.8.6 Additional Examples

Refer to Section 10.2.3 for actual analysis and design of a major glue-laminated girder. In addition, problem 11 in Chapter 9 proposes design of a cantilever-suspended wood bridge. Problems 8, 9, and 10 in Chapter 8 present related aspects of analysis and design of a three-span beam system.

REFERENCES

1. American Institute of Timber Construction, "Timber Construction Standards," AITC 100–69, Fifth Edition, 1969.
2. Wilson, T. R. C., "The Glued-laminated Wooden Arch," Technical Bulletin No. 691, U.S. Department of Agriculture, Washington D. C., October 1939, 123 pp.
3. Timoshenko, S. P., and Young, D. H., "Theory of Structures," Second Edition, McGraw-Hill Book Company, New York, 1964.
4. ICES STRUDL-I and II, MIT Department of Civil Engineering, Cambridge, Massachusetts.
5. Leontovich, V., "Frames and Arches, Condensed Solutions for Structural Analysis," Engineering Societies Monographs, Mc-Graw-Hill Book Company, Inc., New York, 1959.
6. Kleinlogel, A., "Rahmenformeln," W. Ernst und Sohn, Berlin, 1913. Also available in an English version from F.T. Ungar Co.
7. Morris, E. L., "A Modern Concept in Timber Truss Design," Civil Engineering, ASCE, March 1967, pp. 52–53.

8. Luxford, R. F., "Light Wood Trusses," Journal of the Structural Division, ASCE, Vol. 84 No. ST7, Nov. 1958, 48 pages.
9. Gurfinkel, G., "Steel Girders Support Covered Bridge," Engineering News-Record, Vol. 176, No. 11, pp. 191–192, March 18, 1965.
10. Gurfinkel, G., "Design of Uplift-Resistant Bridge Rocker," Civil Engineering Magazine, ASCE, Vol. 37, No. 4, April 1967.
11. Norris, C. H., and Wilbur, J. B., "Elementary Structural Analysis," 2nd Edition, McGraw-Hill Book Company, New York, 1960.
12. Timber Engineering Company, "Typical Designs of Timber Structures," A Reference for Use of Architects and Engineers by Timber Engineering Co., Washington, D. C.
13. Sweet's Catalogue Service, "Architectural Catalogue File," Vol. 2, Structural Systems, Accessories, F. W. Dodge Corporation, New York, a yearly publication.
14. Hansen, H. J., "Design Loads for Wooden Roof Trusses," University of Texas Engineering Experiment Station, Bulletin No. 74.
15. Gordon, Paul, "Details that Cause Failure in Timber Truss Design," Civil Engineering Magazine, ASCE, October 1965, pp. 74–75.
16. "National Design Specification for Stress-Grade Lumber and Its Fastenings," 1977 Edition, National Forest Products Association, Washington, D. C.
17. "Composite Truss of Wood, Wire and Pipe Creates Clear-Span Roof for Hockey Rink," Steel Developments Digest, Vol. 8, No. 1, September 1971. Also see Progressive Architecture, April 1971.
18. Khachaturian, N., and Gurfinkel, G., "Prestressed Concrete," McGraw-Hill Book Co., New York, 1969.
19. Bendtsen, B. A., and Eslyn, W. E., "House Foundation of Treated Wood After 30 Years' Service," U.S.D.A. Forest Products Laboratory, Madison, Wisconsin, August 1968.
20. National Forest Products Association, "The All-Weather Wood Foundation System, Design and Construction Methods," Technical Report No. 7 by NFPA.
21. Terzaghi, K. and Peck, R. B., "Soil Mechanics in Engineering Practice," John Wiley and Sons, N. Y., Second Edition, 1968.
22. Chellis, R. D., "Pile Foundations," McGraw-Hill Book Company, Inc., New York, Second Edition, 1961.
23. American Wood Preservers Institute, "Pile Foundations Know-How," 3rd Edition, October 1969; A series of 14 articles by various authors, available from AWPI, 2600 Virginia Ave., N. W., Washington, D. C. 20037.
24. Dames and Moore, "Timber Foundation Pile Study," published by American Wood Preservers Institute, Washington, D. C., 1966.
25. American Wood Preservers Institute, "Pressure Treated Timber Foundation Piles for Permanent Structures," published by AWPI, Washington, D. C., 1967.
26. Ayers, J. R., and Stoke, R. C., "Timber in Marine Structures," Journal of the Structural Division, ASCE, Vol. 93, No. ST2, Proc. Paper 5168, April 1967, pp. 33–45.
27. ASTM Designation D25–73, "Standard Specification for Round Timber Piles," American Society for Testing and Materials, 1978 Book of ASTM Standards, Part 22, pp. 23–26, 1978.
28. American Association of State Highway Officials, "Standard Specifications for Highway Bridges," Twelfth Edition, 1977, published by AASHTO, 444 N. Capital St., N. W., Washington, D. C., 20004.
29. Forest Products Laboratory, "Wood Handbook," U.S. Department of Agriculture Handbook No. 72, revised August 1974 U.S. Government Printing Office, Washington, D. C.
30. Norum, W. A., "Pole Buildings Go Modern," Journal of the Structural Division, ASCE, Vol. 93, No. ST2, Proc. Paper 5169, April 1967, pp. 47–56.
31. Anderson, L. O., and Smith, W. R., "Houses Can Resist Hurricanes," Forest Products Laboratory, Report FPL 33, Madison, Wis., August, 1965.
32. "How to Build Storm Resistant Structures," by Southern Forest Products Association, 1970.
33. Zornig, H. F., and Sherwood, G. E., "Wood Structures Survive Hurricane Camille's Winds," U.S.D.A. Forest Service Research Paper FPL 123, Forest Products Laboratory, Madison, Wis., October 1969.

34. Patterson, Donald, "Pole Building Design," 6th Edition 1969, Anerican Wood Preservers Institute, 2600 Virginia Ave., N. W., Washington, D. C. 20037.
35. Timoshenko, S. P., and Gere, J. M., "Theory of Elastic Stability," McGraw-Hill Book Co., Inc., New York, Second Edition, 1961.
36. American Institute of Timber Construction, "Timber Construction Manual," John Wiley and Sons, Inc., New York, Second Edition, 1974.
37. American Institute of Timber Construction, "Structural Glued-Laminated Timber," Publication 6.5/Ai, Sweet's Catalogue, 1972.
38. Baker, William A., "Plywood and the Cantilevered Floor Joist System," 1975 by American Plywood Association, Research Report 130, APA, 1119 A Street, Tacoma, Washington, 98401.
39. "Cantilevered In-Line Joint System," American Plywood Association, Tacoma, Washington, 98401.

Appendix Tables

TABLE A2.1.* CLEAR WOOD STRENGTH VALUES UNADJUSTED FOR END USE AND MEASURES OF VARIATION FOR COMMERCIAL SPECIES OF WOOD IN THE UNSEASONED CONDITION (METHOD A).[a]

NOTE – All digits retained in the averages and standard deviations through the units position to permit further computation with minimum round-off error (specific gravity excepted).

Species or Region, or Both	Property																
	Modulus of Rupture[b]			Modulus of Elasticity[c]			Compression Parallel to Grain, crushing strength, max			Shear Strength			Compression, Perpendicular to Grain;[d] Fiber Stress at Proportional Limit		Specific Gravity		
	Avg, psi	Variability Index	Standard Deviation, psi	Avg, 1000 psi	Variability Index	Standard Deviation, 1000 psi	Avg, psi	Variability Index	Standard Deviation, psi	Avg, psi	Variability Index	Standard Deviation, psi	Avg, psi	Standard Deviation, psi	Avg, psi	Variability Index	Standard Deviation
Douglas fir[e]																	
Coast	7665	1.05	1317	1560	1.05	315	3784	1.05	734	904	1.03	131	382	107	0.45	. . .	0.057
Interior West	7713	1.03	1322	1513	1.04	324	3872	1.04	799	936	1.02	137	418	117	0.46	. . .	0.058
Interior North	7438	1.04	1163	1409	1.04	274	3469	1.04	602	947	1.03	126	356	100	0.45	. . .	0.049
Interior South	6784	1.01	908	1162	1.00	200	3113	1.01	489	953	1.00	153	337	94	0.43	. . .	0.045
White fir	5854	1.01	949	1161	1.02	249	2902	1.02	528	756	1.01	78	282	79	0.37	. . .	0.045
Claifornia red fir	5809	1.01	885	1170	1.01	267	2758	1.01	459	767	1.00	146	334	94	0.36	. . .	0.043
Grand fir	5839	1.03	680	1250	1.03	164	2939	1.04	363	739	1.04	97	272	76	0.35	. . .	0.043
Pacific silver fir	6410	1.07	1296	1420	1.05	255	3142	1.06	591	746	1.05	114	225	63	0.39	. . .	0.058
Noble fir	6169	1.07	966	1380	1.08	310	3013	1.08	561	802	1.04	136	274	77	0.37	. . .	0.043
Western hemlock	6637	1.03	1088	1307	1.02	258	3364	1.03	615	864	1.02	105	282	79	0.42	. . .	0.053
Western larch	7652	1.04	1001	1458	1.02	249	3756	1.04	564	869	1.03	85	399	112	0.48	. . .	0.048
Black cottonwood	4890	1.00	951	1083	1.00	197	2200	1.00	360	612	1.00	92	165	46	0.31	. . .	0.034
Southern pine																	
Loblolly	7300	1.08	1199	1402	1.08	321	3511	1.09	612	863	1.05	112	389	109	0.47	1.06	0.057
Longleaf	8538	1.07	1305	1586	1.07	295	4321	1.07	707	1041	1.05	120	479	134	0.54	1.05	0.069
Shortleaf	7435	1.04	1167	1388	1.04	268	3527	1.05	564	905	1.05	125	353	99	0.47	1.05	0.052
Slash	8692	1.09	1127	1532	1.08	295	3823	1.07	547	964	1.05	128	529	148	0.54	1.09	0.062

[a] For tension parallel and perpendicular to grain and modulus of rigidity, see 4.3.
[b] Modulus of rupture values are applicable to material 2 in. (51 mm) in depth.
[c] Modulus of elasticity values are applicable at a ratio of shear span to depth of 14.
[d] All maximum crushing strength perpendicular to grain values are based on standard test data only.
[e] The regional description of Douglas fir is that given on pp. 54–55 of U.S. Forest Service Research Paper FPL 27, "Western Wood Density Survey Report No. 1."

*From Table 1 of ASTM Standard D2555–78, Part 22, ASTM 1978 Book of Standards.

TABLE A2.2.* CLEAR WOOD STRENGTH VALUES UNADJUSTED FOR END USE AND MEASURES OF VARIATION FOR COMMERCIAL SPECIES OF WOOD IN THE UNSEASONED CONDITION (METHOD B) (FOR WOODS GROWN IN THE UNITED STATES).[a]

NOTE – All digits retained in the averages and standard deviations through the units position to permit further computation with minimum round-off error (specific gravity excepted).

NOTE – Values of standard deviation have been calculated using the values for c given in 4.2.

Species (Official Common Tree Names)	Property											
	Modulus of Rupture[b]		Modulus of Elasticity[c]		Compression Parallel to Grain, Crushing Strength, max		Shear Strength		Compression Perpendicular to Grain, Fiber Stress at Proportional Limit		Specific Gravity	
	Avg, psi	Standard Deviation, psi	Avg, 1000 psi	Standard Deviation, 1000 psi	Avg, psi	Standard Deviation, psi	Avg, psi	Standard Deviation, psi	Avg, psi	Standard Deviation, psi	Avg	Standard Deviation
SOFTWOODS												
Basswood, American	4 960	794	1038	228	2220	400	599	84	170	48	0.32	0.032
Baldcypress	6 640	1062	1184	260	3580	644	812	114	403	113	0.43	0.043
Cedar:												
Alaska	6 450	1032	1135	260	3050	549	842	118	349	98	0.42	0.042
Incense	6 220	995	840	185	3150	567	834	117	369	103	0.35	0.035
Port Orford	6 598	860	1297	247	3145	397	842	122	301	71	0.39	0.034
Atlantic white	4 740	758	752	165	2390	430	694	97	244	68	0.31	0.031
Northern white	4 250	680	643	141	1990	358	616	86	234	66	0.29	0.029
Eastern red	7 030	1125	649	143	3570	643	1008	141	700	196	0.46	0.046
Western red	5 184	761	939	223	2774	493	771	115	244	65	0.31	0.027
Fir:												
Balsam	5 517	552	1251	143	2631	283	662	83	187	31.2	0.322	0.025
Subalpine	4 900	664	1052	182	2301	363	696	103	192	44	0.31	0.032
Hemlock:												
Eastern	6 420	1027	1073	236	3080	554	848	119	359	101	0.39	0.039
Mountain	6 270	1003	1038	228	2880	518	933	131	371	104	0.42	0.042
Pine:												
Jack	6 030	965	1068	235	2950	531	754	106	296	83	0.40	0.040
Eastern white	4 930	789	994	219	2440	439	678	95	218	61	0.35	0.035
Lodgepole	5 490	878	1076	237	2610	470	685	96	252	71	0.39	0.039
Monterey	6 625	1060	1420	312	3330	599	875	123	440	123	0.46	0.046
Ponderosa	5 130	821	997	219	2450	441	704	99	282	79	0.39	0.039
Red	5 820	931	1281	282	2730	491	686	96	259	73	0.42	0.042
Sugar	4 893	663	1032	193	2459	386	718	105	214	43	0.34	0.027
Western white	4 688	693	1193	257	2434	406	677	98	192	46	0.35	0.034
Pine, southern yellow:												
Pitch	6 830	1093	1200	264	2950	531	860	120	365	102	0.47	0.047
Pond	7 450	1192	1281	282	3660	659	936	131	441	123	0.51	0.051
Spruce	6 004	1102	1002	286	2835	580	895	136	279	95	0.41	0.041
Sand	7 500	1200	1024	225	3440	619	1143	160	450	126	0.46	0.046
Virginia	7 330	1173	1218	268	3420	616	888	124	390	109	0.46	0.046
Redwood:												
Old growth	7 500	1202	1177	259	4210	758	803	112	424	119	0.39	0.039
Second growth	5 920	947	955	210	3110	560	894	125	269	75	0.34	0.034
Spruce:												
Black	6 118	759	1382	193	2836	417	739	79	242	33.5	0.384	0.028

*From Table 2 of ASTM Standard D2555–78, Part 22 ASTM 1978 Book of Standards.

TABLE A2.2—*CONTINUED*

Table 2 *Continued*

Species (Official Common Tree Names)	Property											
	Modulus of Rupture[b]		Modulus of Elasticity[c]		Compression Parallel to Grain, Crushing Strength, max		Shear Strength		Compression Perpendicular to Grain, Fiber Stress at Proportional Limit		Specific Gravity	
	Avg. psi	Standard Deviation, psi	Avg. 1000 psi	Standard Deviation, 1000 psi	Avg. psi	Standard Deviation, psi	Avg. psi	Standard Deviation, psi	Avg. psi	Standard Deviation, psi	Avg	Standard Deviation
SOFTWOODS—*Continued*												
Engelmann	4 705	692	1029	207	2180	427	637	64	197	50	0.33	0.033
Red	6 003	627	1328	145	2721	313	754	95	262	59.4	0.373	0.025
Sitka	5 660	906	1230	271	2670	481	757	106	279	78	0.38	0.038
White	4 995	878	1141	265	2349	439	636	68	210	51.3	0.328	0.034
Tamarack	7 170	1147	1236	272	3480	626	863	121	389	109	0.49	0.049
HARDWOODS												
Alder, red	6 540	1044	1167	257	2960	484	770	108	250	70	0.38	0.038
Ash:												
Black	6 000	960	1043	229	2300	414	861	120	347	97	0.45	0.045
Green	9 460	1514	1400	308	4200	756	1261	176	734	206	0.53	0.053
White	9 500	1520	1436	316	3990	718	1354	190	667	187	0.54	0.054
Aspen:												
Bigtooth	5 400	864	1120	246	2500	450	732	102	206	58	0.36	0.036
Quaking	5 130	821	860	189	2140	385	656	92	181	51	0.35	0.035
Beech, American	8 570	1371	1381	304	3550	639	1288	180	544	152	0.57	0.057
Birch:												
Paper	6 380	1021	1170	257	2360	425	836	117	273	76	0.48	0.048
Sweet	9 390	1502	1650	363	3740	673	1245	174	473	132	0.60	0.060
Yellow	8 260	1322	1504	331	3380	608	1106	155	428	120	0.55	0.055
Cottonwood:												
Eastern	5 260	842	1013	223	2280	410	682	95	196	55	0.37	0.037
Elm:												
American	7 190	1150	1114	245	2910	524	1002	140	355	99	0.46	0.046
Rock	9 490	1518	1194	263	3780	680	1274	178	610	171	0.57	0.057
Slippery	8 010	1282	1232	271	3320	598	1106	155	415	116	0.49	0.049
Hackberry	6 480	1037	954	210	2650	477	1070	150	399	112	0.49	0.049
Hickory:												
Pecan	9 770	1563	1367	301	3990	718	1482	207	777	218	0.61	0.061
Water	10 740	1718	1563	344	4660	839	1440	202	881	247	0.63	0.063
Mockernut	11 080	1773	1574	346	4480	806	1277	179	812	227	0.64	0.064
Pignut	11 740	1878	1652	363	4810	866	1370	192	923	258	0.67	0.067
Shagbark	11 020	1763	1566	344	4580	824	1520	213	843	236	0.64	0.064
Shellbark	10 530	1685	1343	295	3920	706	1186	166	808	226	0.63	0.063
Bitternut	10 280	1645	1399	308	4570	823	1237	173	799	224	0.62	0.062
Nutmeg	9 060	1450	1289	284	3980	716	1032	144	760	213	0.56	0.056
Magnolia:												
Cucumbertree	7 420	1187	1565	344	3140	565	991	139	330	92	0.44	0.044
Southern magnolia	6 780	1085	1106	243	2700	486	1044	146	462	129	0.46	0.046

TABLE A2.2—*CONTINUED*

Table 2 *Continued*

Species (Official Common Tree Names)	Property											
	Modulus of Rupture[b]		Modulus of Elasticity[c]		Compression Parallel to Grain, Crushing Strength, max		Shear Strength		Compression Perpendicular to Grain, Fiber Stress at Proportional Limit		Specific Gravity	
	Avg, psi	Standard Deviation, psi	Avg, 1000 psi	Standard Deviation, 1000 psi	Avg, psi	Standard Deviation, psi	Avg, psi	Standard Deviation, psi	Avg, psi	Standard Deviation, psi	Avg	Standard Deviation
HARDWOODS – *Continued*												
Maple:												
Bigleaf	7 390	1182	1095	241	3240	583	1108	155	449	126	0.44	0.044
Black	7 920	1267	1328	292	3270	589	1128	158	601	168	0.52	0.052
Sugar	9 420	1507	1546	340	4020	724	1465	205	645	181	0.57	0.057
Red	7 690	1230	1386	305	3280	590	1151	161	405	113	0.50	0.050
Silver	5 820	931	943	207	2490	448	1053	147	369	103	0.44	0.044
Oak, red:												
Black	8 220	1315	1182	260	3470	625	1222	171	706	198	0.56	0.056
Cherrybark	10 850	1736	1790	394	4620	832	1321	185	765	214	0.60	0.060
Northern red	8 300	1328	1353	298	3440	619	1214	170	614	172	0.56	0.056
Southern red	6 920	1107	1141	251	3030	545	934	131	547	153	0.53	0.053
Laurel	7 940	1270	1393	306	3170	571	1182	165	573	160	0.56	0.056
Pin	8 330	1333	1318	290	3680	662	1293	181	715	200	0.58	0.058
Scarlet	10 420	1667	1476	325	4090	736	1411	198	834	234	0.61	0.061
Water	8 910	1426	1552	341	3740	673	1240	174	620	174	0.56	0.056
Willow	7 400	1184	1286	283	3000	540	1184	166	611	171	0.55	0.055
Oak, white:												
Chestnut	8 030	1285	1372	302	3520	634	1212	170	532	149	0.58	0.058
Live	11 930	1909	1575	346	5430	977	2210	309	2039	571	0.81	0.081
Post	8 080	1293	1086	239	3480	626	1278	179	855	239	0.60	0.060
Swamp chestnut	8 480	1357	1350	297	3540	637	1262	177	573	160	0.60	0.060
White	8 300	1328	1246	274	3560	641	1249	175	671	188	0.60	0.060
Bur	7 180	1149	877	193	3290	592	1354	190	677	190	0.60	0.060
Overcup	8 000	1280	1146	252	3370	607	1315	184	539	151	0.56	0.056
Swamp white	9 860	1578	1593	350	4360	785	1296	181	764	214	0.64	0.064
Poplar, balsam	3 860	618	748	165	1690	304	504	71	136	38	0.30	0.030
Sycamore, American	6 470	1035	1065	234	2920	526	996	139	365	102	0.46	0.046
Sweetgum	7 110	1138	1201	264	3040	547	992	139	367	103	0.46	0.046
Tanoak	10 470	1675	1550	341	4650	837	...	...	...	...	0.58	0.058
Tupelo:												
Black	7 040	1126	1031	227	3040	547	1098	154	485	136	0.47	0.047
Water	7 300	1168	1052	231	3370	607	1194	167	480	134	0.46	0.046
Yellow-poplar	5 950	952	1222	269	2660	479	792	111	269	75	0.40	0.040

[a] For tension parallel and perpendicular to grain and modulus of rigidity, see 4.3.
[b] Modulus of rupture values are applicable to material 2 in. (51 mm) in depth.
[c] Modulus of elasticity values are applicable at a ratio of shear span to depth of 14.

TABLE A.2.3. SCHOLTEN'S NOMOGRAPHS.

The Hankinson formula may be solved graphically through use of the charts on this page.

Figure C-1. Bearing Strength of Wood at Angles to the Grain (Hankinson Formula).

The compressive strength of wood depends on the direction of the grain with respect to the direction of the applied load. It is highest parallel to the grain, and lowest perpendicular to the grain. The variation in strength, at angles between parallel and perpendicular, is determined by the Hankinson formula. The Scholten nomographs, shown here, are a graphical solution of this formula which is —

$$F_n = \frac{F_c F_{c\perp}}{F_c \sin^2\theta + F_{c\perp} \cos^2\theta}$$

F_c = Unit stress in compression parallel to the grain.

$F_{c\perp}$ = Unit stress in compression perpendicular to the grain.

θ = Angle between the direction of grain and direction of load normal to the face considered.

F_n = Unit compressive stress at inclination θ with the direction of grain.

The difference between the two charts is in scale, the one on the right to units of 1000 pounds, and the one on the left to units of 100 pounds. These units may be applied to allowable lumber stresses in pounds per square inch, or to total loads in the case of bolts, timber connectors or lag screws.

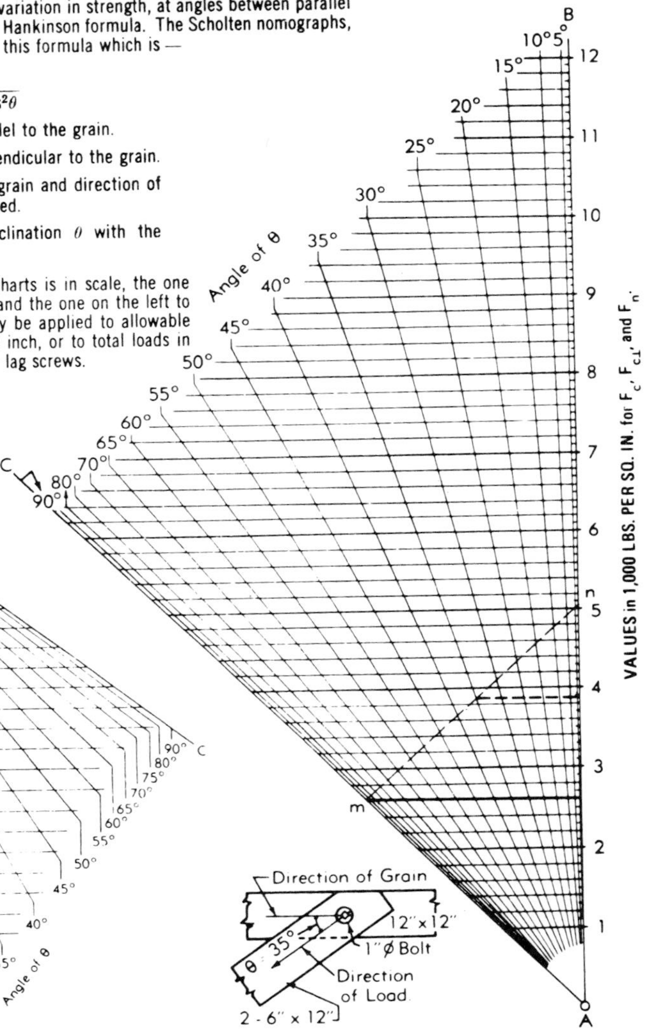

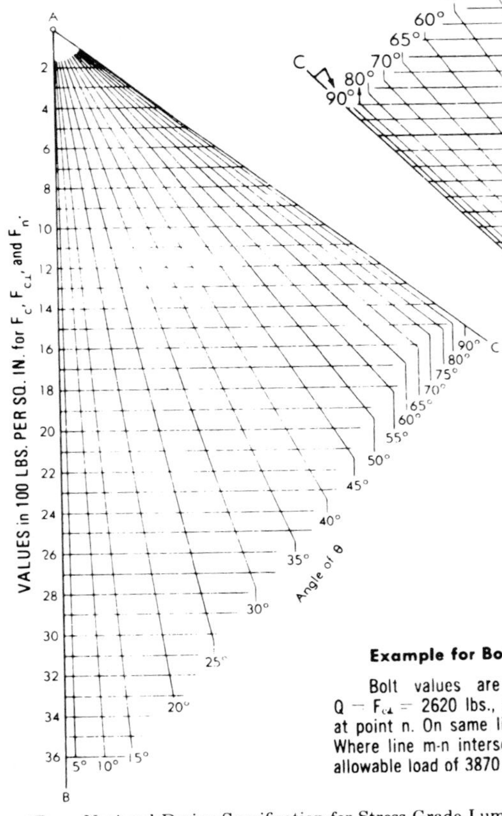

Example for Bolted Joint

Bolt values are obtained from Table 12. Assume P = F_c' = 5030 lbs., Q = $F_{c\perp}$ = 2620 lbs., and θ = 35°. On line A-B, in chart at right, locate 5030 lbs. at point n. On same line A-B, locate 2620 lbs. and project to point m on line A-C. Where line m-n intersects the radial line for 35°, project to line A-B and read the allowable load of 3870 lbs for the 35° angle of load to grain.

†From National Design Specification for Stress-Grade Lumber and Its Fastenings, 1971 Edition by NFPA

TABLE A2.4

APPENDIX M—PROPERTIES OF STRUCTURAL LUMBER

TABLE M-2. Sectional Properties of Standard Dressed (S4S) Sizes

Nominal size b(inches)d	Standard dressed size (S4S) b(inches)d	Area of Section A	Moment of inertia I	Section modulus S	Weight in pounds per linear foot of piece when weight of wood per cubic foot equals:					
					25 lb.	30 lb.	35 lb.	40 lb.	45 lb.	50 lb.
1 x 3	3/4 x 2-1/2	1.875	0.977	0.781	0.326	0.391	0.456	0.521	0.586	0.651
1 x 4	3/4 x 3-1/2	2.625	2.680	1.531	0.456	0.547	0.638	0.729	0.820	0.911
1 x 6	3/4 x 5-1/2	4.125	10.398	3.781	0.716	0.859	1.003	1.146	1.289	1.432
1 x 8	3/4 x 7-1/4	5.438	23.817	6.570	0.944	1.133	1.322	1.510	1.699	1.888
1 x 10	3/4 x 9-1/4	6.938	49.466	10.695	1.204	1.445	1.686	1.927	2.168	2.409
1 x 12	3/4 x 11-1/4	8.438	88.989	15.820	1.465	1.758	2.051	2.344	2.637	2.930
2 x 3	1-1/2 x 2-1/2	3.750	1.953	1.563	0.651	0.781	0.911	1.042	1.172	1.302
2 x 4	1-1/2 x 3-1/2	5.250	5.359	3.063	0.911	1.094	1.276	1.458	1.641	1.823
2 x 5	1-1/2 x 4-1/2	6.750	11.391	5.063	1.172	1.406	1.641	1.875	2.109	2.344
2 x 6	1-1/2 x 5-1/2	8.250	20.797	7.563	1.432	1.719	2.005	2.292	2.578	2.865
2 x 8	1-1/2 x 7-1/4	10.875	47.635	13.141	1.888	2.266	2.643	3.021	3.398	3.776
2 x 10	1-1/2 x 9-1/4	13.875	98.932	21.391	2.409	2.891	3.372	3.854	4.336	4.818
2 x 12	1-1/2 x 11-1/4	16.875	177.979	31.641	2.930	3.516	4.102	4.688	5.273	5.859
2 x 14	1-1/2 x 13-1/4	19.875	290.775	43.891	3.451	4.141	4.831	5.521	6.211	6.901
3 x 1	2-1/2 x 3/4	1.875	0.088	0.234	0.326	0.391	0.456	0.521	0.586	0.651
3 x 2	2-1/2 x 1-1/2	3.750	0.703	0.938	0.651	0.781	0.911	1.042	1.172	1.302
3 x 4	2-1/2 x 3-1/2	8.750	8.932	5.104	1.519	1.823	2.127	2.431	2.734	3.038
3 x 5	2-1/2 x 4-1/2	11.250	18.984	8.438	1.953	2.344	2.734	3.125	3.516	3.906
3 x 6	2-1/2 x 5-1/2	13.750	34.661	12.604	2.387	2.865	3.342	3.819	4.297	4.774
3 x 8	2-1/2 x 7-1/4	18.125	79.391	21.901	3.147	3.776	4.405	5.035	5.664	6.293
3 x 10	2-1/2 x 9-1/4	23.125	164.886	35.651	4.015	4.818	5.621	6.424	7.227	8.030
3 x 12	2-1/2 x 11-1/4	28.125	296.631	52.734	4.883	5.859	6.836	7.813	8.789	9.766
3 x 14	2-1/2 x 13-1/4	33.125	484.625	73.151	5.751	6.901	8.051	9.201	10.352	11.502
3 x 16	2-1/2 x 15-1/4	38.125	738.870	96.901	6.619	7.943	9.266	10.590	11.914	13.238
4 x 1	3-1/2 x 3/4	2.625	0.123	0.328	0.456	0.547	0.638	0.729	0.820	0.911
4 x 2	3-1/2 x 1-1/2	5.250	0.984	1.313	0.911	1.094	1.276	1.458	1.641	1.823
4 x 3	3-1/2 x 2-1/2	8.750	4.557	3.646	1.519	1.823	2.127	2.431	2.734	3.038
4 x 4	3-1/2 x 3-1/2	12.250	12.505	7.146	2.127	2.552	2.977	3.403	3.828	4.253
4 x 5	3-1/2 x 4-1/2	15.750	26.578	11.813	2.734	3.281	3.828	4.375	4.922	5.469
4 x 6	3-1/2 x 5-1/2	19.250	48.526	17.646	3.342	4.010	4.679	5.347	6.016	6.684
4 x 8	3-1/2 x 7-1/4	25.375	111.148	30.661	4.405	5.286	6.168	7.049	7.930	8.811
4 x 10	3-1/2 x 9-1/4	32.375	230.840	49.911	5.621	6.745	7.869	8.933	10.117	11.241
4 x 12	3-1/2 x 11-1/4	39.375	415.283	73.828	6.836	8.203	9.570	10.938	12.305	13.672
4 x 14	3-1/2 x 13-1/4	46.375	678.475	102.411	8.047	9.657	11.266	12.877	14.485	16.094
4 x 16	3-1/2 x 15-1/4	53.375	1034.418	135.66	9.267	11.121	12.975	14.828	16.682	18.536
5 x 2	4-1/2 x 1-1/2	6.750	1.266	1.688	1.172	1.406	1.641	1.875	2.109	2.344
5 x 3	4-1/2 x 2-1/2	11.250	5.859	4.688	1.953	2.344	2.734	3.125	3.516	3.906
5 x 4	4-1/2 x 3-1/2	15.750	16.078	9.188	2.734	3.281	3.828	4.375	4.922	5.469
5 x 5	4-1/2 x 4-1/2	20.250	34.172	15.188	3.516	4.219	4.922	5.675	6.328	7.031
6 x 1	5-1/2 x 3/4	4.125	0.193	0.516	0.716	0.859	1.003	1.146	1.289	1.432
6 x 2	5-1/2 x 1-1/2	8.250	1.547	2.063	1.432	1.719	2.005	2.292	2.578	2.865
6 x 3	5-1/2 x 2-1/2	13.750	7.161	5.729	2.387	2.865	3.342	3.819	4.297	4.774
6 x 4	5-1/2 x 3-1/2	19.250	19.651	11.229	3.342	4.010	4.679	5.347	6.016	6.684
6 x 6	5-1/2 x 5-1/2	30.250	76.255	27.729	5.252	6.302	7.352	8.403	9.453	10.503
6 x 8	5-1/2 x 7-1/2	41.250	193.359	51.563	7.161	8.594	10.026	11.458	12.891	14.323
6 x 10	5-1/2 x 9-1/2	52.250	392.963	82.729	9.071	10.885	12.700	14.514	16.328	18.142
6 x 12	5-1/2 x 11-1/2	63.250	697.068	121.229	10.981	13.177	15.373	17.569	19.766	21.962
6 x 14	5-1/2 x 13-1/2	74.250	1127.672	167.063	12.891	15.469	18.047	20.625	23.203	25.781
6 x 16	5-1/2 x 15-1/2	85.250	1706.776	220.229	14.800	17.760	20.720	23.681	26.641	29.601
6 x 18	5-1/2 x 17-1/2	96.250	2456.380	280.729	16.710	20.052	23.394	26.736	30.078	33.420
6 x 20	5-1/2 x 19-1/2	107.250	3398.484	348.563	18.620	22.344	26.068	29.792	33.516	37.240
6 x 22	5-1/2 x 21-1/2	118.250	4555.086	423.729	20.530	24.635	28.741	32.847	36.953	41.059
6 x 24	5-1/2 x 23-1/2	129.250	5948.191	506.229	22.439	26.927	31.415	35.903	40.391	44.878
8 x 1	7-1/4 x 3/4	5.438	0.255	0.680	0.944	1.133	1.322	1.510	1.699	1.888
8 x 2	7-1/4 x 1-1/2	10.875	2.039	2.719	1.888	2.266	2.643	3.021	3.398	3.776
8 x 3	7-1/4 x 2-1/2	18.125	9.440	7.552	3.147	3.776	4.405	5.035	5.664	6.293
8 x 4	7-1/4 x 3-1/2	25.375	25.904	14.803	4.405	5.286	6.168	7.049	7.930	8.811
8 x 6	7-1/2 x 5-1/2	41.250	103.984	37.813	7.161	8.594	10.026	11.458	12.891	14.323
8 x 8	7-1/2 x 7-1/2	56.250	263.672	70.313	9.766	11.719	13.672	15.625	17.578	19.531
8 x 10	7-1/2 x 9-1/2	71.250	535.859	112.813	12.370	14.844	17.318	19.792	22.266	24.740
8 x 12	7-1/2 x 11-1/2	86.250	950.547	165.313	14.974	17.969	20.964	23.958	26.953	29.948
8 x 14	7-1/2 x 13-1/2	101.250	1537.734	227.813	17.578	21.094	24.609	28.125	31.641	35.156
8 x 16	7-1/2 x 15-1/2	116.250	2327.422	300.313	20.182	24.219	28.255	32.292	36.328	40.365
8 x 18	7-1/2 x 17-1/2	131.250	3349.609	382.813	22.786	27.344	31.901	36.458	41.016	45.573
8 x 20	7-1/2 x 19-1/2	146.250	4634.297	475.313	25.391	30.469	35.547	40.625	45.703	50.781
8 x 22	7-1/2 x 21-1/2	161.250	6211.484	577.813	27.995	33.594	39.193	44.792	50.391	55.990
8 x 24	7-1/2 x 23-1/2	176.250	8111.172	690.313	30.599	36.719	42.839	48.958	55.078	61.198

TABLE A2.4—*CONTINUED*

APPENDIX M—PROPERTIES OF STRUCTURAL LUMBER

TABLE M-2. Sectional Properties of Standard Dressed (S4S) Sizes (Cont'd)

Nominal size b(inches)d	Standard dressed size (S4S) b(inches)d	Area of Section A	Moment of inertia I	Section modulus S	Weight in pounds per linear foot of peice when weight of wood per cubic foot equals:					
					25 lb.	30 lb.	35 lb.	40 lb.	45 lb.	50 lb.
10 x 1	9-1/4 x 3/4	6.938	0.325	0.867	1.204	1.445	1.686	1.927	2.168	2.409
10 x 2	9-1/4 x 1-1/2	13.875	2.602	3.469	2.409	2.891	3.372	3.854	4.336	4.818
10 x 3	9-1/4 x 2-1/2	23.125	12.044	9.635	4.015	4.818	5.621	6.424	7.227	8.030
10 x 4	9-1/4 x 3-1/2	32.375	33.049	18.885	5.621	6.745	7.869	8.993	10.117	11.241
10 x 6	9-1/2 x 5-1/2	52.250	131.714	47.896	9.071	10.885	12.700	14.514	16.328	18.142
10 x 8	9-1/2 x 7-1/2	71.250	333.984	89.063	12.370	14.844	17.318	19.792	22.266	24.740
10 x 10	9-1/2 x 9-1/2	90.250	678.755	142.896	15.668	18.802	21.936	25.069	28.203	31.337
10 x 12	9-1/2 x 11-1/2	109.250	1204.026	209.396	18.967	22.760	26.554	30.347	34.141	37.934
10 x 14	9-1/2 x 13-1/2	128.250	1947.797	288.563	22.266	26.719	31.172	35.625	40.078	44.531
10 x 16	9-1/2 x 15-1/2	147.250	2948.068	380.396	25.564	30.677	35.790	40.903	46.016	51.128
10 x 18	9-1/2 x 17-1/2	166.250	4242.836	484.896	28.863	34.635	40.408	46.181	51.953	57.726
10 x 20	9-1/2 x 19-1/2	185.250	5870.109	602.063	32.161	38.594	45.026	51.458	57.891	64.323
10 x 22	9-1/2 x 21-1/2	204.250	7867.879	731.896	35.460	42.552	49.644	56.736	63.828	70.920
10 x 24	9-1/2 x 23-1/2	223.250	10274.148	874.396	38.759	46.510	54.262	62.014	69.766	77.517
12 x 1	11-1/4 x 3/4	8.438	0.396	1.055	1.465	1.758	2.051	2.344	2.637	2.930
12 x 2	11-1/4 x 1-1/2	16.875	3.164	4.219	2.930	3.516	4.102	4.688	5.273	5.859
12 x 3	11-1/4 x 2-1/2	28.125	14.648	11.719	4.883	5.859	6.836	7.813	8.789	9.766
12 x 4	11-1/4 x 3-1/2	39.375	40.195	22.969	6.836	8.203	9.570	10.938	12.305	13.672
12 x 6	11-1/2 x 5-1/2	63.250	159.443	57.979	10.981	13.177	15.373	17.569	19.766	21.962
12 x 8	11-1/2 x 7-1/2	86.250	404.297	107.813	14.974	17.969	20.964	23.958	26.953	29.948
12 x 10	11-1/2 x 9-1/2	109.250	821.651	172.979	18.967	22.760	26.554	30.347	34.141	37.934
12 x 12	11-1/2 x 11-1/2	132.250	1457.505	253.479	22.960	27.552	32.144	36.736	41.328	45.920
12 x 14	11-1/2 x 13-1/2	155.250	2357.859	349.313	26.953	32.344	37.734	43.125	48.516	53.906
12 x 16	11-1/2 x 15-1/2	178.250	3568.713	460.479	30.946	37.135	43.325	49.514	55.703	61.892
12 x 18	11-1/2 x 17-1/2	201.250	5136.066	586.979	34.939	41.927	48.915	55.903	62.891	69.878
12 x 20	11-1/2 x 19-1/2	224.250	7105.922	728.813	38.932	46.719	54.505	62.292	70.078	77.865
12 x 22	11-1/2 x 21-1/2	247.250	9524.273	885.979	42.925	51.510	60.095	68.681	77.266	85.851
12 x 24	11-1/2 x 23-1/2	270.250	12437.129	1058.479	46.918	56.302	65.686	75.069	84.453	93.837
14 x 2	13-1/4 x 1-1/2	19.875	3.727	4.969	3.451	4.141	4.831	5.521	6.211	6.901
14 x 3	13-1/4 x 2-1/2	33.125	17.253	13.802	5.751	6.901	8.051	9.201	10.352	11.502
14 x 4	13-1/4 x 3-1/2	46.375	47.34	27.052	8.047	9.657	11.266	12.877	14.485	16.094
14 x 6	13-1/2 x 5-1/2	74.250	187.172	68.063	12.891	15.469	18.047	20.625	23.203	25.781
14 x 8	13-1/2 x 7-1/2	101.250	474.609	126.563	17.578	21.094	24.609	28.125	31.641	35.156
14 x 10	13-1/2 x 9-1/2	128.250	964.547	203.063	22.266	26.719	31.172	35.625	40.078	44.531
14 x 12	13-1/2 x 11-1/2	155.250	1710.984	297.563	26.953	32.344	37.734	43.125	48.516	53.906
14 x 16	13-1/2 x 15-1/2	209.250	4189.359	540.563	36.328	43.594	50.859	58.125	65.391	72.656
14 x 18	13-1/2 x 17-1/2	236.250	6029.297	689.063	41.016	49.219	57.422	65.625	73.828	82.031
14 x 20	13-1/2 x 19-1/2	263.250	8341.734	855.563	45.703	54.844	63.984	73.125	82.266	91.406
14 x 22	13-1/2 x 21-1/2	290.250	11180.672	1040.063	50.391	60.469	70.547	80.625	90.703	100.781
14 x 24	13-1/2 x 23-1/2	317.250	14600.109	1242.563	55.078	66.094	77.109	88.125	99.141	110.156
16 x 3	15-1/4 x 2-1/2	38.125	19.857	15.885	6.619	7.944	9.267	10.592	11.915	13.240
16 x 4	15-1/4 x 3-1/2	53.375	54.487	31.135	9.267	11.121	12.975	14.828	16.682	18.536
16 x 6	15-1/2 x 5-1/2	85.250	214.901	78.146	14.800	17.760	20.720	23.681	26.641	29.601
16 x 8	15-1/2 x 7-1/2	116.250	544.922	145.313	20.182	24.219	28.255	32.292	36.328	40.365
16 x 10	15-1/2 x 9-1/2	147.250	1107.443	233.146	25.564	30.677	35.790	40.903	46.016	51.128
16 x 12	15-1/2 x 11-1/2	178.250	1964.463	341.646	30.946	37.135	43.325	49.514	55.703	61.892
16 x 14	15-1/2 x 13-1/2	209.250	3177.984	470.813	36.328	43.594	50.859	58.125	65.391	72.656
16 x 16	15-1/2 x 15-1/2	240.250	4810.004	620.646	41.710	50.052	58.394	66.736	75.078	83.420
16 x 18	15-1/2 x 17-1/2	271.250	6922.523	791.146	47.092	56.510	65.929	75.347	84.766	94.184
16 x 20	15-1/2 x 19-1/2	302.250	9577.547	982.313	52.474	62.969	73.464	83.958	94.453	104.948
16 x 22	15-1/2 x 21-1/2	333.250	12837.066	1194.146	57.856	69.427	80.998	92.569	104.141	115.712
16 x 24	15-1/2 x 23-1/2	364.250	16763.086	1426.646	63.238	75.885	88.533	101.181	113.828	126.476
18 x 6	17-1/2 x 5-1/2	96.250	242.630	88.229	16.710	20.052	23.394	26.736	30.078	33.420
18 x 8	17-1/2 x 7-1/2	131.250	615.234	164.063	22.786	27.344	31.901	36.458	41.016	45.573
18 x 10	17-1/2 x 9-1/2	166.250	1250.338	263.229	28.863	34.635	40.408	46.181	51.953	57.726
18 x 12	17-1/2 x 11-1/2	201.250	2217.943	385.729	34.939	41.927	48.915	55.903	62.891	69.878
18 x 14	17-1/2 x 13-1/2	236.250	3588.047	531.563	41.016	49.219	57.422	65.625	73.828	82.031
18 x 16	17-1/2 x 15-1/2	271.250	5430.648	700.729	47.092	56.510	65.929	75.347	84.766	94.184
18 x 18	17-1/2 x 17-1/2	306.250	7815.754	893.229	53.168	63.802	74.436	85.069	95.703	106.337
18 x 20	17-1/2 x 19-1/2	341.250	10813.359	1109.063	59.245	71.094	82.943	94.792	106.641	118.490
18 x 22	17-1/2 x 21-1/2	376.250	14493.461	1348.229	65.321	78.385	91.450	104.514	117.578	130.642
18 x 24	17-1/2 x 23-1/2	411.250	18926.066	1610.729	71.398	85.677	99.957	114.236	128.516	142.795

TABLE A2.4*—*CONTINUED*

APPENDIX M—PROPERTIES OF STRUCTURAL LUMBER

TABLE M-2. Sectional Properties of Standard Dressed (S4S) Sizes (Cont'd)

Nominal size b(inches)d	Standard dressed size (S4S) b(inches)d	Area of Section A	Moment of inertia I	Section modulus S	Weight in pounds per linear foot of piece when weight of wood per cubic foot equals:					
					25 lb.	30 lb.	35 lb.	40 lb.	45 lb.	50 lb.
20 x 6	19-1/2 x 5-1/2	107.250	270.359	98.313	18.620	22.344	26.068	29.792	33.516	37.240
20 x 8	19-1/2 x 7-1/2	146.250	685.547	182.813	25.391	30.469	35.547	40.625	45.703	50.781
20 x 10	19-1/2 x 9-1/2	185.250	1393.234	293.313	32.161	38.594	45.026	51.458	57.891	64.323
20 x 12	19-1/2 x 11-1/2	224.250	2471.422	429.813	38.932	46.719	54.505	62.292	70.078	77.865
20 x 14	19-1/2 x 13-1/2	263.250	3998.109	592.313	45.703	54.844	63.984	73.125	82.266	91.406
20 x 16	19-1/2 x 15-1/2	302.250	6051.297	780.813	52.474	62.969	73.464	83.958	94.453	104.948
20 x 18	19-1/2 x 17-1/2	341.250	8708.984	995.313	59.245	71.094	82.943	94.792	106.641	118.490
20 x 20	19-1/2 x 19-1/2	380.250	12049.172	1235.813	66.016	79.219	92.422	105.625	118.828	132.031
20 x 22	19-1/2 x 21-1/2	419.250	16149.859	1502.313	72.786	87.344	101.901	116.458	131.016	145.573
20 x 24	19-1/2 x 23-1/2	458.250	21089.047	1794.813	79.557	95.469	111.380	127.292	243.203	159.115
22 x 6	21-1/2 x 5-1/2	118.250	298.088	108.396	20.530	24.635	28.741	32.847	36.953	41.059
22 x 8	21-1/2 x 7-1/2	161.250	755.859	201.563	27.995	33.594	39.193	44.792	50.391	55.990
22 x 10	21-1/2 x 9-1/2	204.250	1536.130	323.396	35.460	42.552	49.644	56.736	63.828	70.920
22 x 12	21-1/2 x 11-1/2	247.250	2724.901	473.896	42.925	51.510	60.095	68.681	77.266	85.851
22 x 14	21-1/2 x 13-1/2	290.250	4408.172	653.063	50.391	60.469	70.547	80.625	90.703	100.781
22 x 16	21-1/2 x 15-1/2	333.250	6671.941	860.896	57.856	69.427	80.998	92.569	104.141	115.712
22 x 18	21-1/2 x 17-1/2	376.250	9602.211	1097.396	65.321	78.385	91.450	104.514	117.578	130.642
22 x 20	21-1/2 x 19-1/2	419.250	13284.984	1362.563	72.786	87.344	101.901	116.458	131.016	145.573
22 x 22	21-1/2 x 21-1/2	462.250	17806.254	1656.396	80.252	96.302	112.352	128.403	144.453	160.503
22 x 24	21-1/2 x 23-1/2	505.250	23252.023	1978.896	87.717	105.260	122.804	140.347	157.891	175.434
24 x 6	23-1/2 x 5-1/2	129.250	325.818	118.479	22.439	26.927	31.415	35.903	40.391	44.878
24 x 8	23-1/2 x 7-1/2	176.250	826.172	220.313	30.599	36.719	42.839	48.958	55.078	61.198
24 x 10	23-1/2 x 9-1/2	223.250	1679.026	353.479	38.759	46.510	54.262	62.014	69.766	77.517
24 x 12	23-1/2 x 11-1/2	270.250	2978.380	517.979	46.918	56.302	65.686	75.069	84.453	93.837
24 x 14	23-1/2 x 13-1/2	317.250	4818.234	713.813	55.078	66.094	77.109	88.125	99.141	110.156
24 x 16	23.1/2 x 15-1/2	364.250	7292.586	940.979	63.238	75.885	88.533	101.181	113.828	126.476
24 x 18	23-1/2 x 17-1/2	411.250	10495.441	1199.479	71.398	85.677	99.957	114.236	128.516	142.795
24 x 20	23-1/2 x 19-1/2	458.250	14520.797	1489.313	79.557	95.469	111.380	127.292	143.203	159.115
24 x 22	23-1/2 x 21-1/2	505.250	19462.648	1810.479	87.717	105.260	122.804	140.347	157.891	175.434
24 x 24	24-1/2 x 23-1/2	552.250	25415.004	2162.979	95.877	115.052	134.227	153.403	172.578	191.753

*From National Design Specification, 1977 Edition by NFPA.

TABLE A2.5.* DESIGN VALUES FOR VISUALLY GRADED STRUCTURAL LUMBER.

Design values for a new Southern Pine species sub-group, Virginia Pine-Pond Pine, and a correction of compression perpendicular stresses for Eastern Hemlock. Also included is a list of those agencies certified by the American Lumber Standards Committee Board of Review for inspection and grading under the grading rules mentioned above.

(Design values listed are for normal loading conditions. See other provisions in the footnotes and in the National Design Specification for adjustments of tabulated values.)

Species and commercial grade	Size classification	Design values in pounds per square inch							Grading rules agency
		Extreme fiber in bending "F_b"		Tension parallel to grain "F_t"	Horizontal shear "F_v"	Compression perpendicular to grain "$F_{c\perp}$"	Compression parallel to grain "F_c"	Modulus of elasticity "E"	
		Single-member uses	Repetitive-member uses						
ASPEN (Surfaced dry or surfaced green. Used at 19% max. m.c.)									
Select Structural		1300	1500	775	60	185	850	1,100,000	
No. 1	2" to 4"	1100	1300	650	60	185	675	1,100,000	
No. 2	thick	925	1050	525	60	185	550	1,000,000	
No. 3	2" to 4"	500	575	300	60	185	325	900,000	
Appearance	wide	1100	1300	650	60	185	825	1,100,000	NELMA
Stud		500	575	300	60	185	325	900,000	NHPMA
									WWPA
Construction	2" to 4"	650	750	400	60	185	625	900,000	
Standard	thick	375	425	225	60	185	500	900,000	(See footnotes
Utility	4" wide	175	200	100	60	185	325	900,000	1 through 11)
Select Structural		1150	1300	750	60	185	750	1,100,000	
No. 1	2" to 4"	950	1100	650	60	185	675	1,100,000	
No. 2	thick	775	900	425	60	185	575	1,000,000	
No. 3	5" and	450	525	250	60	185	375	900,000	
Appearance	wider	950	1100	650	60	185	825	1,100,000	
Stud		450	525	250	60	185	375	900,000	
BALSAM FIR (Surfaced dry or surfaced green. Used at 19% max. m.c.)									
Select Structural	2" to 4"	1750	2000	1000	70	185	1350	1,500,000	
No. 1	thick	1450	1700	850	70	185	1050	1,500,000	
No. 2	2" to 4"	1200	1400	700	70	185	850	1,300,000	
No. 3	wide	675	775	400	70	185	525	1,200,000	
Appearance		1450	1700	850	70	185	1250	1,500,000	NELMA
Stud		675	775	400	70	185	525	1,200,000	NHPMA
									(see footnores
Construction	2" to 4"	875	1000	525	70	185	950	1,200,000	1 through 12)
Standard	thick	500	575	275	70	185	775	1,200,000	
Utility	4" wide	225	275	125	70	185	525	1,200,000	
Select Structural	2" to 4"	1500	1700	1000	70	185	1200	1,500,000	
No. 1	thick	1250	1450	850	70	185	1050	1,500,000	
No. 2	5" and	1050	1200	550	70	185	900	1,300,000	
No. 3	wider	600	700	325	70	185	575	1,200,000	
Appearance		1250	1450	850	70	185	1250	1,500,000	
Stud		600	700	325	70	185	575	1,200,000	
Select Structural	Beams and	1350	—	900	65	185	950	1,400,000	
No. 1	Stringers	1100	—	750	65	185	800	1,400,000	
Select Structural	Posts and	1250	—	825	65	185	1000	1,400,000	
No. 1	Timbers	1000	—	675	65	185	875	1,400,000	
Select	Decking	—	1650	—	—	—	—	1,500,000	NELMA
Commercial		—	1400	—	—	—	—	1,300,000	(see footnotes
									1 through 12)
BLACK COTTONWOOD (Surfaced dry or surfaced green. Used at 19% max. m.c.)									
Select Structural		1000	1200	600	50	100	725	1,200,000	
No. 1	2" to 3"	875	1000	500	50	100	575	1,200,000	
No. 2	thick	725	825	425	50	100	450	1,100,000	
No. 3	2" to 4"	400	450	225	50	100	275	900,000	
Appearance	wide	875	1000	500	50	100	700	1,200,000	NLGA
Stud		400	450	225	50	100	275	900,000	
Construction	2" to 4"	525	600	300	50	100	525	900,000	(A Canadian
Standard	thick	300	325	175	50	100	425	900,000	agency. See
Utility	4" wide	150	150	75	50	100	275	900,000	footnotes 1
									through 12
Select Structural		875	1000	575	50	100	650	1,200,000	and 15
No. 1	2" to 4"	750	875	500	50	100	575	1,200,000	and 16)
No. 2	thick	625	700	325	50	100	475	1,100,000	
No. 3	5" and	350	425	175	50	100	300	900,000	
Appearance	wider	750	875	500	50	100	700	1,200,000	
Stud		350	425	175	50	100	300	900,000	

***From National Design Specification, April 1980 Supplement by NFPA.**

TABLE A2.5—*CONTINUED*

DESIGN VALUES FOR VISUALLY GRADED STRUCTURAL LUMBER

(Design values listed are for normal loading conditions. See other provisions in the footnotes and in the National Design Specification for adjustments of tabulated values.)

Species and commercial grade	Size classification	Design values in pounds per square inch: Extreme fiber in bending "F_b": Single-member uses	Extreme fiber in bending "F_b": Repetitive member uses	Tension parallel to grain "F_t"	Horizontal shear "F_v"	Compression perpendicular to grain "$F_{c\perp}$"	Compression parallel to grain "F_c"	Modulus of elasticity "E"	Grading rules agency
CALIFORNIA REDWOOD (Surfaced dry or surfaced green. Used at 19% max. m.c.)									
Clear Heart Structural	4" & less thick, any width	2300	2650	1550	145	425	2150	1,400,000	
Clear Structural		2300	2650	1550	145	425	2150	1,400,000	
Select Structural	2" to 4" thick 2" to 4" wide	2050	2350	1200	80	425	1750	1,400,000	
Select Structural, Open grain		1600	1850	950	80	270	1300	1,100,000	
No. 1		1700	1950	975	80	425	1400	1,400,000	
No. 1, Open grain		1350	1550	775	80	270	1050	1,100,000	
No. 2		1400	1600	800	80	425	1100	1,250,000	
No. 2, Open grain		1100	1250	625	80	270	825	1,000,000	
No. 3		800	900	475	80	425	675	1,100,000	
No. 3, Open grain		625	725	375	80	270	500	900,000	
Stud		625	725	375	80	270	500	900,000	
Construction	2" to 4" thick 4" wide	825	950	475	80	270	925	900,000	
Standard		450	525	250	80	270	775	900,000	
Utility		225	250	125	80	270	500	900,000	RIS (see footnotes 1 through 6, 8 through 10 and 12)
Select Structural	2" to 4" thick 5" and wider	1750	2000	1150	80	425	1550	1,400,000	
Select Structural, Open grain		1400	1600	925	80	270	1150	1,100,000	
No. 1		1500	1700	975	80	425	1400	1,400,000	
No. 1, Open grain		1150	1350	775	80	270	1050	1,100,000	
No. 2		1200	1400	650	80	425	1200	1,250,000	
No. 2, Open grain		950	1100	500	80	270	875	1,000,000	
No. 3		700	800	375	80	425	725	1,100,000	
No. 3, Open grain		550	650	350	80	270	525	900,000	
Stud		700	800	375	80	425	725	1,100,000	
Clear Heart Structural or Clear Structural	5" by 5" and larger	1850	—	1250	135	425	1650	1,300,000	
Select Structural		1400	—	950	95	425	1200	1,300,000	
No. 1		1200	—	800	95	425	1050	1,300,000	
No. 2		975	—	650	95	425	900	1,100,000	
No. 3		550	—	375	95	425	550	1,000,000	
Select Decking, Close grain	Decking 2" thick 6" and wider	1850	2150	—	—	—	—	1,400,000	RIS (see footnotes 1, 2, 8 and 9)
Select Decking		1450	1700	—	—	—	—	1,100,000	
Commercial Decking		1200	1350	—	—	—	—	1,000,000	
COAST SITKA SPRUCE (Surfaced dry or surfaced green. Used at 19% max. m.c.)									
Select Structural	2" to 3" thick 2" to 4" wide	1500	1700	875	65	290	1100	1,700,000	
No. 1		1250	1450	750	65	290	875	1,700,000	
No. 2		1050	1200	625	65	290	700	1,500,000	
No. 3		575	675	350	65	290	425	1,300,000	
Appearance		1250	1450	725	65	290	1050	1,700,000	
Stud		575	675	350	65	290	425	1,300,000	
Construction	2" to 4" thick 4" wide	750	875	450	65	290	800	1,300,000	
Standard		425	500	250	65	290	650	1,300,000	
Utility		200	225	125	65	290	425	1,300,000	NLGA
Select Structural	2" to 4" thick 5" and wider	1300	1500	850	65	290	975	1,700,000	(A Canadian agency. See footnotes 1 through 12 and 15 and 16)
No. 1		1100	1250	725	65	290	875	1,700,000	
No. 2		900	1050	475	65	290	750	1,500,000	
No. 3		525	600	275	65	290	475	1,300,000	
Appearance		1100	1250	725	65	290	1050	1,700,000	
Stud		525	600	275	65	290	475	1,300,000	
Select Structural	Beams and Stringers	1150	—	675	60	290	775	1,500,000	
No. 1		950	—	475	60	290	650	1,500,000	
Select Structural	Posts and Timbers	1100	—	725	60	290	825	1,500,000	
No. 1		875	—	575	60	290	725	1,500,000	
Select	Decking	1250	1450	—	—	290	—	1,700,000	
Commercial		1050	1200	—	—	290	—	1,500,000	
COAST SPECIES (Surfaced dry or surfaced green. Used at 19% max. m.c.)									
Select Structural	2" to 3" thick 2" to 4" wide	1500	1700	875	65	235	1100	1,500,000	
No. 1		1250	1450	750	65	235	875	1,500,000	
No. 2		1050	1200	625	65	235	700	1,400,000	
No. 3		575	675	350	65	235	425	1,200,000	
Appearance		1250	1450	725	65	235	1050	1,500,000	
Stud		575	675	350	65	235	425	1,200,000	
Construction	2" to 4" thick 4" wide	750	875	450	65	235	800	1,200,000	NLGA
Standard		425	500	250	65	235	650	1,200,000	
Utility		200	225	125	65	235	425	1,200,000	
Select Structural	2" to 4" thick 5" and wider	1300	1500	850	65	235	975	1,500,000	(A Canadian agency. See footnotes 1 through 12 and 15 and 16)
No. 1		1100	1250	725	65	235	875	1,500,000	
No. 2		900	1050	475	65	235	750	1,400,000	
No. 3		525	600	275	65	235	475	1,200,000	
Appearance		1100	1250	725	65	235	1050	1,500,000	
Stud		525	600	275	65	235	475	1,200,000	
Select	Decking	1250	1450	—	—	235	—	1,500,000	
Commercial		1050	1200	—	—	235	—	1,400,000	

TABLE A2.5—*CONTINUED*

DESIGN VALUES FOR VISUALLY GRADED STRUCTURAL LUMBER

(Design values listed are for normal loading conditions. See other provisions in the footnotes and in the National Design Specification for adjustments of tabulated values.)

Species and commercial grade	Size classification	Design values in pounds per square inch: Extreme fiber in bending "F_b" — Single-member uses	Extreme fiber in bending "F_b" — Repetitive member uses	Tension parallel to grain "F_t"	Horizontal shear "F_v"	Compression perpendicular to grain "$F_{c\perp}$"	Compression parallel to grain "F_c"	Modulus of elasticity "E"	Grading rules agency
COTTONWOOD (Surfaced dry or surfaced green. Used at 19% max. m.c.)									
Stud	2" to 3" thick 2" to 4" wide	525	600	300	65	195	350	1,000,000	NHPMA (see footnotes 1 through 12)
Construction	2" to 4" thick 4" wide	675	775	400	65	195	650	1,000,000	
Standard		375	425	225	65	195	525	1,000,000	
Utility		175	200	100	65	195	350	1,000,000	
DOUGLAS FIR-LARCH (Surfaced dry or surfaced green. Used at 19% max. m.c.)									
Dense Select Structural	2" to 4" thick 2" to 4" wide	2450	2800	1400	95	455	1850	1,900,000	WCLIB WWPA (see footnotes 1 through 12)
Select Structural		2100	2400	1200	95	385	1600	1,800,000	
Dense No. 1		2050	2400	1200	95	455	1450	1,900,000	
No. 1		1750	2050	1050	95	385	1250	1,800,000	
Dense No. 2		1700	1950	1000	95	455	1150	1,700,000	
No. 2		1450	1650	850	95	385	1000	1,700,000	
No. 3		800	925	475	95	385	600	1,500,000	
Appearance		1750	2050	1050	95	385	1500	1,800,000	
Stud		800	925	475	95	385	600	1,500,000	
Construction	2" to 4" thick 4" wide	1050	1200	625	95	385	1150	1,500,000	
Standard		600	675	350	95	385	925	1,500,000	
Utility		275	325	175	95	385	600	1,500,000	
Dense Select Structural	2" to 4" thick 5" and wider	2100	2400	1400	95	455	1650	1,900,000	
Select Structural		1800	2050	1200	95	385	1400	1,800,000	
Dense No. 1		1800	2050	1200	95	455	1450	1,900,000	
No. 1		1500	1750	1000	95	385	1250	1,800,000	
Dense No. 2		1450	1700	775	95	455	1250	1,700,000	
No. 2		1250	1450	650	95	385	1050	1,700,000	
No. 3		725	850	375	95	385	675	1,500,000	
Appearance		1500	1750	1000	95	385	1500	1,800,000	
Stud		725	850	375	95	385	675	1,500,000	
Dense Select Structural	Beams and Stringers	1900	—	1100	85	455	1300	1,700,000	WCLIB (see footnotes 1 through 12)
Select Structural		1600	—	950	85	385	1100	1,600,000	
Dense No. 1		1550	—	775	85	455	1100	1,700,000	
No. 1		1300	—	675	85	385	925	1,600,000	
Dense Select Structural	Posts and Timbers	1750	—	1150	85	455	1350	1,700,000	
Select Structural		1500	—	1000	85	385	1150	1,600,000	
Dense No. 1		1400	—	950	85	455	1200	1,700,000	
No. 1		1200	—	825	85	385	1000	1,600,000	
Select Dex	Decking	1750	2000	—	—	385	—	1,800,000	
Commercial Dex		1450	1650	—	—	385	—	1,700,000	
Dense Select Structural	Beams and Stringers	1900	—	1250	85	455	1300	1,700,000	WWPA (see footnotes 1 through 13)
Select Structural		1600	—	1050	85	385	1100	1,600,000	
Dense No. 1		1550	—	1050	85	455	1100	1,700,000	
No. 1		1350	—	900	85	385	925	1,600,000	
Dense Select Structural	Posts and Timbers	1750	—	1150	85	455	1350	1,700,000	
Select Structural		1500	—	1000	85	385	1150	1,600,000	
Dense No. 1		1400	—	950	85	455	1200	1,700,000	
No. 1		1200	—	825	85	385	1000	1,600,000	
Selected Decking	Decking	—	2000	—	—	—	—	1,800,000	
Commercial Decking		—	1650	—	—	—	—	1,700,000	
Selected Decking	Decking	—	2150	(Surfaced at 15% max. m.c. and used at 15% max. m.c.)			—	1,900,000	
Commercial Decking		—	1800				—	1,700,000	
DOUGLAS FIR-LARCH (NORTH) (Surfaced dry or surfaced green. Used at 19% max. m.c.)									
Select Structural	2" to 3" thick 2" to 4" wide	2100	2400	1200	95	385	1550	1,800,000	NLGA (A Canadian agency. See footnotes 1 through 12 and 15 and 16)
No. 1		1750	2050	1050	95	385	1250	1,800,000	
No. 2		1450	1650	850	95	385	1000	1,700,000	
No. 3		800	925	475	95	385	600	1,500,000	
Appearance		1750	2050	1050	95	385	1500	1,800,000	
Stud		800	925	475	95	385	600	1,500,000	
Construction	2" to 4" thick 4" wide	1050	1200	625	95	385	1150	1,500,000	
Standard		600	675	350	95	385	925	1,500,000	
Utility		275	325	175	95	385	600	1,500,000	
Select Structural	2" to 4" thick 5" and wider	1800	2050	1200	95	385	1400	1,800,000	
No. 1		1500	1750	1000	95	385	1250	1,800,000	
No. 2		1250	1450	650	95	385	1050	1,700,000	
No. 3		725	850	375	95	385	675	1,500,000	
Appearance		1500	1750	1000	95	385	1500	1,800,000	
Stud		725	850	375	95	385	675	1,500,000	
Select Structural	Beams and Stringers	1600	—	950	85	385	1100	1,600,000	
No. 1		1300	—	675	85	385	925	1,600,000	
Select Structural	Posts and Timbers	1500	—	1000	85	385	1150	1,600,000	
No. 1		1200	—	825	85	385	1000	1,600,000	
Select	Decking	1750	2000	—	—	385	—	1,800,000	
Commercial		1450	1650	—	—	385	—	1,700,000	

TABLE A2.5—*CONTINUED*

DESIGN VALUES FOR VISUALLY GRADED STRUCTURAL LUMBER
(Design values listed are for normal loading conditions. See other provisions in the footnotes and in the National Design Specification for adjustments of tabulated values.)

Species and commercial grade	Size classification	Design values in pounds per square inch: Extreme fiber in bending "F_b": Single-member uses	Extreme fiber in bending "F_b": Repetitive-member uses	Tension parallel to grain "F_t"	Horizontal shear "F_v"	Compression perpendicular to grain "$F_{c\perp}$"	Compression parallel to grain "F_c"	Modulus of elasticity "E"	Grading rules agency
DOUGLAS FIR SOUTH (Surfaced dry or surfaced green. Used at 19% max. m.c.)									
Select Structural	2" to 4" thick 2" to 4" wide	2000	2300	1150	90	335	1400	1,400,000	WWPA (see footnotes 1 through 13)
No. 1		1700	1950	975	90	335	1150	1,400,000	
No. 2		1400	1600	825	90	335	900	1,300,000	
No. 3		775	875	450	90	335	550	1,100,000	
Appearance		1700	1950	975	90	335	1350	1,400,000	
Stud		775	875	450	90	335	550	1,100,000	
Construction	2" to 4" thick 4" wide	1000	1150	600	90	335	1000	1,100,000	
Standard		550	650	325	90	335	850	1,100,000	
Utility		275	300	150	90	335	550	1,100,000	
Select Structural	2" to 4" thick 5" and wider	1700	1950	1150	90	335	1250	1,400,000	
No. 1		1450	1650	975	90	335	1150	1,400,000	
No. 2		1200	1350	625	90	335	950	1,300,000	
No. 3		700	800	350	90	335	600	1,100,000	
Appearance		1450	1650	975	90	335	1350	1,400,000	
Stud		700	800	350	90	335	600	1,100,000	
Select Structural	Beams and Stringers	1550	—	1050	85	335	1000	1,200,000	
No. 1		1300	—	850	85	335	850	1,200,000	
Select Structural	Posts and Timbers	1400	—	950	85	335	1050	1,200,000	
No. 1		1150	—	775	85	335	925	1,200,000	
Selected Decking	Decking	—	1900	—	—	—	—	1,400,000	
Commercial Decking		—	1600	—	—	—	—	1,300,000	
Selected Decking	Decking	—	2050	(Surfaced at 15% max. m.c. and used at 15% max. m.c.)			—	1,500,000	
Commercial Decking		—	1750				—	1,300,000	
EASTERN HEMLOCK (Surfaced dry or surfaced green. Used at 19% max. m.c.)									
Select Structrual	2" to 4" thick 2" to 4" wide	1750	2050	1050	85	360	1350	1,200,000	NELMA (see footnotes 1 through 12)
No. 1		1500	1750	875	85	360	1050	1,200,000	
No. 2		1250	1450	725	85	360	850	1,100,000	
No. 3		675	800	400	85	360	525	1,000,000	
Appearance		1500	1750	875	85	360	1250	1,200,000	
Stud		675	800	400	85	360	525	1,000,000	
Construction	2" to 4" thick 4" wide	900	1050	525	85	360	950	1,000,000	
Standard		500	575	300	85	360	800	1,000,000	
Utility		250	275	150	85	360	525	1,000,000	
Select Structural	2" to 4" thick 5" and wider	1550	1750	1000	85	360	1200	1,200,000	
No. 1		1300	1500	875	85	360	1050	1,200,000	
No. 2		1050	1250	550	85	360	900	1,100,000	
No. 3		625	700	325	85	360	575	1,000,000	
Appearance		1300	1500	875	85	360	1250	1,200,000	
Stud		625	700	325	85	360	575	1,000,000	
Select Structural	Beams and Stringers	1350	—	925	80	360	950	1,200,000	
No. 1		1150	—	775	80	360	800	1,200,000	
Select Structural	Posts and Timbers	1250	—	850	80	360	1000	1,200,000	
No. 1		1050	—	700	80	360	875	1,200,000	
EASTERN HEMLOCK – TAMARACK (Surfaced dry or surfaced green. Used at 19% max. m.c.)									
Select Structural	2" to 4" thick 2" to 4" wide	1800	2050	1050	85	365	1350	1,300,000	NELMA NHPMA (see footnotes 1 through 12)
No. 1		1500	1750	900	85	365	1050	1,300,000	
No. 2		1250	1450	725	85	365	850	1,100,000	
No. 3		700	800	400	85	365	525	1,000,000	
Appearance		1300	1500	900	85	365	1300	1,300,000	
Stud		700	800	400	85	365	525	1,000,000	
Construction	2" to 4" thick 4" wide	900	1050	525	85	365	975	1,000,000	
Standard		500	575	300	85	365	800	1,000,000	
Utility		250	275	150	85	365	525	1,000,000	
Select Structural	2" to 4" thick 5" and wider	1550	1750	1050	85	365	1200	1,300,000	
No. 1		1300	1500	875	85	365	1050	1,300,000	
No. 2		1050	1200	575	85	365	900	1,100,000	
No. 3		625	725	325	85	365	575	1,000,000	
Appearance		1300	1500	875	85	365	1300	1,300,000	
Stud		625	725	325	85	365	575	1,000,000	
Select Structural	Beams and Stringers	1400	—	925	80	365	950	1,200,000	
No. 1		1150	—	775	80	365	800	1,200,000	
Select Structural	Posts and Timbers	1300	—	875	80	365	1000	1,200,000	
No. 1		1050	—	700	80	365	875	1,200,000	
Select	Decking	1500	1700	—	—	—	—	1,300,000	NELMA (see footnotes 1 thru 12)
Commercial		1250	1450	—	—	—	—	1,100,000	

TABLE A2.5—*CONTINUED*

DESIGN VALUES FOR VISUALLY GRADED STRUCTURAL LUMBER

(Design values listed are for normal loading conditions. See other provisions in the footnotes and in the National Design Specification for adjustments of tabulated values.)

Species and commercial grade	Size classification	Design values in pounds per square inch: Extreme fiber in bending "F_b": Single-member uses	Extreme fiber in bending "F_b": Repetitive-member uses	Tension parallel to grain "F_t"	Horizontal shear "F_v"	Compression perpendicular to grain "$F_{c\perp}$"	Compression parallel to grain "F_c"	Modulus of elasticity "E"	Grading rules agency
EASTERN HEMLOCK – TAMARACK (NORTH) (Surfaced dry or surfaced green. Used at 19% max. m.c.)									
Select Structural	2" to 3" thick 2" to 4" wide	1800	2050	1050	85	365	1350	1,300,000	NLGA (A Canadian agency. See footnotes 1 through 12 and 15 and 16)
No. 1		1500	1750	900	85	365	1050	1,300,000	
No. 2		1250	1450	725	85	365	850	1,100,000	
No. 3		700	800	400	85	365	525	1,000,000	
Appearance		1500	1750	900	85	365	1300	1,300,000	
Stud		700	800	400	85	365	525	1,000,000	
Construction	2" to 4" thick 4" wide	900	1050	525	85	365	975	1,000,000	
Standard		500	575	300	85	365	800	1,000,000	
Utility		250	275	150	85	365	525	1,000,000	
Select Structural	2" to 4" thick 5" and wider	1550	1750	1050	85	365	1200	1,300,000	
No. 1		1300	1500	875	85	365	1050	1,300,000	
No. 2		1050	1200	575	85	365	900	1,100,000	
No. 3		625	725	325	85	365	575	1,000,000	
Appearance		1300	1500	875	85	365	1300	1,300,000	
Stud		625	725	325	85	365	575	1,000,000	
Select Structural	Beams and Stringers	1450	——	850	85	365	950	1,300,000	
No. 1		1200	——	600	85	365	800	1,300,000	
Select Structural	Posts and Timbers	1350	——	900	85	365	1000	1,300,000	
No. 1		1100	——	725	85	365	875	1,300,000	
Select	Decking	1500	1700	——	——	365	——	1,300,000	
Commercial		1250	1450	——	——	365	——	1,100,000	
EASTERN SOFTWOODS (Surfaced dry or surfaced green, Used at 19% max. m.c.)									
Select Structural	2" to 4" thick 2" to 4" wide	1350	1550	800	70	185	1050	1,200,000	NHPMA (see footnotes 1 through 12)
No. 1		1150	1350	675	70	185	825	1,200,000	
No. 2		950	1100	550	70	185	650	1,100,000	
No. 3		525	600	300	70	185	400	1,000,000	
Stud	2" to 4" thick 2" to 4" wide	525	600	300	70	185	400	1,000,000	NELMA NHPMA (see footnotes 1 through 12)
Construction	2" to 4" thick 4" wide	700	800	400	70	185	750	1,000,000	NHPMA (see footnotes 1 through 12)
Standard		375	450	225	70	185	625	1,000,000	
Utility		175	200	100	70	185	400	1,000,000	
Select Structural	2" to 4" thick 5" and wider	1150	1350	775	70	185	925	1,200,000	
No. 1		1000	1150	675	70	185	825	1,200,000	
No. 2		825	950	425	70	185	700	1,100,000	
No. 3		475	550	250	70	185	450	1,000,000	
Appearance		1000	1150	675	70	185	1000	1,200,000	
Stud	2" to 4" thick 5" and wider	475	550	250	70	185	450	1,000,000	NELMA NHPMA (see footnotes 1 through 12)
EASTERN SPRUCE (Surfaced dry or surfaced green. Used at 19% max. m.c.)									
Select Structural	2" to 4" thick 2" to 4" wide	1400	1600	800	70	230	1050	1,500,000	NELMA NHPMA (see footnotes 1 through 12)
No. 1		1200	1350	700	70	230	825	1,500,000	
No. 2		975	1100	575	70	230	650	1,400,000	
No. 3		550	625	325	70	230	400	1,200,000	
Appearance		1200	1350	700	70	230	1000	1,500,000	
Stud		550	625	325	70	230	400	1,200,000	
Construction	2" to 4" thick 4" wide	700	800	400	70	230	750	1,200,000	
Standard		400	450	225	70	230	625	1,200,000	
Utility		175	225	100	70	230	400	1,200,000	
Select Structural	2" to 4" thick 5" and wider	1200	1350	800	70	230	925	1,500,000	
No. 1		1000	1150	675	70	230	825	1,500,000	
No. 2		825	950	425	70	230	700	1,400,000	
No. 3		475	550	250	70	230	450	1,200,000	
Appearance		1000	1150	675	70	230	1000	1,500,000	
Stud		475	550	250	70	230	450	1,200,000	
Select Structural	Beams and Stringers	1050	——	725	65	230	750	1,400,000	
No. 1		900	——	600	65	230	625	1,400,000	
Select Structural	Posts and Timbers	1000	——	675	65	230	775	1,400,000	
No. 1		800	——	550	65	230	675	1,400,000	
Select	Decking	——	1300	——	——	——	——	1,500,000	NELMA (see notes 1 –12)
Commercial		——	1100	——	——	——	——	1,400,000	

TABLE A2.5—*CONTINUED*

DESIGN VALUES FOR VISUALLY GRADED STRUCTURAL LUMBER

(Design values listed are for normal loading conditions. See other provisions in the footnotes and in the National Design Specification for adjustments of tabulated values.)

Species and commercial grade	Size classification	Design values in pounds per square inch: Extreme fiber in bending "F_b" — Single-member uses	Extreme fiber in bending "F_b" — Repetitive-member uses	Tension parallel to grain "F_t"	Horizontal shear "F_v"	Compression perpendicular to grain "$F_{c\perp}$"	Compression parallel to grain "F_c"	Modulus of elasticity "E"	Grading rules agency
EASTERN WHITE PINE (Surfaced dry or surfaced green. Used at 19% max. m.c.)									
Select Structural	2" to 4" thick 2" to 4" wide	1350	1550	800	70	220	1050	1,200,000	NELMA NHPMA (see footnotes 1 through 12)
No. 1		1150	1350	675	70	220	850	1,200,000	
No. 2		950	1100	550	70	220	675	1,100,000	
No. 3		525	600	300	70	220	400	1,000,000	
Appearance		1150	1350	675	70	220	1000	1,200,000	
Stud		525	600	300	70	220	400	1,000,000	
Construction	2" to 4" thick 4" wide	700	800	400	70	220	750	1,000,000	
Standard		375	450	225	70	220	625	1,000,000	
Utility		175	200	100	70	220	400	1,000,000	
Select Structural	2" to 4" thick 5" and wider	1150	1350	775	70	220	950	1,200,000	
No. 1		1000	1150	675	70	220	850	1,200,000	
No. 2		825	950	425	70	220	700	1,100,000	
No. 3		475	550	250	70	220	450	1,000,000	
Appearance		1000	1150	675	70	220	1000	1,200,000	
Stud		475	550	250	70	220	450	1,000,000	
Select Structural	Beams and Stringers	1050	—	700	65	220	675	1,100,000	
No. 1		875	—	600	65	220	575	1,100,000	
Select Structural	Posts and Timbers	975	—	650	65	220	725	1,100,000	
No. 1		800	—	525	65	220	625	1,100,000	
Select	Decking	1150	1300	—	—	—	—	1,200,000	NELMA (see footnotes 1 thru 12)
Commercial		950	1100	—	—	—	—	1,100,000	
EASTERN WHITE PINE (NORTH) (Surfaced dry or surfaced green. Used at 19% max. m.c.)									
Select Structural	2" to 3" thick 2" to 4" wide	1350	1550	800	65	220	1050	1,200,000	NLGA (A Canadian agency. See footnotes 1 through 12 and 15 and 16)
No. 1		1150	1350	675	65	220	850	1,200,000	
No. 2		950	1100	550	65	220	675	1,100,000	
No. 3		525	600	300	65	220	400	1,000,000	
Appearance		1150	1350	675	65	220	1000	1,200,000	
Stud		525	600	300	65	220	400	1,000,000	
Construction	2" to 4" thick 4" wide	700	800	400	65	220	750	1,000,000	
Standard		375	450	225	65	220	625	1,000,000	
Utility		175	200	100	65	220	400	1,000,000	
Select Structural	2" to 4" thick 5" and wider	1150	1350	775	65	220	950	1,200,000	
No. 1		1000	1150	675	65	220	850	1,200,000	
No. 2		825	950	425	65	220	700	1,100,000	
No. 3		475	550	250	65	220	450	1,000,000	
Appearance		1000	1150	675	65	220	1000	1,200,000	
Stud		475	550	250	65	220	450	1,000,000	
Select	Decking	900	1050	—	—	220	—	1,200,000	
Commercial		775	875	—	—	220	—	1,100,000	
EASTERN WOODS (Surfaced dry or surfaced green. Used at 19% max. m.c.)									
Select Structural	2" to 4" thick 2" to 4" wide	1300	1500	775	60	185	850	1,100,000	NELMA NHPMA (see footnotes 1 through 12)
No. 1		1100	1300	650	60	185	675	1,100,000	
No. 2		925	1050	525	60	185	550	1,000,000	
No. 3		500	575	300	60	185	325	900,000	
Appearance		1100	1300	650	60	185	825	1,100,000	
Stud		500	575	300	60	185	325	900,000	
Construction	2" to 4" thick 4" wide	650	750	400	60	185	625	900,000	
Standard		375	425	225	60	185	500	900,000	
Utility		175	200	100	60	185	325	900,000	
Select Structural	2" to 4" thick 5" and wider	1150	1300	750	60	185	750	1,100,000	NHPMA (see footnotes 1 through 12)
No. 1		950	1100	650	60	185	675	1,100,000	
No. 2		775	900	425	60	185	575	1,000,000	
No. 3		450	525	250	60	185	375	900,000	
Appearance		950	1100	650	60	185	825	1,100,000	
Stud		450	525	250	60	185	375	900,000	

TABLE A2.5—*CONTINUED*

DESIGN VALUES FOR VISUALLY GRADED STRUCTURAL LUMBER

(Design values listed are for normal loading conditions. See other provisions in the footnotes and in the National Design Specification for adjustments of tabulated values.)

Species and commercial grade	Size classification	Design values in pounds per square inch							Grading rules agency
		Extreme fiber in bending "F_b"		Tension parallel to grain "F_t"	Horizontal shear "F_v"	Compression perpendicular to grain "$F_{c\perp}$"	Compression parallel to grain "F_c"	Modulus of elasticity "E"	
		Single-member uses	Repetitive-member uses						
ENGELMANN SPRUCE–ALPINE FIR (ENGELMANN SPRUCE–LODGEPOLE PINE) (Surfaced dry or surfaced green. Used at 19% max. m.c.)									
Select Structural		1350	1550	800	70	195	950	1,300,000	
No. 1	2" to 4"	1150	1350	675	70	195	750	1,300,000	
No. 2	thick	950	1100	550	70	195	600	1,100,000	
No. 3	2" to 4"	525	600	300	70	195	375	1,000,000	
Appearance	wide	1150	1350	675	70	195	900	1,300,000	
Stud		525	600	300	70	195	375	1,000,000	
Construction	2" to 4"	700	800	400	70	195	675	1,000,000	
Standard	thick	375	450	225	70	195	550	1,000,000	
Utility	4" wide	175	200	100	70	195	375	1,000,000	
Select Structural		1200	1350	775	70	195	850	1,300,000	
No. 1	2" to 4"	1000	1150	675	70	195	750	1,300,000	WWPA
No. 2	thick	825	950	425	70	195	625	1,100,000	(see footnotes
No. 3	5" and	475	550	250	70	195	400	1,000,000	1 through 13)
Appearance	wider	1000	1150	675	70	195	900	1,300,000	
Stud		475	550	250	70	195	400	1,000,000	
Select Structural	Beams and	1050	——	700	65	195	675	1,100,000	
No. 1	Stringers	875	——	600	65	195	550	1,100,000	
Select Structural	Posts and	975	——	650	65	195	700	1,100,000	
No. 1	Timbers	800	——	525	65	195	625	1,100,000	
Selected Decking	Decking	——	1300	——	——	——	——	1,300,000	
Commercial Decking		——	1100	——	——	——	——	1,100,000	
Selected Decking	Decking	——	1400	(Surfaced at 15% max. m.c. and			——	1,300,000	
Commercial Decking		——	1200	used at 15% max. m.c.)			——	1,200,000	
HEM-FIR (Surfaced dry or surfaced green. Used at 19% max. m.c.)									
Select Structural		1650	1900	975	75	245	1300	1,500,000	
No. 1	2" to 4"	1400	1600	825	75	245	1050	1,500,000	
No. 2	thick	1150	1350	675	75	245	825	1,400,000	
No. 3	2" to 4"	650	725	375	75	245	500	1,200,000	
Appearance	wide	1400	1600	825	75	245	1250	1,500,000	
Stud		650	725	375	75	245	500	1,200,000	
Construction	2" to 4"	825	975	500	75	245	925	1,200,000	WCLIB
Standard	thick	475	550	275	75	245	775	1,200,000	WWPA
Utility	4" wide	225	250	125	75	245	500	1,200,000	
Select Structural		1400	1650	950	75	245	1150	1,500,000	(see footnotes
No. 1	2" to 4"	1200	1400	800	75	245	1050	1,500,000	1 through 12)
No. 2	thick	1000	1150	525	75	245	875	1,400,000	
No. 3	5" and	575	675	300	75	245	550	1,200,000	
Appearance	wider	1200	1400	800	75	245	1250	1,500,000	
Stud		575	675	300	75	245	550	1,200,000	
Select Structural	Beams and	1300	——	750	70	245	925	1,300,000	
No. 1	Stringers	1050	——	525	70	245	750	1,300,000	
Select Structural	Posts and	1200	——	800	70	245	975	1,300,000	WCLIB
No. 1	Timbers	975	——	650	70	245	850	1,300,000	
Select Dex	Decking	1400	1600	——	——	245	——	1,500,000	(see footnotes
Commercial Dex		1150	1350	——	——	245	——	1,400,000	1 through 12))
Select Structural	Beams and	1250	——	850	70	245	925	1,300,000	
No. 1	Stringers	1050	——	725	70	245	775	1,300,000	
Select Structural	Posts and	1200	——	800	70	245	975	1,300,000	WWPA
No. 1	Timbers	950	——	650	70	245	850	1,300,000	
Selected Decking	Decking	——	1600	——	——	——	——	1,500,000	
Commercial Decking		——	1350	——	——	——	——	1,400,000	(see footnotes
Selected Decking	Decking	——	1700	(Surfaced at 15% max. m.c. and			——	1,600,000	1 through 13)
Commercial Decking		——	1450	used at 15% max. m.c.)			——	1,400,000	

TABLE A2.5—*CONTINUED*

DESIGN VALUES FOR VISUALLY GRADED STRUCTURAL LUMBER

(Design values listed are for normal loading conditions. See other provisions in the footnotes and in the National Design Specification for adjustments of tabulated values.)

Species and commercial grade	Size classification	Design values in pounds per square inch: Extreme fiber in bending "F_b" — Single-member uses	Extreme fiber in bending "F_b" — Repetitive-member uses	Tension parallel to grain "F_t"	Horizontal shear "F_v"	Compression perpendicular to grain "$F_{c\perp}$"	Compression parallel to grain "F_c"	Modulus of elasticity "E"	Grading rules agency
HEM-FIR (NORTH) (Surfaced dry or surfaced green. Used at 19% max. m.c.)									
Select Structural	2" to 3" thick 2" to 4" wide	1600	1800	925	75	235	1300	1,500,000	NLGA (A Canadian agency. See footnotes 1 through 12 and 15 and 16)
No. 1		1350	1550	800	75	235	1050	1,500,000	
No. 2		1100	1300	650	75	235	800	1,400,000	
No. 3		625	700	350	75	235	500	1,200,000	
Appearance		1350	1550	800	75	235	1250	1,500,000	
Stud		625	700	350	75	235	500	1,200,000	
Construction	2" to 4" thick 4" wide	800	925	475	75	235	925	1,200,000	
Standard		450	525	275	75	235	775	1,200,000	
Utility		225	250	125	75	235	500	1,200,000	
Select Structural	2" to 4" thick 5" and wider	1350	1550	900	75	235	1150	1,500,000	
No. 1		1150	1350	775	75	235	1050	1,500,000	
No. 2		950	1100	500	75	235	850	1,400,000	
No. 3		550	650	300	75	235	550	1,200,000	
Appearance		1150	1350	775	75	235	1250	1,500,000	
Stud		550	650	300	75	235	550	1,200,000	
Select Structural	Beams and Stringers	1250	––	725	70	235	900	1,300,000	
No. 1		1000	––	500	70	235	750	1,300,000	
Select Structural	Posts and Timbers	1150	––	775	70	235	950	1,300,000	
No. 1		925	––	625	70	235	850	1,300,000	
Select	Decking	1350	1500	––	––	235	––	1,500,000	
Commercial		1100	1300	––	––	235	––	1,400,000	
IDAHO WHITE PINE (Surfaced dry or surfaced green. Used at 19% max. m.c.)									
Select Structural	2" to 4" thick 2" to 4" wide	1350	1550	775	70	190	1100	1,400,000	WWPA (see footnotes 1 through 13)
No. 1		1150	1300	650	70	190	875	1,400,000	
No. 2		925	1050	550	70	190	675	1,300,000	
No. 3		525	600	300	70	190	425	1,200,000	
Appearance		1150	1300	650	70	190	1050	1,400,000	
Stud		525	600	300	70	190	425	1,200,000	
Construction	2" to 4" thick 4" wide	675	775	400	70	190	775	1,200,000	
Standard		375	425	225	70	190	650	1,200,000	
Utility		175	200	100	70	190	425	1,200,000	
Select Structural	2" to 4" thick 5" and wider	1150	1300	775	70	190	950	1,400,000	
No. 1		975	1100	650	70	190	875	1,400,000	
No. 2		800	925	425	70	190	725	1,300,000	
No. 3		475	550	250	70	190	450	1,200,000	
Appearance		975	1100	650	70	190	1050	1,400,000	
Stud		475	550	250	70	190	450	1,200,000	
Select Structural	Beams and Stringers	1000	––	700	65	190	775	1,300,000	
No. 1		850	––	575	65	190	650	1,300,000	
Select Structural	Posts and Timbers	950	––	650	65	190	800	1,300,000	
No. 1		775	––	525	65	190	700	1,300,000	
Selected Decking	Decking	––	1300	––	––	––	––	1,400,000	
Commercial Decking		––	1050	––	––	––	––	1,300,000	
Selected Decking	Decking	––	1400	(Surfaced at 15% max. m.c. and used at 15% max. m.c.)			––	1,500,000	
Commercial Decking		––	1150				––	1,400,000	
LODGEPOLE PINE (Surfaced dry or surfaced green. Used at 19% max. m.c.)									
Select Structural	2" to 4" thick 2" to 4" wide	1500	1750	875	70	250	1150	1,300,000	WWPA (see footnotes 1 through 13)
No. 1		1300	1500	750	70	250	900	1,300,000	
No. 2		1050	1200	625	70	250	700	1,200,000	
No. 3		600	675	350	70	250	425	1,000,000	
Appearance		1300	1500	750	70	250	1050	1,300,000	
Stud		600	675	350	70	250	425	1,000,000	
Construction	2" to 4" thick 4" wide	775	875	450	70	250	800	1,000,000	
Standard		425	500	250	70	250	675	1,000,000	
Utility		200	225	125	70	250	425	1,000,000	
Select Structural	2" to 4" thick 5" and wider	1300	1500	875	70	250	1000	1,300,000	
No. 1		1100	1300	750	70	250	900	1,300,000	
No. 2		925	1050	475	70	250	750	1,200,000	
No. 3		525	625	275	70	250	475	1,000,000	
Appearance		1100	1300	750	70	250	1050	1,300,000	
Stud		525	625	275	70	250	475	1,000,000	
Select Structural	Beams and Stringers	1150	––	775	65	250	800	1,100,000	
No. 1		975	––	650	65	250	675	1,100,000	
Select Structural	Posts and Timbers	1100	––	725	65	250	850	1,100,000	
No. 1		875	––	600	65	250	725	1,100,000	
Selected Decking	Decking	––	1450	––	––	––	––	1,300,000	
Commercial Decking		––	1200	––	––	––	––	1,200,000	
Selected Decking	Decking	––	1550	(Surfaced at 15% max. m.c. and used at 15% max. m.c.)			––	1,400,000	
Commercial Decking		––	1300				––	1,200,000	

TABLE A2.5—*CONTINUED*

DESIGN VALUES FOR VISUALLY GRADED STRUCTURAL LUMBER

(Design values listed are for normal loading conditions. See other provisions in the footnotes and in the National Design Specification for adjustments of tabulated values.)

Species and commercial grade	Size classification	Design values in pounds per square inch							Grading rules agency
		Extreme fiber in bending "F_b"		Tension parallel to grain "F_t"	Horizontal shear "F_v"	Compression perpendicular to grain "$F_{c\perp}$"	Compression parallel to grain "F_c"	Modulus of elasticity "E"	
		Single-member uses	Repetitive-member uses						
MOUNTAIN HEMLOCK (Surfaced dry or surfaced green. Used at 19% max. m.c.)									
Select Structural		1750	2000	1000	95	370	1250	1,300,000	
No. 1	2" to 4"	1450	1700	850	95	370	1000	1,300,000	
No. 2	thick	1200	1400	700	95	370	775	1,100,000	
No. 3	2" to 4"	675	775	400	95	370	475	1,000,000	
Appearance	wide	1450	1700	850	95	370	1200	1,300,000	
Stud		675	775	400	95	370	475	1,000,000	WCLIB
Construction	2" to 4"	875	1000	525	95	370	900	1,000,000	WWPA
Standard	thick	500	575	275	95	370	725	1,000,000	
Utility	4" wide	225	275	125	95	370	475	1,000,000	
Select Structural		1500	1700	1000	95	370	1100	1,300,000	
No. 1	2" to 4"	1250	1450	850	95	370	1000	1,300,000	(see footnotes
No. 2	thick	1050	1200	550	95	370	825	1,100,000	1 through 12)
No. 3	5" and	625	700	325	95	370	525	1,000,000	
Appearance	wider	1250	1450	850	95	370	1200	1,300,000	
Stud		625	700	325	95	370	525	1,000,000	
Select Structural	Beams and	1350	——	775	85	370	875	1,100,000	
No. 1	Stringers	1100	——	550	85	370	725	1,100,000	
Select Structural	Posts and	1250	——	825	85	370	925	1,100,000	WCLIB
No. 1	Timbers	1000	——	675	85	370	800	1,100,000	
Select Dex	Decking	1450	1650	——	——	370	——	1,300,000	(see footnotes
Commercial Dex		1200	1400	——	——	370	——	1,100,000	1 through 12)
Select Structural	Beams and	1350	——	900	90	370	875	1,100,000	
No. 1	Stringers	1100	——	750	90	370	750	1,100,000	
Select Structural	Posts and	1250	——	825	90	370	925	1,100,000	WWPA
No. 1	Timbers	1000	——	675	90	370	800	1,100,000	
Selected Decking	Decking	——	1650	——	——	——	——	1,300,000	
Commercial Decking		——	1400	——	——	——	——	1,100,000	(see footnotes
Selected Decking	Decking	——	1800	(Surfaced at 15% max. m.c. and			——	1,300,000	1 through 13)
Commercial Decking		——	1500	used at 15% max. m.c.)			——	1,200,000	
MOUNTAIN HEMLOCK—HEM-FIR (Surfaced dry or surfaced green. Used at 19% max. m.c.)									
Select Structural		1650	1900	975	75	245	1250	1,300,000	
No. 1	2" to 4"	1400	1600	825	75	245	1000	1,300,000	
No. 2	thick	1150	1350	675	75	245	775	1,100,000	
No. 3	2" to 4"	650	725	375	75	245	475	1,000,000	
Appearance	wide	1400	1600	825	75	245	1200	1,300,000	
Stud		650	725	375	75	245	475	1,000,000	
Construction	2" to 4"	825	975	500	75	245	900	1,000,000	
Standard	thick	475	550	275	75	245	725	1,000,000	
Utility	4" wide	225	250	125	75	245	475	1,000,000	WWPA
Select Structural	2" to 4"	1400	1650	950	75	245	1100	1,300,000	
No. 1	thick	1200	1400	800	75	245	1000	1,300,000	
No. 2	5" and	1000	1150	525	75	245	825	1,100,000	(see footnotes
No. 3	wider	575	675	300	75	245	525	1,000,000	1 through 13)
Appearance		1200	1400	800	75	245	1200	1,300,000	
Stud		575	675	300	75	245	525	1,000,000	
Select Structural	Beams and	1250	——	850	70	245	875	1,100,000	
No. 1	Stringers	1050	——	725	70	245	750	1,100,000	
Select Structural	Posts and	1200	——	800	70	245	925	1,100,000	
No. 1	Timbers	950	——	650	70	245	800	1,100,000	
Selected Decking	Decking	——	1600	——	——	——	——	1,300,000	
Commercial Decking		——	1350	——	——	——	——	1,100,000	
Selected Decking	Decking	——	1700	(Surfaced at 15% max. m.c. and			——	1,300,000	
Commercial Decking		——	1450	used at 15% max. m.c.)			——	1,200,000	
NORTHERN ASPEN (Surfaced dry or surfaced green. Used at 19% max. m.c.)									
Select Structural		1300	1500	750	60	195	850	1,400,000	
No. 1	2" to 3"	1100	1250	650	60	195	675	1,400,000	
No. 2	thick	900	1050	525	60	195	525	1,200,000	
No. 3	2" to 4"	500	575	275	60	195	325	1,100,000	
Appearance	wide	1100	1250	650	60	195	800	1,400,000	
Stud		500	575	275	60	195	325	1,100,000	NLGA
Construction	2" to 4"	650	750	375	60	195	600	1,100,000	(A Canadian
Standard	thick	350	425	200	60	195	500	1,100,000	agency. See
Utility	4" wide	175	200	100	60	195	325	1,100,000	footnotes 1
Select Structural		1100	1250	725	60	195	750	1,400,000	through 12
No. 1	2" to 4"	950	1100	625	60	195	675	1,400,000	and 15
No. 2	thick	775	900	400	60	195	575	1,200,000	and 16)
No. 3	5" and	450	525	250	60	195	350	1,100,000	
Appearance	wider	950	1100	625	60	195	800	1,400,000	
Stud		450	525	250	60	195	350	1,100,000	

TABLE A2.5—*CONTINUED*

DESIGN VALUES FOR VISUALLY GRADED STRUCTURAL LUMBER

(Design values listed are for normal loading conditions. See other provisions in the footnotes and in the National Design Specification for adjustments of tabulated values.)

Species and commercial grade	Size classification	Design values in pounds per square inch: Extreme fiber in bending "F_b": Single-member uses	Repetitive-member uses	Tension parallel to grain "F_t"	Horizontal shear "F_v"	Compression perpendicular to grain "$F_{c\perp}$"	Compression parallel to grain "F_c"	Modulus of elasticity "E"	Grading rules agency
NORTHERN PINE (Surfaced dry or surfaced green. Used at 19% max. m.c.)									
Select Structural	2" to 4" thick 2" to 4" wide	1650	1850	950	70	280	1200	1,400,000	NELMA NHPMA (see footnotes) 1 through 12)
No. 1		1400	1600	825	70	280	975	1,400,000	
No. 2		1150	1300	675	70	280	775	1,300,000	
No. 3		625	725	375	70	280	475	1,100,000	
Appearance		1200	1400	800	70	280	1150	1,400,000	
Stud		625	725	375	70	280	475	1,100,000	
Construction	2" to 4" thick 4" wide	825	950	475	70	280	875	1,100,000	
Standard		450	525	275	70	280	725	1,100,000	
Utility		225	250	125	70	280	475	1,100,000	
Select Structural	2" to 4" thick 5" and wider	1400	1600	950	70	280	1100	1,400,000	
No. 1		1200	1400	800	70	280	975	1,400,000	
No. 2		950	1100	525	70	280	825	1,300,000	
No. 3		575	650	300	70	280	525	1,100,000	
Appearance		1200	1400	800	70	280	1150	1,400,000	
Stud		575	650	300	70	280	525	1,100,000	
Select Structural	Beams and Stringers	1250	——	850	65	280	850	1,300,000	
No. 1		1050	——	700	65	280	725	1,300,000	
Select Structural	Posts and Timbers	1150	——	800	65	280	900	1,300,000	
No. 1		950	——	650	65	280	800	1,300,000	
Select	Decking	1350	1550	——	——	——	——	1,400,000	NELMA (see footnotes 1 thru 12)
Commercial		1150	1300	——	——	——	——	1,300,000	
NORTHERN SPECIES (Surfaced dry or surfaced green. Used at 19% max. m.c.)									
Select Structural	2" to 3" thick 2" to 4" wide	1350	1550	775	65	220	1050	1,100,000	NLGA (A Canadian agency. See footnotes 1 through 12 and 15 and 16)
No. 1		1150	1300	675	65	220	825	1,100,000	
No. 2		925	1050	550	65	220	650	1,000,000	
No. 3		525	600	300	65	220	400	900,000	
Appearance		1150	1300	675	65	220	975	1,100,000	
Stud		525	600	300	65	220	400	900,000	
Construction	2" to 4" thick 4" wide	675	775	400	65	220	750	900,000	
Standard		375	425	225	65	220	600	900,000	
Utility		175	200	100	65	220	400	900,000	
Select Structural	2" to 4" thick 5" and wider	1150	1300	750	65	220	900	1,100,000	
No. 1		975	1150	650	65	220	825	1,100,000	
No. 2		800	925	425	65	220	675	1,000,000	
No. 3		475	550	250	65	220	425	900,000	
Appearance		975	1150	650	65	220	975	1,100,000	
Stud		475	550	250	65	220	425	900,000	
Select	Decking	900	1050	——	——	220	——	1,100,000	
Commercial		775	875	——	——	220	——	1,000,000	
NORTHERN WHITE CEDAR (Surfaced dry or surfaced green. Used at 19% max. m.c.)									
Select Structural	2" to 4" thick 2" to 4" wide	1150	1350	700	65	205	875	800,000	NELMA (see footnotes 1 through 12)
No. 1		1000	1150	600	65	205	675	800,000	
No. 2		825	950	500	65	205	550	700,000	
No. 3		450	525	275	65	205	325	600,000	
Appearance		850	1000	575	65	205	825	800,000	
Stud		450	525	275	65	205	325	600,000	
Construction	2" to 4" thick 4" wide	600	675	350	65	205	625	600,000	
Standard		325	375	200	65	205	500	600,000	
Utility		150	175	100	65	205	325	600,000	
Select Structural	2" to 4" thick 5" and wider	1000	1150	675	65	205	775	800,000	
No. 1		850	1000	575	65	205	675	800,000	
No. 2		700	825	375	65	205	575	700,000	
No. 3		425	475	225	65	205	375	600,000	
Appearance		850	1000	575	65	205	825	800,000	
Stud		425	475	225	65	205	375	600,000	
Select Structural	Beams and Stringers	900	——	600	60	205	600	700,000	
No. 1		750	——	500	60	205	500	700,000	
Select Structural	Posts and Timbers	850	——	575	60	205	650	700,000	
No. 1		675	——	450	60	205	550	700,000	
Select	Decking	975	1100	——	——	——	——	800,000	
Commercial		825	950	——	——	——	——	700,000	

TABLE A2.5—*CONTINUED*

DESIGN VALUES FOR VISUALLY GRADED STRUCTURAL LUMBER

(Design values listed are for normal loading conditions. See other provisions in the footnotes and in the National Design Specification for adjustments of tabulated values.)

Species and commercial grade	Size classification	Design values in pounds per square inch: Extreme fiber in bending "F_b" — Single-member uses	Extreme fiber in bending "F_b" — Repetitive-member uses	Tension parallel to grain "F_t"	Horizontal shear "F_v"	Compression perpendicular to grain "$F_{c\perp}$"	Compression parallel to grain "F_c"	Modulus of elasticity "E"	Grading rules agency
PONDEROSA PINE (Surfaced dry or surfaced green. Used at 19% max. m.c.)									
Select Structural		1400	1650	825	70	235	1050	1,200,000	
No. 1	2" to 3"	1200	1400	700	70	235	850	1,200,000	
No. 2	thick	1000	1150	575	70	235	675	1,100,000	
No. 3	2" to 4"	550	625	325	70	235	400	1,000,000	
Appearance	wide	1200	1400	700	70	235	1000	1,200,000	
Stud		550	625	325	70	235	400	1,000,000	
Construction	2" to 4"	725	825	425	70	235	775	1,000,000	NLGA
Standard	thick	400	450	225	70	235	625	1,000,000	
Utility	4" wide	200	225	100	70	235	400	1,000,000	
Select Structural	2" to 4"	1200	1400	825	70	235	950	1,200,000	
No. 1	thick	1050	1200	700	70	235	850	1,200,000	(A Canadian
No. 2	5" and	850	975	450	70	235	700	1,100,000	agency. See
No. 3	wider	500	575	250	70	235	450	1,000,000	footnotes 1
Appearance		1050	1200	700	70	235	1000	1,200,000	through 12 and
Stud		500	575	250	70	235	450	1,000,000	15 and 16)
Select Structural	Beams and	1100	—	725	65	235	750	1,100,000	
No. 1	Stringers	925	—	500	65	235	625	1,100,000	
Select Structural	Posts and	1000	—	675	65	235	800	1,100,000	
No. 1	Timbers	825	—	550	65	235	700	1,100,000	
Select	Decking	1200	1450	—	—	235	—	1,300,000	
Commercial		1000	1250	—	—	235	—	1,100,000	
PONDEROSA PINE–SUGAR PINE (PONDEROSA PINE–LODGEPOLE PINE) (Surfaced dry or surfaced green. Used at 19% max. m.c.)									
Select Structural		1400	1650	825	70	235	1050	1,200,000	
No. 1	2" to 4"	1200	1400	700	70	235	850	1,200,000	
No. 2	thick	1000	1150	575	70	235	675	1,100,000	
No. 3	2" to 4"	550	625	325	70	235	400	1,000,000	
Appearance	wide	1200	1400	700	70	235	1000	1,200,000	
Stud		550	625	325	70	235	400	1,000,000	
Construction	2" to 4"	725	825	425	70	235	775	1,000,000	
Standard	thick	400	450	225	70	235	625	1,000,000	
Utility	4" wide	200	225	100	70	235	400	1,000,000	
Select Structural	2" to 4"	1200	1400	825	70	235	950	1,200,000	WWPA
No. 1	thick	1050	1200	700	70	235	850	1,200,000	
No. 2	5" and	850	975	450	70	235	700	1,100,000	
No. 3	wider	500	575	250	70	235	450	1,000,000	(see footnotes
Appearance		1050	1200	700	70	235	1000	1,200,000	1 through 13)
Stud		500	575	250	70	235	450	1,000,000	
Select Structural	Beams and	1100	—	725	65	235	750	1,100,000	
No. 1	Stringers	925	—	625	65	235	625	1,100,000	
Select Structural	Posts and	1000	—	675	65	235	800	1,100,000	
No. 1	Timbers	825	—	550	65	235	700	1,100,000	
Selected Decking	Decking	—	1350	—	—	—	—	1,200,000	
Commercial Decking		—	1150	—	—	—	—	1,100,000	
Selected Decking	Decking	—	1450	(Surfaced at 15% max. m.c. and used at 15% max. m.c.)			—	1,300,000	
Commercial Decking		—	1250				—	1,100,000	
RED PINE (Surfaced dry or surfaced green. Used at 19% max. m.c.)									
Select Structural		1400	1600	800	70	280	1050	1,300,000	
No. 1	2" to 3"	1200	1350	700	70	280	825	1,300,000	
No. 2	thick	975	1100	575	70	280	650	1,200,000	
No. 3	2" to 4"	525	625	325	70	280	400	1,000,000	
Appearance	wide	1200	1350	700	70	280	975	1,300,000	
Stud		525	625	325	70	280	400	1,000,000	
Construction	2" to 4"	700	800	400	70	280	750	1,000,000	
Standard	thick	400	450	225	70	280	600	1,000,000	
Utility	4" wide	175	225	100	70	280	400	1,000,000	NLGA
Select Structural	2" to 4"	1200	1350	775	70	280	900	1,300,000	(A Canadian
No. 1	thick	1000	1150	675	70	280	825	1,300,000	agency. See
No. 2	5" and	825	950	425	70	280	675	1,200,000	footnotes 1
No. 3	wider	500	550	250	70	280	425	1,000,000	through 12 and
Appearance		1000	1150	675	70	280	975	1,300,000	15 and 16)
Stud		500	550	250	70	280	425	1,000,000	
Select Structural	Beams and	1050	—	625	65	280	725	1,100,000	
No. 1	Stringers	875	—	450	65	280	600	1,100,000	
Select Structural	Posts and	1000	—	675	65	280	775	1,100,000	
No. 1	Timbers	800	—	550	65	280	675	1,100,000	
Select	Decking	1150	1350	—	—	280	—	1,300,000	
Commercial		975	1100	—	—	280	—	1,200,000	

TABLE A2.5—*CONTINUED*

DESIGN VALUES FOR VISUALLY GRADED STRUCTURAL LUMBER

(Design values listed are for normal loading conditions. See other provisions in the footnotes and in the National Design Specification for adjustments of tabulated values.)

Species and commercial grade	Size classification	Design values in pounds per square inch: Extreme fiber in bending "F_b" — Single-member uses	Extreme fiber in bending "F_b" — Repetitive-member uses	Tension parallel to grain "F_t"	Horizontal shear "F_v"	Compression perpendicular to grain "$F_{c\perp}$"	Compression parallel to grain "F_c"	Modulus of elasticity "E"	Grading rules agency
SITKA SPRUCE (Surfaced dry or surfaced green. Used at 19% max. m.c.)									
Select Structural	2" to 4" thick 2" to 4" wide	1550	1800	925	75	280	1150	1,500,000	WCLIB (see footnotes 1 through 12)
No. 1		1350	1550	775	75	280	925	1,500,000	
No. 2		1100	1250	650	75	280	725	1,300,000	
No. 3		600	700	350	75	280	450	1,200,000	
Appearance		1350	1550	750	75	280	1100	1,500,000	
Stud		600	700	350	75	280	450	1,200,000	
Construction	2" to 4" thick 4" wide	800	925	475	75	280	825	1,200,000	
Standard		450	500	250	75	280	675	1,200,000	
Utility		200	250	125	75	280	450	1,200,000	
Select Structural	2" to 4" thick 5" and wider	1350	1550	900	75	280	1000	1,500,000	
No. 1		1150	1300	775	75	280	925	1,500,000	
No. 2		925	1050	500	75	280	775	1,300,000	
No. 3		525	600	275	75	280	500	1,200,000	
Appearance		1150	1300	750	75	280	1100	1,500,000	
Stud		525	600	275	75	280	500	1,200,000	
Select Structural	Beams and Stringers	1200	—	675	70	280	825	1,300,000	
No. 1		1000	—	500	70	280	675	1,300,000	
Select Structural	Posts and Timbers	1150	—	750	70	280	875	1,300,000	
No. 1		925	—	600	70	280	750	1,300,000	
Select Dex	Decking	1300	1500	—	—	280	—	1,500,000	
Commercial Dex		1100	1250	—	—	280	—	1,300,000	
SOUTHERN PINE (Surfaced at 15% maximum moisture content, K.D. Used at 15% max. m.c.)									
Select Structural	2" to 4" thick 2" to 4" wide	2150	2500	1250	105	405	1800	1,800,000	SPIB (see footnotes 1, 3, 4, 5, 6, 12, 18 and 19)
Dense Select Structural		2500	2900	1500	105	475	2100	1,900,000	
No. 1		1850	2100	1050	105	405	1450	1,800,000	
No. 1 Dense		2150	2450	1250	105	475	1700	1,900,000	
No. 2		1550	1750	900	95	405	1150	1,600,000	
No. 2 Dense		1800	2050	1050	95	475	1350	1,700,000	
No. 3		850	975	500	95	405	675	1,500,000	
No. 3 Dense		1000	1150	575	95	475	800	1,500,000	
Stud		850	975	500	95	405	675	1,500,000	
Construction	2" to 4" thick 4" wide	1100	1250	650	105	405	1300	1,500,000	
Standard		625	725	375	95	405	1050	1,500,000	
Utility		275	300	175	95	405	675	1,500,000	
Select Structural	2" to 4" thick 5" and wider	1850	2150	1200	95	405	1600	1,800,000	
Dense Select Structural		2200	2500	1450	95	475	1850	1,900,000	
No. 1		1600	1850	1050	95	405	1450	1,800,000	
No. 1 Dense		1850	2150	1250	95	475	1700	1,900,000	
No. 2		1300	1500	675	95	405	1200	1,600,000	
No. 2 Dense		1550	1750	800	95	475	1400	1,700,000	
No. 3		750	875	400	95	405	725	1,500,000	
No. 3 Dense		875	1000	450	95	475	850	1,500,000	
Stud		800	900	400	95	405	725	1,500,000	
Dense Standard Decking	2" to 4" thick 2" and wider Decking	2150	2450	—	—	475	—	1,900,000	
Select Decking		1550	1750	—	—	405	—	1,600,000	
Dense Select Decking		1800	2050	—	—	475	—	1,700,000	
Commercial Decking		1550	1750	—	—	405	—	1,600,000	
Dense Commercial Decking		1800	2050	—	—	475	—	1,700,000	
Dense Structural 86	2" to 4" thick	2800	3250	1900	165	475	2300	1,900,000	
Dense Structural 72		2400	2750	1600	135	475	1950	1,900,000	
Dense Structural 65		2150	2450	1450	125	475	1750	1,900,000	

TABLE A2.5—*CONTINUED*

DESIGN VALUES FOR VISUALLY GRADED STRUCTURAL LUMBER

(Design values listed are for normal loading conditions. See other provisions in the footnotes and in the National Design Specification for adjustments of tabulated values.)

Species and commercial grade	Size classification	Design values in pounds per square inch							Grading rules agency
		Extreme fiber in bending "F_b"		Tension parallel to grain "F_t"	Horizontal shear "F_v"	Compression perpendicular to grain "$F_{c\perp}$"	Compression parallel to grain "F_c"	Modulus of elasticity "E"	
		Single-member uses	Repetitive-member uses						
SOUTHERN PINE (Surfaced dry. Used at 19% max. m.c.)									
Select Structural	2" to 4" thick 2" to 4" wide	2000	2300	1150	100	405	1550	1,700,000	
Dense Select Structural		2350	2700	1350	100	475	1800	1,800,000	
No. 1		1700	1950	1000	100	405	1250	1,700,000	
No. 1 Dense		2000	2300	1150	100	475	1450	1,800,000	
No. 2		1400	1650	825	90	405	975	1,600,000	
No. 2 Dense		1650	1900	975	90	475	1150	1,600,000	
No. 3		775	900	450	90	405	575	1,400,000	
No. 3 Dense		925	1050	525	90	475	675	1,500,000	
Stud		775	900	450	90	405	575	1,400,000	
Construction	2" to 4" thick 4" wide	1000	1150	600	100	405	1100	1,400,000	
Standard		575	675	350	90	405	900	1,400,000	
Utility		275	300	150	90	405	575	1,400,000	
Select Structural	2" to 4" thick 5" and wider	1750	2000	1150	90	405	1350	1,700,000	SPIB (see footnotes 1,3,4,5,6, 12, 18 and 19)
Dense Select Structural		2050	2350	1300	90	475	1600	1,800,000	
No. 1		1450	1700	975	90	405	1250	1,700,000	
No. 1 Dense		1700	2000	1150	90	475	1450	1,800,000	
No. 2		1200	1400	625	90	405	1000	1,600,000	
No. 2 Dense		1400	1650	725	90	475	1200	1,600,000	
No. 3		700	800	350	90	405	625	1,400,000	
No. 3 Dense		825	925	425	90	475	725	1,500,000	
Stud		725	850	350	90	405	625	1,400,000	
Dense Standard Decking	2" to 4" thick 2" and wider Decking	2000	2300	—	—	475	—	1,800,000	
Select Decking		1400	1650	—	—	405	—	1,600,000	
Dense Select Decking		1650	1900	—	—	475	—	1,600,000	
Commercial Decking		1400	1650	—	—	405	—	1,600,000	
Dense Commercial Decking		1650	1900	—	—	475	—	1,600,000	
Dense Structural 86	2" to 4" thick	2600	3000	1750	155	475	2000	1,800,000	
Dense Structural 72		2200	2550	1450	130	475	1650	1,800,000	
Dense Structural 65		2000	2300	1300	115	475	1500	1,800,000	
SOUTHERN PINE (Surfaced green. Used any condition)									
Select Structural	2½" to 4" thick 2½" to 4" wide	1600	1850	925	95	270	1050	1,500,000	
Dense Select Structural		1850	2150	1100	95	315	1200	1,600,000	
No. 1		1350	1550	800	95	270	825	1,500,000	
No. 1 Dense		1600	1800	925	95	315	950	1,600,000	
No. 2		1150	1300	675	85	270	650	1,400,000	
No. 2 Dense		1350	1500	775	85	315	750	1,400,000	
No. 3		625	725	375	85	270	400	1,200,000	
No. 3 Dense		725	850	425	85	315	450	1,300,000	
Stud		625	725	375	85	270	400	1,200,000	
Construction	2½" to 4" thick 4" wide	825	925	475	95	270	725	1,200,000	
Standard		475	525	275	85	270	600	1,200,000	
Utility		200	250	125	85	270	400	1,200,000	
Select Structural	2½" to 4" thick 5" and wider	1400	1600	900	85	270	900	1,500,000	
Dense Select Structural		1600	1850	1050	85	315	1050	1,600,000	
No. 1		1200	1350	775	85	270	825	1,500,000	
No. 1 Dense		1400	1600	925	85	315	950	1,600,000	
No. 2		975	1100	500	85	270	675	1,400,000	SPIB
No. 2 Dense		1150	1300	600	85	315	800	1,400,000	
No. 3		550	650	300	85	270	425	1,200,000	(see footnotes 1,3,4,5,6, 12, 17,18 and 19)
No. 3 Dense		650	750	350	85	315	475	1,300,000	
Stud		575	675	300	85	270	425	1,200,000	
Dense Standard Decking	2½" to 4" thick 2" and wider Decking	1600	1800	—	—	315	—	1,600,000	
Select Decking		1150	1300	—	—	270	—	1,400,000	
Dense Select Decking		1350	1500	—	—	315	—	1,400,000	
Commercial Decking		1150	1300	—	—	270	—	1,400,000	
Dense Commercial Decking		1350	1500	—	—	315	—	1,400,000	
No. 1 SR	5" and thicker	1350	—	875	110	270	775	1,500,000	
No. 1 Dense SR		1550	—	1050	110	315	925	1,600,000	
No. 2 SR		1100	—	725	95	270	625	1,400,000	
No. 2 Dense SR		1250	—	850	95	315	725	1,400,000	
Dense Structural 86	2½" and thicker	2100	2400	1400	145	315	1300	1,600,000	
Dense Structural 72		1750	2050	1200	120	315	1100	1,600,000	
Dense Structural 65		1600	1800	1050	110	315	1000	1,600,000	

TABLE A2.5—*CONTINUED*

DESIGN VALUES FOR VISUALLY GRADED STRUCTURAL LUMBER

(Design values listed are for normal loading conditions. See other provisions in the footnotes and in the National Design Specification for adjustments of tabulated values.)

Species and commercial grade	Size classification	Design values in pounds per square inch							Grading rules agency
		Extreme fiber in bending "F_b"		Tension parallel to grain "F_t"	Horizontal shear "F_v"	Compression perpendicular to grain "$F_{c\perp}$"	Compression parallel to grain "F_c"	Modulus of elasticity "E"	
		Single-member uses	Repetitive-member uses						
SPRUCE–PINE–FIR (Surfaced dry or surfaced green. Used at 19% max. m.c.)									
Select Structural	2" to 3" thick 2" to 4" wide	1450	1650	850	70	265	1100	1,500,000	NLGA (A Canadian agency. See footnotes 1 through 12 and 15 and 16)
No. 1		1200	1400	725	70	265	875	1,500,000	
No. 2		1000	1150	600	70	265	675	1,300,000	
No. 3		550	650	325	70	265	425	1,200,000	
Appearance		1200	1400	725	70	265	1050	1,500,000	
Stud		550	650	325	70	265	425	1,200,000	
Construction	2" to 4" thick 4" wide	725	850	425	70	265	775	1,200,000	
Standard		400	475	225	70	265	650	1,200,000	
Utility		175	225	100	70	265	425	1,200,000	
Select Structural	2" to 4" thick 5" and wider	1250	1450	825	70	265	975	1,500,000	
No. 1		1050	1200	700	70	265	875	1,500,000	
No. 2		875	1000	450	70	265	725	1,300,000	
No. 3		500	575	275	70	265	450	1,200,000	
Appearance		1050	1200	700	70	265	1050	1,500,000	
Stud		500	575	275	70	265	450	1,200,000	
Select Structural	Beams and Stringers	1100	——	650	65	265	775	1,300,000	
No. 1		900	——	450	65	265	625	1,300,000	
Select Structural	Posts and Timbers	1050	——	700	65	265	800	1,300,000	
No. 1		850	——	550	65	265	700	1,300,000	
Select	Decking	1200	1400	——	——	265	——	1,500,000	
Commercial		1000	1150	——	——	265	——	1,300,000	
WESTERN CEDARS (Surfaced dry or surfaced green. Used at 19% max. m.c.)									
Select Structural	2" to 4" thick 2" to 4" wide	1500	1750	875	75	265	1200	1,100,000	WCLIB WWPA (see footnotes 1 through 12)
No. 1		1300	1500	750	75	265	950	1,100,000	
No. 2		1050	1200	625	75	265	750	1,000,000	
No. 3		600	675	350	75	265	450	900,000	
Appearance		1300	1500	750	75	265	1100	1,100,000	
Stud		600	675	350	75	265	450	900,000	
Construction	2" to 4" thick 4" wide	775	875	450	75	265	850	900,000	
Standard		425	500	250	75	265	700	900,000	
Utility		200	225	125	75	265	450	900,000	
Select Structural	2" to 4" thick 5" and wider	1300	1500	875	75	265	1050	1,100,000	
No. 1		1100	1300	750	75	265	950	1,100,000	
No. 2		925	1050	475	75	265	800	1,000,000	
No. 3		525	625	275	75	265	500	900,000	
Appearance		1100	1300	750	75	265	1100	1,100,000	
Stud		525	625	275	75	265	500	900,000	
Select Structural	Beams and Stringers	1150	——	675	70	265	875	1,000,000	WCLIB (see footnotes 1 through 12)
No. 1		975	——	475	70	265	725	1,000,000	
Select Structural	Posts and Timbers	1100	——	725	70	265	925	1,000,000	
No. 1		875	——	600	70	265	800	1,000,000	
Select Dex	Decking	1250	1450	——	——	265	——	1,100,000	
Commercial Dex		1050	1200	——	——	265	——	1,000,000	
Select Structural	Beams and Stringers	1150	——	775	70	265	875	1,000,000	WWPA (see footnotes 1 through 13)
No. 1		975	——	650	70	265	725	1,000,000	
Select Structural	Posts and Timbers	1100	——	725	70	265	925	1,000,000	
No. 1		875	——	600	70	265	800	1,000,000	
Selected Decking	Decking	——	1450	——	——	——	——	1,100,000	
Commercial Decking		——	1200	——	——	——	——	1,000,000	
Selected Decking	Decking	——	1550	(Surfaced at 15% max. m.c. and used at 15% max. m.c.)			——	1,100,000	
Commercial Decking		——	1300				——	1,000,000	
WESTERN CEDARS (NORTH) (Surfaced dry or surfaced green. Used at 19% max. m.c.)									
Select Structural	2" to 3" thick 2" to 4" wide	1450	1700	850	70	265	1200	1,100,000	NLGA (A Canadian agency. See footnotes 1 through 12 and 15 and 16)
No. 1		1250	1450	725	70	265	950	1,100,000	
No. 2		1000	1200	600	70	265	750	1,000,000	
No. 3		575	650	325	70	265	450	900,000	
Appearance		1250	1450	725	70	265	1100	1,100,000	
Stud		575	650	325	70	265	450	900,000	
Construction	2" to 4" thick 4" wide	750	850	425	70	265	850	900,000	
Standard		425	475	250	70	265	700	900,000	
Utility		200	225	125	70	265	450	900,000	
Select Structural	2" to 4" thick 5" and wider	1250	1450	825	70	265	1050	1,100,000	
No. 1		1050	1250	725	70	265	950	1,100,000	
No. 2		875	1000	475	70	265	800	1,000,000	
No. 3		525	600	275	70	265	500	900,000	
Appearance		1050	1250	725	70	265	1100	1,100,000	
Stud		525	600	275	70	265	500	900,000	
Select Structural	Beams and Stringers	1150	——	675	65	265	850	1,000,000	
No. 1		925	——	475	65	265	700	1,000,000	
Select Structural	Posts and Timbers	1050	——	700	65	265	900	1,000,000	
No. 1		875	——	575	65	265	800	1,000,000	
Select	Decking	1200	1400	——	——	265	——	1,100,000	
Commercial		1050	1200	——	——	265	——	1,000,000	

TABLE A2.5—*CONTINUED*

DESIGN VALUES FOR VISUALLY GRADED STRUCTURAL LUMBER

(Design values listed are for normal loading conditions. See other provisions in the footnotes and in the National Design Specification for adjustments of tabulated values.)

Species and commercial grade	Size classification	Design values in pounds per square inch: Extreme fiber in bending "F_b": Single-member uses	Repetitive-member uses	Tension parallel to grain "F_t"	Horizontal shear "F_v"	Compression perpendicular to grain "$F_{c\perp}$"	Compression parallel to grain "F_c"	Modulus of elasticity "E"	Grading rules agency
WESTERN HEMLOCK (Surfaced dry or surfaced green. Used at 19% max. m.c.)									
Select Structural	2" to 4" thick 2" to 4" wide	1800	2100	1050	90	280	1450	1,600,000	
No. 1		1550	1800	900	90	280	1150	1,600,000	
No. 2		1300	1450	750	90	280	900	1,400,000	
No. 3		700	800	425	90	280	550	1,300,000	
Appearance		1550	1800	900	90	280	1350	1,600,000	
Stud		700	800	425	90	280	550	1,300,000	WCLIB
Construction	2" to 4" thick 4" wide	925	1050	550	90	280	1050	1,300,000	WWPA
Standard		525	600	300	90	280	850	1,300,000	
Utility		250	275	150	90	280	550	1,300,000	
Select Structural	2" to 4" thick 5" and wider	1550	1800	1050	90	280	1300	1,600,000	
No. 1		1350	1550	900	90	280	1150	1,600,000	(see footnotes 1 through 12)
No. 2		1100	1250	575	90	280	975	1,400,000	
No. 3		650	750	325	90	280	625	1,300,000	
Appearance		1350	1550	900	90	280	1350	1,600,000	
Stud		650	750	325	90	280	625	1,300,000	
Select Structural	Beams and Stringers	1400	——	825	85	280	1000	1,400,000	
No. 1		1150	——	575	85	280	850	1,400,000	
Select Structural	Posts and Timbers	1300	——	875	85	280	1100	1,400,000	WCLIB
No. 1		1050	——	700	85	280	950	1,400,000	
Select Dex	Decking	1500	1750	——	——	280	——	1,600,000	(see footnotes 1 through 12)
Commercial Dex		1300	1450	——	——	280	——	1,400,000	
Select Structural	Beams and Stringers	1400	——	950	85	280	1000	1,400,000	
No. 1		1150	——	775	85	280	850	1,400,000	
Select Structural	Posts and Timbers	1300	——	875	85	280	1100	1,400,000	WWPA
No. 1		1050	——	700	85	280	950	1,400,000	
Selected Decking	Decking	——	1750	——	——	——	——	1,600,000	(see footnotes 1 through 13)
Commercial Decking			1450	——	——	——	——	1,400,000	
Selected Decking	Decking	——	1900	(Surfaced at 15% max. m.c. and used at 15% max. m.c.)			——	1,700,000	
Commercial Decking		——	1600				——	1,500,000	
WESTERN HEMLOCK (NORTH) (Surfaced dry or surfaced green. Used at 19% max. m.c.)									
Select Structural	2" to 3" thick 2" to 4" wide	1800	2100	1050	75	280	1450	1,600,000	
No. 1		1550	1800	900	75	280	1150	1,600,000	
No. 2		1300	1450	750	75	280	900	1,400,000	
No. 3		700	800	425	75	280	550	1,300,000	
Appearance		1550	1800	900	75	280	1350	1,600,000	
Stud		700	800	425	75	280	550	1,300,000	
Construction	2" to 4" thick 4" wide	925	1050	550	75	280	1050	1,300,000	NLGA
Standard		525	600	300	75	280	850	1,300,000	
Utility		250	275	150	75	280	550	1,300,000	(A Canadian agency. See footnotes 1 through 12 and 15 and 16)
Select Structural	2" to 4" thick 5" and wider	1550	1800	1050	75	280	1300	1,600,000	
No. 1		1350	1550	900	75	280	1150	1,600,000	
No. 2		1100	1250	575	75	280	975	1,400,000	
No. 3		650	750	325	75	280	625	1,300,000	
Appearance		1350	1550	900	75	280	1350	1,600,000	
Stud		650	750	325	75	280	625	1,300,000	
Select Structural	Beams and Stringers	1400	——	825	70	280	1000	1,400,000	
No. 1		1150	——	575	70	280	850	1,400,000	
Select Structural	Posts and Timbers	1300	——	875	70	280	1100	1,400,000	
No. 1		1050	——	700	70	280	950	1,400,000	
Select	Decking	1500	1750	——	——	280	——	1,600,000	
Commercial		1300	1450	——	——	280	——	1,400,000	
WESTERN WHITE PINE (Surfaced dry or surfaced green. Used at 19% max. m.c.)									
Select Structural	2" to 3" thick 2" to 4" wide	1350	1550	775	65	235	1100	1,400,000	
No. 1		1150	1300	675	65	235	875	1,400,000	
No. 2		925	1050	550	65	235	675	1,300,000	
No. 3		525	600	300	65	235	425	1,200,000	
Appearance		1150	1300	675	65	235	1050	1,400,000	
Stud		525	600	300	65	235	425	1,200,000	
Construction	2" to 4" thick 4" wide	675	775	400	65	235	775	1,200,000	
Standard		375	425	225	65	235	650	1,200,000	NLGA
Utility		175	200	100	65	235	425	1,200,000	
Select Structural	2" to 4" thick 5" and wider	1150	1300	750	65	235	975	1,400,000	
No. 1		975	1150	650	65	235	875	1,400,000	(A Canadian agency. See footnotes 1 through 12 and 15 and 16)
No. 2		800	925	425	65	235	725	1,300,000	
No. 3		475	550	250	65	235	450	1,200,000	
Appearance		975	1150	650	65	235	1050	1,400,000	
Stud		475	550	250	65	235	450	1,200,000	
Select Structural	Beams and Stringers	1050	——	600	60	235	775	1,300,000	
No. 1		850	——	425	60	235	625	1,300,000	
Select Structural	Posts and Timbers	975	——	650	60	235	800	1,300,000	
No. 1		775	——	525	60	235	700	1,300,000	
Select	Decking	1100	1300	——	——	235	——	1,400,000	
Commercial		925	1050	——	——	235	——	1,300,000	

TABLE A2.5—*CONTINUED*

DESIGN VALUES FOR VISUALLY GRADED STRUCTURAL LUMBER

(Design values listed are for normal loading conditions. See other provisions in the footnotes and in the National Design Specification for adjustments of tabulated values.)

Species and commercial grade	Size classification	Design values in pounds per square inch: Extreme fiber in bending "F_b" — Single-member uses	Extreme fiber in bending "F_b" — Repetitive-member uses	Tension parallel to grain "F_t"	Horizontal shear "F_v"	Compression perpendicular to grain "$F_{c\perp}$"	Compression parallel to grain "F_c"	Modulus of elasticity "E"	Grading rules agency
WHITE WOODS (WESTERN WOODS) (Surfaced dry or surfaced green. Used at 19% max. m.c.)									
Select Structural	2" to 4" thick 2" to 4" wide	1350	1550	775	70	190	950	1,100,000	WWPA (see footnotes 1 through 13)
No. 1		1150	1300	650	70	190	750	1,100,000	
No. 2		925	1050	550	70	190	600	1,000,000	
No. 3		525	600	300	70	190	375	900,000	
Appearance		1150	1300	650	70	190	900	1,100,000	
Stud		525	600	300	70	190	375	900,000	
Construction	2" to 4" thick 4" wide	675	775	400	70	190	675	900,000	
Standard		375	425	225	70	190	550	900,000	
Utility		175	200	100	70	190	375	900,000	
Select Structural	2" to 4" thick 5" and wider	1150	1300	775	70	190	850	1,100,000	
No. 1		975	1100	650	70	190	750	1,100,000	
No. 2		800	925	425	70	190	625	1,000,000	
No. 3		475	550	250	70	190	400	900,000	
Appearance		975	1100	650	70	190	900	1,100,000	
Stud		475	550	250	70	190	400	900,000	
Select Structural	Beams and Stringers	1000	—	700	65	190	675	1,000,000	
No. 1		850	—	575	65	190	550	1,000,000	
Select Structural	Posts and Timbers	950	—	650	65	190	700	1,000,000	
No. 1		775	—	525	65	190	625	1,000,000	
Selected Decking	Decking	—	1300	—	—	—	—	1,100,000	
Commercial Decking		—	1050	—	—	—	—	1,000,000	
Selected Decking	Decking	—	1400	(Surfaced at 15% max. m.c. and used at 15% max. m.c.)			—	1,100,000	
Commercial Decking		—	1150				—	1,000,000	
YELLOW-POPLAR (Surfaced dry or surfaced green. Used at 19% max. m.c.)									
Select Structural	2" to 3" thick 2" to 4" wide	1500	1700	875	80	270	1050	1,500,000	NHPMA (see footnotes 1 through 12)
No. 1		1250	1450	750	80	270	825	1,500,000	
No. 2		1050	1200	625	75	270	650	1,300,000	
No. 3		575	675	350	75	270	400	1,200,000	
Stud		575	675	350	75	270	400	1,200,000	
Construction	2" to 4" thick 4" wide	750	875	450	80	270	750	1,200,000	
Standard		425	500	250	75	270	625	1,200,000	
Utility		200	225	125	75	270	400	1,200,000	
Select Structural	2" to 4" thick 5" and wider	1300	1500	850	75	270	925	1,500,000	
No. 1		1100	1250	725	75	270	825	1,500,000	
No. 2		900	1050	475	75	270	700	1,300,000	
No. 3		525	600	275	75	270	425	1,200,000	
Appearance		1100	1250	725	75	270	1000	1,500,000	
Stud		525	600	275	75	270	425	1,200,000	

Table 4A Footnotes Applicable to VISUALLY GRADED LUMBER

1. Grading rules agencies listed in Tables 4A and 4B include the following:

NELMA	—	Northeastern Lumber Manufacturers Association, Inc. 4 Fundy Road, Falmouth, Maine 04105
NHPMA	—	Northern Hardwood and Pine Manufacturers Association, Inc. Northern Bldg, Green Bay, Wisconsin 54301
NLGA	—	National Lumber Grades Authority (Canada) P.O. Box 97 Ganges, B.C., Canada VDS 1EO
RIS	—	Redwood Inspection Service One Lombard St., San Francisco, California 94111
SPIB	—	Southern Pine Inspection Bureau 4709 Scenic Highway, Pensacola, Florida 32504
WCLIB	—	West Coast Lumber Inspection Bureau 6980 SW Varnes Rd., PO Box 23145 Portland, Oregon 97223
WWPA	—	Western Wood Products Association 1500 Yeon Building, Portland, Oregon 97204

2. The design values herein are applicable to lumber that will be used under dry conditions such as in most covered structures. For 2" to 4" thick lumber the DRY surfaced size shall be used. In calculating design values, the natural gain in strength and stiffness that occurs as lumber dries has been taken into consideration as well as the reduction in size that occurs when unseasoned lumber shrinks. The gain in load carrying capacity due to increased strength and stiffness resulting from drying more than offsets the design effect of size reductions due to shrinkage. For 5" and thicker lumber, the surfaced sizes also may be used because design values have been adjusted to compensate for any loss in size by shrinkage which may occur.

3. Tabulated tension parallel to grain values for all species for 5" and wider, 2" to 4" thick (and 2½" to 4" thick) size classifications apply to 5" and 6" widths only, for grades of Select Structural, No. 1, No. 2, No. 3, Appearance and Stud, (including dense grades). For lumber wider than 6" in these grades, the tabulated "F_t" values shall be multiplied by the following factors:

Grade (2" to 4" thick, 5" and wider) (2½" to 4" thick, 5" and wider) (Includes "Dense" grades)	Multiply tabulated "F_t" values by: 5" & 6" wide	8" wide	10" and wider
Select Structural	1.00	0.90	0.80
No. 1, No. 2, No. 3 and Appearance	1.00	0.80	0.60
Stud	1.00	—	—

4. Design values for all species of Stud grade in 5" and wider size classifications apply to 5" and 6" widths only.

5. Values for "F_b", "F_t", and "F_c" for all species of the grades of Construction, Standard and Utility apply only to 4" widths. Design values for 2" and 3" widths of these grades are available from the grading rules agencies (see Note 1).

6. The values in Table 4A for dimension lumber 2" to 4" in thickness are based on edgewise use. When such lumber is used flatwise, the design values for extreme fiber in bending for all species may be multiplied by the following factors:

Dimension lumber used flatwise — Width	Thickness: 2"	3"	4"
2" to 4"	1.10	1.04	1.00
5" and wider	1.22	1.16	1.11

TABLE A2.5—*CONTINUED*

7. The design values in Table 4A for extreme fiber in bending for decking may be increased by 10 percent for 2" thick decking and by 4 percent for 3" thick decking. (Not applicable to California Redwood and Southern Pine.)

8. When 2" to 4" thick lumber is manufactured at a maximum moisture content of 15 percent and used in a condition where the moisture content does not exceed 15 percent, the design values for surfaced dry or surfaced green lumber shown in Table 4A may be multiplied by the following factors. (For Southern Pine use tabulated design values without adjustment):

2" to 4" thick lumber manufactured and used at 15 percent maximum moisture content (MC 15)					
Extreme fiber in bending "F_b"	Tension parallel to grain "F_t"	Horizontal shear "F_v"	Compression perpendicular to grain "$F_{c\perp}$"	Compression* parallel to grain "F_c"	Modulus* of elasticity "E"
1.08	1.08	1.05	1.00	1.17	1.05
			*For Redwood use only	1.15	1.04

9. When 2" to 4" thick lumber is designed for use where the moisture content will exceed 19 percent for an extended period of time, the design values shown herein shall be multiplied by the following factors, except that for Southern Pine footnote 18 applies:

2" to 4" thick lumber used where moisture content will exceed 19%					
Extreme fiber in bending in "F_b"	Tension parallel to grain "F_t"	Horizontal shear "F_v"	Compression perpendicular to grain "$F_{c\perp}$"	Compression parallel to grain "F_c"	Modulus of elasticity "E"
0.86	0.84	0.97	0.67	0.70	0.97

10. When lumber 5" and thicker is designed for use where the moisture content will exceed 19 percent for an extended period of time, the design values shown in Table 4A (except those for Southern Pine) shall be multiplied by the following factors:

5" and thicker lumber used where moisture content will exceed 19%					
Extreme fiber in bending in "F_b"	Tension parallel to grain "F_t"	Horizontal shear "F_v"	Compression perpendicular to grain "$F_{c\perp}$"	Compression parallel to grain "F_c"	Modulus of elasticity "E"
1.00	1.00	1.00	0.67	0.91	1.00

11. Specific horizontal shear values may be established by use of the following table when length of split, or size of check or shake is known and no increase in them is anticipated. For California Redwood, Southern Pine, Virginia Pine-Pond Pine, or Yellow-Poplar, the provisions in this Footnote apply only to those grades having the following Fv Value: 75 psi - California Redwood, 95 psi - Southern Pine (KD), 90 psi - Southern Pine (S-Dry), 85 psi - Southern Pine (S-Green), 95 psi - Virginia Pine-Pond Pine (KD), 90 psi - Virginia Pine-Pond Pine (S-Dry), 85 psi - Virginia Pine-Pond Pine (S-Green), and 75 psi - Yellow-Poplar.

Shear Stress Modification Factor					
When length of split on wide face is:	Multiply tabulated "F_v" value by: (nominal 2" lumber)	When length of split on wide face is:	Multiply tabulated "F_v" value by: (3" and thicker lumber)	When size of shake* is:	Multiply tabulated "F_v" value by: (3" and thicker lumber)
no split	2.00	no split	2.00	no shake	2.00
½ x wide face	1.67	½ x narrow face	1.67	1/6 x narrow face	1.67
¾ x wide face	1.50	1 x narrow face	1.33	1/3 x narrow face	1.33
1 x wide face	1.33	1½ x narrow face or more	1.00	1/2 x narrow face or more	1.00
1½ x wide face or more	1.00			*Shake is measured at the end between lines enclosing the shake and parallel to the wide face.	

12. Stress rated boards of nominal 1", 1¼" and 1½" thickness, 2" and wider, of most species, are permitted the design values shown for Select Structural, No. 1, No. 2, No. 3, Construction, Standard, Utility, Appearance, Clear Heart Structural and Clear Structural grades as shown in the 2" to 4" thick categories herein, when graded in accordance with the stress rated board provisions in the applicable grading rules. Information on stress rated board grades applicable to the various species is available from the respective grading rules agencies. Information on additional design values may also be available from the respective grading agencies

13. When Decking graded to WWPA rules is surfaced at 15 percent maximum moisture content and used where the moisture content will exceed 15 percent for an extended period of time, the tabulated design values for Decking surfaced at 15 percent maximum moisture content shall be multiplied by the following factors: Extreme Fiber in Bending "F_b" -0.79; Modulus of Elasticity "E" -0.92.

14. To obtain a recommended design value for Spruce Pine, multiply the appropriate design value for Virginia Pine-Pond Pine by the corresponding conversion factor shown below and round to the nearest 100,000 psi for modulus of elasticity; to the next lower multiple of 5 psi for horizontal shear and compression perpendicular to grain; to the next lower multiple of 50 psi for bending, tension parallel to grain and compression parallel to grain if 1000 psi or greater, 25 psi otherwise.

Conversion Factors for Determining Design Values for Spruce Pine							
Design Category	Extreme fiber in bending "F_b": Single member uses	Extreme fiber in bending "F_b": Repetitive member uses	Tension parallel to grain "F_t"	Horizontal Shear "F_v"	Compression perpendicular to grain "$F_{c\perp}$"	Compression parallel to grain "F_c"	Modulus of elasticity "E"
Conversion Factor	.784	.784	.784	.766	.965	.682	.807

15. National Lumber Grades Authority is the Canadian rules writing agency responsible for preparation, maintenance and dissemination of a uniform softwood lumber grading rule for all Canadian species.

16. For species graded to NLGA rules, values shown in Table 4A for Select Structural, No. 1, No. 2, No. 3 and Stud grades are not applicable to 3" x 4" and 4" x 4" sizes.

17. Repetitive member design values for extreme fiber in bending for Southern Pine grades of Dense Structural 86, 72 and 65 apply to 2" to 4" thicknesses only.

18. When 2" to 4" thick Southern Pine lumber is surfaced dry or at 15 percent maximum moisture content (KD) and is designed for use where the moisture content will exceed 19 percent for an extended period of time, the design values in Table 4A for the corresponding grades of 2½" to 4" thick surfaced green Southern Pine lumber shall be used. The net green size may be used in such designs.

19. When 2" to 4" thick Southern Pine lumber is surfaced dry or at 15 percent maximum moisture content (KD) and is designed for use under dry conditions, such as in most covered structures, the net DRY size shall be used in design. For other sizes and conditions of use, the net green size may be used in design.

TABLE A2.5.a.

DESIGN VALUES FOR MACHINE STRESS RATED STRUCTURAL LUMBER

(Design values listed are for normal loading conditions.[8] See footnotes, and other provisions in the National Design Specification, for adjustments of tabulated values.[10])

Grade designation[12]	Grading rules agency (see footnotes 1,2,3,4)	Size classification	Design values in pounds per square inch[11]: Extreme fiber in bending "F_b"[9], Single-member uses	Extreme fiber in bending "F_b"[9], Repetitive-member uses	Tension parallel to grain "F_t"	Compression parallel to grain "F_c"	Modulus of elasticity "E"
900f-1.0E	3,4	Machine rated lumber 2" thick or less All widths	900	1050	350	725	1,000,000
1200f-1.2E	1,2,3,4		1200	1400	600	950	1,200,000
1350f-1.3E	2,4		1350	1550	750	1075[13]	1,300,000
1450f-1.3E	1,3,4		1450	1650	800	1150	1,300,000
1500f-1.3E	2		1500	1750	900	1200	1,300,000
1500f-1.4E	1,2,3,4		1500	1750	900	1200	1,400,000
1650f-1.4E	2		1650	1900	1020	1320	1,400,000
1650f-1.5E	1,2,3,4		1650	1900	1020	1320	1,500,000
1800f-1.6E	1,2,3,4		1800	2050	1175	1450	1,600,000
1950f-1.5E	2		1950	2250	1375	1550	1,500,000
1950f-1.7E	1,2,4		1950	2250	1375	1550	1,700,000
2100f-1.8E	1,2,3,4		2100	2400	1575	1700	1,800,000
2250f-1.6E	2		2250	2600	1750	1800	1,600,000
2250f-1.9E	1,2,4		2250	2600	1750	1800	1,900,000
2400f-1.7E	2		2400	2750	1925	1925	1,700,000
2400f-2.0E	1,2,3,4		2400	2750	1925	1925	2,000,000
2550f-2.1E	1,2,4		2550	2950	2050	2050	2,100,000
2700f-2.2E	1,2,3,4		2700	3100	2150	2150	2,200,000
2850f-2.3E	2		2850	3300	2300	2300	2,300,000
3000f-2.4E	1,2		3000	3450	2400	2400	2,400,000
3150f-2.5E	2		3150	3600	2500	2500	2,500,000
3300f-2.6E	2		3300	3800	2650	2650	2,600,000
900f-1.0E	1,2,3	See footnote 5	900	1050	350	725	1,000,000
900f-1.2E	1,2,3		900	1050	350	725	1,200,000
1200f-1.5E	1,2,3		1200	1400	600	950	1,500,000
1350f-1.8E	1,2		1350	1550	750	1075	1,800,000
1500f-1.8E	3		1500	1750	900	1200	1,800,000
1800f-2.1E	1,2,3		1800	2050	1175	1450	2,100,000

Table 4B. Footnotes Applicable to MACHINE STRESS RATED LUMBER

1. NLGA grading rules, see Footnote 1, Table 4A.
2. SPIB grading rules, see Footnote 1, Table 4A.
3. WCLIB grading rules, see Footnote 1, Table 4A.
4. WWPA grading rules, see Footnote 1, Table 4A.
5. Size classifications for these grades are:
   NLGA - Machine Rated Lumber; 2" thick or less, all widths.
   SPIB - Machine Rated Lumber; 2" thick or less, all widths.
   WCLIB - Machine Rated Joists; 2" thick or less, 6" and wider.
6. Pine includes Idaho White Pine, Lodgepole Pine, Ponderosa Pine or Sugar Pine.
7. Cedar includes Incense or Western Red Cedar.
8. Stresses apply for lumber used at 19 percent maximum moisture content.
9. Tabulated extreme fiber in bending values "F_b" are applicable to lumber loaded on edge. When loaded flatwise, these values may be increased by multiplying by the following factors:

Nominal width (in)	3"	4"	5"	6"	8"	10"	12"	14"
Factor	1.06	1.10	1.12	1.15	1.19	1.22	1.25	1.28

10. Footnotes 1, 2, 9, 11 and 19 to Table 4A apply also to Machine Stress Rated Lumber.
11. Design values for horizontal shear "F_v" and compression perpendicular to grain "$F_{c\perp}$" for lumber used under dry conditions are the same as the values listed in Table 4A for No. 2 visually graded lumber of the appropriate species. For "Mixed Species" graded under WCLIB grading rules, F_v= 70 psi and $F_{c\perp}$ = 190 psi.
12. For any given value of fiber stress in bending, "Fb", the average modulus of elasticity, "E", may vary depending upon species, timber source and other variables. The "E" values included in the "f-E" grade designations in Table 4B are those usually associated with each "Fb" level. Grade stamps may show higher or lower "E" values (in increments of 100,000 psi) if machine rating indicates the assignment is appropriate. When an "E" value associated with a designated "f" level is lower or higher than those listed in Table 4B, the tabulated "F_b", "F_t", and "F_c" values associated with the designated "f" value are applicable. The "E" value for design shall be that associated with the "E" value on the grade stamp.
13. When graded under WWPA grading rules, value shall be 1100 psi.
14. When 2" to 4" thick lumber is designed for use where the moisture content will exceed 19 percent for an extended period of time, the design values shown herein shall be multiplied by the following factors.

2" to 4" thick lumber used where moisture content will exceed 19%: Extreme fiber in bending in "F_b"	Tension parallel to grain "F_t"	Horizontal shear "F_v"	Compression perpendicular to grain "$F_{c\perp}$"	Compression parallel to grain "F_c"	Modulus of elasticity "E"
0.86	0.84	0.97	0.67	0.70	0.97

TABLE A3.1.*

DESIGN VALUES FOR STRUCTURAL GLUED LAMINATED SOFTWOOD TIMBERS: MEMBERS STRESSED PRIMARILY IN BENDING[1, 2, 3, 4, 14]

Design values are for normal load duration and dry conditions of use. See footnotes, and other provisions in the National Design Specification for Wood Construction, for adjustments of calculated values.

Combination Symbol[6]	Species Outer Laminations/ Core Laminations[7]	Design values in pounds per square inch													
		BENDING ABOUT X-X AXIS						BENDING ABOUT Y-Y AXIS				AXIALLY LOADED			
		Loaded Perpendicular to Wide Faces of Laminations						Loaded Parallel to the Wide Faces of the Laminations							
		Extreme Fiber in Bending[5]		Compression Perp. to Grain		Horizontal Shear[12]	Modulus of Elasticity	Extreme Fiber in Bending[5]	Compression Perpendicular to Grain Side Faces	Horizontal Shear	Modulus of Elasticity	Tension Parallel To Grain	Compression Parallel To Grain	Modulus Of Elasticity	
		Tension Zone Stressed In Tension F bxx	Compression Zone Stressed[8] In Tension F bxx	Tension Face[11,12] F c⊥xx	Compression Face[11,12] F c⊥xx	Fvxx	Exx	F byy	F c⊥yy	F vyy	E yy	F_t	F c	E	
VISUALLY GRADED WESTERN SPECIES															
16F-V1	DF/WW	1600	800	385[11,12]	385[12]	140	1,300,000	950	190	130	1,100,000	675	975	1,100,000	
16F-V2	HF/HF	1600	800	385[12]	245[12]	155	1,400,000	1250	245	135	1,300,000	875	1300	1,300,000	
16F-V3	DF/DF	1600	800	385[11,12]	385[12]	165	1,500,000	1450	385	145	1,500,000	950	1550	1,500,000	
16F-V4[9]	DF/N3WW	1600	800	450	385[12]	90[12]	1,500,000	900	190	130	1,300,000	650	600	1,300,000	
16F-V5[9]	DF/N3DF	1600	800	450	385[12]	90[12]	1,600,000	1000	270	135	1,500,000	750	875	1,500,000	
16F-V6[10]	DF/DF	1600	1600	385[11,12]	385[12]	165	1,500,000	1450	385	145	1,400,000	950	1550	1,500,000	
16F-V7[10]	HF/HF	1600	1600	245[12]	245[12]	155	1,400,000	1200	245	135	1,300,000	850	1350	1,300,000	
20F-V1	DF/WW	2000	1000	450	385[12]	140	1,400,000	1000	190	130	1,200,000	750	1000	1,200,000	
20F-V2	HF/HF	2000	1000	385[12]	245[12]	155	1,500,000	1200	245	135	1,400,000	975	1350	1,400,000	
20F-V3	DF/DF	2000	1000	450	385[12]	165	1,600,000	1450	385	145	1,500,000	1000	1550	1,500,000	
20F-V4	DF/DF	2000	1000	410[11,12]	385[12]	165	1,600,000	1450	385	145	1,600,000	1000	1550	1,600,000	
20F-V5[9]	DF/N3WW	2000	1000	450	385[12]	90[12]	1,600,000	1000	190	135	1,300,000	750	725	1,300,000	
20F-V6[9]	DF/N3DF	2000	1000	450	385[12]	90[12]	1,600,000	1000	270	135	1,500,000	775	900	1,500,000	
20F-V7[10]	DF/DF	2000	2000	450	450	165	1,600,000	1450	385	145	1,600,000	1000	1600	1,600,000	
20F-V8[10]	DF/DF	2000	2000	410[11,12]	410[11,12]	165	1,700,000	1450	385	145	1,600,000	1000	1600	1,600,000	
20F-V9[10]	HF/HF	2000	2000	385[12]	385[12]	155	1,500,000	1400	245	135	1,400,000	975	1400	1,400,000	
22F-V1	DF/WW	2200	1100	450	385[12]	140	1,600,000	1050	190	130	1,300,000	850	1100	1,300,000	
22F-V2	HF/HF	2200	1100	385[12]	385[12]	155	1,500,000	1250	245	135	1,400,000	950	1350	1,400,000	
22F-V3	DF/DF	2200	1100	450	385[12]	165	1,700,000	1450	385	145	1,600,000	1050	1500	1,600,000	
22F-V4	DF/DF	2200	1100	410[11,12]	385[12]	165	1,700,000	1450	385	145	1,600,000	1000	1550	1,600,000	
22F-V5[9]	DF/N3WW	2200	1100	450	385[12]	90[12]	1,600,000	1100	190	135	1,400,000	800	725	1,400,000	
22F-V6[9]	DF/N3DF	2200	1100	450	385[12]	90[12]	1,700,000	1250	270	135	1,600,000	900	925	1,600,000	
22F-V7[10]	DF/DF	2200	2200	450	450	165	1,800,000	1450	385	145	1,600,000	1100	1650	1,600,000	
22F-V8[10]	DF/DF	2200	2200	410[11,12]	410[11,12]	165	1,700,000	1450	385	145	1,600,000	1050	1650	1,600,000	
22F-V9[10]	HF/HF	2200	2200	385[12]	385[12]	155	1,500,000	1250	245	135	1,400,000	975	1400	1,400,000	
24F-V1	DF/WW	2400	1200	450	450	140	1,700,000	1250	190	135	1,400,000	1000	1300	1,400,000	
24F-V2	HF/HF	2400	1200	385[12]	385[12]	155	1,500,000	1250	245	135	1,400,000	950	1300	1,400,000	
24F-V3	DF/DF	2400	1200	450	385[12]	165	1,700,000	1500	385	145	1,600,000	1100	1600	1,600,000	
24F-V4	DF/DF	2400	1200	450	450	165	1,800,000	1500	385	145	1,600,000	1150	1650	1,600,000	
24F-V5	DF/HF	2400	1200	450	450	155	1,700,000	1350	245	140	1,500,000	1100	1450	1,500,000	
24F-V6[9]	DF/N3WW	2400	1200	450	385[12]	90[12]	1,700,000	1200	190	140	1,500,000	950	800	1,500,000	
24F-V7[9]	DF/N3DF	2400	1200	450	385[12]	90[12]	1,700,000	1250	270	135	1,600,000	900	950	1,600,000	
24F-V8[10]	DF/DF	2400	2400	450	450	165	1,800,000	1450	385	145	1,600,000	1100	1650	1,600,000	
24F-V9[10]	HF/HF	2400	2400	385[12]	385[12]	155	1,500,000	1500	245	135	1,400,000	1000	1450	1,400,000	
24F-V10[10]	DF/HF	2400	2400	450	450	155	1,800,000	1400	245	140	1,600,000	1150	1600	1,600,000	
Wet-use factors		0.8	0.8	0.667	0.667	0.875	0.833	0.8	0.667	0.875	0.833	0.8	0.73	0.833	

From National Design Specification, April 1980 supplement.

TABLE A3.1—*CONTINUED*

DESIGN VALUES FOR STRUCTURAL GLUED LAMINATED SOFTWOOD TIMBERS: MEMBERS STRESSED PRIMARILY IN BENDING[1,2,3,4,14]

Design values are for normal load duration and dry conditions of use. See footnotes, and other provisions in the National Design Specification for Wood Construction, for adjustments of calculated values.

Combination Symbol[6]	Species Outer Laminations/ Core Laminations[7]	Design values in pounds per square inch												
		BENDING ABOUT X-X AXIS						BENDING ABOUT Y-Y AXIS				AXIALLY LOADED		
		Loaded Perpendicular to Wide Faces of Laminations						Loaded Parallel to the Wide Faces of the Laminations						
		Extreme Fiber in Bending[5]		Compression Perp. to Grain		Horizontal Shear[12]	Modulus of Elasticity	Extreme Fiber in Bending[5]	Compression Perpendicular to Grain Side Faces	Horizontal Shear	Modulus of Elasticity	Tension Parallel To Grain	Compression Parallel To Grain	Modulus of Elasticity
		Tension Zone Stressed In Tension F bxx	Compression Zone Stressed[8] In Tension F bxx	Tension Face[11,12] F c⊥xx	Compression Face[11,12] F c⊥xx	Fvxx	Exx	F byy	F c⊥yy	F vyy	E yy	F_t	F c	E
E-RATED WESTERN SPECIES														
16F-E1	WW/WW	1600	800	190[12]	190[12]	140	1,300,000	1050	190	125	1,200,000	725	925	1,200,000
16F-E2[13]	HF/HF	1600	800	245[12]	245[12]	155	1,400,000	1250	245	135	1,300,000	825	1200	1,300,000
16F-E3	DF/DF	1600	800	385[12]	385[12]	165	1,600,000	1450	385	145	1,500,000	975	1600	1,500,000
16F-E4[9]	DF/N3WW	1600	800	385[12]	385[12]	90[12]	1,600,000	900	190	130	1,300,000	675	675	1,300,000
16F-E5[9]	DF/N3DF	1600	800	385[12]	385[12]	90[12]	1,600,000	1050	270	135	1,500,000	700	900	1,500,000
16F-E6[10]	DF/DF	1600	1600	385[12]	385[12]	165	1,600,000	1500	385	145	1,500,000	1000	1600	1,500,000
16F-E7[10,13]	HF/HF	1600	1600	245[12]	245[12]	155	1,400,000	1250	245	135	1,300,000	850	1150	1,300,000
20F-E1	WW/WW	2000	1000	190[12]	190[12]	140	1,600,000	1100	190	125	1,300,000	800	1050	1,300,000
20F-E2[13]	HF/HF	2000	1000	385[12]	385[12]	155	1,600,000	1400	245	135	1,400,000	925	1550	1,400,000
20F-E3	DF/DF	2000	1000	385[12]	385[12]	165	1,700,000	1550	385	145	1,600,000	1050	1650	1,600,000
20F-E4[9]	DF/N3WW	2000	1000	450	385[12]	90[12]	1,600,000	1100	190	130	1,400,000	800	700	1,400,000
20F-E5[9]	DF/N3DF	2000	1000	385[12]	385[12]	90[12]	1,700,000	1300	270	135	1,600,000	825	975	1,600,000
20F-E6[10]	DF/DF	2000	2000	385[12]	385[12]	165	1,700,000	1600	385	145	1,600,000	1150	1650	1,600,000
20F-E7[10,13]	HF/HF	2000	2000	385[12]	385[12]	155	1,600,000	1500	245	135	1,400,000	1050	1550	1,400,000
22F-E1	DF/DF	2200	1100	450	385[12]	165	1,700,000	1550	385	145	1,600,000	1050	1600	1,600,000
22F-E2[13]	HF/HF	2200	1100	385[12]	385[12]	155	1,600,000	1400	245	135	1,400,000	950	1400	1,400,000
22F-E3[9]	DF/N3WW	2200	1100	450	385[12]	90[12]	1,700,000	1250	190	135	1,400,000	825	750	1,400,000
22F-E4[9]	DF/N3DF	2200	1100	450	385[12]	90[12]	1,800,000	1350	270	135	1,600,000	950	950	1,600,000
22F-E5[10]	DF/DF	2200	2200	450	450	165	1,700,000	1650	385	145	1,600,000	1100	1650	1,600,000
22F-E6[10,13]	HF/HF	2200	2200	385[12]	385[12]	155	1,700,000	1550	245	135	1,500,000	1050	1500	1,500,000
24F-E1	DF/DF	2400	1200	450	450	165	1,800,000	1550	385	145	1,600,000	1100	1600	1,600,000
24F-E2[13]	HF/HF	2400	1200	385[12]	385[12]	155	1,700,000	1450	245	135	1,500,000	1000	1400	1,500,000
24F-E3	DF/HF	2400	1200	450	385[12]	155	1,800,000	1500	245	135	1,500,000	1050	1550	1,500,000
24F-E4	DF/DF	2400	1200	450	450	165	1,800,000	1650	385	145	1,700,000	1100	1700	1,700,000
24F-E5	DF/DF	2400	1200	450	385[12]	165	1,800,000	1650	385	145	1,600,000	1100	1550	1,600,000
24F-E6[13]	HF/WW	2400	1200	385[12]	385[12]	140	1,800,000	1250	190	130	1,400,000	925	1350	1,400,000
24F-E7[9]	DF/N3WW	2400	1200	450	450	90[12]	1,900,000	1400	190	135	1,600,000	975	875	1,600,000
24F-E8[9]	DF/N3DF	2400	1200	450	450	90[12]	1,900,000	1400	270	135	1,700,000	1000	1050	1,700,000
24F-E9[9,13]	HF/N3HF	2400	1200	385[12]	385[12]	90[12]	1,800,000	1350	245	135	1,600,000	950	825	1,600,000
24F-E10[9]	DF/DF	2400	2400	450	450	165	1,900,000	1850	385	145	1,700,000	1300	1750	1,700,000
24F-E11[9,13]	HF/HF	2400	2400	385[12]	385[12]	155	1,800,000	1600	245	135	1,500,000	1150	1550	1,500,000
24F-E12[9]	DF/HF	2400	2400	450	450	155	1,900,000	1750	245	135	1,600,000	1200	1600	1,600,000
24F-E13[9]	DF/DF	2400	2400	450	450	165	1,800,000	1950	385	145	1,700,000	1250	1700	1,700,000
Wet-use factors		0.8	0.8	0.667	0.667	0.875	0.833	0.8	0.667	0.875	0.833	0.8	0.73	0.833

TABLE A3.1—*CONTINUED*

DESIGN VALUES FOR STRUCTURAL GLUED LAMINATED SOFTWOOD TIMBERS: MEMBERS STRESSED PRIMARILY IN BENDING[1,2,3,4,14]

Design values are for normal load duration and dry conditions of use. See footnotes, and other provisions in the National Design Specification for Wood Construction, for adjustments of calculated values.

Combination Symbol[6]	Species Outer Laminations/ Core Laminations[7]	Design values in pounds per square inch												
		BENDING ABOUT X-X AXIS						BENDING ABOUT Y-Y AXIS				AXIALLY LOADED		
		Loaded Perpendicular to Wide Faces of Laminations						Loaded Parallel to the Wide Faces of the Laminations						
		Extreme Fiber in Bending[5]		Compression Perp. to Grain		Horizontal Shear[12]	Modulus of Elasticity	Extreme Fiber in Bending[5]	Compression Perpendicular to Grain Side faces	Horizontal Shear	Modulus of Elasticity	Tension Parallel To Grain	Compression Parallel To Grain	Modulus of Elasticity
		Tension Zone Stressed In Tension F bxx	Compression Zone Stressed[8] In Tension F bxx	Tension Face[11,12] F c⊥xx	Compression Face[11,12] F c⊥xx	Fvxx	Exx	F byy	F c⊥yy	F vyy	E yy	F_t	F c	E
VISUALLY GRADED SOUTHERN PINE														
16F-V1	SP/SP	1600	800	385[11,12]	385[12]	200	1,400,000	1450	385	175	1,300,000	950	1450	1,300,000
16F-V2	SP/SP	1600	800	385[11,12]	385[12]	200	1,400,000	1600	385	175	1,400,000	1000	1550	1,400,000
16F-V3	SP/SP	1600	800	450	450	200	1,400,000	1450	385	175	1,300,000	975	1450	1,300,000
16F-V5[9]	SP/SP	1600	1600	385[11,12]	385[12]	200	1,400,000	1600	385	175	1,400,000	1000	1550	1,400,000
16F-V4[10]	SP/SP	1600	800	385[11,12]	385[12]	90[12]	1,300,000	975	270	150	1,200,000	650	950	1,200,000
20F-V1	SP/SP	2000	1000	450	385[12]	200	1,500,000	1450	385	175	1,400,000	1000	1450	1,400,000
20F-V2	SP/SP	2000	1000	450	385[12]	200	1,600,000	1450	385	175	1,400,000	1050	1550	1,400,000
20F-V3	SP/SP	2000	1000	385[11,12]	385[12]	200	1,400,000	1600	385	175	1,400,000	1000	1500	1,400,000
20F-V4[9]	SP/SP	2000	1000	450	385[12]	90[12]	1,500,000	1100	270	150	1,300,000	725	950	1,300,000
20F-V5[10]	SP/SP	2000	2000	450	450	200	1,600,000	1450	385	175	1,400,000	1050	1550	1,400,000
22F-V1	SP/SP	2200	1100	450	450	200	1,600,000	1600	385	175	1,500,000	1050	1650	1,500,000
22F-V2	SP/SP	2200	1100	385[11,12]	385[12]	200	1,400,000	1600	385	175	1,400,000	1000	1500	1,400,000
22F-V3	SP/SP	2200	1100	450	385[12]	200	1,600,000	1500	385	175	1,400,000	1050	1500	1,400,000
22F-V4[9]	SP/SP	2200	1100	450	385[12]	90[12]	1,600,000	1250	270	155	1,400,000	825	1000	1,400,000
22F-V5[10]	SP/SP	2200	2200	450	450	200	1,600,000	1600	385	175	1,500,000	1050	1600	1,500,000
24F-V1	SP/SP	2400	1200	450	385[12]	200	1,700,000	1500	385	175	1,500,000	1100	1350	1,500,000
24F-V2	SP/SP	2400	1200	450	450	200	1,700,000	1600	385	175	1,500,000	1100	1600	1,500,000
24F-V3	SP/SP	2400	1200	450	450	200	1,800,000	1600	385	175	1,600,000	1150	1700	1,600,000
24F-V4[9]	SP/SP	2400	1200	450	385[12]	90[12]	1,700,000	1250	270	155	1,400,000	850	1050	1,400,000
24F-V5[10]	SP/SP	2400	2400	450	450	200	1,700,000	1600	385	175	1,500,000	1150	1700	1,500,000
E-RATED SOUTHERN PINE														
16F-E1	SP/SP	1600	800	385[12]	385[12]	200	1,600,000	1550	385	175	1,500,000	1050	1600	1,500,000
16F-E2[9]	SP/SP	1600	800	385[12]	385[12]	90[12]	1,600,000	950	270	145	1,300,000	700	1050	1,300,000
16F-E3[10]	SP/SP	1600	1600	385[12]	385[12]	200	1,600,000	1700	385	175	1,500,000	1100	1650	1,500,000
20F-E1	SP/SP	2000	1000	385[12]	385[12]	200	1,700,000	1600	385	175	1,500,000	1050	1600	1,500,000
20F-E2[9]	SP/SP	2000	1000	450	385[12]	90[12]	1,600,000	1100	270	150	1,400,000	750	1000	1,400,000
20F-E3[10]	SP/SP	2000	2000	385[12]	385[12]	200	1,700,000	1800	385	175	1,500,000	1150	1700	1,500,000
22F-E1	SP/SP	2200	1100	450	385[12]	200	1,700,000	1600	385	175	1,500,000	1050	1650	1,500,000
22F-E2[9]	SP/SP	2200	1100	450	385[12]	90[12]	1,600,000	1250	270	155	1,400,000	850	1050	1,400,000
22F-E3[10]	SP/SP	2200	2200	450	450	200	1,700,000	1750	385	175	1,500,000	1150	1650	1,500,000
24F-E1	SP/SP	2400	1200	450	385[12]	200	1,800,000	1600	385	175	1,600,000	1100	1750	1,600,000
24F-E2	SP/SP	2400	1200	450	450	200	1,900,000	1700	385	175	1,600,000	1150	1700	1,600,000
24F-E3[9]	SP/SP	2400	1200	450	450	90[12]	1,800,000	1300	270	155	1,500,000	950	1100	1,500,000
24F-E4[10]	SP/SP	2400	2400	450	450	200	1,800,000	2000	385	175	1,600,000	1250	1750	1,600,000
Wet-use factors		0.8	0.8	0.667	0.667	0.875	0.833	0.8	0.667	0.875	0.833	0.8	0.73	0.833

TABLE A3.1—*CONTINUED*

Footnotes

Applicable to STRUCTURAL GLUED LAMINATED SOFTWOOD TIMBERS:
MEMBERS STRESSED PRIMARILY IN BENDING

1. Design values in this table are based on combinations conforming to "AITC 117-79 – DESIGN, Standard Specifications for Structural Glued Laminated Timber for Softwood Species", by American Institute of Timber Construction, and manufactured in accordance with Department of Commerce Voluntary Product Standard PS56-73, Structural Glued Laminated Timber.

2. The combinations in this table are intended primarily for members stressed in bending due to loads applied perpendicular to the wide faces of the laminations. Design values are tabulated, however, for loading both perpendicular and parallel to the wide faces of the laminations, and for axial loading. For combinations applicable to members loaded primarily axially or parallel to the wide faces of the laminations, see Table 5B.

3. Design values in this table are applicable to members having 4 or more laminations. For members having 2 or 3 laminations, see Table 5B.

4. When moisture content in service will be 16 percent or more, tabulated design values shall be multiplied by the modification factor for wet service conditions, as given in the bottom line of this table.

5. The tabulated design values in bending are applicable to members 12" or less in depth. For members greater than 12" in depth, the requirements of Section 5.3.4 of the National Design Specification apply.

6. The 22F and 24F combinations for members 15" and less in depth may not be readily available and the designer should check on availability prior to specifying. The 16F and 20F combinations are generally available for members 15" and less in depth.

7. The symbols used for species are DF = Douglas Fir-Larch, HF = Hem-Fir, WW = Western Woods or Canadian softwood species, and SP = Southern Pine (N3 refers to No. 3 structural joists and planks or structural light framing grade). For design values for California Redwood, see AITC 117-79 – DESIGN.

8. Design values in this column are for extreme fiber stress in bending when the member is loaded such that the compression zone laminations are subjected to tensile stresses. For more information, see AITC 117-79 – DESIGN. The values in this column may be increased 200 psi where end joint spacing restrictions are applied to the compression zone when stressed in tension.

9. These combinations are intended for straight or slightly cambered members for dry use and industrial appearance grade, because they may contain wane. If wane is omitted these restrictions do not apply.

10 These combinations are balanced and are intended for members continuous or cantilevered over supports and stressed equally in both positive and negative bending.

11. For bending members greater than 15" in depth, these design values for compression perpendicular to grain are 450 psi on the tension face.

12. These design values may be increased in accordance with AITC 117-79 – DESIGN when member conforms with special construction requirements therein. For more information see AITC 117-79 – DESIGN.

13. For these combinations manufacturers may substitute E-rated Douglas Fir-Larch laminations that are 200,000 psi higher in modulus of elasticity than the specified E-rated Hem-Fir, with no change in design values.

14. For fastener design, the appropriate timber connector load group, lag bolt and driven fastener load group, and bolt design value can be classified by the design value for compression perpendicular to grain, as shown in the following table:

Species classification for fastener design			
Compression Perpendicular to Grain Design Value Fc⊥ psi	Timber Connector Load Grouping (NDS Table 8.1A.)	Lag Screw and Driven Fastener Load Grouping (NDS Table 8.1A.)	Bolt Design Values in NDS Table 8.5A
410 or Greater	A	II	Column 3
326 - 410*	B	II	Column 3
325	C	III	Column 3
245 - 324	C	III	Column 8
191 - 244	C	III	Column 12
190 - or less	D	IV	Column 12

*For Fc⊥ = 385 psi for Hem-Fir, use timber connector Group C or driven fastener Group III.

TABLE A3.2.*

DESIGN VALUES FOR STRUCTURAL GLUED LAMINATED SOFTWOOD TIMBERS: MEMBERS STRESSED PRIMARILY IN AXIAL TENSION OR COMPRESSION

(or loaded in bending parallel to the wide face of laminations)[1,2,3,11]

Design values are for normal load duration and dry conditions of use. See footnotes, and other provisions in the National Design Specification for Wood Construction for adjustments of calculated values

Combination Symbol	Species[4]	Modulus of Elasticity	Compression[10] Perpendicular To Grain	Design values in pounds per square inch											
				AXIALLY LOADED			BENDING ABOUT Y-Y AXIS Loaded Parallel to Wide Faces of Laminations						BENDING ABOUT X-X AXIS Loaded Perpendicular to Wide Faces of Laminations		
				Tension Parallel To Grain	Compression Parallel To Grain		Extreme Fiber in Bending[5,6]			Horizontal Shear[7]			Extreme Fiber In Bending[6]		Horizontal Shear[7]
			10	2 or More Laminations	4 or More Laminations	2 or 3 Laminations	4 or More Laminations	3 Laminations	2 Laminations	4 or More Laminations	3 Laminations	2 Laminations	2 Laminations to 15" deep[8]	4 or More Laminations[9]	2 or More Laminations
		E	$F_{c\perp}$	F_t	F_c	F_c	F_{byy}	F_{byy}	F_{byy}	F_{vyy}	F_{vyy}	F_{vyy}	F_{bxx}	F_{bxx}	F_{vxx}
VISUALLY GRADED WESTERN SPECIES															
A-1	DF	1,500,000	385[10]	900	1550	1200	1450	1250	1000	145	135	125	1250	1500	165
A-2	DF	1,700,000	385[10]	1250	1900	1600	1800	1600	1300	145	135	125	1700	2000	165
A-3	DF	1,800,000	450	1450	2300	1850	2100	1850	1550	145	135	125	2000	2300	165
A-4	DF	1,900,000	410[10]	1400	2100	1900	2200	2000	1650	145	135	125	1900	2200	165
A-5	DF	2,000,000	450	1600	2400	2100	2400	2100	1800	145	135	125	2200	2400	165
A-6	DF	1,400,000	270	350	875	550	550	550	550	120	115	105	450	—	140
A-7	DF	1,500,000	385	900	1550	700	1450	1250	1000	145	135	125	1000	—	165
A-8	DF	1,600,000	385[10]	1000	1550	1150	1600	1550	1300	145	135	125	1350	1600	165
A-9	DF	1,800,000	450	1150	1800	1350	1850	1800	1500	145	135	125	1600	1850	165
A-10	DF	1,800,000	385[10]	1300	1950	1450	1950	1750	1500	145	135	125	1750	2100	165
A-11	DF	2,000,000	450	1500	2300	1700	2300	2100	1750	145	135	125	2100	2400	165
A-12	DF	1,800,000	385[10]	1400	1950	1650	2100	1950	1650	145	135	125	1900	2200	165
A-13	DF	2,000,000	450	1600	2300	1950	2400	2300	1950	145	135	125	2200	2400	165
A-14	HF	1,300,000	245[10]	800	1100	975	1200	1050	850	135	130	115	1100	1300	155
A-15	HF	1,400,000	245[10]	1050	1350	1300	1500	1350	1100	135	130	115	1450	1700	155
A-16	HF	1,600,000	245[10]	1200	1500	1450	1750	1550	1300	135	130	115	1600	1900	155
A-17	HF	1,700,000	385	1400	1750	1700	2000	1850	1550	135	130	115	1900	2200	155
A-18	HF	1,300,000	245	425	900	575	700	700	700	135	130	115	575	—	155
A-19	HF	1,400,000	245[10]	850	1300	975	1350	1300	1100	135	130	115	1150	1350	155
A-20	HF	1,600,000	245[10]	975	1450	1250	1550	1500	1250	135	130	115	1350	1550	155
A-21	HF	1,600,000	245[10]	1100	1450	1350	1750	1650	1400	135	130	115	1500	1750	155
A-22	WW	1,000,000	190	525	850	675	800	700	550	120	115	105	725	850	140
A-23	WW	1,000,000	190	275	625	450	450	450	450	120	115	105	400	—	140
A-24	WW	1,100,000	190	550	900	700	900	875	725	120	115	105	775	900	140
A-25	WW	1,200,000	190	650	1000	875	1050	1000	850	120	115	105	875	1050	140
A-26	WW	1,200,000	190	750	1000	1000	1150	1100	925	120	115	105	1000	1150	140
E-RATED WESTERN SPECIES															
A-27	DF	1,800,000	385	900	1750	1200	1450	1250	1000	145	135	125	1250	1500	165
A-28	DF	2,000,000	450	1100	2000	1400	1450	1250	1000	145	135	125	1500	1750	165
A-29	DF	2,200,000	450	1250	2300	1550	1650	1400	1150	145	135	125	1700	2000	165
A-30	DF	1,800,000	385	1550	2100	1700	2400	2400	2100	145	135	125	1800	2100	165
A-31	DF	2,000,000	450	1800	2400	1900	2400	2400	2400	145	135	125	2100	2400	165
A-32	DF	2,200,000	450	1800	2400	2100	2400	2400	2400	145	135	125	2300	2400	165
A-33	HF	1,500,000	245	800	1050	950	1200	1050	850	135	130	115	1100	1300	155
A-34	HF	1,800,000	385	900	1300	1200	1450	1250	1000	135	130	115	1250	1500	155
A-35	HF	2,000,000	385	1100	1550	1400	1450	1250	1000	135	130	115	1500	1750	155
A-36	HF	1,500,000	245	1200	1450	1300	2100	1900	1700	135	130	115	1400	1650	155
A-37	HF	1,800,000	385	1550	1950	1700	2400	2400	2100	135	130	115	1800	2100	155
A-38	HF	2,000,000	385	1800	2400	1900	2400	2400	2400	135	130	115	2100	2400	155
A-39	WW	1,500,000	190	800	1200	950	1200	1050	850	120	115	105	1100	1300	140
A-40	WW	1,800,000	190	900	1500	1200	1450	1250	1000	120	115	105	1250	1500	140
A-41	WW	2,000,000	190	1100	1750	1400	1450	1250	1000	120	115	105	1500	1750	140
A-42	WW	1,500,000	190	1200	1550	1300	2100	1900	1700	120	115	105	1400	1650	140
A-43	WW	1,800,000	190	1550	1950	1700	2400	2400	2100	120	115	105	1800	2100	140
A-44	WW	2,000,000	190	1800	2200	1900	2400	2400	2400	120	115	105	2100	2400	140
Wet-use factors		0.833	0.667	0.8	0.73	0.73	0.8	0.8	0.8	0.875	0.875	0.875	0.8	0.875	0.875

*From National Design Specification, April 1980 Supplement.

TABLE A3.2—*CONTINUED*

DESIGN VALUES FOR STRUCTURAL GLUED LAMINATED SOFTWOOD TIMBERS: MEMBERS STRESSED PRIMARILY IN AXIAL TENSION OR COMPRESSION

(or loaded in bending parallel to the wide face of laminations)[1,2,3,11]

Design values are for normal load duration and dry conditions of use. See footnotes, and other provisions in the National Design Specification for Wood Construction for adjustments of calculated values.

Combination Symbol	Species[4]	Modulus of Elasticity E	Compression[10] Perpendicular To Grain $F_{c\perp}$	Axially Loaded: Tension Parallel To Grain, 2 or More Laminations F_t	Axially Loaded: Compression Parallel To Grain, 4 or More Laminations F_c	Axially Loaded: Compression Parallel To Grain, 2 or 3 Laminations F_c	Bending About Y-Y Axis (Loaded Parallel to Wide Faces of Laminations): Extreme Fiber in Bending[5,6], 4 or More Laminations F_{byy}	Extreme Fiber in Bending[5,6], 3 Laminations F_{byy}	Extreme Fiber in Bending[5,6], 2 Laminations F_{byy}	Horizontal Shear[7], 4 or More Laminations F_{vyy}	Horizontal Shear[7], 3 Laminations F_{vyy}	Horizontal Shear[7], 2 Laminations F_{vyy}	Bending About X-X Axis (Loaded Perpendicular to Wide Faces of Laminations): Extreme Fiber In Bending[6], 2 Laminations to 15" deep[8] F_{bxx}	Extreme Fiber In Bending[6], 4 or More Laminations[9] F_{bxx}	Horizontal Shear[7], 2 or More Laminations F_{vxx}
				Design values in pounds per square inch											
VISUALLY GRADED SOUTHERN PINE															
A-45	SP	1,100,000	270	325	850	550	550	550	550	120	115	105	450	—	140
A-46	SP	1,300,000	385	900	1500	675	1450	1250	1000	175	165	150	1000	—	200
A-47	SP	1,400,000	385[10]	1200	1900	1150	1750	1550	1300	175	165	150	1400	1600	200
A-48	SP	1,700,000	450	1400	2200	1350	2000	1800	1500	175	165	150	1600	1900	200
A-49	SP	1,700,000	385[10]	1350	2100	1450	1950	1750	1500	175	165	150	1800	2100	200
A-50	SP	1,900,000	450	1550	2300	1700	2300	2100	1750	175	165	150	2100	2400	200
A-51	SP	1,700,000	385[10]	1300	1900	1600	2100	1950	1650	175	165	150	1750	2100	200
A-52	SP	1,900,000	450[10]	1500	2200	1850	2400	2300	1950	175	165	150	2100	2400	200
E-RATED SOUTHERN PINE															
A-53	SP	1,800,000	385	900	1900	1200	1450	1250	1000	175	165	150	1250	1500	200
A-54	SP	2,000,000	450	1100	2300	1400	1450	1250	1000	175	165	150	1500	1750	200
A-55	SP	2,200,000	450	1250	2400	1550	1650	1400	1150	175	165	150	1700	2000	200
A-56	SP	1,800,000	385	1550	1850	1700	2400	2400	2100	175	165	150	1800	2100	200
A-57	SP	2,000,000	450	1800	2400	1900	2400	2400	2400	175	165	150	2100	2400	200
A-58	SP	2,200,000	450	1800	2400	2100	2400	2400	2400	175	165	150	2300	2400	200
Wet-use factors		0.833	0.667	0.8	0.73	0.73	0.8	0.8	0.8	0.875	0.875	0.875	0.8	0.875	0.875

Footnotes

Applicable to STRUCTURAL GLUED LAMINATED SOFTWOOD TIMBERS: MEMBERS STRESSED PRIMARILY IN AXIAL TENSION OR COMPRESSION

1. Design values in this table are based on combinations conforming to "AITC 117-79—DESIGN, Standard Specifications for Structural Glued Laminated Timber of Softwood Species", by American Institute of Timber Construction, and manufactured in accordance with Department of Commerce Voluntary Product Standard PS56-73, Structural Glued Laminated Timber.

2. The combinations in this table are intended primarily for members loaded either axially or in bending with the loads acting parallel to the wide faces of the laminations (bending about Y-Y axis). Design values for bending due to load applied perpendicular to the wide faces of the laminations (bending about X-X axis) are also included, although the combinations in Table 5A are usually better suited for this condition of loading.

3. When moisture content in service will be 16 percent or more, tabulated design values shall be multiplied by the modification factor for wet service conditions, as given in the bottom line of this table.

4. The symbols used for species are DF = Douglas Fir-Larch, HF = Hem-Fir, WW = Western Woods and Canadian softwood species, and SP = Southern Pine. For design values for California Redwood, see AITC 117-79—DESIGN.

5. The values of F_{byy} in Table 5B are based on members 12" in depth (bending about Y-Y axis). When member depth is less than 12", the values of F_{byy} may be multiplied by the following factors:

Depth	Multiply tabulated F_{byy} by:
10.75"	1.01
8.75"	1.04
6.75"	1.07
5.125"	1.10
3.125"	1.16

6. The tabulated design values in bending are applicable to members 12" or less in depth. For members greater than 12" in depth, the requirements of Section 5.3.4 of the National Design Specification apply to F_{byy} and F_{bxx}.

7. The design values in horizontal shear in Table 5B are based on members that do not contain wane.

8. The design values in bending about the X-X axis in this column are for members up to 15" in depth without tension laminations.

9. The design values in bending about the X-X axis in this column are for members having special tension laminations, and apply to members having 4 or more laminations. When these values are used in design and the member is specified by combination symbol, the design should also specify the required design value in bending.

10. These design values may be increased in accordance with AITC 117-79—DESIGN when member conforms with special construction requirements therein. For more information see AITC 117-79—DESIGN.

11. Footnote 14 of Table 5A applies also to Table 5B.

TABLE A3.3.*

DESIGN VALUES FOR STRUCTURAL GLUED LAMINATED HARDWOOE

Design values are for normal load duration and dry conditions[3] of use. See footnotes, and other provisions in Specification for Wood Construction, for adjustments of calculated values.

Pounds per square inch

PART A

SPECIES	Multiply the appropriate stress module in part B by the factors below to determine design value for			
	Extreme fiber in bending "F_b" or tension parallel to grain "F_t"	Compression parallel to grain F_c	Modulus of elasticity E	Hor sl
	Factor	Factor	Factor	
Hickory, true and pecan	3.85	3.05	1.80	
Beech, American	3.05	2.45	1.70	
Birch, sweet and yellow	3.05	2.45	1.90	
Elm, rock	3.05	2.45	1.40	
Maple, black and sugar (hard maple)	3.05	2.45	1.70	
Ash, commercial white	2.80	2.20	1.70	
Oak, commercial red and white	2.80	2.05	1.60	
Elm, American and slippery (white or soft elm)	2.20	1.60	1.40	
Sweetgum (red or sap gum)	2.20	1.60	1.40	
Tupelo, black (blackgum)	2.20	1.60	1.20	
Tupelo, water	2.20	1.60	1.30	
Ash, black	2.00	1.30	1.30	
Yellow poplar	2.00	1.45	1.50	
Cottonwood, Eastern	1.55	1.20	1.20	
Modification factor for wet service conditions	0.80	0.73	0.833	(

PART B — Values for use in computing design values with the factors of Part A together with limitations required to permit the use of such stresses

Combination symbol	Ratio of size of maximum permitted knot to finished width of lamination	Number of laminations	Extreme fiber in bending[4]			Tension parallel to grain[4]		Compression parallel to grain[4]		Modulus of elasticity
			Stress module	Maximum grain slope		Stress module	Maximum grain slope	Stress module	Maximum grain slope	Stress module
				Outer lams.[5]	Core					
A	0.1	4 to 14	800	1:16	1:8	500	1:16	970	1:15	1,000,000
		15 or more	800	1:16	1:8	500	1:16	970	1:15	
B	0.2	4 to 14	770	1:16	1:8	500	1:16	920	1:15	1,000,000
		15 or more	800	1:16	1:8	500	1:16	930	1:15	
C	0.3	4 to 14	600	1:12	1:8	450	1:15	860	1:14	900,000
		15 or more	660	1:12	1:8	450	1:16	870	1:14	
D	0.4	4 to 14	450	1:8	1:8	350	1:10	780	1:12	800,000
		15 or more	520	1:8	1:8	350	1:12	810	1:12	
E	0.5	4 to 14	300	1:8	1:8	300	1:8	690	1:10	800,000
		15 or more	380	1:8	1:8	300	1:8	730	1:10	

Footnotes Applicable to STRUCTURAL GLUED LAMINATED HARDWOOD TIMBERS

1. Standard Specifications for Hardwood Glued Laminated Timbers, AITC 119-76, by American Institute of Timber Construction, applies.

2. The design values in bending obtained from this Table apply when the wide faces of the laminations are normal to the direction of the load. They also apply when the loading is parallel to the wide faces of the laminations, provided certain additional restrictions, given in the applicable specification indicated in Note 1, are applied.

3. For wet service conditions, where moisture content of the member will be 16 percent or more, multiply design value by the appropriate modification factor from the bottom of Part A of this table.

4. Stress modules apply only when laminations have a slope of grain no steeper than the values listed, as required in the specification indicated in Note 1. For tension and compression parallel to grain values to apply, all laminations must conform to slope of grain requirements. As an alternative to specifying more restrictive slope of grain limitations in order to use tabulated values for "secondary" stresses, such as when a member is subject to reversal of axial load or to combined flexure and axial load, consult American Institute of Timber Construction for advice on "secondary stresses appropriate for the slope of grain required for the principal stress."

5. For bending, outer laminations means each outer 10 percent of the depth of the member, measured from each face at any cross section of the member as finally installed.

*From National Design Specification, April 1980 Supplement.

Table A.3.4 GLUED-LAMINATED MEMBERS

Unit Properties of Sections*

Symbol Identification
A = Area = in²
S = Section Modulus = in³
I = Moment of Inertia = in⁴
C_f = Size Effect Factor

To determine weight per lineal foot (in pounds) divide the area by 4.

¾ Inch Lamination Thickness †

C_f	No. Lams	Depth	3⅛″ A	3⅛″ S	3⅛″ I	5⅛″ A	5⅛″ S	5⅛″ I	6¾″ A	6¾″ S	6¾″ I	8¾″ A	8¾″ S	8¾″ I	10¾″ A	10¾″ S	10¾″ I	12¼″ A	12¼″ S	12¼″ I	14¼″ A	14¼″ S	14¼″ I
1.000	5	3¾	12	8	14	19	12	23	25	16	30	33	21	39	40	25	47	46	29	54	53	33	63
1.000	6	4½	14	11	24	23	17	39	30	23	51	39	30	67	48	36	82	55	41	93	64	48	108
1.000	7	5¼	16	14	38	27	24	62	35	31	81	46	40	106	56	49	130	64	56	148	75	66	172
1.000	8	6	19	19	56	31	31	92	41	41	122	53	53	158	65	65	194	74	74	221	86	86	257
1.000	9	6¾	21	24	80	35	39	131	46	51	173	59	67	224	73	82	276	83	93	314	96	108	365
1.000	10	7½	23	29	110	38	48	180	51	63	237	66	82	308	81	101	378	92	115	431	107	134	501
1.000	11	8¼	26	36	146	42	58	240	56	77	316	72	99	409	89	122	503	101	139	573	118	162	667
1.000	12	9	28	42	190	46	69	311	61	91	410	79	118	532	97	145	653	110	165	744	128	192	866
1.000	13	9¾	31	50	241	50	81	396	66	107	521	85	139	676	105	170	830	119	194	946	139	226	1,101
1.000	14	10½	33	57	302	54	94	494	71	124	651	92	161	844	113	198	1,037	129	225	1,182	150	262	1,375
1.000	15	11¼	35	66	371	58	108	608	76	142	801	98	185	1,038	121	227	1,276	138	258	1,454	160	301	1,691
1.000	16	12	38	75	450	62	123	738	81	162	972	105	210	1,260	129	258	1,548	147	294	1,764	171	342	2,052
0.993	17	12¾	40	85	540	65	139	885	86	183	1,166	112	237	1,511	137	291	1,857	156	332	2,116	182	386	2,461
0.987	18	13½	42	95	641	69	156	1,051	91	205	1,384	118	266	1,794	145	327	2,204	165	372	2,512	192	433	2,922
0.981	19	14¼	45	106	754	73	174	1,236	96	229	1,628	125	296	2,110	153	364	2,592	175	415	2,954	203	482	3,436
0.976	20	15	47	117	879	77	192	1,441	101	253	1,898	131	328	2,461	161	403	3,023	184	459	3,445	214	534	4,008
0.970	21	15¾	49	129	1,017	81	212	1,669	106	279	2,198	138	362	2,849	169	445	3,500	193	507	3,988	224	589	4,640
0.965	22	16½	52	142	1,170	85	233	1,919	111	306	2,527	144	397	3,276	177	488	4,024	202	556	4,586	235	647	5,334
0.961	23	17¼	54	155	1,337	88	254	2,192	116	335	2,887	151	434	3,743	185	533	4,598	211	608	5,240	246	707	6,095
0.956	24	18	56	169	1,519	92	277	2,491	122	365	3,281	158	473	4,253	194	581	5,225	221	662	5,954	257	770	6,926
0.952	25	18¾	59	183	1,717	96	300	2,815	127	396	3,708	164	513	4,807	202	630	5,905	230	718	6,729	267	835	7,828
0.948	26	19½	61	198	1,931	100	325	3,167	132	428	4,171	171	555	5,407	210	681	6,643	239	776	7,569	278	903	8,805
0.944	27	20¼	63	214	2,162	104	350	3,546	137	461	4,671	177	598	6,055	218	735	7,439	248	837	8,477	289	974	9,861
0.940	28	21	66	230	2,412	108	377	3,955	142	496	5,209	184	643	6,753	226	790	8,296	257	900	9,454	299	1,047	10,997
0.936	29	21¾	68	246	2,680	112	404	4,394	147	532	5,788	190	690	7,503	234	848	9,217	266	966	10,504	310	1,124	12,218
0.933	30	22½	70	264	2,966	115	432	4,865	152	570	6,407	197	738	8,306	242	907	10,204	276	1,034	11,628	321	1,202	13,526
0.929	31	23¼	73	282	3,273	119	462	5,368	157	608	7,070	203	788	9,164	250	969	11,259	285	1,104	12,830	331	1,284	14,925
0.926	32	24	75	300	3,600	123	492	5,904	162	648	7,776	210	840	10,080	258	1,032	12,384	294	1,176	14,112	342	1,368	16,416
0.923	33	24¾	77	319	3,948	127	523	6,475	167	689	8,528	217	893	11,055	266	1,098	13,582	303	1,251	15,477	353	1,455	18,004
0.920	34	25½	80	339	4,318	131	555	7,082	172	732	9,327	223	948	12,091	274	1,165	14,854	312	1,328	16,927	363	1,544	19,690

TABLE A.3.4—*CONTINUED.*

¾ Inch Lamination Thickness †

C_f	No. Lams	Depth	3⅛″ A	3⅛″ S	3⅛″ I	5⅛″ A	5⅛″ S	5⅛″ I	6¾″ A	6¾″ S	6¾″ I	8¾″ A	8¾″ S	8¾″ I	10¾″ A	10¾″ S	10¾″ I	12¼″ A	12¼″ S	12¼″ I	14¼″ A	14¼″ S	14¼″ I
0.917	35	26¼	82	359	4,710	135	589	7,725	177	775	10,174	230	1,005	13,189	282	1,235	16,204	322	1,407	18,465	374	1,637	21,479
0.914	36	27	84	380	5,126	139	623	8,406	182	820	11,072	236	1,063	14,352	290	1,306	17,633	331	1,488	20,093	385	1,731	23,374
0.911	37	27¾	87	401	5,565	142	658	9,126	187	866	12,020	243	1,123	15,582	298	1,380	19,143	340	1,572	21,814	395	1,829	25,376
0.908	38	28½	89	423	6,028	146	694	9,887	192	914	13,021	249	1,185	16,880	306	1,455	20,738	349	1,658	23,631	406	1,929	27,490
0.906	39	29¼	91	446	6,517	150	731	10,688	197	963	14,077	256	1,248	18,248	314	1,533	22,418	358	1,747	25,547	417	2,032	29,717
0.903	40	30	94	469	7,031	154	769	11,531	203	1,013	15,188	263	1,313	19,688	323	1,613	24,188	368	1,838	27,563	428	2,138	32,063
0.901	41	30¾	96	493	7,572	158	808	12,418	208	1,064	16,355	269	1,379	21,201	331	1,694	26,047	377	1,931	29,682	438	2,246	34,528
0.898	42	31½	98	517	8,140	161	848	13,349	213	1,116	17,581	276	1,447	22,791	339	1,778	28,000	386	2,026	31,907	449	2,357	37,116
0.896	43	32¼	101	542	8,735	165	888	14,325	218	1,170	18,867	282	1,517	24,458	347	1,864	30,048	395	2,124	34,241	460	2,470	39,831
0.894	44	33	103	567	9,359	169	930	15,348	223	1,225	20,215	289	1,588	26,204	355	1,951	32,194	404	2,223	36,686	470	2,586	42,675
0.892	45	33¾	106	593	10,011	173	973	16,419	228	1,282	21,624	295	1,661	28,032	363	2,041	34,439	413	2,326	39,244	481	2,705	45,652
0.889	46	34½	108	620	10,693	177	1,017	17,538	233	1,339	23,098	302	1,736	29,942	371	2,133	36,786	423	2,430	41,919	492	2,827	48,763
0.887	47	35¼	110	647	11,406	181	1,061	18,706	238	1,398	24,638	308	1,812	31,938	379	2,226	39,238	432	2,537	44,713	502	2,951	52,013
0.885	48	36	113	675	12,150	185	1,107	19,926	243	1,458	26,244	315	1,890	34,020	387	2,322	41,796	441	2,646	47,628	513	3,078	55,404
0.883	49	36¾	115	703	12,925	188	1,154	21,198	248	1,519	27,919	322	1,970	36,191	395	2,420	44,463	450	2,757	50,667	524	3,208	58,939
0.881	50	37½	117	732	13,733	192	1,201	22,522	253	1,582	29,663	328	2,051	38,452	403	2,520	47,241	459	2,871	53,833	534	3,340	62,622
0.879	51	38¼	120	762	14,574	196	1,250	23,901	258	1,646	31,479	335	2,134	40,806	411	2,621	50,133	469	2,987	57,128	545	3,475	66,455
0.877	52	39	122	792	15,448	200	1,299	25,334	263	1,711	33,367	341	2,218	43,253	419	2,725	53,140	478	3,105	60,555	556	3,612	70,441
0.875	53	39¾	124	823	16,356	204	1,350	26,824	268	1,778	35,329	348	2,304	45,797	427	2,831	56,265	487	3,226	64,116	566	3,753	74,584
0.874	54	40½	127	854	17,300	208	1,401	28,371	273	1,845	37,367	354	2,392	48,439	435	2,939	59,510	496	3,349	67,814	577	3,896	78,886
0.872	55	41¼	129	886	18,279	211	1,453	29,977	278	1,914	39,482	361	2,482	51,180	443	3,049	62,878	505	3,474	71,652	588	4,041	83,350
0.870	56	42	131	919	19,294	215	1,507	31,642	284	1,985	41,675	368	2,573	54,023	452	3,161	66,371	515	3,602	75,632	599	4,190	87,980
0.868	57	42¾	134	952	20,346	219	1,561	33,367	289	2,056	43,947	374	2,665	56,969	460	3,274	69,990	524	3,731	79,756	609	4,341	92,777
0.867	58	43½	136	986	21,436	223	1,616	35,154	294	2,129	46,301	381	2,760	60,020	468	3,390	73,739	533	3,863	84,028	620	4,494	97,747
0.865	59	44¼	138	1,020	22,564	227	1,673	37,004	299	2,203	48,737	387	2,856	63,178	476	3,508	77,619	542	3,998	88,449	631	4,650	102,890
0.863	60	45	141	1,055	23,731	231	1,730	38,918	304	2,278	51,258	394	2,953	66,445	484	3,628	81,633	551	4,134	93,023	641	4,809	108,211
0.862	61	45¾	143	1,090	24,937	235	1,788	40,896	309	2,355	53,864	400	3,052	69,823	492	3,750	85,783	560	4,273	97,753	652	4,971	113,712
0.860	62	46½	145	1,126	26,184	238	1,847	42,941	314	2,433	56,556	407	3,153	73,314	500	3,874	90,071	570	4,415	102,639	663	5,135	119,397
0.859	63	47¼	148	1,163	27,471	242	1,907	45,052	319	2,512	59,337	413	3,256	76,919	508	4,000	94,500	579	4,558	107,686	673	5,302	125,268
0.857	64	48	150	1,200	28,800	246	1,968	47,232	324	2,592	62,208	420	3,360	80,640	516	4,128	99,072	588	4,704	112,896	684	5,472	131,328
0.856	65	48¾	152	1,238	30,171	250	2,030	49,481	329	2,674	65,170	427	3,466	84,479	524	4,258	103,789	597	4,852	118,271	695	5,644	137,581
0.854	66	49½	155	1,276	31,585	254	2,093	51,800	334	2,757	68,224	433	3,573	88,439	532	4,390	108,653	606	5,003	123,814	705	5,819	144,029
0.853	67	50¼	157	1,315	33,043	258	2,157	54,190	339	2,841	71,373	440	3,682	92,520	540	4,524	113,667	616	5,155	129,528	716	5,997	150,675
0.852	68	51	159	1,355	34,545	261	2,222	56,653	344	2,926	74,816	446	3,793	96,725	548	4,660	118,833	625	5,310	135,415	727	6,177	157,523
0.850	69	51¾	162	1,395	36,091	265	2,288	59,189	349	3,013	77,597	453	3,906	101,055	556	4,798	124,153	634	5,468	141,477	737	6,360	164,575
0.849	70	52½	164	1,436	37,683	269	2,354	61,800	354	3,101	81,396	459	4,020	105,513	564	4,938	129,630	643	5,627	147,718	748	6,546	171,835
0.847	71	53¼	166	1,477	39,321	273	2,422	64,487	359	3,190	84,934	466	4,135	110,100	572	5,080	135,265	652	5,789	154,139	759	6,735	179,305
0.846	72	54	169	1,519	41,006	277	2,491	67,250	365	3,281	88,574	473	4,253	114,818	581	5,225	141,062	662	5,954	160,745	770	6,926	186,989
0.845	73	54¾	171	1,561	42,739	281	2,560	70,091	370	3,372	92,316	479	4,372	119,668	589	5,371	147,021	671	6,120	167,536	780	7,119	194,888
0.844	74	55½	173	1,604	44,519	284	2,631	73,012	375	3,465	96,162	486	4,492	124,654	597	5,519	153,146	680	6,289	174,515	791	7,316	203,008

TABLE A.3.4—*CONTINUED.*

0.842	75	56¼	176	1,648	46,349	288	2,703	76,012	380	3,560	100,113	492	4,614	129,776	605	5,669	159,439	689	6,460	181,686	802	7,515	211,350
0.841	76	57	178	1,692	48,227	292	2,775	79,093	385	3,655	104,171	499	4,738	135,037	613	5,821	165,902	698	6,633	189,051	812	7,716	219,917
0.840	77	57¾	181	1,737	50,156	296	2,849	82,256	390	3,752	108,337	505	4,864	140,437	621	5,975	172,537	707	6,809	196,612	823	7,921	228,712
0.839	78	58½	183	1,782	52,136	300	2,923	85,503	395	3,850	112,613	512	4,991	145,980	629	6,132	179,347	717	6,987	204,373	834	8,128	237,739
0.837	79	59¼	185	1,828	54,167	304	2,999	88,834	400	3,949	117,001	518	5,120	151,667	637	6,290	186,334	726	7,167	212,334	844	8,338	247,001
0.836	80	60	188	1,875	56,250	308	3,075	92,250	405	4,050	121,500	525	5,250	157,500	645	6,450	193,500	735	7,350	220,500	855	8,550	256,500
0.835	81	60¾	190	1,922	58,386	311	3,152	95,753	410	4,152	126,113	532	5,382	163,480	653	6,612	200,847	744	7,535	228,873	866	8,765	266,240
0.834	82	61½	192	1,970	60,575	315	3,231	99,343	415	4,255	130,842	538	5,516	169,610	661	6,777	208,378	753	7,722	237,454	876	8,983	276,222
0.833	83	62¼	195	2,018	62,818	319	3,310	103,022	420	4,360	135,688	545	5,651	175,892	669	6,943	216,095	763	7,912	246,248	887	9,203	286,452
0.832	84	63	197	2,067	65,116	323	3,390	106,791	425	4,465	140,651	551	5,788	182,326	677	7,111	224,000	772	8,103	255,256	898	9,426	296,931
0.831	85	63¾	199	2,117	67,470	327	3,471	110,651	430	4,572	145,735	558	5,927	188,915	685	7,282	232,096	781	8,298	264,482	908	9,652	307,662
0.830	86	64½	202	2,167	69,880	331	3,554	114,602	435	4,680	150,939	564	6,067	195,662	693	7,454	240,385	790	8,494	273,927	919	9,881	318,649
0.829	87	65¼	204	2,218	72,345	334	3,637	118,646	440	4,790	156,266	571	6,209	202,567	701	7,628	248,868	799	8,693	283,594	930	10,112	329,895
0.827	88	66	206	2,269	74,869	338	3,721	122,785	446	4,901	161,717	578	6,353	209,633	710	7,805	257,549	809	8,894	293,486	941	10,346	341,402
0.826	89	66¾	209	2,321	77,450	342	3,806	127,018	451	5,013	167,293	584	6,498	216,861	718	7,983	266,429	818	9,097	303,605	951	10,582	353,173

*From Koppers Unit Laminated Structures, Publication W-516, April 1971.

†Data for 1 1/2 inch lamination thickness may be obtained by entering the table with double the number of laminations or with the actual depth of the section.

TABLE A4.1*

Classification of Species

Group 1	Group 2		Group 3	Group 4	Group 5 [a]
Apitong [b][c]	Cedar, Port Orford	Maple, Black	Alder, Red	Aspen	Basswood
Beech, American	Cypress	Mengkulang [b]	Birch, Paper	Bigtooth	Fir, Balsam
Birch	Douglas Fir 2[d]	Meranti, Red [b][e]	Cedar, Alaska	Quaking	Poplar, Balsam
Sweet	Fir	Mersawa [b]	Fir, Subalpine	Cativo	
Yellow	California Red	Pine	Hemlock, Eastern	Cedar	
Douglas Fir 1[d]	Grand	Pond	Maple, Bigleaf	Incense	
Kapur [b]	Noble	Red	Pine	Western Red	
Keruing [b][c]	Pacific Silver	Virginia	Jack	Cottonwood	
Larch, Western	White	Western White	Lodgepole	Eastern	
Maple, Sugar	Hemlock, Western	Spruce	Ponderosa	Black (Western Poplar)	
Pine	Lauan	Red	Spruce		
Caribbean	Almon	Sitka	Redwood	Pine	
Ocote	Bagtikan	Sweetgum	Spruce	Eastern White	
Pine, Southern	Mayapis	Tamarack	Black	Sugar	
Loblolly	Red Lauan	Yellow-poplar	Engelmann		
Longleaf	Tangile		White		
Shortleaf	White Lauan				
Slash					
Tanoak					

(a) Design stresses for Group 5 not assigned.

(b) Each of these names represents a trade group of woods consisting of a number of closely related species.

(c) Species from the genus Dipterocarpus are marketed collectively: Apitong if originating in the Philippines; Keruing if originating in Malaysia or Indonesia.

(d) Douglas fir from trees grown in the states of Washington, Oregon, California, Idaho, Montana, Wyoming, and the Canadian Provinces of Alberta and British Columbia shall be classed as Douglas fir No. 1. Douglas fir from trees grown in the states of Nevada, Utah, Colorado, Arizona and New Mexico shall be classed as Douglas fir No. 2.

(e) Red Meranti shall be limited to species having a specific gravity of 0.41 or more based on green volume and oven dry weight.

*From Plywood Design Specification, 1978 by APA.

TABLE A.4.2*

Veneer Grades

Grade	Description
N	Smooth surface "natural finish" veneer. Select, all heartwood or all sapwood. Free of open defects. Allows not more than 6 repairs, wood only, per 4x8 panel, made parallel to grain and well matched for grain and color.
A	Smooth, paintable. Not more than 18 neatly made repairs, boat, sled, or router type, and parallel to grain, permitted. May be used for natural finish in less demanding applications.
B	Solid surface. Shims, circular repair plugs and tight knots to 1 inch across grain permitted. Some minor splits permitted.
C Plugged	Improved C veneer with splits limited to 1/8 inch width and knotholes and borer holes limited to 1/4 x 1/2 inch. Admits some broken grain. Synthetic repairs permitted.
C	Tight knots to 1-1/2 inch. Knotholes to 1 inch across grain and some to 1-1/2 inch if total width of knots and knotholes is within specified limits. Synthetic or wood repairs. Discoloration and sanding defects that do not impair strength permitted. Limited splits allowed. Stitching permitted.
D	Knots and knotholes to 2-1/2 inch width across grain and 1/2 inch larger within specified limits. Limited splits allowed. Stitching permitted. Limited to Interior grades of plywood.

*From Plywood Design Specification, 1978 by APA.

Table A.4.3
STRENGTH AND RIGIDITY*†

Plywood is a superior structural material. Its cross-laminated construction provides two-way strength and rigidity in each panel. This along with the panel's large size accounts for its advantages as a bracing material.

A comparison of strength and rigidity in framed wall applications is shown in the accompanying U.S. Forest Products Laboratory table. It clearly points up plywood's superiority in diaphragm applications. In walls or roofs plywood sheathing often eliminates the need for additional bracing.

STRENGTH AND RIGIDITY OF FRAME WALLS		RELATIVE RIGIDITY	RELATIVE STRENGTH
1" x 8" DIAGONAL SHEATHING	(8d nails, 2 per stud crossing)	1.0	1.0
25/32" FIBERBOARD	(8d nails, spaced 3" at all vertical edges; 5-1/2" to 6" elsewhere.)	1.6	1.6
HORIZONTAL SHEATHING	(1 x 8 sheathing; 1 x 4 let in braces; 8d nails, 2 per stud crossing.)	1.5	1.7
1/4" PLYWOOD NAILED	(6d nails spaced 5" at edges, 10" elsewhere.)	2.0	2.2
1/4" PLYWOOD GLUED TO FRAME		3.7	3.1

Loads are applied laterally in the plane of the test panel.

Table A.4.4
FASTENERS*†

NAILS — Ultimate Lateral Loads in Douglas fir lumber (lb. per common nail) (1) (2)

Plywood Thickness	6d	8d	10d	16d
5/16"	275	305		
3/8"	275	340		
1/2"		350	425	
5/8"		350	425	445
3/4"			410	445

(1) Assume 3/8" edge distance.
(2) For galvanized casing nails, multiply tabulated values by 0.6.

SCREWS — Ultimate Withdrawal Loads (2) (lb. per screw)

Penetration into Plywood (1)	Size of Wood Screw 6	7	8	9	10	12
3/8"	150	165	180	195		
1/2"	200	220	240	260	280	
5/8"	250	275	300	325	350	395
3/4"	300	330	360	390	420	475

(1) Screws driven perpendicular to face of panel.
(2) Values may be increased one-fourth for sheet metal screws.

STAPLES — Ultimate Loads (1) (2) (lb. per staple)

Penetration Into Lumber	Lateral Load	Withdrawal Load
3/4"	160	100
1"	180	150
1-1/4"	200	200
1-1/2"	220	

(1) Values are for 3/8" and thicker plywood, 16 gage galvanized staples with 7/16" crown, driven into Douglas fir lumber.
(2) Some plastic coated staples may provide higher values.

BOLTS — Ultimate Loads at listed angle between plywood face grain and load direction (lb. per bolt) (1) (2)

Plywood Thickness	1/2" Bolt 0°	1/2" Bolt 45°	1/2" Bolt 90°	3/4" Bolt 0°	3/4" Bolt 45°	3/4" Bolt 90°
5/16"			2400			
1/2"	4740	6200	4570	7800	6720	6900
5/8"	5400	5400	4320	8070	7720	6900
3/4"	5280	3500	5630	7920	7270	8650
13/16"						8820

(1) Values are for plywood gussets on both sides of 2-1/2" Douglas fir lumber with bolt loaded in double shear.
(2) To develop loads tabulated, plywood end distance is six times bolt diameter, "D" for 0 to 45 degrees, and 8D for 90 degrees; edge distance is 3D for 0, and 6D for 45 degrees and 90 degrees. Reduce loads proportionately for reduced end distances. Minimum end distance is 3D for 0 and 45 degrees, 4D for 90 degrees.

*Values shown are based on the physical properties of Douglas fir.
†From "Plywood Properties and Grades" by APA.

Table A.4.5. BENDING*†

Simple curves are easy to form with plywood. A continuous rounded bracing produces best results. When application calls for abrupt curvatures, fasten panel end to shorter radius first. The following radii have been found to be appropriate minimums for mill-run panels of thicknesses shown bent dry. Shorter radii can be developed by selection for bending of areas free of knots and short grain, and/or by wetting or steaming. (Exterior glue recommended). Panels to be glued should be re-dried first. An occasional panel may develop localized fractures at thes radii. To achieve these radii with narrow strips, select for straight grain and freedom from knots.

MINIMUM BENDING RADIUS (ft.)		
Panel Thickness	Across Grain	Parallel to Grain
1/4″	2′	5′
5/16″	2′	6′
3/8″	3′	8′
1/2″	6′	12′
5/8	8′	16′
3/4″	12′	20′

*Values shown are based on the physical properties of Douglas fir.
†From "Plywood Properties and Grades" by APA.

TABLE A.4.6

Guide to Appearance Grades of Plywood[1]

SPECIFIC GRADES AND THICKNESSES MAY BE IN LOCALLY LIMITED SUPPLY. SEE YOUR DEALER BEFORE SPECIFYING.

Type	Grade Designation (2)	Description and Most Common Uses	Typical (3) Grade-trademarks	Veneer Grade: Face	Veneer Grade: Inner Plies	Veneer Grade: Back	Most Common Thicknesses (inch)					
Interior Type	N-N, N-A N-B INT-APA	Cabinet quality. For natural finish furniture, cabinet doors, built-ins, etc. Special order items.	N-N G1 INT-APA PS1-74 000 / N-A G2 INT-APA PS1-74 000	N	C	N,A, or B						3/4
	N-D-INT-APA	For natural finish paneling. Special order item.	N-D G2 INT-APA PS1-74 000	N	D	D	1/4					
	A-A INT-APA	For applications with both sides on view, built-ins, cabinets, furniture, partitions. Smooth face; suitable for painting.	A-A G1 INT-APA PS1-74 000	A	D	A	1/4		3/8	1/2	5/8	3/4
	A-B INT-APA	Use where appearance of one side is less important but where two solid surfaces are necessary.	A-B G1 INT-APA PS1-74 000	A	D	B	1/4		3/8	1/2	5/8	3/4
	A-D INT-APA	Use where appearance of only one side is important. Paneling, built-ins, shelving, partitions, flow racks.	A-D GROUP 1 INTERIOR PS 1-74 000 (APA)	A	D	D	1/4		3/8	1/2	5/8	3/4
	B-B INT-APA	Utility panel with two solid sides. Permits circular plugs.	B-B G2 INT-APA PS1-74 000	B	D	B	1/4		3/8	1/2	5/8	3/4
	B-D INT-APA	Utility panel with one solid side. Good for backing, sides of built-ins, industry shelving, slip sheets, separator boards, bins.	B-D GROUP 2 INTERIOR PS 1-74 000 (APA)	B	D	D	1/4		3/8	1/2	5/8	3/4
	DECORATIVE PANELS—APA	Rough-sawn, brushed, grooved, or striated faces. For paneling, interior accent walls, built-ins, counter facing, displays, exhibits.	DECORATIVE B-D G1 INT-APA PS1-74	C or btr.	D	D		5/16	3/8	1/2	5/8	
	PLYRON INT-APA	Hardboard face on both sides. For counter tops, shelving, cabinet doors, flooring. Faces tempered, untempered, smooth, or screened.	PLYRON INT-APA 000		C & D					1/2	5/8	3/4
Exterior Type	A-A EXT-APA	Use where appearance of both sides is important. Fences, built-ins, signs, boats, cabinets, commercial refrigerators, shipping containers, tote boxes, tanks, ducts. (4)	A-A G1 EXT-APA PS1-74 000	A	C	A	1/4		3/8	1/2	5/8	3/4
	A-B EXT-APA	Use where the appearance of one side is less important. (4)	A-B G1 EXT-APA PS1-74 000	A	C	B	1/4		3/8	1/2	5/8	3/4
	A-C EXT-APA	Use where the appearance of only one side is important. Soffits, fences, structural uses, boxcar and truck lining, farm buildings. Tanks, trays, commercial refrigerators. (4)	A-C GROUP 1 EXTERIOR PS 1-74 000 (APA)	A	C	C	1/4		3/8	1/2	5/8	3/4
	B-B EXT-APA	Utility panel with solid faces. (4)	B-B G2 EXT-APA PS1-74 000	B	C	B	1/4		3/8	1/2	5/8	3/4
	B-C EXT-APA	Utility panel for farm service and work buildings, boxcar and truck lining, containers, tanks, agricultural equipment. Also as base for exterior coatings for walls, roofs. (4)	B-C GROUP 2 EXTERIOR PS 1-74 000 (APA)	B	C	C	1/4		3/8	1/2	5/8	3/4
	HDO EXT-APA	High Density Overlay plywood. Has a hard, semi-opaque resin-fiber overlay both faces. Abrasion resistant. For concrete forms, cabinets, counter tops, signs, tanks. (4)	HDO A-A G1 EXT-APA PS1-74	A or B	C or C plgd	A or B			3/8	1/2	5/8	3/4
	MDO EXT-APA	Medium Density Overlay with smooth, opaque, resin-fiber overlay one or both panel faces. Highly recommended for siding and other outdoor applications, built-ins, signs, displays. Ideal base for paint. (4)(6)	MDO B-B G2 EXT-APA PS1-74 000	B	C	B or C			3/8	1/2	5/8	3/4
	303 SIDING EXT-APA	Proprietary plywood products for exterior siding, fencing, etc. Special surface treatment such as V-groove, channel groove, striated, brushed, rough-sawn and texture-embossed MDO. Stud spacing (Span Index) and face grade classification indicated on grade stamp.	303 SIDING GROUP 1 EXTERIOR PS 1-74 000 (APA)	(5)	C	C			3/8	1/2	5/8	
	T 1-11 EXT-APA	Special 303 panel having grooves 1/4" deep, 3/8" wide, spaced 4" or 8" o.c. Other spacing optional. Edges shiplapped. Available unsanded, textured and MDO.	303 SIDING 6 S/W T 1-11 GROUP 2 16 oc SPAN EXTERIOR PS 1-74 000 (APA)	C or btr.	C	C				19/32	5/8	
	PLYRON EXT-APA	Hardboard faces both sides, tempered, smooth or screened.	PLYRON EXT-APA 000		C					1/2	5/8	3/4
	MARINE EXT-APA	Ideal for boat hulls. Made only with Douglas fir or western larch. Special solid jointed core construction. Subject to special limitations on core gaps and number of face repairs. Also available with HDO or MDO faces.	MARINE A-A EXT-APA PS1-74 000	A or B	B	A or B	1/4		3/8	1/2	5/8	3/4

(1) Sanded both sides except where decorative or other surfaces specified.
(2) Can be manufactured in Group 1, 2, 3, 4 or 5.
(3) The species groups, Identification Indexes and Span Indexes shown in the typical grade-trademarks are examples only. See "Group," "Identification Index" and "Span Index" for explanations and availability.
(4) Can also be manufactured in Structural I (all plies limited to Group 1 species) and Structural II (all plies limited to Group 1, 2, or 3 species).
(5) C or better for 5 plies. C Plugged or better for 3 plies.
(6) Also available as a 303 siding.

TABLE A.4.7

Guide to Engineered Grades of Plywood

SPECIFIC GRADES AND THICKNESSES MAY BE IN LOCALLY LIMITED SUPPLY. SEE YOUR DEALER BEFORE SPECIFYING.

Type	Grade Designation	Description and Most Common Uses	Typical Grade-trademarks (1)	Veneer Grade: Face	Veneer Grade: Inner Plies	Veneer Grade: Back	Most Common Thicknesses (inch)				
Interior Type	C-D INT-APA	For wall and roof sheathing, subflooring, industrial uses such as pallets. Most commonly available with exterior glue (CDX). Specify exterior glue where construction delays are anticipated and for treated-wood foundations. (7)	C-D 32/16 APA INTERIOR PS 1-74 000; C-D 24/0 APA INTERIOR PS 1-74 000 EXTERIOR GLUE	C	D	D	5/16	3/8	1/2	5/8	3/4
Interior Type	STRUCTURAL I C-D INT-APA and STRUCTURAL II C-D INT-APA	Unsanded structural grades where plywood strength properties are of maximum importance: structural diaphragms, box beams, gusset plates, stressed-skin panels, containers, pallet bins. Made only with exterior glue. See (6) for species group requirements. Structural I more commonly available. (7)	STRUCTURAL I C-D 24/0 APA INTERIOR PS 1-74 000 EXTERIOR GLUE	C(3)	D(3)	D(3)	5/16	3/8	1/2	5/8	3/4
Interior Type	STURD-I-FLOOR INT-APA	For combination subfloor-underlayment. Provides smooth surface for application of resilient floor covering. Possesses high concentrated- and impact-load resistance during construction and occupancy. Manufactured with exterior glue only. Touch-sanded. Available square edge or tongue-and-groove. (7)	STURD-I-FLOOR 24oc T&G 23/32 INCH INTERIOR 000 EXTERIOR GLUE NRB-108	C Plugged	(4)	D				19/32 5/8	23/32 3/4
Interior Type	STURD-I-FLOOR 48 O.C. (2-4-1) INT-APA	For combination subfloor-underlayment on 32- and 48-inch spans. Provides smooth surface for application of resilient floor coverings. Possesses high concentrated- and impact-load resistance during construction and occupancy. Manufactured with exterior glue only. Unsanded or touch-sanded. Available square edge or tongue-and-groove. (7)	STURD-I-FLOOR 48oc 2-4-1 T&G 1 1/8 INCH INTERIOR 000 EXTERIOR GLUE NRB-108	C Plugged	C(5) & D	D	1-1/8				
Interior Type	UNDERLAYMENT INT-APA	For application over structural subfloor. Provides smooth surface for application of resilient floor coverings. Touch-sanded. Also available with exterior glue. (2)(6)	UNDERLAYMENT GROUP 1 APA INTERIOR PS 1-74 000	C Plugged	C(5) & D	D		3/8	1/2	19/32 5/8	23/32 3/4
Interior Type	C-D PLUGGED INT-APA	For built-ins, wall and ceiling tile backing, cable reels, walkways, separator boards. Not a substitute for Underlayment or Sturd-I-Floor as it lacks their indentation resistance. Touch-sanded. Also made with exterior glue. (2) (6)	C-D PLUGGED GROUP 2 APA INTERIOR PS 1-74 000	C Plugged	D	D		3/8	1/2	19/32 5/8	23/32 3/4
Exterior Type	C-C EXT-APA	Unsanded grade with waterproof bond for subflooring and roof decking, siding on service and farm buildings, crating, pallets, pallet bins, cable reels, treated-wood foundations. (7)	C-C 42/20 APA EXTERIOR PS 1-74 000	C	C	C	5/16	3/8	1/2	5/8	3/4
Exterior Type	STRUCTURAL I C-C EXT-APA and STRUCTURAL II C-C EXT-APA	For engineered applications in construction and industry where full Exterior type panels are required. Unsanded. See (6) for species group requirements. (7)	STRUCTURAL I C-C 32/16 APA EXTERIOR PS 1-74 000	C	C	C	5/16	3/8	1/2	5/8	3/4
Exterior Type	STURD-I-FLOOR EXT-APA	For combination subfloor-underlayment under resilient floor coverings where severe moisture conditions may be present, as in balcony decks. Possesses high concentrated- and impact-load resistance during construction and occupancy. Touch-sanded. Available square edge or tongue-and-groove. (7)	STURD-I-FLOOR 20oc 5/8 INCH EXTERIOR 000 NRB-108	C Plugged	C(5)	C				19/32 5/8	23/32 3/4
Exterior Type	UNDERLAYMENT C-C PLUGGED EXT-APA	For application over structural subfloor. Provides smooth surface for application of resilient floor coverings where severe moisture conditions may be present. Touch-sanded. (2)(6)	UNDERLAYMENT C-C PLUGGED GROUP 2 APA EXTERIOR PS 1-74 000	C Plugged	C(5)	C		3/8	1/2	19/32 5/8	23/32 3/4
Exterior Type	C-C PLUGGED EXT-APA	For use as tile backing where severe moisture conditions exist. For refrigerated or controlled atmosphere rooms, pallet fruit bins, tanks, box car and truck floors and linings, open soffits. Touch-sanded. (2)(6)	C-C PLUGGED GROUP 2 APA EXTERIOR PS 1-74 000	C Plugged	C	C		3/8	1/2	19/32 5/8	23/32 3/4
Exterior Type	B-B PLYFORM CLASS I & CLASS II EXT-APA	Concrete form grades with high reuse factor. Sanded both sides. Mill-oiled unless otherwise specified. Special restrictions on species. Available in HDO and Structural I. Class I most commonly available. (8)	B-B PLYFORM CLASS I APA EXTERIOR PS 1-74 000	B	C	B				5/8	3/4

(1) The species groups, Identification Indexes and Span Indexes shown in the typical grade-trademarks are examples only. See "Group," "Identification Index" and "Span Index" for explanations and availability.
(2) Can be manufactured in Group 1, 2, 3, 4, or 5.
(3) Special improved grade for structural panels.
(4) Special veneer construction to resist indentation from concentrated loads, or other solid wood-base materials.
(5) Special construction to resist indentation from concentrated loads.
(6) Can also be manufactured in Structural I (all plies limited to Group 1 species) and Structural II (all plies limited to Group 1, 2, or 3 species).
(7) Specify by Identification Index for sheathing and Span Index for Sturd-I-Floor panels.
(8) Made only from certain wood-species to conform to APA specifications.

TABLE A.4.8*

Guide to Identification Indexes[1]

Thickness (Inch)	C-D INT-APA C-C EXT-APA		
	Group 1 & Structural I	Group 2[2] or 3 & Structural II[2]	Group 4[3]
5/16	20/0	16/0	12/0
3/8	24/0	20/0	16/0
1/2	32/16	24/0	24/0
5/8	42/20	32/16	30/12
3/4	48/24	42/20	36/16
7/8	-------	48/24	42/20

(1) Sheathing/subflooring is ordinarily ordered by Identification Index. Check local availability if a specific combination of thickness and Identification Index is desired. Some panels, for example Structural II of all thicknesses, and panels with Identification Indexes of 30/12 and 36/16, may be difficult to obtain.

(2) Panels with Group 2 outer plies and special thickness and construction requirements, or Structural II panels with Group 1 faces, may carry the Identification Index numbers shown for Group 1 panels.

(3) Panels made with Group 4 outer plies may carry the Identification Index numbers shown for Group 3 panels when they conform to special thickness and construction requirements detailed in PS 1.

*From Plywood Specification and Grade Guide by APA.

TABLE A.4.9*

EFFECTIVE SECTION PROPERTIES FOR PLYWOOD

Face Plies of Different Species Group from Inner Plies (Includes all Product Standard Grades except those noted in Table 2.)

(1) Nominal Thickness (in.)	(2) Approximate Weight (psf)	(3) Effective Thickness for Shear (in.)	Stress Applied Parallel to Face Grain: (4) A Area (in.²/ft)	(5) I Moment of Inertia (in.⁴/ft)	(6) KS Eff. Section Modulus (in.³/ft)	(7) Ib/Q Rolling Shear Constant (in.²/ft)	Stress Applied Perpendicular to Face Grain: (8) A Area (in.²/ft)	(9) I Moment of Inertia (in.⁴/ft)	(10) KS Eff. Section Modulus (in.³/ft)	(11) Ib/Q Rolling Shear Constant (in.²/ft)
UNSANDED PANELS										
5/16-U	1.0	0.283	1.914	0.025	0.124	2.568	0.660	0.001	0.023	–
3/8-U	1.1	0.293	1.866	0.041	0.162	3.108	0.799	0.002	0.033	–
1/2-U	1.5	0.316	2.500	0.086	0.247	4.189	1.076	0.005	0.057	2.585
5/8-U	1.8	0.336	2.951	0.155	0.379	5.270	1.354	0.011	0.095	3.252
3/4-U	2.2	0.467	3.403	0.243	0.501	6.823	1.632	0.036	0.232	3.717
7/8-U	2.6	0.757	4.109	0.344	0.681	7.174	2.925	0.162	0.542	5.097
1-U	3.0	0.859	3.916	0.493	0.859	9.244	3.611	0.210	0.660	6.997
1-1/8-U	3.3	0.877	4.725	0.676	1.047	9.960	3.079	0.288	0.768	8.679
SANDED PANELS										
1/4-S	0.8	0.278	1.307	0.009	0.067	2.182	0.681	0.001	0.018	–
3/8-S	1.1	0.294	1.307	0.027	0.125	3.389	1.181	0.004	0.053	–
1/2-S	1.5	0.450	1.947	0.077	0.266	4.834	1.281	0.018	0.150	3.099
5/8-S	1.8	0.572	2.475	0.129	0.356	6.442	1.627	0.045	0.234	3.922
3/4-S	2.2	0.589	2.884	0.197	0.452	7.881	2.104	0.093	0.387	4.842
7/8-S	2.6	0.608	2.942	0.278	0.547	8.225	3.199	0.157	0.542	5.698
1-S	3.0	0.846	3.776	0.423	0.730	8.882	3.537	0.253	0.744	7.644
1-1/8-S	3.3	0.865	3.854	0.548	0.840	9.883	3.673	0.360	0.918	9.032
TOUCH-SANDED PANELS										
1/2-T	1.5	0.346	2.698	0.083	0.271	4.252	1.159	0.006	0.061	2.746
19/32-T	1.7	0.491	2.618	0.123	0.337	5.403	1.610	0.019	0.150	3.220
5/8-T	1.8	0.497	2.728	0.141	0.364	5.719	1.715	0.023	0.170	3.419
23/32-T	2.1	0.503	3.181	0.196	0.447	6.600	2.014	0.035	0.226	3.659
3/4-T	2.2	0.509	3.297	0.220	0.477	6.917	2.125	0.041	0.251	3.847
2-4-1 1-1/8-T	3.3	0.855	4.592	0.653	0.995	9.933	4.120	0.283	0.763	7.452

Table 2. Structural I, II and Marine

(1) Nominal Thickness (in.)	(2) Approximate Weight (psf)	(3) Effective Thickness for Shear (in.)	Stress Applied Parallel to Face Grain: (4) A Area (in.²/ft)	(5) I Moment of Inertia (in.⁴/ft)	(6) KS Eff. Section Modulus (in.³/ft)	(7) Ib/Q Rolling Shear Constant (in.²/ft)	Stress Applied Perpendicular to Face Grain: (8) A Area (in.²/ft)	(9) I Moment of Inertia (in.⁴/ft)	(10) KS Eff. Section Modulus (in.³/ft)	(11) Ib/Q Rolling Shear Constant (in.²/ft)
UNSANDED PANELS										
5/16-U	1.0	0.356	2.375	0.025	0.144	2.567	1.188	0.002	0.029	–
3/8-U	1.1	0.371	2.226	0.041	0.195	3.107	1.438	0.003	0.043	–
1/2-U	1.5	0.543	2.906	0.091	0.318	4.497	2.325	0.017	0.145	2.574
5/8-U	1.8	0.715	3.464	0.157	0.437	5.993	2.925	0.052	0.267	3.238
3/4-U	2.2	0.747	4.406	0.247	0.573	7.046	2.938	0.086	0.369	3.697
7/8-U	2.6	0.776	4.388	0.346	0.690	6.948	3.510	0.192	0.584	5.086
1-U	3.0	1.088	5.200	0.529	0.922	8.512	6.500	0.366	0.970	6.986
1-1/8-U	3.3	1.119	6.654	0.751	1.164	9.061	5.542	0.503	1.131	8.675
SANDED PANELS										
1/4-S	0.8	0.342	1.680	0.013	0.092	2.172	1.226	0.001	0.027	–
3/8-S	1.1	0.373	1.680	0.038	0.177	3.382	2.120	0.007	0.078	–
1/2-S	1.5	0.545	1.947	0.078	0.271	4.816	2.305	0.030	0.217	3.076
5/8-S	1.8	0.717	3.112	0.131	0.361	6.526	2.929	0.077	0.343	3.887
3/4-S	2.2	0.748	3.848	0.202	0.464	7.926	3.787	0.162	0.570	4.812
7/8-S	2.6	0.778	3.952	0.288	0.569	7.539	5.759	0.275	0.798	5.671
1-S	3.0	1.091	5.215	0.479	0.827	7.978	6.367	0.445	1.098	7.639
1-1/8-S	3.3	1.121	5.593	0.623	0.955	8.840	6.611	0.634	1.356	9.031
TOUCH-SANDED PANELS										
1/2-T	1.5	0.543	2.698	0.084	0.282	4.580	2.486	0.020	0.162	2.720
19/32-T	1.7	0.707	3.127	0.124	0.349	6.094	2.899	0.050	0.259	3.183
5/8-T	1.8	0.715	3.267	0.144	0.378	6.552	3.086	0.060	0.293	3.383
23/32-T	2.1	0.739	4.059	0.201	0.469	6.971	3.625	0.078	0.350	3.596
3/4-T	2.2	0.746	4.209	0.226	0.503	7.379	3.825	0.092	0.388	3.786

*From Plywood Design Specification, 1978 by APA.

TABLE A.4.10*

Allowable Stresses for Plywood.

Conforming to U.S. Product Standard PS 1-74 for Construction and Industrial Plywood. Normal Load Basis in PSI.

TYPE OF STRESS		SPECIES GROUP of FACE PLY	GRADE STRESS LEVEL *				
			S-1		*S-2*		*S-3*
			WET	DRY	WET	DRY	DRY ONLY
EXTREME FIBER STRESS IN BENDING (F_b) TENSION IN PLANE OF PLIES (F_t) Face Grain Parallel or Perpendicular to Span (At 45° to Face Grain Use 1/6 F_t)	F_b & F_t	1	1430	2000	1190	1650	1650
		2, 3	980	1400	820	1200	1200
		4	940	1330	780	1110	1110
COMPRESSION IN PLANE OF PLIES. (F_c). Parallel or Perpendicular to Face Grain (At 45° to Face Grain Use 1/3 F_c)	F_c	1	970	1640	900	1540	1540
		2	730	1200	680	1100	1100
		3	610	1060	580	990	990
		4	610	1000	580	950	950
SHEAR IN PLANE PERPENDICULAR TO PLIES Parallel or Perpendicular to Face Grain (At 45° to Face Grain Use 2 F_v)	F_v	1	205	250	205	250	210
		2,3	160	185	160	185	160
		4	145	175	145	175	155
SHEAR, ROLLING, IN THE PLANE OF PLIES Parallel or Perpendicular to Face Grain (At 45° to Face Grain Use 1 1/3 F_s)	F_s	MARINE & STRUCTURAL I	63	75	63	75	
		STRUCTURAL II, STURD-I-FLOOR and 2-4-1	49	56	49	56	55
		ALL OTHER	44	53	44	53	48
MODULUS OF RIGIDITY Shear in Plane Perpendicular to Plies	G	1	70,000	90,000	70,000	90,000	82,000
		2	60,000	75,000	60,000	75,000	68,000
		3	50,000	60,000	50,000	60,000	55,000
		4	45,000	50,000	45,000	50,000	45,000
BEARING (ON FACE) Perpendicular to Plane of Plies	$F_{c\perp}$	1	210	340	210	340	340
		2,3	135	210	135	210	210
		4	105	160	105	160	160
MODULUS OF ELASTICITY IN BENDING IN PLANE OF PLIES. Face Grain Parallel or Perpendicular to Span	E	1	1,500,000	1,800,000	1,500,000	1,800,000	1,800,000
		2	1,300,000	1,500,000	1,300,000	1,500,000	1,500,000
		3	1,100,000	1,200,000	1,100,000	1,200,000	1,200,000
		4	900,000	1,000,000	900,000	1,000,000	1,000,000

* See page 14 for Guide.
To qualify for stress level S-1, gluelines must be exterior and only veneer grades N, A, and C are allowed in either face or back.
For stress level S-2, gluelines must be exterior and veneer grade B, C-plugged and D are allowed on the face or back.
Stress level S-3 includes all panels with interior or intermediate glue lines.

*From Plywood Design Specification, 1978 by APA.

Table A.4.11 PLYWOOD PANEL CONSTRUCTIONS USED IN CALCULATING SECTION PROPERTIES†

Other constructions in common use produce higher Section-Property values

Plywood Thickness (inches)	No. of Plys	Veneer Thickness (inches): For properties parallel to face grain			For properties perpendicular to face grain		
		Faces	Centers	Crossbands	Faces	Centers	Crossbands
Unsanded Panels							
5/16 U	3	1/10 0.100	–	1 @ 1/10 0.100	1/10 0.100	–	1 @ 1/10 0.100
3/8 U	3	1/10 0.100	–	1 @ 3/16 0.175	1/8 0.125	–	1 @ 1/8 0.125
1/2 U*	3* or 4*	1/8 0.125	–	1 @ 1/4 0.250	1/6 0.167	–	1 @ 1/6 0.167
1/2 U	5	1/10 0.100	1 @ 1/10 0.100	2 @ 1/10 0.100	1/10 0.100	1 @ 1/10 0.100	2 @ 1/10 0.100
5/8 U*	3* or 4*	7/32 0.208	–	1 @ 7/32 0.208	7/32 0.208	–	1 @ 7/32 0.208
5/8 U	5	1/10 0.100	1 @ 3/16 0.182	2 @ 1/8 0.121	1/8 0.125	1 @ 1/8 0.125	2 @ 1/8 0.125
3/4 U	5	1/10 0.100	1 @ 3/16 0.183	2 @ 3/16 0.183	1/8 0.125	1 @ 1/8 0.125	2 @ 3/16 0.188
13/16 U	5	1/10 0.100	1 @ 3/16 0.184	2 @ 7/32 0.214	1/8 0.125	1 @ 3/16 0.188	2 @ 3/16 0.188
7/8 U	7	1/10 0.100	2 @ 3/16 0.188	3 @ 1/10 0.100	1/8 0.125	2 @ 1/8 0.125	3 @ 1/8 0.125
1 U	7	1/10 0.100	2 @ 1/8 0.123	3 @ 3/16 0.185	1/8 0.125	2 @ 1/10 0.098	3 @ 3/16 0.184
1-1/8 U	7	1/10 0.100	2 @ 3/16 0.185	3 @ 3/16 0.185	1/8 0.125	2 @ 7/32 0.219	3 @ 1/6 0.146
Sanded Panels**							
1/4 S	3	1/10 0.070	–	1 @ 1/10 0.100	1/10 0.070	–	1 @ 1/10 0.100
3/8 S	3	1/10 0.070	–	1 @ 7/32 0.235	1/8 0.100	–	1 @ 3/16 0.175
1/2 S	5	1/10 0.070	1 @ 1/8 0.120	2 @ 1/8 0.120	1/8 0.100	1 @ 1/10 0.100	2 @ 1/10 0.100
5/8 S	5	1/10 0.070	1 @ 1/8 0.121	2 @ 3/16 0.182	1/8 0.100	1 @ 3/16 0.182	2 @ 1/8 0.121
3/4 S	5	1/10 0.070	1 @ 3/16 0.183	2 @ 7/32 0.214	1/8 0.100	1 @ 3/16 0.183	2 @ 3/16 0.183
7/8 S	7	1/10 0.070	2 @ 1/10 0.096	3 @ 3/16 0.181	1/8 0.100	2 @ 1/10 0.096	3 @ 1/6 0.161
1 S	7	1/10 0.070	2 @ 1/6 0.160	3 @ 3/16 0.180	1/8 0.100	2 @ 1/8 0.123	3 @ 3/16 0.185
1-1/8 S***	7	1/10 0.070	2 @ 3/16 0.179	3 @ 7/32 0.209	1/8 0.100	2 @ 3/16 0.185	3 @ 3/16 0.185

*Allowed in STANDARD sheathing grades only
**Includes Touch-Sanded
***for 1-1/8 2-4-1 use the following

1-1/8 2-4-1	7	1/10 0.070	2 @ 3/16 0.179	3 @ 7/32 0.209	3/16 0.158	2 @ 1/8 0.125	3 @ 3/16 0.187

†From "Plywood Design Specification" by APA.

TABLE A.4.12. ROLLING SHEAR† (See example in Section 10.3.1.7)

A and y′ for Computing Q_s *

Plywood Thickness (in.)	Structural I, or any Group-4 Panel				All Other Panels			
	Face Grain ‖ to stringers		Face Grain ⊥ to stringers		Face Grain ‖ to stringers		Face Grain ⊥ to stringers	
	Area (in.²)	y′ (in.)	Area (in.²)	y′ (in.)	Area (in.²)	y′ (in.)	Area (in.²)	y′ (in.)
Unsanded Panels								
5/16	4.800	0.050	4.800	0.150	4.800	0.050	2.400	0.150
3/8	4.800	.050	6.000	.188	4.800	.050	3.000	.188
1/2	9.600	.150	9.600	.250	7.200	.117	4.800	.250
5/8	13.543	.219	12.000	.312	9.171	.175	6.000	.312
3/4	13.600	.260	18.000	.375	9.200	.205	9.000	.375
Sanded Panels								
1/4	3.360	0.035	4.800	0.120	3.360	0.035	2.400	0.120
3/8	3.360	.035	8.400	.188	3.360	.035	4.200	.188
1/2	9.120	.171	9.600	.250	6.240	.134	4.800	.250
5/8	9.180	.211	11.657	.312	6.270	.164	5.829	.312
3/4	12.144	.281	17.600	.375	7.752	.228	8.800	.375
2·4·1	...	...	...	...	20.553	.476	26.916	.562

* Note: Area based on 48″ wide panel. For other widths, use a proportionate area.

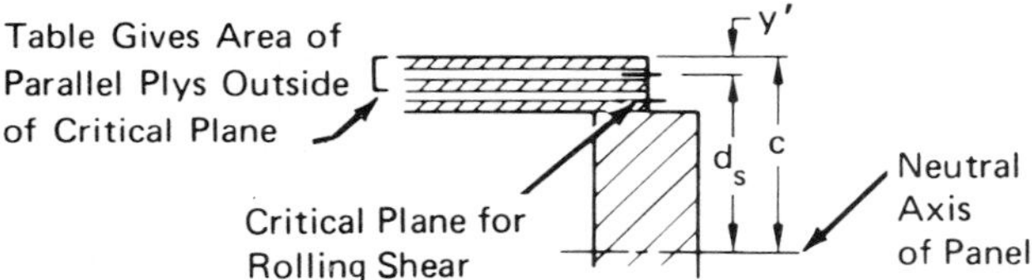

With this table, the designer can figure the Q of the plys outside the critical plane by simply multiplying two figures. One of these figures is given directly in the table. It represents the "transformed area" of the required plys, A, for a 4-ft-wide panel.

The distance (d_s) by which this area must be multiplied is not presented directly, but is easy to obtain. It is the distance from the centroid of the transformed area to the neutral axis of the panel. To obtain it, subtract from the total distance between N.A. and outside of panel, c, the distance between outside of panel and centroid of transformed area, y′. In equation form,

$$d_s = c - y'$$

†From "Plywood Design Specification, Supplement No. 3" by APA.

Table A.4.13 DESIGN OF PLYWOOD DIAPHRAGMS.†

(a)—Recommended Shear in pounds per foot for horizontal plywood Diaphragns for Wind or Seismic Loading.

(Plywood and framing assumed already designed for perpendicular loads)

Plywood grade (c)	Common nail size	Minimum nail penetration in framing (in.)	Minimum nominal plywood thickness (in.)	Minimum nominal width of framing member (in.)	Blocked diaphragms: Nail spacing at diaphragm boundaries (all Cases) and continuous panel edges parrallel to load (in.) (Cases 3 and 4) (a)				Unblocked diaphragms: Nails spaced 6" max. at supported edges (a)	
					6	4	2-1/2	2	Load perpendicular to unblocked edges and continuous panel joints (Case 1)	All other configurations (Cases 2, 3 & 4)
					Nail spacing at other plywood panel edges (in.)					
					6	6	4	3		
STRUCTURAL I INT-DFPA or EXT-DFPA	6d	1-1/4	5/16	2 3	185 210	250 280	375 420	420 475	165 185	125 140
	8d	1-1/2	3/8	2 3	270 300	360 400	530 600	600 675	240 265	180 200
	10d	1-5/8	1/2	2 3	320 360	425 480	640(b) 720	730(b) 820	285 320	215 240
C-C EXT-DFPA STRUCTURAL II INT-DFPA, STANDARD C-D INT-DFPA, sheathing and other DFPA grades except Species Group 5	6d	1-1/4	5/16	2 3	170 190	225 250	335 380	380 430	150 170	110 125
			3/8	2 3	185 210	250 280	375 420	420 475	165 185	125 140
	8d	1-1/2	3/8	2 3	240 270	320 360	480 540	545 610	215 240	160 180
			1/2	2 3	270 300	360 400	530 600	600 675	240 265	180 200
	10d	1-5/8	1/2	2 3	290 325	385 430	575(b) 650	655(b) 735	255 290	190 215
			5/8	2 3	320 360	425 480	640(b) 720	730(b) 820	285 320	215 240

(a) Space nails 12 in. on center along intermediate framing members.
(b) Reduce tabulated allowable shears 10 per cent when boundary members provide less than 3-inch nominal nailing surface.
(c) All recommendations based on the use of DFPA grade-trademarked plywood.
Notes: Design for diaphragm stresses depends on direction of continuous panel joints with reference to load, not on direction of long dimensions of plywood sheet. Continuous framing may be in either direction for blocked diaphragms.

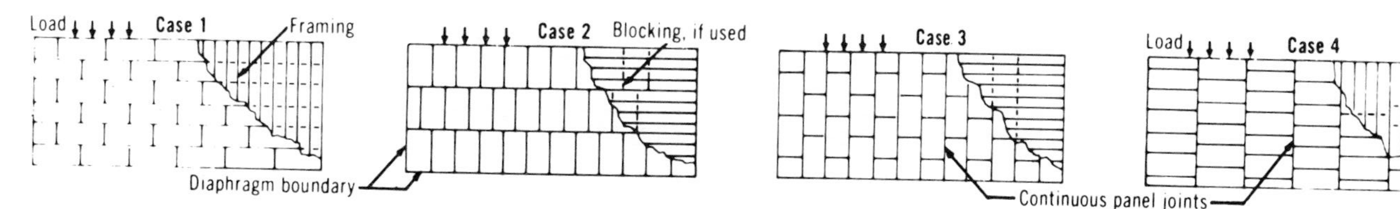

TABLE A4.13.—*CONTINUED*

(b)–Recommended Shear in Pounds Per Foot for Plywood Shear Walls for Wind or Seismic Loading (a)

Plywood grade	Minimum nominal plywood thickness (in.)	Minimum nail penetration in framing (in.)	Plywood applied direct to framing					Plywood applied over 1/2″ gypsum sheathing				
			Nail size (common or galvanized box)	Nail spacing at plywood panel edges (in.)				Nail size (common or galvanized box)	Nail spacing at plywood panel edges			
				6	4	2½	2		6	4	2½	2
STRUCTURAL I INT-DFPA or EXT-DFPA	5/16 or 1/4	1-1/4	6d	200	300	450	510	8d	200	300	450	510
	3/8	1-1/2	8d	280	430	640	730	10d	280	430	640(d)	730(d)
	1/2	1-5/8	10d	340	510	770(d)	870(d)	—	—	—	—	—
C-C EXT-DFPA STRUCTURAL II INT-DFPA STANDARD C-D INT-DFPA DFPA Panel siding and other DFPA grades (c)	5/16 or 1/4 (b)	1-1/4	6d	180	270	400	450	8d	180	270	400	450
	3/8	1-1/2	8d	260	380	570	640	10d	260	380	570(d)	640(d)
	1/2	1-5/8	10d	310	460	690(d)	770(d)	—	—	—	—	—
			Nail size (galvanized casing)					Nail size (galvanized casing)				
DFPA Plywood Panel Siding (c)	5/16 (b)	1-1/4	6d	140	210	320	360	8d	140	210	320	360
	3/8	1-1/2	8d	160	240	360	410	10d	160	240	360	410

(a) All panel edges backed with 2-inch nominal or wider framing. Plywood installed either horizontally or vertically. Space nails at 12 in. on center along intermediate framing members.

(b) 3/8″ minimum recommended when applied direct to framing as exterior siding.

(c) Except Group 5 species.

(d) Reduce tabulated allowable shears 10% when boundary members provide less than 3-inch nominal nailing surface.

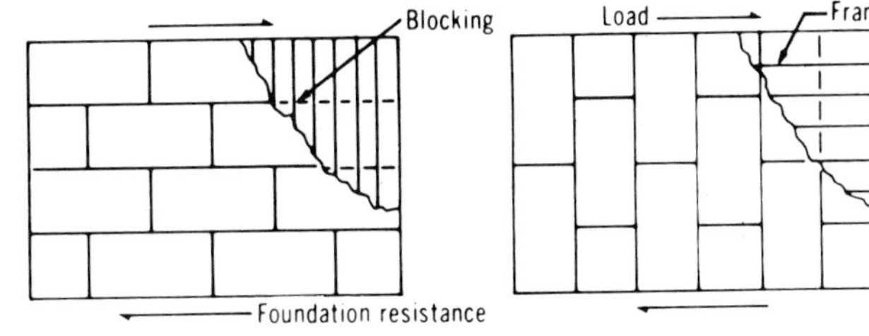

†From "Plywood Diaphragm Construction" by APA.

TABLE A.4.14*

Butt Joints – Tension and Flexure

Plywood Thickness (inches)	Length of Splice Plate (inches)	Maximum Stress (psi)			
		All STRUCT. I Grades	Group 1	Group 2 and Group 3	Group 4
1/4 5/16 3/8 Sanded 3/8 Unsanded	6 8 10 12	1500	1200	1000	900
1/2	14	1500	1000	950	900
5/8 & 3/4	16	1200	800	750	700

*From Plywood Design Specification, 1978 by APA.

TABLE A.4.15*

Key to Identification Index, Span Index and Species Group

For panels with "Index" as across top, and thickness as at left,
use stress for species group given in table.

THICKNESS (IN)	IDENTIFICATION INDEX[1] (C D and C C grades)							
	12/0	16/0	20/0	24/0	32/16	42/20	48/24	
					Span Index (STURD-I-FLOOR grades)			
					16 o.c.	20 o.c.	24 o.c.	48 o.c.
5/16	4	3	1					
3/8		4	3	1				
1/2				4	1			
5/8[2]					3	1		
3/4[2]						3	1	
7/8						4	3	
1-1/8								1

(1) 30/12-5/8″ and 36/16-3/4″ panels also sometimes available. Check your local supplier for availability. Use Group 4 stresses.

(2) For STURD-I-FLOOR grades, 5/8″ may be 19/32″, and 3/4″ may be 23/32″.

*From Plywood Design Specification, 1978 by APA

TABLE A.4.16

GUIDE TO USE OF ALLOWABLE STRESS AND SECTION PROPERTIES TABLES

	Plywood Grade	Description and Use	Typical Trademarks	Veneer Grade			Common Thicknesses	Grade Stress Level (Table 3)	Species Group	Section Property Table
				Face	Back	Inner				
INTERIOR OR PROTECTED APPLICATIONS	APA RATED SHEATHING EXP 1 or 2(3)	Unsanded sheathing grade for wall, roof, subflooring, and industrial applications such as pallets and for for engineering design, with proper stresses. Manufactured with intermediate and exterior glue (1). For permanent exposure to weather or moisture only Exterior type plywood is suitable.	APA RATED SHEATHING 32/16 1/2 INCH SIZED FOR SPACING EXPOSURE 1 000 PS 1-74 C-D INT/EXT GLUE NRB-108	*C*	*D*	*D*	5/16, 3/8, 1/2, 5/8, 3/4	*S-3* (1)	See "Key to Span Rating"	Table 1 (unsanded)
	APA STRUCTURAL I RATED SHEATHING EXP 1 or APA STRUCTURAL II(2) RATED SHEATHING EXP 1	Plywood grades to use where strength properties are of maximum importance, such as plywood-lumber components. Made with exterior glue only. STRUCTURAL I is made from all Group 1 woods. STRUCTURAL II allows Group 3 woods.	APA RATED SHEATHING STRUCTURAL I 24/0 3/8 INCH SIZED FOR SPACING EXPOSURE 1 000 PS 1-74 C-D INT/EXT GLUE NRB-108	*C*	*D*	*D*	5/16, 3/8, 1/2, 5/8, 3/4	*S-2*	Structural I use Group 1 Structural II See "Key to Span Rating"	Table 2 (unsanded) Table 1 (unsanded)
	APA RATED STURD-I-FLOOR EXP 1 or 2(3)	For combination subfloor-underlayment. Provides smooth surface for application of resilient floor coverings. Possesses high concentrated and impact load resistance during construction and occupancy. Manufactured with intermediate and exterior glue. Touch-sanded (4). Available with tongue and groove.(5)	APA RATED STURD-I-FLOOR 20 oc 5/8 INCH SIZED FOR SPACING EXPOSURE 1 000 INT/EXT GLUE NRB-108 FHA-UM-64	*C* plugged	*D*	*C* & *D*	19/32, 5/8, 23/32, 3/4, 1-1/8 (2-4-1)	*S-3* (1)	See "Key to Span Rating"	Table 1 (touch-sanded)
	APA UNDERLAYMENT INT	For underlayment under resilient floor coverings. Available with exterior glue. Touch-sanded. Available with tongue and groove.(5)	APA UNDERLAYMENT GROUP 1 INTERIOR 000 PS 1-74 EXTERIOR GLUE	*C* plugged	*D*	*C* & *D*	1/2, 19/32, 5/8, 23/32, 3/4	*S-3* (1)	As Specified	Table 1 (touch-sanded)
	APA C-D PLUGGED INT	For built-ins, wall and ceiling tile backing, NOT for underlayment. Available with exterior glue. Touch-sanded.(5)	APA C-D PLUGGED GROUP 2 INTERIOR 000 PS 1-74 EXTERIOR GLUE	*C* plugged	*D*	*D*	1/2, 19/32, 5/8, 23/32, 3/4	*S-3* (1)	As Specified	Table 1 (touch-sanded)
	APA APPEARANCE GRADES INT	Generally applied where a high quality surface is required. Includes APA N-N, N-A, N-B, N-D, A-A, A-B, A-D, B-B, and B-D INT grades.(5)	APA A-D GROUP 1 INTERIOR 000 PS 1-74 EXTERIOR GLUE	*B* or better	*D* or better	*C* & *D*	1/4, 3/8, 1/2, 5/8, 3/4	*S-3* (1)	As Specified	Table 1 (sanded)

GUIDE TO USE OF ALLOWABLE STRESS AND SECTION PROPERTIES TABLES (Continued)

	Plywood Grade	Description and Use	Typical Trademarks	Veneer Grade Face	Veneer Grade Back	Veneer Grade Inner	Common Thicknesses	Grade Stress Level (Table 3)	Species Group	Section Property Table
EXTERIOR APPLICATIONS	APA RATED SHEATHING EXT[3]	Unsanded sheathing grade with waterproof glue bond for wall, roof, subfloor and industrial applications such as pallet bins.	APA RATED SHEATHING 48/24 3/4 INCH SIZED FOR SPACING EXTERIOR 000	*C*	*C*	*C*	5/16, 3/8, 1/2, 5/8, 3/4	*S-1*	See "Key to Span Rating"	Table 1 (unsanded)
EXTERIOR APPLICATIONS	APA STRUCTURAL I RATED SHEATHING EXT or APA STRUCTURAL II[2] RATED SHEATHING EXT	"Structural" is a modifier for this unsanded sheathing grade. For engineered applications in construction and industry where full exterior-type panels are required. STRUCTURAL I is made from Group 1 woods only.	APA RATED SHEATHING STRUCTURAL I 24/0 3/8 INCH SIZED FOR SPACING EXTERIOR 000	*C*	*C*	*C*	5/16, 3/8, 1/2, 5/8, 3/4	*S-1*	Structural I use Group 1 Structural II See "Key to Span Rating"	Table 2 (unsanded) Table 1 (unsanded)
EXTERIOR APPLICATIONS	APA RATED STURD-I-FLOOR EXT[3]	For combination subfloor-underlayment for resilient floor coverings where severe moisture conditions may be present, as in balcony decks. Possesses high concentrated and impact load resistance during construction and occupancy. Touch-sanded (4). Available with tongue and groove.(5)	APA RATED STURD-I-FLOOR 20 oc 19/32 INCH SIZED FOR SPACING EXTERIOR 000	*C* plugged	*C*	*C*	19/32, 5/8, 23/32, 3/4,	*S-2*	See "Key to Span Rating"	Table 1 (touch-sanded)
EXTERIOR APPLICATIONS	APA UNDERLAYMENT EXT and APA C-C PLUGGED EXT	Underlayment for floor under resilient floor coverings where severe moisture conditions may exist. Also for controlled atmosphere rooms and many industrial applications. Touch-sanded. Available with tongue and groove.(5)	APA C-D PLUGGED GROUP 2 INTERIOR 000	*C* plugged	*C*	*C*	1/2, 19/32, 5/8, 23/32, 3/4	*S-2*	As Specified	Table 1 (touch-sanded)
EXTERIOR APPLICATIONS	APA B-B PLYFORM CLASS I or II[2]	Concrete-form grade with high reuse factor. Sanded both sides, mill-oiled unless otherwise specified. Available in HDO. For refined design information on this special-use panel see APA publication "Plywood for Concrete Forming" (form V345). Design using values from this specification will result in a conservative design.(5)	APA PLYFORM B-B CLASS I EXTERIOR 000	*B*	*B*	*C*	5/8, 3/4	*S-2*	Class I use Group 1; Class II use Group 3	Table 1 (sanded)
EXTERIOR APPLICATIONS	APA MARINE EXT	Superior Exterior-type plywood made only with Douglas Fir or Western Larch. Special solid-core construction. Available with MDO or HDO face. Ideal for boat hull construction.	MARINE A-A EXT APA PS 1 74 000	*A* or *B*	*A* or *B*	*B*	1/4, 3/8, 1/2, 5/8, 3/4	*A* face & back use *S-1*; *B* face or back use *S-2*	Group 1	Table 2 (sanded)
EXTERIOR APPLICATIONS	APA APPEARANCE GRADES EXT	Generally applied where a high quality surface is required. Includes APA A-A, A-B, A-C, B-B, B-C, HDO and MDO EXT.(5)	APA A-C GROUP 1 EXTERIOR 000	*B* or better	*C* or better	*C*	1/4, 3/8, 1/2, 5/8, 3/4	*A* or *C* face and back use *S-1*; *B* face or back use *S-2*	As Specified	Table 1 (sanded)

(1) When exterior glue is specified, i.e. "Interior with exterior glue" or "Exposure 1," stress level 2 (S-2) should be used.
(2) Check local suppliers for availability of STRUCTURAL II and PLYFORM Class II grades.
(3) Properties and stresses apply only to APA RATED STURD-I-FLOOR and APA RATED SHEATHING manufactured entirely with veneers.
(4) APA RATED STURD-I-FLOOR 2-4-1 may be produced unsanded.
(5) May be modified to STRUCTURAL I. For such designation use Group 1 stresses and Table 2 section properties.

TABLE A6.1a*

BOLT DESIGN VALUES

Design values, in pounds, on one bolt loaded at both ends (double shear)* for following species

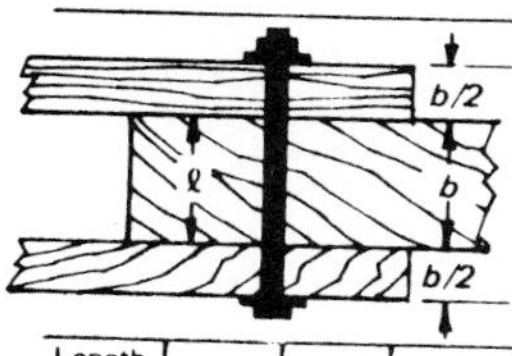

Length of bolt in main member ℓ	Diameter of bolt D	ℓ/D	Projected area of bolt $A=\ell \times D$	1 DOUGLAS FIR LARCH (Dense), SOUTHERN PINE (Dense) Parallel to grain P	1 Perpendicular to grain Q	2 ASH, Commercial White, HICKORY Parallel to grain P	2 Perpendicular to grain Q	3 CALIFORNIA REDWOOD (Close grain), DOUGLAS FIR-LARCH, SOUTHERN PINE, SOUTHERN CYPRESS Parallel to grain P	3 Perpendicular to grain Q	4 BEECH, BIRCH, Sweet & Yellow, MAPLE, Black & Sugar Parallel to grain P	4 Perpendicular to grain Q	5 OAK, Red & White Parallel to grain P	5 Perpendicular to grain Q	6 DOUGLAS FIR SOUTH Parallel to grain P	6 Perpendicular to grain Q
1-1/2	1/2	3.00	.750	1100	500	1080	780	940	430	900	480	830	650	870	370
	5/8	2.40	.938	1380	570	1360	880	1180	490	1130	540	1050	730	1090	420
	3/4	2.00	1.125	1660	630	1630	980	1420	540	1360	600	1260	820	1310	470
	7/8	1.71	1.313	1940	700	1910	1080	1660	600	1590	670	1470	900	1530	520
	1	1.50	1.500	2220	760	2180	1170	1890	650	1820	730	1690	980	1750	570
2	1/2	4.00	1.000	1370	670	1340	1040	1170	570	1120	640	1040	870	1080	500
	5/8	3.20	1.250	1820	760	1790	1170	1550	650	1490	720	1380	980	1440	560
	3/4	2.67	1.500	2210	840	2170	1300	1890	720	1810	810	1680	1090	1740	630
	7/8	2.29	1.750	2580	930	2540	1430	2200	790	2120	890	1960	1200	2040	690
	1	2.00	2.000	2960	1010	2910	1570	2520	870	2420	970	2250	1310	2330	750
2-1/2	1/2	5.00	1.250	1480	840	1450	1290	1260	720	1210	800	1120	1080	1170	620
	5/8	4.00	1.563	2140	950	2100	1460	1820	810	1750	910	1620	1220	1690	710
	3/4	3.33	1.875	2710	1060	2660	1630	2310	900	2220	1010	2060	1360	2140	790
	7/8	2.86	2.188	3210	1160	3160	1790	2740	990	2630	1110	2440	1500	2530	860
	1	2.50	2.500	3680	1270	3620	1960	3150	1080	3020	1210	2800	1640	2910	940
3	1/2	6.00	1.500	1490	1010	1460	1460	1270	860	1220	930	1130	1130	1180	750
	5/8	4.80	1.875	2290	1140	2260	1760	1960	970	1880	1090	1740	1470	1810	850
	3/4	4.00	2.250	3080	1270	3020	1950	2630	1080	2520	1210	2340	1630	2430	940
	7/8	3.43	2.625	3770	1390	3710	2150	3220	1190	3090	1330	2870	1800	2980	1040
	1	3.00	3.000	4390	1520	4320	2350	3750	1300	3600	1450	3340	1960	3460	1130
3-1/2	1/2	7.00	1.750	1490	1140	1460	1460	1270	980	1220	970	1130	1130	1180	870
	5/8	5.60	2.188	2320	1330	2280	2020	1980	1130	1900	1250	1760	1690	1830	990
	3/4	4.67	2.625	3280	1480	3220	2280	2800	1260	2690	1410	2440	1910	2590	1100
	7/8	4.00	3.063	4190	1630	4120	2510	3580	1390	3430	1550	3180	2100	3300	1210
	1	3.50	3.500	5000	1770	4920	2740	4270	1520	4100	1690	3800	2290	3950	1320
4	1/2	8.00	2.000	1490	1180	1460	1460	1270	1010	1220	960	1130	1130	1180	960
	5/8	6.40	2.500	2330	1510	2290	2180	1990	1290	1910	1350	1770	1770	1840	1130
	3/4	5.33	3.000	3340	1690	3280	2590	2850	1440	2740	1600	2540	2170	2630	1260
	7/8	4.57	3.500	4450	1860	4370	2870	3800	1590	3650	1770	3380	2400	3510	1380
	1	4.00	4.000	5470	2030	5380	3130	4670	1730	4480	1940	4160	2620	4320	1510
4-1/2	5/8	7.20	2.813	2330	1640	2290	2230	1990	1400	1910	1380	1770	1770	1840	1270
	3/4	6.00	3.375	3350	1900	3300	2820	2860	1620	2750	1750	2550	2360	2650	1410
	7/8	5.14	3.938	4530	2090	4460	3220	3870	1790	3720	1990	3450	2690	3580	1560
	1	4.50	4.500	5770	2280	5680	3520	4930	1950	4730	2180	4390	2940	4560	1700
	1-1/4	3.60	5.625	7980	2670	7850	4120	6820	2280	6540	2550	6070	3450	6300	1990
5-1/2	5/8	8.80	3.438	2330	1650	2290	2150	1990	1410	1910	1330	1770	1770	1840	1380
	3/4	7.33	4.125	3350	2200	3290	2980	2860	1880	2740	1840	2540	2490	2640	1720
	7/8	6.29	4.813	4570	2550	4490	3710	3900	2180	3740	2300	3470	3110	3600	1900
	1	5.50	5.500	5930	2790	5830	4260	5070	2380	4860	2640	4510	3560	4680	2080
	1-1/4	4.40	6.875	8940	3260	8790	5040	7640	2790	7330	3120	6800	4210	7060	2430
7-1/2	5/8	12.00	4.688	2330	1480	2290	1870	1990	1260	1910	1150	1770	1560	1840	1290
	3/4	10.00	5.625	3350	2130	3290	2710	2860	1820	2750	1670	2550	2260	2640	1800
	7/8	8.57	6.563	4560	2840	4480	3720	3890	2430	3740	2300	3460	3110	3600	2360
	1	7.50	7.500	5950	3550	5850	4760	5080	3030	4880	2950	4520	3980	4700	2800
	1-1/4	6.00	9.375	9310	4450	9150	6620	7950	3800	7630	4090	7080	5530	7350	3310
9-1/2	3/4	12.67	7.125	3350	1920	3290	2420	2860	1640	2740	1500	2550	2020	2640	1700
	7/8	10.86	8.313	4570	2660	4490	3360	3900	2270	3750	2080	3470	2810	3610	2260
	1	9.50	9.500	5950	3460	5850	4460	5080	2960	4880	2760	4520	3730	4700	2900
	1-1/4	7.60	11.875	9300	5210	9140	6960	7950	4450	7620	4300	7070	5820	7340	4140
	1-1/2	6.33	14.250	13410	6480	13190	9380	11460	5530	10990	5800	10200	7840	10590	4830
11-1/2	7/8	13.14	10.062	4560	1980	4490	3050	3900	2060	3750	2300	3470	2550	3600	2170
	1	11.50	11.500	5950	3240	5850	4080	5080	2770	4880	2530	4520	3420	4690	2780
	1-1/4	9.20	14.375	9300	5110	9150	6610	7950	4360	7630	4090	7070	5520	7340	4270
	1-1/2	7.67	17.250	13410	7200	13180	9570	11450	6150	10990	5920	10190	8000	10580	5740
13-1/2	1	13.50	13.500	5960	2410	5850	3730	5100	2530	4880	2830	4520	3120	4710	2680
	1-1/4	10.80	16.875	9300	4860	9140	6150	7950	4160	7620	3800	7070	5140	7340	4130
	1-1/2	9.00	20.250	13400	7070	13180	9190	11450	6040	10990	5680	10190	7680	10580	5920

*Three (3) member joint.

*From National Design Specification, 1977 by NFPA.

TABLE A6.1a—CONTINUED

BOLT DESIGN VALUES

Design values, in pounds, on one bolt loaded at both ends (double shear)* for following species

				7 SWEETGUM & TUPELO		8 EASTERN HEMLOCK-TAMARACK, CALIFORNIA REDWOOD (Open grain), HEM-FIR, WESTERN HEMLOCK		9 MOUNTAIN HEMLOCK, WESTERN CEDARS, NORTHERN PINE		10 SPRUCE-PINE-FIR, SITKA SPRUCE, YELLOW POPLAR, EASTERN SPRUCE, LODGEPOLE PINE		11 RED PINE, WESTERN WHITE PINE, PONDEROSA PINE-SUGAR PINE, EASTERN WHITE PINE, BALSAM FIR, IDAHO WHITE PINE		12 ASPEN, EASTERN COTTONWOOD, ENGELMANN SPRUCE-ALPINE FIR, NORTHERN WHITE CEDAR	
Length of bolt in main member ℓ	Diameter of bolt D	ℓ/D	Projected area of bolt $A=\ell \times D$	Parallel to grain P	Perpendicular to grain Q	Parallel to grain P	Perpendicular to grain Q	Parallel to grain P	Perpendicular to grain Q	Parallel to grain P	Perpendicular to grain Q	Parallel to grain P	Perpendicular to grain Q	Parallel to grain P	Perpendicular to grain Q
1-1/2	1/2	3.00	.750	810	410	800	280	750	300	680	280	630	190	530	210
	5/8	2.40	.938	1010	460	1000	310	930	340	850	320	790	210	660	230
	3/4	2.00	1.125	1220	510	1200	350	1120	370	1020	350	950	240	800	260
	7/8	1.71	1.313	1420	560	1400	380	1310	410	1190	390	1110	260	930	290
	1	1.50	1.500	1620	620	1600	420	1490	450	1360	420	1260	290	1060	310
2	1/2	4.00	1.000	1050	540	1040	370	990	400	900	370	840	250	700	280
	5/8	3.20	1.250	1350	610	1330	410	1240	450	1130	420	1050	290	890	310
	3/4	2.67	1.500	1620	680	1600	460	1490	500	1360	470	1260	320	1060	350
	7/8	2.29	1.750	1890	750	1870	510	1740	550	1580	520	1480	350	1240	380
	1	2.00	2.000	2160	820	2130	550	1990	600	1810	560	1690	380	1420	420
2-1/2	1/2	5.00	1.250	1190	680	1180	460	1190	500	1080	470	1010	320	840	340
	5/8	4.00	1.563	1650	770	1620	520	1550	560	1410	530	1310	360	1100	390
	3/4	3.33	1.875	2020	850	1990	580	1870	620	1700	590	1580	400	1330	430
	7/8	2.86	2.188	2360	940	2330	630	2180	690	1980	650	1840	440	1550	480
	1	2.50	2.500	2700	1030	2670	690	2490	750	2260	700	2110	480	1770	520
3	1/2	6.00	1.500	1230	810	1210	550	1280	590	1160	560	1080	380	910	410
	5/8	4.80	1.875	1840	920	1810	620	1800	670	1640	630	1530	430	1280	470
	3/4	4.00	2.250	2370	1030	2340	690	2230	750	2030	700	1890	480	1590	520
	7/8	3.43	2.625	2820	1130	2790	760	2610	820	2380	780	2210	520	1860	570
	1	3.00	3.000	3240	1230	3200	830	2990	900	2710	850	2530	570	2120	620
3-1/2	1/2	7.00	1.750	1230	920	1210	640	1280	690	1160	650	1080	440	910	480
	5/8	5.60	2.188	1910	1070	1890	730	1970	780	1790	740	1670	500	1400	550
	3/4	4.67	2.625	2610	1200	2580	810	2540	870	2310	820	2150	560	1800	610
	7/8	4.00	3.063	3220	1320	3180	890	3030	960	2760	900	2570	610	2160	670
	1	3.50	3.500	3760	1440	3710	970	3480	1050	3170	990	2950	670	2480	730
4	1/2	8.00	2.000	1230	960	1210	700	1280	790	1160	750	1080	500	910	550
	5/8	6.40	2.500	1920	1230	1900	830	2000	900	1820	840	1690	570	1420	620
	3/4	5.33	3.000	2720	1370	2690	920	2770	1000	2520	940	2340	630	1970	690
	7/8	4.57	3.500	3530	1500	3480	1020	3400	1100	3090	1030	2880	700	2420	760
	1	4.00	4.000	4210	1640	4150	1110	3960	1200	3600	1130	3360	760	2820	830
4-1/2	5/8	7.20	2.813	1920	1320	1890	930	2000	1010	1810	950	1690	640	1420	700
	3/4	6.00	3.375	2770	1540	2730	1040	2880	1120	2610	1060	2440	710	2050	780
	7/8	5.14	3.938	3680	1690	3630	1140	3700	1240	3360	1160	3130	790	2630	860
	1	4.50	4.500	4560	1850	4500	1250	4380	1350	3990	1270	3710	860	3120	940
	1-1/4	3.60	5.625	6020	2160	5940	1460	5590	1580	5080	1490	4740	1000	3980	1100
5-1/2	5/8	8.80	3.438	1920	1330	1900	1010	1990	1190	1810	1110	1690	750	1420	820
	3/4	7.33	4.125	2760	1780	2730	1270	2870	1370	2610	1290	2430	870	2040	950
	7/8	6.29	4.813	3770	2070	3720	1400	3920	1510	3560	1420	3320	960	2790	1050
	1	5.50	5.500	4890	2260	4820	1520	5000	1650	4550	1550	4240	1050	3560	1150
	1-1/4	4.40	6.875	7040	2640	6940	1780	6730	1930	6110	1820	5700	1230	4780	1340
7-1/2	5/8	12.00	4.688	1920	1200	1890	950	2000	1150	1820	1090	1690	730	1420	800
	3/4	10.00	5.625	2770	1720	2730	1320	2880	1590	2620	1500	2440	1010	2050	1110
	7/8	8.57	6.563	3770	2300	3720	1730	3910	2010	3560	1900	3310	1280	2780	1400
	1	7.50	7.500	4910	2870	4850	2060	5110	2250	4650	2110	4330	1430	3640	1560
	1-1/4	6.00	9.375	7680	3600	7580	2430	7990	2940	7260	2760	6770	1870	5680	2040
9-1/2	3/4	12.67	7.125	2760	1550	2720	1250	2870	1520	2610	1430	2430	970	2040	1060
	7/8	10.86	8.313	3770	2150	3720	1660	3910	2010	3560	1890	3320	1280	2780	1400
	1	9.50	9.500	4920	2800	4850	2130	5110	2560	4650	2410	4330	1630	3640	1780
	1-1/4	7.60	11.875	7680	4220	7570	3040	7980	4480	7250	4210	6750	2850	5680	3110
	1-1/2	6.33	14.250	11060	5240	10910	3540	11520	3830	10470	3610	9760	2440	8190	2660
11-1/2	7/8	13.14	10.062	3770	1950	3700	1590	3930	1710	3570	1610	3330	1090	2790	1190
	1	11.50	11.500	4920	2620	4860	2040	5120	2490	4650	2340	4330	1580	3640	1730
	1-1/4	9.20	14.375	7680	4130	7570	3130	7970	3750	7240	3530	6750	2380	3670	2610
	1-1/2	7.67	17.250	11020	5830	10930	4210	11470	4640	10430	4370	9720	2950	8160	3230
13-1/2	1	13.50	13.500	4920	2400	4850	1970	5130	2410	4670	2270	4350	1530	3650	1680
	1-1/4	10.80	16.875	7700	3940	7590	3030	7990	3680	7270	3460	6770	2340	5690	2560
	1-1/2	9.00	20.250	11070	5720	10930	4340	11510	5150	10460	4850	9750	3280	8190	3580

*Three (3) member joint.

TABLE A6.1b. MULTIPLE-MEMBER BOLTED JOINT

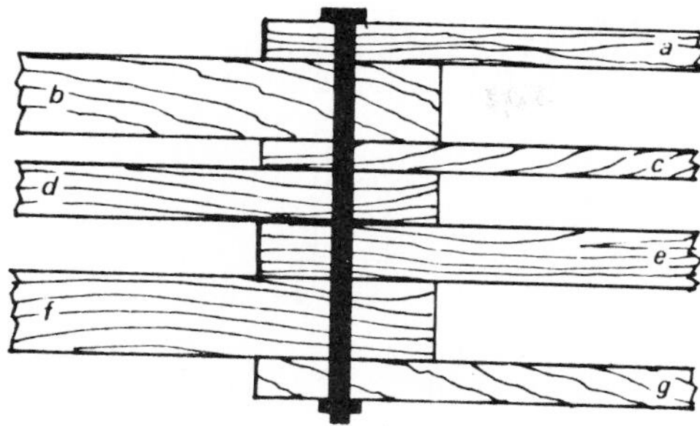

8.5.8.5 For a joint with four or more members not of equal thickness (see Figure 8.5H), the design value may be determined from the design values for the individual shear planes, as follows:

(a) Resolve the multi-member joint into the maximum number of contigous 3-member joints, as in Figure 8.5H.

(b) For each such 3-member joint, determine the applicable design load in accordance with standard procedures, and assign one-half of the load to each shear plane in the joint.

(c) For those shear planes to which two different design loads have been assigned, the design value shall be the lesser of the two loads.

(d) For assemblies in which the load is shared equally among the members, or in which the distribution of load among members is indeterminate, the design value for the multi-member joint shall be the least design value for any one shear plane times the number of shear planes in the joint.

(e) For assemblies in which the load on each member is known, the bolt design value for any member in the joint shall be the sum of the individual bolt design values for each of the two shear planes acting on that member, as determined in (a) through (c), above.

a-b-c

b-c-d

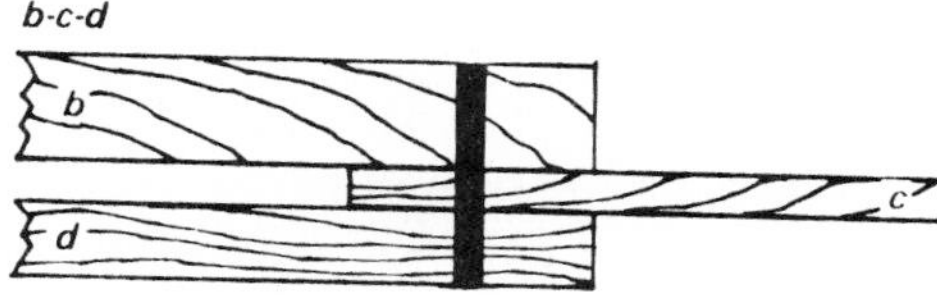

c-d-e

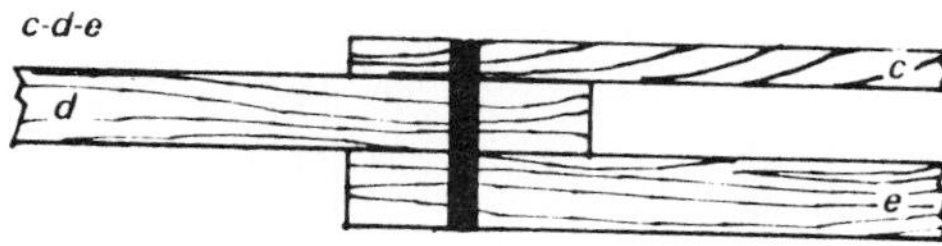

d-e-f

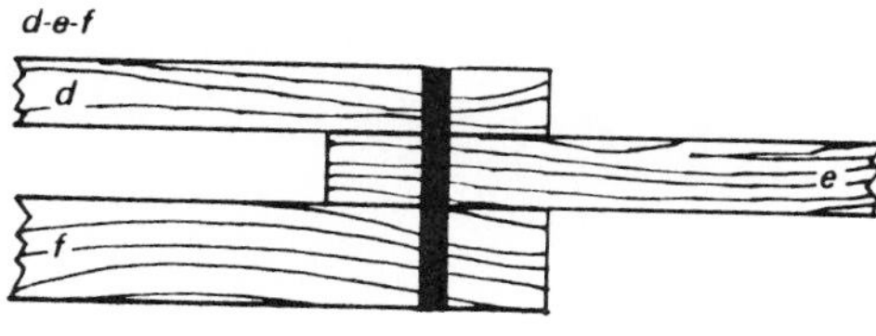

e-f-g

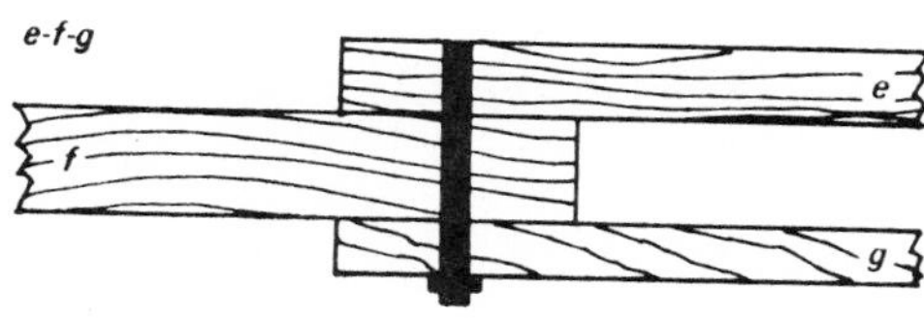

Figure 8.5H

TABLE A6.2a*

TYPICAL DIMENSIONS OF STANDARD LAG SCREWS FOR WOOD

[All dimensions in inches]

D = Nominal diameter.
Ds = *D* = Diameter of shank.
Dr = Diameter at root of thread.
W = Width of head across flats.
H = Height of head.
L = Nominal length.
S = Length of shank.
T = Length of thread.
E = Length of tapered tip.
N = Number of threads per inch.

Nominal length *L* in inches*	Item	Dimensions of lag screw with nominal diameter *D* of—												
		3/16	1/4	5/16	3/8	7/16	1/2	9/16	5/8	3/4	7/8	1	1-1/8	1-1/4
All lengths	*Ds=D*	0.190	0.250	0.3125	0.375	0.4375	0.500	0.5625	0.625	0.750	0.875	1.000	1.125	1.250
	Dr	0.120	0.173	0.227	0.265	0.328	0.371	0.435	0.471	0.579	0.683	0.780	0.887	1.012
	E	5/32	3/16	1/4	1/4	9/32	5/16	3/8	3/8	7/16	1/2	9/16	5/8	3/4
	H	9/64	11/64	13/64	1/4	19/64	21/64	3/8	27/64	1/2	19/32	21/32	3/4	27/32
	W	9/32	3/8	1/2	9/16	5/8	3/4	7/8	15/16	1-1/8	1-5/16	1-1/2	1-11/16	1-7/8
	N	11	10	9	7	7	6	6	5	4-1/2	4	3-1/2	3-1/4	3-1/4
1	*S*	1/4	1/4	1/4	1/4	1/4	1/4	...	...	...	...	...	...	...
	T	3/4	3/4	3/4	3/4	3/4	3/4	...	...	...	...	...	...	...
	T-E	19/32	9/16	1/2	1/2	15/32	7/16	...	...	...	...	...	...	...
1½	*S*	3/8	3/8	3/8	3/8	3/8	3/8	...	...	...	...	...	...	...
	T	1-1/8	1-1/8	1-1/8	1-1/8	1-1/8	1-1/8	...	...	...	...	...	...	...
	T-E	31/32	15/16	7/8	7/8	27/32	13/16	...	...	...	...	...	...	...
2	*S*	1/2	1/2	1/2	1/2	1/2	1/2	1/2	1/2	...	...	...	...	...
	T	1-1/2	1-1/2	1-1/2	1-1/2	1-1/2	1-1/2	1-1/2	1-1/2	...	...	...	...	...
	T-E	1-11/32	1-5/16	1-1/4	1-1/4	17/32	1-3/16	1-1/8	1-1/8	...	...	...	...	...
2½	*S*	1	1	7/8	7/8	3/4	3/4	3/4	3/4	...	...	...	...	...
	T	1-1/2	1-1/2	1-5/8	1-5/8	1-3/4	1-3/4	1-3/4	1-3/4	...	...	...	...	...
	T-E	1-11/16	1-5/16	1-3/8	1-3/8	1-15/32	1-7/16	1-3/8	1-3/8	...	...	...	...	...
3	*S*	1	1	1	1	1	1	1	1	1	1	1	...	...
	T	2	2	2	2	2	2	2	2	2	2	2	...	...
	T-E	1-27/32	1-13/16	1-3/4	1-3/4	1-23/32	1-11/16	1-5/8	1-5/8	1-9/16	1-1/2	1-7/16	...	...
4	*S*	1-1/2	1-1/2	1-1/2	1-1/2	1-1/2	1-1/2	1-1/2	1-1/2	1-1/2	1-1/2	1-1/2	1-1/2	1-1/2
	T	2-1/2	2-1/2	2-1/2	2-1/2	2-1/2	2-1/2	2-1/2	2-1/2	2-1/2	2-1/2	2-1/2	2-1/2	2-1/2
	T-E	2-11/32	2-5/16	2-1/4	2-1/4	2-7/32	2-3/16	2-1/8	2-1/8	2-1/16	2	1-15/16	1-7/8	1-3/4
5	*S*	2	2	2	2	2	2	2	2	2	2	2	2	2
	T	3	3	3	3	3	3	3	3	3	3	3	3	3
	T-E	2-27/32	2-13/16	2-3/4	2-3/4	2-23/32	2-11/16	2-5/8	2-5/8	2-9/16	2-1/2	2-7/16	2-3/8	2-1/4
6	*S*	2-1/2	2-1/2	2-1/2	2-1/2	2-1/2	2-1/2	2-1/2	2-1/2	2-1/2	2-1/2	2-1/2	2-1/2	2-1/2
	T	3-1/2	3-1/2	3-1/2	3-1/2	3-1/2	3-1/2	3-1/2	3-1/2	3-1/2	3-1/2	3-1/2	3-1/2	3-1/2
	T-E	3-11/32	3-5/16	3-1/4	3-1/4	3-7/32	3-3/16	3-1/8	3-1/8	3-1/16	3	2-15/16	2-7/8	2-3/4
7	*S*	3	3	3	3	3	3	3	3	3	3	3	3	3
	T	4	4	4	4	4	4	4	4	4	4	4	4	4
	T-E	3-27/32	3-13/16	3-3/4	3-3/4	3-23/32	3-11/16	3-5/8	3-5/8	3-9/16	3-1/2	3-7/16	3-3/8	3-1/4
8	*S*	3-1/2	3-1/2	3-1/2	3-1/2	3-1/2	3-1/2	3-1/2	3-1/2	3-1/2	3-1/2	3-1/2	3-1/2	3-1/2
	T	4-1/2	4-1/2	4-1/2	4-1/2	4-1/2	4-1/2	4-1/2	4-1/2	4-1/2	4-1/2	4-1/2	4-1/2	4-1/2
	T-E	4-11/16	4-5/16	4-1/4	4-1/4	4-7/32	4-3/16	4-1/8	4-1/8	4-1/16	4	3-15/16	3-7/8	3-3/4
9	*S*	4	4	4	4	4	4	4	4	4	4	4	4	4
	T	5	5	5	5	5	5	5	5	5	5	5	5	5
	T-E	4-27/32	4-13/16	4-3/4	4-3/4	4-23/32	4-11/16	4-5/8	4-5/8	4-9/16	4-1/2	4-7/16	4-3/8	4-1/4
10	*S*	4-3/4	4-3/4	4-3/4	4-3/4	4-3/4	4-3/4	4-3/4	4-3/4	4-3/4	4-3/4	4-3/4	4-3/4	4-3/4
	T	5-1/4	5-1/4	5-1/4	5-1/4	5-1/4	5-1/4	5-1/4	5-1/4	5-1/4	5-1/4	5-1/4	5-1/4	5-1/4
	T-E	5-3/32	5-1/16	5	5	4-31/32	4-15/16	4-7/8	4-7/8	4-13/16	4-3/4	4-11/16	4-5/8	4-1/2
11	*S*	5-1/2	5-1/2	5-1/2	5-1/2	5-1/2	5-1/2	5-1/2	5-1/2	5-1/2	5-1/2	5-1/2	5-1/2	5-1/2
	T	5-1/2	5-1/2	5-1/2	5-1/2	5-1/2	5-1/2	5-1/2	5-1/2	5-1/2	5-1/2	5-1/2	5-1/2	5-1/2
	T-E	5-11/32	5-9/32	5-1/4	5-1/4	5-7/32	5-3/16	5-1/8	5-1/8	5-1/16	5	4-15/16	4-7/8	4-3/4
12	*S*	6	6	6	6	6	6	6	6	6	6	6	6	6
	T	6	6	6	6	6	6	6	6	6	6	6	6	6
	T-E	5-27/32	5-13/16	5-3/4	5-3/4	5-23/32	5-11/16	5-5/8	5-5/8	5-9/16	5-1/2	5-7/16	5-3/8	5-1/4

*Length of thread *T* on intervening bolt lengths is the same as that of the next shorter length listed. The length of thread *T* on standard lag screw lengths *L* in excess of 12 inches is equal to 1/2 the lag screw length, *L*/2.

*From National Design Specification, 1977 by NFPA.

TABLE A6.2b. LAG SCREWS—WITHDRAWAL DESIGN VALUES.

Normal load duration, dry service conditions

Design values for load in withdrawal in pounds per inch of penetration of threaded part into side grain of member holding point.

D = the shank diameter in inches.
G = specific gravity of the wood based on weight and volume when oven-dry.

Specific gravity G	Lag screw diameter D											
	1/4	5/16	3/8	7/16	1/2	9/16	5/8	3/4	7/8	1	1-1/8	1-1/4
	0.250	0.3125	0.375	0.4375	0.500	0.5625	0.625	0.750	0.875	1.000	1.125	1.250
0.75	413	489	560	629	695	759	822	942	1058	1169	1277	1382
0.68	357	422	484	543	600	656	709	813	913	1009	1103	1193
0.67	349	413	473	531	587	641	694	796	893	987	1078	1167
0.66	341	403	463	519	574	627	678	778	873	965	1054	1141
0.62	311	367	421	473	523	571	618	708	795	879	960	1039
0.55	260	307	352	395	437	477	516	592	664	734	802	868
0.54	253	299	342	384	425	464	502	576	646	714	780	844
0.51	232	274	314	353	390	426	461	528	593	656	716	775
0.49	218	258	296	332	367	401	434	498	559	617	674	730
0.48	212	250	287	322	356	389	421	482	542	599	654	708
0.47	205	242	278	312	345	377	408	467	525	580	634	686
0.46	199	235	269	302	334	365	395	453	508	562	613	664
0.45	192	227	260	292	323	353	382	438	492	543	594	642
0.44	186	220	252	283	312	341	369	423	475	525	574	621
0.43	179	212	243	273	302	330	357	409	459	508	554	600
0.42	173	205	235	264	291	318	344	395	443	490	535	579
0.41	167	198	226	254	281	307	332	381	428	473	516	559
0.40	161	190	218	245	271	296	320	367	412	455	497	538
0.39	155	183	210	236	261	285	308	353	397	438	479	518
0.38	149	176	202	227	251	274	296	340	381	422	461	498
0.37	143	169	194	218	241	263	285	326	367	405	443	479
0.36	137	163	186	209	231	253	273	313	352	389	425	460
0.35	132	156	179	200	222	242	262	300	337	373	407	441
0.33	121	143	164	184	203	222	240	275	309	341	373	403
0.31	110	130	149	167	185	202	218	250	281	311	339	367

TABLE A6.3. LAG SCREWS. LATERAL LOAD DESIGN VALUES WITH WOOD SIDE PIECES.*

Normal load duration, dry service conditions

Thickness of side member (inches)	Length of lag screw (inches)	Diameter of lag screw shank (inches)	GROUP I		GROUP II		GROUP III		GROUP IV	
			Total lateral load per lag screw in single shear (pounds)		Total lateral load per lag screw in single shear (pounds)		Total lateral load per lag screw in single shear (pounds)		Total lateral load per lag screw in single shear (pounds)	
			Parallel to grain	Perpendicular to grain	Parallel to grain	Perpendicular to grain	Parallel to grain	Perpendicular to grain	Parallel to grain	Perpendicular to grain
1½"	4"	1/4	200	190	170	170	130	120	100	100
		5/16	280	230	210	180	150	130	120	100
		3/8	320	240	240	180	170	130	140	100
		7/16	350	250	270	190	190	140	150	110
		1/2	390	250	290	190	210	140	170	110
		5/8	470	280	360	210	260	150	200	120
	5"	1/4	230	220	200	190	180	170	160	150
		5/16	330	280	290	250	230	200	190	160
		3/8	430	330	370	280	260	200	210	160
		7/16	540	380	400	290	290	210	230	160
		1/2	580	380	440	280	310	200	250	160
		5/8	710	420	530	320	380	230	310	180
	6"	1/4	270	260	230	220	200	200	180	180
		5/16	380	320	330	280	290	250	260	220
		3/8	480	370	420	320	360	280	290	220
		7/16	590	420	510	360	400	280	320	230
		1/2	700	460	600	390	430	280	340	220
		5/8	860	510	710	430	510	310	410	250
	7"	1/4	280	270	240	230	210	210	190	180
		5/16	400	340	350	300	310	270	280	230
		3/8	520	400	450	340	410	310	360	270
		7/16	640	460	560	390	500	350	420	290
		1/2	760	500	660	430	560	360	450	290
		5/8	910	550	790	470	640	380	510	310
2½"	6"	3/8	450	340	370	280	270	200	210	160
		7/16	570	400	430	310	310	220	250	180
		1/2	620	410	470	310	340	220	270	180
		5/8	730	440	550	330	390	240	320	190
		3/4	820	450	620	340	440	240	360	200
		7/8	930	490	710	370	500	260	400	210
		1	1040	520	790	390	560	280	450	230
	7"	3/8	500	380	430	330	370	280	300	220
		7/16	660	470	570	410	420	300	340	240
		1/2	830	540	650	420	460	300	370	240
		5/8	1000	600	750	450	540	320	430	260
		3/4	1110	610	840	460	600	330	480	270
		7/8	1260	660	950	500	680	360	550	290
		1	1420	710	1070	540	770	380	620	310
	8"	3/8	550	420	480	360	430	320	380	290
		7/16	720	510	630	440	550	390	440	310
		1/2	890	580	770	500	600	390	480	310
		5/8	1230	740	970	580	700	420	560	340
		3/4	1430	790	1080	600	780	430	620	340
		7/8	1590	830	1200	630	860	450	690	360
		1	1800	900	1360	680	970	490	780	390
	9"	3/8	600	460	520	390	460	350	410	310
		7/16	790	560	680	480	610	430	540	380
		1/2	970	630	840	540	750	490	600	390
		5/8	1310	790	1130	680	860	520	690	420
		3/4	1670	920	1340	740	960	530	770	420
		7/8	1920	1000	1450	760	1040	540	830	430
		1	2170	1090	1640	820	1180	590	940	470

*From National Design Specification, 1977 by NFPA.

TABLE A6.4. LAG SCREWS—LATERAL LOAD DESIGN VALUES WITH 1/2″ METAL SIDE PIECES.

Normal load duration, dry service conditions

Length of lag screw (inches)	Diameter of lag screw shank (inches)	GROUP I Total lateral load per lag screw in single shear (pounds)		GROUP II Total lateral load per lag screw in single shear (pounds)		GROUP III Total lateral load per lag screw in single shear (pounds)		GROUP IV Total lateral load per lag screw in single shear (pounds)	
		Parallel to grain	Perpendicular to grain	Parallel to grain	Perpendicular to grain	Parallel to grain	Perpendicular to grain	Parallel to grain	Perpendicular to grain
3″	1/4	240	185	210	160	155	120	125	100
	5/16	355	240	265	180	190	130	155	105
	3/8	420	255	320	195	230	140	180	110
	7/16	485	275	370	210	265	150	210	120
	1/2	550	285	415	215	295	155	240	125
	5/8	645	310	490	235	350	170	280	135
4″	1/4•	275	210	235	185	210	165	190	145
	5/16	410	280	355	240	290	200	235	160
	3/8	570	345	480	290	345	210	275	165
	7/16	750	425	575	320	405	230	320	180
	1/2	830	430	625	325	450	235	360	185
	5/8	975	465	740	355	530	255	425	205
5″	5/16	435	295	375	255	335	230	300	205
	3/8	615	375	535	325	470	295	375	230
	7/16	820	465	710	405	535	350	430	245
	1/2	1045	540	850	440	610	315	490	255
	5/8	1330	635	1005	480	720	345	580	280
	3/4	1580	695	1190	525	855	375	690	305
6″	5/16•	445	305	400	270	345	235	305	205
	3/8	630	385	545	330	490	300	430	260
	7/16	850	480	735	415	660	375	545	310
	1/2	1100	570	945	490	770	400	615	320
	5/8	1640	790	1250	600	900	430	720	345
	3/4	1970	865	1480	650	1060	460	850	370
7″	3/8•	645	390	555	340	500	305	440	270
	7/16	865	490	750	425	670	380	590	335
	1/2	1120	580	970	505	865	450	745	385
	5/8	1700	820	1460	700	1020	490	900	430
	3/4	2360	1040	2030	890	1290	570	1040	460
8″	7/16•	875	500	760	430	680	385	600	340
	1/2	1140	590	985	510	880	455	775	400
	5/8	1750	840	1500	720	1325	635	1070	560
	3/4	2475	1090	2130	935	1550	680	1250	555
	7/8	3280	1365	2720	1130	1950	810	1560	715
9″	1/2•	1150	600	990	515	885	460	780	405
	5/8	1770	850	1510	725	1360	650	1200	575
	3/4	2520	1110	2160	950	1780	785	1435	630
	7/8	3350	1390	2880	1200	2060	855	1660	690
10″	5/8•	1800	865	1540	740	1380	660	1220	585
	3/4	2540	1120	2190	965	1970	865	1625	715
	7/8	3420	1420	2960	1230	2340	970	1890	785
	1	4420	1770	3710	1485	2660	1065	2140	855
11″	3/4•	2580	1130	2220	970	2000	880	1765	780
	7/8	3450	1430	2990	1240	2600	1080	2100	870
	1	4500	1800	3880	1550	2970	1190	2370	950
12″	7/8	3470	1440	3000	1250	2690	1120	2320	965
	1	4520	1810	3900	1560	3290	1320	2630	1050
	1-1/8	5660	2260	4900	1960	3570	1430	2870	1150
13″	7/8•	3500	1455	3030	1260	2710	1130	2390	990
	1	4550	1820	3930	1570	3520	1410	2890	1155
	1-1/8	5700	2280	4920	1970	3920	1570	3120	1250
14″	1	4570	1830	3950	1570	3530	1410	3120	1240
	1-1/8	5740	2300	4950	1980	4380	1750	3500	1400
	1-1/4	7020	2800	6060	2420	4830	1930	3910	1560
15″	1	4580	1830	3960	1580	3550	1420	3130	1250
	1-1/8	5770	2310	4980	1990	4460	1790	3820	1530
	1-1/4	7070	2830	6110	2450	5250	2100	4180	1670
16″	1•	4600	1840	3960	1580	3550	1420	3130	1250
	1-1/8•	5800	2320	5000	2000	4470	1790	3950	1580
	1-1/4•	7120	2850	6150	2460	5500	2200	4520	1810

•Greater lengths do not provide higher loads.

TABLE A.6.5*

GROUPING OF SPECIES FOR FASTENING DESIGN[1]

Grouping for timber connector loads		Grouping for lag screw, drift bolt, nail, spike, wood screw and metal plate connector loads		
Connector load group*	Species of wood	Group	Species of wood	Specific gravity** (G)
Group A	Ash, Commercial White Beech Birch, Sweet & Yellow Douglas Fir-Larch (Dense)*** Hickory & Pecan Maple, Black & Sugar Oak, Red & White Southern Pine (Dense)	Group I	Ash, Commercial White Beech Birch, Sweet & Yellow Hickory & Pecan Maple, Black & Sugar Oak, Red & White	0.62 0.68 0.66 0.75 0.66 0.67
Group B	Douglas Fir-Larch*** Southern Pine (Med. grain) Sweetgum & Tupelo	Group II	Douglas Fir - Larch*** Southern Pine Sweetgum & Tupelo	0.51 0.55 0.54
Group C	California Redwood (Close grain) Douglas Fir, South Eastern Hemlock-Tamarack*** Eastern Spruce Hem - Fir*** Lodgepole Pine Mountain Hemlock Northern Aspen Northern Pine Ponderosa Pine**** Ponderosa Pine-Sugar Pine Red Pine**** Sitka Spruce Southern Cypress Spruce-Pine-Fir Western Hemlock Yellow Poplar	Group III	California Redwood (Close grain) Douglas Fir, South Eastern Hemlock-Tamarack*** Eastern Spruce Hem - Fir*** Lodgepole Pine Mountain Hemlock Northern Aspen Northern Pine Ponderosa Pine**** Ponderosa Pine-Sugar Pine Red Pine**** Sitka Spruce Southern Cypress Spruce-Pine-Fir Western Hemlock Yellow Poplar	0.42 0.48 0.46 0.43 0.42 0.44 0.47 0.42 0.46 0.49 0.42 0.42 0.43 0.48 0.42 0.48 0.46
Group D	Aspen Balsam Fir Black Cottonwood California Redwood (Open grain) Coast Sitka Spruce Cottonwood, Eastern Eastern White Pine*** Engelmann Spruce - Alpine Fir Idaho White Pine Northern White Cedar Western Cedars*** Western White Pine	Group IV	Aspen Balsam Fir Black Cottonwood California Redwood (Open grain) Coast Sitka Spruce Coast Species Cottonwood, Eastern Eastern White Pine*** Eastern Woods Engelmann Spruce - Alpine Fir Idaho White Pine Northern Species Northern White Cedar West Coast Woods (Mixed Species) Western Cedars*** Western White Pine White Woods (Western Woods)	0.40 0.38 0.33 0.37 0.39 0.39 0.41 0.38 0.38 0.36 0.40 0.35 0.31 0.35 0.35 0.40 0.35

*When stress graded.

**Based on weight and volume when oven-dry.

***Also applies when species name includes the designation "North".

****Applies when graded to NLGA rules.

Note: Coarse grain Southern Pine, as used in some glued laminated timber combinations, is in Group C.

*From National Design Specification, 1977 by NFPA.

1. For Glulam members see footnote 14 of TABLE A.3.1.

Table A.6.6 Allowable loads for one SPLIT RING and Bolt in single shear†

(The allowable loads below are for normal loading conditions. See other provisions of NDS for adjustments of these tabulated allowable loads)

Split-Ring diam. (inches)	Bolt diam. (inches)	Number of faces of piece with connectors on same bolt	Thickness (net) of lumber (inches)	Loaded parallel to grain (0°): Edge distance min. (inches)	Loaded parallel to grain (0°): Allowable load per connector unit and bolt (pounds)				Loaded perpendicular to grain (90°): Edge distance (inches)		Loaded perpendicular to grain (90°): Allowable load per connector unit and bolt (pounds)			
					Group A woods	Group B woods	Group C woods	Group D woods	Unloaded edge min.	Loaded-edge (See Fig. 12)	Group A woods	Group B woods	Group C woods	Group D woods
2½	½	1	1 min.	1¾	2630	2270	1900	1640	1¾	1¾ min.	1580	1350	1130	970
										2¾ or more	1900	1620	1350	1160
			1½ or more	1¾	3160	2730	2290	1960	1¾	1¾ min.	1900	1620	1350	1160
										2¾ or more	2280	1940	1620	1390
		2	1½ min.	1¾	2430	2100	1760	1510	1¾	1¾ min.	1460	1250	1040	890
										2¾ or more	1750	1500	1250	1070
			2 or more	1¾	3160	2730	2290	1960	1¾	1¾ min.	1900	1620	1350	1160
										2¾ or more	2280	1940	1620	1390
4	¾	1	1 min.	2¾	4090	3510	2920	2520	2¾	2¾ min.	2370	2030	1700	1470
										3¾ or more	2840	2440	2040	1760
			1½ or more	2¾	6020	5160	4280	3710	2¾	2¾ min.	3490	2990	2490	2150
										3¾ or more	4180	3590	2990	2580
		2	1½ min.	2¾	4110	3520	2940	2540	2¾	2¾ min.	2480	2040	1700	1470
										3¾ or more	2980	2450	2040	1760
			2	2¾	4950	4250	3540	3050	2¾	2¾ min.	2870	2470	2050	1770
										3¾ or more	3440	2960	2460	2120
			2½	2¾	5830	5000	4160	3600	2¾	2¾ min.	3380	2900	2410	2080
										3¾ or more	4050	3480	2890	2500
			3 or more	2¾	6140	5260	4380	3790	2¾	2¾ min.	3560	3050	2540	2190
										3¾ or more	4270	3660	3050	2630

†From National Design Specification for Stress-Grade Lumber and Its Fastening, 1977 Edition by NFPA.

Table A.6.7 Allowable loads for one TOOTHED-RING and bolt in single shear†

(The allowable loads below are for normal loading conditions. See other provisions of NDS for adjustments of these tabulated allowable loads)

Toothed-Ring diam. (inches)	Bolt diam. (inches)	Number of faces of piece with connectors on same bolt	Thickness (net) of lumber (inches)	Loaded parallel to grain (0°): Edge distance min. (inches)	Loaded parallel to grain (0°): Allowable load per connector unit and bolt (pounds): Group A woods	Group B woods	Group C woods	Group D woods	Loaded perpendicular to grain (90°): Edge distance (inches): Unloaded-edge min.	Loaded-edge (See Fig. 12)	Loaded perpendicular to grain (90°): Allowable load per connector unit and bolt (pounds): Group A woods	Group B woods	Group C woods	Group D woods
2⅝	⅝	1	1 min.	1¾	1820	1650	1490	1290	1¾	1¾ min.	1210	1100	990	860
										2½ or more	1390	1260	1140	990
			1½ or more	1¾	2270	2030	1850	1610	1¾	1¾ min.	1510	1370	1240	1070
										2½ or more	1730	1570	1430	1230
		2	1½ min.	1¾	1750	1580	1420	1240	1¾	1¾ min.	1160	1050	960	820
										2½ or more	1330	1210	1100	940
			2	1¾	2010	1830	1640	1420	1¾	1¾ min.	1340	1220	1090	950
										2½ or more	1540	1400	1250	1090
			2½ or more	1¾	2270	2030	1850	1610	1¾	1¾ min.	1510	1370	1240	1070
										2½ or more	1730	1570	1430	1230
4	¾	1	1 min.	2¾	2840	2590	2330	2020	2¾	2¾ min.	1900	1720	1550	1340
										3¾ or more	2280	2060	1860	1610
			1½ or more	2¾	3480	3160	2850	2460	2¾	2¾ min.	2310	2110	1900	1640
										3¾ or more	2770	2530	2280	1970
		2	1½ min.	2¾	2780	2520	2270	1960	2¾	2¾ min.	1840	1680	1510	1310
										3¾ or more	2210	2020	1810	1570
			2	2¾	3070	2790	2520	2180	2¾	2¾ min.	2050	1860	1680	1450
										3¾ or more	2460	2240	2020	1740
			2½	2¾	3400	3090	2790	2410	2¾	2¾ min.	2260	2060	1860	1610
										3¾ or more	2710	2470	2230	1930
			3 or more	2¾	3700	3360	3030	2620	2¾	2¾ min.	2460	2240	2020	1750
										3¾ or more	2950	2690	2420	2100

†From National Design Specification for Stress-Grade Lumber and Its Fastening, 1977 Edition by NFPA.

Table A.6.8 Allowable loads for one SHEAR-PLATE unit and bolt in single shear†

(The allowable loads below are for normal loading conditions. See other provisions of NDS for adjustments of these tabulated allowable loads)

Shear-plate diam. (inches)	Bolt diam. (inches)	Number of faces of piece with connectors on same bolt	Thickness (net) of lumber (inches)	Loaded parallel to grain (0°)					Loaded perpendicular to grain (90°)					
				Edge distance min. (inches)	Allowable load per connector unit and bolt (pounds)				Edge distance (inches)		Allowable load p.r connector unit and bolt (pounds)			
					Group A woods	Group B woods	Group C woods	Group D woods	Unloaded-edge min.	Loaded-edge (See Fig. 12)	Group A woods	Group B woods	Group C woods	Group D woods
2⅝	¾	1	1½ min.	1¾	3110[1]	2760	2220	2010	1¾	1¾ min.	1810	1550	1290	1110
										2¾ or more	2170	1860	1550	1330
		2	1½ min.	1¾	2420	2080	1730	1500	1¾	1¾ min.	1410	1210	1010	870
										2¾ or more	1690	1450	1210	1040
			2	1¾	3190[1]	2730	2270	1960	1¾	1¾ min.	1850	1590	1320	1140
										2¾ or more	2220	1910	1580	1370
			2½ or more	1¾	3330[1]	2860	2380	2060	1¾	1¾ min.	1940	1660	1380	1200
										2¾ or more	2320	1990	1650	1440
4	¾ or ⅞	1	1½ min.	2¾	4370	3750	3130	2700	2¾	2¾ min.	2540	2180	1810	1550
										3¾ or more	3040	2620	2170	1860
			1¾ or more	2¾	5090[1]	4360	3640	3140	2¾	2¾ min.	2950	2530	2110	1810
										3¾ or more	3540	3040	2530	2200
		2	1¾ min.	2¾	3390	2910	2420	2090	2¾	2¾ min.	1970	1680	1400	1250
										3¾ or more	2360	2020	1680	1410
			2	2¾	3790	3240	2700	2330	2¾	2¾ min.	2200	1880	1570	1360
										3¾ or more	2640	2260	1880	1630
			2½	2¾	4310	3690	3080	2660	2¾	2¾ min.	2500	2140	1780	1540
										3¾ or more	3000	2550	2140	1850
			3	2¾	4830	4140	3450	2980	2¾	2¾ min.	2800	2400	2000	1720
										3¾ or more	3360	2880	2400	2060
			3½ or more	2¾	5030[1]	4320	3600	3110	2¾	2¾ min.	2920	2500	2090	1800
										3¾ or more	3500	3000	2510	2160

NOTES

1. Loads followed by "1" in the above table, exceed those permitted by Note 3, but are needed for proper determination of loads for other angles of load to grain. Note 3 limitations apply in all cases.

2. For metal side plates, tabulated loads apply except that, for 4" shear plates, the parallel-to-grain (not perpendicular) loads for wood side plates shall be increased 18, 11, 5 and 0 percent for groups A, B, C and D woods, respectively, but loads shall not exceed those permitted by Note 3.

3. The allowable loads for all loadings, except wind, shall not exceed 2900 lbs. for 2⅝" shear plates; 4970 lbs. and 6760 lbs. for 4" shear plates with ¾" and ⅞" bolts, respectively; or, for wind loading, shall not exceed 3870 lbs., 6630 lbs. and 9020 lbs., respectively. If bolt threads are in bearing on the shear plate, reduce the preceding values by one-ninth.

4. Metal side plates, when used, shall be designed in accordance with accepted metal practices. For steel, the following unit stresses, in pounds per square inch, are suggested for all loadings except wind: net section in tension, 20,000; shear, 12,500; double-shear bearing, 28,125; single-shear bearing, 22,500; for wind, these values may be increased one-third; if bolt threads are in bearing, reduce the preceding shear and bearing values by one-ninth.

†From National Design Specification for Stress-Grade Lumber and Its Fastening, 1977 Edition by NFPA.

A.6.9 DESIGN AND LOAD DATA FOR CONNECTORS†

FACTORS AFFECTING DESIGN DATA—SPACINGS, DISTANCES, LOADS

ANGLE OF LOAD TO GRAIN

The angle of load to grain is the angle between the resultant load exerted by the connector acting on the member and the longitudinal axis of the member. See figure #1.

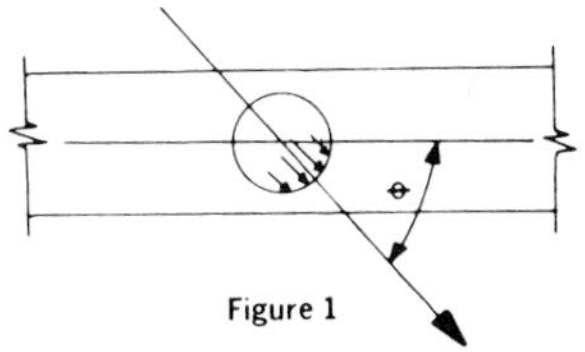

Figure 1

ANGLE OF AXIS TO GRAIN

The angle of axis to grain is the angle of connector axis formed by a line joining the centers of two adjacent connectors located in the same face of a member in a joint and the longitudinal axis of the member. See figure #2.

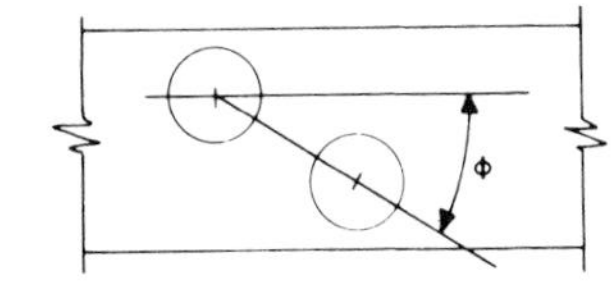

Figure 2

USE OF DATA AND CHARTS

General—The charts in A.6.10.a through A.6.14b have been arranged with two facing pages devoted to each size of each type of the major connectors. On the left hand page are the load chart and size data as well as required increases or decreases for various conditions and on the right hand page are the spacing, end distance and edge distance charts.

Load Charts—The load charts show the allowable normal loads for one connector unit and bolt in single shear. The connector unit consists of one split ring or toothed ring, a pair of shear plates, or a single shear plate used with a steel sideplate.

The charts are broken vertically into the three species groups. Select the group from page 4 according to the species of lumber specified and use the portion of the chart applying to this group. Within each group there are several curves, each representing a thicknesses of lumber and the number of loaded faces. Use the curve conforming to the condition existing in the joint. Each curve is plotted according to the Hankinson Formula with load in pounds and angle of load to grain as the variables. Select the proper angle at bottom or top of chart, proceed vertically to the selected curve and proceed horizontally to read the alloable normal load. Lumber thichnesses less than those shown on the load data charts for the corresponding number of loaded faces are not recommended.

For more than one connector unit, multiply the connector load by the number of units.

Connector Data—The data given cover dimensions of the connectors, minimum lumber sizes, recommended bolt and bolt hole diameters, recommended washer sizes and similar self explanatory information.

A.6.9 (cont.) DESIGN AND LOAD DATA FOR CONNECTORS†

Lag screws of the same diameter as the recommended bolt sizes may be used instead of bolts in accordance with provisions set forth in the "National Design Specification for Stress-Grade Lumber and Its Fastenings." (1)

Connector Specifications—The design data and other information in the publication are based on the use of standard connectors and tools. It is recommended that the specifications be closely followed to insure the use of products for which the data were prepared.

Adjustments for Loading Durations

The allowable connector loads given on the charts are for loadings of normal duration. Normal duration of loading contemplates the joint being fully loaded for approximately ten years, either continuously or cumulatively, by the maximum allowable normal design load shown on the charts and/or that 90% of this load is applied throughout the remainder of the life of the structure. Factors are given with the data on each connector which, applied to the load values in the charts, give the allowable loads for other durations of loading from those of impact character to those applied permanently.

Permanent loading, for which a factor of 90% is given, contemplates the connector will be fully loaded to that percentage of the normal design values on the chart applied permanently or for many years.

Information on design with combinations of loads of different durations is given in National Design Specification for Stress-Grade Lumber and Its Fastenings.

Decreases for Moisture Content Condition—The expected condition of lumber when fabricated and used should be determined and the required decreases if necessary made in accordance with the tabulated data with each connector.

Design of Eccentric Joints and of Beams Supported by Fastenings—Eccentric connector joints and beams supported by connectors shall be designed so that H in the following formula does not exceed the allowable unit stresses in horizontal shear

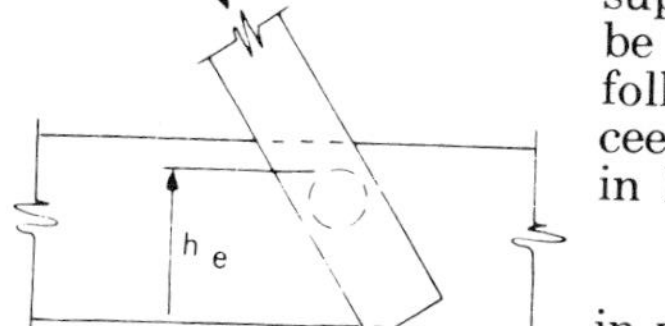

FIGURE 3·

$$H = \frac{V}{bh.}$$

in which

$h.$ (with connectors) = the depth of the member less the distance from the unloaded edge of the member to the nearest edge of the nearest connector.

Maximum Permissible Loads on Shear Plates—Due to the type of test failure, there are limits beyond which shear plates should not be loaded. These limits vary with the bolt size. These limits are given with the shear plate data.

Spacing Charts—Spacing is the distance between centers of connectors measured along a line joining their centers. "R" in figure #4 is the spacing between the rings shown.

On the right hand page of the data for each connector, is the spacing chart. Each chart has five elliptical curves representing recommended spacing for full load at the particular angle of load to grain noted on the curve. For intermediate angles of load, straight line interpolation may be used. If the spacing for full load is desired, select the proper angle of load to grain curve and find where it intersects the radial lines representing angle of axis to grain, the distance from the point to the lower left hand corner is the spacing. It is probably more convenient, however, in laying out this spacing to use the parallel to

A.6.9 (cont.) DESIGN AND LOAD DATA FOR CONNECTORS†

grain and perpendicular to grain components or measurements of the spacing. The parallel to grain component may be read at the bottom of the chart by projecting downward from the point on the curve. The perpendicular component of the spacing may be read at the left hand side of the chart by projecting horizontally from the point on the curve.

The sixth curve on the chart is a quarter-circle. This curve represents the spacing for 50% of full load for any angle of load to grain and also the minimum spacing permissible. For percentages between 50% and 100% of full load for an angle of load to grain interpolate radially on a straight line between the 50% curve and the curve corresponding to the proper angle of load to grain.

Reductions in load for edge distance and end distance are not additive to spacing reductions but are coincident.

End Distance Charts—End distance is the distance measured parallel to grain from the center of a connector to the square cut end of the member. If the end of the member is not square cut, the end distance shall be taken as the distance from any point on the center half of the connector diameter drawn perpendicular to the center line of the piece to the nearest point on the end of the member measured parallel to grain. The distance measured perpendicular to the end cut to the center of the connector shall never be less than the required edge distance. Figure #5 demonstrates end distance measurement (A).

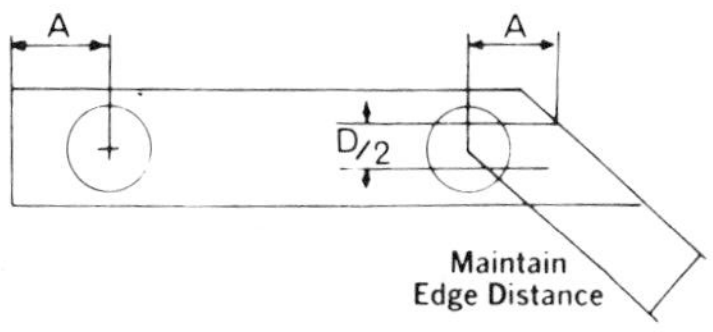

FIGURE 5

On the same page with the spacing chart, the end distance chart will be found. This chart is divided into two sections, depending on whether the member is in tension or compression. If in tension, project vertically on the chart from the end distance to the curve, then horizontally to get the percentage of full load allowable. This process can be reversed of course by going from percent of full load required to spacing required. In compression there is an additional variable of angle of load to grain. The operation is the same except the curve for the proper angle of load to grain should be used. For intermediate angles interpolate between the curves on a straight line. On some of the charts the curves are cut off at the right hand side of the chart. The end distance dimension marking this cut off is the minimum permissible and gives full load for an angle of load to grain of zero degrees.

Reductions in load for edge distance or spacing are not additive to end distance reductions but are coincident.

Edge Distance Charts—Edge distance is the distance from the edge of the member to the center of the connector closest to the edge of the member measured perpendicular to the edge. The loaded edge distance is the edge distance measured from the edge toward which the load induced by the connector acts. The unloaded edge distance is the edge distance measured from the edge away from which the load induced by the connector acts. Figure #4 shows a typical measurement of edge distance with "B" being the unloaded edge distance and "C" the loaded edge distance.

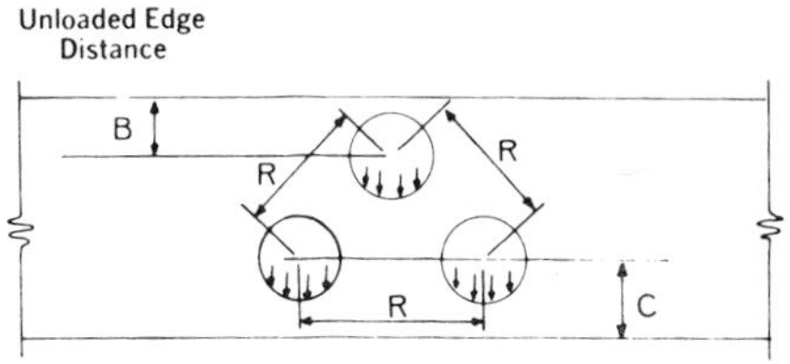

FIGURE 4

A.6.9 DESIGN AND LOAD DATA FOR CONNECTORS†

On the same page with the spacing and end distance charts, the edge distance chart will be found. For unloaded edge distance the standard and minimum edge distance is the dimension at the right hand edge of the chart. For loaded edge distance there is a variation according to angle of load to grain. Select the proper curve for the desired angle and then a given edge distance projected vertically to the curve and then horizontally to the side will give the percentage of full load allowable. The upper right hand corner represents the standard and minimum loaded edge distance for zero degrees angle of load to grain. For intermediate angles, interpolate on a straight line.

Effect on Connector Loads of Grade of Lumber and Species—The various commercial stress grades of lumber are affected by characteristics which have no effect on the strength values of connectors. Therefore, no variation in connector loads is permissible for variations in assigned stresses to lumber.

Connector loads do vary depending on the species of lumber with which they are used. The species have been classified in groups A, B, C, and D. Loads for species not listed may be obtained on request from the Timber Engineering Company.

†From Design Manual for Teco Timber Connector Construction by Timber Engineering Co.

Table A.6.10.a
Design and Load Data for 2½″ Split-Ring Connectors†

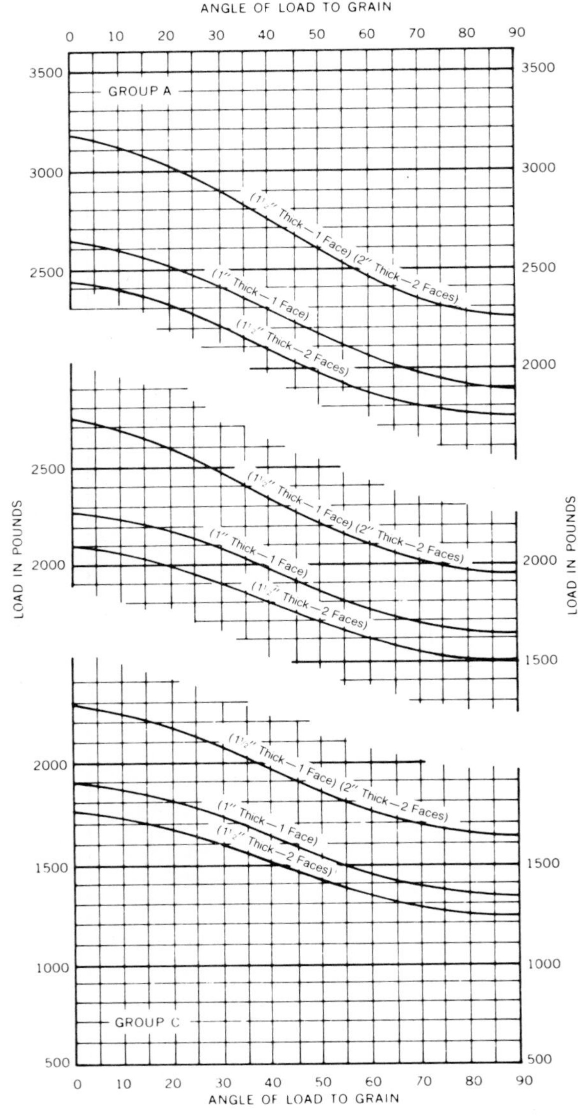

LOAD CHART
FOR NORMAL LOADING
ONE 2½″ SPLIT RING AND BOLT IN SINGLE SHEAR

2½″ SPLIT RING DATA

Split Ring—Dimensions	
Inside Diameter at center when closed	2½″
Inside diameter at center when installed	2.54″
Thickness of ring at center	0.163″
Thickness of ring at edge	0.123″
Depth	¾″
Lumber, Minimum dimensions allowed	
Width	3½″
Thickness, rings in one face	1″
Thickness, rings opposite in both faces	1½″
Bolt, diameter	½″
Bolt hole, diameter	9/16″
Projected Area for portion of one ring within a member, square inches	1.10
Washer, minimum	
Round, Cast or Malleable Iron, diameter	2⅛″
Square Plate	
Length of Side	2″
Thickness	⅛″
(For trussed rafters and similar light construction standard wrought washers may be used.)	

PERCENTAGES FOR DURATION OF MAXIMUM LOAD

Two Months Loading, as for snow	115%
Seven Days Loading	125%
Wind or Earthquake Loading	133⅓%
Impact Loading	200%
Permanent Loading	90%

DECREASES FOR MOISTURE CONTENT CONDITIONS

Condition when Fabricated	Seasoned	Unseasoned	Unseasoned
Condition when Used	Seasoned	Seasoned	Unseasoned or Wet
Shear Plates	0%	20%	33%

†Courtesy Timber Engineering Company

Table A.6.10.b Design and Load Data for 2½″ Split-Ring Connectors†

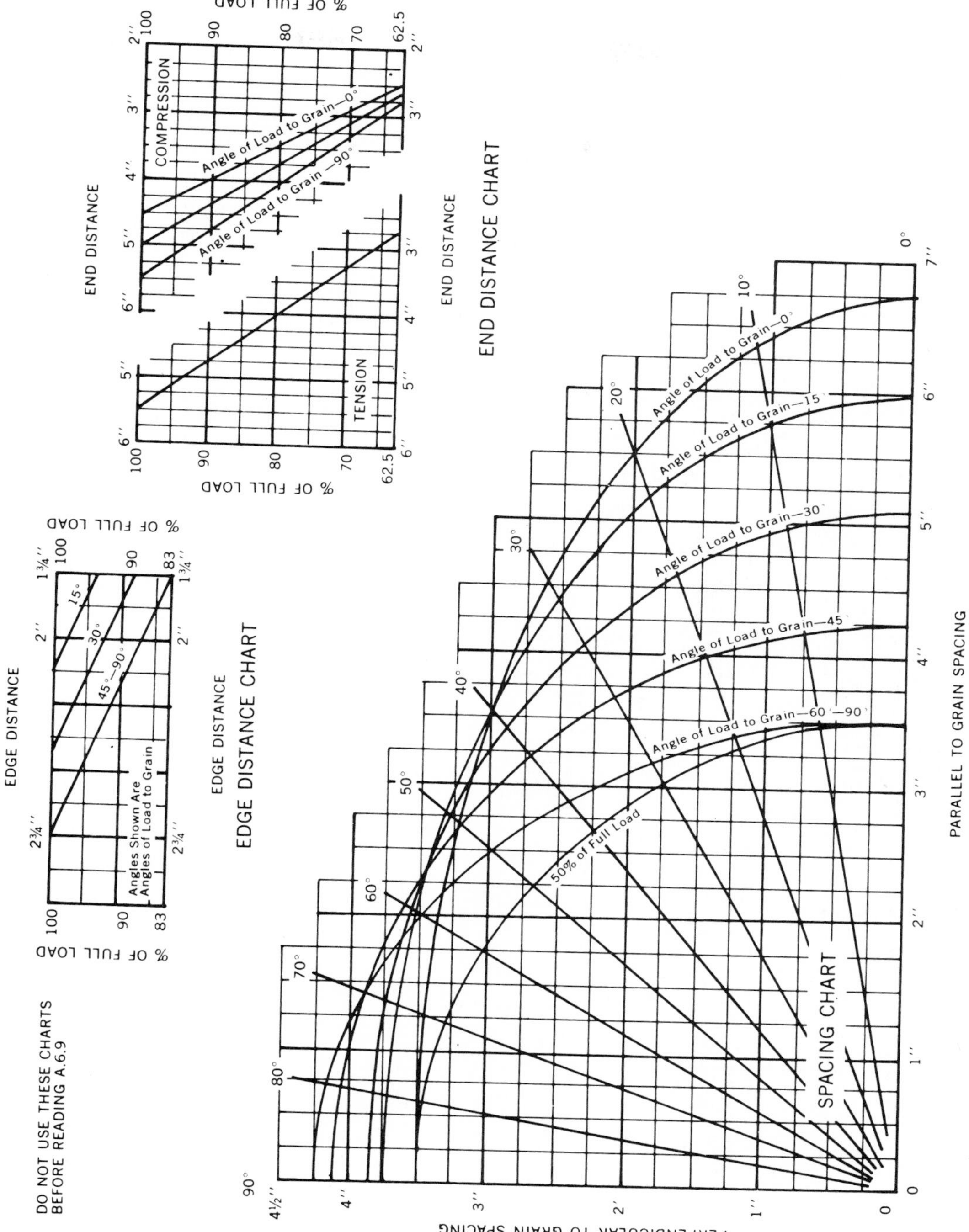

Table A.6.11.a Design and Load Data for 4″ Split-Ring Connectors†

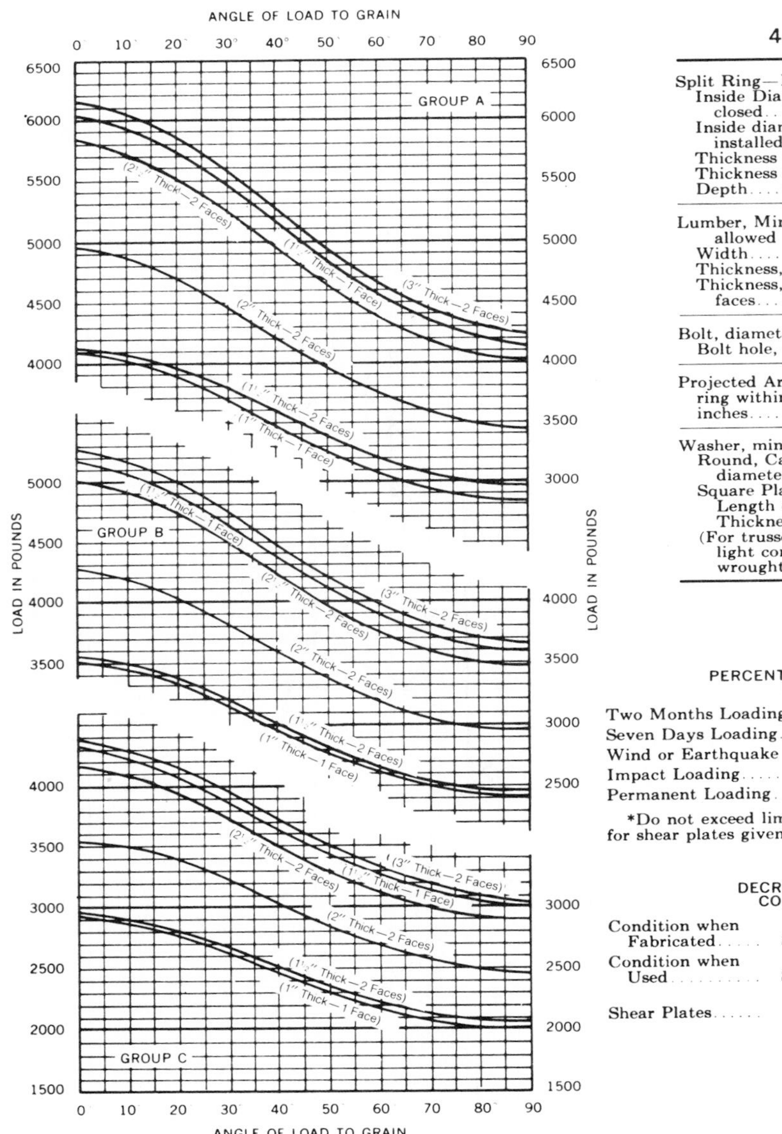

LOAD CHART
FOR NORMAL LOADING
ONE 4″ SPLIT RING AND BOLT IN SINGLE SHEAR

4″ SPLIT RING DATA

Split Ring—Dimensions	
Inside Diameter at center when closed	4″
Inside diameter at center when installed	4.06″
Thickness of ring at center	0.193″
Thickness of ring at edge	0.133″
Depth	1″
Lumber, Minimum dimensions allowed	
Width	5½″
Thickness, rings in one face	1″
Thickness, rings opposite in both faces	1½″
Bolt, diameter	¾″
Bolt hole, diameter	13/16″
Projected Area for portion of one ring within a member, square inches	2.24
Washer, minimum	
Round, Cast or Malleable Iron, diameter	3″
Square Plate	
Length of Side	3″
Thickness	3/16″
(For trussed rafters and similar light construction standard wrought washers may be used.)	

PERCENTAGES FOR DURATION OF MAXIMUM LOAD

Two Months Loading, as for snow	*115%
Seven Days Loading	*125%
Wind or Earthquake Loading	*133⅓%
Impact Loading	*200%
Permanent Loading	90%

*Do not exceed limitations for maximum allowable loads for shear plates given elsewhere on this page.

DECREASES FOR MOISTURE CONTENT CONDITIONS

Condition when Fabricated	Seasoned	Unseasoned	Unseasoned
Condition when Used	Seasoned	Seasoned	Unseasoned or Wet
Shear Plates	0%	20%	33%

†Courtesy Timber Engineering Company

Table A.6.11.b Design and Load Data for 4″ Split-Ring Connectors

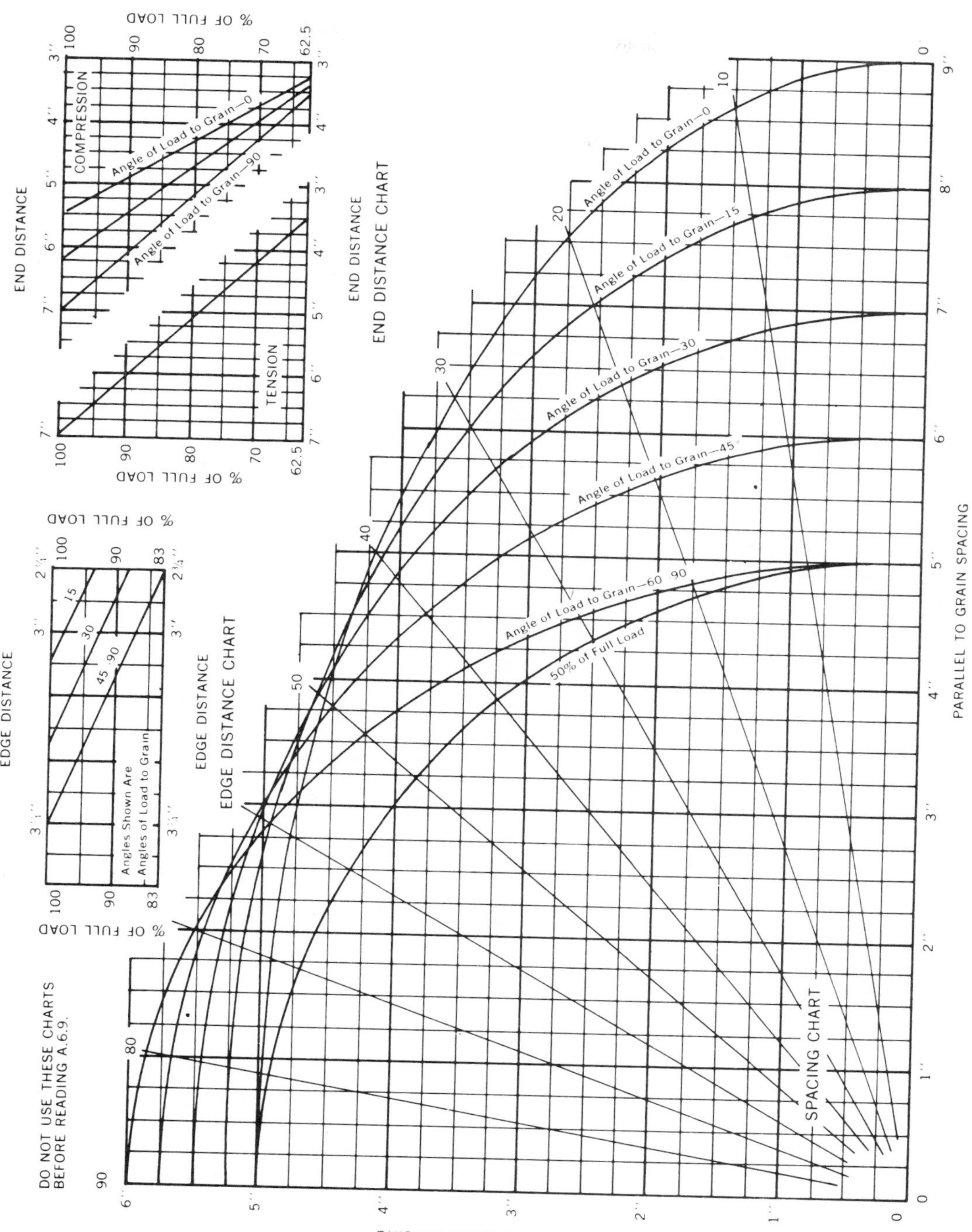

Table A.6.12.a

Design and Load Data for 2⅝″ Shear-Plate Connectors†

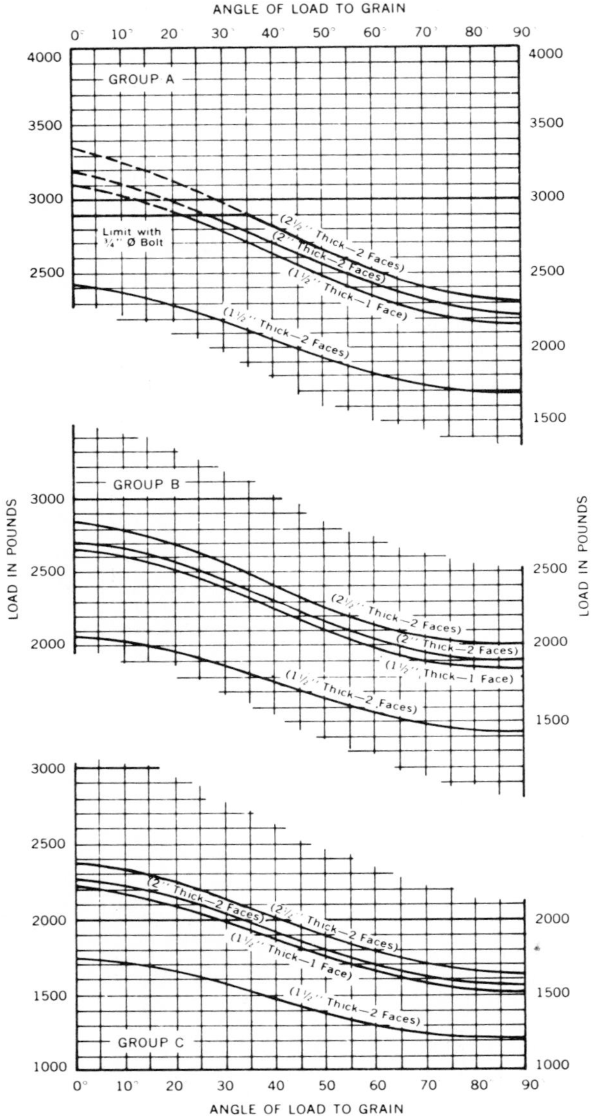

LOAD CHART
FOR NORMAL LOADING
ONE 2⅝″ SHEAR-PLATE UNIT AND BOLT IN SINGLE SHEAR

2⅝″ SHEAR PLATE DATA

	Pressed Steel Reg.	Pressed Steel Lt. Ga.
Shear Plates, Dimensions		
Material		
Diameter of plate	2.62″	2.62″
Diameter of bolt hole	.81″	.81″
Depth of plate	.42″	.35″
Lumber, minimum dimensions		
Face, width	3½″	3½″
Thickness, plates in one face only	1½″	1½″
Thickness, plates opposite in both faces	1½″	1½″
Steel Shapes or Straps (Thickness required when used with shear plates) Thickness of steel side plates shall be determined in accordance with A.I.S.C. recommendations.		
Hole, diameter in steel straps or shapes	13/16″	13/16″
Bolt, diameter	3/4″	3/4″
Bolt Hole, diameter in timber	13/16″	13/16″
Washers, standard, timber or timber connections only		
Round, cast or malleable iron, diameter	3″	3″
Square Plate		
Length of side	3″	3″
Thickness	1/4″	1/4″
(For trussed rafters and other light structures standard wrought washers may be used.)		
Projected Area, for one shear plate, square inches	1.18	1.00

PERCENTAGES FOR DURATION OF MAXIMUM LOAD

Two Months Loading, as for snow	*115%
Seven Days Loading	*125%
Wind or Earthquake Loading	*133⅓%
Impact Loading	*200%
Permanent Loading	90%

*Do not exceed limitations for maximum allowable loads for shear plates given elsewhere on this page.

DECREASES FOR MOISTURE CONTENT CONDITIONS

Condition when Fabricated	Seasoned	Unseasoned	Unseasoned
Condition when Used	Seasoned	Seasoned	Unseasoned or Wet
Shear Plates	0%	20%	33%

MAXIMUM PERMISSIBLE LOADS ON SHEAR PLATES

The allowable loads for all loadings except wind shall not exceed 2900 lbs. for 2⅝″ shear plates with ¾″ bolts. The allowable wind load shall not exceed 3870 lbs. If bolt threads bear on the shear plate, reduce the preceding values by one-ninth.

†Courtesy Timber Engineering Company

Table A.6.12.b Design and Load Data for $2\frac{5}{8}''$ Shear-Plate Connectors

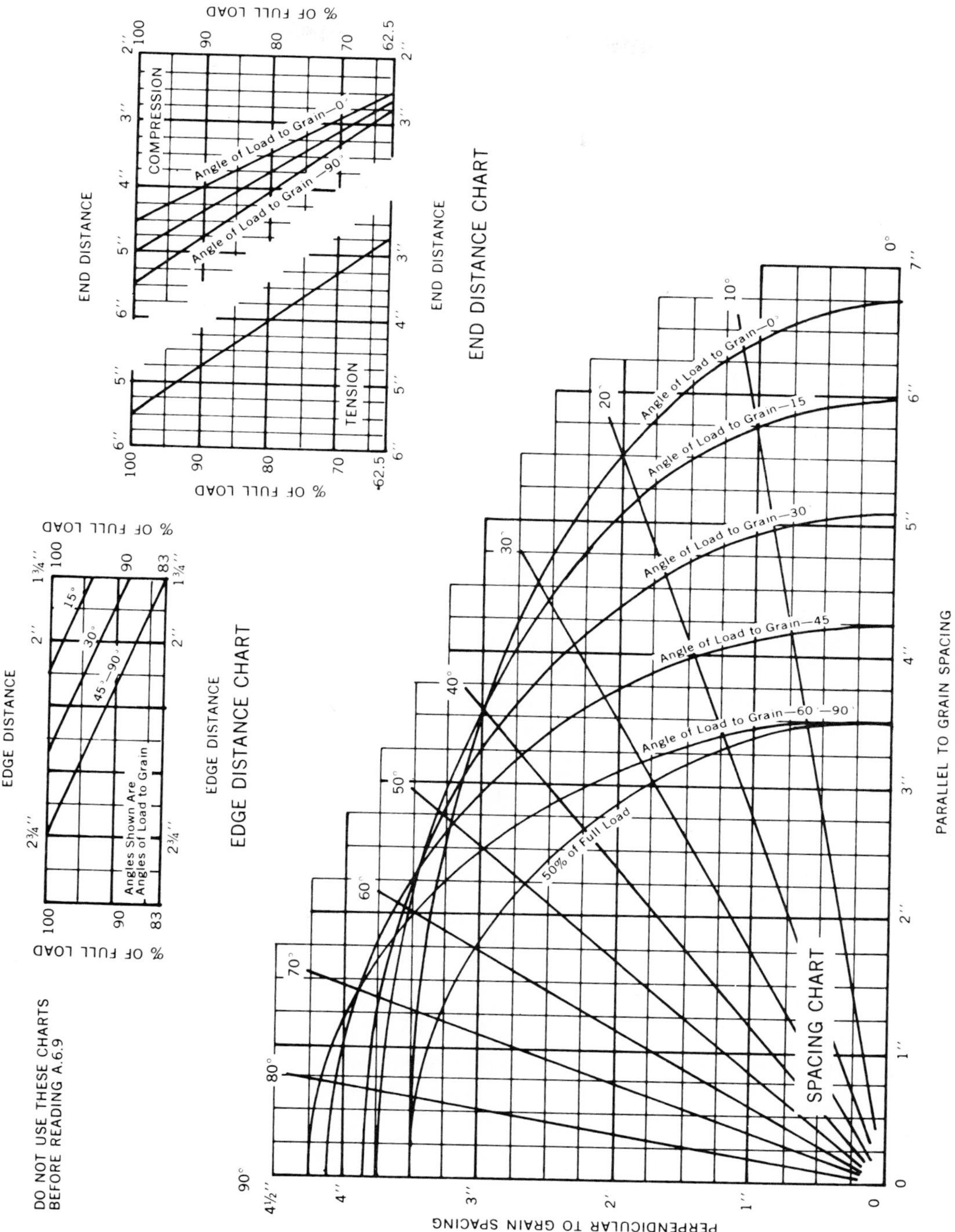

Table A.6.13.a
Design and Load Data for 4″ Shear Plate Connectors †
(Wood to Wood)

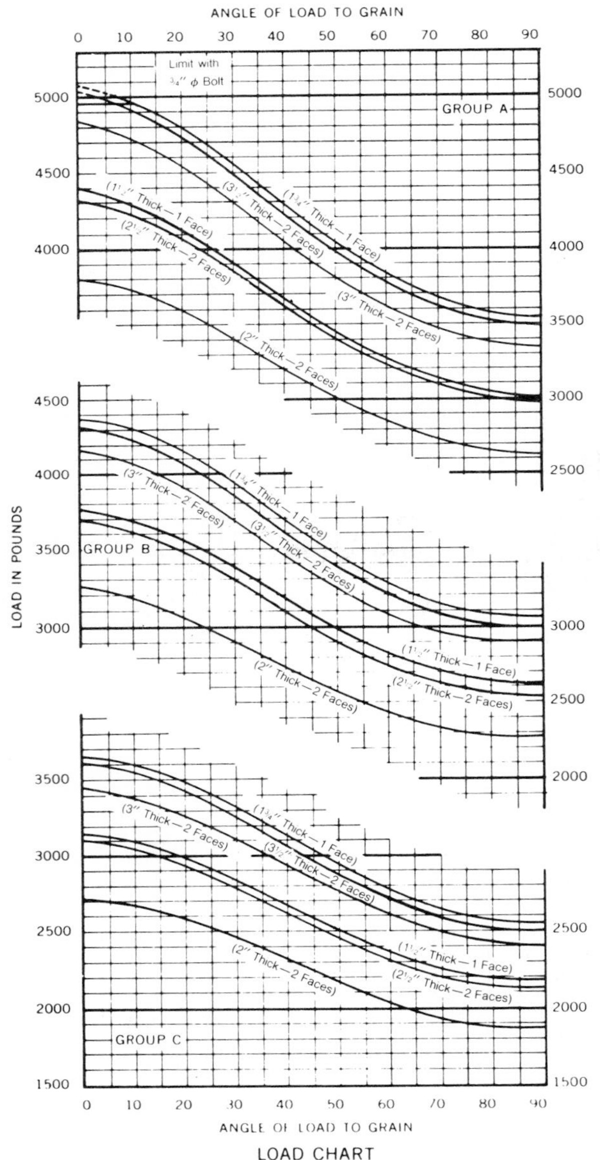

LOAD CHART
FOR NORMAL LOADING
ONE 4″ SHEAR-PLATE UNIT AND BOLT IN SINGLE SHEAR

4″ SHEAR PLATE DATA

Shear Plates, Dimensions		
Material	Malleable Iron	Malleable Iron
Diameter of plate	4.03″	4.03″
Diameter of bolt hole	.81″	.94″
Depth of plate	.64″	.64″
Lumber, minimum dimensions		
Face, width	5½″	5½″
Thickness, plates in one face only	1½″	1½″
Bolt, diameter	3/4″	7/8″
Bolt Hole, diameter in timber	13/16″	15/16″
Washers, standard, timber or timber connections only		
Round, cast or malleable iron, diameter	3″	3½″
Square Plate		
Length of side	3″	3″
Thickness	1/4″	1/4″
(For trussed rafters and other light structures standard wrought washers may be used.)		
Projected Area, for one shear plate, square inches	2.58	2.58

PERCENTAGES FOR DURATION OF MAXIMUM LOAD

Two Months Loading, as for snow	*115%
Seven Days Loading	*125%
Wind or Earthquake Loading	*133⅓%
Impact Loading	*200%
Permanent Loading	90%

*Do not exceed limitations for maximum allowable loads for shear plates given elsewhere on this page.

DECREASES FOR MOISTURE CONTENT CONDITIONS

Condition when Fabricated	Seasoned	Unseasoned	Unseasoned
Condition when Used	Seasoned	Seasoned	Unseasoned or Wet
Shear Plates	0%	20%	33%

MAXIMUM PERMISSIBLE LOADS ON SHEAR PLATES

The allowable loads for all loadings except wind shall not exceed 4970 lbs. for 4″ shear plates with ¾″ bolts and 6760 lbs. for 4″ shear plates with ⅞″ bolts. The allowable wind loads shall not exceed 6630 lbs. when used with a ¾″ bolt and 9020 lbs. when used with a ⅞″ bolt. If bolt threads bear on the shear plate, reduce the preceding value by one-ninth.

†Courtesy Timber Engineering Company

Table A.6.13.b Design and Load Data for 4″ Shear Plate Connectors (Wood to Wood)

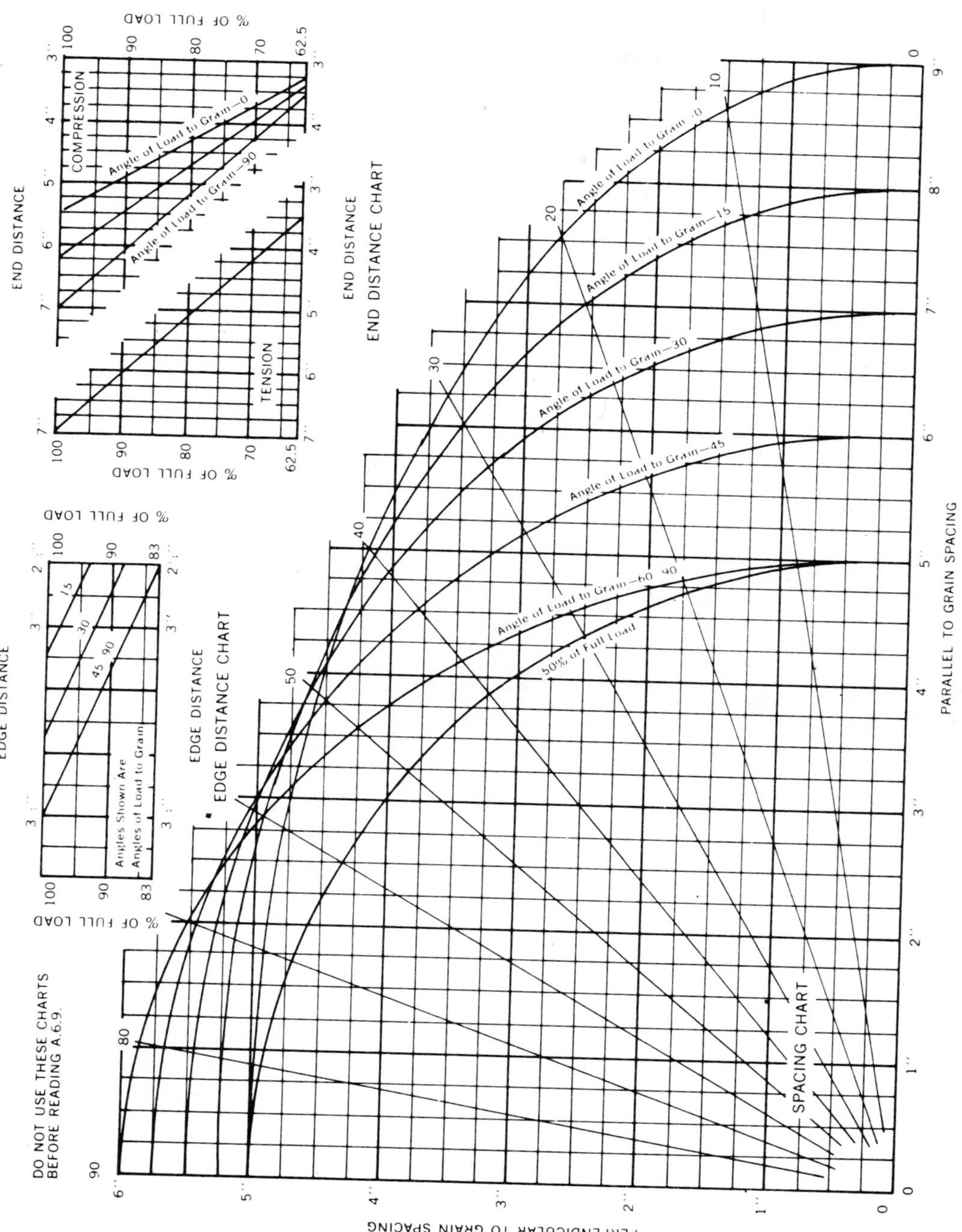

Table A.6.14.a
Design and Load Data for 4″ Shear-Plate Connectors †
(Wood to Steel)

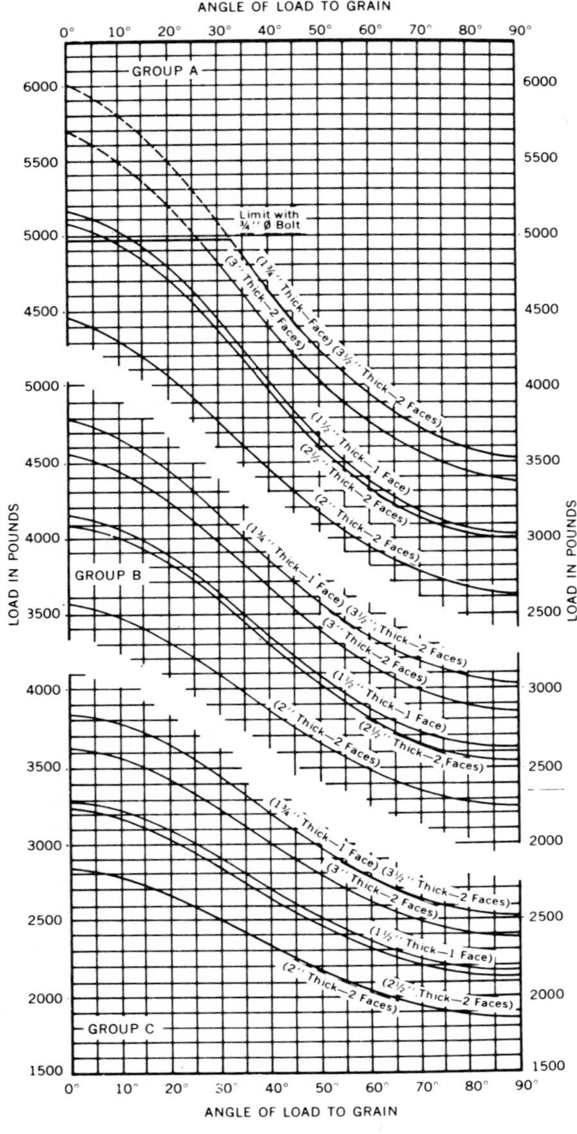

LOAD CHART
FOR NORMAL LOADING
ONE 4″ SHEAR-PLATE UNIT AND BOLT IN SINGLE SHEAR

4″ SHEAR PLATE DATA

Shear Plates, Dimensions		
Material	Malleable Iron	Malleable Iron
Diameter of plate	4.03″	4.03″
Diameter of bolt hole	.81″	.94″
Depth of plate	.64″	.64″
Lumber, minimum dimensions		
Face, width	5½″	5½″
Thickness, plates in one face only	1½″	1½″
Steel Shapes or Straps (Thickness required when used with shear plates) Thickness of steel side plates shall be determined in accordance with A.I.S.C. recommendations.		
Hole, diameter in steel straps or shapes	13/16″	15/16″
Bolt, diameter	3/4″	7/8″
Bolt Hole, diameter in timber	13/16″	15/16″
Projected Area, for one shear plate, square inches	2.58	2.58

PERCENTAGES FOR DURATION OF MAXIMUM LOAD

Two Months Loading, as for snow	*115%
Seven Days Loading	*125%
Wind or Earthquake Loading	*133⅓%
Impact Loading	*200%
Permanent Loading	90%

*Do not exceed limitations for maximum allowable loads for shear plates given elsewhere on this page.

DECREASES FOR MOISTURE CONTENT CONDITIONS

Condition when Fabricated	Seasoned	Unseasoned	Unseasoned
Condition when Used	Seasoned	Seasoned	Unseasoned or Wet
Shear Plates	0%	20%	33%

MAXIMUM PERMISSIBLE LOADS ON SHEAR PLATES

The allowable loads for all loadings except wind shall not exceed 4970 lbs. for 4″ shear plates with ¾″ bolts and 6760 lbs. for 4″ shear plates with ⅞″ bolts. The allowable wind loads shall not exceed 6630 lbs. when used with a ¾″ bolt and 9020 lbs. when used with a ⅞″ bolt. If bolt threads bear on the shear plate, reduce the preceding value by one-ninth.

†Courtesy Timber Engineering Company

Table A.6.14.b Design and Load Data for 4″ Shear-Plate Connectors (Wood to Steel)

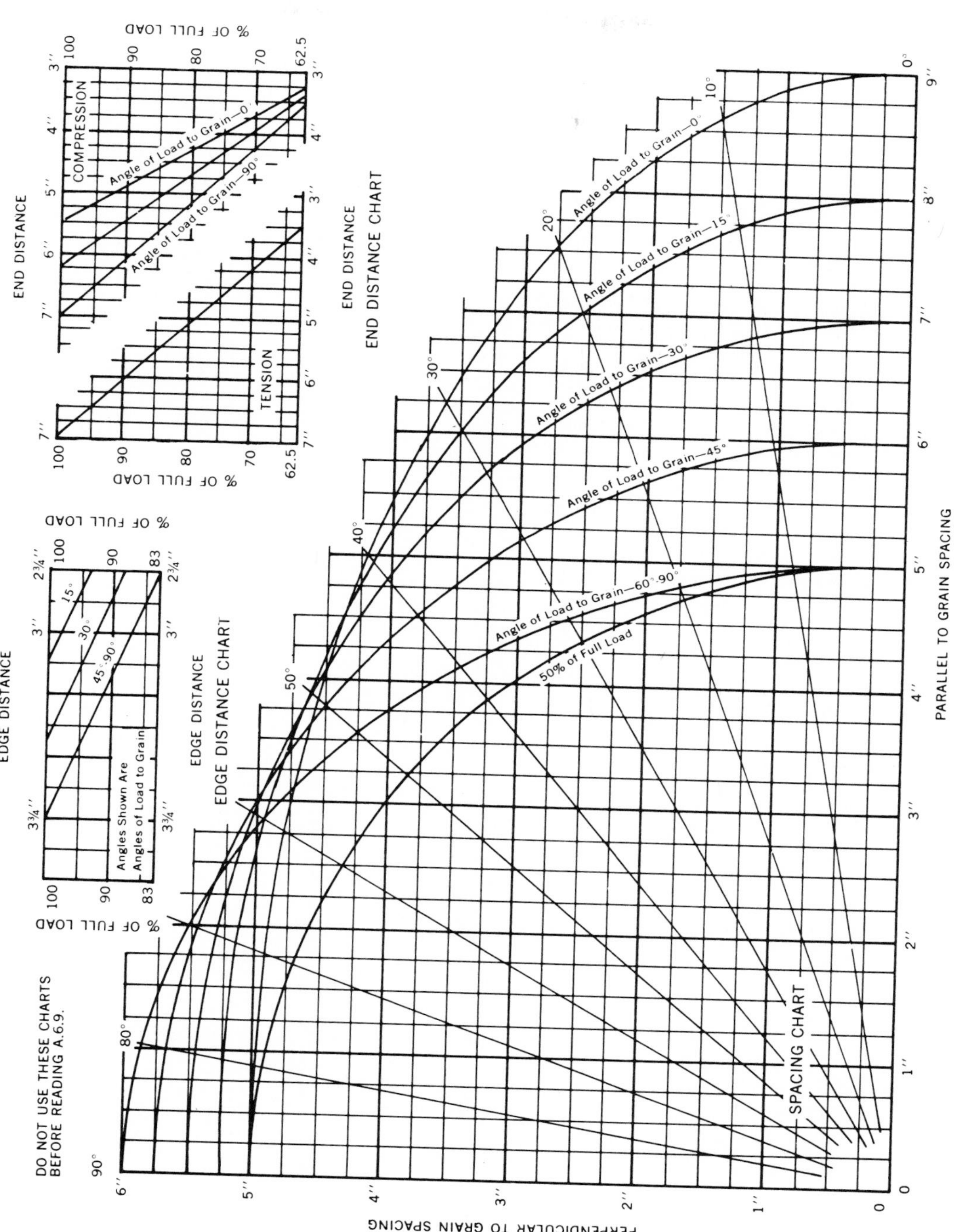

Table A.6.15 Projected Area of Connectors and Bolts†

(For Use In Determining Net Sections)

Connectors		Bolt Diam.	Placement of Connectors	Total Projected Area in Square Inches of Connectors & Bolts in Lumber Thickness of				
No.	Size			1½″	2½″	3½″	5½″	7½″
SPLIT RINGS 1	2½	½ ½	One Face Two Faces	1.71 2.60	2.27 3.16	2.84 3.73	3.89 4.78	5.02 5.91
2	4	¾ ¾	One Face Two Faces	3.01 4.85	3.86 5.66	4.64 6.47	6.16 8.00	7.79 9.62
SHEAR PLATES 1	2⅝	¾ ¾	One Face Two Faces	2.00 2.81	2.81 2.68	3.62 4.43	4.14 5.96	6.77 7.58
1	2⅝ LG	¾ ¾	One Face Two Faces	1.87 2.56	2.68 3.38	3.50 4.19	5.02 5.71	6.65 7.34
2	4	¾ ¾	One Face Two Faces	3.24 —	4.05 6.11	4.87 6.93	6.39 8.45	8.01 10.07
2-A	4	⅞ ⅞	One Face Two Faces	3.33 —	4.27 6.25	5.21 7.19	6.97 8.95	8.84 10.82

†Courtesy Timber Engineering Company

TABLE A.6.16*

Connectors in End Grain

8.4.15.1 When connectors are installed in a surface that is not parallel to the general direction of the grain of the member, such as the end of a square-cut member, or the sloping surface of a member cut at an angle to its axis, or the surface of a glued laminated timber cut at an angle to the direction of the laminations, design values shall be determined in accordance with 8.4.15.2 to 8.4.15.4. (See Figures 8.4F through 8.4L.)

8.4.15.2 The following definitions and notations apply to 8.4.15.1 to 8.5.15.4.

Side-grain surface means a surface parallel to the general direction of the wood fibers ($\alpha = 0°$), such as the top, bottom and sides of a straight member.

Sloping surface means a surface cut at an angle, α, other than 0° or 90° to the general direction of the wood fibers.

Square-cut surface means a surface perpendicular to the general direction of the wood fibers ($\alpha = 90°$).

Axis of cut defines the direction of a sloping surface relative to the general direction of the wood fibers. For a sloping cut symmetrical about one of the major axes of the member, as in Figures 8.4F, 8.4J, 8.4K and 8.4L, the axis of cut is parallel to a major axis. For an asymmetrical sloping surface (i.e. one that slopes relative to both major axes of the member), the axis of cut is the direction of a line defining the intersection of the sloping surface with any plane that is both normal to the sloping surface and also is aligned with the general direction of the wood fibers. (See Figures 8.4G and 8.4F.)

α = the least angle formed between a sloping surface and the general direction of the wood fibers (i.e., the acute angle between the axis of cut and the general direction of the fibers. Sometimes called the slope of the cut.) (See Figures 8.4F through 8.4L.)

θ = the angle between the direction of applied load and the axis of cut of a sloping surface, measured in the plane of the sloping surface. (See Figure 8.4L.)

P = design value for a connector unit in a side-grain surface, when loaded in a parallel-to-grain direction ($\alpha = 0°$, $\theta = 0°$).

Q = design value for a connector unit in a side-grain surface, when loaded in a perpendicular-to-grain direction ($\alpha = 0°$, $\theta = 90°$).

Q_{90} = design value for a connector unit in a square-cut surface, when loaded in any direction in the plane of the surface ($\alpha = 90°$).

$P\alpha$ = design value for a connector unit in a sloping surface, when loaded in a direction parallel to the axis of cut ($0° < \alpha < 90°$, $\theta = 0°$).

$Q\alpha$ = design value for a connector unit in a sloping surface, when loaded in a direction perpendicular to the axis of cut ($0° < \alpha < 90°$, $\theta = 90°$).

$N\alpha$ = design value for a connector unit in a sloping surface, when direction of load is at an angle θ from the axis of cut.

8.4.15.3 For connectors installed in square-cut or sloping surfaces, design values shall be determined from the following applications of the Hankinson formula:

(a) Square-cut surface; loaded in any direction ($\alpha = 90°$). (See Figure 8.4H.)

$$Q_{90} = 0.60\ Q$$

(b) Sloping surface, loaded parallel to axis of cut ($0° < \alpha < 90°$, $\theta = 0°$). See Figure 8.4J.)

$$P\alpha = \frac{PQ_{90}}{P \sin^2 \alpha + Q_{90} \cos^2 \alpha}$$

(c) Sloping surface; loaded perpendicular to axis of cut ($0° < \alpha < 90°$, $\theta = 90°$). (See Figure 8.4K.)

$$Q\alpha = \frac{Q\ Q_{90}}{Q \sin^2 \alpha + Q_{90} \cos^2 \alpha}$$

(d) Sloping surface; loaded at angle θ to axis of cut ($0° < \alpha < 90°$, $0° < \theta < 90°$). (See Figure 8.4L.)

$$N\alpha = \frac{P\alpha\ Q\alpha}{P\alpha \sin^2 \theta + Q\alpha \sin^2 \theta}$$

*From National Design Specification, 1977 by NFPA.

TABLE A.6.16—*CONTINUED*

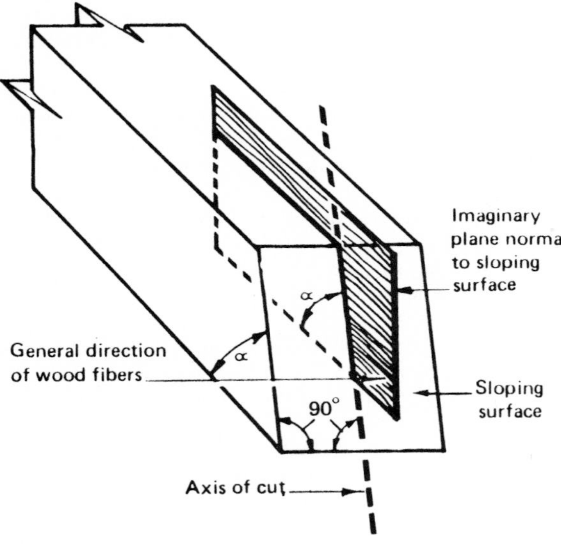

Figure 8.4F. Axis of cut for symetrical sloping end; cut at angle α.

8.4.15.4 The provisions for edge distance, end distance and spacing given in 8.4.12, 8.4.13 and 8.4.14, for connectors in side-grain surfaces, shall apply to connectors in square-cut surfaces and sloping surfaces, as follows:

(a) Square-cut surface, loaded in any direction—apply provisions for perpendicular-to-grain loading.

(b) Sloping surface with α from 45° to 90°, loaded in any direction apply provisions for perpendicular-to-grain loading.

(c) Sloping surface with α less than 45°, loaded parallel to axis of cut—apply provisions for parallel-to-grain loading.

(d) Sloping surface with α less than 45°, loaded perpendicular to axis to cut—apply provisions for perpendicular-to-grain loading.

(e) Sloping surface with α less than 45°, loaded at angle θ to axis of cut—apply provisions for members loaded at angles to grain other than 0° and 90°.

8.4.15.5 The provisions of 3.4.7 shall apply to the design of connector joints in end-grain surfaces.

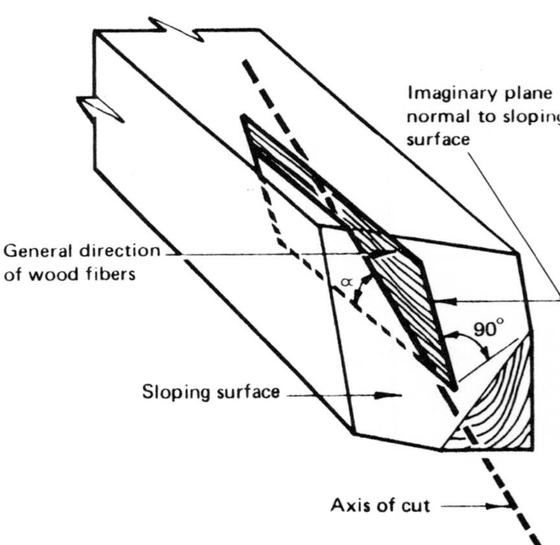

Figure 8.4G. Axis of cut for asymetrical compound sloped end cut.

TABLE A.6.16—*CONTINUED*

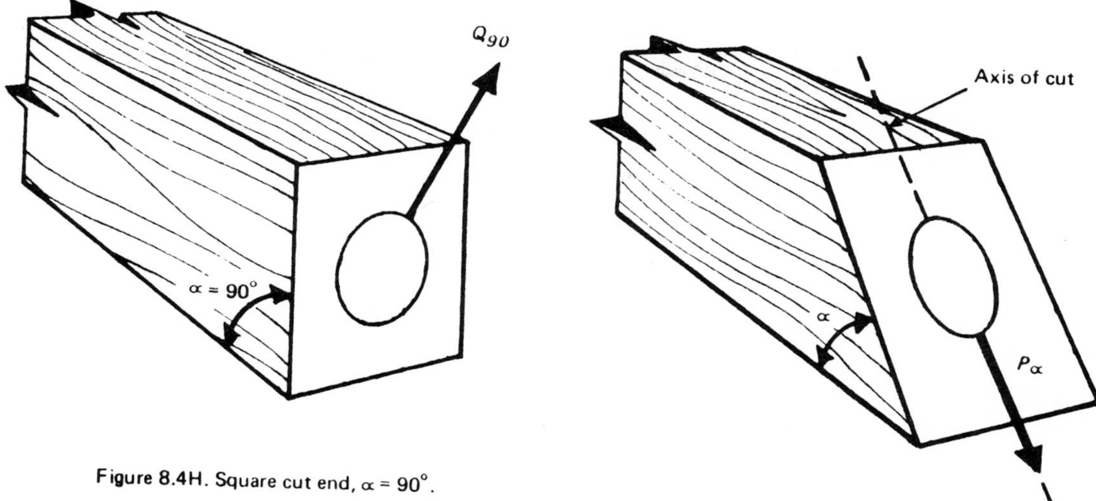

Figure 8.4H. Square cut end, α = 90°.

Figure 8.4J. Load parallel to axis of cut; θ = 0°.

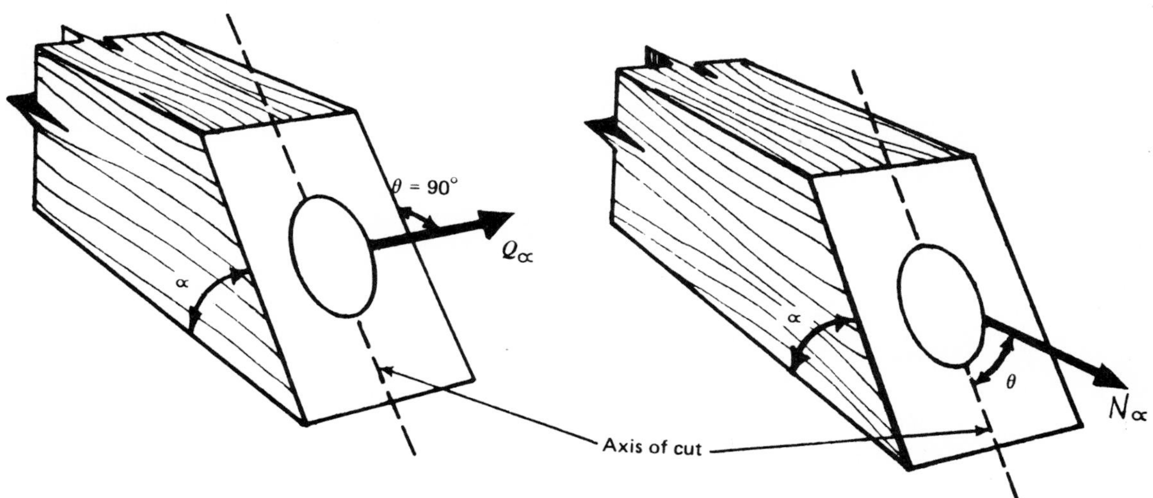

Figure 8.4K. Load perpendicular to axis of cut; θ = 90°.

Figure 8.4L. Load at angle θ to axis of cut.

TABLE A.6.17*

NAIL AND SPIKE SIZES

Pennyweight	Length Inches	Wire diameter, inches			
		Box nails	Common wire nails	Threaded hardened-steel nails	Common wire spikes
6*d*	2	0.099	0.113	0.120	–
8*d*	2½	0.113	0.131	0.120	–
10*d*	3	0.128	0.148	0.135	0.192
12*d*	3¼	0.128	0.148	0.135	0.192
16*d*	3½	0.135	0.162	0.148	0.207
20*d*	4	0.148	0.192	0.177	0.225
30*d*	4½	0.148	0.207	0.177	0.244
40*d*	5	0.162	0.225	0.177	0.263
50*d*	5½	–	0.244	0.177	0.283
60*d*	6	–	0.263	0.177	0.283
70*d*	7	–	–	0.207	–
80*d*	8	–	–	0.207	–
90*d*	9	–	–	0.207	–
5/16	7	–	–	–	0.312
3/8	8½	–	–	–	0.375

*From National Design Specification, 1977 by NFPA.

TABLE A.6.18. NAILS AND SPIKES—WITHDRAWAL DESIGN VALUES.*

Normal load duration

Design values in withdrawal in pounds per inch of penetration into side grain of member holding point.

d = pennyweight of nail or spike. *G* = specific gravity of the wood, based on weight and volume when oven-dry.

Specific gravity *G*	Penny wt. / Diam.	Size of common nail 6*d* 0.113	8*d* 0.131	10*d* 0.148	12*d* 0.148	16*d* 0.162	20*d* 0.192	30*d* 0.207	40*d* 0.225	50*d* 0.244	60*d* 0.263	Size of threaded nail* 30*d* 0.177	40*d* 0.177	50*d* 0.177	60*d* 0.177	70*d* 0.207	80*d* 0.207	90*d* 0.207
0.75		76	88	99	99	109	129	139	151	164	177	129	129	129	129	151	151	151
0.68		59	69	78	78	85	101	109	118	128	138	101	101	101	101	118	118	118
0.67		57	66	75	75	82	97	105	114	124	133	97	97	97	97	114	114	114
0.66		55	64	72	72	79	94	101	110	119	128	94	94	94	94	110	110	110
0.62		47	55	62	62	68	80	86	94	102	110	80	80	80	80	94	94	94
0.55		35	41	46	46	50	59	64	70	76	81	59	59	59	59	70	70	70
0.54		33	39	44	44	48	57	61	67	72	78	57	57	57	57	67	67	67
0.51		29	34	38	38	42	49	53	58	63	67	49	49	49	49	58	58	58
0.49		26	30	34	34	38	45	48	52	57	61	45	45	45	45	52	52	52
0.48		25	29	33	33	36	42	46	50	54	58	42	42	42	42	50	50	50
0.47		24	27	31	31	34	40	43	47	51	55	40	40	40	40	47	47	47
0.46		22	26	29	29	32	38	41	45	48	52	38	38	38	38	45	45	45
0.45		21	25	28	28	30	36	39	42	46	49	36	36	36	36	42	42	42
0.44		20	23	26	26	29	34	37	40	43	47	34	34	34	34	40	40	40
0.43		19	22	25	25	27	32	35	38	41	44	32	32	32	32	38	38	38
0.42		18	21	23	23	26	30	33	35	38	41	30	30	30	30	35	35	35
0.41		17	19	22	22	24	29	31	33	36	39	29	29	29	29	33	33	33
0.40		16	18	21	21	23	27	29	31	34	37	27	27	27	27	31	31	31
0.39		15	17	19	19	21	25	27	29	32	34	25	25	25	25	29	29	29
0.38		14	16	18	18	20	24	25	28	30	32	24	24	24	24	28	28	28
0.37		13	15	17	17	19	22	24	26	28	30	22	22	22	22	26	26	26
0.36		12	14	16	16	17	21	22	24	26	28	21	21	21	21	24	24	24
0.35		11	13	15	15	16	19	21	23	24	26	19	19	19	19	23	23	23
0.33		10	11	13	13	14	17	18	19	21	23	17	17	17	17	19	19	19
0.31		8	10	11	11	12	14	15	17	18	19	14	14	14	14	17	17	17

*Loads for threaded, hardened steel nails, in 6*d* to 20*d* sizes, are the same as for common nails.

Specific gravity *G*	Penny wt. / Diam.	Size of box nail 6*d* 0.099	8*d* 0.113	10*d* 0.128	12*d* 0.128	16*d* 0.135	20*d* 0.148	30*d* 0.148	40*d* 0.162	Size of common spike 10*d* 0.192	12*d* 0.192	16*d* 0.207	20*d* 0.225	30*d* 0.244	40*d* 0.263	50*d* 0.283	60*d* 0.283	5/16" 0.312	3/8" 0.375
0.75		67	78	86	86	91	99	99	109	129	129	139	151	164	177	190	190	210	252
0.68		52	59	67	67	71	78	78	85	101	101	109	118	128	138	149	149	164	197
0.67		50	57	65	65	68	75	75	82	97	97	105	114	124	133	144	144	158	190
0.66		48	55	63	63	66	72	72	79	94	94	101	110	119	128	138	138	152	183
0.62		41	47	53	53	56	62	62	68	80	80	86	94	102	110	118	118	130	157
0.55		31	35	40	40	42	46	46	50	59	59	64	70	76	81	88	88	97	116
0.54		29	33	38	38	40	44	44	48	57	57	61	67	72	78	84	84	92	111
0.51		25	29	33	33	35	38	38	42	49	49	53	58	63	67	73	73	80	96
0.49		23	26	30	30	31	34	34	38	45	45	48	52	57	61	66	66	72	87
0.48		22	25	28	28	30	33	33	36	42	42	46	50	54	58	62	62	69	83
0.47		21	24	27	27	28	31	31	34	40	40	43	47	51	55	59	59	65	78
0.46		20	22	25	25	27	29	29	32	38	38	41	45	48	52	56	56	62	74
0.45		19	21	24	24	25	28	28	30	36	36	39	42	46	49	53	53	58	70
0.44		18	20	23	23	24	26	26	29	34	34	37	40	43	47	50	50	55	66
0.43		17	19	21	21	23	25	25	27	32	32	35	38	41	44	47	47	52	63
0.42		16	18	20	20	21	23	23	26	30	30	33	35	38	41	45	45	49	59
0.41		15	17	19	19	20	22	22	24	29	29	31	33	36	39	42	42	46	56
0.40		14	16	18	18	19	21	21	23	27	27	29	31	34	37	40	40	44	52
0.39		13	15	17	17	18	19	19	21	25	25	27	29	32	34	37	37	41	49
0.38		12	14	16	16	17	18	18	20	24	24	25	28	30	32	35	35	38	46
0.37		11	13	15	15	16	17	17	19	22	22	24	26	28	30	33	33	36	43
0.36		11	12	14	14	14	16	16	17	21	21	22	24	26	28	30	30	33	40
0.35		10	11	13	13	14	15	15	16	19	19	21	23	24	26	28	28	31	38
0.33		9	10	11	11	12	13	13	14	17	17	18	19	21	23	24	24	27	32
0.31		7	8	9	9	10	11	11	12	14	14	15	17	18	19	21	21	23	28

*From National Design Specification, 1977 by NFPA.

TABLE A.6.19. NAILS AND SPIKES—LATERAL LOAD DESIGN VALUES.*

Normal load duration

Design values for lateral loads (single shear) for nails and spikes penetrating not less than 10 diameters in Group I species, 11 diameters in Group II species, 13 in diameters in Group III species, and 14 diameters in Group IV species, into the member holding the point. Nail size in pennyweight. Diameters and lengths in inches. Loads in pounds.

BOX NAILS

Penny weight	6*d*	8*d*	10*d*	12*d*	16*d*	20*d*	30*d*	40*d*
Length	2	2½	3	3¼	3½	4	4½	5
Diameter	0.099	0.113	0.128	0.128	0.135	0.148	0.148	0.162
10 Diameters	0.99	1.13	1.28	1.28	1.35	1.48	1.48	1.62
11 Diameters	1.09	1.24	1.41	1.41	1.49	1.63	1.63	1.78
13 Diameters	1.29	1.47	1.66	1.66	1.76	1.92	1.92	2.11
14 Diameters	1.39	1.58	1.79	1.79	1.89	2.07	2.07	2.27
Species group I	64	77	93	93	101	116	116	133
Species group II	51	63	76	76	82	94	94	108
Species group III	42	51	62	62	67	77	77	88
Species group IV	34	41	49	49	54	61	61	70

COMMON WIRE NAILS

Penny weight	6*d*	8*d*	10*d*	12*d*	16*d*	20*d*	30*d*	40*d*	50*d*	60*d*
Length	2	2½	3	3¼	3½	4	4½	5	5½	6
Diameter	0.113	0.131	0.148	0.148	0.162	0.192	0.207	0.225	0.244	0.263
10 Diameters	1.13	1.31	1.48	1.48	1.62	1.92	2.07	2.25	2.44	2.63
11 Diameters	1.24	1.44	1.63	1.63	1.78	2.11	2.28	2.48	2.68	2.89
13 Diameters	1.47	1.70	1.92	1.92	2.11	2.50	2.69	2.93	3.17	3.42
14 Diameters	1.58	1.83	2.07	2.07	2.27	2.69	2.90	3.15	3.42	3.68
Species group I	77	97	116	116	133	172	192	218	246	275
Species group II	63	78	94	94	108	139	155	176	199	223
Species group III	51	64	77	77	88	114	127	144	163	182
Species group IV	41	51	61	61	70	91	102	115	130	146

THREADED HARDENED STEEL NAILS AND SPIKES

Penny weight	6*d*	8*d*	10*d*	12*d*	16*d*	20*d*	30*d*	40*d*	50*d*	60*d*	70*d*	80*d*	90*d*
Length	2	2½	3	3¼	3½	4	4½	5	5½	6	7	8	9
Diameter	0.120	0.120	0.135	0.135	0.148	0.177	0.177	0.177	0.177	0.177	0.207	0.207	0.207
10 Diameters	1.20	1.20	1.35	1.35	1.48	1.77	1.77	1.77	1.77	1.77	2.07	2.07	2.07
11 Diameters	1.32	1.32	1.49	1.49	1.63	1.95	1.95	1.95	1.95	1.95	2.28	2.28	2.28
13 Diameters	1.56	1.56	1.76	1.76	1.92	2.30	2.30	2.30	2.30	2.30	2.69	2.69	2.69
14 Diameters	1.68	1.68	1.89	1.89	2.07	2.48	2.48	2.48	2.48	2.48	2.90	2.90	2.90
Species group I	77	97	116	116	133	172	172	172	172	172	218	218	218
Species group II	63	78	94	94	108	139	139	139	139	139	176	176	176
Species group III	51	64	77	77	88	114	114	114	114	114	144	144	144
Species group IV	41	51	61	61	70	91	91	91	91	91	115	115	115

COMMON WIRE SPIKES

Penny weight	10*d*	12*d*	16*d*	20*d*	30*d*	40*d*	50*d*	60*d*	5/16"	3/8"
Length	3	3¼	3½	4	4½	5	5½	6	7	8½
Diameter	0.192	0.192	0.207	0.225	0.244	0.263	0.283	0.283	0.312	0.375
10 Diameters	1.92	1.92	2.07	2.25	2.44	2.63	2.83	2.83	3.12	3.75
11 Diameters	2.11	2.11	2.28	2.48	2.68	2.89	3.11	3.11	3.43	4.13
13 Diameters	2.50	2.50	2.69	2.93	3.17	3.42	3.68	3.68	4.06	4.88
14 Diameters	2.69	2.69	2.90	3.15	3.42	3.68	3.96	3.96	4.37	5.25
Species group I	172	172	192	218	246	275	307	307	356	468
Species group II	139	139	155	176	199	223	248	248	288	379
Species group III	114	114	127	144	163	182	203	203	235	310
Species group IV	91	91	102	115	130	146	163	163	188	248

*From National Design Specification, 1977 by NFPA.

TABLE A.6.20. WOOD SCREWS—WITHDRAWAL DESIGN VALUES.*

Normal load duration

Design values in withdrawal in pounds per inch of penetration of threaded part into side grain of member holding point.

g = gauge of screw. D = shank diameter in inches. G = specific gravity of the wood, based on weight and volume when oven-dry.

Specific gravity G	Screw size										
	g = 6	7	8	9	10	12	14	16	18	20	24
	D = 0.138	0.151	0.164	0.177	0.190	0.216	0.242	0.268	0.294	0.320	0.372
0.75	220	241	262	283	304	345	387	428	470	511	594
0.68	181	198	215	232	250	284	318	352	386	420	489
0.67	176	193	209	226	242	275	309	342	375	408	474
0.66	171	187	203	219	235	267	299	332	364	396	460
0.62	151	165	179	193	207	236	264	293	321	349	406
0.55	119	130	141	152	163	186	208	230	253	275	320
0.54	114	125	136	147	157	179	200	222	243	265	308
0.51	102	112	121	131	140	160	179	198	217	236	275
0.49	94	103	112	121	130	147	165	183	200	218	254
0.48	90	99	107	116	124	141	158	175	192	209	243
0.47	87	95	103	111	119	136	152	168	184	201	233
0.46	83	91	99	106	114	130	145	161	177	192	224
0.45	79	87	94	102	109	124	139	154	169	184	214
0.44	76	83	90	97	104	119	133	147	162	176	205
0.43	72	79	86	93	100	113	127	141	154	168	195
0.42	69	76	82	89	95	108	121	134	147	160	186
0.41	66	72	78	85	91	103	116	128	140	153	178
0.40	63	69	75	80	86	98	110	122	134	145	169
0.39	60	65	71	76	82	93	105	116	127	138	161
0.38	57	62	67	73	78	89	99	110	121	131	153
0.37	54	59	64	69	74	84	94	104	114	124	145
0.36	51	56	60	65	70	80	89	99	108	118	137
0.35	48	53	57	62	66	75	84	93	102	111	129
0.33	43	47	51	55	59	67	75	83	91	99	115
0.31	38	41	45	48	52	59	66	73	80	87	102

Approximately two-thirds of the length of a standard wood screw is threaded.

*From National Design Specification, 1977 by NFPA.

TABLE A.6.21. WOOD SCREWS—LATERAL LOAD DESIGN VALUES.*

Normal load duration
Design values for lateral loads (shear) in pounds for screws embedded to approximately 7 times the shank diameter into the member holding the point. For less penetration, reduce loads in proportion. Penetration should not be less than 4 times the shank diameter

Species group	SIZE OF SCREW											
	g =	6	7	8	9	10	12	14	16	18	20	24
	D =	0.138	0.151	0.164	0.177	0.190	0.216	0.242	0.268	0.294	0.320	0.372
	$7D$ =	0.966	1.057	1.148	1.239	1.330	1.512	1.694	1.876	2.058	2.240	2.604
	$4D$ =	0.552	0.604	0.656	0.708	0.760	0.864	0.968	1.072	1.176	1.280	1.488
Group I	=	91	109	129	150	173	224	281	345	415	492	664
Group II . . .	=	75	90	106	124	143	185	232	284	342	406	548
Group III . .	=	62	74	87	101	117	151	190	233	280	332	448
Group IV . .	=	48	58	68	79	91	118	148	181	218	258	349

*From National Design Specification, 1977 by NFPA.

Table A.6.22 DESIGN AND LOAD DATA FOR TECO CONNECTORS†

TECO Spike-Grids

TECO SPIKE-GRID.			
Type	Flat	Single Curve	Circular
Size, square	4⅛″	4⅛″	3¼″
Total depth of grids, maximum	1″	1.38″	1.20″
Diameter of bolt hole	1.06″	1.06″	1.33″
Weight, per 100 grids, lbs	50	75	26
LUMBER DIMENSIONS, minimum recommended for installation of flat grids			
Face width			
Thickness	5½″	5½″	5½″
Grids one face only	1½″	1½″	1½″
Grids opposite in both faces	2½″	………	2½″
Minimum diameter of pile for curved grids	………	10″	………
BOLT, diameter	¾″ or 1″	¾″ or 1″	¾″ or 1″
BOLT HOLE, diameter in timber	13/16″ or 1-1/16″	13/16″ or 1-1/16″	13/16″ or 1-1/16″
WASHERS			
Round, cast or malleable iron	Standard Size for Bolt Diameter Used.		
Square plate	3″x3″x¼″ Punched for Bolt Diameter Used.		
SPACING OF GRIDS, minimum, center to center			
0°–30° angle of load to grain			
Spacing parallel to grain	7″	7″	7″
Spacing perpendicular to grain	5½″	5½″	5½″
30°–90° angle of load to grain			
Spacing parallel or perpendicular to grain	5½″	5½″	5½″
END DISTANCES, center of grid to end of piece (tension or compression members)			
Standard	7″	7″	7″
Minimum, reduce loads 15%	5″	5″	5″
EDGE DISTANCES, center of grid to edge of piece			
Load applied at any angle to grain			
Standard	3¾″	3¾″	3¾″
Minimum, reduce loads 15%	2¾″	2¾″	2¾″
PROJECTED AREA for portion of one grid within member, square inches	2.06	2.06	1.95

SPIKE-GRID SPECIFICATIONS

Spike-grids shall be TECO spike-grids as manufactured by the Timber Engineering Company. Spike-grid timber connectors shall be manufactured according to A. S. T. M. Standard Specifications A 47-33, Grade 35018, for malleable iron castings. They shall consist of four rows of opposing spikes forming a 4⅛″ square grid with 16 teeth which are held in place by fillets. Fillets for the flat and circular grid in cross section shall be diamond shaped. Fillets for the single curve grids shall be increased in depth to allow for curvature.

WIND AND EARTHQUAKE LOADS

For wind or earthquake loads alone or a combination of wind or earthquake with dead or live loads or both, the safe loads on spike-grids may be taken as 120% of the Design Loads provided the resulting size and number of connectors is not less than required for the dead and live loads alone.

IMPACT

When using Design Loads, the load on a spike-grid due to a force producing impact shall be taken as 115% of the sum of the force as a static load and the load due to its impact.

LOADS IN RELATION TO DISTANCES AND SPACINGS

Standard Design Loads are for standard distances and spacings. Standard and minimum distances and spacings with load reduction factors are given in the table. Loads for end and edge distances and spacings intermediate of standard and minimum may be determined by interpolation.

DESIGN LOADS* FOR ONE SPIKE-GRID AND BOLT IN SINGLE SHEAR

GROUP A			GROUP B			GROUP C		
Type of Grid	Bolt Diameter	Allowable Load	Type of Grid	Bolt Diameter	Allowable Load	Type of Grid	Bolt Diameter	Allowable Load
FLAT	¾″	3900#	FLAT	¾″	3500#	FLAT	¾″	3000#
	1″	4200#		1″	3800#		1″	3300#
SINGLE CURVE	¾″	4200#	SINGLE CURVE	¾″	3800#	SINGLE CURVE	¾″	3200#
	1″	4500#		1″	4100#		1″	3500#
CIRCULAR	¾″	3500#	CIRCULAR	¾″	3100#	CIRCULAR	¾″	2600#
	1″	3800#		1″	3400#		1″	2900#

* Allowable loads on spike-grids same for all angles of load to grain.

†Courtesy Timber Engineering Company

Table A.6.23 DESIGN AND LOAD DATA FOR TECO CONNECTORS†

TECO Clamping Plates

	Plain	Flanged
CLAMPING-PLATES.		
Type	Plain	Flanged
Length of plate	5¼"	8"
Width of plate	5¼"	5
Depth of flange	....	2"
Length of teeth	.68"	.72"
Diameter of bolt hole	1.12"	1.12"
Weight per 100 pieces, lbs	65	200
LUMBER DIMENSIONS, minimum required for installation of plates		
Face width	6½"	6½"
Thickness, plates in one face only	1½"	2½"
Thickness, plates opposite on both faces	2½"	5½"
BOLT diameter, minimum	¾"	¾"
BOLT HOLE, diameter in timber	13/16"	13/16"
WASHERS		
Round, cast or malleable iron, diameter	3"	3"
Square plate:		
Length of side	3"	3"
Thickness	¼"	¼"
SPACING OF CLAMPING-PLATES		
Minimum center-to-center		
Parallel to grain	6"	8½"
Perpendicular to grain	6"	5½"
END DISTANCES		
Center of plate to end of piece		
Plain type	5"	....
Flanged type:		
Toothed side	....	5"
Flanged side	....	2½"
EDGE DISTANCES		
Center of plate to edge of piece		
Plain type	3¼"	....
Flanged type:		
Toothed side	....	3¼"
Flanged side	....	4"

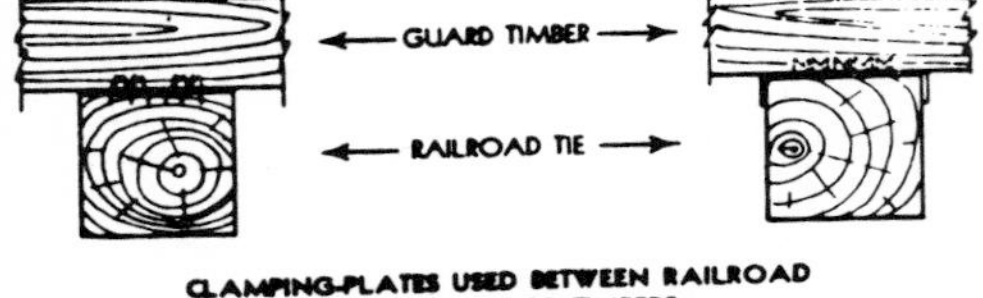

CLAMPING-PLATES USED BETWEEN RAILROAD TIES AND GUARD TIMBERS

Plates may be used with a bolt through each joint or at every third or fourth joint.

CLAMPING-PLATE SPECIFICATIONS

Clamping-plates shall be TECO Clamping-plates as manufactured by the Timber Engineering Co., Washington, D. C.

Clamping-plate timber connectors shall be stamped cold from mild steel conforming to A. S. T. M. Standard Specifications for carbon steel A 17-29, Type A, Grade 1.

Plain Clamping-Plate—Plain clamping-plates shall consist of a square steel plate with a central bolt hole and twelve teeth projecting from each face. The flat sides of the teeth on each face shall be parallel to each other and arranged with the flat sides of the teeth on opposite faces of the plate at right angles to each other.

Flanged Clamping-Plate—Flanged clamping-plates shall consist of a rectangular steel plate with a central bolt hole around which fourteen teeth shall project from one face with their flat sides parallel to the long edges of the plate. The metal near the two narrow ends of the plate shall be bent at right angles to the plate and in a direction opposite to the teeth to form flanges.

WIND AND EARTHQUAKE LOADS

For wind or earthquake loads alone or a combination of wind or earthquake with dead or live loads or both, the safe loads on clamping-plates may be taken as 116 per cent of the Design Loads provided the resulting number of connectors is not less than required for the dead and live loads alone.

IMPACT

When using Design Loads, the load on a clamping-plate due to a force producing impact shall be taken as 115 per cent of the sum of the force as a static load and the load due to its impact.

LOADS IN RELATION TO DISTANCES AND SPACINGS

Design Loads are for standard distances and spacings. Spacings of plates and end and edge distances less than the minimum specified are not recommended.

CONDITION OF LUMBER

Tabulated loads apply to the plain clamping-plate (where the teeth rather than the flange determine the load capacity) used in seasoned lumber. Loads for the plain clamping-plates in green lumber should not exceed 60 per cent of those shown. Tabulated loads apply to the flanged clamping-plate in either seasoned or green lumber.

†Courtesy Timber Engineering Company

DESIGN LOADS PER CLAMPING-PLATE

Species	Clamping-Plate Size (inches)	Type of Joint Connection	Angle of Load to grain 90°
GROUP A	5¼x5¼	¾" Through Bolt	3,400#
		No Through Bolt*	2,460#
	5x8	Teeth Only (No Through Bolt)	2,920#
		One Flange Only (No Through Bolt)	1,610#
GROUP B	5¼x5¼	¾" Through Bolt	3,080#
		No Through Bolt	2,230#
	5x8	Teeth Only (No Through Bolt)	2,660#
		One Flange Only (No Through Bolt)	1,610#
GROUP C	5¼x5¼	¾" Through Bolt	2,770#
		No Through Bolt	2,010#
	5x8	Teeth Only (No Through Bolt)	2,390#
		One Flange Only (No Through Bolt)	1,610#

* Joint members held in contact by bolts outside joint area.

Table A.6.24 DESIGN AND LOAD DATA FOR TECO PILE CAP CONNECTORS†

ALLOWABLE DESIGN WIND UPLIFT LOADS — CONNECTOR TO TIMBER PILE

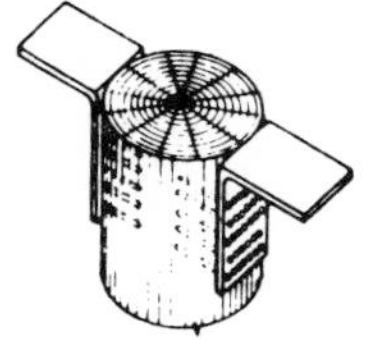

PILE SPECIES	TWO PLATES (MIN.)	ADDITIONAL PLATES
southern pine	20,000#	10,000# per plate
west coast douglas fir	20,000#	10,000# per plate
oak	20,000#	10,000# per plate

Allowable design loads established by full scale joint tests for lateral resistance of nails in pile, and net section tensile strength of plates. Adjustments may be made for other durations of applied load as outlined in appendix H of National Design Specification for Stress Grade Lumber and Its Fastenings.

DEVELOPMENT OF CONNECTOR DESIGN VALUE IN CONCRETE CAP

TEN-CON DATA

TECO TEN-CON connectors (type A) are formed in the shape of a right angle, with one leg having specially stamped truncated circular holes to receive the special nails for attachment to the pile. The other angular leg acts to form a bond with the concrete pile cap.

TEN-CON connectors are precision manufactured from ¼" hot rolled SAE 1020 carbon steel, with a yield stress of 35,00 psi. Overall dimensions are: width=4", length=15" (7½" leg + 1½" bend + 6" leg). A total of 33 nail holes are provided in the 7½" leg. TECO GRIP-LAM nails, 3½" long, are packed with the connectors.

TEN-CON SPECIFICATIONS

Uplift resisting piles shall be anchored to pile caps with TECO TEN-CON (type A) piling connectors as manufactured by Timber Engineering Company, Washington, D. C. Connectors to be made of ¼" thick SAE 1020 hot rolled steel. Connectors are to be attached to piles with TECO 3½" GRIP-LAM nails. All nail holes are to be filled unless otherwise specified on the drawings.

GRIP-LAM NAILS

Manufactured of heat-treated steel, TECO GRIP-LAM nails carry four times the load capacity of an ordinary nail and eliminate tearing of beam fibres.

c + d

d/2 c d/2

d

Critical section assumed by ACI code. section 1207.

Plane of failure assumed when cone is pulled out of concrete.

Bottom re-bars in concrete cap.

NOTE: In addition to shear in plain concrete, stirrups or other re-inforcing may be used to develop pull-out strength.

45°

Bottom of cap

$V_c = \frac{V}{b_o d}$ = unit shear stress in concrete

V = total load

d = depth to reinforcing steel

c = pile diameter

$b_o = \pi (c + d)$ = circumference at critical section

EXAMPLE: Given:

V = 20,000# (allowable uplift load for 2 connector plates to pile)

$d = 7\frac{1}{2}''$ $c = 10\frac{1}{2}''$

$b_o = \pi (c + d)\ \frac{1}{2}$

$= 3.14 \times 18 \times 0.5 = 28.2''$

Solve:

$V_c = \frac{V}{b_o d} = \frac{20{,}000}{28.2 \times 7.5} = 94.2$ psi

V_c allowable for concrete = 100 psi.

NOTE ABOUT b_o: Where only two plates are used, b_o might be assumed to be equivalent to one half a full circumference, or some other fraction thereof. The quantity b_o should be established through experience and engineering judgment.

†Courtesy Timber Engineering Company

TEN-CON FLAT PLATE CONNECTORS

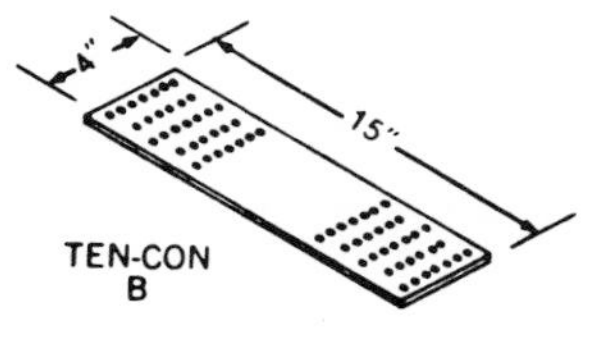

WIND ANCHOR

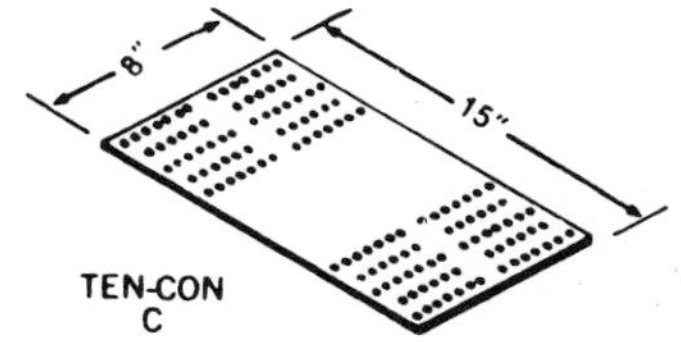

BEAM SPLICE

TABLE A.6.25. WOOD SIDE PLATE MODIFICATION FACTORS FOR CONNECTOR, BOLT AND LATERALLY LOADED LAG-SCREW JOINTS.*

A_1/A_2	A_1 (in^2)†	Number of fasteners in a row										
		2	3	4	5	6	7	8	9	10	11	12
	<12	1.00	0.92	0.84	0.76	0.68	0.61	0.55	0.49	0.43	0.38	0.34
	12 – 19	1.00	0.95	0.88	0.82	0.75	0.68	0.62	0.57	0.52	0.48	0.43
	>19 – 28	1.00	0.97	0.93	0.88	0.82	0.77	0.71	0.67	0.63	0.59	0.55
0.5	>28 – 40	1.00	0.98	0.96	0.92	0.87	0.83	0.79	0.75	0.71	0.69	0.66
•	>40 – 64	1.00	1.00	0.97	0.94	0.90	0.86	0.83	0.79	0.76	0.74	0.72
‡	>64	1.00	1.00	0.98	0.95	0.91	0.88	0.85	0.82	0.80	0.78	0.76
	<12	1.00	0.97	0.92	0.85	0.78	0.71	0.65	0.59	0.54	0.49	0.44
	12 – 19	1.00	0.98	0.94	0.89	0.84	0.78	0.72	0.66	0.61	0.56	0.51
	>19 – 28	1.00	1.00	0.97	0.93	0.89	0.85	0.80	0.76	0.72	0.68	0.64
1.0	>28 – 40	1.00	1.00	0.99	0.96	0.92	0.89	0.86	0.83	0.80	0.78	0.75
•	>40 – 64	1.00	1.00	1.00	0.97	0.94	0.91	0.88	0.85	0.84	0.82	0.80
‡	>64	1.00	1.00	1.00	0.99	0.96	0.93	0.91	0.88	0.87	0.86	0.85

Notes: 1. A_1 = cross-sectional area of main member(s) before boring or grooving.
2. A_2 – sum of the cross-sectional areas of side members before boring or grooving.

•When A_1/A_2 exceeds 1.0, use A_2/A_1.

†When A_1/A_2 exceeds 1.0, use A_2 instead of A_1.

‡For A_1/A_2 between 0 and 1.0, interpolate or extrapolate from the tabulated values.

*From National Design Specification, 1977 by NFPA.

TABLE A.6.26. METAL SIDE PLATE MODIFICATION FACTORS FOR CONNECTOR, BOLT AND LATERALLY LOADED LAG-SCREW JOINTS.*

A_1/A_2	A_1 (in^2)	Number of fasteners in a row										
		2	3	4	5	6	7	8	9	10	11	12
2–12	25 – 39	1.00	0.94	0.87	0.80	0.73	0.67	0.61	0.56	0.51	0.46	0.42
	40 – 64	1.00	0.96	0.92	0.87	0.81	0.75	0.70	0.66	0.62	0.58	0.55
	65 – 119	1.00	0.98	0.95	0.91	0.87	0.82	0.78	0.75	0.72	0.69	0.66
	120 – 199	1.00	0.99	0.97	0.95	0.92	0.89	0.86	0.84	0.81	0.79	0.78
12–18	40 – 64	1.00	0.98	0.94	0.90	0.85	0.80	0.75	0.70	0.67	0.62	0.58
	65 – 119	1.00	0.99	0.96	0.93	0.90	0.86	0.82	0.79	0.75	0.72	0.69
	120 – 199	1.00	1.00	0.98	0.96	0.94	0.92	0.89	0.86	0.83	0.80	0.78
	200	1.00	1.00	1.00	0.98	0.97	0.95	0.93	0.91	0.90	0.88	0.87
18–24	40 – 64	1.00	1.00	0.96	0.93	0.89	0.84	0.79	0.74	0.69	0.64	0.59
	65 – 119	1.00	1.00	0.97	0.94	0.92	0.89	0.86	0.83	0.80	0.76	0.73
	120 – 199	1.00	1.00	0.99	0.98	0.96	0.94	0.92	0.90	0.88	0.86	0.85
	200	1.00	1.00	1.00	1.00	0.98	0.96	0.95	0.93	0.92	0.92	0.91
24–30	40 – 64	1.00	0.98	0.94	0.90	0.85	0.80	0.74	0.69	0.65	0.61	0.58
	65 – 119	1.00	0.99	0.97	0.93	0.90	0.86	0.82	0.79	0.76	0.73	0.71
	120 – 199	1.00	1.00	0.98	0.96	0.94	0.92	0.89	0.87	0.85	0.83	0.81
	200	1.00	1.00	0.99	0.98	0.97	0.95	0.93	0.92	0.90	0.89	0.89
30–35	40 – 64	1.00	0.96	0.92	0.86	0.80	0.74	0.68	0.64	0.60	0.57	0.55
	65 – 119	1.00	0.98	0.95	0.90	0.86	0.81	0.76	0.72	0.68	0.65	0.62
	120 – 199	1.00	0.99	0.97	0.95	0.92	0.88	0.85	0.82	0.80	0.78	0.77
	200	1.00	1.00	0.98	0.97	0.95	0.93	0.90	0.89	0.87	0.86	0.85
35–42	40 – 64	1.00	0.95	0.89	0.82	0.75	0.69	0.63	0.58	0.53	0.49	0.46
	65 – 119	1.00	0.97	0.93	0.88	0.82	0.77	0.71	0.67	0.63	0.59	0.56
	120 – 199	1.00	0.98	0.96	0.93	0.89	0.85	0.81	0.78	0.76	0.73	0.71
	200	1.00	0.99	0.98	0.96	0.93	0.90	0.87	0.84	0.82	0.80	0.78

Notes: 1. A_1 = Cross-sectional area of main member before boring or grooving.
2. A_2 = Sum of cross-sectional areas of metal side plates before drilling.

*From National Design Specification, 1977 by NFPA.

TABLE A.6.27. FASTENER LOAD MODIFICATION FACTORS FOR MOISTURE CONTENT.*

Type of fastener	Condition of wood[1]		Factor
	At time of fabrication	In service	
Timber connectors[2]	Dry	Dry	1.0
	Partially seasoned[3]	Dry	See Note 3
	Wet	Dry	0.8
	Dry or wet	Partially seasoned or wet	0.67
Bolts or lag screws	Dry	Dry	1.0
	Partially seasoned[3] or wet	Dry	See Table 8.1C
	Dry or wet	Exposed to weather	0.75
	Dry or wet	Wet	0.67
Drift bolts or pins - Laterally loaded	Dry or wet	Dry	1.0
	Dry or wet	Partially seasoned or wet, or subject to wetting and drying	0.70
Wire nails and spikes			
–Withdrawal loads	Dry	Dry	1.0
	Partially seasoned or wet	Will remain wet	1.0
	Partially seasoned or wet	Dry	0.25
	Dry	Subject to wetting and drying	0.25
–Lateral loads	Dry	Dry	1.0
	Partially seasoned or wet	Dry or wet	0.75
	Dry	Partially seasoned or wet	0.75
Threaded, hardened steel nails	Dry or wet	Dry or wet	1.0
Wood screws	Dry or wet	Dry	1.0
	Dry or wet	Exposed to weather	0.75
	Dry or wet	Wet	0.67
Metal plate connectors	Dry	Dry	1.0
	Partially seasoned or wet	Dry or wet	0.8

1. Condition of wood definitions applicable to fasteners are

"Dry" wood has a moisture content of 19 percent or less.

"Wet" wood has a moisture content at or above the fiber saturation point (approximately 30 percent).

"Partially seasoned" wood, for the purposes of Table 8.1B, has a moisture content greater than 19 percent but less than the fiber saturation point (approximately 30 percent).

"Exposed to weather" implies that the wood may vary in moisture content from dry to partially seasoned, but is not expected to reach the fiber saturation point at times when the joint is under full design load.

"Subject to wetting and drying" implies that the wood may vary in moisture content from dry to partially seasoned or wet, or vice versa, with consequent effects on the tightness of the joint.

2. For timber connectors, moisture content limitations apply to a depth of 3/4 inch from the surface of the wood.

3. When timber connectors, bolts or laterally loaded lag screws are installed in wood that is partially seasoned at the time of fabrication but that will be dry before full design load is applied, proportional intermediate values may be used.

*From National Design Specification, 1977 by NFPA.

TABLE A.6.28. MODIFICATION FACTORS FOR LATERALLY LOADED BOLTS AND LAG SCREWS IN TIMBER SEASONED IN PLACE.*

Factors apply when wood is at or above the fiber saturation point (wet) at time of fabrication but dries to a moisture content of 19 percent or less (dry) before full design load is applied. For wood partially seasoned when fabricated, adjusted intermediate values may be used.

Arrangement of bolts or lag screws	Type of splice plate	Modification factor
–One fastener only, or –Two or more fasteners placed in a single line parallel to grain, or –Fasteners placed in two or more lines parallel to grain with separate splice plates for each line	Wood or metal	1.0
–All other arrangements	Wood or metal	0.4

*From National Design Specification, 1977 by NFPA.

TABLE A.7.1. DESIGN AID FOR THE DETERMINATION OF LATERAL STABILITY. (After P. F. Barber.)

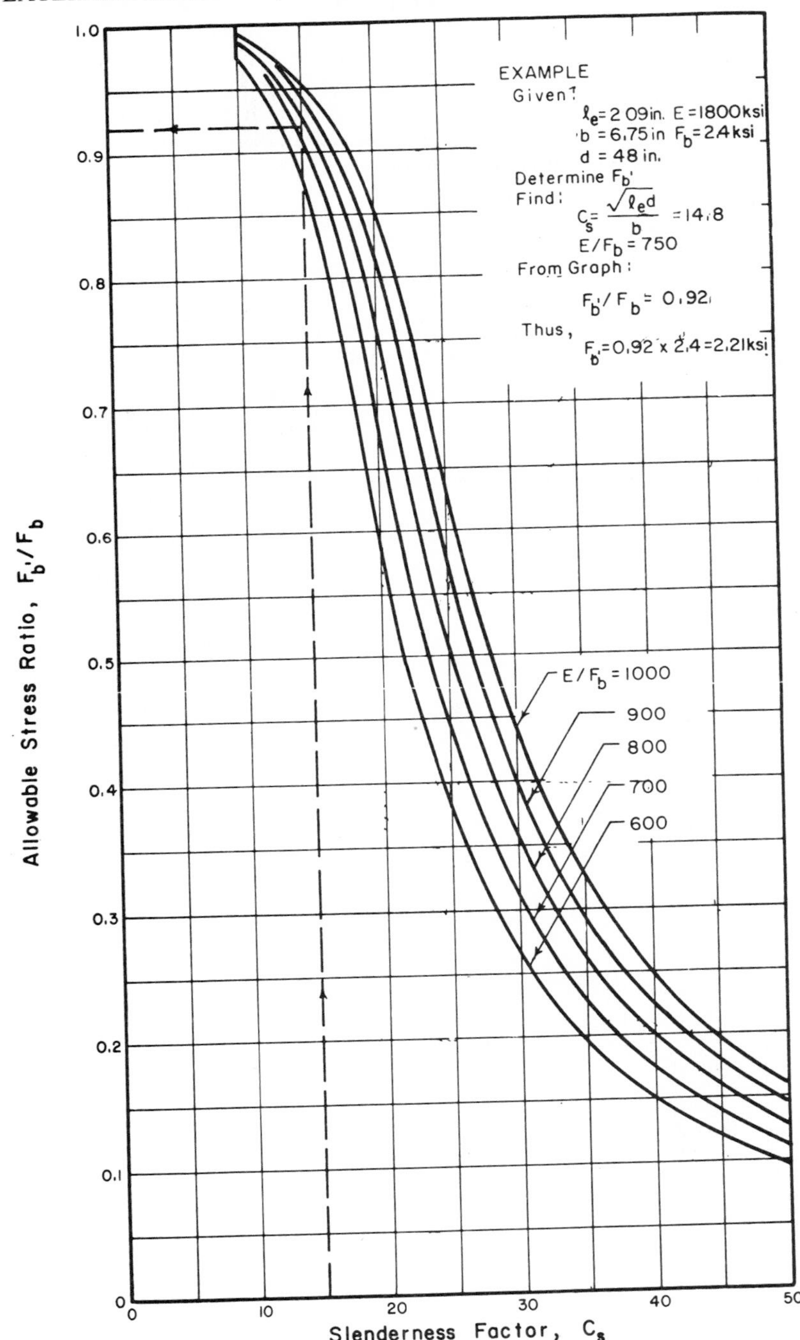

A. 9.1—Typical Sections of Composite Wood-Concrete Slabs. From T. K. May's Report, see Ref. 6, Chapter 9.

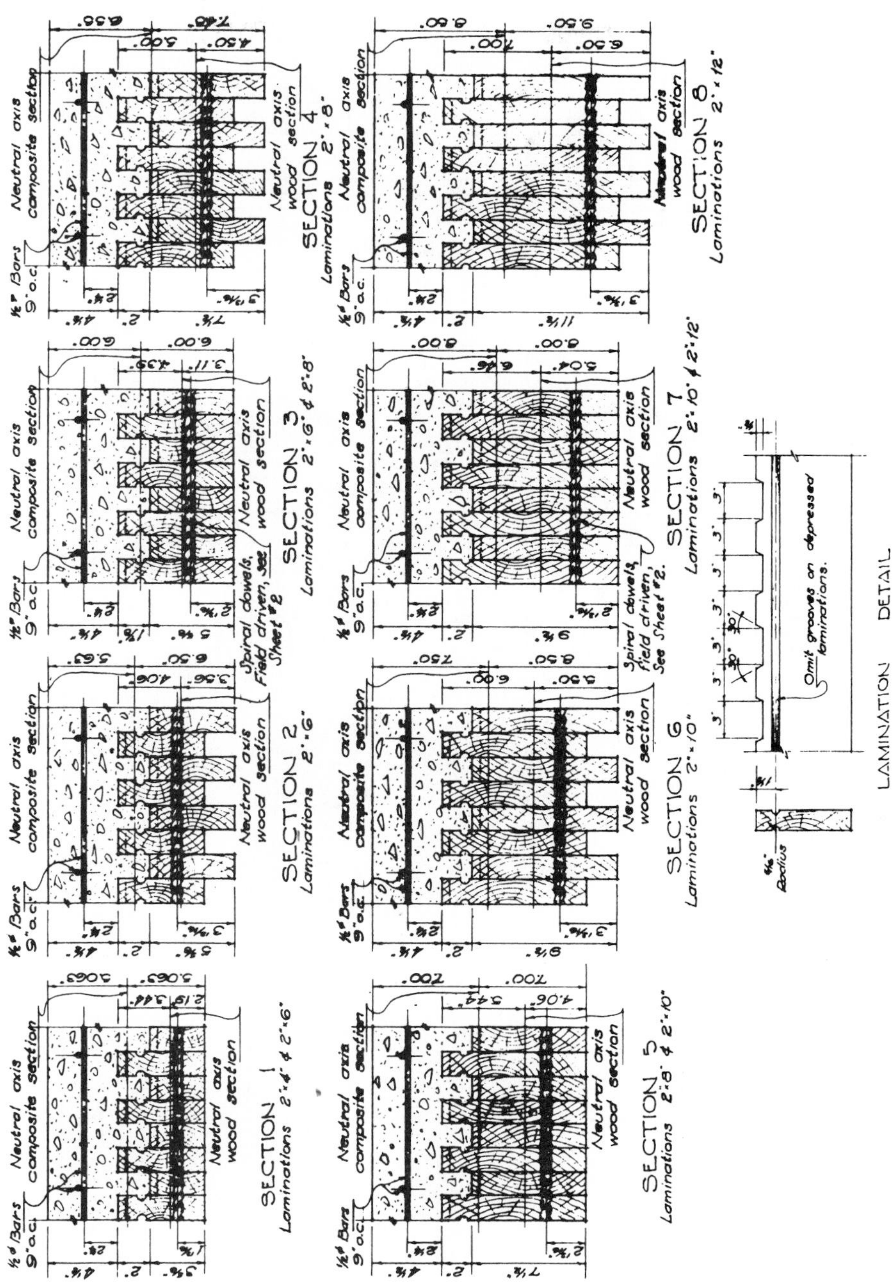

A. 9.2—Initial Proportioning of Section for Composite Wood-Concrete Composite Slab. Use in relation to Table A.9.1. From T. K. May's Report.

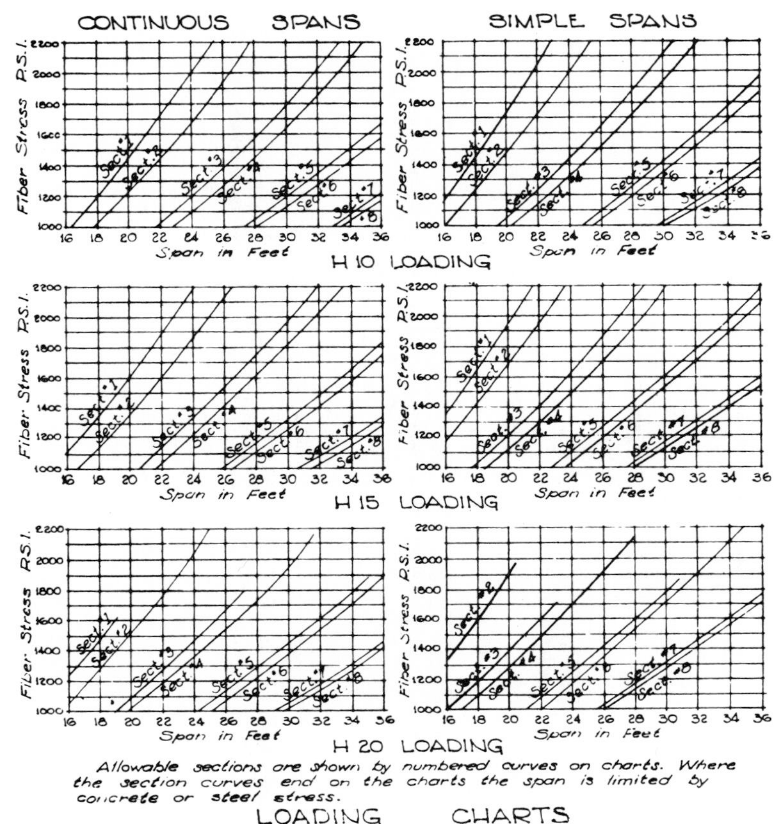

TABLE A.11.1. END-BEARING PILES—SPECIFIED TIP CIRCUMFERENCES WITH MINIMUM BUTT CIRCUMFERENCES.*

Tip Circumference, in. (mm) / Required Minimum Length, ft (m)	16 (406)	19 (483)	22 (559)	25 (635)	28 (711)	31 (787)	35 (889)	38 (965)
	Minimum Circumferences 3 ft from Butt in. (mm)							
20 (6.1)	22.0 (559)	24.0 (610)	27.0 (686)	30.0 (762)	33.0 (838)	36.0 (914)	40.0 (1016)	43.0 (1092)
30 (9.1)	23.5 (597)	26.5 (673)	29.5 (749)	32.5 (826)	35.5 (902)	38.5 (978)	42.5 (1080)	45.5 (1156)
40 (12.2)	26.0 (660)	29.0 (737)	32.0 (813)	35.0 (889)	38.0 (965)	41.0 (1041)	45.0 (1143)	48.0 (1219)
50 (15.2)	28.5 (724)	31.5 (800)	34.5 (876)	37.5 (953)	40.5 (1029)	43.5 (1105)	47.5 (1206)	50.5 (1283)
60 (18.3)	31.0 (787)	34.0 (864)	37.0 (940)	40.0 (1016)	43.0 (1092)	46.0 (1168)	50.0 (1270)	53.0 (1346)
70 (21.3)	33.5 (851)	36.5 (927)	39.5 (1003)	42.5 (1080)	45.5 (1156)	48.5 (1232)	52.5 (1334)	55.5 (1410)
80 (24.4)	36.0 (914)	39.0 (991)	42.0 (1067)	45.0 (1143)	48.0 (1219)	51.0 (1295)	55.0 (1397)	58.0 (1473)
90 (27.4)	38.6 (980)	41.6 (1057)	44.6 (1133)	47.6 (1209)	50.6 (1285)	53.6 (1361)	57.6 (1463)	60.5 (1537)
100 (30.5)	41.0 (1041)	44.0 (1118)	47.0 (1194)	50.0 (1270)	53.0 (1346)	56.0 (1422)	60.0 (1524)	
110 (33.5)	43.6 (1107)	46.6 (1184)	49.6 (1260)	52.6 (1336)	55.6 (1412)	61.0 (1549)		
120 (36.6)	46.0 (1168)	49.0 (1245)	52.0 (1321)	55.0 (1397)	58.0 (1473)			

*From ASTM Designation D25–73, Part 22, 1978.

TABLE A.11.2. FRICTION PILES—SPECIFIED BUTT CIRCUMFERENCES WITH MINIMUM TIP CIRCUMFERENCES.*

NOTE—Where the taper applied to the butt circumferences calculate to a circumference at the tip of less than 16 in. (406 mm), the individual values have been increased to 16 in. (406 mm) to assure a minimum of 5-in. (127-nm) tip for purposes of driving.

Required Minimum Circumference, in. (mm) 3 ft (914 mm) from Butt	22 (559)	25 (635)	28 (711)	31 (787)	35 (889)	38 (965)	41 (1041)	44 (1118)	47 (1194)	50 (1270)	57 (1448)
Length, ft (m)	Minimum Tip Circumferences in. (mm)										
20 (6.1)	16.0 (406)	16.0 (406)	16.0 (406)	18.0 (457)	22.0 (559)	25.0 (635)	28.0 (711)				
30 (9.1)	16.0 (406)	16.0 (406)	16.0 (406)	16.0 (406)	19.0 (483)	22.0 (559)	25.0 (635)	28.0 (711)			
40 (12.2)				16.0 (406)	17.0 (432)	20.0 (508)	23.0 (584)	26.0 (660)	29.0 (737)		
50 (15.2)					16.0 (406)	17.0 (432)	19.0 (483)	22.0 (559)	25.0 (635)	28.0 (711)	
60 (18.3)						16.0 (406)	16.0 (406)	18.6 (472)	21.6 (549)	24.6 (625)	31.6 (803)
70 (21.3)						16.0 (406)	16.0 (406)	16.0 (406)	16.2 (411)	19.2 (488)	26.2 (665)
80 (24.4)							16.0 (406)	16.0 (406)	16.0 (406)	16.0 (406)	21.8 (554)
90 (27.4)							16.0 (406)	16.0 (406)	16.0 (406)	16.0 (406)	19.5 (495)
100 (30.5)							16.0 (406)	16.0 (406)	16.0 (406)	16.0 (406)	18.0 (457)
110 (33.5)										16.0 (406)	16.0 (406)
120 (36.6)											16.0 (406)

*From ASTM Designation D25–73, Part 22, 1978

TABLE A.11.3. DESIGN VALUES FOR TREATED ROUND TIMBER PILES.*

Design values for normal load duration and wet conditions of use, pounds per square inch

Species	Compression parallel to grain F_c	Extreme fiber in bending F_b	Horizontal shear F_v	Compression perpendicular to grain $F_{c\perp}$	Modulus of elasticity E
Pacific Coast Douglas Fir[1]	1250	2450	115	230	1,500,000
Southern Pine[2]	1200	2400	110	250	1,500,000
Red Oak[3]	1100	2450	135	350	1,250,000
Red Pine[4]	900	1900	85	155	1,280,000

1. Pacific Coast Douglas Fir values apply to this species as defined in ASTM Designation D1760-76, Standard Specification for Pressure Treatment of Timber Products. For fastener design, use Douglas Fir-Larch design values.

2. Southern Pine values apply to Longleaf, Slash, Loblolly and Shortleaf Pines.

3. Red Oak values apply to Northern and Southern Red Oak.

4. Red Pine values apply to Red Pine grown in the United States. For fastener design, use Northern Pine design values.

*From National Design Specification, 1977 by NFPA.

Table A.11.4 DECKING SIZES, WEIGHTS AND COVERAGE FACTORS†

Species	Size, in. Nominal	Size, in. Actual	Weight psf	Coverage Factor(1) bd. ft./sq. ft.
SOLID HEAVY TIMBER DECKING				
Douglas fir/larch	4 x 6	3½ x 5¼	9.3	4.57
	3 x 6	2½ x 5¼	6.6	3.43
Hem-fir	4 x 6	3½ x 5¼	8.6	4.57
	3 x 6	2½ x 5¼	6.2	3.43
Southern pine	4 x 6	3½ x 5¼	10.5	4.57
	3 x 6	2½ x 5¼	7.5	3.43
Western red cedar	4 x 6	3½ x 5¼	6.7	4.57
	3 x 6	2½ x 5¼	4.8	3.43
Western white spruce	4 x 6	3½ x 5¼	8.3	4.57
	3 x 6	2½ x 5¼	5.9	3.43
GLUED LAMINATED TIMBER DECKING(2)				
Douglas fir/larch	5 x 6	3 13/16 x 5⅜	10.5	5.58
	4 x 6	3 1/16 x 5⅜	8.5	4.46
	3 x 6S	2⅝ x 5⅜	7.0	3.35
	3 x 6	2¼ x 5⅜	6.5	3.35
Idaho white pine	5 x 6	3 13/16 x 5⅜	9.5	5.58
	4 x 6	3 1/16 x 5⅜	7.6	4.46
	3 x 6S	2⅝ x 5⅜	5.8	3.35
	3 x 6	2¼ x 5⅜	5.0	3.35
Inland red cedar	5 x 6	3 13/16 x 5⅜	7.5	5.58
	4 x 6	3 1/16 x 5⅜	7.0	4.46
	3 x 6S	2⅝ x 5⅜	6.5	3.35
	3 x 6	2¼ x 5⅜	4.5	3.35
Ponderosa pine	3 x 8	2 11/32 x 7 15/32	6.0	3.30
Southern pine	5 x 6	3 11/16 x 6⅜	10.5	5.58
	4 x 6	3 x 5⅜	8.5	4.46
	3 x 6	2¼ x 5⅜	6.5	3.35
West coast hemlock	4 x 6	3 x 5 15/32	7.9	4.40
	4 x 6S	2¾ x 5 15/32	7.3	4.40
	3 x 6	2¼ x 5 15/32	5.9	3.30

(1) To estimate board feet of decking required, multiply square feet of area to be covered by the coverage factor. Add for jobsite trimming and waste for irregular areas.
(2) Laminated decking sizes may vary between manufacturers and the designer should check with the supplier to determine actual sizes and load carrying capacities.
†From "Structural Glued-Laminated Timber," Publication 6.5/Ai, by AITC, 1972.

Table A.11.5 SOLID DECKING [1] [2] †

PART 2—SIMPLE SPAN—END JOINTS OVER SUPPORTS

SPECIES	GRADE	E[3] VALUE	SIZE[1]	SPAN 8′ 1/180	8′ 1/240	9′ 1/180	9′ 1/240	10′ 1/180	10′ 1/240	11′ 1/180	11′ 1/240	12′ 1/180	12′ 1/240	13′ 1/180	13′ 1/240	14′ 1/180	14′ 1/240	15′ 1/180	15′ 1/240	16′ 1/180	16′ 1/240	17′ 1/180	17′ 1/240	18 1/180	18 1/240
DOUGLAS FIR/LARCH	**SELECT**	**1.80**	**4x6**							**172**	**129**	**132**	**99**	**104**	**78**	**83**	**62**	**68**	**51**	**56**	**42**	**47**	**35**	**39**	**29**
	COMM.	1.70	4x6							162	122	125	94	98	74	79	59	64	48	53	40	44	33	37	28
	SELECT	**1.80**	**3x6**	**163**	**122**	**114**	**86**	**83**	**62**	**63**	**47**	**48**	**36**	**38**	**28**	**30**	**23**	**25**	**19**	**20**	**15**	**17**	**13**	**14**	**11**
	COMM.	1.70	3x6	154	115	108	81	79	59	59	44	46	34	36	27	29	22	23	17	19	14	16	12	13	10
HEM-FIR	**SELECT**	**1.50**	**4x6**					**191**	**143**	**143**	**107**	**110**	**83**	**87**	**65**	**69**	**52**	**56**	**42**	**47**	**35**	**39**	**29**	**33**	**25**
	COMM.	1.40	4x6					178	133	134	100	103	77	81	61	65	49	53	40	43	33	36	27	30	23
	SELECT	**1.50**	**3x6**	**136**	**102**	**95**	**71**	**69**	**52**	**52**	**39**	**40**	**30**	**32**	**24**	**25**	**19**	**21**	**15**	**17**	**13**	**14**	**11**		
	COMM.	1.40	3x6	127	95	84	67	65	49	49	37	38	28	30	22	24	18	19	14	16	12	13	10		
SOUTHERN PINE	**SELECT & COMM.**	**1.60**	**4x6**							**153**	**115**	**118**	**88**	**93**	**69**	**74**	**56**	**60**	**45**	**50**	**37**	**41**	**31**	**35**	**26**
	SELECT & COMM.	1.60	3x6	145	109	102	76	74	56	56	42	43	32	34	25	27	20	22	16	18	14	15	11		
WESTERN RED CEDAR	**SELECT**	**1.10**	**4x6**			**192**	**144**	**140**	**105**	**105**	**79**	**81**	**61**	**64**	**48**	**51**	**38**	**41**	**31**	**34**	**26**	**28**	**21**	**24**	**18**
	COMM.	1.00	4x6			174	131	127	95	95	72	74	55	58	43	46	35	38	28	31	23	26	19	22	16
	SELECT	**1.10**	**3x6**	**99**	**75**	**70**	**52**	**51**	**38**	**38**	**29**	**29**	**22**	**23**	**17**	**19**	**14**	**15**	**11**						
	COMM.	1.00	3x6	90	68	64	48	46	35	35	26	27	20	21	16	17	13	14	10						
WESTERN WHITE SPRUCE	**SELECT & COMM.**	**1.32**	**4x6**					(160) [5]	**126**	**126**	**94**	**97**	**73**	**76**	**57**	**61**	**46**	**50**	**37**	**41**	**31**	**34**	**26**	**29**	**22**
	SELECT & COMM.	1.32	3x6	119	90	84	63	61	46	46	34	35	27	28	21	22	17	18	14	15	11				

Table A.11.5 continued

PART 3—CONTROLLED RANDOM LAYUP CONTINUOUS OVER 3 OR MORE EQUAL SPANS

				10′	11′	12′	13′	14′	15′	16′	17′	18′	19′	20′
DOUGLAS FIR/LARCH	SELECT	1.80	4x6			200 / 150	157 / 118	126 / 94	102 / 77	84 / 63	70 / 53	59 / 44	50 / 38	43 / 32
	COMM.	1.70	4x6			189 / 142	148 / 111	119 / 89	97 / 72	80 / 60	66 / 50	56 / 42	48 / 36	41 / 31
	SELECT	1.80	3x6	126 / 94	94 / 71	73 / 55	57 / 43	46 / 34	37 / 28	31 / 23	26 / 19	22 / 16	18 / 14	16 / 12
	COMM.	1.70	3x6	119 / 89	89 / 67	69 / 52	54 / 41	43 / 32	35 / 26	29 / 22	24 / 18	20 / 15	17 / 13	15 / 11
HEM-FIR	SELECT	1.50	4x6			166 / 125	131 / 98	105 / 79	85 / 64	70 / 53	59 / 44	49 / 37	42 / 31	36 / 27
	COMM.	1.40	4x6		(179) / 151	(150) / 117	122 / 92	98 / 73	80 / 60	66 / 49	55 / 41	46 / 35	39 / 29	34 / 25
	SELECT	1.50	3x6	105 / 79	79 / 59	61 / 46	48 / 36	38 / 29	31 / 23	26 / 19	21 / 16	18 / 13	15 / 11	13 / 10
	COMM.	1.40	3x6	98 / 73	73 / 55	57 / 42	45 / 33	36 / 27	29 / 22	24 / 18	20 / 15	17 / 13	14 / 11	
SOUTHERN PINE	SELECT & COMM.	1.60	4x6		(194) / 173	(163) / 133	(139) / 105	112 / 84	91 / 68	75 / 56	62 / 47	53 / 39	45 / 34	38 / 29
	SELECT & COMM.	1.60	3x6	112 / 84	84 / 63	65 / 49	51 / 38	41 / 31	33 / 25	27 / 20	23 / 17	19 / 14	16 / 12	14 / 10
WESTERN RED CEDAR	SELECT	1.10	4x6		158 / 119	122 / 92	96 / 72	77 / 58	63 / 47	52 / 39	43 / 32	36 / 27	31 / 23	26 / 20
	COMM.	1.00	4x6	192 / 144	144 / 108	111 / 83	87 / 65	70 / 52	57 / 43	47 / 35	39 / 29	33 / 25	28 / 21	24 / 18
	SELECT	1.10	3x6	77 / 58	58 / 43	44 / 33	35 / 26	28 / 21	23 / 17	19 / 14	16 / 12	13 / 10		
	COMM.	1.00	3x6	70 / 52	52 / 39	40 / 30	32 / 24	25 / 19	21 / 16	17 / 13	14 / 11			
WESTERN WHITE SPRUCE	SELECT	1.32	4x6	(207) / 190	(171) / 143	(143) / 110	115 / 86	92 / 69	75 / 56	62 / 46	52 / 39	43 / 33	37 / 28	32 / 24
	COMM.	1.32	4x6	(160)	(132)	(111) / 110	(94) / 86	(81) / 69	(71) / 56	62 / 46	52 / 39	43 / 33	37 / 28	32 / 24
	SELECT & COMM.	1.32	3x6	(81) ⑥ / 69	(67) ⑦ / 52	53 / 40	42 / 31	34 / 25	27 / 21	23 / 17	19 / 14	16 / 12	13 / 10	

Part 2 & 3 Footnotes

① Allowable uniformly distributed total roof load in pounds per square foot of roof surface for flat roofs (up to 3/12 pitch), consisting of live load and dead load including weight of deck (see Table 1). Loads are for dry condition of use.

② Loads shown in parentheses ( ) are controlled by bending. All others controlled by deflection. Allowable unit stresses in bending are based on a 15% increase for 2-month load duration, as for snow.

③ Modulus of elasticity, times 10^{-6} psi.

④ Sizes shown are nominal sizes. See Table A.11.4 for actual sizes.

⑤ Value shown is for Commercial grade. Select grade is controlled by deflection at 168 psf.

⑥ Value shown is for Commercial grade. Select grade is controlled by deflection at 92 psf.

⑦ Value shown is for Commercial grade. Select grade is controlled by deflection at 69 psf.

† From "Structural Glued-Laminated Timber," Publication 6.5/Ai, by AITC, 1972.

Table A.11.6 LAMINATED DECKING ① ⑨ †

PART 4—SIMPLE SPAN—END JOINTS OVER SUPPORTS

SPECIES	E③ VALUE	SIZE	SPAN 8′ 1/180	8′ 1/240	9′ 1/180	9′ 1/240	10′ 1/180	10′ 1/240	11′ 1/180	11′ 1/240	12′ 1/180	12′ 1/240	13′ 1/180	13′ 1/240	14′ 1/180	14′ 1/240	15′ 1/180	15′ 1/240	16′ 1/180	16′ 1/240	17′ 1/180	17′ 1/240	18′ 1/180	18′ 1/240
DOUGLAS FIR/LARCH④	**1.80**	**5x6 ⑥ 3¹¹⁄₁₆"x5⅜"**									166	125	131	98	105	79	85	64	70	53	59	44	49	37
DOUGLAS FIR/LARCH④	1.80	4x6 (H) 3¹⁄₁₆"x5⅜"					149	112	112	84	87	65	68	51	54	41	44	33	36	27	30	23	26	19
DOUGLAS FIR/LARCH	**1.80**	**3x6S ⑥ 2⅝"x5⅜"**					95	71	71	53	55	41	43	32	34	26	28	21	23	17				
DOUGLAS FIR/LARCH④	1.80	3x6 (H) 2¼"x5⅜"	116	87	81	61	59	44	44	33	34	26	27	20	22	16								
IDAHO WHITE PINE⑥	**1.50**	**5x6 ⑥ 3¹¹⁄₁₆"x5⅜"**									138	104	109	82	87	66	71	53	58	44	49	37	41	31
IDAHO WHITE PINE⑤	1.50	4x6 (H) 3¹⁄₁₆"x5⅜"					124	93	93	70	72	54	57	43	46	34	37	28	30	23	25	19	21	16
IDAHO WHITE PINE⑥	**1.50**	**3x6S ⑥ 2⅝"x5⅜"**	154	115	108	81	80	59	59	44	46	34	36	27	29	22	23	18	19	14				
IDAHO WHITE PINE⑤	1.50	3x6 (H) 2¼"x5⅜"	96	72	68	51	49	37	37	28	28	21	22	17	18	14								
INLAND RED CEDAR	**1.20**	**5x6 ⑥ 3¹¹⁄₁₆"x5⅜"**									111	83	87	65	70	52	57	43	47	35	39	29	33	25
INLAND RED CEDAR	1.20	4x6 (H) 3¹⁄₁₆"x5⅜"					100	75	75	56	58	43	45	34	36	27	30	22	24	18				
INLAND RED CEDAR	**1.20**	**3x6S ⑥ 2⅝"x5⅜"**	123	92	86	65	63	47	47	36	36	27	29	22	23	17								
INLAND RED CEDAR	1.20	3x6 (H) 2¼"x5⅜"	77	58	54	41	39	30	30	22	23	17												
PONDEROSA PINE	**1.25**	**3x8 ⑦ 2¹¹⁄₃₂"x7¹⁵⁄₃₂"**	110	82	77	57	56	42	42	31	32	24	25	19	20	15								
SOUTHERN PINE	1.80	5x6 3¹¹⁄₁₆"x5⅜"													94	71	77	58	63	47	53	40	44	33
SOUTHERN PINE	**1.80**	**4x6 3"x5⅜"**									81	61	64	48	51	38	42.	31	34	26	29	21	24	18
SOUTHERN PINE	1.80	3x6 2¼"x5⅜"	115	86	81	61	59	44	44	33	34	26	27	20	21	16	17	13	14	11				
WEST COAST HEMLOCK	**1.75**	**4x6 3"x5¹⁵⁄₃₂"**					164	123	123	92	95	71	75	56	60	45	49	36	40	30	33	25	28	21
WEST COAST HEMLOCK	1.75	4x6S 2¾"x5¹⁵⁄₃₂"					126	95	95	71	73	55	58	43	46	35	37	28	31	23	26	19	22	16
WEST COAST HEMLOCK	**1.75**	**3x6 2¼"x5¹⁵⁄₃₂"**	135	101	95	71	69	52	52	39	40	30	32	24	25	19	21	15	17	13				

Table A.11.6 continued

PART 5—CONTROLLED RANDOM LAYUP CONTINUOUS OVER 3 OR MORE EQUAL SPANS(2)

			10′	11′	12′	13′	14′	15′	16′	17′	18′	19′	20′
DOUGLAS FIR/LARCH(4)	**1.80**	**5x6 (6) 3 13/16″x5 3/8″**					**177 / 133**	**144 / 108**	**113 / 89**	**99 / 74**	**83 / 62**	**71 / 53**	**61 / 46**
DOUGLAS FIR/LARCH(4)	1.80	4x6 (6) 3 1/16″x5 3/8″			146 / 110	115 / 86	92 / 69	75 / 56	62 / 46	51 / 39	43 / 32	37 / 28	32 / 24
DOUGLAS FIR/LARCH	**1.80**	**3x6S (6) 2 5/8″x5 3/8″**	**160 / 120**	**120 / 90**	**92 / 69**	**73 / 54**	**58 / 44**	**47 / 36**	**39 / 29**	**32 / 24**	**27 / 20**		
DOUGLAS FIR/LARCH(4)	1.80	3x6 (6) 2 1/4″x5 3/8″	100 / 75	75 / 56	58 / 43	45 / 34	36 / 27	30 / 22	24 / 18				
IDAHO WHITE PINE(5)	**1.50**	**5x6 (6) 3 13/16″x5 3/8″**					**148 / 111**	**120 / 90**	**99 / 74**	**82 / 62**	**69 / 52**	**59 / 44**	**51 / 38**
IDAHO WHITE PINE(5)	1.50	4x6 (6) 3 1/16″x5 3/8″			122 / 91	96 / 72	77 / 58	62 / 47	51 / 39	43 / 32	36 / 27	31 / 23	26 / 20
IDAHO WHITE PINE(5)	**1.50**	**3x6S (6) 2 5/8″x5 3/8″**	**133 / 100**	**100 / 75**	**77 / 58**	**61 / 46**	**48 / 36**	**40 / 30**	**32 / 24**	**27 / 20**	**23 / 17**		
IDAHO WHITE PINE(5)	1.50	3x6 (6) 2 1/4″x5 3/8″	83 / 62	62 / 47	48 / 36	38 / 28	30 / 23	25 / 18	20 / 15				
INLAND RED CEDAR	**1.20**	**5x6 (6) 3 13/16″x5 3/8″**					**118 / 88**	**96 / 72**	**79 / 59**	**66 / 49**	**56 / 42**	**47 / 35**	**40 / 30**
INLAND RED CEDAR	1.20	4x6 (6) 3 1/16″x5 3/8″			98 / 73	77 / 58	61 / 46	50 / 37	41 / 31	34 / 26	29 / 22	25 / 18	21 / 16
INLAND RED CEDAR	**1.20**	**3x6S (6) 2 5/8″x5 3/8″**	**106 / 80**	**80 / 60**	**62 / 46**	**48 / 36**	**39 / 29**	**32 / 24**	**26 / 20**				
INLAND RED CEDAR	1.20	3x6 (6) 2 1/4″x5 3/8″	67 / 50	50 / 38	38 / 29	30 / 23	24 / 18	20 / 15					
PONDEROSA PINE	**1.25**	**3x8 (7) 2 11/32″x7 15/32″**	**79 / 59**	**59 / 44**	**45 / 34**	**36 / 27**	**28 / 21**	**23 / 17**					
SOUTHERN PINE	1.80	5x6 (8) 3 11/16″x5 3/8″					160 / 120	130 / 97	107 / 80	89 / 67	75 / 56	64 / 48	55 / 41
SOUTHERN PINE	**1.80**	**4x6 (8) 3″x5 3/8″**			**137 / 103**	**108 / 81**	**86 / 65**	**70 / 53**	**58 / 43**	**48 / 36**	**41 / 30**	**35 / 26**	**30 / 22**
SOUTHERN PINE	1.80	3x6 2 1/4″x5 3/8″	100 / 75	75 / 56	58 / 43	45 / 34	36 / 27	30 / 22	24 / 18				
WEST COAST HEMLOCK	**1.75**	**4x6 3″x5 15/32″**			**137 / 103**	**108 / 81**	**86 / 65**	**70 / 53**	**58 / 43**	**48 / 36**	**41 / 30**	**35 / 26**	**30 / 22**
WEST COAST HEMLOCK	1.75	4x6S 2 3/4″x5 15/32″			106 / 79	83 / 62	67 / 50	54 / 41	45 / 33	37 / 28	31 / 23	27 / 20	23 / 17
WEST COAST HEMLOCK	**1.75**	**3x6 2 1/4″x5 15/32″**	**100 / 75**	**75 / 56**	**58 / 43**	**45 / 34**	**36 / 27**	**30 / 22**	**24 / 18**				

Part 4 & 5 Footnotes

(1) Allowable uniformly distributed total roof load in pounds per square foot of roof surface for flat roofs, consisting of live load and dead load including weight of desk (see Table A.11.4). These roof loads do not exceed bending loads allowable under recognized bending formulas. Loads are for dry condition of use.

(2) Total loads for controlled random layup are 12% higher than recognized by I.C.B.O. Research Recommendations.

(3) Modulus of elasticity, times 10^{-6} psi.

(4) Southern Pine also available at same load values.

(5) Inland White Fir also available at same load values.

(6) Also available in 8″ nominal width (7 3/8″ net). Coverage factors: 5x8-5.43, 4x8-4.34, 3x8S and 3x8-3.26.

(7) Also available in 10″ and 12″ nominal widths (9 15/32″ and 11 15/32″ net). Coverage factor 3.20 for each width.

(8) Douglas Fir/Larch also available at same load values.

(9) All load values assume installation conforming to manufacturer's recommendations.

† From "Structural Glued-Laminated Timber," Publication 6.5/Ai, by AITC, 1972.

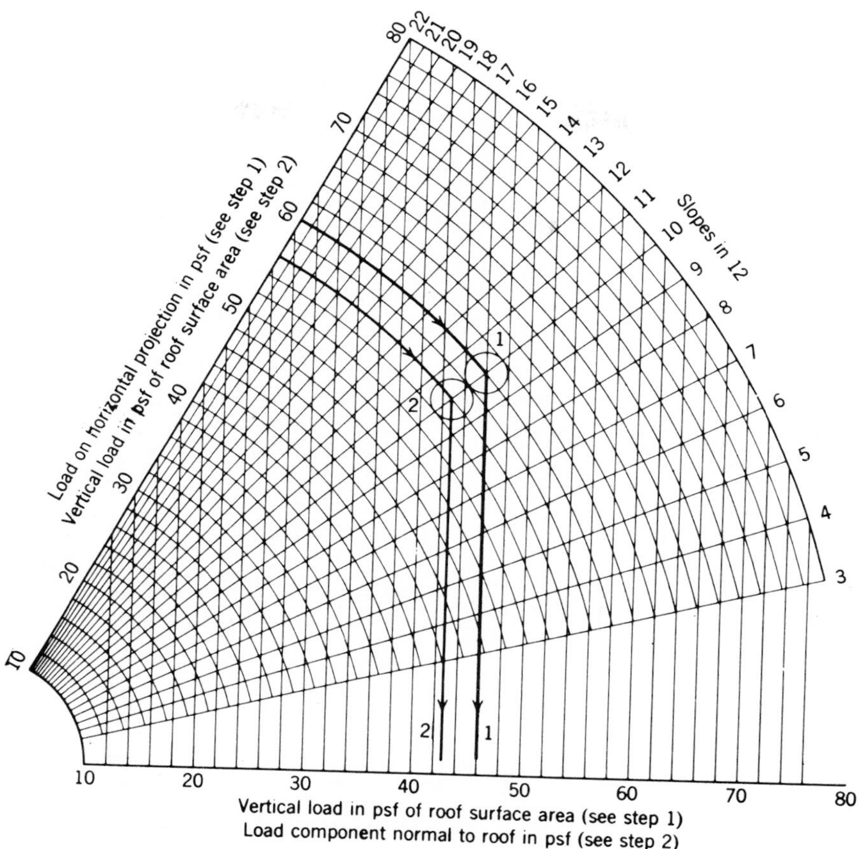

TABLE A.11.7 LOAD CONVERSION. Example: 60 psf live load and 10 psf dead load on 10 in 12 slope. Step 1:60 psf live load on horizontal projection = 46 psf of roof surface area vertical load on 10 in 12 roof slope. Step 2:10 psf of roof surface area dead load plus 46 psf of roof surface area live load = 56 psf of roof surface area combined load acting vertically; 56 psf of roof surface area vertical total load = 43 psf normal to roof causing bending and deflection. From "Timber Construction Manual," by AITC, John Wiley and Sons, New York, Second Edition, 1974.

TABLE A.11.8.[1] CANTILEVER-SUSPENDED JOISTS.

16 in. Spacing

Joist Lumber	House Width						
	22′	24′	26′	28′	30′	32′	34′
	Joist Size						
Douglas fir-larch							
No. 1	2x6*	2x6*	2x8	2x8*	2x8*	2x10	2x10
No. 2	2x6*	2x8	2x8	2x8*	2x8*	2x10	2x10
No. 3	2x8	2x8*	2x10	2x10	2x10*	2x12	2x12*
Southern pine							
No. 1	2x6*	2x6*	2x8	2x8*	2x8*	2x10	2x10
No. 2 MG KD†	2x6*	2x8	2x8	2x8*	2x8*	2x10	2x10
No. 2	2x8	2x8	2x8*	2x8*	2x10	2x10	2x10*
No. 3	2x8	2x8*	2x10	2x10*	2x10*	2x12	2x12*
Hem-fir							
No. 1	2x8	2x8	2x8	2x8*	2x8*	2x10	2x10*
No. 2	2x8	2x8	2x8*	2x8*	2x10	2x10	2x10*
No. 3	2x10	2x10	2x10*	2x12	2x12	2x12*	—

24 in. Spacing

Joist Lumber	House Width						
	22′	24′	26′	28′	30′	32′	34′
	Joist Size						
Douglas fir-larch							
No. 1	2x8	2x8	2x10	2x10	2x10	2x10*	2x12
No. 2	2x8	2x8*	2x10	2x10	2x10	2x10*	2x12
No. 3	2x10	2x10*	2x12	2x12	2x12*	—	—
Southern pine							
No. 1	2x8	2x8	2x10	2x10	2x10	2x10*	2x12
No. 2 MG KD†	2x8	2x8	2x10	2x10	2x10	2x10*	2x12
No. 2	2x8*	2x10	2x10	2x10*	2x12	2x12	—
No. 3	2x10	2x10*	2x12	2x12*	2x12*	—	—
Hem-fir							
No. 1	2x8	2x10	2x10	2x10	2x12	2x12	—
No. 2	2x8*	2x10	2x10	2x10*	2x12	2x12	—
No. 3	2x12	2x12	2x12*	—	—	—	—

*Joist size is a reduction from HUD-MPS requirement for simple spans, as shown in the NFPA Span Tables for Joists and Rafters, adopted as the standard for conventional floor construction. (See MPS 4900.1, Appendix E).

†Medium grain, kiln dried.

1. From APA's "Cantilevered In-Line Joist System."

TABLE A.11.9.* CANTILEVER-SUSPENDED JOISTS CUTTING SCHEDULE.

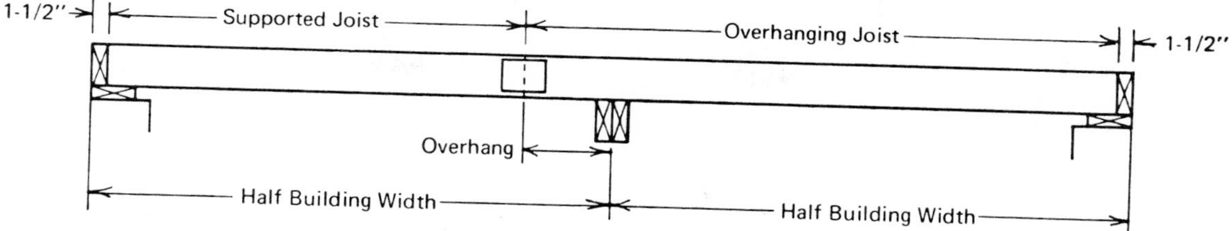

Building Width	Overhanging Joist	Supported Joist	Overhang
22' - 0"	12' - 0"	9' - 9"	1' - 1-1/2"
24' - 0"	14' - 0"	9' - 9"	2' - 1-1/2"
26' - 0"	15' - 9"	10' - 0"	2' - 10-1/2"
28' - 0"	16' - 0"	11' - 9"	2' - 1-1/2"
30' - 0"	17' - 9"	12' - 0"	2' - 10-1/2"
32' - 0"	18' - 0"	13' - 9"	2' - 1-1/2"
34' - 0"	20' - 0"	13' - 9"	3' - 1-1/2"

*From APA's Cantilevered In-Line Joist System."

Name Index

Subject Index